Solution and Surface Polymerization

Functional and Modified Polymeric Materials Two-Volume Set

Eli Ruckenstein
Hangquan Li
Chong Cheng

Volumes in the Set:

Concentrated Emulsion Polymerization (ISBN: 9780367134556)

Solution and Surface Polymerization (ISBN: 9780367134563)

Solution and Surface Polymerization

Eli Ruckenstein
Hangquan Li
Chong Cheng

CRC Press is an imprint of the
Taylor & Francis Group, an informa business

CRC Press
Taylor & Francis Group
6000 Broken Sound Parkway NW, Suite 300
Boca Raton, FL 33487-2742

Printed on acid-free paper

International Standard Book Number-13: 978-0-3671-3456-3 (Hardback)

Visit the Taylor & Francis Web site at
http://www.taylorandfrancis.com

and the CRC Press Web site at
http://www.crcpress.com

Contents

Preface

With the current annual worldwide production of over 400 million tons, polymers are indispensable materials that have been broadly used in everyday life and in almost every industry. As the chemical process to convert small molecule precursors (i.e., monomers) to polymers, polymerization can be performed in homogeneous systems in bulk or solution, as well as heterogeneous systems in emulsion or suspension, at interface, or on surface. Although polymers have versatile mechanical properties, and are commonly produced for general applications such as plastic, rubber and fiber, their structures can be tailored to target new and enhanced properties and applications. Accordingly, the preparation and comprehensive studies of functional and modified polymeric materials represent an increasingly important research direction in polymer field. This book set of two volumes summarizes the research work performed by Professor Eli Ruckenstein and co-workers on functional and modified polymeric materials over the past decades. Volume 2 (this book) presents the research studies mainly using solution or surface polymerization in material synthesis. The corresponding polymer-based materials include polymers with special functional groups, degradable and de-cross-linkable polymers, pervaporation membranes, soluble conductive polymers, composites with conductive surfaces, and so on. Volume 1 depicts the research studies using emulsion polymerization, especially concentrated emulsion polymerization, as the synthetic method. The corresponding polymeric materials include conductive polymer composites, core-shell latex particles, enzyme/catalyst/herbicide carriers, and toughened and compatibilized plastics.

There are six chapters in this book (Volume 2), each containing research papers by Professor Eli Ruckenstein and his co-workers on a specific topic or on miscellaneous topics. Chapter 1 describes the synthesis of well-defined polymers having specific functional groups by living ionic polymerization in solution. With excellent living characteristics, anionic polymerization is utilized as the major polymerization technique, while cationic polymerization is employed as the complementary polymerization technique in some cases. By using different synthetic pathways, a variety of functionalities, such as acetal, carboxylic, hydroxyl, glycidyl, isobutoxyethyl and vinyl groups, are introduced to polymers. With well-controlled structures and functionalities, these polymers are novel model polymers. They can be further modified to form other types of functional polymers, as well as grafted copolymers and crosslinked polymer networks.

Chapter 2 reports the preparation of several types of degradable and de-cross-linkable polymers. Each of these polymers has a different set of cleavable functionalities and is associated with a different preparation method. Because commonly used polymers result in significant environmental concerns, the development of degradable and de-cross-linkable polymers represents a critical research direction to dissipate such concerns.

Chapter 3 reports systematic studies in the preparation and assessment of semi- and interpenetrating polymer network (IPN) pervaporation membranes. As an important energy efficient separation technique, pervaporation has critical industrial applications, but commonly used polymer pervaporation membranes often lack the required comprehensive material properties. With tuneable multiple polymeric components for targeting superb comprehensive properties, semi-IPN and IPN membranes are highly promising materials for pervaporation applications.

Chapter 4 presents several approaches to preparation of soluble conducting polymers. Because typical conducting polymers have low processability that restricts broad large-scale applications, it is critically important to develop soluble conducting polymers that can be readily processed via solutions. Doping, grafting, and copolymerization strategies are utilized to incorporate solubility-enhancing structures with conducting polymer chains, resulting in soluble conducting polymers.

Chapter 5 focuses on the preparation approaches for composite materials with conductive polymer surfaces. Because conductive polymers generally would not possess appropriate processability and mechanical properties, composite materials with conducting polymers on the surface of non-conductive base materials may possess improved comprehensive properties and applicability. The approaches presented in this chapter can allow the preparation of different types of conductive polymer layers, with the use of different types of base materials, leading to a variety of surface-conducting composite materials.

Chapter 6 depicts research studies on miscellaneous topics, including coordination polymerization, dendritic polymers and the corresponding organic-inorganic hybrid materials, and plastics toughening using non-emulsion approaches. Although these studies are not systematic, they are still significant and innovative.

This book set will be of considerable value to material scientists in universities and industry, and to graduate students. In particular, this volume set will be of interest to researchers working in solution and surface polymerization, ionic polymerization, degradable and de-cross-linkable polymers, pervaporation membranes, conductive polymers and composites, etc.

Authors

Eli Ruckenstein is a Distinguished Professor at the State University of New York (SUNY) at Buffalo. He has published more than 1,000 papers in numerous areas of engineering science and has received a large number of awards from the American Chemical Society and the American Institute of Chemical Engineers. Dr. Ruckenstein has also received the Founders Gold Medal Award from the National Academy of Engineering and the National Medal of Science from President Clinton. He is a member of the National Academy of Engineering and of the American Academy of Art and Sciences.

Hangquan Li received a PhD in polymer science and engineering from Beijing University of Chemical Technology in 1990. He was a visiting scientist at the State University of New York (SUNY) at Buffalo, working with Dr. Eli Ruckenstein from 1993 to 1996. He has been a Professor at Beijing University of Chemical Technology since 1996. He has published over 100 papers, mainly on polymer research.

Chong Cheng received a PhD in chemistry (polymer) from City University of New York in 2003. He currently is an Associate Professor in the Department of Chemical and Biological Engineering at the State University of New York (SUNY) at Buffalo. He has published over 70 papers on polymer synthesis and characterization, as well as biomedical applications of polymers.

1 Functional Polymers by Living Ionic Polymerization

CONTENTS

Living polymerizations can exert superior control to macromolecular structures as compared to conventional chain polymerizations. With the absence or greatly suppressed chain-breaking reactions, living polymerizations with fast propagation relative to initiation generally lead to well-defined polymers with predetermined number-average molecular weight (M_n) and narrow polydispersity index (PDI; also called molecular weight dispersity, Đ). Polymers with complexed structures, such as block copolymers and graft copolymers, can also be prepared by living polymerizations.

Living ionic polymerizations include living anionic polymerization, which was first reported by Swartz in 1956, and living cationic polymerization, which was first reported by several research teams in the early 1980s. Living anionic polymerization is particularly important in polymer science, because not only commercial thermoplastic elastomers but also narrowly dispersed polymer standards are produced by this technique. For both living anionic and living cationic polymerizations, a major restriction is their limited functional group tolerance because the ionic reactive centers can react with a broad variety of functional groups. Therefore, it is important to develop strategies for the preparation of functional polymers by living ionic polymerization.

This chapter collects ten papers on the synthesis of functional polymers by living ionic polymerizations. In these studies, living anionic polymerization was used as the major polymerization approach, and in some cases, living cationic polymerization was also employed. Four key strategies to access functional polymers are described in these papers.

The first strategy is selective living anionic polymerization of asymmetrical divinyl (or multivinyl) monomers (Sections 1.1 through 1.4). For each of the asymmetrical divinyl monomers, it has one vinyl group (V_A; as methacryloyl or styryl group) that is polymerizable under the living anionic polymerization conditions and the other vinyl group (V_B; as alkene group) that is non-polymerizable under the same conditions but can perform functional group transformation (FGT). Thus, well-defined functional polymers carrying quantitative V_B groups were formed by selective living anionic polymerization, and they could be further converted to other types of functional polymers with bromide, hydroxyl or other functionalities through FGT of V_B groups. This innovative selective polymerization strategy has also inspired the development of selective living radical polymerization of asymmetrical divinyl monomers.

The second strategy employs living anionic polymerization of acetal-protected vinyl monomers (Sections 1.5 and 1.6). Innovative synthesis of acetal-protected vinyl monomers was developed at first, then acetal-protected vinyl polymers were prepared by living anionic polymerization of these monomers, and finally well-defined functional polymers carrying carboxylic or hydroxyl functionalities were obtained after facile deprotection of these polymers under slightly acidic conditions.

The third strategy is based on living anionic polymerization of glycidyl-functionalized monomers (such as glycidyl methacrylate), in which glycidyl functionalities are intact under the polymerization conditions (Section 1.7). The resulting well-defined glycidyl-functionalized polymers were then used as backbone precursor polymers to react with strong living anions for grafting-onto reactions. Various well-defined copolymers with grafted structures were obtained.

The fourth strategy utilizes living anionic polymerization of inimers, each consisting of a monomer functionality for anionic polymerization and an initiator group for cationic polymerization (Sections 1.8 through 1.10). Anionic polymerization of such inimers yielded well-defined polyinitiators that can initiate living cationic polymerization for further synthesis of graft copolymers. These studies elegantly provided one of the first examples of the synthesis of complexed polymers by strategic combination of different types of living polymerizations, and such tandem living polymerization approaches have been further adopted in many other studies.

1.1 Novel Monodisperse Functional (Co)polymers Based on the Selective Living Anionic Polymerization of a New Bifunctional Monomer, *trans, trans*-1-Methacryloyloxy-2,4-hexadiene*

Hongmin Zhang and Eli Ruckenstein

Chemical Engineering Department, State University of New York at Buffalo, Amherst, New York 14260

ABSTRACT A new bifunctional monomer, *trans*, *trans*-1-methacryloyloxy-2,4-hexadiene (MAHE), was prepared through the reaction between the sodium salt of 2,4-hexadien-1-ol and methacryloyl chloride. Using 1,1-diphenylhexyllithium (DPHL) as initiator, in tetrahydrofuran (THF), in the presence of LiCl ($[LiCl]/[DPHL]_0 = 3$), at −70°C, the anionic polymerization of MAHE proceeded quantitatively, generating monodisperse polymers ($M_w/M_n = 1.04–1.05$) with well-controlled molecular weights. Under similar conditions, well-defined block copolymers with controlled molecular weights and compositions as well as very narrow molecular weight distributions ($M_w/M_n = 1.04–1.06$) were prepared by the anionic block copolymerization of MAHE and methyl methacrylate (MMA), regardless of the polymerization sequence (MAHE followed by MMA or vice versa). Similarly, the copolymers resulting from the random copolymerization of these two monomers possessed monodispersity

* *Macromolecules* 2001, 34, 3587–3593.

($M_w/M_n = 1.04–1.05$), regardless of the feed amount ratio. Further, block copolymers of styrene (St) and MAHE with various compositions were obtained by the anionic block copolymerization of MAHE to a 1,1-diphenylethylene-capped anionic living poly(St). In both the homopolymerization of MAHE and its copolymerization with MMA or St, its 2,4-hexadienyl side group remained unreacted. This side group was further reacted with bromine or osmium tetroxide. Using an excess amount of bromine, at room temperature, the bromination efficiency reached 95%, generating a new functional polymer, poly(2,3,4,5-tetrabromohexyl methacrylate). The osmylation was carried out by reacting the (co)polymer with an excess of *N*-methylmorpholine *N*-oxide, in the presence of a small amount of osmium tetroxide as catalyst, at room temperature. This osmylation reaction changed poly(MAHE) to a new functional polymer, poly(2,3,4,5-tetrahydroxyhexyl methacrylate): with an extremely high water solubility. Well-defined amphiphilic block copolymers were also obtained via the osmylation after block copolymerizations.

1.1.1 INTRODUCTION

The properties and performances of a functional polymer depend mainly on the kind and number of functional groups. For instance, the presence of halogen atoms can greatly increase the flame retardancy of the polymer,[1] and the introduction of hydroxyl groups can improve its hydrophilicity. It is well known that the polymers of alkyl methacrylates are insoluble in alcohol and water, because of their poor hydrophilicity. However, poly(2-hydroxyethyl methacrylate) [poly(HEMA)], which possesses one hydroxyl group in each repeating unit, is soluble in alcohol but insoluble in water.[2] When an additional hydroxyl group was introduced into the side chain, the poly(2,3-dihydroxypropyl methacrylate) [poly(DHPMA)] thus obtained became water-soluble.[3] It is expected that an increasing hydrophilicity of methacrylate polymers can be achieved by a further increase of the hydroxyl groups in the side chains.

The molecular weight distribution (MWD) is also a very important factor which can affect the properties and the performances of functional polymers. For a polymer to possess uniform properties, its MWD should be narrowly controlled, and even an effort should be made to obtain monodisperse functional polymers ($M_w/M_n \leq 1.05$). The present paper presents an approach to several novel monodisperse functional homo- and copolymers via the selective living anionic polymerization of a new bifunctional monomer followed by additional reactions.

It is well known that the methacrylate monomers are much more reactive than the conjugated dienes in an anionic polymerization, and for this reason, the diblock copolymer of these two monomers can only be obtained via a polymerization sequence of diene followed by methacrylate.[4] The new bifunctional monomer, *trans, trans*-1-methacryloyloxy-2,4-hexadiene (MAHE, **1** in Scheme 1.1.1), possesses both a methacryloyl type and a conjugated diene type polymerizable group. The reactivity of the former group is greater than that of the latter. In addition, the steric hindrance due to the presence of a methyl and an ester group at the two ends of the diene further reduces its reactivity. Therefore, it is expected that the methacryloyl type C=C bond

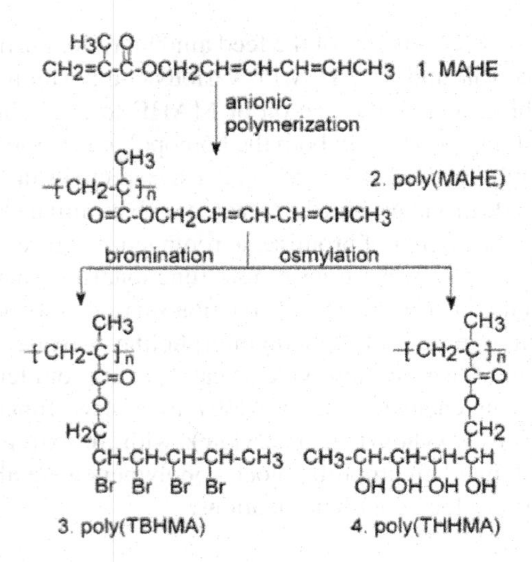

SCHEME 1.1.1

can be selectively polymerized under suitable conditions to generate a functional polymer with a reactive dienyl group in each of its repeating units (**2** in Scheme 1.1.1). Furthermore, because the diene side groups can undergo many kinds of reactions, such as bromination, osmylation, Diels–Alder reaction, etc., a number of novel functional polymers can be obtained by reactions after the polymerization. As shown in Scheme 1.1.1, the bromination of poly(MAHE) can introduce four bromine atoms into each of its repeating units (**3** in Scheme 1.1.1). On the other hand, its reaction with *N*-methylmorpholine *N*-oxide (NMO) in the presence of osmium tetroxide as catalyst generates a functional polymer (**4** in Scheme 1.1.1) that possesses very high water solubility because of the presence of four hydroxyl groups in each of its repeating units. In addition, a number of novel functional copolymers, such as amphiphilic copolymers, can also be obtained by the copolymerization of MAHE with other monomers, such as St, MMA, etc., followed by additional reactions.

1.1.2 EXPERIMENTAL SECTION

1.1.2.1 Materials

THF was dried with CaH_2 under reflux for more than 24 h, distilled, and distilled again from a solution of 1,1-diphenylhexyllithium (DPHL) just before use. Toluene was washed with concentrated sulfuric acid and then with water, dried over $MgSO_4$, distilled over CaH_2, and distilled again from a DPHL solution before use. Hexane was first dried and distilled over CaH_2 and then distilled from a solution of *n*-BuLi. Methyl methacrylate (MMA, Aldrich, 99%) was dried over CaH_2 with magnetic stirring under reduced pressure and vacuum-distilled twice over CaH_2. Styrene (St, 99%, Aldrich) was washed with a 10% aqueous NaOH solution and then with water, dried overnight with $MgSO_4$, distilled over CaH_2, and finally distilled in the presence of phenylmagnesium chloride just before use. 1,1-Diphenylethylene

(DPE, Aldrich, 97%) was distilled over CaH_2 and then distilled in the presence of DPHL under reduced pressure. Lithium chloride (Aldrich, 99.99%) was dried at 120°C for 24 h and dissolved in THF.[5] *n*-BuLi (Aldrich, 1.6 M solution in hexane) was diluted with purified hexane. Methacryloyl chloride (Acros, 97%), *trans, trans*-2,4-hexadien-1-ol (Aldrich, 97%), bromine (Aldrich, 99.5%), osmium tetroxide (Aldrich, 4.0 wt% aqueous solution), and (NMO, Aldrich, 97%) were used as received.

1.1.2.2 Synthesis of MAHE

MAHE was prepared through the reaction between the sodium salt of *trans, trans*-2,4-hexadien-1-ol (HDO) and methacryloyl chloride. A 1000 mL round-bottom flask equipped with a condenser, a paddle stirrer, and a dropping funnel with a pressure-equalization arm was degassed and heated under reduced pressure. To this flask, 400 mL of CaH_2-purified toluene was added, followed by the addition of sodium hydride (Aldrich; 60% dispersion in mineral oil; 10.4 g, 0.26 mol). Under the protection of nitrogen, 100 mL of toluene solution of HDO (25 g, 0.25 mol) was dropwise added in 1.5 h. The reaction mixture was stirred for 1 h at room temperature to generate the sodium salt of HDO. The reaction system was cooled to 0°C, and methacryloyl chloride (24 g, 0.22 mol) was dropwise added in 1.0 h. After about 1.5 h, the reaction was terminated by carefully adding an aqueous solution of sodium hydroxide (1.0 M, 100 mL). The reaction mixture was washed three times with water, evaporated to remove toluene, and distilled under reduced pressure (bp: 42°C/0.5 Torr; yield: 74% based on the amount of methacryloyl chloride employed). The monomer thus obtained was distilled twice over CaH_2 and finally distilled in the presence of $Al(C_2H_5)_3$[6] prior to polymerization. As shown later in Figure 1.1.3A, the chemical shifts and their intensities in the ^1H NMR spectrum of the prepared MAHE are consistent with its molecular structure.

1.1.2.3 Anionic (Co)Polymerizations

All polymerizations, namely, the anionic homopolymerization of MAHE, its random and block copolymerizations with MMA, and its block copolymerization with St, were carried out in a round-bottom flask under argon with magnetic stirring.

1.1.2.4 Homopolymerization

The anionic homopolymerization of MAHE was performed in THF, at −70°C, in the presence of LiCl ($[LiCl]/[DPHL]_0 = 3$).[5] Before the monomer addition, the initiator DPHL was first prepared in situ. After THF, DPE, and a THF solution of LiCl were added with dry syringes, the flask was cooled to −40°C and *n*-BuLi (in hexane) was added. The reaction between *n*-BuLi and DPE was allowed to last 15 min. Then, the system was cooled to −70°C, and the polymerization reaction was induced by the addition of prechilled MAHE to the above system. About 1 h later, the polymerization was quenched with a small amount of methanol (ca. 1 mL). Usually, the polymerization solution was divided into three parts. Two parts were further used for the bromination and osmylation, and the third part was treated with a trace amount of inhibitor BHT, evaporated to dryness, and vacuum-dried for 24 h at room temperature.

1.1.2.5 Copolymerization with MMA

The anionic block copolymerization of MAHE and MMA was carried out by using the monomer addition sequence of MAHE followed by MMA or vice versa, in THF, in the presence of LiCl ($[LiCl]/[DPHL]_0 = 3$), at −70°C. The polymerization times for MMA and MAHE were 30 and 50 min, respectively. The random copolymerization of the two monomers was carried out under the same conditions, except that the two monomers were added at the same time. The initiator preparation and the polymer purification were carried out in ways similar to those employed for the homopolymer of MAHE.

1.1.2.6 Block Copolymerization with St

The block copolymer of MAHE and St was prepared using the polymerization sequence of St, followed by MAHE. The living poly(St) was first prepared by the anionic polymerization of St, which was carried out in a mixture of toluene and THF (3:1 by volume), at −55°C, using *n*-BuLi as the initiator. After toluene, THF, and a hexane solution of *n*-BuLi were introduced into a flask kept at −55°C, the polymerization was started by adding prechilled St. While the polymerization was proceeding, THF, DPE ($mol_{DPE}/mol_{n\text{-BuLi}} = 1.4$), and a THF solution of LiCl ($mol_{LiCl}/mol_{n\text{-BuLi}} = 3$) were introduced into another flask, to which a hexane solution of *n*-BuLi was dropwise added until the red color of DPHL appeared, to remove the impurities. Then, this mixture was immediately introduced into the living poly(St) solution. The color of the system changed instantaneously from yellow to deep red, implying that the living end of poly(St) was rapidly capped by DPE. This reaction was allowed to last 20 min to ensure a complete transformation of the living site. Subsequently, prechilled MAHE was added and its polymerization allowed to last 40 min. After the termination with methanol (ca. 1 mL), the bromination, osmylation, and the polymer purification were carried out in ways similar to those for the homopoly(MAHE).

1.1.2.7 Bromination

The reaction between the diene side chains of the resulting (co)polymers and bromine was carried out by directly employing the polymerization solution in the next bromination step. The (co)polymer was allowed to react with an excess amount of bromine at room temperature, in a dark box, under nitrogen, with magnetic stirring. After a selected time, the polymer was precipitated by pouring the reaction mixture into methanol, washed three times with methanol, immersed in methanol overnight, and finally vacuum-dried at 45°C for 24 h.

1.1.2.8 Osmylation

The osmylation reactions of the homopolymer of MAHE and of its block copolymer with MMA or St were carried out with magnetic stirring, under nitrogen, at room temperature. Upon the (co)polymerization, a solution containing a certain amount of (co)polymer was transferred into a nitrogen-protected flask, to which THF, acetone, methanol, and NMO were sequentially added. Then, the system was degassed and reprotected with nitrogen, and the reaction was started by adding osmium tetroxide (4.0 wt% aqueous solution) with a syringe. After a certain time, the resulting polymer was precipitated by pouring the reaction solution into a mixture of hexane and ethanol (1:1 by volume). Then, the polymer was redissolved in hot methanol (for homopolymer) or pyridine (for block copolymer), and the precipitation into the above mixture was twice repeated. The (co)polymer thus obtained was kept in the above mixture overnight, washed with ethanol, vacuum-dried at 40°C for 24 h, and finally freeze-dried from a 1,4-dioxane solution containing a small amount of water.

1.1.2.9 Measurements

^{1}H NMR spectra were recorded in $CDCl_3$, D_2O, or pyridine-d_5 on an Inova-400 spectrometer. M_n and M_w/M_n of the (co)polymers were determined by gel permeation chromatography (GPC) on the basis of a polystyrene calibration curve. The GPC measurements were carried out using THF as solvent, at 30°C, with a 1.0 mL/min flow rate and a 1.0 cm/min chart speed. Three polystyrene gel columns (Waters, 7.8 × 300 mm: one HR 5E, part no. 44228; one linear, part no. 10681; and one HR 4E, part no. 44240) were used, which were connected to a Waters 515 precision pump. FT-IR spectra were recorded using KBr pellets on a Perkin-Elmer 1760-X spectrometer. The elemental analysis was carried out by Atlantic Microlab, Inc.

1.1.3 RESULTS AND DISCUSSION

1.1.3.1 Anionic Polymerization of MAHE

Monomer purification played an important role in the anionic polymerization of MAHE. Calcium hydride-purified MAHE was not pure enough for the anionic polymerization. Even though a high initiator concentration was employed ($[DPHL]_0$ = 50 mM, $[MAHE]_0$ = 0.6 M), no polymer could be obtained because of the presence of a trace amount of unreacted alcohol (HDO) during the monomer preparation procedure (see Experimental Section). To remove this impurity, CaH_2-purified MAHE was further distilled in the presence of triethylaluminum,[6] and the monomer thus obtained was employed in the polymerization immediately. Using DPHL as the initiator, in the presence of LiCl ($[LiCl]/[DPHL]_0 = 3$), in THF, at −70°C, the anionic polymerization of MAHE proceeded smoothly, generating poly(MAHE) quantitatively. As shown in Table 1.1.1, monodisperse polymers (M_w/M_n = 1.04–1.05) with well-controlled molecular weights were obtained in every case. Even when the initiator concentration was very low ($[DPHL]_0$ = 1.0 mM, PMAHE-5 in Table 1.1.1), the resulting large molecular weight poly(MAHE) still possessed monodispersity ($M_n = 8.55 \times 10^4$, $M_w/M_n = 1.05$). As shown in Figure 1.1.1A, its GPC chromatogram exhibits a sharp, symmetrical peak, which is even narrower than that of the poly(St) standard (Figure 1.1.1C, M_n = 17 500, $M_w/M_n < 1.06$). Very few polar functional polymers were reported to possess such a narrow MWD.

TABLE 1.1.1
Selective Anionic Homopolymerization of MAHE[a]

	$[DPHL]_0$(mM)	$[MAHE]_0$(M)	$10^{-4}M_n$		M_w/M_n[b]
			calcd	obsd[b]	
PMAHE-1	10.0	0.48	0.82	0.81	1.05
PMAHE-2	6.7	0.60	1.51	1.51	1.04
PMAHE-3	4.0	0.48	2.02	1.94	1.04
PMAHE-4	2.0	0.48	4.01	4.04	1.04
PMAHE-5	1.0	0.48	8.00	8.55	1.05

[a] The initiator DPHL was prepared in situ before the monomer addition via the reaction between *n*-BuLi and an excess amount of DPE ($[DPE]/[n\text{-BuLi}]_0 = 1.4$) at −40°C for about 15 min. The polymerization was carried out in THF, in the presence of LiCl ($[LiCl]/[DPHL]_0 = 3$), at −70°C, for 45–60 min. The monomer conversion was 100% in each case.

[b] Determined by GPC.

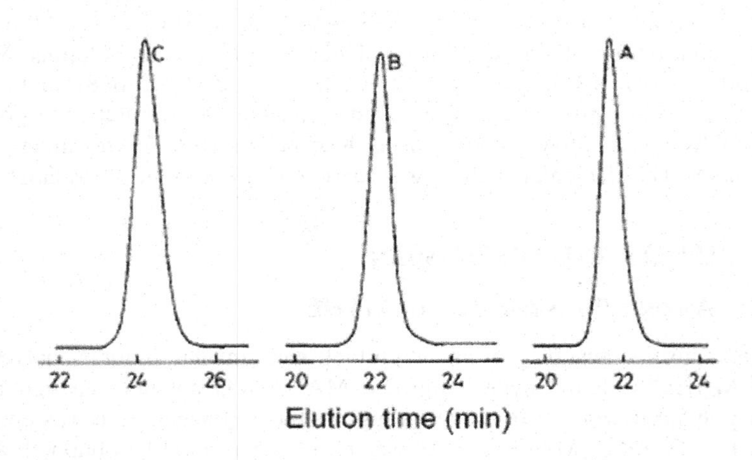

FIGURE 1.1.1 GPC traces of poly(MAHE) (A; PMAHE-5 in Table 1.1.1, M_n = 85,500, M_w/M_n = 1.05), poly(MAHE-*co*-MMA) (B; random-4 in Table 1.1.3, M_n = 66,900, M_w/M_n = 1.04), and polystyrene standard (C; from Pressure Chemical, lot no. 41220, M_n = 17,500, M_w/M_n < 1.06).

To verify the living nature of the anionic polymerization of MAHE, two successive monomer addition experiments were carried out. The polymerization conditions and the GPC results are presented in Figure 1.1.2. In the first stage, poly(MAHE) with a very narrow MWD (peak A, M_n = 12,000, M_w/M_n = 1.04) was obtained. After the same amount of MAHE was added to the system, the GPC peak shifted toward the high molecular weight, but the MWD remained narrow (peak B, M_n = 24,000, M_w/M_n = 1.05), and no precursor polymer remained. The above results clearly indicate that the anionic polymerization of MAHE proceeded in a living manner.

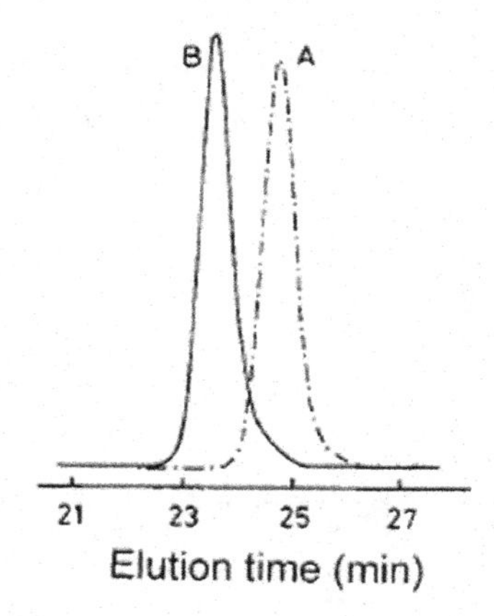

FIGURE 1.1.2 GPC traces of poly(MAHE)s obtained by a two times of addition of the same amount of monomer for anionic polymerization (A) First-time monomer addition, $M_n = 12{,}000$, $M_w/M_n = 1.04$ ($[DPHL]_0 = 5.19$ mM, $[MAHE]_0 = 0.344$ M, [LiCl] = 15.6 mM, at –70°C). (B) Second-time monomer addition, $M_n = 24{,}000$, $M_w/M_n = 1.05$.

Parts A and B of Figure 1.1.3 present the ^1H NMR spectra of MAHE and its polymer, respectively. After polymerization, peaks a, b, and b′ in Figure 1.1.3A due to α-CH_3 and H_2C= (b and b′) of the methacryloyl group disappeared completely, and two new absorptions corresponding to $-CH_3$ (peak a in Figure 1.1.3B) and $-CH_2-$ of poly(MAHE) main chain emerged at 0.75–1.10 and 1.88 ppm, respectively. On the other hand, peaks c–h due to the 2,4-hexadienyl ester group remained unchanged after polymerization. Furthermore, as shown later in Table 1.1.8, the resulting poly(MAHE) is soluble in a number of solvents, such as benzene, chloroform, THF, 1,4-dioxane, acetone, etc. Therefore, one can conclude that the methacryloyl type C=C bond of MAHE was selectively polymerized, that no cross-linking and any other side reaction occurred during polymerization, and that the polymer thus obtained is a functional polymer possessing a conjugated diene group in each of its repeating units.

1.1.3.2 Anionic Copolymerization of MAHE and MMA

Both the anionic block and random copolymerizations of MAHE and MMA were carried out under conditions similar to those for its homopolymerization. For the block copolymerization, a two-step sequential monomer addition, namely, MAHE followed by MMA or vice versa, was employed. As shown in Table 1.1.2, regardless of the polymerization sequence, well-defined block copolymers were obtained in both cases. The molecular weight could be controlled at each step, the MWD of either the precursor or the final copolymer possessed monodispersity (M_w/M_n = 1.04–1.06), and the composition determined by ^1H NMR was in good agreement with that designed. As an example, Figure 1.1.4A presents the GPC chromatograms of poly(MAHE-b-MMA) (peak A-b; block-3 in Table 1.1.2) and of its living poly(MAHE) precursor (peak A-a; $M_n = 16{,}800$, $M_w/M_n = 1.05$). After the second

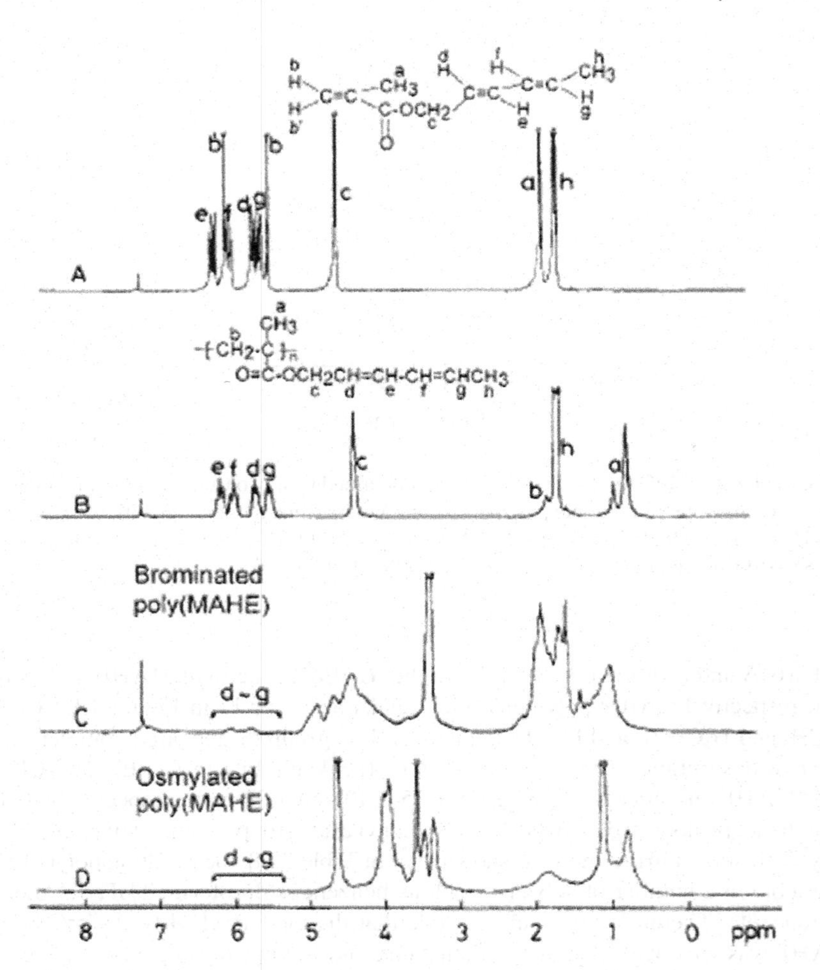

FIGURE 1.1.3 ^{1}H NMR spectra of MAHE (A; in $CDCl_3$), poly(MAHE) (B; in $CDCl_3$; PMAHE-2 in Table 1.1.1), brominated poly(MAHE) (C; in $CDCl_3$; Br-PMAHE-2′, see Tables 1.1.1 and 1.1.5) and osmylated poly(MAHE) (D; in D_2O; o-PMAHE-2, see Tables 1.1.1 and 1.1.6).

step polymerization of MMA from the living end of poly(MAHE), the M_n of the resulting copolymer increased (M_n = 39,500, M_w/M_n = 1.04), while the monodispersity remained unchanged. This result further confirmed the living nature of the anionic polymerization of MAHE.

Similar to the block copolymerization, the random copolymerization of MAHE and MMA also proceeded smoothly to generate their copolymers. As shown in Table 1.1.3, no matter how the feed amount ratio of the two monomers changed

TABLE 1.1.2
Anionic Block Copolymerization of MAHE and MMA[a]

	Polymerization Sequence	$[DPHL]_0$ (mM)	$[M1]_0$ (M)	$[M2]_0$ (M)	First Step		Second Step		W_{MAHE}/W_{MMA} [c]
					$10^{-4}M_n$ [b]	M_w/M_n [b]	$10^{-4}M_n$ [b]	M_w/M_n [b]	
block-1	MAHE→MMA	8.4	0.30	1.50	0.64 (0.62)	1.06	2.56(2.41)	1.05	27/73 (25/75)
block-2	MAHE→MMA	4.2	0.30	0.50	1.32 (1.21)	1.04	2.60 (2.40)	1.06	51/49 (50/50)
block-3	MAHE→MMA	3.3	0.30	0.75	1.68 (1.54)	1.05	3.95 (3.81)	1.04	39/61 (40/60)
block-4	MMA→MAHE	6.3	0.50	0.30	1.08 (0.82)	1.05	1.96 (1.61)	1.05	48/52 (50/50)
block-5	MMA→MAHE	3.3	0.33	0.60	1.12 (1.03)	1.05	4.56 (4.05)	1.04	26/74 (25/75)

[a] The copolymerization was carried out by a sequential monomer addition of MAHE followed by MMA or vice versa, in THF, in the presence of LiCl ($[LiCl]/[DPHL]_0 = 3$), at −70°C. The polymerization times for MAHE and MMA were 50 and 30 min, respectively. The polymer yield was quantitative in each case.

[b] Determined by GPC. The data in parentheses are the calculated M_n.

[c] The weight ratio of MAHE and MMA segments in the resulting copolymer, which was determined by ^1H NMR. The data in parentheses are the calculated weight amount ratios of the two monomers.

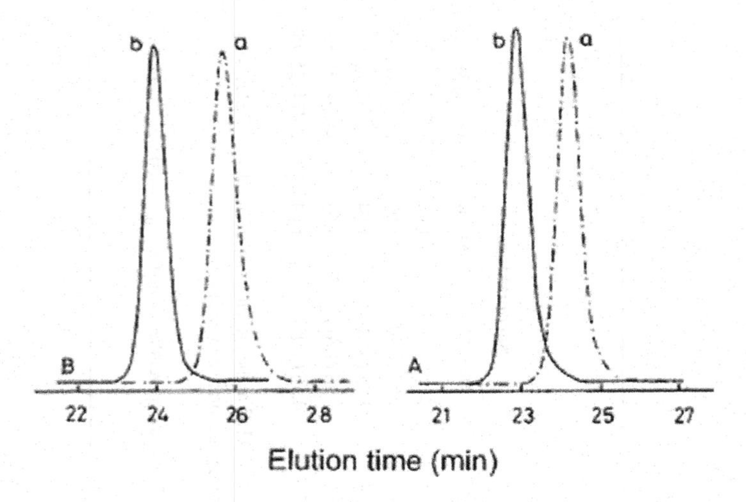

FIGURE 1.1.4 GPC traces of block copolymers and their precursors. A-a: living poly(MAHE) (see block-3 in Table 1.1.2, M_n = 16,800, M_w/M_n = 1.05). A-b: block copolymer, poly(MAHE-*b*-MMA) (block-3 in Table 1.1.2, M_n = 39,500, M_w/M_n = 1.04) obtained via the successive anionic polymerization of MMA from A-a. B-a: living poly(St) (see StMA-1 in Table 1.1.4, M_n = 4900, M_w/M_n = 1.05). B-b: block copolymer, poly(St-*b*-MAHE) (StMA-1 in Table 1.1.4, M_n = 19,500, M_w/M_n = 1.04) obtained via the anionic polymerization of MAHE from B-a.

(W_{MAHE}/W_{MMA} = 30/70, 40/60, 50/50, and 67/33), the monomer conversion was 100%, and any of the copolymers possessed monodispersity (M_w/M_n = 1.04–1.05) as well as well-controlled composition. As illustrated in Figure 1.1.1B, the GPC chromatogram of a copolymer (random-4 in Table 1.1.3, M_n = 66,900, M_w/M_n = 1.04) exhibits a single, sharp peak, which is as narrow as that of homopoly(MAHE) (peak A).

TABLE 1.1.3
Anionic Copolymerization of MAHE with MMA[a]

	$[DPHL]_0$ (mM)	$[MAHE]_0$ (M)	$[MMA]_0$ (M)	$10^{-4}M_n$ (calcd)	$10^{-4}M_n$ (obsd)[b]	M_w/M_n[b]	W_{MAHE}/W_{MMA}[c]
random-1	6.8	0.24	0.60	1.49	1.59	1.05	40/60 (40/60)
random-2	4.0	0.30	0.50	2.52	2.81	1.05	52/48 (50/50)
random-3	4.0	0.18	0.70	2.52	2.95	1.04	33/67 (30/70)
random-4	1.3	0.30	0.25	5.80	6.69	1.04	65/35 (67/33)

[a] The anionic copolymerization was carried out in THF, in the presence of LiCl ($[LiCl]/[DPHL]_0$ = 3), at −78°C, for 50 min. The polymer yield was quantitative in each case.
[b] Determined by GPC.
[c] Weight ratio of MAHE and MMA units in the copolymer, determined by ^1H NMR. The data in parentheses are the feed amount ratios of the two monomers.

1.1.3.3 Anionic Block Copolymerization of St and MAHE

In contrast to the block copolymerization of MAHE with MMA, the diblock copolymer of St and MAHE can only be obtained by using a one-way polymerization sequence of St followed by MAHE, due to the large difference between the reactivities of the two monomers.[4d] The anionic polymerization of St was carried out using *n*-BuLi as the initiator, in a mixture of toluene and THF (3:1 by volume), at −55°C, for 50 min. Then, a THF solution of LiCl ($mol_{LiCl}/mol_{n\text{-BuLi}} = 3$) and of DPE ($mol_{DPE}/mol_{n\text{-BuLi}} = 1.4$) was added to change the living end to a bulkier and less reactive diphenyl carbanion.[7] Subsequently, the anionic polymerization of MAHE proceeded from this DPE-capped living site by adding MAHE to the above system. As shown in Figure 1.1.4B and Table 1.1.4, a new single peak (B-b) due to the block copolymer (StMA-1 in Table1.1.4) appeared in the higher molecular weight side after the polymerization of MAHE, and the peak (B-a) of its living poly(St) precursor (M_n = 4900, M_w/M_n = 1.05) disappeared. The molecular weight of the resulting copolymer (M_n = 19,500, M_w/M_n = 1.04) increased and was close to the designed value [M_n(calcd) = 20,100], whereas the MWD remained very narrow. Consequently, a pure diblock copolymer, poly(St-*b*-MAHE), free of its precursor polymers, was obtained. In addition, the weight ratios of St and MAHE segments in the block copolymers determined by ^{1}H NMR were very close to the feed amount ratios of the two monomers (Table 1.1.4).

1.1.3.4 Polymer Reactions

The diene compounds possess multiple and useful reactivities. Just as St, the dienes, especially 1,3-butadiene and isoprene, constitute some of the most important nonpolar monomers, and their polymerizations have been extensively investigated.[8] Besides their polymerization capability, they can also undergo many other reactions. The addition reaction of a diene with a halogen or a halogen hydride produces halogenated compounds.[9] The osmylation of a diene with osmium tetroxide generates hydroxylated products. Another most widely known reaction is the Diels–Alder reaction between conjugated diene and unlimited number of dienophiles, such as maleic anhydride, acrolein, alkyl propenoate, cyanoethene, etc.[10]

Because of its very reactive properties, dienyl groups have been recently introduced at the ends of the common polymers, such as poly(MMA), to prepare end-functional polymers or macromonomers.[11] The present paper intends to develop an approach for new functional polymers with diene side groups and further to other new functional polymers. Obviously, numerous monodisperse functional polymers can be obtained through the reactions between the anionically prepared (co)polymers of MAHE with other reagents. As examples, two kinds of reactions are reported in what follows.

1.1.3.5 Bromination

The addition between a conjugated diene and the same molar amount of bromine usually generates both 1,2- and 1,4-addition products.[9] If an excess amount of bromine is employed, the remaining single C=C bond can further react with Br_2 to generate tetrabrominated products. As shown in Table 1.1.5, the bromination

TABLE 1.1.4
Anionic Block Copolymerization of St with MAHE

	Living Poly(St)[a]				Block Copolymer[b]				
	$[n\text{-BuLi}]_0$ (mM)	$[St]_0$ (M)	$10^{-4}M_n$[c]	M_w/M_n[c]	$[\text{Living PSt}]_0$ (mM)	$[MAHE]_0$ (M)	$10^{-4}M_n$[c]	M_w/M_n[c]	W_{St}/W_{MAHE}[d]
StMA-1	13	0.60	0.49 (0.49)	1.05	5.0	0.45	1.95 (2.01)	1.04	26/74 (25/75)
StMA-2	10	0.96	1.10 (1.00)	1.04	4.4	0.27	2.01 (2.02)	1.05	48/52 (50/50)
StMA-3	6.7	0.96	1.68 (1.50)	1.06	3.3	0.10	2.24 (2.01)	1.07	76/24 (75/25)

[a] The anionic block copolymerization was carried out using a polymerization sequence of St followed by MAHE. The anionic polymerization of St was performed using *n*-BuLi as initiator, in a mixture of toluene and THF (3/1 by volume), at –55°C, for about 50 min.

[b] After the polymerization of St, a THF solution of DPE ($[DPE]/[\text{living PSt}]_0 = 1.4$) and LiCl ($[LiCl]/[\text{living PSt}]_0 = 3$) was added. In this manner, the living poly(St) solution was diluted, and its living end was capped with DPE. Then, MAHE was added and polymerized at –70°C for 40 min. The copolymer yield was 100% in each case.

[c] Determined by GPC. The data in parentheses are the calculated M_n.

[d] The weight ratio of St and MAHE segments in the copolymer, determined by ^1H NMR. The data in parentheses are the feed amount ratios of the two monomers.

TABLE 1.1.5
Bromination of the (Co)polymers[a]

	Precursor/g[b]	C═C (mmol)[c]	Br_2(mmol)	Time (h)	Functionality (%)[d]
Br-PMAHE-2	PMAHE-2/1.0	12	36	20	87
Br-PMAHE-2′	PMAHE-2/0.5	6	36	75	95
Br-random-3	random-3/1.0	3.6	20	75	92
Br-block-2	block-2/1.0	6	29	75	90
Br-StMA-2	StMA-2/1.0	6	39	54	93

[a] A polymerization solution was directly employed and was diluted to 5% (w/v) with THF. The reaction was carried out by adding bromine with a dry syringe under nitrogen with magnetic stirring.
[b] See Tables 1.1.1 through 1.1.4.
[c] The molar amount of single double bonds, which is twice that of the conjugated diene side chains.
[d] Determined by ^{1}H NMR.

of the (co)polymers of MAHE was carried out by reacting the (co)polymer with a large excess amount of bromine ($mol_{Br_2}/mol_{C=C}$ = 3–6.5) with the goal to obtain poly(2,3,4,5-tetrabromohexyl methacrylate) [poly(TBHMA), **3** in Scheme 1.1.1] or its copolymer. The polymerization solution was directly used for the reaction with Br_2 at room temperature. Because the atomic weight of bromine is large (80) and the number of atoms added to a single repeating unit can be up to 4, the polymer weight was greatly increased after reaction. For instance, after 1.0 g of PMAHE-2 (Table 1.1.5) was reacted with 2.9 g of bromine for 20 h, 2.6 g of adduct was recovered. The functionality calculated for this polymer yield was 89%, which was close to that (87%) determined by ^{1}H NMR. Regardless if the homopolymer of MAHE or its copolymer with St or MMA was used for bromination, it was found that the functionality easily reached about 90% but hardly reached 100%. Figure 1.1.3C presents the ^{1}H NMR spectrum of a brominated poly(MAHE) (Br-PMAHE-2′ in Table 1.1.5), which was obtained by reacting 0.5 g of PMAHE-2 ($mol_{C=C}$ = 6 mmol) with 2.9 g of Br_2 (36 mmol) for 75 h. Comparing Figure 1.1.3C with the spectrum of its poly(MAHE) precursor (Figure 1.1.3B), one can observe that the conjugated diene (peaks d–g) are no longer present after bromination. However, a small fraction (5%) of single C═C bonds (5.75–6.14 ppm, Figure 1.1.3C) remained in the reaction product. The functionality calculated according to the intensity ratio was 95%.

Parts A, B, and C of Figure 1.1.5 depict the FT-IR spectra of MAHE, poly(MAHE), and brominated poly(MAHE), respectively. In the spectrum of the monomer (A), the absorption of the olefinic bond conjugated with the carbonyl group appeared at 1637 cm^{-1} (A-b). The olefinic bond stretching vibration of the conjugated diene produces two C═C stretching bands because of the absence of a center of symmetry. One absorption is located at 1662 cm^{-1} (A-a), and the other one (A-a′, 1630 cm^{-1}) is overlapped with that (A-b) of the methacryloyl type C═C bond. After selective polymerization, the absorption (b) of the methacryloyl type C═C bonds disappeared, while those (a and a′, Figure 1.1.5B) corresponding to the diene groups remained unchanged. After this sample was subjected to bromination (Figure 1.1.5C), the two peaks of the diene group disappeared almost completely (functionality 95%).

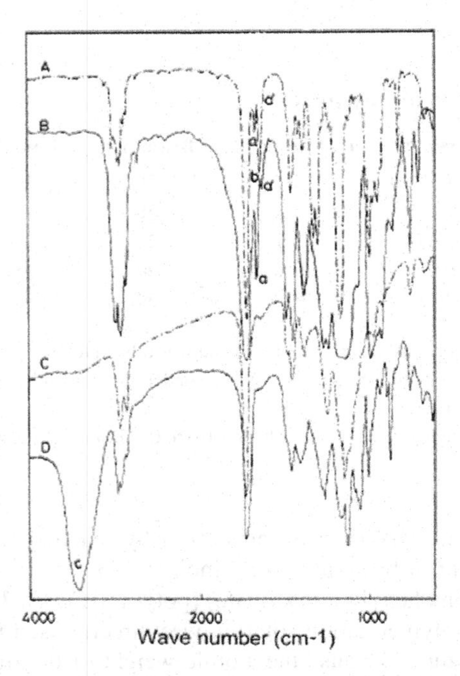

FIGURE 1.1.5 FT-IR spectra of MAHE (A), PMAHE-2 (B; Table 1.1.1), and its brominated (C; Table 1.1.5) and osmylated (D; Table 1.1.6) products.

1.1.3.6 Osmylation

We have demonstrated that the osmylation of the C=C bond-containing polymer constitutes a perfect approach to functional polymers with excellent hydrophilicities.[3b] During this process, the C=C bond reacted with NMO in the presence of a trace amount of osmium tetroxide as catalyst. The dihydroxylation of C=C bond proceeded smoothly without cross-linking or any other side reaction, and the functionality reached 100% under suitable conditions. However, the polymers previously subjected to osmylation contained only a single C=C bond in each repeating unit, which was changed to two hydroxyl groups. In the present paper, the osmylation of poly(MAHE), which contains a diene group in each of its repeating units (two single C=C bonds), was carried out to prepare the new functional polymer, poly(2,3,4,5-tetrahydroxyhexyl methacrylate) [poly(THHMA), **4** in Scheme 1.1.1].

The osmylation was carried out with magnetic stirring, under nitrogen, at room temperature. The polymerization solution was directly subjected to reaction without polymer separation. Compared to the molar amount of C=C bonds, an excess of NMO ($mol_{NMO}mol_{C=C}$ = 1.5–1.8) and a small amount of OsO_4 ($mol_{OsO_4}/mol_{C=C}$ = 1.3–1.8/100) were employed (Table 1.1.6). Figure 1.1.3D presents the ^{1}H NMR spectrum of an osmylated poly(MAHE) (o-PMAHE-2 in Table 1.1.6). Compared to its poly(MAHE) precursor (Figure 1.1.3B), peaks d–g corresponding to the diene side groups of poly(MAHE) disappeared completely after osmylation. The FT-IR

TABLE 1.1.6
Osmylation of the (Co)polymers[a]

	Precursor[b] (g)	C=C (mmol)[c]	NMO (mmol)[d]	OsO_4[e] (mL)	Solvents (by vol [mL])[f] THF/toluene/acetone/CH_3OH	Time (h)	Functionality[g] (%)
o-PMAHE-2	PMAHE-2/0.5	6	10	0.7	10/0/3/7	24	100[h]
o-PMAHE-3	PMAHE-3/1.0	12	18	1.5	16/0/3/10	48	100[h]
o-block-5	block-5/0.53	4.8	8	0.5	10/0/4/4	36	100
o-StMA-1	StMA-1/0.5	4.6	7	0.5	8/2/3/4	28	100
o-StMA-2	StMA-2/0.53	3.2	6	0.5	10/2/2/4	26	100

[a] The reaction was carried out with magnetic stirring, under nitrogen, at room temperature. The polymer yield was 100% in each case.
[b] A polymerization solution containing a certain amount of (co)polymer was directly employed.
[c] Molar amount of single C=C bonds, which is twice that of the conjugated diene side chains of poly(MAHE) units.
[d] Molar amount of *N*-methylmorpholine *N*-oxide (NMO).
[e] A 4.0 wt% of aqueous solution, which contained 1.3–1.8 mol% of OsO_4 compared to the molar amount of C=C bonds in MAHE segments, was employed.
[f] To a polymerization solution, THF, acetone, and methanol were sequentially added.
[g] Determined by ^{1}H NMR.
[h] The quantitative functionality was also confirmed by elemental analysis.

analysis also confirmed the quantitative hydroxylation. Figure 1.1.5D presents the FT-IR spectrum of o-PMAHE-2. After osmylation, peaks a and a′ (Figure 1.1.5B) due to the conjugated diene C=C double bond of poly(MAHE) disappeared, and a strong, broad band, D-c, corresponding to the hydroxyl groups emerged. For the osmylated block copolymers, similar NMR and FT-IR results were obtained.

As shown in Table 1.1.7, the elemental analysis results of homopoly(MAHE) and its osmylated product are consistent with the calculated values. The calculations

TABLE 1.1.7
Elemental Analyses of Poly(MAHE) and Its Osmylated Product

		C (%)		H (%)		N (%)	
		Theory	Found	Theory	Found	Theory	Found
PMAHE-3[a]	$C_{10}H_{14}O_2$	72.24	72.70	8.51	8.57		
o-PMAHE-3[b]	$C_{10}H_{18}O_6 \cdot 0.2H_2O$[c]	50.49	50.51	7.81	7.87	0.00	0.00

[a] See Table 1.1.2.
[b] See Table 1.1.6.
[c] Each repeating unit contains 0.2 water molecule.

indicated that each repeating unit of poly(THHMA) contained 0.2 water molecule. On the other hand, even though a nitrogen-containing compound, NMO, was employed in the osmylation process, the elemental analysis did not indicate the presence of nitrogen in the resulting polymer. Consequently, the reagents were completely removed during the purification process, and a very pure poly(THHMA) was obtained. The combination of ^{1}H NMR, FT-IR, and the elemental analysis clearly indicates that this osmylation procedure is not only suitable for single C=C bond-containing polymer but also suitable for the polymers with conjugated diene side groups.

1.1.3.7 Solubility

Poly(MAHE) and its osmylated product poly(THHMA) exhibit different solubilities. For instance, as shown in Table 1.1.8, PMAHE-3 is soluble in benzene, chloroform, DMF, pyridine, 1,4-dioxane, THF, and acetone but insoluble in ethanol, methanol, and water. For poly(THHMA) (o-PMAHE-3 in Table 1.1.8), water became the best solvent, and the dissolution was completed in only a few seconds. Ethanol and methanol are poor solvents because their polarity is less than that of water. Poly(THHMA) is insoluble in ethanol and can only partially dissolve in methanol at room temperature, although heating can accelerate its dissolution rate. For comparison, solubility experiments for poly(HEMA) and poly(DHPMA) were also carried out. Compared to the four hydroxyl groups in each repeating unit of poly(THHMA), poly(HEMA) and poly(DHPMA) possess only one and two hydroxyl groups in each of their repeating units, respectively. As shown in Table 1.1.8, poly(HEMA) is soluble in ethanol and methanol but insoluble in water. While poly(DHPMA) is insoluble in ethanol, methanol is its best solvent and it can dissolve slowly in water. It is clear that the hydrophilicity of the polymers increases as the hydroxyl number in the side chain increases, and poly(THHMA) possesses the highest water solubility. In addition, as shown in Table 1.1.8, the osmylated block copolymers possess quite different solubilities compared to those of their precursors. They are soluble only in DMF and pyridine and become swollen in water.

1.1.4 CONCLUSION

The new bifunctional monomer, *trans, trans*-1-methacryloyloxy-2,4-hexadiene (MAHE), underwent selective living anionic polymerization without cross-linking and any other side reaction, generating monodisperse polymers ($M_w/M_n = 1.04$–1.05) with well-controlled molecular weights. The 2,4-hexadienyl side groups of the resulting polymers could further react with bromine. This bromination introduced four bromine atoms into each of the repeating units of poly(MAHE). The diene side groups of poly(MAHE) could also react with *N*-methyl-morpholine *N*-oxide in the presence of a trace amount of osmium tetroxide as catalyst. This osmylation changed poly(MAHE) to a new functional polymer, poly(2,3,4,5-tetrahydroxyhexyl methacrylate), which possessed extremely high water solubility because of the presence of four hydroxyl groups in each of its repeating units. Under suitable conditions,

TABLE 1.1.8
Solubilities of the (Co)polymers[a]

	Hexane	Bz	$CHCl_3$	DMF	Pyridine	1,4-Dioxane	THF	Acetone	C_2H_5OH	CH_3OH	H_2O
PMAHE-3	P	S	S	S	S	S	S	S	I	I	I
o-PMAHE-3	I	I	I	S	D	I	I	I	I	D, P[b]	S[c]
PDHPMA[d]	I	I	I	S	S	I	I	I	I	S	S
PHEMA[e]	I	I	I	S	S	I	I	I	S	S	I
Br-PMAHE-2′	I	S	S	S	S	C	S	P	I	I	I
block-4	I	S	S	S	S	S	S	S	I	I	I
o-block-4	I	I	I	S	D	I	I	I	I	D	SW
StMA-2	I	S	S	S	S	S	S	S	I	I	I
o-StMA-2	I	I	I	S	S	I	I	I	I	I	SW

[a] The experiment was carried out at room temperature (23°C). The amounts of polymer and the solvent were 0.03 g and 1.0 mL, respectively. S = soluble; D = dissolved slowly; P = partially soluble; C = cloudy; SW = swollen; I = insoluble.
[b] Only partially dissolved after 24 h.
[c] Dissolved completely in a few seconds.
[d] Poly(2,3-dihydroxypropyl methacrylate) prepared in a previous paper [ref 3b].
[e] Poly(2-hydroxyethyl methacrylate); see ref 2.

well-defined block and random copolymers of MAHE with MMA and the block copolymer of MAHE with St were also prepared. These copolymers possessed controlled molecular weights and compositions as well as monodispersity. The osmylation after the block copolymerization generated well-defined novel amphiphilic block copolymers.

REFERENCES

1. (a) Green, J. *J. Fire Sci.* **1996**, *14*, 426. (b) Camps, M.; Jebri, A.; Dronet, J. C. *J. Fire Sci.* **1996**, *14*, 251.
2. (a) Hirao, A.; Kata, H.; Yamaguchi, K.; Nakahama, S. *Macromolecules* **1986**, *19*, 1294. (b) Ruckenstein, E.; Zhang, H. M. *J. Polym. Sci., Polym. Chem.* **1998**, *36*, 1865.
3. (a) Mori, H.; Hirao, A.; Nakahama, S. *Macromolecules* **1994**, *27*, 35. (b) Zhang, H. M.; Ruckenstein, E. *Macromolecules* **2000**, *33*, 4738.
4. (a) Noshay, A.; McGrath, J. E. *Block Copolymerization*; Academic Press: New York, 1977. (b) Van Beylen, M.; Bywater, S.; Smets, G.; Szwarc, M.; Worsfold, D. J. *Adv. Polym. Sci.* **1988**, *86*, 87. (c) Morton, M. *Anionic Polymerization: Principle and Practice*; Academic Press: New York, 1983. (d) Hsieh, H. L.; Quirk, R. P. *Anionic Polymerization*; Marcel Dekker: New York, 1996.
5. Fayt, R.; Forte, R.; Jacobs, C.; Jerome, R.; Ouhadi, T.; Teyssie, Ph.; Varshney, S. K. *Macromolecules* **1987**, *20*, 1442.
6. Allen, R. D.; Long, T. E.; McGrath, J. E. *Polym. Bull.* **1986**, *15*, 127.
7. (a) Schulz, G.; Hocher, H. *Angew. Chem., Int. Ed. Engl.* **1980**, *19*, 219. (b) Ruckenstein, E.; Zhang, H. M. *Macromolecules* **1998**, *31*, 9127.
8. For examples: (a) Kobatake, S.; Harword, H. J.; Quirk, R. P.; Priddy, D. B. *Macromolecules* **1999**, *32*, 10. (b) Pispas, S.; Allorio, S.; Hadjichristidis, N.; Mays, J. W. *Macromolecules* **1996**, *29*, 2903.
9. Vollhardt, K. P. C.; Schore, N. E. *Organic Chemistry*; Freeman, W. H. and Company: New York, 1994; p. 515.
10. (a) Diels, O.; Alder, K. Ann. **1928**, *460*, 98. (b) Diels, O.; Alder, K. *Ber.* **1929**, *62*, 2081. (c) Solomons, T. W. G. *Organic Chemistry*; John Wiley & Sons: New York, 1992; p. 501.
11. (a) Mizawa, T.; Takenaka, K.; Shiomi, T. *J. Polym. Sci., Polym. Chem.* **1999**, *37*, 3464. (b) Mizawa, T.; Takenaka, K.; Shiomi, T. *J. Polym. Sci., Polym. Chem.* **2000**, *38*, 237.

1.2 A Novel Successive Route to Well-Defined Water-Soluble Poly(2,3-dihydroxypropyl methacrylate) and Amphiphilic Block Copolymers Based on an Osmylation Reaction*

Hongmin Zhang and Eli Ruckenstein
Chemical Engineering Department, State University of New York at Buffalo, Amherst, New York 14260

ABSTRACT A novel two-step successive route to well-defined water-soluble poly(2,3-dihydroxypropyl methacrylate) [poly(DHPMA)] and the amphiphilic block copolymers of DHPMA with styrene (St) or methyl methacrylate (MMA) was developed by combining the living anionic polymerization method with an osmylation reaction. Homopolymers of allyl methacrylate (AMA) with various molecular weights (M_n = 5900–170, 000) and narrow molecular weight distributions (MWD; M_w/M_n = 1.06–1.16) were first prepared by the anionic polymerization of AMA using 1,1-diphenylhexyllithium (DPHL) as the initiator, in THF, in the presence of LiCl ([LiCl]/[DPHL]$_0$ = 3), at −70°C. Under similar conditions, the anionic block copolymerization of AMA with MMA generated well-defined di- and triblock copolymers with controlled molecular weights and compositions as well as narrow MWDs (M_w/M_n = 1.05–1.16). Further, block copolymers of St and AMA with various compositions were prepared by performing the anionic block copolymerization of AMA from a 1,1-diphenylethylene-capped anionic living poly(St). In both the homopolymerization of AMA and its block copolymerization with MMA or St, its allyl side group remained unreacted.

* *Macromolecules* 2000, 33, 4738–4744.

Without the (co)polymer separation from the (co)polymerization solution, the allyl groups of the (co)polymer were directly reacted with an excess of *N*-methylmorpholine *N*-oxide (NMO), in the presence of a small amount of osmium tetroxide as catalyst, at room temperature. This osmylation procedure changed the poly(AMA) component to poly(DHPMA). During this process, no cross-linking or polymer chain damage occurred, and well-defined water-soluble poly(DHPMA) and amphiphilic block copolymers were obtained. The molecular weight, molecular architecture, and the copolymer composition affect the hydrophilicity of the osmylated (co)polymers, and all these factors can be accurately controlled during the (co)polymerization step.

1.2.1 INTRODUCTION

A number of amphiphilic block copolymers have been prepared via the living anionic polymerization technique.[1] In these copolymers, poly(methacrylic acid) and poly(2-hydroxyethyl methacrylate) [poly(HEMA)] were most usually used as hydrophilic segments,[2] although they are not water-soluble. Mori et al.[3] reported an anionic synthetic method of poly(2,3-dihydroxypropyl methacrylate) [poly(DHPMA)] using a three-step approach. First, a protected monomer, (2,2-dimethyl-1,3-dioxalan-4-yl) methyl methacrylate, was prepared through the reaction between isopropylidene glycerol and methacryloyl chloride. Then, the protected monomer was subjected to anionic polymerization. Finally, the acetal protecting group was eliminated by hydrolysis to obtain poly(DHPMA). Compared to poly(HEMA), poly(DHPMA) is more hydrophilic and even water-soluble, because of the presence of two hydroxyl groups in each of its repeating units. In addition, the block copolymer containing poly(DHPMA) as hydrophilic segment exhibits interesting surface properties.[3b]

In the present paper, a more convenient and successive route to well-defined poly(DHPMA) and further to several kinds of amphiphilic block copolymers containing poly(DHPMA) hydrophilic segments was developed. As an example, the approach to the amphiphilic block copolymer consisting of polystyrene [poly(St)] and poly(DHPMA) is presented in Scheme 1.2.1. The anionic living poly(St) was reacted with 1,1-diphenylethylene (DPE) to obtain a DPE-capped living poly(St) (**2** in Scheme 1.2.1), from which the anionic block copolymerization of allyl methacrylate (AMA) was performed, generating a block copolymer, poly(St-b-AMA) (**3** in Scheme 1.2.1). During this process, the allyl side group of AMA remained unreacted. Without polymer separation from the polymerization solution, the allyl side groups were directly reacted with *N*-methylmorpholine *N*-oxide (NMO) in the presence of a trace amount of osmium tetroxide. This osmylation changed the poly(AMA) segment to poly(DHPMA), and an amphiphilic block copolymer, poly(St-b-DHPMA) (**4** in Scheme 1.2.1), was thus obtained. This osmylation process was also applied to homopoly(AMA), and in this case, a well-defined water-soluble poly(DHPMA) was prepared. Furthermore, the block copolymerization of AMA and MMA, followed by the successive osmylation, generated several di- and triblock amphiphilic copolymers, such as poly(DHPMA-*b*-MMA), poly(MMA-*b*-DHPMA-*b*-MMA), and poly(DHPMA-*b*-MMA-*b*-DHPMA) This (co) polymer synthetic route has the following advantages: (i) all the monomers employed are commercially available; (ii) only two synthetic steps are involved and can be performed

n-Bu{ CH2-CH}nCH2-CH⁻ Li⁺ —DPE / LiCl→ poly(St)-CH2-C⁻ Li⁺

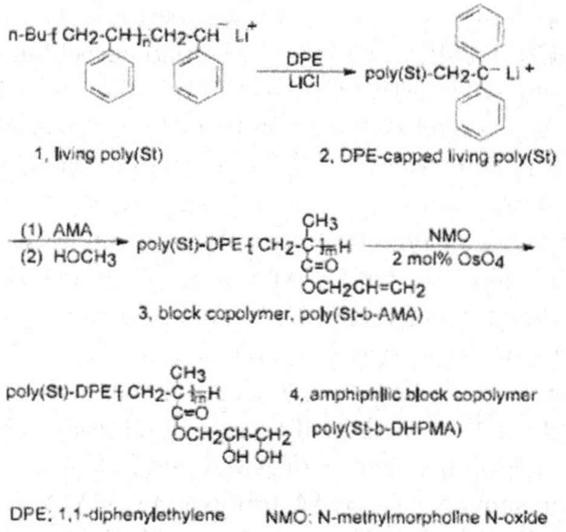

1, living poly(St)

2, DPE-capped living poly(St)

(1) AMA (2) HOCH3 → poly(St)-DPE{ CH2-C }mH (CH3; C=O; OCH2CH=CH2) —NMO / 2 mol% OsO4→

3, block copolymer, poly(St-b-AMA)

poly(St)-DPE{ CH2-C }mH (CH3; C=O; OCH2CH-CH2; OH OH)

4, amphiphilic block copolymer poly(St-b-DHPMA)

DPE: 1,1-diphenylethylene NMO: N-methylmorpholine N-oxide

SCHEME 1.2.1

successively; (iii) this is a general method which is expected to be suitable for a number of polymers containing C=C bonds in their side groups, such as 1,2-poly(butadiene),[4] poly[4-(vinylphenyl)-1-butene],[5] etc.

1.2.2 EXPERIMENTAL SECTION

1.2.2.1 Materials

Tetrahydrofuran (THF) was dried with CaH_2 under reflux for more than 24 h, distilled, and distilled again from a solution of 1,1-diphenylhexyllithium (DPHL) just before use. Toluene was washed with concentrated sulfuric acid and then with water, dried over $MgSO_4$, distilled over CaH_2, and distilled again from a DPHL solution before use. Hexane was first dried and distilled over CaH_2 and then distilled from a solution of *n*-BuLi. Methyl methacrylate (MMA; Aldrich, 99%) and allyl methacrylate (AMA; Aldrich, 98%) were dried over CaH_2 with magnetic stirring under reduced pressure, vacuum distilled over CaH_2, and finally distilled in the presence of Al(*sec*-Bu)$_3$.[6] Styrene (St; 99%, Aldrich) was washed with 10% aqueous NaOH solution and then with water, dried overnight with $MgSO_4$, distilled over CaH_2, and finally distilled in the presence of phenylmagnesium chloride just before use. 1,1-Diphenylethylene (DPE; Aldrich, 97%) was distilled over CaH_2 and then distilled in the presence of DPHL under reduced pressure. Lithium chloride (Aldrich, 99.99%) was dried at 120°C for 24 h and dissolved in THF.[7] *n*-BuLi (Aldrich, 1.6 M solution in hexane) was diluted with purified hexane. Osmium tetroxide (Aldrich, 2.5 wt% in 2-methyl-2-propanol), *N*-methylmorpholine *N*-oxide (NMO; Aldrich, 97%), and benzoic anhydride (Aldrich, 90%) were used as received.

1.2.2.2 Polymerization

All polymerizations, namely, the anionic homopolymerization of AMA and its block copolymerization with MMA or St, were carried out in a round-bottom flask under an overpressure of argon with magnetic stirring.

The anionic homopolymerization of AMA was performed in THF, at −70°C, in the presence of LiCl ($[LiCl]/[DPHL]_0 = 3$).[8] Before monomer addition, the initiator DPHL was first prepared in situ. After THF, DPE, and a THF solution of LiCl were added with dry syringes, the flask was cooled to −40°C, and *n*-BuLi (in hexane) was added. The reaction between *n*-BuLi and DPE was allowed to last 20 min. Then, the system was cooled to −70°C, and the polymerization reaction was induced by the addition of prechilled AMA to the above system. One hour later, the polymerization system was quenched with a small amount of methanol (ca. 1 mL). After a certain amount of the polymerization solution was taken out and directly subjected to the osmylation reaction, the remaining part of the polymerization solution was poured into a mixture of water and methanol (4:1 by volume) to precipitate the polymer. The polymer was washed with methanol, vacuum-dried overnight at 40°C, and finally freeze-dried from its benzene solution.

The anionic block copolymerization of AMA and MMA was carried out by using the monomer addition sequence of AMA followed by MMA, or vice versa, in THF, in the presence of LiCl ($[LiCl]/[DPHL]_0 = 3$), at −70°C. For the preparation of triblock copolymers of AMA and MMA, a three-step sequential monomer addition, namely, AMA → MMA → AMA or MMA → AMA → MMA, was adopted. The polymerization times for MMA and AMA were 40 and 60 min, respectively. The initiator preparation and the polymer purification were performed in the ways similar to those employed for the homopolymer of AMA.

The block copolymer of AMA and St was prepared using a polymerization sequence of St, followed by AMA. The living poly(St) was first prepared by the anionic polymerization of St, which was carried out in a mixture of toluene and THF (1:1 by volume), at −60°C, using *n*-BuLi as the initiator. After toluene, THF, and a hexane solution of *n*-BuLi were introduced into a flask kept at −60°C, the polymerization was started by adding prechilled St to the above system. While the polymerization was proceeding, a certain amount of THF, DPE ($mol_{DPE}/mol_{n\text{-}BuLi} = 1.2$), and a THF solution of LiCl ($mol_{LiCl}/mol_{n\text{-}BuLi} = 3$) were introduced into another flask, to which a hexane solution of *n*-BuLi was dropwise added until the red color of DPHL appeared, to remove impurities. Then, this mixture was immediately introduced into the living poly(St) solution. The color of the system changed instantaneously from yellow to deep red, implying that the living end of poly(St) was rapidly capped by DPE. This reaction was allowed to last 20 min to ensure a complete transformation of the living site. Subsequently, the prechilled AMA was added, and its polymerization lasted 1 h. After the termination with methanol (ca. 1 mL), a fraction of the polymerization solution was directly subjected to osmylation. For the remaining part, the precipitation and purification of the polymer were carried out in the ways similar to those for homopoly(AMA).

1.2.2.3 Osmylation

The osmylation reactions of the homopolymer of AMA and of its block copolymer with MMA or St were carried out with magnetic stirring, under the protection of nitrogen, at room temperature, for 24 h. Upon the (co)polymerization, a solution containing a certain amount of (co)polymer was transferred into a nitrogen-protected flask, to which acetone, methanol, water, and NMO (solid) were sequentially added. Then, the system was degassed and reprotected with nitrogen, and the reaction was started by adding osmium tetroxide (2.5 wt% solution in 2-methyl-2-propanol) with a syringe. As the reaction was proceeding, the system became turbid gradually. To keep the reaction proceeding homogeneously, a small amount of methanol was added. After 24 h, the resulting (co)polymer was precipitated by pouring the reaction solution into a mixture of hexane and ethanol (1:1 by volume). The (co)polymer was kept in this mixture overnight, washed with ethanol, vacuum-dried at 40°C for 24 h, and finally freeze-dried from a 1,4-dioxane solution containing a small amount of methanol.

1.2.2.4 Reprotection of the Hydroxylated (Co)polymer

To verify by GPC whether the osmylation affected the main chain structures of the (co)polymers and/or whether the cross-linking occurred during the process, the osmylated products were reprotected by reacting the formed hydroxyl groups with benzoic anhydride. A small amount of osmylated PAMA-2 (Table 1.2.1; 0.35 g) was dissolved in 12 mL of pyridine, to which 4.2 g of benzoic anhydride was added, and the reaction was allowed to last 24 h at room temperature. Then, the polymer was precipitated into methanol, washed with methanol, and vacuum-dried overnight.

1.2.2.5 Measurements

^1H and ^{13}C NMR spectra were recorded in $CDCl_3$, CD_3OD, or a mixture of CD_3OD with THF-d_8 on an Inova-500 spectrometer. M_n and M_w/M_n of the (co)polymer were determined by gel permeation chromatography (GPC) on the basis of a polystyrene calibration curve. The GPC measurements were carried out using THF as solvent, at 30°C, with a 1.0 mL/min flow rate and a 1.0 cm/min chart speed. Three polystyrene gel columns (Waters, 7.8 × 300 mm; one HR 5E, part no. 44228, one Linear, part no. 10681, and one HR 4E, part no. 44240) were used, which were connected to a Waters 515 precision pump. FT-IR spectra were recorded using KBr tablets on a Perkin-Elmer 1760-X spectrometer. The elemental analysis was carried out by Atlantic Microlab, Inc.

TABLE 1.2.1
Anionic Homopolymerization of AMA[a]

			$10^{-4} M_n$		
	$[DPHL]_0$ (mM)	$[AMA]_0$ (M)	calcd	obsd[b]	M_w/M_n[b]
PAMA-1	15.7	0.660	0.55	0.59	1.12
PAMA-2	10.0	0.794	1.02	0.98	1.06
PAMA-3	2.8	0.743	3.34	3.78	1.08
PAMA-4	0.7	0.743	13.4	17.0	1.16

[a] The polymerization was carried out in THF, in the presence of LiCl ($[LiCl]/[DPHL]_0 = 3$), at −70°C, for 1 h. The monomer conversion was 100% in each case.

[b] Determined by GPC.

1.2.3 RESULTS AND DISCUSSION

1.2.3.1 Anionic Homopolymerization of AMA and Its Block Copolymerization with MMA

The anionic homopolymerization of AMA has been systematically investigated in a previous paper.[8] Using bulky DPHL as initiator, in the presence of LiCl, in THF, at a low temperature (–30°C to –70°C), AMA could undergo living anionic polymerization smoothly, generating a uniform size functional polymer with a reactive allyl group in each repeating unit. To investigate the effect of molecular weight on the hydrophilicity of the osmylated poly(AMA), in the present paper, several homopolymers of AMA with different molecular weights (M_n = 6000–170 000) were prepared using DPHL as the initiator, in THF, in the presence of LiCl ([LiCl]/[DPHL]$_0$ = 3), at –70°C. As shown in Table 1.2.1, monomer conversion was 100% in each case, molecular weight was well-controlled, and molecular weight distribution (MWD) was very narrow (M/M_n = 1.06–1.16).

The di- and triblock copolymers of AMA and MMA were prepared under conditions similar to those for homopoly(AMA). For diblock copolymers, a two-step sequential monomer addition, namely, MMA followed by AMA, or vice versa, was employed. As shown in Table 1.2.2, regardless of the polymerization sequence, well-defined block copolymers were obtained in both cases. Molecular weight could be controlled at each step, the MWD of either the precursor or the final copolymer was very narrow (M/M_n = 1.05–1.12), and their composition determined by ^{1}H NMR was in good agreement with that designed. As an example, Figure 1.2.1A presents the GPC chromatograms of poly(AMA-b-MMA) (peak A-b; AbM-1 in Table 1.2.2) and its precursor [peak A-a; poly(AMA), M_n = 11, 200, M_w/M_n = 1.06]. After the second step polymerization of MMA from the living end of poly(AMA), the M_n was almost doubled increased (M_n = 22, 300, M_w/M_n = 1.05), while the MWD remained very narrow. The above results confirmed the living nature of the anionic polymerization of AMA.

To investigate the effects of molecular architectures of the block copolymers on the hydrophilicity of their osmylated products, ABA and BAB type triblock copolymers of AMA and MMA were also prepared using a three-step sequential anionic polymerization technique. As shown in Table 1.2.3, the weight ratios of AMA and MMA for both MbAbM and AbMbA were designed to be W_{AMA}/W_{MMA} = 50/50. However, the former triblock copolymer possesses one poly(AMA) segment in the middle of the polymer chain. On the other hand, the latter triblock copolymer has two poly(AMA) segments at the two sides of the polymer chain. Therefore, after osmylation, the hydrophilic poly(DHPMA) segments will be located in different places, and this might have a contribution to their hydrophilicities. As shown in Table 1.2.3 and Figure 1.2.2, despite a three-step polymerization procedure, the block copolymers thus obtained were almost free of their precursors, their molecular weights and compositions were well-controlled, and their MWDs were narrow (M_w/M_n = 1.14–1.16).

TABLE 1.2.2
Preparation of Diblock Copolymers of AMA (A) and MMA (M) via Their Anionic Copolymerization[a]

					First Step			Second Step				
	Polymerization Sequence	$[DPHL]_0$ (mM)	$[M1]_0$ (M)	$[M2]_0$ (M)	$10^{-4}M_n$ calcd	$10^{-4}M_n$ obsd[b]	M_w/M_n[b]	$10^{-4}M_n$ calcd	$10^{-4}M_n$ obsd[b]	M_w/M_n[b]	W_{AMA}/W_{MMA}[c] calcd	W_{AMA}/W_{MMA}[c] obsd[d]
MbA-1	MMA → AMA	5.00	0.500	0.400	1.02	1.04	1.05	2.03	2.23	1.06	50/50	47/53
MbA-2	MMA → AMA	5.00	0.330	0.530	0.68	0.72	1.08	2.06	2.39	1.11	67/33	66/34
AbM-1	AMA → MMA	5.00	0.400	0.500	1.03	1.12	1.06	2.03	2.23	1.05	50/50	49/51
AbM-2	AMA → MMA	4.00	0.270	0.670	0.87	0.93	1.07	2.55	2.88	1.12	33/67	35/65

[a] The copolymerization was carried out by means of a sequential monomer addition of MMA followed by AMA, or vice versa, in THF, in the presence of LiCl ($[LiCl]/[DPHL]_0 = 3$), at −70°C. The polymerization times for MMA and AMA were 40 and 60 min, respectively. The polymer yield was quantitative in each case.
[b] Determined by GPC.
[c] The weight ratio of AMA and MMA segments in the resulting copolymer.
[d] Determined by ^{1}H NMR.

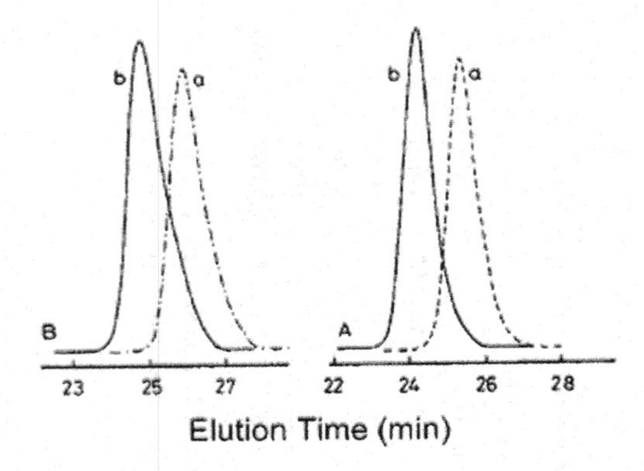

FIGURE 1.2.1 GPC traces of diblock copolymers and their precursors. A-a: living poly(AMA) (see AbM-1 in Table 1.2.2, $M_n = 11,200$, $M_w/M_n = 1.06$). A-b: block copolymer, poly(AMA-*b*-MMA) (AbM-1 in Table 1.2.2, $M_n = 22,300$, $M_w/M_n = 1.05$) obtained via the successive anionic polymerization of MMA from A-a. B-a: living poly(St) (see SbA-1 in Table 1.2.4, $M_n = 6800$, $M_w/M_n = 1.07$). B-b: block copolymer, poly(St-b-AMA) (see SbA-1 in Table 1.2.4, $M_n = 14,100$, $M_w/M_n = 1.09$) obtained via the anionic polymerization of AMA from B-a.

TABLE 1.2.3
Preparation of ABA and BAB Triblock Copolymers of AMA and MMA by a Three-Step Sequential Anionic Copolymerization of the Two Monomers[a]

	Polymerization Sequence (weight [g])[b]	DPHL (mmol)	$10^{-4}M_n$		M_w/M_n[c]	W_{AMA}/W_{MMA}[d]	
			calcd	obsd[c]		calcd	obsd[e]
MbAbM	MMA(2.0) → AMA(4.0) → MMA(2.0)	0.35	2.31	2.54	1.16	50/50	50/50
AbMbA	AMA(2.0) → MMA(4.0) → AMA(2.0)	0.30	2.69	2.92	1.14	50/50	48/52

[a] The anionic copolymerization was carried out using a three-step sequential monomer addition, in THF, in the presence of LiCl ($mol_{LiCl}/mol_{DPHL} = 3$), at −70°C. The total volume of the polymerization system was 60 mL. The polymerization times for MMA and AMA were 40 and 60 min, respectively. The copolymer yield was 100% in each case.

[b] The data in the parentheses are the weights of monomers added to the polymerization system sequentially.

[c] Determined by GPC.

[d] The weight ratio of AMA and MMA segments in the resulting copolymer.

[e] Determined by ^{1}H NMR.

1.2.3.2 Anionic Block Copolymerization of St and AMA

A block copolymer of St and AMA was prepared using the polymerization sequence of St followed by AMA. The anionic polymerization of St was carried out in a mixture of toluene and THF (1:1 by volume), at −60°C, for 1 h. However,

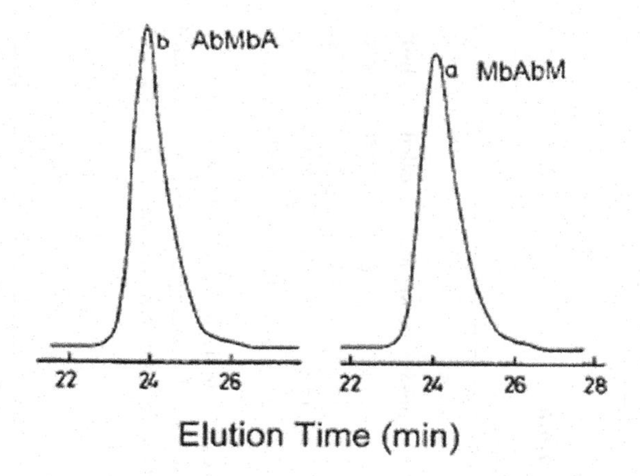

FIGURE 1.2.2 GPC traces of triblock copolymers: (a) poly(MMA-b-AMA-b-MMA) (MbAbM in Table 1.2.3, M_n = 25, 400, M_w/M_n = 1.16). (b) poly(AMA-b-MMA-b-AMA) (AbMbA in Table 1.2.3, M_n = 29, 200, M_w/M_n = 1.14).

the second step living polymerization of AMA proceeded better in a polar solvent (THF) and in the presence of LiCl.[8] In addition, the living site of poly(St) is too reactive and may attack the ester group of AMA, causing unwanted side reactions. Considering the above factors, a THF solution of LiCl ($mol_{LiCl}/mol_{n\text{-BuLi}}$ = 3) and of DPE ($mol_{DPE}/mol_{n\text{-BuLi}}$ = 1.2) was added after the anionic polymerization of St. A rapid addition reaction between the living site of poly(St) and DPE occurred, generating a DPE-capped living poly(St) containing a bulkier and less reactive end carbanion (**2** in Scheme 1.2.1), whose structure is similar to that of DPHL.[9] Subsequently, the anionic polymerization of AMA proceeded from the living site of the DPE-capped poly(St) by adding AMA to the above system. As shown in Figure 1.2.1B and Table 1.2.4 (SbA-1), a new single peak (B-b) corresponding to the block copolymer appeared in the higher molecular weight side after the polymerization of AMA, and the peak (B-a) of its living poly(St) precursor (M_n = 6800, M_w/M_n = 1.07) disappeared. The molecular weight of the block copolymer (M_n = 14, 100, M_w/M_n = 1.09) is about twice as large as that of poly(St) and almost equal to the designed value, and its MWD is almost as narrow as that of its precursor. Consequently, a pure diblock copolymer, poly(St-*b*-AMA), free of homopolymers, was obtained. In addition, the weight ratios of St and AMA segments in the block copolymers determined by ^{1}H NMR were very close to the feed amount ratios of the two monomers (Table 1.2.4).

1.2.3.3 Osmylation of (Co)polymers

As well-known, the reaction of small olefin molecules with osmium tetroxide has been the most reliable method for cis-dihydroxylation of the C=C bond.[10] However, when OsO_4 is used stoichiometrically, its high cost greatly limits its application in organic synthesis. In addition, the tedious separation of the main product from the

TABLE 1.2.4
Preparation of Diblock Copolymers of St (S) and AMA (A) via Their Anionic Copolymerization

	Living Poly(St)[a]					Block Copolymer[b]						
	$[n\text{-BuLi}]_0$ (mM)	$[\text{St}]_0$ (M)	$10^{-3}M_n$ calcd	$10^{-3}M_n$ obsd[c]	M_w/M_n[c]	$[\text{Living PSt}]_0$ (mM)	$[\text{AMA}]_0$ (M)	$10^{-3}M_n$ calcd	$10^{-3}M_n$ obsd[c]	M_w/M_n[c]	W_{St}/W_{MA}[d] calcd	W_{St}/W_{MA}[d] obsd[e]
SbA-1	15.0	0.962	6.73	6.80	1.07	7.50	0.400	13.5	14.1	1.09	50/50	51/49
SbA-2	20.0	0.641	3.39	4.15	1.12	8.57	0.680	13.4	13.8	1.13	25/75	26/74
SbA-3	12.0	1.15	10.0	10.7	1.08	7.50	0.200	13.4	13.5	1.09	75/25	77/23
SbA-4	7.5	0.962	13.4	15.0	1.09	3.75	0.400	26.8	29.0	1.10	50/50	52/48

[a] The living poly(St) was first prepared by the anionic polymerization of St, which was carried out in a mixture of toluene and THF (1:1 by volume), at −60°C, for 1 h.

[b] After the polymerization of St, a THF solution of DPE ($[\text{DPE}]/[\text{living PSt}]_0 = 1.2$) and LiCl ($[\text{LiCl}]/[\text{living PSt}]_0 = 3$) was added. The living end of poly(St) was reacted with DPE for 20 min at −60°C, and this was followed by the introduction of AMA. The polymerization of AMA lasted 1 h. The copolymer yield was 100% in each case.

[c] Determined by GPC.

[d] The weight ratio of St and AMA segments in the copolymer.

[e] Determined by ^{1}H NMR.

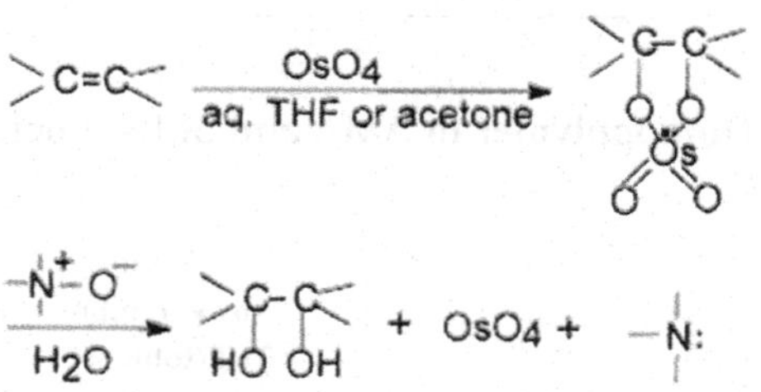

OsO_4
aq. THF or acetone

H_2O
HO OH
+ OsO_4 +

SCHEME 1.2.2

byproducts is also a shortcoming of this reaction. VanRheenen et al.[11] improved the osmylation process by combining osmium tetroxide with a tertiary amine *N*-oxide, such as *N*-methylmorpholine *N*-oxide (NMO) (Scheme 1.2.2). Using this method, the amount of OsO_4 can be reduced to 1 mol% compared to the molar amount of C=C bonds.

In the present paper, this osmylation process was applied to the dihydroxylation of homopoly(AMA) and the block copolymer of AMA with MMA or St, with the goal to prepare well-defined water-soluble poly(DHPMA) and hydrophilic-hydrophobic block copolymers.

The osmylation was carried out with magnetic stirring, under the protection of nitrogen, at room temperature, for 24 h. Because the homopolymerization and block copolymerization were carried out in THF, the polymerization solution was directly subjected to osmylation without polymer separation. As shown in Table 1.2.5, to a certain amount of polymerization solution, acetone, methanol, water, and NMO were sequentially added under the protection of nitrogen. The methanol was needed to dissolve the solid NMO. For the regeneration of OsO_4, a small amount of water was necessary (Scheme 1.2.2). In addition, the presence of acetone increased the compatibility between methanol, water, THF, and toluene. Compared to the molar amount of C=C bonds, an excess of NMO ($mol_{NMO}/mol_{C=C}$ = 1.2–1.5) and a small amount of OsO_4 ($mol_{OsO_4}/mol_{C=C}$ = 2/100) were employed. As soon as the alcohol solution of OsO_4 was introduced into the above system, a light-red color appeared at once, indicating a rapid formation of a complex between the C=C bond and OsO_4 (Scheme 1.2.2). As the reaction proceeded, the system became gradually turbid, indicating that the osmylated product has a different solubility than its precursor. To keep the reaction proceeding homogeneously, methanol was dropwise added during the osmylation process. About 24 h later, the resulting polymer was precipitated by pouring the reaction solution into a mixture of hexane and ethanol (1:1 by volume). The regenerated OsO_4, the excess of NMO, and the byproduct *N*-methylmorpholine were easily removed, because they are soluble in ethanol.

Figure 1.2.3A and B depicts the ^1H NMR spectra of poly(AMA) (PAMA-2 in Table 1.2.1) and of its osmylated product (Tables 1.2.5 and 1.2.6). After osmylation, the peaks c, d, and e corresponding to the allyl side groups of poly(AMA) disappeared completely, and the spectrum of the osmylated product became consistent with the

TABLE 1.2.5
Osmylation of the Homopolymer of AMA and of Its Block Copolymer with MMA or St[a]

	(Co)polymer[b] Weight (g)	AMA (mmol)[c]	NMO (mmol)[d]	OsO_4[e] (mL)	Solvents (by vol, mL)[f] THF/toluene/ acetone/ CH_3OH/H_2O	CH_3OH[g] (mL)	Functionality[h] (%)
PAMA-1	0.50	4.0	5.6	0.8	5.5/0/2.0/2.0/0.5	0.5	100
PAMA-2	0.50	4.0	6.0	0.8	6.0/0/4.0/2.0/0.5	1.0	100
PAMA-4	0.50	4.0	6.0	0.8	8.0/0/5.0/1.0/0.5	4.0	100
MbA-1	1.0	4.0	4.8	0.8	8.5/0/3.0/1.5/0.5	1.5	100
MbA-2	1.0	5.3	8.0	1.1	8.5/0/3.0/2.0/0.5	2.0	100
AbM-2	1.0	2.6	3.9	0.6	8.5/0/4.0/1.0/0.5	0.5	100
MbAbM	1.0	4.0	5.0	0.8	8.5/0/3.0/1.5/0.5	1.5	100
AbMbA	1.0	4.0	6.0	0.8	8.5/0/3.0/1.5/0.5	1.5	100
SbA-1	1.0	4.0	6.0	0.8	7.0/2.0/2.0/1.0/0.5	1.0	100
SbA-2	1.0	6.0	9.0	1.2	6.6/1.6/3.0/1.5/0.5	4.0	100
SbA-3	1.0	2.0	3.0	0.4	6.4/2.6/2.0/1.0/0.5	0.5	100
SbA-4	1.0	4.0	6.0	0.8	7.0/2.0/2.0/1.0/0.5	0.5	100

[a] The reaction was carried out with magnetic stirring, under the protection of nitrogen, at room temperature, for 24 h. The polymer yield was 100% in each case.
[b] A polymerization solution containing a certain amount of (co)polymer was directly employed.
[c] Molar amount of AMA component in the (co)polymer.
[d] Molar amount of *N*-methylmorpholine *N*-oxide (NMO).
[e] A 2.5 wt% solution in 2-methyl-2-propanol, which contained 2 mol% of OsO_4 compared to the molar amount of C=C bonds in AMA segments, was employed.
[f] To a polymerization solution, acetone, methanol, and water were sequentially added.
[g] To keep the reaction proceeding homogeneously, methanol was dropwise added during the osmylation process.
[h] Determined by ^{1}H NMR.

molecular structure of poly(DHPMA). The osmylated homopolymer was also identified by ^{13}C NMR [o-PAMA-2 (Tables 1.2.5 and 1.2.6) in CD_3OD, δ 178.3, 178.1, 177.9 (C=O), 69.7, 66.3, 63.1 (CH_2CHCH_2), 54.7 (CH_2), 45.2, 44.9 (–C–), 18.6, 16.5 (CH_3)]. The complete disappearance of the absorptions (δ 131.8, 119.0, 65.8) due to the allyl group and the appearance of the absorptions (δ 69.7, 66.3, 63.1) corresponding to the dihydroxypropyl indicate that after osmylation a complete change from poly(AMA) to poly(DHPMA) took place.

The ^{1}H and ^{13}C NMR results are also supported by the FT-IR measurements. Parts A and B of Figure 1.2.4 present the FT-IR spectra of PAMA-2 and its osmylated product, respectively. After osmylation, the peak A-a due to the C=C double bond of poly(AMA) disappeared completely, and the two strong bands, B-b and B-c

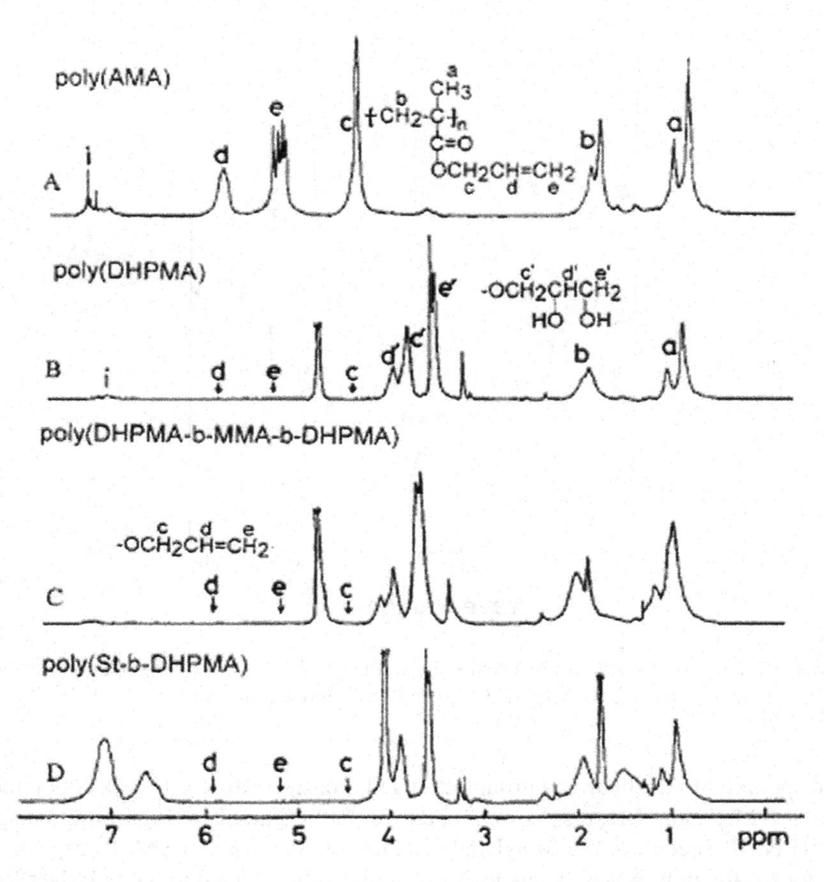

FIGURE 1.2.3 ^{1}H NMR spectra of PAMA-2 (A; in $CDCl_3$; Table 1.2.1, M_n = 9800, M_w/M_n = 1.06), osmylated PAMA-2 (B; in CD_3-OD), osmylated triblock copolymer AbMbA [C; Table 1.2.3; in a mixture of CD_3OD and THF-d_8 (3:1 by weight)], and osmylated diblock copolymer SbA-1 (D; Table 1.2.4; in a mixture of CD_3OD and THF-d_8 (1:3 by weight)]. Peak i: phenyl groups of the initiator DPHL.

TABLE 1.2.6
Elemental Analyses of Poly(AMA) and Its Osmylated Product

		C (%)		H (%)		N (%)	
		Theory	Found	Theory	Found	Theory	Found
PAMA-2[a]	$C_7H_{10}O_2$	66.64	67.07	8.01	8.01		
o-PAMA-2[b]	$C_7H_{12}O_4 \cdot {}^1/_5H_2O$[c]	51.34	51.38	7.63	7.48	0.00	0.00

[a] See Table 1.2.1.
[b] See Table 1.2.5.
[c] Each repeating unit contains $\frac{1}{5}$ water molecule (see ref 3a).

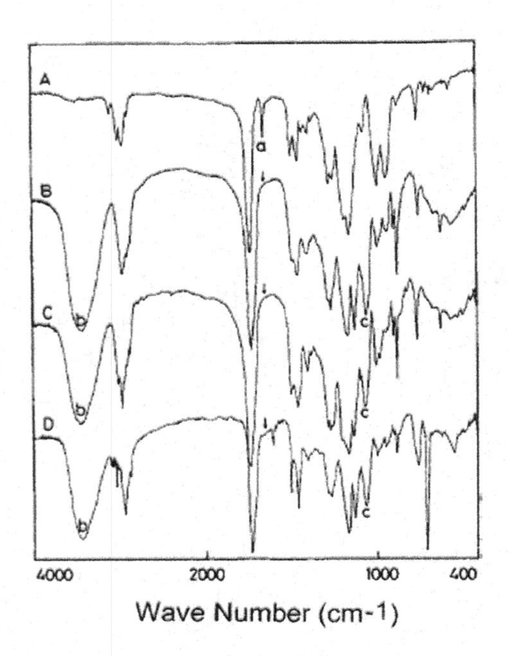

FIGURE 1.2.4 FT-IR spectra of PAMA-2 (A, Table 1.2.1) and of the osmylation products of PAMA-2 (B, Table 1.2.5), MbA-1 (C, Tables 1.2.2 and 1.2.5), and SbA-1 (D, Tables 1.2.4 and 1.2.5).

corresponding to the hydroxyl groups emerged. For the osmylated block copolymers, similar NMR and FT-IR results were obtained. Parts C and D of Figure 1.2.3 display the ^{1}H NMR spectra of the osmylated products of AbMbA and SbA-1, respectively. Obviously, the poly(AMA) segments were completely changed to the poly(DHPMA) ones. As shown in Figure 1.2.4C and D, the formation of the diol side groups in these block copolymers is confirmed by the appearance of the strong absorptions of C-b, C-c and D-b, D-c.

As shown in Table 1.2.6, the elemental analysis results of homopoly(AMA) and its osmylated product are consistent with the calculated values. The calculations have indicated that each repeating unit of poly(DHPMA) contained $^1/_5$ water molecule. It is of interest to notice that when poly(DHPMA) was prepared by polymerizing a protected monomer followed by deprotection,[3a] the same amount of water per repeating unit was found. On the other hand, even though a nitrogen-containing compound, NMO, was employed in the osmylation process, the elemental analysis did not indicate the presence of nitrogen in the resulting polymer. Consequently, the reagents were completely removed during the purification process, and very pure poly(DHPMA) was obtained.

To verify by GPC whether cross-linking and/or chain damage occurred during osmylation, the dihydroxylate of poly(AMA) was reprotected by reacting the formed hydroxyl groups with benzoic anhydride (see Experimental Section). As shown in Figure 1.2.5b, the GPC chromatogram of the polymer thus obtained exhibits a single peak, and its MWD ($M_n = 20,500$, $M_w/M_n = 1.08$) is as narrow as that of its precursor

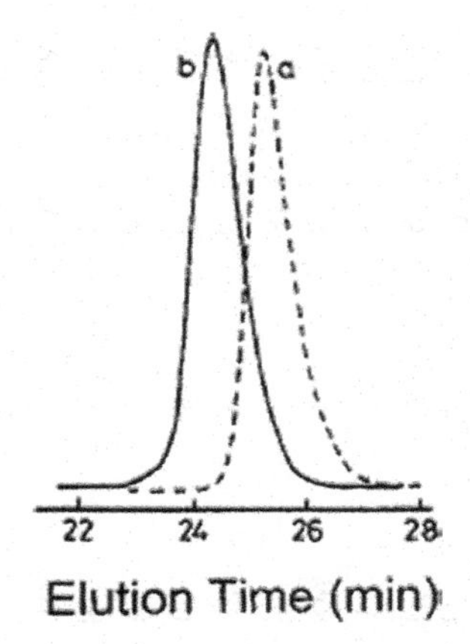

FIGURE 1.2.5 GPC traces of PAMA-2 (a; Table 1.2.1, $M_n = 9800$, $M_w/M_n = 1.06$) and of the reprotected product (b; $M_n = 20,500$, $M_w/M_n = 1.08$) of its dihydroxylate.

(PAMA-2 in Table 1.2.1, $M_n = 9800$, $M_w/M_n = 1.06$). Consequently, no crosslinking occurred during osmylation, and this process did not affect the chain structure of the polymer.

1.2.3.4 Solubilities of the (Co)polymers before and after Osmylation

The poly(AMA) and its osmylated product exhibit different solubilities. For instance, as shown in Table 1.2.7, PAMA-2 is soluble in benzene, chloroform, DMF, pyridine, 1,4-dioxane, THF, and acetone but insoluble in hexane, ethanol, methanol, and water. However, its osmylated product is soluble in DMF, pyridine, methanol, and water but insoluble in any of the other solvents tested. The molecular weight affects the solubility of the dihydroxylate of poly(AMA). When the molecular weight is small, such as PAMA-1 ($M_n = 5900$) and PAMA-2 ($M_n = 9800$), their osmylated products dissolved in methanol more rapidly than in water. However, the dissolution of dihydroxylate of PAMA-4, which has a large molecular weight ($M_n = 170,000$), is much faster in water than in methanol. For comparison, solubility experiments for poly(HEMA) were also carried out. This polymer possesses only one hydroxyl in each of its repeating units. As shown in Table 1.2.7, in contrast to poly(DHPMA), it is soluble in ethanol and methanol but insoluble in water. As expected, the molecular architectures of the triblock copolymers, poly(DHPMA-*b*-MMA-*b*-DHPMA) and poly(MMA-*b*-DHPMA-**b**-MMA)., affected their hydrophilicity. The former is completely soluble in methanol and partially soluble in water. In contrast, the latter triblock copolymer, which possesses a poly(DHPMA) hydrophilic segment in the middle of the polymer chain, is only swollen by water and dissolved very slowly in methanol. The composition of the block copolymers has also an effect on their solubility. For instance, o-SbA-1 ($W_{St}/W_{AMA} = 51/49$) is soluble in THF and 1,4-dioxane but only swollen in methanol and water; o-SbA-2 ($W_{St}/W_{AMA} = 26/74$) is partially soluble in water and methanol but insoluble in THF and 1,4-dioxane; o-SbA-3 ($W_{St}/W_{AMA} = 77/23$) is insoluble in polar solvents but soluble in benzene and $CHCl_3$. Consequently,

TABLE 1.2.7
Solubilities of the (Co)polymers before and after Osmylation[a]

No.[b]	Hexane	Bz	$CHCl_3$	DMF	Pyridine	1,4-Dioxane	THF	Acetone	C_2H_5OH	CH_3OH	H_2O
PAMA-2	I	S	S	S	S	S	S	S	I	I	I
o-PAMA-2	I	I	I	S	S	I	I	I	I	S	S
o-PAMA-1	I	I	I	S	S	I	I	I	I	S	S
o-PAMA-4	I	I	I	S	S	I	I	I	I	D	S
PHEMA[c]	I	I	I	S	S	I	I	I	S	S	I
MbA-1	I	S	S	S	S	S	S	S	I	I	I
o-MbA-1	I	I	SW	S	S	I	I	SW	I	S	SW
o-MbA-2	I	I	I	S	S	I	I	I	I	S	SW
o-AbM-2	I	I	SW	S	S	I	I	SW	I	S	SW
o-MbAbM	I	I	SW	S	S	I	I	I	I	D	SW
o-AbMbA	I	I	I	S	S	I	I	I	I	S	P
SbA-1	I	S	S	S	S	S	S	S	I	I	I
o-SbA-1	I	I	SW	S	S	D	S	SW	I	SW	SW
o-SbA-2	I	I	I	S	S	I	I	I	I	P	P
o-SbA-3	I	D	D	S	S	D	S	I	I	I	SW
o-SbA-4	I	SW	SW	S	S	SW	SW	I	I	SW	SW

[a] The experiment was carried out at room temperature (23°C). The amounts of polymer and the solvent were 0.03 g and 1.0 mL, respectively. S = soluble; D = dissolved slowly; P = partially soluble; W = wetting; SW = swollen; I = insoluble.
[b] o-: the (co)polymers after osmylation.
[c] Poly(2-hydroxyethyl methacrylate) prepared in a previous paper.[2d]

the molecular weight, molecular architecture, and the copolymer composition can affect the hydrophilicity of the osmylated (co)polymers, and all these factors can be accurately controlled in the (co)polymerization step.

REFERENCES

1. (a) Noshay, A.; McGrath, J. E. *Block Copolymerization;* Academic Press: New York, 1977. (b) Van Beylen, M.; Bywater, S.; Smets, G.; Szwarc, M.; Worsfold, D. J. *Adv. Polym. Sci.* **1988**, *86*, 87. (c) Rempp, P.; Franta, E.; Herz, J. E. *Adv. Polym. Sci.* **1988**, *86*, 145. (d) Morton, M. *Anionic Polymerization: Principle and Practice*; Academic Press: New York, 1983. (e) Hsieh, H. L.; Quirk, R. P. *Anionic Polymerization*; Marcel Dekker: New York, 1996. (f) Pitsikalis, M.; Pispas, S.; Mays, J. W.; Hadjichristidis, N. *Adv. Polym. Sci.* **1998**, *135*, 1.
2. For examples: (a) Hirao, A.; Kata, H.; Yamaguchi, K.; Nakahama, S. *Macromolecules* **1986**, *19*, 1294. (b) Ramireddy, C.; Tuzar, Z.; Prochazka, K.; Webber, S. E.; Munk, P. *Macromolecules* **1992**, *25*, 2541. (c) Zhang, H. M.; Ruckenstein, E. *Macromolecules* **1998**, *31*, 7575; **1998**, *31*, 9127. (d) Ruckenstein, E.; Zhang, H. M. *J. Polym. Sci., Polym. Chem.* **1998**, *36*, 1865.
3. (a) Mori, H.; Hirao, A.; Nakahama, S. *Macromolecules* **1994**, *27*, 35. (b) Mori, H.; Hirao, A.; Nakahama, S. *Macromolecules* **1994**, *27*, 4093.
4. (a) Chung, T. C.; Raate, M.; Berluche, E.; Schulz, D. N. *Macromolecules* **1988**, *21*, 1903. (b) Mao, G.; Wang, J.; Clingman, S. R.; Ober, C. K.; Chen, J. T.; Thomas, E. L. *Macromolecules* **1997**, *30*, 2556.
5. Zhang, H. M.; Ruckenstein, E. *Macromolecules* **1999**, *32*, 5495.
6. Allen, R. D.; Long, T. E.; McGrath, J. E. *Polym. Bull.* **1986**, *15*, 127.
7. Fayt, R.; Forte, R.; Jacobs, C.; Jerome, R.; Ouhadi, T.; Teyssie, Ph.; Varshney, S. K. *Macromolecules* **1987**, *20*, 1442.
8. Zhang, H. M.; Ruckenstein, E. *J. Polym. Sci., Polym. Chem.* **1997**, *35*, 2901.
9. Schulz, G.; Hocher, H. *Angew. Chem., Int. Ed. Engl.* **1980**, *19*, 219.
10. Gundstone, F. D. In *Advances in Organic Chemistry*; Raphael, R. A., Taylor, E. C., Wynberg, H., Eds.; Interscience Publishers: New York, 1960; Vol. 1, p. 110.
11. VanRheenen, V.; Kelly, R. C.; Cha, D. Y. *Tetrahedron Lett.* **1976**, 23, 1973.

1.3 Well-Defined Graft Copolymers Based on the Selective Living Anionic Polymerization of the Bifunctional Monomer 4-(Vinylphenyl)-1-butene*

Eli Ruckenstein and Hongmin Zhang

Chemical Engineering Department, State University of New York at Buffalo, Amherst, New York 14260

ABSTRACT The anionic copolymerization of the bifunctional monomer 4-(vinylphenyl)-1-butene (VSt) with styrene (St) was carried out in a mixture of toluene and tetrahydrofuran (THF), using *n*-BuLi as initiator, at −40°C. During this process, the styrene type C=C bond of VSt was selectively polymerized, and the other one in its butenyl moiety remained unchanged. The molecular weight and the composition of the resulting copolymer could be well-controlled, and its molecular weight distribution was very narrow (M_w/M_n = 1.03–1.04). The unreacted butenyl groups of the VSt units were further reacted with chlorodimethylsilane, generating another functional copolymer with reactive chlorodimethylbutylsilyl side chains. This functional polymer was used as a backbone polymer and reacted with the anionic living polymers of St, isoprene (Is), or methyl methacrylate (MMA). This coupling reaction resulted in the formation of well-defined graft (co)polymers with a poly(St) backbone and either poly(St) or poly(Is) or poly(MMA) side chains.

1.3.1 INTRODUCTION

The "grafting onto" method has been often employed for the preparation of graft (co)polymers by the anionic polymerization technique.[1] This method is based on the coupling reaction of the electrophilic groups attached to the backbone polymer and

* *Macromolecules* 1999, 32, 6082–6087.

the propagating site of an anionic living polymer.[2] The key point of this method is to generate a backbone polymer that possesses suitable reactive groups. The scope of the present paper is to present a novel route to functional polymers with electrophilic groups and further to well-defined graft (co)polymers, based on the selective living anionic polymerization of a new bifunctional monomer.

Recently, we have investigated the anionic homopolymerization of a novel bifunctional monomer 4-(vinylphenyl)-1-butene (VSt) and its block copolymerization with styrene (St).[3] Under suitable conditions, the styrene type C=C bond of VSt could be selectively polymerized, generating a uniform-size functional polymer with a reactive butenyl group in each of its repeating units. The further reaction of the butenyl side chains of homopoly(VSt) or the block copolymer poly(VSt-*b*-St) with 9-borabicyclo[3.3.1]nonane, followed by the addition of NaOH and H_2O_2, generated a new uniform-size hydrophilic functional polymer poly(4-hydroxybutylstyrene) or an amphiphilic block copolymer containing both poly(4-hydroxybutylstyrene) hydrophilic and poly(St) hydrophobic segments.

In the present paper, we will further demonstrate the advantages of the selective living polymerization of the novel bifunctional monomer by employing the resulting functional polymer in the preparation of well-defined graft (co)polymers. First, the anionic copolymerization of VSt and St was carried out (Scheme 1.3.1). Similar to the homo- and block copolymerizations, the resulting statistical copolymer, poly(St-*co*-VSt) (**1** in Scheme 1.3.1), possesses an unreacted C=C bond in each of its VSt units. Subsequently, this functional side chain was reacted with chlorodimethylsilane, producing another new functional copolymer, poly(St-*co*-StSiCl) (**2** in Scheme 1.3.1), with reactive chlorosilyl side chains. The further coupling reaction of these side chains with the anionic living polymers of either St, or isoprene (Is), or methyl methacrylate (MMA) generated well-defined graft (co)polymers (**3** in Scheme 1.3.1) with a poly(St) backbone and either poly(St) or poly(Is) or poly(MMA) side chains.

1.3.2 EXPERIMENTAL SECTION

Materials. Tetrahydrofuran (THF) was dried with CaH_2 under reflux for more than 24 h, distilled, and distilled again from a solution of sodium naphthalenide just before use. Toluene and benzene were washed with concentrated sulfuric acid and then with water, dried with $MgSO_4$, distilled over CaH_2, and finally distilled from a *n*-BuLi solution. Hexane was first dried and distilled over CaH_2 and then distilled from a solution of *n*-BuLi. Vinylbenzyl chloride (VBC; Aldrich, 97%) was dried with CaH_2 and distilled under reduced pressure. Styrene (St; Aldrich, 99%) and isoprene (Is; Aldrich, 99%) were washed with 10 wt% aqueous NaOH solution and then with water, dried overnight with $MgSO_4$, distilled over CaH_2, and finally distilled in the presence of phenylmagnesium chloride prior to polymerization. Methyl methacrylate (MMA; Aldrich, 99%) was washed and purified in a way similar to that used for St and Is, except the final distillation was carried out in the presence of triisobutylaluminum.[4] 1,1-Diphenylethylene (DPE; Aldrich, 97%) was distilled over CaH_2 and then distilled in the presence of 1,1-diphenylhexyllithium (DPHL) under reduced

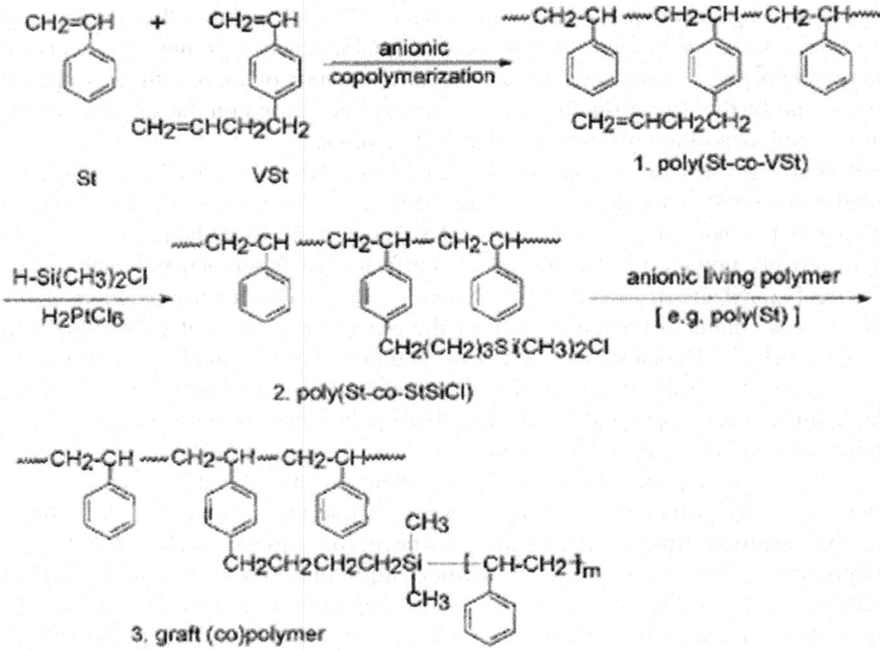

CH_2=CH + CH_2=CH → anionic copolymerization → ~CH_2-CH~CH_2-CH~CH_2-CH~

CH_2=$CHCH_2CH_2$ CH_2=$CHCH_2CH_2$

St   VSt   1. poly(St-co-VSt)

H-Si$(CH_3)_2$Cl / H_2PtCl_6 → ~CH_2-CH~CH_2-CH~CH_2-CH~ → anionic living polymer [e.g. poly(St)]

$CH_2(CH_2)_3Si(CH_3)_2Cl$

2. poly(St-co-StSiCl)

~CH_2-CH~CH_2-CH~CH_2-CH~

CH_3

$CH_2CH_2CH_2CH_2$Si—[CH-CH_2]$_m$

CH_3

3. graft (co)polymer

SCHEME 1.3.1

pressure. Lithium chloride (99.99%, Aldrich) was dried at 120°C for 24 h and dissolved in purified THF.[5] Hydrogen hexachloroplatinate (Aldrich) was dissolved in propanol (5 wt%), and this solution was directly used. Chlorodimethylsilane (98%, Aldrich) was distilled under the protection of N_2 before use. *n*-BuLi (Aldrich, 1.6 M solution in hexane) was diluted with purified hexane. Allylmagnesium chloride (AMC; Aldrich, 2.0 M solution in THF) was used as received. The bifunctional monomer, 4-vinylphenyl-1-butene (VSt), was prepared via the coupling reaction between VBC and AMC[3] and distilled in the presence of phenylmagnesium chloride under reduced pressure prior to polymerization.

Preparation of the Backbone Copolymer of VSt and St [Poly(St-*co*-VSt)]. All the polymerizations were carried out in a round-bottom glass flask, under an overpressure of argon, with magnetic stirring, at selected temperatures. Poly(St-*co*-VSt) was prepared via the anionic copolymerization of VSt and St, which was performed in a mixture of toluene and THF (2:1 by volume), at −40°C, using *n*-BuLi as initiator. After the solvents were added with dry syringes, the flask was placed in a bath kept at −40°C, and the initiator solution (*n*-BuLi in hexane) was added. The polymerization was induced by introducing a prechilled mixture of VSt and St into the above initiator solution. After 50 min, the system was quenched by adding a small amount of methanol, and the copolymer was precipitated by pouring the polymerization solution into a large amount of methanol. The copolymer thus obtained was washed with methanol and vacuum-dried at 40°C for more than 24 h.

Preparation of the Backbone Copolymer Containing Chlorosilyl Reactive Side Chains [Poly(St-*co*-StSiCl] by the Hydrosilylation of the C=C Functional Groups of VSt Units in Poly[St-*co*-VSt]. In a round-bottom flask, equipped with a condenser and a magnetic stirring bar and protected with nitrogen, a benzene solution of poly(St-*co*-VSt) and a trace amount of propanol solution of H_2PtCl_6 (5 wt%) were introduced. After the temperature was raised to 60°C, chlorodimethylsilane was dropwise added carefully with a syringe in about 20 min. Then, this reaction was allowed to last more than 16 h at 80°C. Subsequently, the benzene solution was freeze-dried for 12 h, followed by vacuum-drying at 60°C for an additional 10 h. The copolymer thus obtained was dissolved in purified toluene, and the solution was employed in the next coupling reaction.

Preparation of Anionic Living Side Chain Polymers of St, Is, or MMA. The living poly(St), poly(Is), and poly(MMA) were prepared by the anionic living polymerizations of the corresponding monomers. The anionic polymerization of St was carried out under conditions similar to those for the preparation of poly(St-*co*-VSt). Upon the completion of polymerization, a small amount of solution (ca. 0.2 mL) was taken out for a GPC measurement. Without termination, this living polymer solution was used in the next coupling reaction step.

The anionic polymerization of Is was performed in benzene, at room temperature, using *n*-BuLi as initiator. To promote the dissociation of the initiator and/or the propagating site and to accelerate the polymerization reaction, a trace amount of polar solvent THF ($[THF]/[n\text{-BuLi}]_0 = 1.2–1.4$) was added. The polymerization was induced by adding the monomer to a mixture of benzene, THF, and initiator. After 70 min, the living poly(Is) solution was used for its coupling with poly(St-*co*-StSiCl).

The anionic polymerization of MMA was carried out using DPHL as initiator, in THF, at −78°C, in the presence of LiCl ($[LiCl]/[DPHL]_0 = 1.2$). After THF, DPE, and a THF solution of LiCl were added with dry syringes, the flask was cooled to −40°C, and *n*-BuLi (in hexane) was added. The deep red color of DPHL appeared at once, and the reaction between *n*-BuLi and DPE was allowed to continue for 15 min. The polymerization was started by the addition of the prechilled MMA to the above system, and the reaction was allowed to last 20 min. This THF solution of the living poly(MMA) was used for the preparation of the graft copolymer.

Preparation of the Graft (Co)polymer via the Coupling of Poly(St-*co*-StSiCl) with an Anionic Living Polymer of St, Is, or MMA. The graft (co)polymers were prepared through the coupling reaction between the chlorosilyl side chains of poly(St-*co*-StSiCl) and the anionic living polymer of St, Is, or MMA. After the anionic polymerization of a monomer, a toluene solution of poly(St-*co*-StSiCl) was introduced with a dry syringe into the living polymer solution. The reaction temperatures for living poly(St), poly(Is), and poly(MMA) were −40, 23, and −50°C, respectively. After the reaction lasted 1.5 h, the polymer was precipitated by pouring the reaction mixture into a large amount of methanol, washed with methanol, and vacuum-dried at 40°C for 24 h.

Measurements. ^1H NMR spectra were recorded in $CDCl_3$ on an INOVA-500 spectrometer. M_n and M_w/M_n of the polymer were determined by gel permeation chromatography (GPC), on the basis of a polystyrene calibration curve, and by vapor pressure osmometry (VPO). The VPO measurements were performed in toluene, at 45°C, using an UIC SA 070 vapor pressure osmometer. The GPC measurements were carried out using THF as solvent, at 30°C, at a 1.0 mL/min flow rate and a 1.0 cm/min chart speed. Three polystyrene gel columns (Waters, 7.8 × 300 mm; one HR 5E, part no. 44228, one Linear, part no. 10681 and one HR 4E, part no. 44240) were used, which were connected to a Waters 515 precision pump.

1.3.3 RESULTS AND DISCUSSION

Preparation of the Backbone Copolymer, Poly(St-*co*-VSt), by the Anionic Copolymerization of VSt and St. As shown in Scheme 1.3.1, the functional units of the backbone copolymer, poly(VSt-*co*-St), will be turned into grafting points after hydrosilylation. If the copolymerization of VSt and St can proceed smoothly in a living manner, the grafting number can be controlled by the feed amount ratio of the two monomers.

In a previous paper,[3] the effects of the initiator, temperature, and solvent on the selective anionic polymerization of VSt were investigated and the optimum conditions determined. Using *n*-BuLi as initiator, in a mixture of toluene and THF (5:1–2:1 by volume), at −40°C, VSt could undergo anionic polymerization in a living manner without cross-linking or any other side reaction. The polymer thus obtained possesses a controlled molecular weight and a very narrow molecular weight distribution (M_w/M_n = 1.03–1.05). The quantitative presence of the unreacted 1-butene type C=C double bond was verified by ^{1}H NMR and FT-IR. The block copolymerization of VSt and St could also proceed smoothly either in the polymerization sequence VSt followed by St, or vice versa, generating a well-defined block copolymer with a controlled molecular weight and composition and a very narrow molecular weight distribution (M_w/M_n = 1.03–1.07). The above results indicate that the living site of VSt or St can initiate quickly the living anionic polymerization of the other monomer. Therefore, the copolymerization of these two monomers can be also expected to proceed smoothly.

The copolymerization of VSt and St was carried out under conditions similar to those for the homopolymerization of VSt and its block copolymerization with St (see Experimental Section), except the two monomers were added simultaneously to the initiator solution. As shown in Table 1.3.1, the monomer conversion is quantitative in every case, the molecular weight determined by GPC is in good agreement with that calculated, and the MWD of the copolymer is very narrow (M_w/M_n = 1.03–1.04). As illustrated in Figure 1.3.1b (SVS-2 in Table 1.3.1, $M_n = 1.39 \times 10^4$, M_w/M_n = 1.03), the GPC chromatogram exhibits a very sharp and symmetrical peak. The above results indicate that the anionic copolymerization of VSt and St proceeded in a living manner and that a uniform-size copolymer, poly(St-*co*-VSt), was obtained.

Figure 1.3.2A depicts the ^{1}H NMR spectrum of the copolymer, SVS-2 (Table 1.3.1). The absorptions corresponding to the –CH=CH_2 groups of VSt units are quantitatively detectable (peak g, =CH_2; peak f, –CH=). The weight content (25.6%) of VSt

TABLE 1.3.1
Anionic Copolymerizaiton of VSt and St[a]

No.	$[n\text{-BuLi}]_0$ (mM)	$[VSt]_0$ (M)	$[St]_0$ (M)	$10^{-3}M_n$ (calcd)	$10^{-3}M_n$[b] (obsd)	M_w/M_n[b]
SVS-1	18.8	0.441	0.685	7.53	8.08	1.04
SVS-2	7.8	0.160	0.739	13.2	13.9	1.03
SVS-3	6.5	0.168	1.09	21.6	22.9	1.03

[a] The anionic copolymerization was carried out in a mixture of toluene and THF (2:1 by volume) at −40°C for 50 min. The polymer yield was 100% in each case.
[b] Determined by GPC on the basis of a standard poly(St) calibration curve.

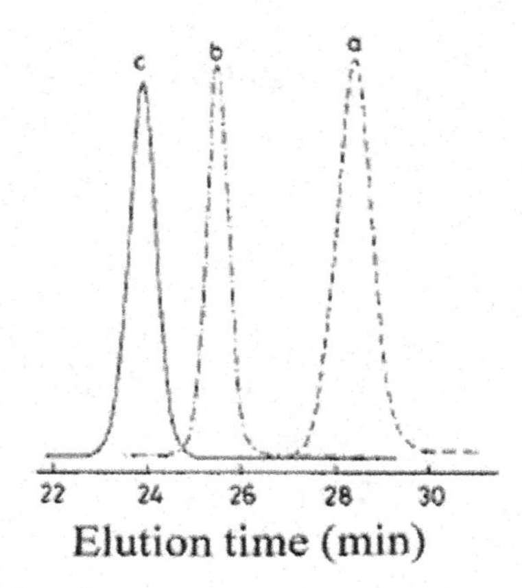

FIGURE 1.3.1 GPC traces of the graft copolymer (c; GP-5 in Table 1.3.4, $M_{n(GPC)} = 3.23 \times 10^4$, $M_{n(VPO)} = 4.01 \times 10^4$, $M_w/M_n = 1.05$) and its precursors: a, poly(Is) (PIs-1 in Table 1.3.3, $M_n = 1060$, $M_w/M_n = 1.11$); b, poly(St-*co*-VSt) (SVS-2 in Table 1.3.1, $M_n = 1.39 \times 10^4$, $M_w/M_n = 1.03$).

units based on the intensities of $=CH_2$ and of the phenyl groups of St units is close to its feed amount (24.7%). Furthermore, the resulting copolymer is soluble in common organic solvents, such as benzene, chloroform, THF, 1,4-dioxane, etc. Therefore, one can conclude that the composition of the copolymer was well-controlled, that no crosslinking reaction occurred during the copolymerization, and that the C=C functional side chains of VSt units remained unreacted.

Hydrosilylation of the C=C Functional Groups of Poly(St-*co*-VSt). Chlorosilyl groups possess high reactivity with nucleophilic reagents, including the anionic living polymers.[6] On the other hand, the reaction of chlorodimethylsilane with C=C bond in the presence of Pt-based catalyst has been frequently used in the preparation of compounds containing reactive chlorosilyl groups.[7] In the present paper,

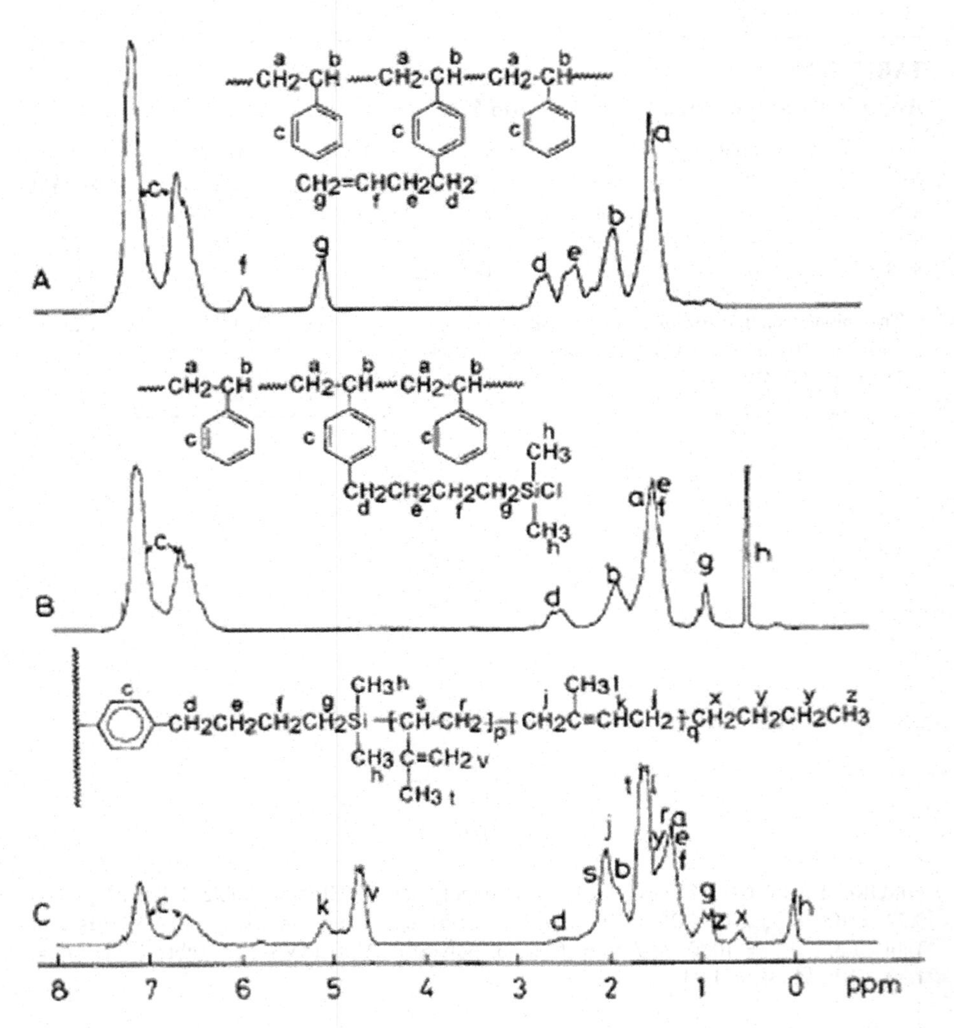

FIGURE 1.3.2 ^{1}H NMR spectra of poly(St-*co*-VSt) (A; SVS-2 in Table 1.3.1), its hydrosilylation product (B; SVSiCl-2 in Table 1.3.2), and the graft copolymer (C; GP-5 in Table 1.3.4) prepared by the coupling reaction between B and the living poly(Is) (PIs-1 in Table 1.3.3).

this hydrosilylation reaction is employed for the preparation of a new functional copolymer containing reactive chlorosilyl side chains, which is further used as the backbone polymer for the preparation of graft (co)polymers.

For an easy purification of the resulting polymer by freeze-drying after hydrosilylation, the purified benzene was selected as solvent. The reaction was carried out using an excess molar amount of chlorodimethylsilane compared to the C=C bonds, at 80°C, for more than 16 h (see Table 1.3.2 and Experimental Section). As soon as the chlorodimethylsilane was dropwise added to the benzene solution of poly(St-*co*-VSt) and of the catalyst H_2PtCl_6, the system started to boil vigorously and to darken gradually, indicating that the addition reaction between C=C

TABLE 1.3.2
Hydrosilylation of the C=C Functional Side Chains of Poly(St-*co*-VSt)[a]

No.	Poly(St-*co*-VSt), wt%[b]	C=C,[c] (mmol)	$(CH_3)_2SiHCl$ (mmol)	*t* (h)	Functionality[d] (%)
SVSiCl-1	SVS-1 5.5	11.1	18.0	24	100
SVSiCl-2	SVS-2 6.4	5.38	16.0	16	100

[a] The reaction was started at 60°C by carefully adding chlorodimethylsilane to a benzene solution of poly(St-*co*-VSt) and of a trace amount of H_2PtCl_6 (0.1 mL 5.0 wt% propanol solution). Then, this reaction was allowed to last more than 16 h at 80°C.
[b] Benzene solution.
[c] Molar amount of C=C bonds in poly(St-*co*-VSt).
[d] Determined by ^{1}H NMR.

and Si–H occurred. Because of the proton sensitivity of the chlorosilyl group, the resulting copolymer could not be purified by the common method, namely, the precipitation in methanol. Instead, the reaction mixture was directly freeze-dried, followed by the vacuum-drying at 60°C. After a small amount of sample was taken out under the protection of nitrogen for a NMR measurement, this product was dissolved in purified toluene, and the solution was used in the next coupling reaction step.

Figure 1.3.2B presents the ^{1}H NMR spectrum of the hydrosilylation product, poly(St-*co*-StSiCl) (SVSiCl-2 in Table 1.3.2). Comparing this spectrum with that of its precursor poly(St-*co*-VSt), the absorptions due to $-CH{=}CH_2$ (peaks g and f in Figure 1.3.2A) disappeared completely, and the peaks corresponding to the chlorodimethylsilyl groups (peak h in Figure 1.3.2B) emerged quantitatively. Therefore, the butenyl side chains of VSt units of poly(St-*co*-VSt) were completely changed to chlorodimethyl butylsilyl groups.

Preparation of the Graft (Co)polymers with Poly(St) or Poly(Is) Side Chains. The graft (co)polymers were prepared through a two-step process. A living side chain polymer was first prepared, followed by its coupling with the chlorosilyl groups of poly(St-*co*-StSiCl). The anionic polymerization of St was carried out in a mixture of toluene and THF ([THF] = 2.1 M) at −40°C, and that of Is was performed in benzene in the presence of a trace amount of THF ($[THF]/[n\text{-BuLi}]_0$ = 1.2–1.4) at room temperature (see Experimental Section). As shown in Table 1.3.3, the living poly(St) or poly(Is) thus prepared possesses a controlled molecular weight and a narrow MWD (M_w/M_n = 1.08–1.12). To a living polymer solution of St or Is, a toluene solution of poly(St-*co*-StSiCl) was added, and the coupling reaction was allowed to last 1.5 h at the corresponding polymerization temperature of St or Is.

As illustrated in Figure 1.3.1, a very sharp and single peak due to the graft copolymer emerges after the coupling reaction, in the short elution time region (peak c, GP-5 in Table 1.3.4), with a higher molecular weight ($M_{n(GPC)} = 3.23 \times 10^4$, $M_{n(VPO)} = 4.01 \times 10^4$, $M_w/M_n = 1.05$) than those of its precursor polymers, namely, the side chain poly(Is) (peak a, PIs-1 in Table 1.3.3, $M_n = 1060$, $M_w/M_n = 1.11$) and the backbone polymer SVS-2 (Table 1.3.1, $M_n = 1.39 \times 10^4$, $M_w/M_n = 1.03$). In addition, it is clear from the GPC chromatograms (Figure 1.3.1) that the graft copolymer

TABLE 1.3.3
Preparation of the Anionic Living Side Chain Polymers[a]

No.	Initiator (mM)	$[M]_0$ (M)	Solvent	Additive	Temp (°C)	Time (h)	M_n (calcd)	M_n (obsd)[b]	M_w/M_n[b]
PSt-1	*n*-BuLi 52.7	0.456	toluene	[THF] = 2.1 M	−40	50	960	920	1.12
PSt-2	*n*-BuLi 26.4	0.456	toluene	[THF] = 2.1 M	−40	60	1850	1880	1.08
PSt-3	*n*-BuLi 52.7	0.456	toluene	[THF] = 2.1 M	−40	50	960	960	1.11
PSt-4	*n*-BuLi 52.7	0.628	toluene	[THF] = 2.1 M	−40	60	1300	1260	1.10
PIs-1	*n*-BuLi 52.7	0.735	benzene	[THF] = 63 mM	23	70	1000	1060	1.11
PIs-2	*n*-BuLi 52.7	1.27	benzene	[THF] = 79 mM	23	70	1700	1760	1.11
PIs-3	*n*-BuLi 37.4	1.21	benzene	[THF] = 52 mM	23	70	2260	2340	1.12
PMMA-1	DPHL[c] 52.7	0.600	THF	[LiCl] = 63 mM	−78	20	1200	1290	1.10
PMMA-2	DPHL[c] 33.3	0.513	THF	[LiCl] = 40 mM	−78	20	1600	1740	1.11

[a] The polymerization was started by adding the monomer to the mixture of solvent, additive, and initiator.
[b] Determined by GPC.
[c] 1,1-Diphenylhexyllithium (DPHL) was prepared in situ, before the monomer addition, by the reaction between *n*-BuLi and 1,1-diphenylethylene, at −40°C, for about 15 min.

TABLE 1.3.4
Preparation of the Graft Copolymers (GP)[a]

No.	Backbone[b]	-$Si(CH_3)_2Cl$ (mmol)[c]	Living Polymer[d] (mmol)	$10^{-4}M_n$ (calcd)	Graft Copolymer $10^{-4}M_{n(VPO)}$[e]	$10^{-4}M_{n(GPC)}$[f]	M_w/M_n[f]	Peak no.
GP-1	SVSiCl-1	1.06	PSt-1 1.58	3.14		2.22[g]	1.13[g]	Double[h]
GP-2	SVSiCl-1	1.06	PSt-2 0.84	4.57	4.74	3.19	1.08	Single
GP-3	SVSiCl-2	1.10	PSt-3 1.06	3.48	3.80	2.51	1.08	Single
GP-4	SVSiCl-2	1.10	PSt-4 1.06	4.10	3.98	3.07	1.06	Single
GP-5	SVSiCl-2	1.10	PIs-1 1.10	3.78	4.01	3.23	1.05	Single
GP-6	SVSiCl-2	1.10	PIs-2 0.94	5.03	4.86	3.49	1.09	Single
GP-7	SVSiCl-2	0.88	PIs-3 0.67	5.28	5.11	4.02	1.12	Single
GP-8	SVSiCl-1	1.06	PMMA-1 1.58	4.08		2.83[g]	1.15[g]	Double[h]
GP-9	SVSiCl-1	1.06	PMMA-2 1.06	5.23		3.16	1.10	Single

[a] The coupling reaction was carried out by introducing a toluene solution of poly(St-*co*-StSiCl) into a living anionic polymer solution of St, Is, or MMA, and the reaction was allowed to last 1.5 h.
[b] See Tables 1.3.1 and 1.3.2.
[c] Molar amount of the chlorosilyl groups of the backbone copolymer.
[d] Molar amount of the living side chain polymer, which is equal to that of the initiator (see Table 1.3.3 and the Experimental Section).
[e] Determined by VPO.
[f] Determined by GPC.
[g] The results only correspond to the graft polymer, not including the excess of side chain polymers.
[h] Double peaks because of excess side chain polymers.

obtained was free of its precursor polymers. As shown in Figure 1.3.2C, besides the absorptions (a to h) corresponding to the backbone polymer (see Figure 1.3.2B), the peaks due to poly(Is) side chains (peaks j, k, l, r, s, t, v, x, y, and z) are also present. All the above results indicate that a pure graft copolymer with a narrow MWD and designed backbone and side chain lengths was obtained.

To obtain a pure graft copolymer that is free of its precursors, it is important to control the mole ratio of the chlorosilyl group of the backbone and the anionic living polymer. As shown in Table 1.3.4, if the amount of the latter is larger than that of the former, the excess side chain polymer will remain in the reaction system. For instance, when an excess living poly(St) was employed (GP-1 in Table 1.3.4), its red color was maintained during the whole polymerization process, and the GPC chromatogram exhibited double peaks belonging to the produced graft polymer and the excess side chain poly(St), respectively. When the amount of living polymer was less than that of the chlorosilyl group (GP-2 to GP-7), the color of the living polymer disappeared at once upon the introduction of the toluene solution of the backbone into the reaction system, and each of the graft (co)polymers thus obtained exhibited a single, symmetrical peak. In addition, the molecular weights of the graft (co)polymers determined by VPO are close to those calculated, but much larger than those determined by GPC, because the graft (co)polymers possess smaller hydrodynamic volumes compared to the corresponding linear polymers.

Preparation of the Graft Copolymer with Poly(MMA) Side Chains. A graft copolymer with poly(MMA) side chains was prepared using a procedure similar to that for the graft (co)polymers with poly(St) or poly(Is) side chains. The anionic polymerization of MMA was carried out using the bulky initiator DPHL, in a polar solvent THF, at a low temperature −78°C. To stabilize the propagating site and to avoid side reactions, LiCl ($[LiCl]/[DPHL]_0 = 1.2$) was employed as additive.[5] After the polymerization lasted 20 min, the temperature was raised to −50°C, and the coupling reaction was started by introducing a toluene solution of poly(St-*co*-StSiCl) into the above system. Before quenching, a small amount of solution was taken out for a GPC measurement. As shown in Figure 1.3.3, the formation of the graft copolymer (GP-8 in Table 1.3.4) is confirmed by the emergence of a new peak (C-b), with a larger molecular weight ($M_{n(GPC)} = 2.83 \times 10^4$, $M_w/M_n = 1.15$) than its precursor polymers, namely, the side chain poly(MMA) (A; PMMA-1 in Table 1.3.3, $M_n = 1290$, $M_w/M_n = 1.10$) and the backbone polymer (B; SVS-1 in Table 1.3.1, $M_n = 8080$, $M_w/M_n = 1.04$). The graft copolymer (peak C-b) is accompanied by the excess of poly(MMA) side chain (peak C-a). In contrast to the graft (co)polymers with poly(St) or poly(Is) side chains, this graft copolymer is unstable in proton-containing media, such as methanol or water. Upon the coupling reaction, the product was precipitated in methanol and kept in methanol for about 12 h. As well-known, this constitutes a general purification method for the common polymers, particularly for poly(St)-based (co)polymers. Surprisingly, the GPC result of the polymer thus treated was different from those of the graft copolymers involving poly(St) or poly(Is) side chains. As shown in Figure 1.3.3D, the graft copolymer (D-c) almost completely decomposed back to its precursor polymers (D-a and D-b). To examine this decomposition, the THF solution of the graft copolymer with poly(MMA) side chains (GP-9 in Table 1.3.4) was poured into either water or a dilute HCl aqueous solution. The decomposition took place in both media, and particularly when the acidic solution was used, the decomposition completed almost instantaneously. Therefore, the graft

D-b
D-a
D-c
B
A
C-b
C-a
22
24
26
28
30
Elution time (min)

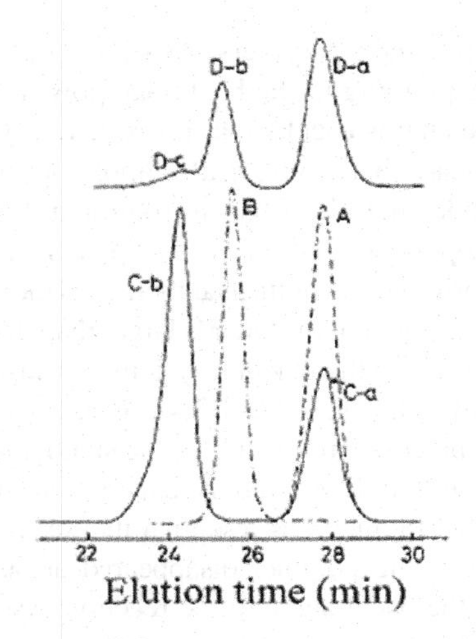

FIGURE 1.3.3 GPC traces of the graft copolymer (C-b; GP-8 in Table 1.3.4) containing poly(MMA) side chains, its precursors (A and B), and its decomposed product (D). A: Living poly(MMA) (PMMA-1 in Table 1.3.3, $M_n = 1290$, $M_w/M_n = 1.10$). B: Poly(St-*co*-VSt) (SVS-1 in Table 1.3.1, $M_n = 8080$, $M_w/M_n = 1.04$). C-b: Graft copolymer ($M_{n(GPC)} = 2.83 \times 10^4$, $M_w/M_n = 1.15$) prepared by the coupling reaction between A and the hydrosilylation product of B (SVSiCl-1 in Table 1.3.2). This graft copolymer is accompanied by the excess of poly(MMA) (C-a). D-a and D-b: Decomposed products from the graft copolymer (D-c).

copolymer with poly(MMA) side chains is stable in organic solvents, such as THF, toluene, etc., but decomposes in the polar solvents containing an active proton, such as alcohol, water, etc. For comparison, the graft copolymers with poly(St) or poly(Is) side chains were also subjected to the same treatment, but no change in the molecular structures was detected even in an acidic environment. As shown in Scheme 1.3.2, the presence of the strong electron-withdrawing carbonyl group of the last MMA unit makes the Si–C linkage (indicated by an arrow) very weak, and for this reason, it is easily fractured in proton-containing media.

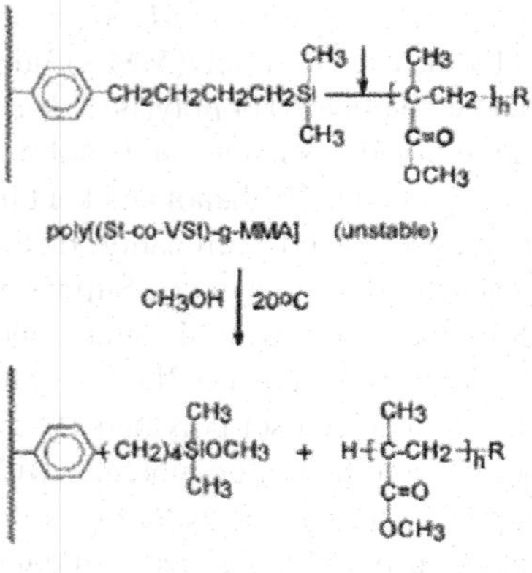

CH3
CH3
CH2CH2CH2CH2Si
C-CH2
R
CH3
C=O
OCH3
poly[(St-co-VSt)-g-MMA]
(unstable)
CH3OH
20oC
CH3
CH3
CH2)4SiOCH3
+
H
C-CH2
R
CH3
C=O
OCH3

SCHEME 1.3.2

1.3.4 CONCLUSIONS

Using *n*-BuLi as initiator, in a mixture of toluene and THF (2:1), at −40°C, the anionic copolymerization of the bifunctional monomer 4-(vinylphenyl)-1-butene (VSt) and styrene (St) could proceed smoothly, generating a copolymer with controlled molecular weight and composition, and very narrow molecular weight distribution (M_w/M_n = 1.03–1.04). The butenyl type C=C double bonds of VSt remained unreacted during the copolymerization process and could be further reacted with chlorodimethylsilane in the presence of the catalyst H_2PtCl_6, generating another new functional copolymer with a reactive chlorosilyl group in each of its VSt units. Subsequently, graft copolymers with poly(St), poly(Is), or poly(MMA) side chains were prepared by the coupling reactions between the chlorosilyl groups of the above copolymer and the anionic living polymers of St, Is, or MMA, respectively. The graft copolymer thus obtained was free of its precursor polymers and possessed a very narrow dispersity (M_w/M_n = 1.05–1.15). The molecular weights of both its backbone and side chains, and hence the total molecular weight of the graft (co)polymer as well as its composition, could be well controlled. In addition, the stability of the graft copolymer with poly(MMA) side chains was different from those of the graft (co)polymers with poly(St) or poly(Is) side chains. The former graft copolymer could be easily decomposed by alcoholysis or hydrolysis at room temperature.

REFERENCES AND NOTES

1. (a) Hsieh, H. L.; Quirk, R. P. *Anionic Polymerization*; Marcel Dekker: New York, 1996. (b) Rempp, P.; Franta, E.; Herz, J. E. *Adv. Polym. Sci.* **1988**, *86*, 145. (c) Norton, M. *Anionic Polymerization: Principles and Practice*; Academic Press: New York, 1983. (d) Van Beylen, M.; Bywater, S.; Smets, G.; Szwarc, M.; Worsfold, D. J. *Adv. Polym. Sci.* **1988**, *86*, 87. (e) Pitsikalis, M.; Pispas, S.; Mays, J. W.; Hadjichristidis, N. *Adv. Polym. Sci.* **1998**, *135*, 1. (f) Kennedy, J. P.; Marechal, E. *Carbocationic Polymerizations*; John Wiley & Sons: New York, 1982; p. 411.
2. (a) Gallot, Y.; Grubisic, Z.; Rempp, P.; Benoit, H. *J. Polym. Sci.* **1968**, *C22*, 527. (b) Ishizu, K.; Fukutomi, T.; Kakurai, T.; Noguchi, T. *Polym. J.* **1973**, *4*, 105. (c) Pitsikalis, M.; Moodward, J.; Mays, J. W.; Hadjichristidis, N. *Macromolecules* **1997**, *30*, 5384. (d) Se, K.; Yamazaki, H.; Shibamoto, T.; Takano, A.; Fujimoto, T. *Macromolecules* **1997**, *30*, 1570. (e) Takaki, M.; Asami, R.; Mizuno, M. *Macromolecules* **1977**, *10*, 845. (f) Zhang, H. M.; Ruckenstein, E. *Macromolecules* **1998**, *31*, 4753. (g) Ruckenstein, E.; Zhang, H. M. *J. Polym. Sci., Polym. Chem.* **1999**, *37*, 105.
3. Zhang, H. M.; Ruckenstein, E. *Macromolecules* **1999**, *32*, 5495.
4. Allen, R. D.; Long. T. E.; McGrath, J. E. *Polym. Bull.* **1986**, *15*, 127.
5. Fayt, R.; Forte, R.; Jacobs, C.; Jerome, R.; Quhadi, T.; Teyssie, Ph.; Varshney, S. K. *Macromolecules* **1987**, *20*, 1442.
6. (a) Kazama, H.; Tezuka, Y.; Imai, K. *Polym. J.* **1987**, *19*, 1091. (b) Roovers, J.; Toporowski, P.; Martin, J. *Macromolecules* **1989**, *22*, 1897.
7. (a) Kawakami, Y.; Miki, Y.; Tsuda, T.; Murthy, R. A. N.; Yamashita, Y. *Polym. J.* **1982**, *14*, 913. (b) Cameron, G. G.; Chisholm, M. S. *Polymer* **1985**, *26*, 437.

1.4 Selective Living Anionic Polymerization of a Novel Bifunctional Monomer 4-(Vinylphenyl)-1-butene and the Preparation of Uniform Size Functional Polymers and Amphiphilic Block Copolymers*

Hongmin Zhang and Eli Ruckenstein

Chemical Engineering Department, State University of New York at Buffalo, Amherst, New York 14260

ABSTRACT A novel bifunctional monomer, namely, 4-(vinylphenyl)-1-butene (VSt), was prepared by the coupling reaction between vinylbenzyl chloride and allylmagnesium chloride with a high yield (93%) and monomer purity. This monomer contains both a styrene type and a 1-butene type C=C double bond. The former double bond can be selectively polymerized by anionic polymerization to generate a polymer with a polystyrene [poly(St)] backbone and functional butenyl side chains. The effects of the initiator, temperature, and solvent were investigated, and the optimum conditions for the selective living anionic polymerization of this monomer were determined. Using *n*-BuLi as initiator, in a mixture of toluene and tetrahydrofuran (5:1–2:1), at −40°C, VSt could undergo anionic polymerization in a living manner without cross-linking and any other side reactions. The polymer thus obtained possesses a controlled molecular weight and a very narrow molecular weight distribution (M_w/M_n = 1.03–1.05). The quantitative

* *Macromolecules* 1999, 32, 5495–5500.

presence of the unreacted 1-butene type C=C double bonds was verified by ^{1}H NMR and FT-IR. The block copolymerization of VSt and St could also proceed smoothly in the polymerization sequence VSt followed by St, or vice versa, to generate a well-defined block copolymer with a controlled molecular weight and composition and a very narrow molecular weight distribution (M_w/M_n = 1.03–1.07). The C=C double bonds of the side chains of poly(VSt) were further reacted first with 9-borabicyclo[3.3.1]nonane (9-BBN), followed by the addition of sodium hydroxide and hydrogen peroxide. This procedure generated a uniform size hydrophilic functional polymer, poly(4-hydroxybutylstyrene), without destroying the main-chain structure of the polymer. This hydroxylation method was also applied to the block copolymer of VSt and St, and a well-defined amphiphilic block copolymer, containing both the hydrophilic poly(4-hydroxybutylstyrene) and the hydrophobic poly(St) segments, was obtained.

1.4.1 INTRODUCTION

Tremendous progress was achieved in the living polymerizations over the past two decades.[1–6] These well-established techniques allow to control the primary structure of the polymers, e.g., the molecular weight, molecular weight distribution (MWD), sequence distribution, stereoregularity, chain-end structures, and branching. The incorporation of functional groups into the polymer chains can greatly improve their properties, such as hydrophilicity, hydrophobicity, biocompatibility, adhesion, etc., thus providing polymers with additional useful characteristics. Therefore, the application of living polymerization methods to functional monomers is a route to high-performance polymers with well-defined molecular architectures.

Well-defined functional polymers have been prepared by combining a protection method with the living anionic polymerization technique.[7–9] The functional group of the monomer was first masked with a suitable group. Then, this protected monomer was subjected to anionic polymerization. Finally, the protecting group was removed under suitable conditions. Compared to this three-step synthetic procedure, the selective living polymerization method of a bifunctional monomer suggested in this paper is a more convenient route to functional polymers. If a monomer contains two functional groups with different polymerizibilities, one of them can be selectively polymerized under certain conditions, using a living polymerization method, without the interference of the other functional group. This one-step living polymerization process can generate a uniform size functional polymer with a functional group in each of its repeating units. For instance, the bifunctional monomer 4-(vinylphenyl)-1-butene (VSt, 1 in Scheme 1.4.1), prepared via the coupling reaction between vinylbenzyl chloride (VBC) and allylmagnesium chloride, contains two C=C double bonds: one is directly connected to the benzene ring, and the other one is separated from the benzene ring by an ethyl group. The latter bond cannot undergo anionic polymerization because ethyl is an electron-donating group. In contrast, the

$CH_2{=}CH{-}C_6H_4{-}CH_2Cl$ $\xrightarrow[-\,MgCl_2]{CH_2{=}CHCH_2MgCl}$ $CH_2{=}CH{-}C_6H_4{-}CH_2CH_2CH{=}CH_2$

VBC 1. VSt

$\xrightarrow[\text{polymerization}]{\text{anionic}}$ $-\!\!\!\left[CH_2{-}CH\right]_n$ ($C_6H_4CH_2CH_2CH{=}CH_2$) $\xrightarrow[2)\ NaOH,\ H_2O_2]{1)\ 9\text{-}BBN}$ $-\!\!\!\left[CH_2{-}CH\right]_n$ ($C_6H_4CH_2CH_2CH_2CH_2OH$)

2. poly(VSt) 3. poly(StOH)

SCHEME 1.4.1

former can be subjected to anionic polymerization because the benzene ring stabilizes the propagating carbanion. Therefore, if the polymerization conditions are carefully selected, the selective anionic polymerization of VSt can generate a functional polymer with a controlled molecular weight, a narrow MWD, and a C=C double bond in each of its repeating units [poly(VSt), 2 in Scheme 1.4.1]. The latter double bonds can be further reacted with numerous reagents, such as 9-bora-bicyclo[3.3.1]nonane (9-BBN),[10] dimethylchlorosilane [$HSi(CH_3)_2Cl$],[11] RCOOOH, etc., to generate other new uniform size functional polymers. In the present paper, the new monomer VSt was first prepared. Then, the effects of the initiator, solvent, and temperature on the selective anionic polymerization of VSt were investigated, and the optimum conditions to obtain a uniform size functional poly(VSt) were determined. Further, the resulting poly(VSt) was reacted with 9-BBN, and this was followed by the addition of sodium hydroxide and hydrogen peroxide. This hydroxylation process generated a new hydrophilic polymer, poly(4-hydroxybutylstyrene) [poly(StOH), 3 in Scheme 1.4.1]. The anionic block copolymerization of VSt and St was also carried out, and a well-defined block copolymer with a controlled molecular weight and composition and a very narrow MWD was obtained. The hydroxylation of the block copolymer generated another uniform size block copolymer with poly(St) hydrophobic and poly(StOH) hydrophilic segments.

1.4.2 EXPERIMENTAL SECTION

Materials. Tetrahydrofuran (THF) was dried with CaH_2 under reflux for more than 24 h, distilled, and distilled again from a solution of sodium naphthalenide just before use. Toluene was washed with concentrated sulfuric acid and then with water, dried with $MgSO_4$, distilled over CaH_2, and finally distilled from a *n*-BuLi solution. Hexane was first dried and distilled over CaH_2 and then distilled from a solution of *n*-BuLi. Vinylbenzyl chloride (VBC; Aldrich, 97%) was dried with CaH_2 and distilled under reduced pressure. Styrene (St; Aldrich, 99%) was washed with 10% aqueous NaOH solution and then with water, dried overnight with $MgSO_4$, distilled over CaH_2, and finally distilled in the presence of phenylmagnesium chloride prior to polymerization. *n*-BuLi (Aldrich, 1.6 M solution in hexane) and *sec*-BuLi (Aldrich, 1.3 M solution in cyclohexane) were diluted with purified hexane. Sodium naphthalenide was prepared by reacting a small excess amount of naphthalene with sodium metal in THF at room temperature. Allylmagnesium chloride (AMC; Aldrich, 2.0 M solution in THF), 9-borabicyclo[3.3.1]nonane (9-BBN; Aldrich, 0.5 M solution in THF), and hydrogen peroxide (H_2O_2; Fisher, 30 wt% aqueous solution) were used as received.

Synthesis of 4-(Vinylphenyl)-1-butene (VSt). VSt was prepared through the coupling reaction between VBC and AMC. A 1000 mL round-bottom flask, equipped with a condenser, a paddle stirrer, and a dropping funnel with a pressure-equalization arm, was degassed and heated under reduced pressure. To this flask, 170 mL of THF solution of AMC (0.340 mol) was carefully added under the protection of N_2. Further, this solution was diluted by adding 100 mL of purified THF and cooled to 0°C. Then, 50.0 g (0.318 mol) of VBC was dropwise added in 1.5 h with stirring, and the stirring was continued for an additional 2.5 h. After a saturated aqueous solution of NH_4Cl was added slowly, the mixture was washed three times with water, evaporated to remove the THF, and distilled under reduced pressure (bp: 43°C/0.2 Torr, yield: 93% based on the amount of VBC employed). The monomer thus obtained was twice distilled in the presence of phenylmagnesium chloride prior to polymerization. As shown later in Figure 1.4.3A, the chemical shifts and their intensities in the ^{1}H NMR spectrum of the prepared VSt are consistent with its molecular structure.

Polymerization. The anionic homopolymerization of VSt and its block copolymerization with St were carried out in a round-bottom glass flask, under an overpressure of argon, with magnetic stirring, at selected temperatures. After the solvent(s) (THF, toluene, or both) were added with dry syringes, the flask was placed in a bath kept at a selected temperature, and an initiator (*n*-BuLi in hexane, *sec*-BuLi in cyclohexane, or sodium naphthalenide in THF) was added. The polymerization was induced by adding prechilled VSt. For the block copolymerization of VSt and St, the sequence VSt followed by St, or vice versa, was used. The system was quenched by adding a small amount of methanol, and the (co)polymer was precipitated by pouring the polymerization solution into a large amount of methanol. The (co)polymer thus obtained was washed with methanol and vacuum-dried at 40°C for more than 24 h.

Hydroxylation of the (Co)polymer. The hydroxylation of poly(VSt) or poly(VSt-*b*-St) was carried out under the protection of N_2 in a round-bottom flask containing a magnetic stirring bar. For instance, 1.2 g of poly(VSt) (PVSt-16, Tables 1.4.1 and 1.4.3) was dissolved in 30 mL of purified THF. To this system, 23 mL of THF solution of 9-BBN (0.5 M) was added at room temperature, and the reaction was allowed to last 16 h. After 2.4 mL of aqueous solution of NaOH (5.0 M) was introduced into the system, the flask was cooled to 10°C, and 4.0 mL of aqueous solution of H_2O_2 (30 wt%) was dropwise added carefully with a syringe to avoid the decomposition of H_2O_2 caused by a temperature increase.[10] Then, the stirring was continued for an additional 4 h at room temperature. After the system was concentrated by evaporation to about 20 mL, the polymer was precipitated by pouring the polymer solution into a large amount of a mixture of water and methanol (9:1), washed with water and then with methanol, and finally vacuum-dried at 40°C for 24 h.

Measurements. ^{1}H NMR spectra were recorded in $CDCl_3$ or 1,4-dioxane-d_8 on an INOVA-400 spectrometer. M_n and M_w/M_n of the polymer were determined by gel permeation chromatography (GPC) on the basis of a polystyrene calibration curve. The GPC measurements were carried out using THF as solvent, at 30°C,

TABLE 1.4.1
Anionic Polymerization of VSt

No.	Initiator (mM)	$[VSt]_0$ (M)	Time (min)	Solvent[a]	Temp (°C)	Conv (%)	$10^{-3}M_{k}$[b]	$10^{-3}M_n$[c]	M_w/M_n[c]	f[d] (%)
PVSt-1	*n*-BuLi 21.1	0.523	40	THF	−70	100	3.97	3.82	1.31	100
PVSt-2	*n*-BuLi 6.8	0.723	40	THF	−70	100	16.8	22.1	1.36	76
PVSt-3	Na-Naph[e] 18.8	0.435	40	THF	−70	100	7.37	22.3	1.58	33
PVSt-4	*sec*-BuLi 21.1	0.523	40	Tol	20	[h]				
PVSt-5	*n*-BuLi 21.1	0.523	40	Tol	20	[h]				
PVSt-6	*n*-BuLi 24.3	0.523	40	Tol/THF[g]	20	100	3.45	17.6	1.32	20
PVSt-7	*n*-BuLi 24.3	0.523	40	Tol/THF = 40/1	0	89	3.08	9.63	1.15	32
PVSt-8	*n*-BuLi 24.3	0.523	40	Tol/THF = 40/1	−10	80	2.77	6.89	1.12	40
PVSt-9	*n*-BuLi 24.3	0.523	40	Tol/THF = 40/1	−20	51	1.79	3.89	1.11	46
PVSt-10	*n*-BuLi 24.3	0.523	40	Tol/THF = 40/1	−40	35	1.19	2.11	1.14	54
PVSt-11	*n*-BuLi 24.3	0.523	40	Tol/THF = 10/1	−40	82	2.84	5.08	1.13	56
PVSt-12	*n*-BuLi 20.0	0.497	40	Tol/THF = 5/1	−40	100	3.98	3.92	1.04	100
PVSt-13	*n*-BuLi 20.0	0.497	40	Tol/THF = 2/1	−40	100	3.98	3.88	1.03	100
PVSt-14	*n*-BuLi 20.0	0.497	40	Tol/THF = 1/1	−40	100	3.98	3.79	1.08	100
PVSt-15	*n*-BuLi 20.0	0.497	40	Tol/THF = 1/2	−40	100	3.98	4.12	1.17	100
PVSt-16	*n*-BuLi 7.4	0.516	60	Tol/THF = 2/1	−40	100	11.1	11.4	1.04	97
PVSt-17	*n*-BuLi 7.0	0.647	60	Tol/THF = 2/1	−40	100	14.6	15.2	1.03	96
PVSt-18	*n*-BuLi 6.6	0.651	60	Tol/THF = 2/1	−40	100	15.6	17.1	1.04	91
PVSt-19	*n*-BuLi 6.9	0.966	80	Tol/THF = 2/1	−40	100	22.1	24.8	1.07	89

[a] Tol = toluene; THF = tetrahydrofuran; Tol/THF = the volume ratio of the two solvents.
[b] Calculated M_n.
[c] Determined by GPC on the basis of a polystyrene calibration curve.
[d] Initiator efficiency.
[e] Sodium naphthalenide.
[g] $[THF] = [n\text{-BuLi}]_0 = 24.3$ mM.
[h] Trace.

at a 1.0 mL/min flow rate and a 1.0 cm/min chart speed. Three polystyrene gel columns (Waters, 7.8 × 300 mm; one HR 5E, part no. 44228, one Linear, part no. 10681, and one HR 4E, part no. 44240) were used, which were connected to a Waters 515 precision pump. The FT-IR spectra were recorded on a Perkin-Elmer 1760-X spectrometer using a film prepared from a 1,4-dioxane [for poly(VSt)] or a THF [for poly(StOH)] solution.

1.4.3 RESULTS AND DISCUSSION

Selective Living Anionic Polymerization of VSt. It is well-known that the living anionic polymerization of St can proceed smoothly either in a polar solvent, such as THF, at a low temperature, or in a nonpolar solvent, such as benzene and toluene, at room temperature. Because VSt has a molecular structure similar to that of St, its anionic polymerization was first examined under the optimum conditions determined for St. When the polymerization was carried out in THF, at −70°C, using *n*-BuLi as initiator, poly(VSt) with a broader MWD was obtained quantitatively (PVSt-1 and 2 in Table 1.4.1, M_w/M_n = 1.31–1.36). As shown in Figure 1.4.1a, the GPC chromatogram of the obtained polymer exhibits an unsymmetrical peak with a heavy tailing in the low molecular weight region. When the initiator *n*-BuLi was replaced with sodium naphthalenide, besides the above tailing, the resulting polymer possessed an even broader MWD (M_w/M_n = 1.58) and an uncontrolled molecular weight (PVSt-3). The above results indicate that a deactivation of the propagating sites occurred during the polymerization process. In this polar solvent (THF), the propagating carbanion has a very high reactivity and may deactivate by capturing a proton from the butenyl of VSt, especially from the two protons of the methylene directly connected to the benzene ring.

When the polymerization of VSt was carried out in a nonpolar solvent (toluene), at a high temperature (20°C), using *sec*-BuLi or *n*-BuLi as initiator, only a trace amount of polymer was obtained (PVSt-4 and 5 in Table 1.4.1), which is quite different from that obtained via the anionic polymerization of St. Because of the high temperature

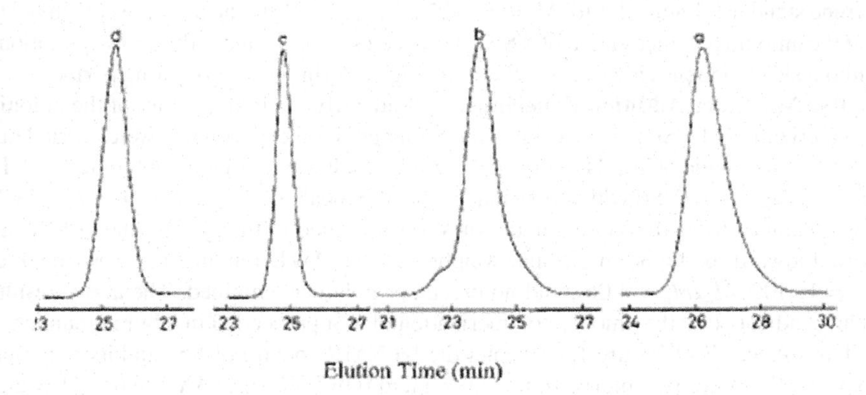

FIGURE 1.4.1 GPC traces of poly(VSt)s. (a) PVSt-1 (Table 1.4.1, M_n = 3820, M_w/M_n = 1.31) prepared in THF at −70°C. (b) PVSt-6 (M_n = 17 600, M_w/M_n = 1.32) prepared in toluene at 20°C in the presence of a trace amount of THF ([THF] = [*n*-BuLi]$_0$ = 24.3 mM). (c) PVSt-17 (M_n = 15 200, M_w/M_n = 1.03) prepared in a mixture of toluene and THF (2:1) at −40°C. (d) Poly(St) standard (Pressure Chemical, lot no. 8b; M_n = 10 000, M_w/M_n < 1.06).

and the low propagating rate in the nonpolar solvent, the side reaction of the initiator with the butenyl protons of VSt was dominant. When, however, a trace amount of THF ([THF] = [*n*-BuLi]$_0$) was added to the latter system, the monomer was polymerized quantitatively (PVSt-6 in Table 1.4.1). Consequently, the introduction of THF into the nonpolar solvent is beneficial because it increases both the polymerization rate and the initiator efficiency. However, the initiator efficiency was still very low (20%), and in the GPC chromatogram of the obtained polymer, a shoulder appeared in the high molecular weight side (Figure 1.4.1b), which most likely represents a cross-linked polymer whose amount increased as the polymerization lasted longer. At the high temperature of 20°C, the C=C double bond of the butenyl of VSt may also participate to some extent in the polymerization, resulting in the cross-linking side reaction.

The effect of the temperature on the selective anionic polymerization of VSt was investigated in toluene from 0 to −40°C (PVSt-7–10), in the presence of a small amount of THF (Tol/THF = 40/1, volume ratio). As shown in Table 1.4.1, as the temperature decreased from 0 to −40°C, the initiator efficiency increased from 32% to 54%, but the monomer conversion decreased from 89% to 35%. This means that the low temperature is beneficial in depressing the side reactions in the initiating period but also decreases the rate of the propagating reaction.

Keeping the temperature at −40°C, the effect of the volume ratio of toluene and THF (Tol/THF) on the polymerization of VSt was investigated (PVSt-11–15 in Table 1.4.1). For the ratios Tol/THF = 10/1, 5/1, 2/1, 1/1 and 1/2, the best results were obtained in the range 5/1 to 2/1. As shown in Table 1.4.1, the monomer conversion was 100% in that range, the determined molecular weight was in good agreement with that calculated, and the MWD was very narrow (M_w/M_n = 1.03 and 1.04). In addition, by changing the feed amounts of the initiator and the monomer, uniform size poly(VSt)s with different molecular weights were prepared (PVSt-16–19 in Table 1.4.1). As shown in Figure 1.4.1c, the GPC chromatogram of the obtained polymer exhibits a very sharp and symmetrical peak (PVSt-17, M_n = 15 200, M_w/M_n = 1.03), comparable to that of the polystyrene standard (Figure 1.4.1d, M_n = 10 000, M_w/M_n < 1.06). The above results indicate that the anionic polymerization of VSt can proceed smoothly under the above conditions without side reactions either during the initiating or during the propagating period.

Two Monomer Addition Experiments. To identify the living nature of the anionic polymerization of VSt, two successive monomer addition experiments were carried out at −40°C, for a ratio of Tol/THF of 2/1. The GPC results are presented in Figure 1.4.2. In the first stage, poly(VSt) with a very narrow MWD (peak a, M_n = 5480, M_w/M_n = 1.03) was obtained. After the same amount of VSt was added to the system, the GPC peak shifted toward the higher molecular weight, but the MWD remained narrow (peak b, M_n = 11 100, M_w/M_n = 1.06), and no precursor polymer remained. The above results clearly indicate that the anionic polymerization of VSt proceeded in a living manner.

Parts A and B of Figure 1.4.3 depict the ^{1}H NMR spectra of VSt and its polymer, respectively. After polymerization, peaks a and b in Figure 1.4.3A due to CH_2= and =CH– of the styrene type vinyl disappeared completely, and two new absorptions corresponding to –CH_2– and –CH– of poly(VSt) main chain emerged at 1.41 and 1.79 ppm, respectively. On the other hand, the peaks f and g due to –CH= and =CH_2 of the butenyl remained unreacted after polymerization. Furthermore, as shown later in Table 1.4.4, the resulting poly(VSt) is soluble in a number of solvents, such as

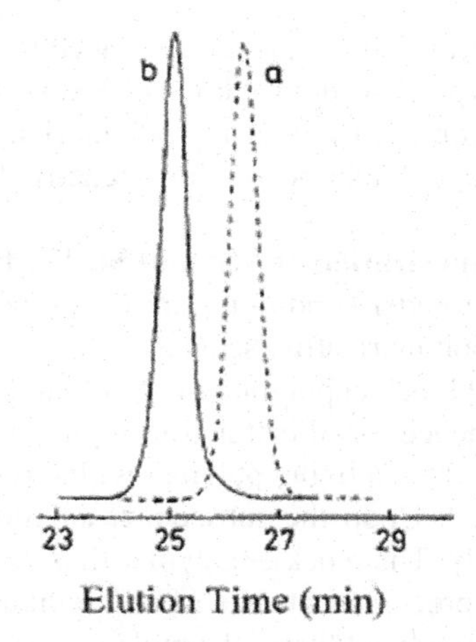

FIGURE 1.4.2 GPC traces of poly(VSt)s obtained by two monomer addition experiments. (a) First monomer addition, $M_n = 5480$, $M_w/M_n = 1.03$ ([n-BuLi]$_0$ = 18.2 mM, [VSt]$_0$ = 0.610 M, Tol/THF = 2/1). (b) Second monomer addition, M_n = 11 100, M_w/M_n = 1.06.

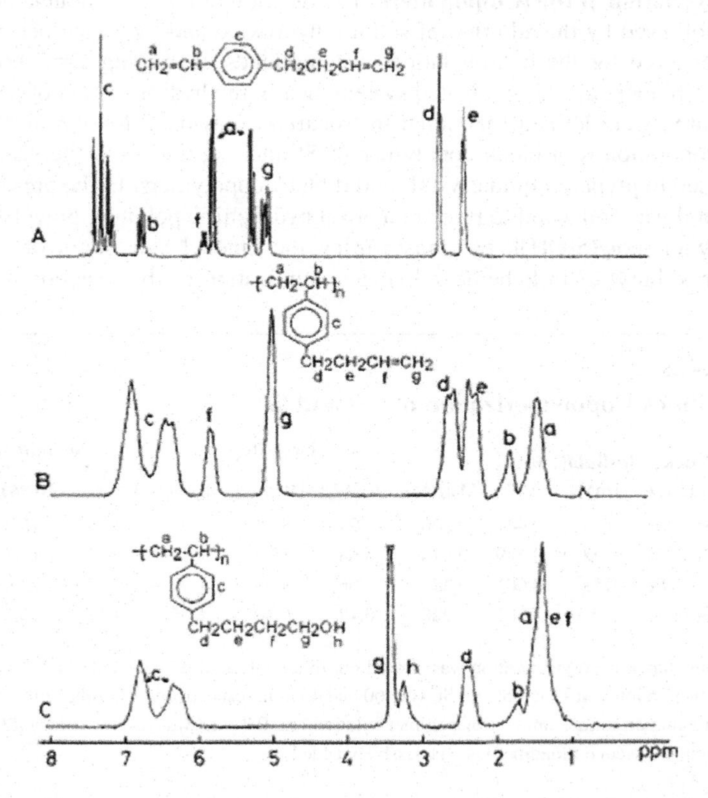

FIGURE 1.4.3 ^{1}H NMR spectra of VSt (A; in $CDCl_3$), poly(VSt) (B; in $CDCl_3$; see PVSt-16 in Table 1.4.1), and its hydroxylate poly(StOH) (C; in 1,4-dioxane-d_8, see PStOH-16 in Table 1.4.3).

hexane, benzene, chloroform, THF, 1,4-dioxane, acetone, etc. Therefore, one can conclude that the styrene type C=C double bond of VSt was selectively polymerized, that no cross-linking reaction occurred during polymerization, and that the polymer thus obtained is a functional polymer possessing a reactive C=C double bond in each of its repeating units.

Anionic Block Copolymerization of VSt and St. The block copolymerization of VSt and St was carried out under conditions similar to those for the homopolymerization of VSt, using the polymerization sequence VSt–St, or vice versa. As shown in Table 1.4.2, well-defined block copolymers were obtained in each case, regardless of the polymerization sequence. As illustrated in Figure 1.4.4, the first step anionic polymerization of VSt generated a living polymer with a very narrow MWD (peak a, M_n = 5920, M_w/M_n = 1.04). Upon the subsequent addition of St, the polymerization continued quantitatively. The block copolymer thus obtained possessed a higher molecular weight than its precursor, and its MWD remained very narrow (peak b, M_n = 19 500, M_w/M_n = 1.03). In addition, the molecular weight and the composition of each of the block copolymers can be well-controlled. Therefore, the anionic block copolymerization of VSt and St not only generated well-defined block copolymer but also demonstrated the living polymerization characteristics of VSt.

Hydroxylation of the (Co)polymers. The reaction of the C=C double bond with 9-BBN, followed by the addition of sodium hydroxide and hydrogen peroxide, has been often used for the hydroxylation of compounds containing C=C bonds. For instance, Chung et al.[10a] investigated systematically the hydroboration of polydienes and prepared well-defined functional hydrocarbon polymers. Mao et al.[10d] applied this hydroboration to a block copolymer of St and isoprene, and the product was further used to prepare liquid crystal–coil diblock copolymers. In the present paper, this method was employed to prepare a novel hydrophilic polymer, poly(4-hydroxybutylstyrene) [poly(StOH)], by transforming the butenyl side chain of poly(VSt) to 4-hydroxybutyl (3 in Scheme 1.4.1). The application of this reaction to a block

TABLE 1.4.2
Anionic Block Copolymerization of VSt and St[a]

					First Step			Second Step		
No.	Block Sequence	$[n\text{-BuLi}]_0$ (mM)	$[M_1]_0$ (M)	$[M_2]_0$(M)	$10^{-3}M_k$[b]	$10^{-3}M_n$[c]	M_w/M_n[c]	$10^{-3}M_k$[b]	$10^{-3}M_n$[c]	M_w/M_n[c]
VbS-1	VSt → St	8.90	0.447	0.566	7.99	8.29	1.04	14.6	15.3	1.04
VbS-2	VSt → St	7.00	0.249	0.876	5.68	5.92	1.04	18.7	19.5	1.03
SbV-1	St → VSt	15.6	0.601	0.435	4.06	4.39	1.06	8.47	7.96	1.07
SbV-2	St → VSt	10.2	0.601	0.220	6.18	6.40	1.05	9.59	10.5	1.06

[a] The anionic block copolymerization was carried out in a mixture of toluene and THF (2:1) at −40°C. The polymerization times for VSt and St were 60 and 40 min, respectively. After the polymerization of the first monomer, a trace amount of polymer solution (ca. 0.2 mL) was taken out for GPC measurement. Then, the second monomer was successively added.

[b] Calculated M_n.

[c] Determined by GPC on the basis of a polystyrene calibration curve.

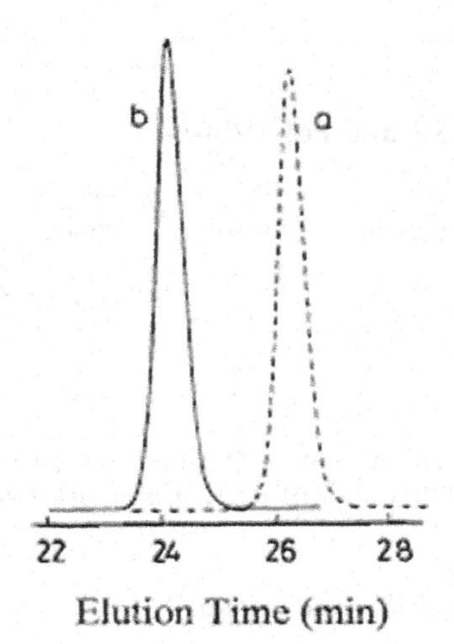

FIGURE 1.4.4 GPC traces of the block copolymer of VSt and St (peak b, VbS-2 in Table 1.4.2, $M_n = 19\ 500$, $M_w/M_n = 1.03$) and its living poly(VSt) precursor (peak a, $M_n = 5920$, $M_w/M_n = 1.04$).

copolymer of VSt and St generated an amphiphilic block copolymer containing both poly(StOH) hydrophilic and poly(St) hydrophobic segments.

As shown in Table 1.4.3, the hydroxylation was carried out in THF at room temperature. The C=C double bonds of the side chains of poly(VSt) were allowed to react with an excess of 9-BBN overnight, and subsequently, aqueous solutions of NaOH and H_2O_2 were added successively. The ^{1}H NMR spectra of the precursor poly(VSt) and its hydroxylate, poly(StOH), are presented in Figure 1.4.3B and C, respectively. As above, the C=C double bond (peaks f and g in Figure 1.4.3B) of the butenyl remained unreacted after the polymerization of VSt. However, upon hydroxylation, these two peaks disappeared completely, and two new peaks emerged at 3.25 and 3.40 ppm (peaks h and g in Figure 1.4.3C), corresponding to –OH and –CH_2O– of the 4-hy-droxybutyl side chain of poly(StOH). This means that the butenyl side chains of poly(VSt) completely changed to 4-hydroxybutyl side chains. This result was also supported by the FT-IR measurements. As shown in Figure 1.4.5, the absorption (A-a; 1620 cm^{-1}) due to the C=C double bonds in poly(VSt) side chains disappeared completely after hydroxylation, and a broad band (B-b; 3100–3600 cm^{-1}) corresponding to the hydroxyl group of poly(StOH) appeared. For the hydroxylation of the block copolymers of VSt and St, similar ^{1}H NMR and FT-IR results were obtained.

To verify whether the main chain of poly(VSt) was not destroyed during hydroxylation, GPC measurements for the resulting polymers were carried out, but no MWD change was observed. For instance, the molecular weight of the block copolymer VbS-1 (Table 1.4.2) increased somewhat after hydroxylation (from $M_n = 15\ 300$ to $M_n = 16\ 200$) due to the addition of water to the C=C double bond of its side chain.

TABLE 1.4.3
Hydroxylation of Poly(VSt) and Poly(VSt-*b*-St)[a]

No.	Polymer[b]	C=C[c] (mmol)	9-BBN[d] (mmol)	NaOH[e] (mmol)	H_2O_2[f] (mmol)	Functionality[g] (%)
PStOH-12	PVSt-12	6.3	13	13	39	100
PStOH-16	PVSt-16	7.6	12	12	36	100
h-VbS-1	VbS-1	5.0	10	10	30	100
h-SbV-1	SbV-1	7.0	14	14	42	100

[a] The reaction between the C=C double bond and 9-BBN was carried out in THF at room temperature with magnetic stirring for 16 h. Then, NaOH and H_2O_2 were added successively, and the reaction was allowed to last 4 h.
[b] 4.0 g/100 mL of THF solution.
[c] The amount of C=C double bond in the (co)polymer.
[d] 0.5 M THF solution.
[e] 5.0 M aqueous solution.
[f] 30 wt% aqueous solution.
[g] Determined by ^1H NMR.

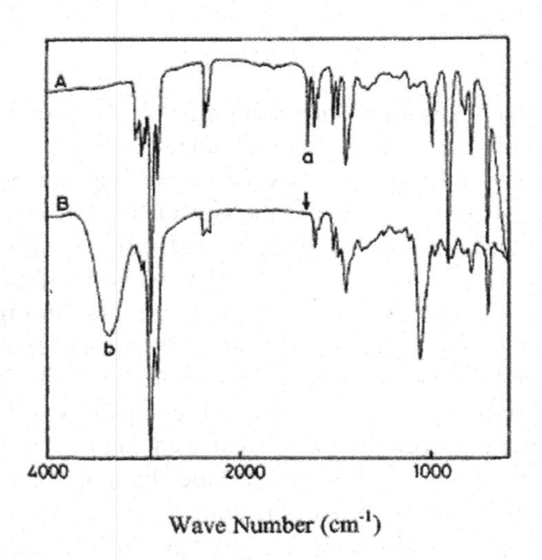

FIGURE 1.4.5 FT-IR spectra of poly(VSt) (A, PVSt-16 in Table 1.4.1) and its hydroxylate poly(StOH) (B, PStOH-16 in Table 1.4.3). Peaks A-a and B-b correspond to C=C double bonds in poly(VSt) side chains and hydroxy groups in poly(StOH) side chains, respectively.

However, its MWD remained unchanged, and the GPC chromatogram still exhibited a single, symmetrical peak (Figure 1.4.6). The above results clearly indicate that the poly(VSt) was changed to another functional polymer poly(StOH) and that this hydroxylation process has not affected the main chain of the polymer. For this reason, poly(StOH) is also a uniform size functional polymer.

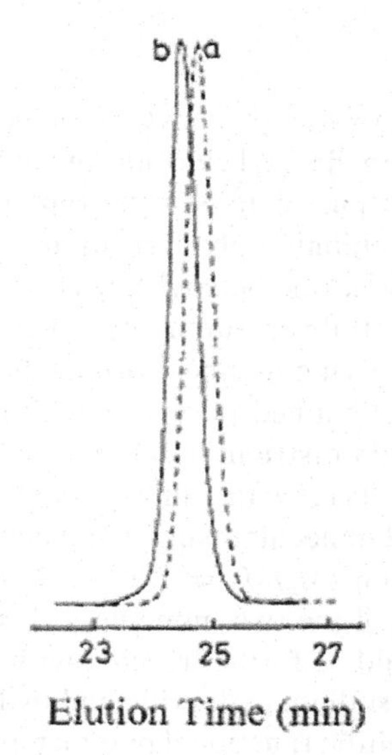

FIGURE 1.4.6 GPC traces of the block copolymer, poly(VSt-*b*-St) (peak a; VbS-1 in Table 1.4.2, M_n = 15 300, M_w/M_n = 1.04) and its hydroxylate (peak b; M_n = 16 200, M_w/M_n = 1.06).

Compared to their precursors, the hydroxylates of poly(VSt) and poly(VSt-*b*-St) exhibit different solubilities. As shown in Table 1.4.4, in contrast to poly(VSt), which is insoluble in methanol and ethanol, its hydroxylate is soluble in these polar solvents but insoluble in nonpolar solvents, such as hexane, benzene, chloroform, etc. On the other hand, because of the coexistence of hydrophilic poly(StOH) and hydrophobic poly(St) segments in the hydroxylate of the block copolymer, the resulting poly-(StOH-*b*-St) is soluble in those solvents that are good solvents for both segments, such as THF, 1,4-dioxane, etc. It is, however, wetted by some nonpolar, such as benzene, and polar, such as alcohols, solvents.

TABLE 1.4.4
Solubility before and after Hydroxylation[a]

No.	Hexane	Benzene	$CHCl_3$	THF	Dioxane	Acetone	CH_3OH	C_2H_5OH	H_2O
PVSt-16	S	S	S	S	S	S	I	I	I
PStOH-16	I	I	I	S	S	S	S	S	I
VbS-1	S	S	S	S	S	S	I	I	I
h-VbS-1	I	W	D	S	S	S	W	W	I

[a] The experiment was carried out at room temperature. The amounts of (co)polymer and solvent were 0.03 g and 1.0 mL, respectively. S = soluble; D = dissolved slowly; W = wetting; I = insoluble. h-VbS-1 = hydroxylate of VbS-1.

1.4.4 CONCLUSIONS

A new bifunctional monomer 4-(vinylphenyl)-1-butene (VSt) was prepared, by the coupling reaction between vinylbenzyl chloride and allylmagnesium chloride, with a high yield (93%) and high purity. Its styrene type C=C double bond could be selectively polymerized by anionic polymerization to generate a polymer with a polystyrene [poly(St)] backbone and butenyl side chains. Using *n*-BuLi as initiator, in a mixture of toluene and tetrahydrofuran (5:1–2:1), at −40°C, VSt could undergo anionic polymerization in a living manner without crosslinking or any other side reaction. The polymer thus obtained possesses a controlled molecular weight, a very narrow molecular weight distribution (M_w/M_n = 1.031.05), and an unreacted C=C double bond in each of its repeating units. A well-defined block copolymer of VSt and St with a controlled molecular weight and composition and a very narrow molecular weight distribution (M_w/M_n = 1.03–1.07) could also be obtained via the block copolymerization of the two monomers. The C=C double bonds of the side chains of poly(VSt) could be further reacted with 9-borabicyclo[3.3.1]nonane, followed by the addition of sodium hydroxide and hydrogen peroxide, to generate another uniform size hydrophilic functional polymer, poly(4-hydroxybutylstyrene). This hydroxylation process could also be applied to the block copolymer of VSt and St, and a well-defined amphiphilic block copolymer containing both the hydrophilic poly(4-hydroxybutylstyrene) and the hydrophobic poly(St) segments could be obtained.

REFERENCES

1. (a) Szwarc, M. *Adv. Polym. Sci.* **1983**, *49*, 1. (b) Van Beylen, M.; Bywater, S.; Smets, G.; Szwarc, M.; Worsfold, D. J. *Adv. Polym. Sci.* **1988**, *86*, 87. (c) Rempp, P.; Franta, E.; Herz, J. E. *Adv. Polym. Sci.* **1988**, *86*, 145. (d) Norton, M. *Anionic Polymerization: Principles and Practice*; Academic Press: New York, 1983. (e) Hsieh, H. L.; Quirk, R. P. *Anionic Polymerization*; Marcel Dekker: New York, 1996.
2. (a) Majoros, I.; Nagy, A.; Kennedy, J. P. *Adv. Polym. Sci.* **1994**, *112*, 1. (b) Sawamoto, M. *Prog. Polym. Sci.* **1991**, *16*, 111. (c) Matyjaszewski, K., Ed. *Cationic Polymerization: Mechanism, Synthesis, and Applications*; Marcel Dekker: New York, 1996.
3. (a) Grubbs, R. H.; Tumas, W. *Science.* **1989**, *243*, 907. (b) Ivin, K., Saegusa, T., Eds. *Ring Opening Polymerization*; Elsevier: New York, 1984.
4. (a) Webster, O. W. *Science.* **1991**, *251*, 887. (b) Webster, O. W.; Hertler, W. R.; Sogah, D. Y.; Farnham, W. B.; RajanBabu, T. V. *J. Am. Chem. Soc.* **1983**, *105*, 5706.
5. (a) Aida, T.; Inoue, S. *Macromolecules.* **1981**, *14*, 1166. (b) Aida, T.; Maekawa, Y.; Asano, S.; Inoue, S. *Macromolecules.* **1988**, 21, 1195.
6. (a) Matyjaszewski, K.; Woodworth, B. E.; Metzner, N. *Macromolecules.* **1998**, *31*, 7999. (b) Otsu, T.; Matsumoto, A. *Adv. Polym. Sci.* **1998**, *136*, 75.
7. (a) Nakahama, S.; Hirao, A. *Prog. Polym. Sci.* **1990**, *15*, 299. (b) Hirao, A.; Ishino, Y.; Nakahama, S. *Macromolecules.* **1988**, *21*, 261. (c) Mori, H.; Hirao, A.; Nakahama, S. *Macromolecules.* **1994**, *27*, 35.
8. (a) Kase, T.; Imahori, M.; Kazama, T.; Isono, Y.; Fujimoto, T. *Macromolecules.* **1991**, *24*, 1714. (b) Kijima, M.; Se, K.; Fujimoto, T. *Polymer.* **1992**, *33*, 2402.
9. (a) Ruckenstein, E.; Zhang, H. M. *J. Polym. Sci., Polym. Chem.* **1998**, *36*, 1865. (b) Zhang, H. M.; Ruckenstein, E. *Macromolecules.* **1998**, *31*, 7575.

10. (a) Chung, T. C.; Raate, M.; Berluche, E.; Schulz, D. N. *Macromolecules.* **1988**, *21*, 1903. (b) Kitayama, T.; Nakagawa, O.; Kishiro, S.; Nishiura, T.; Hatada, K. *Polym. J.* **1993**, *25*, 707. (c) Bayer, U.; Stadler, R. *Macromol. Chem. Phys.* **1994**, *195*, 2709. (d) Mao, G.; Wang, J.; Clingman, S. R.; Ober, C. K.; Chen, J. T.; Thomas, E. L. *Macromolecules* **1997**, *30*, 2556.
11. (a) Kawakami, Y.; Miki, Y.; Tsuda, T.; Murthy, R. A. N.; Yamashita, Y. *Polym. J.* **1982**, *14*, 913. (b) Cameron, G. G.; Chisholm, M. S. *Polymer.* **1985**, *26*, 437. (c) Kazama, H.; Tezuka, Y.; Imai, K. *Polym. J.* **1987**, *19*, 1091. (d) Roovers, J.; Toporowski, P.; Martin, J. *Macromolecules.* **1989**, *22*, 1897.

1.5 Living Anionic Copolymerization of 1-(Alkoxy)ethyl Methacrylates with Polar and/or Nonpolar Monomers and the Preparation of Amphiphilic Block Copolymers Containing Poly(methacrylic acid) Hydrophilic Segments at Higher Temperatures than Usually Employed*

Eli Ruckenstein and Hongmin Zhang

Chemical Engineering Department, State University of New York at Buffalo, Amherst, New York 14260

ABSTRACT The anionic copolymerization of each of the following three novel methacrylate monomers–1-(ethoxy)ethyl methacrylate (EEMA), 1-(butoxy)ethyl methacrylate (BEMA), and 1-(tert-butoxy)ethyl methacrylate

* *Macromolecules* 1998, 31, 9127–9133.

(tBEMA)—with methyl methacrylate (MMA) and/or styrene (St) was carried out. (1) The random copolymerization with MMA proceeded smoothly in tetrahydrofuran (THF), using 1,1-diphenyl-hexyllithium (DPHL) as the initiator, in the presence of LiCl ($[LiCl]/[DPHL]_0 = 1$), at −40°C. The copolymer thus obtained possessed controlled molecular weight and composition, and its molecular weight distribution (MWD) was narrow ($M_w/M_n = 1.08–1.10$). (2) The block copolymer of each of the new monomers with MMA was prepared by the sequential anionic polymerization of the two monomers; the polymerization sequence MMA–new monomer controlled better the molecular weight and led to a narrower MWD than the inverse one. (3) A well-controlled block copolymerization of St with EEMA or with tBEMA was achieved at higher temperatures (≥ −35°C) than usually employed (−78°C), and it should be emphasized that, even at 0°C, a well-defined diblock copolymer consisting of poly(St) and poly(tBEMA) could be obtained. (4) A block copolymer consisting of poly(St) and a random copolymer of MMA and EEMA, poly[St-*b*-(MMA-*co*-EEMA)], was prepared by the sequential monomer addition St–mixture of MMA and EEMA, for various weight ratios of MMA and EEMA. (5) In the preparation of the triblock copolymer with the sequence St, MMA, and EEMA, the molecular weight of the polymer increased step by step and the MWD remained narrow ($M_w/M_n = 1.09$). By changing the polymerization sequence, the hydrophilic segment could be located either in the middle or at the end of the copolymer chain. The protecting group, 1-(alkoxy)ethyl of each of the new monomers, could be easily eliminated after copolymerization, using a mild acidic environment. Thus a copolymer, containing poly(MAA) as hydrophilic segment, with different solubility than its precursor copolymer could be obtained.

1.5.1 INTRODUCTION

The amphiphilic block copolymers containing poly(methacrylic acid) (poly(MAA)) as hydrophilic segment have been widely investigated because of their interesting surface properties.[1] These copolymers exhibit microphase separation in solid state and form spherical micelles in selective solvents. The polar solvents allow the formation of micelles having hydrophobic cores. In addition, the copolymers containing poly(MAA) segments are useful in the preparation of ionomers and microgel particles.[2]

The block copolymers of MAA and polar monomers, such as alkyl methacrylates, could be prepared using the group transfer polymerization (GTP) method,[3] by protecting the carboxyl group of MAA with *tert*-butyl, benzyl, or trimethylsilyl. However, the block copolymer of MAA and a nonpolar monomer, such as styrene (St), cannot be obtained by this method, because GTP cannot be applied to nonpolar monomers. To prepare well-defined block copolymers of MAA and St, the anionic block copolymerization of *tert*-butyl methacrylate and St was carried out in tetrahydrofuran (THF) at −78°C.[4] This was followed by the elimination of the *tert*-butyl side chains of the obtained copolymer. The latter step was usually accomplished by hydrolysis, using either an aqueous hydrochloric acid in 1,4-dioxane at 85°C for about 5 h or *p*-toluenesulfonic acid in wet toluene at 80°C for 4 h.

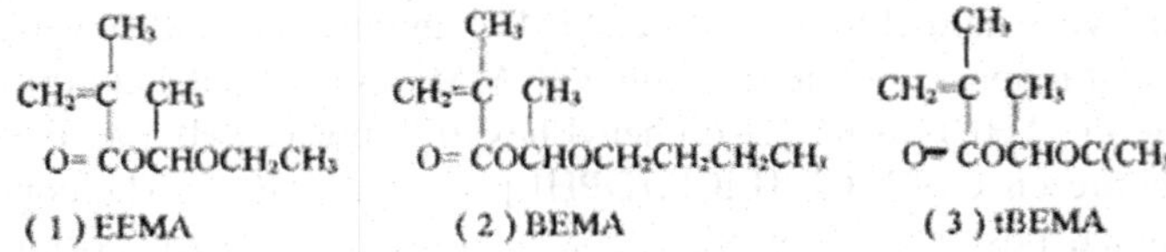

$CH_2{=}C(CH_3)$ – $O{=}COCH(CH_3)OCH_2CH_3$ (1) EEMA

$CH_2{=}C(CH_3)$ – $O{=}COCH(CH_3)OCH_2CH_2CH_2CH_3$ (2) BEMA

$CH_2{=}C(CH_3)$ – $O{=}COCH(CH_3)OC(CH_3)_3$ (3) tBEMA

FIGURE 1.5.1 Molecular structures of 1-(alkoxy)ethyl methacrylates: (1) 1-(ethoxy)ethyl methacrylate (EEMA), (2) 1-(butoxy)ethyl methacrylate (BEMA), and (3) 1-(*tert*-butoxy) ethyl methacrylate (tBEMA).

In a previous paper, we developed a new preparation method for a well-defined homopoly(MAA).[5] Three novel monomers, 1-(ethoxy)ethyl methacrylate (EEMA), 1-(butoxy)ethyl methacrylate (BEMA), and 1-(*tert*-butoxy)ethyl methacrylate (tBEMA) (Figure 1.5.1), were first prepared. These monomers could undergo anionic polymerization at higher temperatures (–40 to 0°C) than the common alkyl methacrylates, and the polymers thus obtained possessed controlled molecular weights and narrow molecular weight distributions (MWD, M_w/M_n = 1.06–1.09). In addition, these polymers were easily hydrolyzed under mild acidic conditions to generate well-defined poly(MAA). This method provides the possibility to prepare, at relatively high temperatures, amphiphilic block copolymers containing poly(MAA) as the hydrophilic segment. The present paper will focus on the syntheses of the following copolymers by the anionic copolymerizations of the new monomers with methyl methacrylate (MMA) and/or St: (1) random and diblock copolymers of EEMA, BEMA, or tBEMA with MMA; (2) diblock copolymers of St with EEMA or tBEMA; (3) diblock copolymer containing a poly(St) block and a random copolymer block of MMA and EEMA; (4) ABC and ACB triblock copolymers of St, MMA, and EEMA. The copolymers prepared under appropriate conditions possessed controlled molecular weights and compositions, and their MWDs were narrow. The ester groups of poly(EEMA), poly(BEMA), or poly(tBEMA) segment in the copolymer could be easily eliminated by hydrolysis in a mild acidic environment to generate a copolymer containing poly(MAA) as hydrophilic segment. Compared to the method based on *tert*-butyl methacrylate, the polymerization temperature is much higher (–35 to 0°C instead of –78°C) and the elimination of the protecting group much easier (mild acidic conditions instead of strong ones and a very short reaction time instead of a very long one).

1.5.2 EXPERIMENTAL SECTION

Materials. Tetrahydrofuran (THF) was dried with CaH_2 under reflux for more than 24 h, distilled, and distilled again from a solution of 1,1-diphenylhexyllithium (DPHL) just before use. Toluene was washed with concentrated sulfuric acid and then with water, dried over $MgSO_4$, distilled over CaH_2, and again distilled from a DPHL solution before use. Hexane was first dried and distilled over CaH_2 and then distilled from a solution of *n*-BuLi. Methyl methacrylate (MMA, Aldrich, 99%) and styrene (St, Aldrich, 99%) were washed with 10% aqueous sodium hydroxide solution and then with water, dried overnight with $MgSO_4$, and distilled over CaH_2. Prior to polymerization, these two monomers were finally distilled in the presence of triisobutylaluminum[6] and benzylmagnesium chloride, respectively. As described in the previous paper,[5] 1-(ethoxy)ethyl methacrylate (EEMA), 1-(butoxy)ethyl methacrylate (BEMA), and 1-(*tert*-butoxy)ethyl methacrylate (tBEMA) were prepared through the reaction of methacrylic acid (MAA) and the corresponding alkyl vinyl ether, namely, ethyl, butyl, and *tert*-butyl vinyl ether, respectively. Prior to polymerization, these monomers were purified by two successive distillations over CaH_2 under reduced pressure. 1,1-Diphenylethylene (DPE, Aldrich, 97%) was distilled over CaH_2 and then distilled again in the presence of DPHL under reduced pressure. Lithium chloride (Aldrich, 99.99%) was dried at 120°C for 24 h and dissolved in THF.[7] *n*-BuLi (Aldrich, 1.6 M solution in hexane) and *sec*-BuLi (Aldrich, 1.3 M solution in cyclohexane) were diluted with purified hexane.

Random and Block Copolymerizations of 1-(Alkoxy)ethyl Methacrylate and MMA. All copolymerizations were carried out with magnetic stirring in a round-bottom glass flask under an overpressure of argon. The random and block copolymerizations of EEMA, BEMA, or tBEMA with MMA were carried out in THF, at a selected temperature, in the presence of LiCl. After THF, DPE, and a THF solution of LiCl were added with dry syringes, the flask was cooled to a selected temperature and *n*-BuLi (in hexane) was added. The deep red color of DPHL appeared at once, and the reaction between *n*-BuLi and DPE was allowed to continue for 15 min. For the random copolymerization, a prechilled mixture of MMA and EEMA or BEMA or tBEMA was added to the above system to induce the copolymerization. In the case of block copolymerization, the two monomers were sequentially added either from MMA to one of the new monomers or vice versa. After a certain time, the system was quenched by adding a small amount of methanol, evaporated to dryness, and then vacuum-dried overnight to obtain the copolymer.

Preparation of the Block Copolymers Containing Poly(St) Segments. Poly(St-*b*-EEMA), Poly(St-*b*-tBEMA), Poly[St-*b*-(MMA-*co*-EEMA)], Poly(St-*b*-MMA-*b*-EEMA), and Poly(St-*b*-EEMA-*b*-MMA). In the preparation of the block copolymers containing poly(St) segments, St was first polymerized via anionic polymerization. This was carried out using *sec*-BuLi as the initiator, in toluene, at 20°C. After toluene and *sec*-BuLi were transferred to the flask placed in a 20°C water bath, St was added to start the polymerization, and the reaction was allowed to last 50 min. Meanwhile, to a THF solution of LiCl ($mol_{LiCl}/mol_{sec\text{-}BuLi} = 1$) and of DPE ($mol_{DPE}/mol_{sec\text{-}BuLi} = 1.2$), *sec*-BuLi was first dropwise added until the red color of DPHL appeared, to remove the impurities. Then, this solution was immediately transferred to the above toluene solution of living poly(St). After the reaction between the living site of poly(St) and DPE was allowed to last 15 min, the system was cooled to a selected temperature and the second monomer was added. For the preparation of the diblock copolymer, poly(St-*b*-EEMA) or poly(St-*b*-tBEMA), the second step polymerization of EEMA (or tBEMA) was carried out at −35°C and also between −20 and +20°C for 50 min. For the preparation of poly[St-*b*-(MMA-*co*-EEMA)], the random copolymerization of MMA and EEMA with the living poly(St) was started by adding a mixture of the two monomers, and the reaction was allowed to last 30 min at −40°C. In the case of the triblock copolymer, poly(St-*b*-MMA-*b*-EEMA), MMA was first polymerized with the living poly(St), at −65°C, for 15 min, and this was followed by the third step, polymerization of EEMA, at −35°C, for 50 min. For poly(St-*b*-EEMA-*b*-MMA), the monomer addition sequence St, EEMA, and MMA was used. After the polymerization of the last monomer, the system was quenched by adding a small amount of methanol, evaporated to dryness, and then vacuum-dried overnight to obtain the copolymer.

Elimination of the Protecting Groups. The protecting groups 1-(ethoxy)ethyl, 1-(butoxy)ethyl, and 1-(*tert*-butoxy)ethyl of 1-(alkoxy)ethyl methacrylate segments in the copolymers were eliminated by hydrolysis in a mild acidic environment. For instance, 3.0 g of vacuum-dried block copolymer of MMA and EEMA (MbE-1 in Tables 1.5.2 and 1.5.6) was redissolved in 30 mL of THF, to which 3.0 mL of HCl aqueous solution (5.0 M) was added with magnetic stirring at room temperature. After 2 min, this mixture was poured into water to precipitate the polymer. The polymer thus obtained was washed with water and vacuum-dried at 40°C for more than 24 h. The block copolymers containing a poly(St) segment were hydrolyzed in a similar way. A mixture of water and methanol ($V_{water}/V_{methanol}$ = 4/1) was used to precipitate the copolymers.

Measurements. ^{1}H NMR spectra were recorded in $CDCl_3$, THF-d_8, or CD_3OD on a VXR-400 spectrometer. M_n and M_w/M_n of the polymer were determined by gel permeation chromatography (GPC) on the basis of a polystyrene calibration curve. The GPC measurements were carried out using THF as solvent, at 30°C, with a 1.0 mL/min flow rate and a 1.0 cm/min chart speed. Three polystyrene gel columns (Waters, 7.8 × 300 mm; one HR 5E, part no. 44228, one Linear, part no. 10681, and one HR 4E, part no. 44240) which were connected to a Waters 515 precision pump were used. After polymerization, a trace of THF solution (ca. 0.1 mL) was taken out, diluted with THF, and injected immediately. The FT-IR spectra were recorded on a Perkin-Elmer 1760-X spectrometer using KBr tablets.

1.5.3 RESULTS AND DISCUSSION

Random Copolymerization of 1-(Alkoxy)ethyl Methacrylate with MMA. LiCl has been often employed in the anionic polymerization of either the common[7] or the functional[8] (meth)acrylates, because it controls the molecular weight and narrows the MWD of the polymer. Therefore, the random copolymerization of MMA and 1-(alkoxy)ethyl methacrylate was carried out in the presence of LiCl, using THF as the solvent. A bulky initiator, DPHL, was first prepared in situ before the monomer addition, via the reaction of *n*-BuLi with DPE ([DPE]/[*n*-BuLi]$_0$ = 1.2), at the selected polymerization temperature, for about 15 min. The copolymerization was induced by adding a prechilled mixture of MMA and EEMA or BEMA or tBEMA, and the reaction was allowed to last 45 min.

In the anionic homopolymerization of EEMA, BEMA, and tBEMA,[5] the optimum results were obtained for a molar ratio [LiCl]/[DPHL]$_0$ of unity, at a temperature around −40°C. A higher molar ratio of [LiCl]/[DPHL]$_0$ = 3 and/or a lower temperature (−80°C), which constitute the optimum conditions for the anionic polymerization of the common alkyl methacrylates, such as MMA, were not suitable for the new monomers. Under the latter conditions, the monomer conversion hardly reached 100%, the molecular weight was out of control, and the MWD was broad.

Because both MMA and one of the new monomers participate in the reaction simultaneously, the random copolymerization was examined not only under the optimum conditions for the new monomer but also under those for MMA. As shown in Table 1.5.1, under the suitable polymerization conditions for MMA (−80 to −60°C, $[LiCl]/[DPHL]_0 = 3$, see ME-1, MB-1, MB-2, MT-1, and MT-2 in Table 1.5.1), copolymers with relatively narrow MWDs ($M_w/M_n = 1.11–1.23$) were obtained quantitatively, but the molecular weights were only fairly controlled. On the other hand, when the copolymerization was carried out under the optimum conditions for the new monomers (−40°C, $[LiCl]/[DPHL]_0 = 1$, see ME-2, MB-3 and MT-3, in Table 1.5.1), the monomer conversion was 100% in every case, the molecular weight of the obtained copolymer was in good agreement with the calculated value, and the MWD was very narrow ($M_w/M_n = 1.08–1.10$). In addition, as proved by ^{1}H NMR, the composition of the copolymer coincided with the feed amounts of the two monomers. For instance, the weight ratio of the two kinds of units in the copolymer MB-3 determined from the peak intensities was $W_{MMA}W_{BEMA} = 48/52$, which is close to the feed amount ratio (50/50). Hence, the composition of the copolymer was also well controlled. The above results indicate that the optimum conditions for the homopolymerization of 1-(alkoxy)ethyl methacrylate are also suitable for their copolymerizations with MMA.

Block Copolymerization of 1-(Alkoxy)ethyl Methacrylate with MMA. The block copolymerization was carried out in THF, using DPHL as the initiator,

TABLE 1.5.1
Anionic Random Copolymerization of 1-(Alkoxy)ethyl Methacrylate (M_1) and MMA[a]

No.	$[M_1]_0$ (M)	$[MMA]_0$ (M)	$[DPHL]_0$ (mM)	[LiCl] (mM)	Temp (°C)	$10^{-3}M_k$[b]	$10^{-3}M_n$[c]	M_w/M_n[c]
ME-1[d]	0.38	0.59	11.8	11.8	−80	10.3	12.7	1.11
ME-2[d]	0.38	0.59	11.8	11.8	−40	10.3	10.3	1.08
MB-1[e]	0.33	0.63	33.3	100	−60	3.98	3.31	1.22
MB-2[e]	0.12	0.71	16.6	33.3	−60	5.86	4.86	1.23
MB-3[e]	0.31	0.59	12.8	12.8	−40	9.35	10.2	1.09
MT-1[f]	0.31	0.90	31.9	100	−60	4.87	6.17	1.12
MT-2[f]	0.08	0.78	16.7	50.0	−60	5.81	7.06	1.18
MT-3[f]	0.30	0.59	11.8	11.8	−40	9.97	9.44	1.10

[a] The initiator, DPHL, was first prepared by the reaction of n-BuLi and DPE ($[DPE]/[n\text{-BuLi}]_0 = 1.2$), at the corresponding polymerization temperature, for 15 min. The copolymerization was carried out in THF, by adding a prechilled mixture of MMA and EEMA or BEMA or tBEMA to the THF solution of DPHL, and the reaction was allowed to last 45 min. The yields of the copolymers were quantitative in all cases.

[b] Calculated number-average molecular weight.

[c] Determined by GPC.

[d, e, f] Random copolymers of MMA with EEMA, MMA with BEMA, and MMA with tBEMA, respectively.

in the presence of LiCl ($[LiCl]/[DPHL]_0 = 1$). The polymerization temperatures for 1-(alkoxy)ethyl methacrylate and MMA were −40 and −65°C, and the polymerization times were 45 and 15 min, respectively. Both 1-(alkoxy)ethyl methacrylate and MMA are polar monomers and possess similar molecular structures. Therefore, their block copolymerization is expected to proceed similarly either in the sequence starting with MMA or vice versa. Indeed, as shown in Table 1.5.2, the block copolymers were obtained quantitatively in both cases, although the MWDs were somewhat different. For the polymerization sequence MMA–new monomer (MbE-1, MbB-1, MbT-1, and MbT-2 in Table 1.5.2), the molecular weight increased after the second step polymerization and was consistent with the designed value, and the MWD remained narrow (M_w/M_n = 1.09–1.12). For the opposite sequence, the MWDs of the obtained block copolymers were somewhat broader (M_w/M_n = 1.17–1.38; EbM-1, EbM-2, BbM-1, TbM-1, and TbM-2 in Table 1.5.2). This is caused by the slow transformation rate of the living site from poly[1-(alkoxy)ethyl methacrylate] to poly(MMA) compared to the propagation rate of MMA, which is very fast.[9]

Block Copolymerization of St and EEMA. The block copolymerization of St and EEMA was carried out using the sequence St–EEMA. For the anionic polymerization of St, toluene and *sec*-BuLi were used as the solvent and the initiator, respectively, because St can undergo smoothly anionic polymerization in a nonpolar solvent at room temperature.[10] However, the second step living polymerization of EEMA is preferably carried out in a polar solvent (THF), in the presence of LiCl.[5,9]

TABLE 1.5.2
Anionic Block Copolymerization of 1-(Alkoxy)Ethyl Methacrylate and MMA[a]

No.	Polymerization Sequence $M_1 \rightarrow M_2$	$[M_1]_0$ (M)	$[M_2]_0$ (M)	$[DPHL]_0$ (mM)	First Step $10^{-3}M_k$[b]	First Step $10^{-3}M_n$[c]	First Step M_w/M_n[c]	Second Step $10^{-3}M_k$[b]	Second Step $10^{-3}M_n$[c]	Second Step M_w/M_n[c]
EbM-1	EEMA → MMA	0.42	0.70	17.7	3.99	4.00	1.11	7.94	7.84	1.18
EbM-2	EEMA → MMA	0.42	0.67	11.4	6.06	5.69	1.07	12.2	12.9	1.17
MbE-1	MMA → EEMA	0.67	0.42	11.4	6.12	6.44	1.06	11.9	12.2	1.09
BbM-1	BEMA → MMA	0.33	0.63	20.5	3.23	3.10	1.15	6.30	6.80	1.38
MbB-1	MMA → BEMA	0.63	0.33	20.5	3.31	3.09	1.18	6.30	6.15	1.12
TbM-1	tBEMA → MMA	0.32	0.63	16.7	3.80	4.10	1.17	7.58	8.88	1.18
TbM-2	tBEMA → MMA	0.34	0.66	10.7	6.15	5.96	1.11	12.3	11.7	1.17
MbT-1	MMA → tBEMA	0.63	0.32	16.7	4.01	4.31	1.11	7.58	7.96	1.09
MbT-2	MMA → tBEMA	0.66	0.34	8.3	8.19	8.51	1.06	15.8	16.0	1.11

[a] The initiator, DPHL, was in situ prepared before the first monomer addition, by the reaction of *n*-BuLi and DPE ($[DPE]/[n\text{-BuLi}]_0$=1.2), in the presence of LiCl ($[LiCl]/[n\text{-BuLi}]_0$=1), at the corresponding polymerization temperature, for 15 min. The polymerization was induced by adding the first monomer (M_1) to the above system. After a selected time, the second monomer was continuously added. The temperatures for 1-(alkoxy)ethyl methacrylate and MMA were −40°C and −65°C, and the polymerization times were 45 and 15 min, respectively. The polymer yield was quantitative for every case.

[b, c] See Table 1.5.1.

Further, the living site of poly(St) is too reactive and may attack the ester group of EEMA, causing unwanted side reactions. Considering the above factors, a THF solution of LiCl ($mol_{LiCl}/mol_{sec\text{-}BuLi} = 1$) and of DPE ($mol_{DPE}/mol_{sec\text{-}BuLi} = 1.2$) was added after the anionic polymerization of St. The reaction between the living site of poly(St) and DPE took place at once, and the color of the system changed from yellow to deep red, implying the formation of a bulkier and less reactive carbanion with a structure similar to DPHL.[11] After 15 min, the system was cooled to −35°C, and the second step copolymerization was started by adding prechilled EEMA to the above system. Upon the addition of the monomer, the color changed from deep red to transparent immediately, indicating the fast transformation of the living site from the DPE capped poly(St) to poly(EEMA). The reaction was allowed to last 50 min at −35°C.

As shown in Figure 1.5.2 and Table 1.5.3 (SE-3), a new sharp and symmetrical peak (b) corresponding to the block copolymer appears after the polymerization of EEMA, in the higher molecular weight side, and the peak (a) of its living poly(St) precursor (M_n = 10 700, M_w/M_n = 1.07) disappears. The molecular weight of the block copolymer (M_n = 20 400, M_w/M_n = 1.08) is about twice as large as that of poly(St), hence almost equal to the designed value, and the MWD is almost as narrow as that of its precursor. These results indicate that the living site of poly(St) initiated the second step, polymerization of EEMA, and that a pure diblock copolymer, poly(St-*b*-EEMA), free of homopolymer, was obtained. In addition, the weight ratio of the two blocks, determined by ^{1}H NMR (for SE-3, W_{St}/W_{EEMA} = 51/49), is very close to the feed amount ratio of St and EEMA (50/50).

Preparation of Diblock Copolymer of St and tBEMA (Poly(St-*b*-tBEMA)) at 0°C. In a previous paper,[5] we found that tBEMA can undergo anionic polymerization smoothly at a higher temperature than EEMA and BEMA, because the bulky *tert*-butyl ester group stabilizes the propagating site. Even at +20°C, this monomer could polymerize and a narrow MWD (M_w/M_n = 1.18) obtained. These results

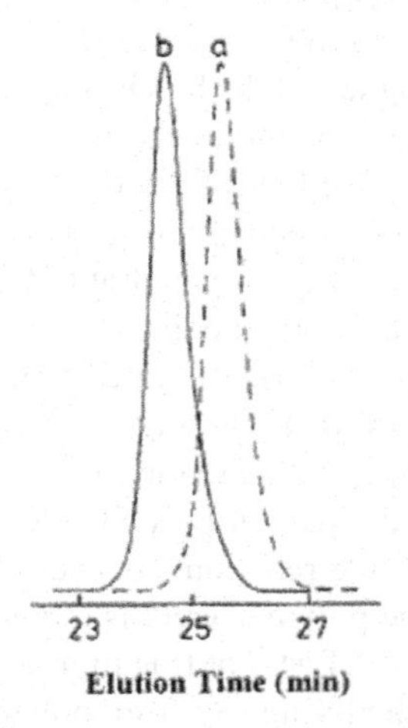

FIGURE 1.5.2 GPC traces of the diblock copolymer of St and EEMA (b, SE-3 in Table 1.5.3, M_n = 20 400, M_w/M_n = 1.08) prepared by the anionic block copolymerization of EEMA from its living poly(St) precursor (a, see SE-3 in Table 1.5.3, M_n = 10 700, M_w/M_n = 1.07).

TABLE 1.5.3
Preparation of Diblock Copolymer of St and EEMA (SE) or St and tBEMA (ST)[a]

	Living Poly(St)				Block Copolymer				
No.	$[sec\text{-BuLi}]_0$/ mM	$10^{-3}M_k$[b]	$10^{-3}M_n$[c]	M_w/M_n[c]	Temp (°C)	$[PSt\text{–}Li^+]_0$[d]/ mM	$10^{-3}M_k$[b]	$10^{-3}M_n$[c]	M_w/M_n[c]
SE-1	31.2	2.62	2.63	1.12	−35	16.6	5.30	5.02	1.14
SE-2	15.6	5.19	5.13	1.07	−35	8.3	10.5	10.2	1.08
SE-3	8.0	10.0	10.7	1.07	−35	4.3	20.3	20.4	1.08
ST-1	16.0	5.06	5.28	1.09	0	9.5	10.2	11.1	1.12
ST-2	16.0	5.06	5.39	1.10	20	9.5	10.2	8.82	1.27[e]
ST-3	8.0	10.1	10.1	1.07	0	4.8	20.2	20.7	1.19
ST-4	8.0	10.1	10.3	1.09	−20	4.8	20.2	19.8	1.17

[a] The anionic polymerization of St was first carried out in toluene, at 20°C, for 50 min ($[St]_0 = 0.77$ M). Then, a THF solution of LiCl ($mol_{LiCl}/mol_{sec\text{-BuLi}} = 1$) and of DPE ($mol_{DPE}/mol_{sec\text{-BuLi}} = 1.2$) was added. After 10 min, the second step polymerization was induced by introducing EEMA or tBEMA into the system ($[EEMA]_0 = 0.28$ M, $[tBEMA]_0 = 0.26$ M), and the reaction was allowed to last an additional 50 min. The polymer yield was quantitative for every case.
[b] Calculated number-average molecular weight.
[c] Determined by GPC on the basis of a poly(St) calibration curve.
[d] Concentration of the living sites of poly(St)s after the addition of the THF solution of LiCl and DPE.
[e] Double peaks (see Figure 1.5.3B-b).

encouraged us to attempt its copolymerization with St at higher temperatures (≥ 0°C). The experiments were carried out by using the procedure employed in the preparation of poly(St-*b*-EEMA), at a different temperature. As shown in Table 1.5.3, for a temperature in the second step of either −20 or 0°C, the polymer yield was quantitative, and the obtained block copolymer possessed a well-controlled molecular weight and a narrow MWD ($M_w/M_n = 1.121.19$; ST-1, ST-3, and ST-4 in Table 1.5.3). As illustrated in Figure 1.5.3A, the chromatogram of the block copolymer obtained at 0°C exhibits a single peak (A-b); the corresponding molecular weight ($M_n = 11\,100$, $M_w/M_n = 1.12$; ST-1 in Table 1.5.3) is about twice that of its living poly(St) precursor ($M_n = 5280$, $M_w/M_n = 1.09$), and the MWD remains narrow. Therefore, the diblock copolymer, poly(St-*b*-tBEMA), can be smoothly prepared at 0°C without side reactions. These results could be relevant from a practical point of view.

Furthermore, the block copolymerization of St and tBEMA was also attempted at room temperature. As shown in Figure 1.5.3B, the main peak of the produced diblock copolymer (peak B-b) is accompanied by a shoulder in the low molecular weight region. This means that a part of poly(St) remained unreacted in the second step. This was not caused by side reactions, because the main peak of the diblock copolymer is not broad and the polymer yield is quantitative. The possible reason is that the polymerization rate of tBEMA at this high temperature is much faster than the rate of transformation of the living site from poly(St) to poly(tBEMA).

Preparation of Poly[St-*b*-(MMA-*co*-EEMA)]. The hydrolysis of the poly(St-*b*-EEMA) or poly(St-*b*-tBEMA) prepared above generates an amphiphilic block copolymer, poly(St-*b*-MAA), in which the hydrophilic segment consists of 100%

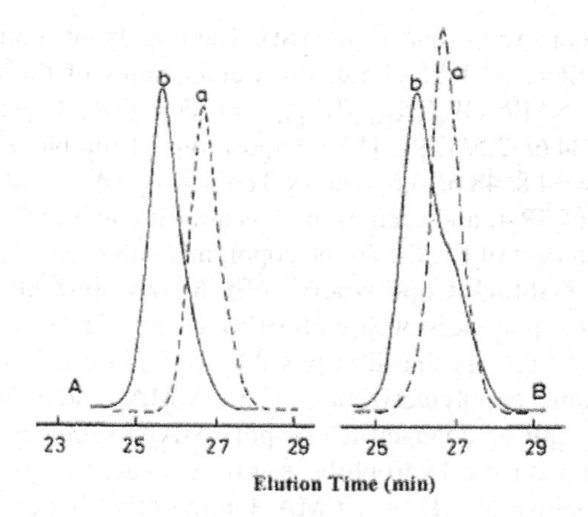

FIGURE 1.5.3 GPC traces of diblock copolymers of St with tBEMA and their precursors: A-a, living poly(St) (see ST-1 in Table 1.5.3, M_n = 5280, M_w/M_n = 1.09); A-b, poly(St-*b*-tBEMA) (ST-1 in Table 1.5.3, M_n = 11 100, M_w/M_n = 1.12) prepared by the anionic block copolymerization of tBEMA at 0°C from A-a; B-a, living poly(St) (see ST-2 in Table 1.5.3, M_n = 5390, M_w/M_n = 1.10); B-b, poly(St-*b*-tBEMA) (ST-2 in Table 1.5.3, M_n = 8820, M_w/M_n = 1.27) prepared by the anionic block copolymerization of tBEMA at 20 °C from B-a.

MAA. It is, however, useful to be able to control the carboxyl content. This can be achieved by the block copolymerization of St with a mixture of EEMA and MMA. The preparation conditions of this copolymer were as those for poly(St-*b*-EEMA), except that the second step, a random copolymerization, was carried out at −40°C for 30 min. As shown in Table 1.5.4, upon the completion of the random copolymerization of EEMA and MMA, the molecular weight became larger than that of the precursor poly(St) and close to the calculated value; the MWD of the block

TABLE 1.5.4
Preparation of Poly[St-*b*-(MMA-*co*-EEMA)][a]

	Living Poly(St)				Block Copolymer					
No.	[*sec*-BuLi]$_0$/ mM	$10^{-3}M_k{}^b$	$10^{-3}M_n{}^c$	$M_w/M_n{}^c$	[PSt−Li$^+$]$_0{}^d$/ mM	[MMA]$_0$	[EEMA]$_0$	$10^{-3}M_k{}^b$	$10^{-3}M_n{}^c$	$M_w/M_n{}^c$
SR-1	25.0	3.26	3.00	1.12	12.5	0.38	0.24	9.33	9.63	1.09
SR-2	15.0	5.34	5.45	1.09	7.5	0.56	0.12	15.3	15.0	1.12
SR-3	15.0	5.34	5.19	1.10	7.5	0.19	0.24	12.9	12.4	1.10

[a] The anionic polymerization of St was first carried out in toluene, at 20 °C, for 50 min ([St]0 = 0.77 M). After a THF solution of LiCl (molLiCl/molsec-BuLi = 1) and of DPE (molDPE/molsec-BuLi = 1.2) was added, the second step polymerization was started by adding a mixture of MMA and EEMA to the above system, and the reaction was allowed to last 30 min at -40 °C. The polymer yield was quantitative for every case.

[b,c,d] See Table 1.5.3.

copolymer was as narrow as that of poly(St). The copolymer composition can also be controlled. For SR-1 in Table 1.5.4, the weight ratios of the three components determined by ^{1}H NMR ($W_S/W_{MMA}/W_{EEMA}$ = 35.8/30.8/33.4) coincide with the theoretical values (34.6/32.5/32.9). The compositions of the block copolymers SR-2 ($W_{st}/W_{MMA}/W_{EEMA}$ = 34.8/ 48.6/16.6) and SR-3 ($W_{st}/W_{MMA}/W_{EEMA}$ = 41.3/19.4/39.3) are different from that of SR-1, and each of them is almost equal to the monomer weight ratios. Hence, the content of EEMA in the copolymer can be controlled accurately.

Preparation of Triblock Copolymers of St, MMA, and EEMA. As indicated above, well-defined copolymers of St with EEMA, and of MMA with EEMA, have been obtained. This suggests that it is possible to prepare triblock copolymers by the sequential anionic copolymerization of St, MMA, and EEMA. Because the poly(EEMA) block can be changed to the poly(MAA) segment, one can obtain a triblock copolymer with the hydrophilic segment located at its end by using the polymerization sequence St–MMA–EEMA. On the other hand, by performing the copolymerization in the sequence St–EEMA–MMA, one can obtain a triblock copolymer with the hydrophilic segment located in the middle of the copolymer chain.

The preparation of the living poly(St) and its reaction with DPE were carried out as described above. The polymerizations of MMA and EEMA that followed were performed at −65 and −35°C, respectively. As shown in Table 1.5.5, for the polymerization sequence St, MMA, and EEMA (SME-1 and SME-2), the three polymerization steps proceeded smoothly. The molecular weight increased step by step and was very close to the calculated value at every stage. As illustrated by the GPC measurements (Figure 1.5.4), the polymer obtained at each stage exhibited a very sharp and symmetrical peak, indicating that no deactivation of the living site occurred and that the final triblock copolymer was free of its precursors. The weight ratios of the three blocks in SME-2 determined by ^{1}H NMRare $W_{St}/W_{MMA}/W_{EEMA}$ = 33.5/32.7/33.8, extremely close to the theoretical values (33.3/33.3/33.3).

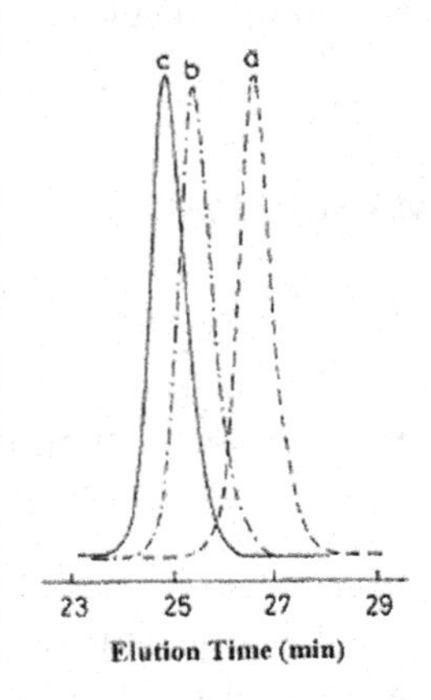

FIGURE 1.5.4 GPC traces of the triblock copolymer of St, MMA, and EEMA and its precursors: a, living poly(St) (see SME-2 in Table 1.5.5, M_n = 5130, M_w/M_n = 1.07); b, living diblock copolymer of St and MMA (see SME-2 in Table 1.5.5, M_n = 10 600, M_w/M_n = 1.09) prepared by the anionic copolymerization of MMA from a; c, triblock copolymer (SME-2 in Table 1.5.5, M_n = 15 500, M_w/M_n = 1.09) prepared by the continuous block copolymerization of EEMA from b.

When the polymerization sequence St–EEMA–MMA was employed, the copolymers were also obtained quantitatively. The second step copolymerization of EEMA from DPE capped living poly(St) proceeded smoothly, just like the preparation of the diblock copolymer, poly(St-*b*-EEMA). However, after the continuous polymerization of MMA, the GPC chromato gram of the obtained polymer exhibited a double-peak distribution. In the low molecular weight range of the main peak corresponding to the triblock copolymer, a shoulder appeared, which can be attributed to the diblock copolymer, poly(St-*b*-EEMA). Because of the presence of this unreacted diblock copolymer, the MWD of the final product was broader (M_w/M_n = 1.27–1.35, SEM-1 and SEM-2 in Table 1.5.5). The too fast polymerization rate of MMA might be the cause for the presence of the unreacted diblock copolymer, poly(St-*b*-EEMA). This does not occur in the polymerization sequence St–MMA–EEMA, because the polymerization rate of EEMA is much slower.

Elimination of the Protecting Groups. The 1-(alkoxy)ethyl groups of poly[1-(alkoxy)ethyl methacrylate)] segments in the copolymers were eliminated through hydrolysis reactions. As soon as a small amount of HCl aqueous solution was added to the THF solution of each copolymer, the reaction completed almost instantaneously. As an example, Figure 1.5.5 depicts the ^{1}H NMR spectra of poly[St-*b*-(MMA-*co*-EEMA)] (A; SR-1 in Tables 1.5.4 and 1.5.6) and its hydrolyzed product (B, in THF-d_8, and C, in CD_3OD). After hydrolysis, the peaks c, d, e, and f due to the protecting group, 1-(ethoxy)ethyl, disappeared completely, and peak j corresponding to the ester group (OCH_3) of MMA units remained unchanged. This means that the ester groups of the EEMA units can be selectively eliminated. The elimination reactions of the protecting groups were also carried out for the other copolymers. Figure 1.5.6 presents the FT-IR spectra of the following hydrolyzed copolymers: the random copolymer of MMA and BEMA (A; see MB-3 in Table 1.5.1); the diblock

TABLE 1.5.5
Preparation of Triblock Copolymer of St (S), MMA (M), and EEMA (E)[a]

	Living Poly(St)				Living Diblock Copolymer				Triblock Copolymer		
No.	[*sec*-BuLi]$_0$/ mM	$10^{-3}M_k$[b]	$10^{-3}M_n$[c]	M_w/M_n[c]	[PSt$^-$Li$^+$]$_0$[d]/ mM	$10^{-3}M_k$[b]	$10^{-3}M_n$[b]	$10^{-3}M_n$[c]	$10^{-3}M_k$[c]	$10^{-3}M_n$[c]	M_w/M_n[c]
SEM-1[e]	26.0	2.62	2.65	1.11	14.2	5.19	5.19	1.11	7.75	7.92	1.10
SEM-2[e]	13.0	5.18	5.13	1.07	7.1	10.3	10.6	1.09	15.4	15.5	1.09
SEM-1[f]	26.0	2.62	2.71	1.10	14.2	5.19	5.44	1.12	7.75	6.77	1.27[g]
SEM-2[f]	13.0	5.18	5.30	1.08	7.1	10.3	10.5	1.11	15.4	13.5	1.35[g]

[a] The anionic polymerization of St was first carried out in toluene, at 20°C, for 50 min ($[St]_0$) = 0.64 M). Before the second-step polymerization, a THF solution of LiCl ($mol_{LiCl}/mol_{sec\text{-}BuLi}$=1) and of DPE ($mol_{DPE}/mol_{sec\text{-}BuLi}$=1.2) was added. For second- and third-step polymerizations of MMA or EEMA, their initial concentrations were 0.36 and 0.23 M, the polymerization times were 15 and 50 min, and the temperatures were –65 and –35°C, respectively. The polymer yield was quantitative for every case.

[b, c, d] See Table 1.5.3.

[e, f] Polymerization sequences were St→MMA→EEMA and St→EEMA→MMA, respectively.

[g] Double peaks.

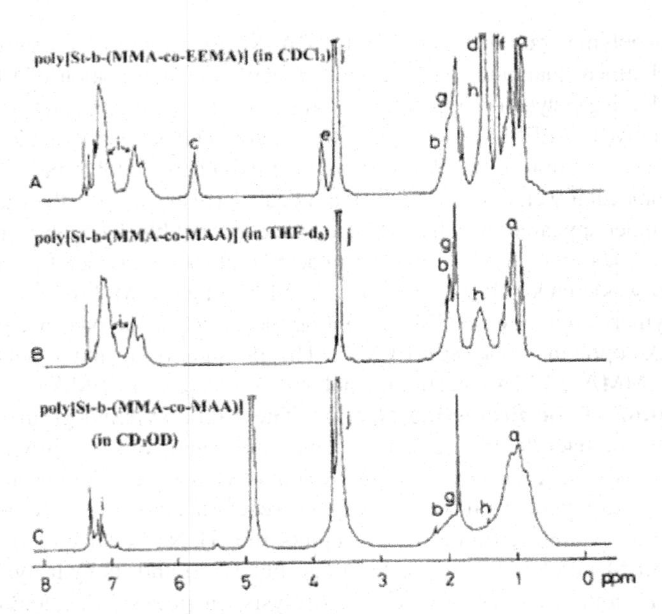

FIGURE 1.5.5 ^{1}H NMR spectra of poly[St-*b*-(MMA-*co*-EEMA)] (A, SR-1 in Table 1.5.4, in $CDCl_3$) and its hydrolyzed copolymer, poly[St-*b*-(MMA-*co*-MAA)] (B, in THF-d_8; C, in CD_3OD): peak a, α-CH_3 of MMA and EEMA (or MAA) units; peak b, –CH_2– in the main chain of MMA and EEMA (or MAA) units; peaks c, d, e, and f, $OCH(CH_3)O$, $OCH(CH_3)O$, CH_2CH_3 and CH_2CH_3 in EEMA units, respectively; peaks g, h, and i, CH, CH_2, and C_6H_5 in poly(St) segment, respectively; peak j, OCH_3 in MMA units.

copolymer of St and EEMA (B; see SE-3 in Table 1.5.3); and the triblock copolymer of St, MMA, and EEMA (C; see SME-2 in Table 1.5.5). Each of them exhibits a broad absorption corresponding to the carboxyl group (peaks A-a, B-a and C-a; 2500–3800 cm^{-1}). All the above results indicate that the protecting group, 1-(alkoxy)-ethyl, is sufficiently stable in a basic environment to allow the smooth anionic copolymerization of the corresponding monomers. In contrast, the protecting group is unstable even under mildly acidic conditions. This allows its much easier elimination than that of the *tert*-butyl group from poly(St-*b*-*tert*-butyl methacrylate). Therefore, it is more advantageous to prepare the hydrophilic MAA segment using the present procedure than the traditional one based on *tert*-butyl methacrylate.

Each hydrolyzed copolymer has a quite different solubility than its precursor (see Table 1.5.6). Before hydrolysis, all the copolymers were soluble in benzene, $CHCl_3$, *N,N*-dimethylformamide, 1,4-dioxane, THF, and acetone but insoluble in hexane and water. In addition, the block copolymers containing poly(St) segments could be dissolved neither in methanol nor in ethanol. However, after hydrolysis, every copolymer became soluble in alcohol, due to the presence of poly(MAA) hydrophilic segment. For instance, poly[St-*b*-(MMA-*co*-EEMA)] (SR-1 in Tables 1.5.4 and 1.5.6) was insoluble in methanol and ethanol, because these are poor solvents for its poly(St) segment. However, its hydrolyzed copolymer (h-SR-1 in Table 1.5.6) could be slowly dissolved in the two solvents, generating transparent solutions. Because the latter copolymer is soluble in both THF and methanol, its ^{1}H NMR measurements were carried out separately in

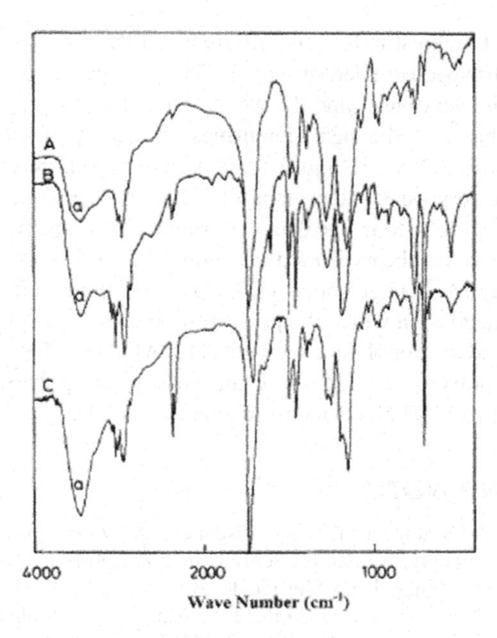

FIGURE 1.5.6 FT-IR spectra (KBr tablets) of the copolymers containing poly(MAA) segments. h-MB-3 (A), h-SE-3 (B), and h-SME-2 (C) were obtained by hydrolysis of MB-3 (Table 1.5.1), SE-3 (Table 1.5.3), and SME-2 (Table 1.5.5), respectively.

TABLE 1.5.6
Solubility Comparison of the Copolymers before and after Hydrolysis[a]

	Hexane	Benzene	$CHCl_3$	DMF[b]	Dioxane	THF	Acetone	CH_3OH	C_2H_5OH	H_2O	NaOH[c]
MB-3	I	S	S	S	S	S	S	D	D	I	I
h-MB-3[d]	I	I	I	S	I	D	D	S	S	I	D
MbE-1	I	S	S	S	S	S	S	D	D	I	I
h-MbE-1[d]	I	I	I	S	I	S	D	S	S	I	C
SE-3	I	S	S	S	S	S	S	I	I	I	I
h-SE-3[d]	I	I	I	S	D	S	I	C	C	I	I
SR-1	I	S	S	S	S	S	S	I	I	I	I
h-SR-1[d]	I	I	C	S	D	S	S	D	D	I	I
SME-2	I	S	S	S	S	S	S	I	I	I	I
h-SME-2[d]	I	I	C	S	D	S	S	C	C	I	I
SEM-1	I	S	S	S	S	S	S	I	I	I	I

[a] The experiment was carried out at room temperature. The amounts of copolymer and the solvent were 0.06 g and 2.0 mL, respectively. S, soluble; D, dissolved slowly; C, cloudy; I, insoluble.
[b] *N,N*-Dimethylformamide.
[c] 5.0 M aqueous solution.
[d] Hydrolyzed products from the corresponding copolymers.

THF-d_8 and CD_3OD. Interestingly, quite different results were obtained. Figure 1.5.5B presents the ^1H NMR spectrum determined in THF-d_8. The peaks corresponding to the three components are very clear, and the weight ratios ($W_{St}/W_{MMA}/W_{EEMA}$ = 42/41/17), calculated on the basis of the peak intensities, are close to the theoretical values (42/40/18). However, in CD_3OD (Figure 1.5.5C), the absorptions corresponding to the poly(St) segment are very weak and even hardly detectable (peaks g, h, and i). The ratios ($W_{St}/W_{MMA}/W_{EEMA}$ = 7.9/56.8/35.3) calculated from the peak intensities are far from the theoretical values. This can be explained as follows. It is well-known that methanol is a good solvent for poly(MAA) but a poor one for poly(St). In methanol, this copolymer is expected to form micelles, in which the poly(St) segments are located inside, forming a hydrophobic core, and the copolymer segments of MMA and MAA are located outside, thus wrapping the poly(St) segments. For this reason, the poly(St) component could hardly be detected by the ^1H NMR measurement in CD_3OD.

REFERENCES AND NOTES

1. (a) Desjardins, A.; Vandeven, T. G. M.; Eisenberg, A. *Macromolecules.* **1992**, *25*, 2412. (b) Bosse, F.; Eisenberg, A.; Xu, R.; Schreiber, H. P. *J. Appi. Polym. Sci.* **1994**, *51*, 521. (c) Gandhi, J. V.; Maher, J. V.; Shaffer, K. A.; Chapman, T. M. *Langmuir.* **1997**, *13*, 1592. (d) May-Colaianni, S. E.; Gandhi, J. V.; Maloy, K. J.; Maher, J. V.; Kuhar, K. A.; Chapman, T. M. *Macromolecules.* **1993**, *26*, 6595.
2. (a) Kutsumizu, S.; Hara, H.; Schlick, S. *Macromolecules.* **1997**, *30*, 2320. (b) Kutsumizu, S.; Schlick, S. *Macromolecules.* **1997**, *30*, 2329. (c) Saunders, B. R.; Crowther, H. M.; Vincent, B. *Macromolecules.* **1997**, *30*, 482.
3. Rannard, S. P.; Billingham, N. C.; Armes, S. P.; Mykytiuk, J. *Eur. Polym. J.* **1993**, *29*, 407.
4. (a) Ramireddy, C.; Tuzar, Z.; Prochazka, K.; Webber, S. E.; Munk, P. *Macromolecules.* **1992**, *25*, 2541. (b) Desjardins, A.; Vandeven, T. G. M.; Eisenberg, A. *Macromolecules.* **1992**, *25*, 2412. (c) Zhang, H. M.; Ishikawa, H.; Ohata, M.; Kazama, T.; Isono, Y.; Fujimoto, T. *Polymer.* **1992**, *33*, 828. (d) Selb, J.; Gollot, Y. Y. In *Development in Block Copolymers*; Goodman, I., Ed.; Elsevier Publishers: London, UK, 1988; Vol. 2. (e) Long, T. E.; Allen, R. D.; McGrath, J. E. In *ACS Symposium Series No. 364*; Benham, J. L., Kinstle, J. F., Eds.; American Chemical Society: Washington, DC, 1988; Chapter 19. (f) Pitsikalis, M.; Woodward, J.; Mays, J. W.; Hadjichristidis, N. *Macromolecules.* **1997**, *30*, 5384.
5. Zhang, H. M.; Ruckenstein, E. *Macromolecules.* **1998**, *31*, 7575.
6. Allen, R. D.; Long, T. E.; McGrath, J. E. *Polym. Bull.* **1986**, *15*, 127.
7. (a) Varshney, S. K.; Hautekeer, J. P.; Fayt, R.; Jerome, R.; Teyssie, Ph. *Macromolecules.* **1990**, *23*, 2618. (b) Fayt, R.; Forte, R.; Jacobs, C.; Jerome, R.; Ouhadi, T.; Teyssie, Ph.; Varshney, S. K. *Macromolecules.* **1987**, *20*, 1442. (c) Kunkel,; Muller, A. H. E.; Janata, M.; Lochman, L. *Makromol. Chem, Macromol. Symp.* **1992**, *60*, 315.
8. Mori, H.; Wakisawa, O.; Hirao, A.; Nakahama, S. *Macromol. Chem. Phys.* **1994**, *195*, 3213. (b) Ruckenstein, E.; Zhang, H. M. *J. Polym. Sci., Polym. Chem.* **1998**, *36*, 1865. (c) Hild, G.; Lamps, J. P.; Rempp, P. *Polymer.* **1993**, *34*, 2875.
9. Hatada, K.; Kitayama, T.; Ute, K. *Prog. Polym. Sci.* **1988**, *10*, 189.
10. (a) Norhay, A.; McGrath, J. E. *Block Copolymerization*; Academic Press: New York, 1977. (b) Van Beylen, M.; Bywater, S.; Smets, G.; Szwarc, M.; Worsfold, D. J. *Adv. Polym. Sci.* **1988**, *86*, 87. (c) Rempp, P.; Franta, E.; Herz, J. *Adv. Polym. Sci.* **1988**, *86*, 145. (d) Norton, M. *Anionic Polymerization: Principles and Practice*; Academic Press: New York, 1983. (e) Hsieh, H. L.; Quirk, R. P. *Anionic Polymerization*; Marcel Dekker: New York, 1996. (f) Pitsikalis, M.; Pispas, S.; Ways, J. W.; Hadjichristidis, N. *Adv. Polym. Sci.* **1998**, *135*, 1.
11. Shulz, G.; Hocher, H. *Angew. Chem., Int. Ed. Engl.* **1980**, *19*, 219.

1.6 A Novel Breakable Cross-linker and pH-Responsive Star-Shaped and Gel Polymers*

Eli Ruckenstein and Hongmin Zhang

Chemical Engineering Department, State University of New York at Buffalo, Amherst, New York 14260

ABSTRACT A bifunctional methacrylate monomer, namely, ethylene glycol di(1-methacryloyloxy)ethyl ether (**1**), was prepared through the addition reaction between ethylene glycol divinyl ether and methacrylic acid. **1** was used as a cross-linker in the preparation of a star-shaped poly(methyl methacrylate) [poly(MMA)], a branched soluble poly(MMA), and a polymer gel. The addition of **1** to an anionically prepared living poly(MMA) solution generated a star-shaped polymer with a central poly(**1**) gel core and several poly(MMA) arms. On the other hand, when MMA and **1** were simultaneously added to a tetrahydrofuran (THF) solution of an anionic initiator, a branched soluble poly(MMA) or a polymer gel was obtained, depending on the amount of **1**. The cross-linking points in the above polymers could be easily broken by hydrolysis under acidic conditions, leading to linear polymers. In contrast to the common polymer gels, the present polymer gel could be broken to soluble polymers in an acidic medium. However, it was just swollen in a basic or a neutral medium. The hydrolyzed product from the star-shaped polymer was a block copolymer consisting of poly(MMA) and poly(methacrylic acid) segments, and those hydrolyzed from the branched polymers and polymer gels were random copolymers of MMA and methacrylic acid. All the hydrolyzed polymers possessed quite different solubilities than those of their precursors.

* *Macromolecules* 1999, 32, 3979–3983.

1.6.1 INTRODUCTION

Divinyl cross-linking reagents (cross-linkers) have been often employed for the preparation of star-shaped (co)polymers. A linear living polymer was first prepared using a living polymerization technique, and this was subsequently followed by the reaction of its living end with a small amount of divinyl compound. For instance, the addition of divinylbenzene (DVB) to an anionic living polystyrene [poly(St)] solution led to the formation of a star-shaped poly(St) with a central poly(DVB) gel core.[1] This method was also extended to cationic,[2] group transfer,[3] and metathesis[4] polymerizations, in which divinyl ethers, divinyl esters, and norbornadiene dimers were used as cross-linkers, respectively. This synthetic technique could be easily carried out by just adding a bifunctional monomer to a completed living polymerization system. However, the arm number of the resulting (co)polymer could hardly be controlled.

In the present paper, we synthesize a novel cross-linker, ethylene glycol di(1-methacryloyloxy)ethyl ether (**1**), whose bonds indicated by arrows in Figure 1.6.1 can be broken in acidic media. This novel cross-linker was employed in the preparation of a star-shaped poly(methyl methacrylate) [poly(MMA)], a branched soluble poly(MMA), and a polymer gel. The star-shaped poly(MMA) was prepared using a traditional method:[1–4] the living poly(MMA) was allowed to react with a small amount of **1** to form a living block copolymer, which had a short segment of **1** attached to the end of the polymer chain; the subsequent intermolecular reactions of the pendant vinyl groups of **1** with the living ends of the polymer chains resulted in a star-shaped polymer with a central poly(**1**) core. On the other hand, the simultaneous introduction of **1** and MMA into a THF solution of an anionic initiator resulted in the formation of a branched soluble poly(MMA) or a polymer gel, depending on the amount of **1** added, because the intermolecular cross-linking can occur as the polymerization proceeds. In contrast to the star-shaped polymers and polymer gels based on the conventional cross-linkers, such as ethylene glycol dimethacrylate, those prepared using **1** could be easily broken by hydrolysis under mild acidic conditions, generating linear polymers. Compared to most polymer gels, the present gels exhibited different responses in the acidic and basic media. In contrast to their swelling in a basic or neutral medium, completely clear solutions of linear polymers could be obtained when an acidic medium was employed. This may be useful for controlled drug release and is relevant to environmental protection.

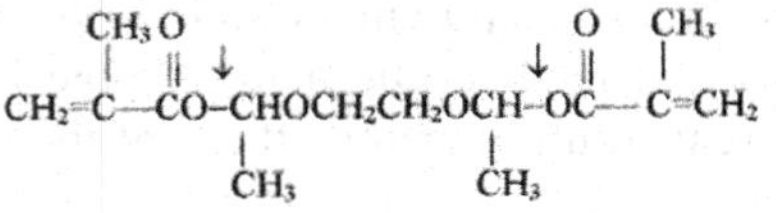

CH_3 O ↓ ↓ O CH_3

$CH_2{=}C{-}CO{-}CHOCH_2CH_2OCH{-}OC{-}C{=}CH_2$

CH_3 CH_3

FIGURE 1.6.1 Molecular structure of ethylene glycol di(1-methacryloyloxy)ethyl ether (**1**). The linkages indicated with arrows can be easily broken under acidic conditions.

1.6.2 EXPERIMENTAL SECTION

Materials. Tetrahydrofuran (THF) was dried with CaH_2 under reflux for more than 24 h, distilled, and distilled again from a solution of 1,1-diphenylhexyllithium (DPHL) just before use. Toluene was washed with concentrated sulfuric acid and then with water, dried with $MgSO_4$, and distilled over CaH_2. Hexane was first dried and distilled over CaH_2 and then distilled from a solution of *n*-BuLi. Methyl methacrylate (MMA, Aldrich, 99%) was washed with a 10% aqueous sodium hydroxide solution and then with water, dried overnight with $MgSO_4$, and distilled twice over CaH_2 prior to polymerization. 1,1-Diphenylethylene (DPE, Aldrich, 97%) was distilled over CaH_2 and then distilled in the presence of DPHL under reduced pressure. Lithium chloride (Aldrich, 99.99%) was dried at 120°C for 24 h and dissolved in THF. *n*-BuLi (Aldrich, 1.6 M solution in hexane) was diluted with purified hexane.

Synthesis of Ethylene Glycol Di(1-methacryloyloxy)ethyl Ether (1). **1** was prepared through the addition reaction between ethylene glycol divinyl ether (EGDE; Aldrich, 97%) and methacrylic acid (MAA; Aldrich, 99.8%) in the presence of a trace amount of the inhibitor 4-*tert*-butylcatechol, under the protection of nitrogen, with magnetic stirring. In a 250 mL round-bottom flask equipped with a condenser and a magnetic stirrer, 25.0 g (0.21 mol) of EGDE and a small amount of 4-*tert*-butylcatechol were introduced. After 4-*tert*-butylcatechol dissolved and the temperature was raised to 70°C, MAA (36.5 g, 0.42 mol) was dropwise added with a syringe in about 20 min. The reaction was allowed to last an additional 6.0 h, and the crude product was distilled under high vacuum. The monomer was dissolved in purified toluene (30% v/v), and this solution was purified with CaH_2 and filtered through a tube filter with reduced ends in a completely sealed apparatus. This purification process was repeated prior to polymerization, and the toluene solution was directly used. The high purity of **1** was confirmed by **1**H NMR ($CDCl_3$): δ 1.42 (d, 6H, $OCH(CH_3)O$), 1.95 (s, 6H,α-CH_3), 3.60–3.83 (m, 4H, OCH_2CH_2O), 5.59 and 6.15 (2s, 4H, CH_2=), 6.00 (m, 2H, $OCH(CH_3)O$).

Polymerization. The anionic polymerization was carried out in THF, in a round-bottom glass flask, under an overpressure of argon, with magnetic stirring, at a selected temperature, in the presence of LiCl.[5] After THF, DPE, and a THF solution of LiCl were added with dry syringes, the flask was cooled to −40°C and *n*-BuLi (in hexane) was added. The deep red color of DPHL appeared at once, and the reaction between *n*-BuLi and DPE was allowed to continue for 15 min. For the preparation of star-shaped polymer, prechilled MMA was first added, and the polymerization was allowed to last 50 min at −78°C. Then, the system was warmed to −50°C, and a toluene solution of **1** was added. After the cross-linking reaction lasted 3 h, the system was quenched with a small amount of methanol, and the polymer was precipitated by pouring the polymerization solution into hexane. Then, the polymer was reprecipitated in ethanol from a benzene solution and vacuum-dried overnight.

In the cases of branched poly(MMA) and polymer gels, a prechilled mixture of MMA and a toluene solution of **1** was added to the initiator solution, and the reaction was allowed to last 2 h at −50°C. The branched poly(MMA) was purified in a way similar to that used for the star-shaped polymer. To purify the polymer gel, hexane containing a small amount of methanol was added, and after 3 h, it was dried under reduced pressure at 50°C for 24 h.

Hydrolysis of the Star-Shaped Polymer, the Branched Polymer, and the Polymer Gel. The hydrolysis of star-shaped or branched poly(MMA) was carried out in acetone, in the presence of a small amount of an aqueous solution of HCl, at room temperature, with magnetic stirring. For instance, 1.2 g of vacuum-dried star-shaped poly(MMA) (SSP-3 in Table 1.6.1) was redissolved in 30 mL of acetone, to which 1.0 mL of HCl aqueous solution (6.0 M) was added. After 20 min, this mixture was poured into hexane to precipitate the polymer. The polymer thus obtained was washed with hexane and vacuum-dried at 40°C for 24 h.

The hydrolysis of the polymer gel was performed either in THF or in acetone, using either acetic acid or an aqueous solution of HCl. For comparison, the reaction was also carried out under basic conditions in the presence of sodium hydroxide. In 30 mL of solvent, 1.0 g of polymer gel and a certain amount of acid were added. The corresponding time to form a completely transparent solution was recorded (Table 1.6.3). The hydrolyzed polymer was purified as described above.

Measurements. ^{1}H NMR spectra were recorded in $CDCl_3$ or CD_3OD on an INOVA-400 spectrometer. M_n and M_w/M_n of the polymer were determined by gel permeation chromatography (GPC) on the basis of a polystyrene calibration curve. The GPC measurements were carried out using THF as solvent, at 30°C, with a 1.0 mL/min flow rate and a 1.0 cm/min chart speed. Three polystyrene gel columns (Waters, 7.8 × 300 mm; one HR 5E, part no. 44228, one Linear, part no. 10681, and one HR 4E, part no. 44240) were used, which were connected to a Waters 515 precision pump. The FT-IR spectra were recorded on a Perkin-Elmer 1760-X spectrometer using KBr tablets.

TABLE 1.6.1
Preparation of Star-Shaped Poly(MMA) (SSP)[a]

		Star[b]		Hydrolyzed Product[c]		
No.	$[1]_0/[DPHL]_0$	M_w	M_n/M_n[d]	M_n (calcd)	M_n (obsd)[d]	M_w/M_n[d]
SSP-1	3	8 420	1.42	4760	4840	1.12
SSP-2	5	10 200	1.72	5100	5460	1.12
SSP-3	8	15 300	2.24	5620	6500	1.15

[a] The initiator, DPHL, was prepared via the reaction of *n*-BuLi with DPE ($[DPE]/[n\text{-}BuLi]_0 = 1.2$), in THF, at −40°C, in the presence of LiCl ($[LiCl]/[n\text{-}BuLi]_0 = 1.2$), for 15 min. The anionic polymerization of MMA was performed by adding prechilled MMA ($[MMA]_0 = 0.667$ M) to the above initiator solution ($[DPHL]_0 = 16.7$ mM), and the reaction was allowed to last 50 min at −78°C. Then, the system was warmed to −50°C, and a toluene solution of **1** was added. This cross-linking reaction was allowed to last an additional 3 h.

[b] Before reprecipitation.

[c] The hydrolysis was carried out after the reprecipitation.

[d] Determined by GPC.

1.6.3 RESULTS AND DISCUSSION

Preparation of Star-Shaped Poly(MMA). The preparation of star-shaped poly(MMA) was carried out in a two-step process,[1–4] namely, the living anionic polymerization of MMA and the reaction of the resulting living polymer with **1**. In the first step, the initiator DPHL was prepared in situ before the monomer addition, via the reaction of *n*-BuLi with DPE ($[DPE]/[n\text{-BuLi}]_0 = 1.2$), in THF, at −40°C, in the presence of LiCl ($[LiCl]/[n\text{-BuLi}]_0 = 1.2$),[5] for about 15 min. The anionic polymerization was induced by adding prechilled MMA ($[MMA]_0 = 0.667$ M) to the above initiator solution ($[DPHL]_0 = 16.7$ mM), and the reaction was allowed to last 50 min at −78°C. The molecular weight (M_n) of the obtained poly(MMA) was in good agreement with the designed value ($M_k = 4230$), and the molecular weight distribution (MWD) was very narrow ($M_w/M_n \leq 1.06$; see Figures 1.6.2A and 1.6.3A).

In the second step, the living poly(MMA) was allowed to react with 1 for $[\mathbf{1}]_0/[DPHL]_0$ ratios[6] of 3, 5, or 8. The cross-linking reaction was allowed to last 3 h at −50°C. The reaction proceeded quantitatively and produced a completely soluble product. As shown in Figure 1.6.2, the resulting polymers possess broad MWDs, and the star-shaped polymers (peak c) are accompanied by low molecular weight polymers (peaks a and b). Peak a corresponds to a higher molecular weight ($M_n \approx 5600$) than that of its living poly(MMA) precursor ($M_n = 4140$), and this can be attributed to a block copolymer of MMA and **1**. The M_n corresponding to peak b is about 14 000, and this fraction is most likely a star-shaped polymer with a low arm number (about 3). The presence of peaks a and b indicates that the cross-linking reaction was incomplete. As the amount of **1** increased, the a and b fractions decreased, being incorporated into the star-shaped polymer with high molecular weight.

It has been reported that the molecular structure of the cross-linker greatly affects the yield of the star-shaped polymer. For instance, when the cationically prepared living polymer of isobutyl vinyl ether (IBVE) was reacted with bisphenol A derived divinyl ether, a star-shaped poly(IBVE) was obtained with high selectivity.[2a] However,

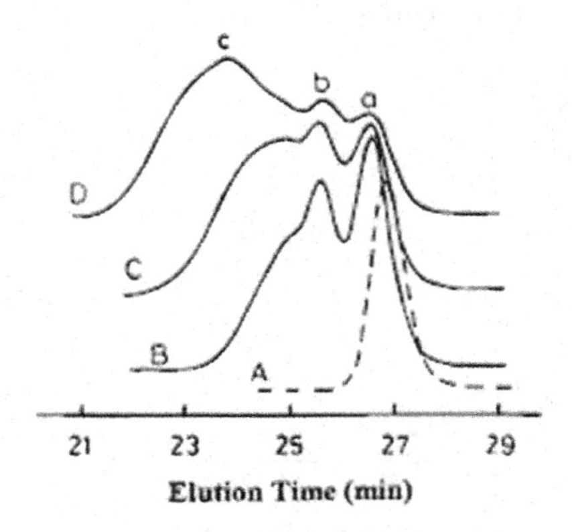

FIGURE 1.6.2 GPC traces of star-shaped polymers and their precursor: (A) living linear poly(MMA) ($M_n = 4140$, $M_w/M_n = 1.06$); (B, C, and D) star-shaped poly(MMA)s (Table 1.6.1; B, sSP-1; C, SSP-2; D, SSP-3) prepared by adding **1** to the above living poly(MMA) solution, for $[\mathbf{1}]_0/[DPHL]_0$ ratios of 3, 5, and 8, respectively.

when this cross-linker with rigid aromatic units was replaced with a flexible divinyl ether, such as di(ethylene glycol) divinyl ether, the resulting star-shaped polymer was accompanied by low molecular weight polymers. Similarly, the cross-linker **1** also possesses a flexible spacer. After the cross-linking proceeds to a certain extent, the flexibility of the cross-linker makes its pendant double bond less exposed and hence less accessible for further reactions with the incoming living chains. The increase of the amount of **1** will generate a larger core, thus providing a larger number of accessible vinyl groups. For this reason, the fraction of star-shaped polymer will increase (Figure 1.6.2D).

Because of the presence of the low molecular weight fractions (peaks a and b), the average molecular weight of the resulting polymer was low. For instance, the M_n of SSP-3 in Table 1.6.1 is 15 300, and according to this value, the calculated arm number of the star-shaped polymer is small (3.6). However, this is not the real value, because the star-shaped polymer is not pure. To obtain a pure star-shaped polymer, reprecipitation was carried out to remove the low molecular weight fractions. For instance, 2.0 g of SSP-3 (Table 1.6.1) was redissolved in 60 mL of benzene, and this solution was poured into 400 mL of ethanol. As shown in Figure 1.6.3B, the reprecipitated polymer exhibits a single GPC peak. According to the molecular weight of the reprecipitated SSP-3 ($M_n = 34\,000$), the calculated arm number of the star-shaped polymer is about 8, which is much higher than that before reprecipitation.

Preparation of the Branched Soluble Poly(MMA) and Polymer Gel. Great attention was accorded to polymer gels, because of their applications in various fields, such as medicine, nutritive and petrochemical industries, agriculture, biotechnology, etc.[7] The synthesis of pH-sensitive polymer gels and their applications in drug delivery have been widely investigated.[8] Because of the presence of ionizable groups in this kind of polymer gels, swelling or deswelling can occur along with pH changes. However, the pH change does not affect or change their chemical composition and molecular structure. In contrast to those polymer gels, a different polymer gel was prepared using the cross-linker **1**, because this insoluble gel can be changed to soluble linear polymers by changing the pH.

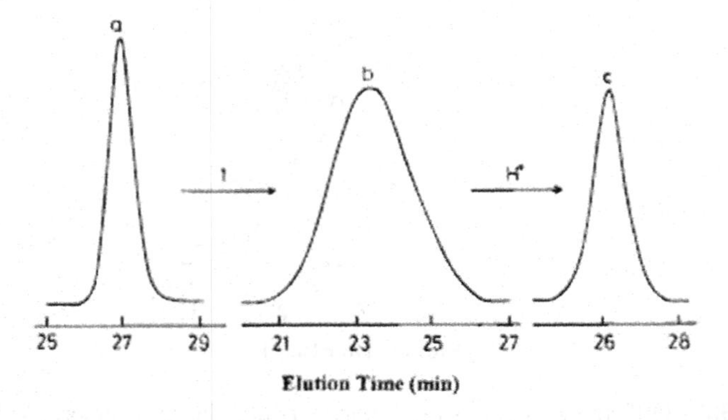

FIGURE 1.6.3 GPC traces of reprecipitated star-shaped poly(MMA) from SSP-3 (peak b, $M_n = 34\,000$, $M_w/M_n = 1.55$), its living poly(MMA) precursor (peak a, $M_n = 4180$, $M_w/M_n = 1.05$), and its hydrolyzed polymer (peak c, $M_n = 6500$, $M_w/M_n = 1.15$).

In contrast to the preparation of the star-shaped poly(MMA), MMA and a toluene solution of **1** were introduced into a THF solution of the initiator (DPHL) at the same time. The reaction was allowed to last 2 h at –50°C. Because both MMA and the cross-linker **1** participate simultaneously in the polymerization, a polymer gel could be obtained through the intermolecular cross-linking reaction.

For a fixed initial concentration of MMA ($[MMA]_0 = 0.588$ M), the characteristics of a polymer gel depend on the concentrations of both the initiator and the cross-linker **1**. Low $[DPHL]_0$ and high $[\mathbf{1}]_0$ are beneficial for the cross-linking reaction. As shown in Table 1.6.2, when $[DPHL]_0 = 14.7$ mM and $[\mathbf{1}]_0 = 29.4$ mM (BP-1), cross-linking occurred to some extent, but the polymer remained soluble. Even when the concentration of **1** was doubled to increase the number of cross-linking points (BP-2), or the concentration of DPHL was reduced to half to increase the molecular weight (BP-3), the polymers still remained soluble, and the magnetic stirring was still possible despite a very high viscosity. The resulting polymers (BP-2 and -3) have higher molecular weights and broader MWDs than BP-1 (Table 1.6.2). Consequently, when the concentration of the cross-linker **1** was sufficiently low and/or the initiator concentration was sufficiently high, the resulting products were soluble branched polymers, probably accompanied by some looped and cyclic polymer chains. For this reason, the MWDs were broad. However, when $[\mathbf{1}]_0$ was taken twice as large and $[DPHL]_0$ was simultaneously reduced to half that for BP-1, the magnetic stirring became impossible after only a few minutes upon the addition of MMA and **1**, and a polymer gel was generated (PG-1 in Table 1.6.2). The further reduction of $[DPHL]_0$ (PG-2 in Table 1.6.2) or the homopolymerization of **1** ($[DPHL]_0 = 7.4$ mM, $[\mathbf{1}]_0 = 0.400$ M) generated polymer gels easily.

Hydrolysis of Star-Shaped Polymers, Branched Polymers, and Polymer Gels. The hydrolysis of the star-shaped or branched polymer was carried out in acetone, at room temperature, in the presence of a small amount of hydrochloric acid (see Experimental Section). Figure 1.6.3B presents the GPC chromatogram of a reprecipitated star-shaped poly(MMA) (peak b), which has, obviously, a larger molecular weight and a broader

TABLE 1.6.2
Preparation of the Branched Soluble Poly(MMA) (BP) and Polymer Gel (PG)[a]

				Before Hydrolysis		After Hydrolysis	
No.	$[DPHL]_0$ (mM)	$[\mathbf{1}]_0$ (mM)	Stirring	M_n[b]	M_w/M_n[b]	M_n[b]	M_w/M_n[b]
BP-1	14.7	29.4	OK	14 100	1.69	9 440	1.40
BP-2	14.7	58.8	OK	33 000	2.43	9 790	1.46
BP-3	7.4	29.4	OK	49 100	3.10	15 100	1.57
PG-1	7.4	58.8	NO	gel		12 900	1.63
PG-2	5.0	58.8	NO	gel		15 900	1.74

[a] The initiator, DPHL, was prepared via the reaction of *n*-BuLi with DPE ($[DPE]/[n\text{-BuLi}]_0 = 1.2$), in THF, at −40°C, in the presence of LiCl ($[LiCl]/[n\text{-BuLi}]_0 = 1.2$), for 15 min. The polymerization was induced by adding a prechilled mixture of MMA ($[MMA]_0 = 0.588$ M) and a toluene solution of **1** to the above initiator system, and the reaction was allowed to last 2 h at −50°C.

[b] Determined by GPC.

MWD ($M_n = 34\,000$, $M_w/M_n = 1.55$) than its living linear poly(MMA) precursor (peak a, $M_n = 4180$, $M_w/M_n = 1.05$). The hydrolysis of this star-shaped polymer resulted in the formation of a polymer with a lower molecular weight and a narrower MWD (peak c, $M_n = 6500$, $M_w/M_n = 1.15$). This hydrolyzed product is most likely a block copolymer consisting of poly(MMA) and poly(methacrylic acid) [poly(MAA)] segments as described below, and for this reason, its molecular weight is larger than that of the linear poly(MMA) precursor (peak a).

To identify the cleavage point, the hydrolyzed product was characterized by ^{1}H NMR and FT-IR. Figure 1.6.4 depicts the ^{1}H NMR spectra of the star-shaped polymer (B) and its hydrolyzed product (C). For comparison, the ^{1}H NMR spectrum of **1** is also included in Figure 1.6.4A. Comparing parts A and B, one can observe that the peaks a and b, corresponding to the α-methyl and $H_2C=$ of **1**, shifted to 0.8–1.1 and 1.8–2.1 ppm after cross-linking and overlapped with the absorptions of the α-CH_3 (B-a) and –CH_2– (B-b) belonging to poly(MMA). On the other hand, the peaks c, d, and e of the side chain of **1** did not change and could be detected in the spectrum of the star-shaped polymer (B). However, these absorptions disappeared completely in the spectrum of the hydrolyzed product (C), indicating that the ester groups of the cross-linker **1** were eliminated to yield the poly(MAA) segment. This result was also confirmed by FT-IR; the hydrolyzed polymer exhibits a broad absorption (2500–3800 cm^{-1}) corresponding to the carboxyl group of poly(MAA) segment. On the other hand, the absorption due to the ester group (–OCH_3) of the poly(MMA) segment is still present quantitatively in the **1**H NMR spectrum of the hydrolyzed polymer (peak f in Figure 1.6.4C). The above results indicate that the hydrolysis

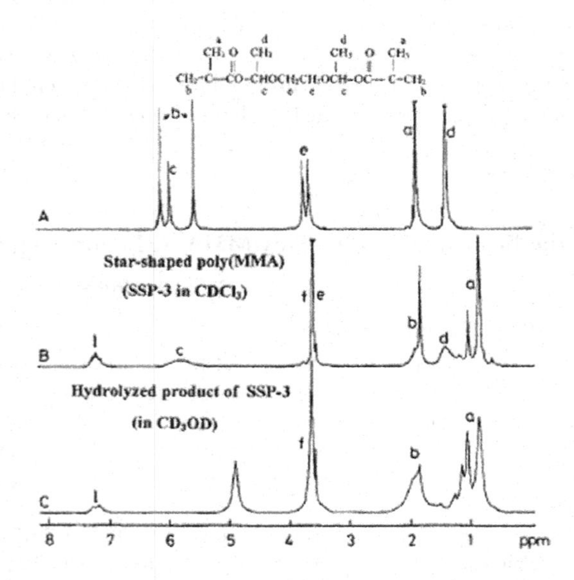

FIGURE 1.6.4 ^{1}H NMR spectra of **1** (A; in $CDCl_3$), star-shaped poly(MMA) (B; SSP-3 in $CDCl_3$; see Table 1.6.1 and Figure 1.6.3B) and its hydrolyzed polymer (C; in CD_3OD, see Table 1.6.1 and Figure 1.6.3C). Absorptions due to poly(MMA) segment: peak a, α-CH_3; peak b, –CH_2– in the main chain; peak f, –OCH_3 in the side chain. I: C_6H_5 of the initiator (DPHL).

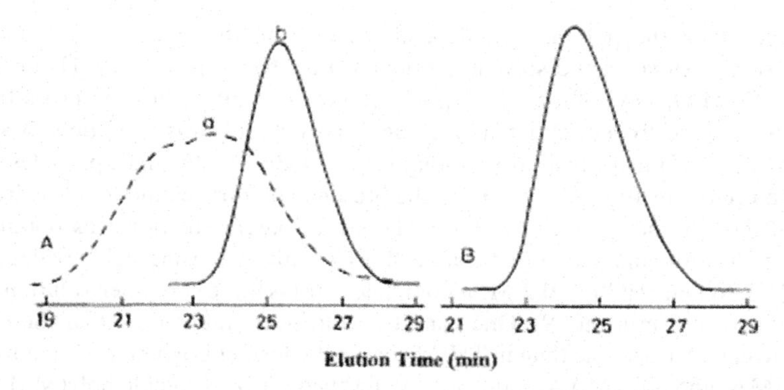

FIGURE 1.6.5 GPC traces of the hydrolyzed polymers from the branched soluble poly(MMA) (A) and polymer gel (B). A-a: BP–2 in Table 1.6.2, $M_n = 33\ 000$, $M_w/M_n = 2.43$. A-b: Hydrolyzed polymer from A-a, $M_n = 9790$, $M_w/M_n = 1.46$. B: Hydrolyzed polymer ($M_n = 12\ 900$, $M_w/M_n = 1.63$) from polymer gel, PG–1 (Table 1.6.2).

reaction selectively fractured the cross-linking points to generate linear block copolymers of MMA and MAA.

Similarly, the hydrolysis of the branched soluble polymer can also generate a linear polymer via the same mechanism. As shown in Figure 1.6.5A, the hydrolyzed polymer possesses a lower molecular weight and a narrower MWD (peak A-b, $M_n = 9790$, $M_w/M_n = 1.46$) than its precursor polymer (peak A-a, BP-2 in Table 1.6.2, $M_n = 33\ 000$, $M_w/M_n = 2.43$). Its FT-IR spectrum also confirmed the presence of the carboxyl groups of poly(MAA) units (2500–3800 cm^{-1}). However, the hydrolyzed product is a random copolymer of MMA and MAA, because MMA and the cross-linker **1** participated in the polymerization simultaneously.

The hydrolysis of the polymer gel is more interesting, because this is a process that changes the insoluble polymer to a soluble linear one. As shown in Table 1.6.3,

TABLE 1.6.3
Response of the Poly(MMA) Gel[a]

No.	Solvent	Catalyst/M	Time	Response
1	Acetone	NaOH/0.42	24 h	Swelling
2	Acetone	H_2O/0.95	24 h	Swelling
3	THF	AA[b]/0.26	36 h	Peptization
4	Acetone	AA[b]/0.26	24 h	Peptization
5	THF	HCl/0.015	20 min	Clear solution
6	Acetone	HCl/0.015	15 min	Clear solution
7	Acetone	HCl/0.05	10 min	Clear solution
8	THF	HCl/0.2	10 min	Clear solution
9	Acetone	HCl/0.2	5 min	Clear solution

[a] The reaction was carried out at 20°C with magnetic stirring (solvent, 30 mL; polymer gel, 1.0 g).
[b] Acetic acid.

the response of the polymer gel depends mainly on the acidity of the medium and to some extent on the solvent employed. In a basic (no. 1, NaOH) or a neutral (no. 2, H_2O) environment, the polymer gel is swollen, but the cross-linking points are not destroyed. When a weak acid, acetic acid, was used (nos. 3 and 4), the polymer gel transformed after a long time (24–36 h) into small particles which dispersed in the medium. However, in the presence of a trace amount of hydrochloric acid (nos. 5 and 6, [HCl] = 0.015 M), a transparent solution was obtained in less than 20 min, and this time became shorter with increasing acid concentration (nos. 7–9). When [HCl] = 0.2 M, a completely transparent polymer solution could be obtained in 5 min (no. 9). One can also note from Table 1.6.3 that the change occurred easier in acetone than in THF, because the former is a better solvent for both poly(MMA) and poly(MAA) segments. As for the branched soluble poly(MMA), the hydrolyzed product of the poly(MMA) gel is also a linear random copolymer of MMA and MAA. Its GPC chromatogram exhibits a single peak (Figure 1.6.5B), and the presence of carboxyl groups in the MAA units was confirmed by FT-IR (2500–3800 cm^{-1}). In addition, the polymer gel prepared via the homopolymerization of **1** could also be hydrolyzed to a linear polymer [poly(MAA)] in methanol in the presence of HCl ([HCl] = 0.2 M) in 20 min.

The hydrolyzed products of star-shaped polymers, branched polymers, and polymer gels possess quite different solubilities compared to those of their precursors (Table 1.6.4). The hydrolyzed star-shaped polymers (h-SSP-1, h-SSP-2,

TABLE 1.6.4
Solubility before and after Hydrolysis[a]

	CH_3OH	Acetone	THF	Dioxane	DMF	$CHCl_3$	Benzene
SSP-1	I	S	S	S	S	S	S
h-SSP-1	D	S	S	D	S	S	I
SSP-2	I	S	S	S	S	S	S
h-SSP-2	S	S	S	D	S	C	I
SSP-3	I	S	S	D	S	S	D
h-SSP-3	S	S	D	D	S	C	I
BP-1	I	S	S	D	S	S	S
h-BP-1	W	S	S	S	S	S	C
BP-2	I	S	D	D	D	S	D
h-BP-2	W	S	S	D	S	C	I
BP-3	I	S	S	D	S	S	D
h-BP-3	W	S	S	D	S	D	C
GP-1	I	I	I	I	I	I	I
h-GP-1	W	S	S	D	S	W	I
GP-2	I	I	I	I	I	I	I
h-GP-2	W	S	S	D	S	W	I

[a] The experiment was carried out at room temperature. The amounts of polymer (or polymer gel) and solvent were 0.03 g and 1.0 mL, respectively. S = soluble; D = dissolved slowly; W = wetting; C = cloudy; I = insoluble; h- = hydrolyzed product from the corresponding polymer or polymer gel.

and h-SSP-3) are soluble in methanol, but insoluble in benzene, due to the presence of the hydrophilic poly(MAA) block. On the other hand, the polymer gel is insoluble in all solvents before hydrolysis. However, its hydrolyzed product is soluble in acetone, THF, 1,4-dioxane, and *N,N*-dimethylformamide (DMF), but insoluble in benzene, and only wetted by methanol due to the presence of the random MAA units.

REFERENCES

1. (a) Zilliox, J. G.; Rempp, P.; Parrod, J. *J. Polym. Sci., Part C* **1968**, *22*, 145. (b) Worsfold, D. J.; Zilliox, J. G.; Rempp, P. *Can. J. Chem.* **1969**, *47*, 3379. (c) Bywater, S. *Adv. Polym. Sci.* **1979**, *30*, 90.
2. (a) Kanaoka, S.; Sawamoto, M.; Higashimura, T. *Macromolecules* **1991**, *24*, 2309, 5741; **1993**, *26*, 254. (b) Kanaoka, S.; Omura, T.; Sawamoto, M.; Higashimura, T. *Macromolecules* **1992**, *25*, 6407.
3. (a) Sogah, D. Y.; Hertler, W. R.; Webster, O. W.; Cohen, G. M. *Macromolecules* **1987**, *20*, 1473. (b) Quirk, R. P.; Ren, T. *Polym. Int.* **1993**, *32*, 205.
4. (a) Bazan, G. C.; Schrock, R. R. *Macromolecules* **1991**, *24*, 817. (b) Saunders, R. S.; Cohen, R. E.; Wong, S. J.; Schrock, R. R. *Macomolecules* **1992**, *25*, 2055.
5. (a) Varshney, S. K.; Hautekeer, J. P.; Fayt, R.; Jerome, R.; Teyssie, Ph. *Macromolecules* **1990**, *23*, 2618. (b) Fayt, R.; Forte, R.; Jacobs, C.; Jerome, R.; Ouhadi, T.; Teyssie, Ph.; Varshney, S. K. *Macromolecules* **1987**, *20*, 1442. (c) Kunkel, D.; Muller, A. H. E.; Janata, M.; Lochman, L. *Makromol. Chem., Macromol. Symp.* **1992**, *60*, 315.
6. The cross-linker **1** added to the living poly(MMA) solution reacts with the living end (P*) of poly(MMA). Because one initiator (DPHL) generates one living poly(MMA) chain in the first step, the living anionic polymerization of MMA, $[\mathbf{1}]_0/[P^*]_0$ is equal to $[\mathbf{1}]_0/[DpHL]_0$.
7. (a) Shibayama, M.; Tanaka, T. *Adv. Polym. Sci.* **1993**, *109*, 1. (b) Samsonov, G. V.; Kuznetsova, N. P. *Adv. Polym. Sci.* **1992**, *104*, 1. (c) Kazanskii, K. S.; Dubrovskii, S. A. *Adv. Polym. Sci.* **1992**, *104*, 97.
8. (a) Siegel, R. A. *Adv. Polym. Sci.* **1993**, *109*, 233. (b) Dong, L. C.; Hoffman, A. S. *J. Control Release* **1991**, *15*, 141.

1.7 Graft, Block–Graft and Star-Shaped Copolymers by an *In Situ* Coupling Reaction*

Eli Ruckenstein and Hongmin Zhang

Chemical Engineering Department, State University of New York at Buffalo, Amherst, New York 14260

ABSTRACT A tetrahydrofuran (THF) solution of the living poly(glycidyl methacrylate) [poly(GMA)] was prepared by the living anionic polymerization of GMA, using 1,1-diphenylhexyllithium (DPHL) as initiator, in the presence of LiCl ($[LiCl]/[DPHL]_0 = 3$), at −45°C. Upon introduction of a benzene solution of living polystyrene [poly(St)] into the above system at −30°C, a very rapid coupling reaction between the epoxy groups of the living poly(GMA) and the propagating sites of the living poly(St) generated a graft copolymer containing a polar backbone and nonpolar side chains. When benzene solutions of living poly(St) and polyisoprene were sequentially transferred to the THF solution of living poly(GMA), a graft copolymer possessing two different nonpolar side chains was obtained. Further, a THF solution of the living block copolymer of methyl methacrylate (MMA) and GMA was prepared by the sequential anionic polymerization of the two monomers, at low temperatures. A block–graft copolymer having a poly(MMA-*b*-GMA) block backbone and poly(St) side chains was synthesized by reacting the living poly(St) with the epoxy groups of the above living block copolymer. In addition, when the poly(GMA) segment in poly(MMA-*b*-GMA) was short, a star-shaped copolymer containing a poly(MMA) arm and several poly(St) arms was obtained. GPC and ^{1}H NMR measurements indicated that all the above copolymers possess high purity, designed molecular architectures, controlled molecular weights and narrow molecular weight distributions ($M_w/M_n = 1.10–1.23$).

1.7.1 INTRODUCTION

The living polymerization techniques provide the possibility to synthesize well-defined graft copolymers.[1–3] Three approach routes have been suggested, namely the "grafting from," "grafting onto," and "grafting through" routes.[4] The first is based

* *Macromolecules* 1998, 31, 4753–4759.

on the sequential polymerizations from the backbone to side chains, with the backbone acting as a macroinitiator for the latter step. In this manner, using a continuous transformation from anionic to cationic polymerization, several well-defined graft and block–graft copolymers, consisting of a polymethacrylate backbone and poly(alkyl vinyl ether) side chains have been recently prepared.[5] The present paper will focus on the preparation of well-defined graft, block–graft, and star-shaped copolymers by a "grafting onto" method. These copolymers contain both polar (polymethacrylate) and nonpolar (polystyrene or/and polyisoprene) components.

The "grafting onto" method by the anionic polymerization technique is based on a nucleophilic reaction of electrophilic groups attached to a backbone polymer (generally prepared by radical polymerization) and propagating sites of a living polymer. The electrophilic groups attached to the backbone can be esters,[6] chloro- or bromomethyls,[7] and dimethylvinylsilyls.[8] The coupling reaction between the epoxy groups and living anionic polymers was also employed in the preparation of a graft copolymer consisting of a polystyrene [poly(St)] backbone and poly(St) side chains, by reacting the living poly(St) with poly(*p*-vinylstyrene oxide) backbone or its styrene copolymer,[9] which was prepared by radical homo- or copolymerization. Generally, the backbone polymer had to be purified carefully by several reprecipitations and freeze-drying under high vacuum. However, even with such precautions, the impurities were not completely removed and usually caused the deactivation of a part of the living polymers. Because of the presence of the deactivated homopolymer, it was difficult to obtain pure graft copolymers.

Recently, the living anionic polymerization of the bifunctional glycidyl methacrylate (GMA) has been achieved by using a bulky initiator with low reactivity, in the presence or absence of LiCl, in a polar solvent, at a low temperature.[10–13] The C=C bonds of GMA were selectively polymerized, and the epoxy groups remained unreacted. In this manner, the functional poly(GMA) and its methyl methacrylate (MMA) copolymer with controlled molecular weights and narrow molecular weight distributions were obtained. This technique will be employed in what follows as a step in the preparation of model copolymers with complex molecular architectures. The scope of the present paper is to synthesize the following graft, block–graft, and star-shaped copolymers by *in situ* coupling reactions. (1) AB_x graft copolymer (**1** in Figure 1.7.1), consisting of a poly(GMA)

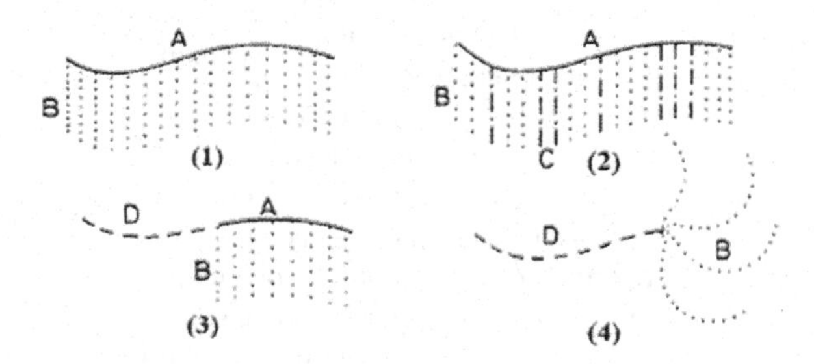

FIGURE 1.7.1 Schematic representation of four model copolymers prepared in the present paper. (1) AB_x graft copolymer; (2) AB_xC_y graft copolymer; (3) DAB_x block–graft copolymer; (4) DaB_x star-shaped copolymer. Components: A, poly(glycidyl methacrylate); B, polystyrene; C, polyisoprene; D, poly(methyl methacrylate); linked with a short poly(glycidyl methacrylate) segment with a few repeating units.

backbone and poly(St) or polyisoprene [poly(Is)] side chains; (2) AB_xC_y graft copolymer (**2** in Figure 1.7.1), consisting of a poly(GMA) backbone and two different side chains, poly(St) and poly(Is); (3) DAB_x block–graft copolymer (**3** in Figure 1.7.1), consisting of a poly(MMA-*b*-GMA) block backbone and poly(St) side chains attached to the poly(GMA) segment of the backbone; (4) DaB_x star-shaped copolymer (**4** in Figure 1.7.1), consisting of a poly(MMA) arm and several poly(St) arms.

As shown in Scheme 1.7.1, a tetrahydrofuran (THF) solution of the living poly(GMA) (**1** in Scheme 1.7.1) and a benzene solution of the living poly(St) were separately prepared at the same time. Then, the coupling reaction between the epoxy groups of the living poly(GMA) and the propagating sites of the living poly(St) was carried out by introducing the solution of the latter into the former. Thus, an AB_x graft copolymer (**2** in Scheme 1.7.1) was prepared. Furthermore, when both the living poly(St) and poly(Is) solutions were sequentially introduced into the solution of living poly(GMA), an AB_xC_y graft copolymer could be produced (**3** in Scheme 1.7.1). The preparation routes of DAB_x block–graft and DaB_x star-shaped copolymers are presented in Scheme 1.7.2. A THF solution of the living block copolymer, poly(MMA-*b*-GMA) (**5** in Scheme 1.7.2), was first prepared by the sequential anionic polymerization of MMA and GMA in THF at low temperature; its reaction with living poly(St) generated a DAB_x graft copolymer (**6** in Scheme 1.7.2). If the polymerization degrees of MMA and St are large and that of GMA (*m*) is small, the poly(GMA) segment is very short and a DaB_x star-shaped copolymer could be obtained (**6** in Scheme 1.7.2).

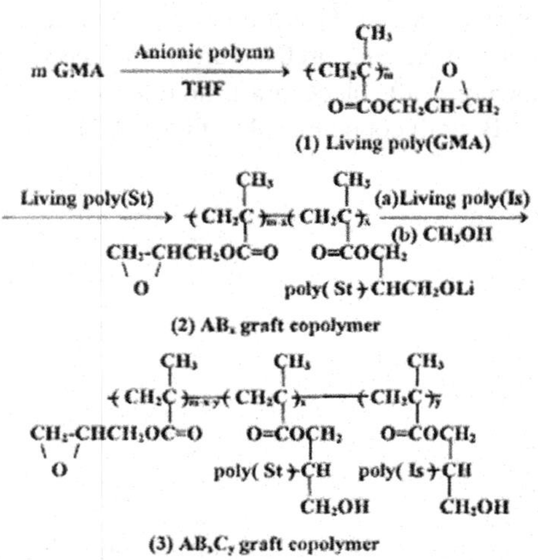

m GMA —Anionic polymn, THF→ (1) Living poly(GMA)

—Living poly(St)→ (2) AB_x graft copolymer —(a) Living poly(Is), (b) CH_3OH→ (3) AB_xC_y graft copolymer

SCHEME 1.7.1

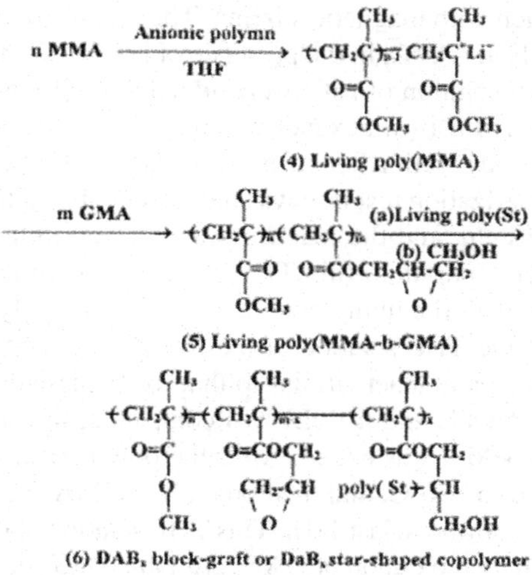

n MMA —Anionic polymn / THF→ $-(CH_2C(CH_3)(C(=O)OCH_3))_{n-1}-CH_2C(CH_3)(C(=O)OCH_3)^- Li^+$

(4) Living poly(MMA)

—m GMA→ $-(CH_2C(CH_3)(C(=O)OCH_3))_n-(CH_2C(CH_3)(C(=O)OCH_2CH(O)CH_2))_m$ —(a) Living poly(St) / (b) CH_3OH→

(5) Living poly(MMA-b-GMA)

$-(CH_2C(CH_3)(C(=O)OCH_3))_n-(CH_2C(CH_3)(C(=O)OCH_2CH(O)CH_2))_{m-x}-(CH_2C(CH_3)(C(=O)OCH_2CH(CH_2OH)-(St)-poly))_x$

(6) DAB_x block-graft or DaB_x star-shaped copolymer

SCHEME 1.7.2

1.7.2 EXPERIMENTAL SECTION

Materials. Tetrahydrofuran (THF) was dried with CaH_2 under reflux for more than 24 h, distilled, and distilled again from a solution of 1,1-diphenylhexyllithium (DPHL) just before use. Benzene was washed with concentrated sulfuric acid and then with water, dried over $MgSO_4$, distilled over CaH_2, and distilled from an *n*-BuLi solution before use. Hexane was first dried and distilled over CaH_2 and then distilled from a solution of *n*-BuLi. Methyl methacrylate (MMA, Aldrich, 99%) was washed with 10% aqueous sodium hydroxide solution and then with water, dried overnight with $MgSO_4$, distilled over CaH_2, and finally distilled in the presence of triisobutylaluminum[14] prior to polymerization. Glycidyl methacrylate (GMA, Aldrich, 97%) was dried over CaH_2 with magnetic stirring under reduced pressure for more than 24 h and vacuum distilled slowly, and the middle fraction (about half volume) was recovered and distilled twice over CaH_2 prior to polymerization. Styrene (St, 99%, Aldrich) and isoprene (Is, 99%, Aldrich) were washed with 10% aqueous NaOH solution and then with water, dried overnight with $MgSO_4$, distilled over CaH_2, and finally distilled in the presence of benzylmagnesium chloride just before use. 1,1-Diphenylethylene (DPE, Aldrich, 97%) was distilled over CaH_2 and then distilled in the presence of DPHL under reduced pressure. Lithium chloride (Aldrich, 99.99%) was dried at 120°C for 24 h and dissolved in THF.[15] *n*-BuLi (Aldrich, 1.6 M solution in hexane) was diluted with purified hexane.

Preparation of the THF Solutions of living Poly(GMA) by the Anionic Polymerization of GMA and of living Block Copolymer of MMA and GMA by the Sequential Anionic Polymerization of the Two Monomers. All polymerizations, namely homopolymerizations of GMA, St, and Is and the block copolymerization of MMA and GMA, were carried out in a round-bottom glass flask under an overpressure of nitrogen with magnetic stirring. The anionic polymerization of GMA was performed in THF, at −45°C, in the presence of LiCl ($[LiCl]/[DPHL]_0 = 3$). After THF, DPE, and a THF solution of LiCl were added with dry syringes, the flask was cooled to −45°C and *n*-BuLi (in hexane) was added. The deep red color of DPHL appeared at once, and the reaction between *n*-BuLi and DPE was allowed to continue for 15 min. The polymerization reaction was induced by the addition of the prechilled GMA to the above system, and the reaction was allowed to last 1.0 h. Without termination and polymer separation, this THF solution was directly used in the next coupling reaction step with the living poly(St) or/and living poly(Is).

The preparation of the THF solution of the living block copolymer of MMA and GMA was carried out in a manner similar to that for living poly(GMA). However, as soon as the reaction of DPE and *n*-BuLi was completed, the solution was cooled to −70°C and prechilled MMA was added. After the polymerization lasted 40 min, the temperature was raised to −50°C, and then prechilled GMA was added and the block copolymerization was allowed to last 1.0 h. This THF solution of the living poly(MMA-*b*-GMA) was used in the preparation of DAB_x graft and DaB_x star-shaped copolymers.

Preparation of the Benzene Solution of Living Poly(St) [or Poly(Is)] by the Anionic Polymerization of St (or Is). The anionic polymerization of St or Is was carried out using *n*-BuLi as initiator, in benzene, in the presence of a small amount of THF ($[THF] = [n\text{-BuLi}]_0$). After benzene, THF and the hexane solution of *n*-BuLi were introduced into a flask placed in a 10°C water bath, the polymerization was started by adding St (or Is) to the above system. After 10 min, the temperature was raised to 25°C and the polymerization reaction was allowed to last 1.0 h. The benzene solution thus obtained was directly employed in the next coupling reaction step.

Syntheses of AB_x and AB_xC_y Graft, DAB_x Block–Graft, and DaB_x Star-Shaped Copolymers by the Coupling Reaction between the Epoxy Groups of Living Poly(GMA) or Poly(MMA-*b*-GMA) with the Living Sites of Poly(St) or/and Poly(Is). The coupling reaction was carried out with vigorous magnetic stirring, at −30°C, for 30 min. For the preparation of AB_x graft copolymer, the benzene solution of living poly(St) was transferred to the THF solution of living poly(GMA). In the case of AB_xC_y graft copolymer, the benzene solutions of both living poly(St) and poly(Is) were sequentially transferred. The DAB_x block–graft copolymer or DaB_x star-shaped copolymer was prepared in the same way as the AB_x graft copolymer, except that the THF solution of living poly(GMA) was replaced by that of living poly(MMA-*b*-GMA). Although the coupling reaction occurred rapidly, it was still allowed to last 30 min to ensure a complete conversion. Then, the reaction was quenched, by adding a small amount of methanol, and the reaction mixture was poured into the methanol to precipitate the polymer. The polymer was reprecipitated by pouring again its THF solution into methanol and was finally vacuum-dried overnight.

Measurements. ^1H NMR spectra were recorded in $CDCl_3$ on a VXR–400 spectrometer. M_n and M_w/M_n of the polymer were determined by gel permeation chromatography (GPC) on the basis of a polystyrene calibration curve. The GPC measurements were carried out using THF as solvent, at 30°C, with a 1.0 mL/min flow rate and a 1.0 cm/min chart speed. Three polystyrene gel columns (Waters, 7.8 × 300 mm; one HR 5E, Part No. 44228, one Linear, Part No. 10681, and one HR 4E, Part No. 44240) were used, which were connected to a Waters 515 precision pump.

1.7.3 RESULTS AND DISCUSSION

Synthesis of AB_x (or AC_y) Graft Copolymer. The THF solution of living poly(GMA) was prepared by the living anionic polymerization according to a reported method.[10–13] The initiator DPHL was first prepared by reacting *n*-BuLi with DPE ($[DPE]/[n\text{-BuLi}]_0 = 1.2$) at −45°C for 15 min. The anionic polymerization of GMA was carried out in THF, at −45°C, in the presence of LiCl[16] ($[LiCl]/[n\text{-BuLi}]_0 = 3$). The initial monomer concentration was fixed at $[GMA]_0 = 0.47$ M in all cases. The polymerization degree ($m = [GMA]_0/[DPHL]_0$) and the theoretical average number molecular weight ($M_n = mM_{GMA} + M_I$, where M_{GMA} and M_I are the molecular weights of GMA and the initiator moiety 1,1-diphenylhexyl, respectively) were controlled by changing the initial concentration of DPHL. For instance, for $[DPHL]_0 = 14.0$ mM, the determined M_n of the obtained poly(GMA) was 4940, which is close to the calculated value ($M_k = 5000$). In addition, the molecular weight distribution of the polymer was very narrow ($M_w/M_n = 1.06$) and the GPC chromatogram exhibited a single sharp peak (peak b in Figure 1.7.2). In the other cases, the determined molecular weights are also in good agreement with those calculated and the molecular weight distributions are narrow ($M_w/M_n = 1.06–1.15$; see Tables 1.7.1 and 1.7.3). These results indicate that the anionic polymerization of GMA proceeded in a living manner. After the polymerization lasted 1.0 h, a small amount of solution (about 0.1 mL) was taken out with a dry syringe for the GPC measurement, and the remaining solution was directly used in the next coupling reaction step.

The benzene solution of living poly(St) or poly(Is) was prepared by a traditional method.[17] The anionic polymerization was carried out in benzene, using *n*-BuLi as initiator, at 25°C. To promote the dissociation of *n*-BuLi and accelerate the polymerization, a small amount of a polar solvent, THF, was added ($[THF] = [n\text{-BuLi}]_0$). Because of the THF, the living poly(St) solution acquired a deep red color and this

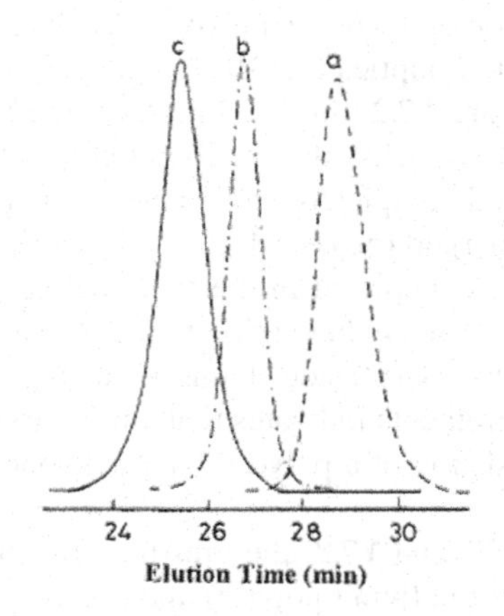

FIGURE 1.7.2 GPC traces of the AB_x graft copolymer and its precursors: (a) living poly(St) ($M_n = 1180$; $M_w/M_n = 1.13$); (b) living poly(GMA) ($M_n = 4940$; $M_w/M_n = 1.06$); (c) graft copolymer prepared by the coupling reaction of **a** and **b** (AB_x-2 in Tables 1.7.1 and 1.7.2; $M_n = 1.21 \times 10^4$; $M_w/M_n = 1.19$).

TABLE 1.7.1
Preparation Conditions of AB_x (or AC_y) Graft Copolymer[a]

	Living Poly(GMA)				Living Poly(St) or Poly(Is)			
No.	M_n[b]	M_w/M_n[b]	m[c]	$mmol_{epoxy}$[d]		M_n[b]	M_w/M_n[b]	mmol[e]
AB_x-1	4050	1.07	26.8	14.1	PSt	1330	1.10	2.22
AB_x-2	4940	1.06	33.1	14.1	PSt	1180	1.13	3.04
AC_y-1	3900	1.09	25.8	14.1	PIs	1430	1.14	2.06

[a] The coupling reaction was carried out by introducing the benzene solution of living poly(St) or poly(Is) into the THF solution of living poly(GMA) with vigorous magnetic stirring. The reaction was allowed to last 30 min at −30°C.
[b] Determined by GPC.
[c] Polymerization degree (m) of poly(GMA).
[d] Molar amount of epoxy groups (equal to the amount of GMA).
[e] Molar amount of living poly(St) or poly(Is) (equal to the amount of n-BuLi used for the preparation of the corresponding living polymer).

allowed us to observe via the color change the end of the next coupling reaction step. The living polymer possesses controlled molecular weight and narrow molecular weight distribution (M_w/M_n = 1.08–1.14; see Tables 1.7.1, 1.7.3 and 1.7.5). After a small amount of sample (about 0.1 mL) was taken out from the system for the GPC measurement, the remaining solution was immediately used in the next coupling reaction step.

The coupling reaction between the epoxy groups of living poly(GMA) and the living sites of poly(St) or poly(Is) was carried out with vigorous magnetic stirring, at −30°C, for 30 min. As soon as the benzene solution of the living poly(St) was introduced into the THF solution of the living poly(GMA), the deep red color of the living poly(St) disappeared, which implies that the coupling reaction took place instantaneously. As shown in Figure 1.7.2, the GPC peaks corresponding to living poly(St) (peak a; M_n = 1180; M_w/M_n = 1.13) and the living poly(GMA) (peak b; M_n = 4940; M_w/M_n = 1.06) disappeared completely and a new single peak due to the formation of the graft copolymer emerged (peak c; $M_n = 1.21 \times 10^4$; M_w/M_n = 1.19; see AB_x-2 in Tables 1.7.1 and 1.7.2). As depicted in the ^{1}H NMR spectrum (I in Figure 1.7.3), besides the peaks (a, b, c, d, and e) belonging to the backbone polymer, poly(GMA), the peaks (f, g, and i) of the poly(St) side chains are also present. The combination of GPC and ^{1}H NMR measurements indicates that a pure graft copolymer, free of precursor polymers and consisting of a poly(GMA) backbone and poly(St) side chains, was obtained.

As shown in Tables 1.7.1 and 1.7.2, the grafting efficiency (f) can be controlled through the molar ratio of the living poly(St) and epoxy groups, $f = mol_{PSt}/mol_{epoxy}$. The graft number in each poly(GMA) backbone is given by $N = fm$, where m is the polymerization degree of GMA. Furthermore, the calculated molecular weight (M_k) of the graft copolymer is given by the expression $M_k = M_{n(poly(GMA))} + NM_{n(PSt)}$. Because of the living polymerization, the real molecular weight is expected to be

TABLE 1.7.2
Characterization of AB_x (or AC_y) Graft Copolymer

No.	f^a (%)	N^b	$10^{-3}M_n$ calcd[c]	$10^{-3}M_n$ obsd[d]	M_w/M_n[d]
AB_x-1	15.7	4.2	9.64	9.32	1.18
AB_x-2	21.6	7.1	13.3	12.1	1.19
AC_y-1	14.6	3.8	9.33	8.2	1.18

[a] Grafting efficiency, $f = \text{mol}_{(\text{PSt or PIs})}/\text{mol}_{\text{epoxy}}$ (see Table 1.7.1).
[b] Graft number in each poly(GMA) chain, $N = fm$.
[c] Calculated with the expression $M_k = M_{n(\text{poly(GMA)})} + NM_{n(\text{PSt or PIs})}$.
[d] Determined by GPC.

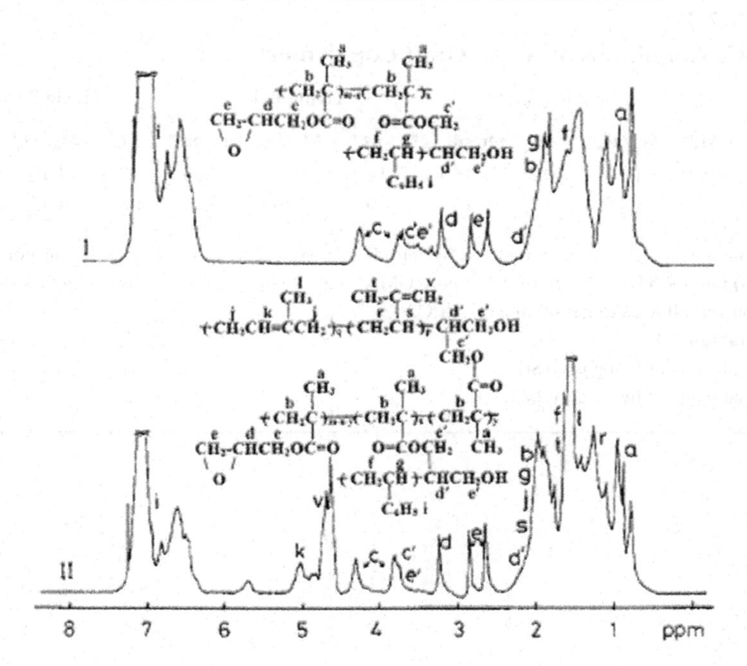

FIGURE 1.7.3 ^{1}H NMR spectra of AB_x (I; AB_x-2 in Tables 1.7.1 and 1.7.2) and AB_xC_y (II; AB_xC_y-1 in Tables 1.7.3 and 1.7.4) graft copolymers.

near the calculated one. As shown in Table 1.7.2, the calculated M_k is close, but somewhat larger than that determined by GPC. It should be emphasized that a polystyrene calibration curve was employed for the determination of the molecular weight by GPC and that polystyrene is a linear molecule. The molecular weights determined by GPC for the branched molecules are, therefore, expected to be smaller than the true ones, because their hydrodynamic volumes are smaller than those of the corresponding linear molecules.

Synthesis of AB_xC_y Graft Copolymer. The "grafting from" method cannot be applied to the synthesis of the graft copolymer possessing two or more different side chains (AB_xC_y graft copolymer). However, the "grafting onto" technique (coupling method) is suitable for the preparation of this kind of copolymer. For the graft copolymer AB_xC_y-1 (Table 1.7.3), three kinds of living polymers were prepared at the same time. As shown in Table 1.7.3 and Figure 1.7.4, the M_n of living poly(St) (peak a in Figure 1.7.4; M_n = 1420; M_w/M_n = 1.12) is almost equal to that of living poly(Is) (peak b in Figure 1.7.4; M_n = 1490; M_w/M_n = 1.10) and their molar amounts are almost the same. The coupling reaction was carried out by sequentially transferring the benzene solutions of the above two living polymers to the THF solution of living poly(GMA) (peak c in Figure 1.7.4; M_n = 5960; M_w/M_n = 1.10). As shown in Figure 1.7.4, after the coupling reaction, a new single peak corresponding to the graft

TABLE 1.7.3
Synthetic Conditions of AB_xC_y Graft Copolymer[a]

	Living Poly(GMA)				Living Poly(St)			Living Poly(Is)		
No.	M_n[b]	M_w/M_n[b]	m[c]	$mmol_{epoxy}$[d]	M_n[b]	M_w/M_n[b]	mmol[e]	M_n[b]	M_w/M_n[b]	mmol[f]
AB_xC_y-1	5960	1.10	40.3	14.1	1420	1.12	2.06	1490	1.10	1.97
AB_xC_y-2	8360	1.15	57.2	14.1	1090	1.14	3.29	1070	1.13	3.35

[a] The coupling reaction was carried out by introducing the benzene solutions of living poly(St) and poly(Is) into the THF solution of living poly(GMA) sequentially, with vigorous magnetic stirring. The reaction was allowed to last 40 min at −30°C.

[b–d] See Table 1.7.1.

[e] Molar amounts of living poly(St).

[f] Molar amounts of living poly(Is).

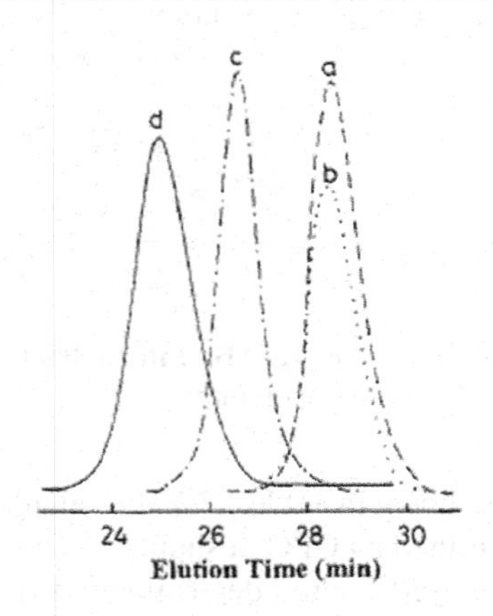

FIGURE 1.7.4 GPC traces of the AB_xC_y graft copolymer and its precursors: (a) living poly(St) (M_n = 1420; M_w/M_n = 1.12); (b) living poly(Is) (M_n = 1490; M_w/M_n = 1.10); (c) living poly(GMA) (M_n = 5960; M_w/M_n = 1.10); (d) graft copolymer obtained after the coupling reaction of **a**, **b**, and **c** (AB_xC_y-1 in Tables 1.7.3 and 1.7.4; $M_n = 1.71 \times 10^4$; M_w/M_n = 1.21).

TABLE 1.7.4
Characterization of AB_xC_y Graft Copolymer

No.	f_{PSt}[a] (%)	N_{PSt}[b] (%)	f_{PIs}[c] (%)	N_{PIs}[d]	$10^{-3}M_n$ calcd[e]	$10^{-3}M_n$ obsd[f]	M_w/M_n[f]
AB_xC_y-1	14.5	5.9	14.0	5.6	22.7	17.1	1.21
AB_xC_y-2	23.3	13.3	23.8	13.6	37.4	27.5	1.23

[a] Grafting efficiency of poly(St), $f_{PSt} = mol_{PSt}/mol_{epoxy}$.
[b] Graft number of poly(St) in each poly(GMA) chain, $N_{PSt} = mf_{PSt}$.
[c] Grafting efficiency of poly(Is), $f_{PIs} = mol_{PIs}/mol_{epoxy}$.
[d] Graft number of poly(Is) in each poly(GMA) chain, $N_{PIs} = mf_{PIs}$.
[e] Calculated with the expression: $M_k = M_{n(poly(GMA))} + N_{PSt}M_{n(PSt)} + N_{PIs}M_{n(PIs)}$.
[f] Determined by GPC.

copolymer appears in the higher molecular weight range (peak d in Figure 1.7.4; $M_n = 1.71 \times 10^4$; $M_w/M_n = 1.21$), and no peak due to the precursor polymers remains. The ^{1}H NMR spectrum of this graft copolymer is presented in part II of Figure 1.7.3 and, compared to part I, contains the additional absorptions j, k, l, r, s, t, and v due to the poly(Is) side chains. Therefore, this graft copolymer consists of a poly(GMA) backbone and both poly(St) and poly(Is) side chains.

The calculated grafting efficiencies and graft numbers of both poly(St) and poly(Is) are listed in Table 1.7.4. For AB_xC_y-1, the grafting efficiency is $f = f_{PSt} + f_{PIs} = 28.6\%$ and the graft number in each poly(GMA) backbone $N = N_{PSt} + N_{PIs} = 11.5$. The calculated molecular weight ($M_k = 2.27 \times 10^4$) is larger than the value ($M_n = 1.71 \times 10^4$) determined by GPC.

Synthesis of DAB_x Block–Graft Copolymer. It has been reported that the carbonyl groups of poly(MMA) can react with the living poly(St) to generate a graft copolymer,[6] although the coupling reaction did not proceed quantitatively and a part of the living poly(St) remained unreacted. However, in the presence of poly(GMA), we have shown that the living poly(St) reacts with the epoxy groups of poly(GMA) and that the poly(MMA) remains unreacted.[18] In other words, the reactivity of epoxy is much higher than that of carbonyl. Therefore, when the living poly(St) is mixed with the block copolymer of MMA and GMA, the former will selectively react with the epoxy groups of the poly(GMA) segment. In this manner, a block–graft copolymer consisting of a poly(MMA-*b*-GMA) block backbone and poly(St) side chains attached to the poly(GMA) segment can be generated.

The block copolymer of MMA and GMA was prepared under conditions similar to those used for the homopolymerization of GMA. However, MMA was first polymerized at –70°C, and GMA was sequentially added at –50°C. One example is given in Figure 1.7.5. The living poly(MMA) (peak b; $M_n = 1780$; $M_w/M_n = 1.09$) was prepared in THF, at a ratio $[MMA]_0/[DPHL]_0 = 0.67/0.04$, in the presence of LiCl ($[LiCl]/[DPHL]_0 = 3$). After GMA was introduced into the above system, the GPC peak shifted to a higher molecular weight position (peak c), and no poly(MMA) precursor remained. This indicates that all the living sites of poly(MMA) were consumed to initiate the polymerization of GMA and the block copolymer thus

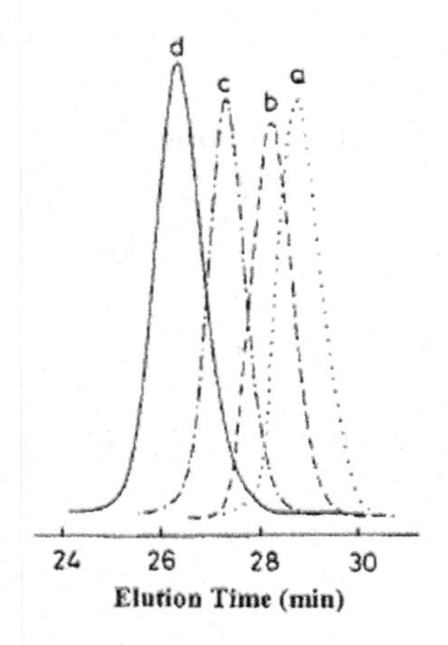

FIGURE 1.7.5 GPC traces of the DAB$_x$ block–graft copolymer and its precursors: (a) living poly(St) (M_n = 1260; M_w/M_n = 1.09); (b) living poly(MMA) (M_n = 1780; M_w/M_n = 1.09); (c) living poly(MMA-*b*-GMA) (M_n = 3400; M_w/M_n = 1.07) prepared by the anionic block copolymerization of GMA from **b**; (d) block–graft copolymer obtained by the coupling reaction of **a** and **c** (DAB$_x$-2 in Tables 1.7.5 and 1.7.6; M_n = 7410; M_w/M_n = 1.12).

obtained had a controlled molecular weight and a narrow molecular weight distribution (M_n = 3400; M_w/M_n = 1.07). For other block copolymers of MMA and GMA (Table 1.7.5), similar results were obtained. As shown by ^{1}H NMR spectrum in part I of Figure 1.7.6, besides the peaks a, b, c, d, and e due to the poly(GMA) segment,

TABLE 1.7.5
Preparation Conditions of DAB$_x$ Block–Graft and DaB$_x$ Star-Shaped Copolymers[a]

	Living Poly(MMA-*b*-GMA)				Living Poly(St)		
No.	M_n[b]	M_w/M_n[b]	$W_{MMA}W_{GMA}$[c]	mmol$_{epoxy}$[d]	M_n[b]	M_w/M_n[b]	mmol[e]
DAB$_x$-1	3670	1.08	50/50	14.1	1750	1.08	3.13
DAB$_x$-2	3400	1.07	50/50	14.1	1260	1.09	3.75
DAB$_x$-3	5830	1.07	50/50	14.1	930	1.13	5.63
DaB$_x$-1	2340	1.09	80/20	3.52	1010	1.10	3.60
DaB$_x$-2	2350	1.10	80/20	3.52	1740	1.09	3.60
DaB$_x$-3	3150	1.08	69/31	6.34	1020	1.10	6.50

[a] The coupling reaction was carried out by introducing the benzene solution of living poly(St) into the THF solution of the living block copolymer of MMA and GMA with vigorous magnetic stirring. The reaction was allowed to last 30 min at −30°C.
[b] Determined by GPC.
[c] Weight percent of poly(MMA) and poly(GMA) segments, respectively.
[d] Molar amounts of epoxy groups.
[e] Molar amounts of living poly(St).

TABLE 1.7.6
Characterization of DAB$_x$ Block–Graft Copolymer

				$10^{-3}M_n$		
No.	**f[a] (%)**	**m[b] (%)**	**N[c]**	**calcd[d]**	**obsd[e]**	**M_w/M_n[e]**
DAB$_x$-1	22.2	12.1	2.7	8.40	9.42	1.18
DAB$_x$-2	26.6	11.4	3.0	7.18	7.41	1.12
DAB$_x$-3	40.0	19.7	7.9	13.2	10.8	1.15

[a] Grafting efficiency, $f = \text{mol}_{PSt}/\text{mol}_{epox}$.
[b] Polymerization degree (m) of poly(GMA) segment.
[c] Graft number in each backbone, $N = fm$.
[d] Calculated with the expression $M_k = M_{n(block)} + NM_{n(PSt)}$.
[e] Determined by GPC.

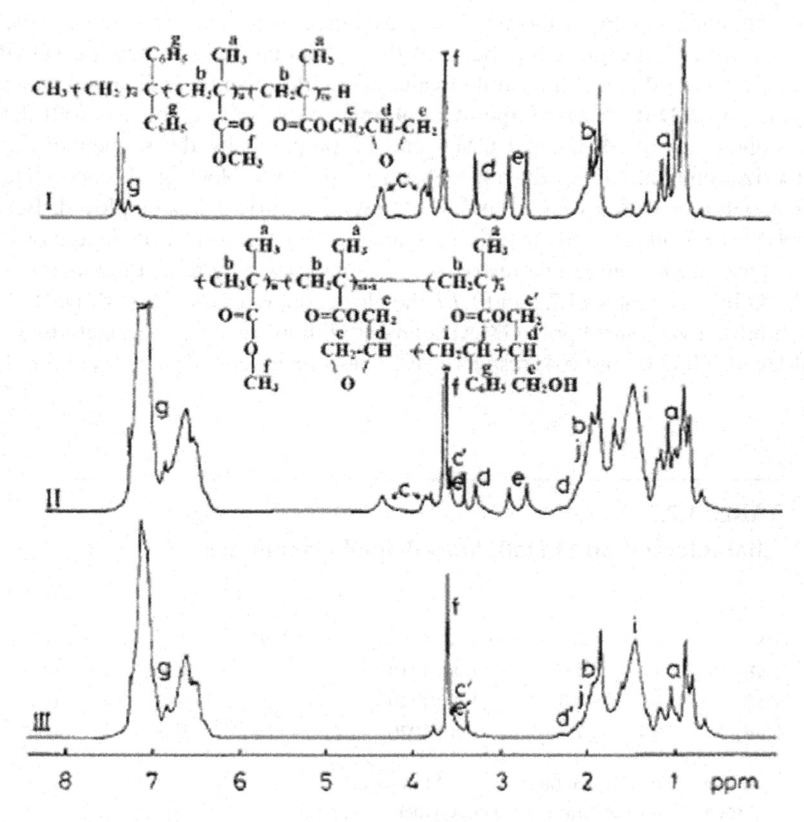

FIGURE 1.7.6 ^{1}H NMR spectra of poly(MMA-*b*-GMA) (I; M_n = 5830; M_w/M_n = 1.07; the weight ratio of two components, W_{MMA}/W_{GMA} = 50/50), DAB$_x$ block–graft copolymer (II; DAB$_x$-1 in Tables 1.7.5 and 1.7.6), and DaB$_x$ star-shaped copolymer (III; DaB$_x$-2 in Tables 1.7.5 and 1.7.7; its structure is similar to that of DAB$_x$-1 in part II).

the absorptions a, b, and f due to the poly(MMA) segment are also present. Without termination and polymer separation, the THF solution of the block copolymer was immediately used in the coupling reaction.

The coupling reaction was carried out in a way similar to that used in the preparation of AB_x graft copolymer (Table 1.7.5). As shown in Figure 1.7.5, the benzene solution of the living poly(St) (peak a; M_n = 1260; M_w/M_n = 1.09) was introduced into the THF solution of poly(MMA-*b*-GMA) (peak c; M_n = 3400; M_w/M_n = 1.07), and the block–graft copolymer thus produced had a high molecular weight and narrow molecular weight distribution (peak d; M_n = 7410; M_w/M_n = 1.12; DAB_x-2 of Table 1.7.6). The ^1H NMR spectrum (part II of Figure 1.7.6) confirms the presence of poly(MMA-*b*-GMA) backbone, and of poly(St) side chains (peaks g, i, and j). The above results indicate that a pure block–graft copolymer consisting of a poly(MMA-*b*-GMA) backbone and poly(St) side chains was obtained.

As shown in Table 1.7.6, when the graft number and the molecular weight are smaller (DAB_x-1 and DAB_x-2), the determined M_n of the block–graft copolymer is close to the calculated value. In this case, the molecular architecture is close to a linear one, and the GPC determination is expected to provide a value near the real one. However, as the graft number and the molecular weight increase (DAB_x-3), Table 1.7.6 shows that the determined value is smaller than the calculated value.

Synthesis of DaB_x Star-Shaped Copolymer. As described above, a well-defined block copolymer of MMA and GMA can be prepared by the sequential anionic polymerization method. For the above synthesis of DAB_x block–graft copolymer, the block copolymer with a weight ratio W_{MMA}/W_{GMA} = 50/50 was employed. Because the polymerization proceeds in a living manner, the polymerization degree of GMA in the block copolymer can be restricted to a small value by reducing the amount of GMA. As listed in Tables 1.7.5 and 1.7.7, the block copolymers in DaB_x-1, DaB_x-2, and DaB_x-3 have a very short poly(GMA) segment with average polymerization degrees of GMA of 3.0, 3.0, and 6.4, respectively. These block copolymers were employed

TABLE 1.7.7
Characterization of DaB_x Star-Shaped Copolymer

				$10^{-3}M_n$		
No.	m[a]	N[b]	$M_{n(PMMA)}/M_{n(PSt)}$[c]	calcd[d]	obsd[e]	M_w/M_n[e]
DaB_x-1	3.0	4.0	1680/1010	5.37	4.40	1.12
DaB_x-2	3.0	4.0	1690/1740	7.57	6.80	1.14
DaB_x-3	6.4	7.4	2010/1020	9.68	7.83	1.10

[a] Polymerization degree (m) of poly(GMA) segment.
[b] Arm number of the star-shaped copolymer, $N = m + 1$.
[c] The molecular weights of poly(MMA) arm and each poly(St) arm, respectively.
[d] Calculated with the expression $M_k = M_{n(block)} + mM_{n(PSt)}$.
[e] Determined by GPC.

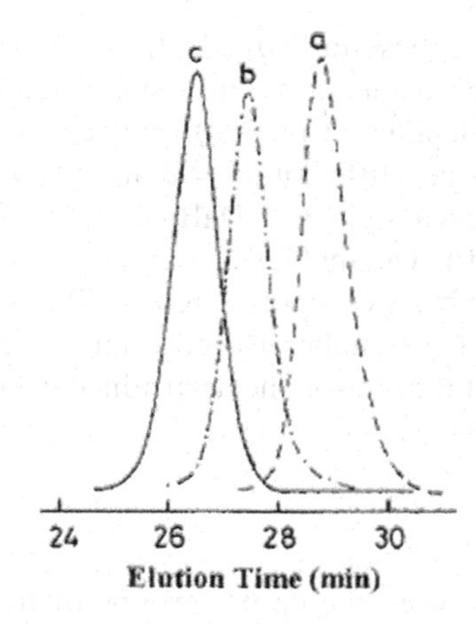

FIGURE 1.7.7 GPC traces of the DaB_x star-shaped copolymer and its precursors: (a) living poly(St) (M_n = 1020; M_w/M_n = 1.10); (b) living poly(MMA-*b*-GMA) (M_n = 3150; M_w/M_n = 1.08); (c) star-shaped copolymer obtained by the coupling reaction of a and b (DaB_x-3 in Tables 1.7.5 and 1.7.7; M_n = 7830; M_w/M_n = 1.10).

for the synthesis of the star-shaped copolymers that contain a poly(MMA) arm and several poly(St) arms (DaB_x star-shaped copolymer, **4** in Figure 1.7.1).

In the coupling reaction, the molar amount of living poly(St) employed was selected to be equal to that of epoxy groups (Table 1.7.5). As illustrated in Figure 1.7.7, when the benzene solution of living poly(St) (peak a; M_n = 1020; M_w/M_n = 1.10; 6.50 mmol) was added to the THF solution of the living poly(MMA-*b*-GMA) (peak b; M_n = 3150; M_w/M_n = 1.08; amount of epoxy, 6.34 mmol), the coupling reaction took place and a star-shaped copolymer with a high molecular weight and narrow molecular weight distribution was generated (peak c; M_n = 7830; M_w/M_n = 1.10; DaB_x–3 in Table 1.7.7). As for the above graft and block–graft copolymers, the single, very sharp GPC peak implies that this star-shaped copolymer is not mixed with its precursor polymers.

Part III of Figure 1.7.6 presents the ^{1}H NMR spectrum of a star-shaped copolymer (DaB_x–2 in Tables 1.7.5 and 1.7.7). The absorptions due to poly(MMA) component (a, b, and f) and poly(St) component (g, i, and j) can be easily detected. However, the absorptions corresponding to the poly(GMA) segment can hardly be detected, because the weight ratio of the three components in this copolymer is W_{MMA}: W_{GMA}: W_{St} = 4:1:14. This copolymer can be considered a star-shaped copolymer that consists of just two components, poly(MMA) and poly(St).

As shown in Tables 1.7.5 and 1.7.7, in the preparation of DaB_x–1 and DaB_x–2, the block copolymers of MMA and GMA employed have almost the same molecular weight, molecular weight distribution, and composition. Therefore, the star-shaped copolymers based on them possess the same arm number, namely a poly(MMA) arm and three poly(St) arms, and the molecular weights of the poly(MMA) arms are almost equal (Table 1.7.7). However, the poly(St) arms in DaB_x–2 have a larger molecular weight ($M_{n(PSt)}$ = 1740) than those in DaB$_x$–1 ($M_{n(PSt)}$ = 1010). This causes a difference in the total molecular weights of the final star-shaped copolymers. Comparing DaB$_x$–1 with DaB$_x$–3, the lengths of poly(St) arms are almost the same. However, their poly(St) arm numbers are different (m = 3 and 6.4, respectively). As shown in Table 1.7.7, the final molecular weight of DaB$_x$–3 (M_n = 7830) is much larger than that of DaB$_x$–1 (M_n = 4400). The molecular weights of the star-shaped copolymers determined by GPC on the basis of a polystyrene calibration curve are smaller than the calculated values. They have smaller hydrodynamic volumes than the corresponding linear polymers, and for this reason the determined values should be smaller than the real ones.

1.7.4 CONCLUSION

The coupling reaction between the epoxy groups of living poly(glycidyl methacrylate) [poly(GMA)] and the propagating sites of living polystyrene [poly(St)] or polyisoprene (poly(Is)) can generate a graft copolymer with a polar backbone and nonpolar side chains. When both the living poly(St) and poly(Is) were reacted with the epoxy groups of living poly(GMA), a graft copolymer possessing two kinds of nonpolar side chains attached to a poly(GMA) backbone was obtained. If the backbone polymer is a block copolymer of MMA and GMA, the coupling reaction with living poly(St) can produce a block–graft copolymer, which contains a poly(MMA-*b*-GMA) block backbone and poly(St) side chains attached to the poly(GMA) segment. In addition, when the poly(GMA) segment is short, a star-shaped copolymer possessing one poly(MMA) arm and several poly(St) arms are obtained.

Because both the backbone and side chain polymers are prepared *in situ* prior to the coupling reaction and the living polymer solutions were directly employed, no impurity was introduced into the system. Both the backbone and side chain polymers prepared separately have controlled molecular weights and narrow molecular weight distributions. On the basis of these well-defined living precursor polymers, model copolymers with designed molecular architectures, controlled molecular weights, and narrow molecular weight distributions can be synthesized.

REFERENCES

1. Rempp, P.; Franta, E.; Herz, J. E. *Adv. Polym. Sci.* **1988**, *86*, 145.
2. Sawamoto, M. *Int. J. Polym. Mater.* **1991**, *15*, 197.
3. Pitsikalis, M.; et al. *Adv. Polym. Sci.* **1998**, *135*, 1.
4. Kennedy, J. P.; Marechal, E. *Carbocationic Polymerization.* John Wiley & Sons: New York, 1982; p 411.
5. Zhang, H. M.; Ruckenstein, E. *Macromolecules* **1998**, *31*, 746; **1998**, *31*, 2977.
6. For example: (a) Gallot, Y.; Rempp, P.; Parrod, J. *Polym. Lett.* **1963**, *1*, 329. (b) Gallot, Y.; Grubisic, Z.; Rempp, P.; Benoit, H. *J. Polym. Sci.* **1968**, *C22*, 527. (c) Ishizu, K.; Fukutomi, T.; Kakurai, T.; Noguchi, T. *Polym. J.* **1973**, *4*, 105; **1975**, *7*, 438.

7. For example: (a) Candau, F.; Afshar-Taromi, F.; Rempp, P. *Polymer* **1977**, *18*, 1252. (b) Ishizu, K.; Fukutomi, T.; Kakurai, T. *Polym. J.* **1975**, *7*, 228. (c) Takaki, M.; Asami, R.; Ishikawa, M. *Macromolecules* **1977**, *10*, 850. (d) Pitsikalis, M.; Woodward, J.; Mays, J. W.; Hadjichristidis, N. *Macromolecules* **1997**, *30*, 5384.
8. Se, K.; Yamazaki, H.; Shibamoto, T.; Takano, A.; Fujimoto, T. *Macromolecules* **1997**, *30*, 1570.
9. Takaki, M.; Asami, R.; Mizuno, M. *Macromolecules* **1977**, *10*, 845.
10. Leemans, L.; Fayt, R.; Teyssie, Ph. *J. Polym. Sci., Polym. Chem.* **1990**, *28*, 1255.
11. Leemans, L.; Fayt, R.; Teyssie, Ph. *J. Polym. Sci., Polym. Chem.* **1990**, *28*, 2187.
12. Hild, G.; Lamps, J. P.; Rempp, P. *Polymer* **1993**, *34*, 2875.
13. Hild, G.; Lamps, J. P. *Polymer* **1995**, *36*, 4841.
14. Allen, R. D.; Long, T. E.; McGrath, J. E. *Polym. Bull.* **1986**, *15*, 127.
15. Varshney, S. K.; Hautekeer, J. P.; Fayt, R.; Jerome, R.; Teyssie, Ph. *Macromolecules* **1990**, *23*, 2618.
16. Fayt, R.; Forte, R.; Jacobs, C.; Jerome, R.; Ouhadi, T.; Teyssie, Ph.; Varshney, S. K. *Macromolecules* **1987**, *20*, 1442.
17. For example: (a) Szwarc, M. *Adv. Polym. Sci.* **1983**, *49*, 1. (b) van Beylen, M.; Bywater, S.; Smets, G.; Szwarc, M.; Worsfold, D. J. *Adv. Polym. Sci.* **1988**, *86*, 87.
18. Ruckenstein, E.; Zhang, H. M. *J. Polym. Sci., Polym. Chem.*, **1999**, *37*, 105.

1.8 Graft Copolymers by Combined Anionic and Cationic Polymerizations Based on the Homopolymerization of a Bifunctional Monomer*

Hongmin Zhang and Eli Ruckenstein

Chemical Engineering Department, State University of New York at Buffalo, Amherst, New York 14260

ABSTRACT The bifunctional monomer 2-(vinyloxy)ethyl methacrylate (VEMA) was polymerized both anionically and cationically. Using 1,1-diphenylhexyllithium (DPHL) as initiator, tetrahydrofuran (THF) as solvent and the low temperature of −60°C, the C=C double bond of the methacryloyl group of VEMA underwent smoothly anionic polymerization, without cross-linking or side reactions. The polymer had a controlled molecular weight and a narrow molecular weight distribution (M_w/M_n = 1.06–1.12). On the other hand, the C=C double bond of the vinyloxy group of VEMA can undergo cationic polymerization. A polymer with controlled molecular weight and narrow molecular weight distribution (M_w/M_n = 1.11–1.13) was prepared using 2-[1-acetoxyethoxy] ethyl methacrylate (**4**)/$EtAlCl_2$ as initiator in the presence of THF, a weak Lewis base. Two methods were employed to prepare graft copolymers.

* *Macromolecules* 1998, 31, 746–752.

(A) The anionically prepared polymer of VEMA was separated from solution after quenching the polymerization and purified by freeze-drying; then the vinyloxy groups of the side chains were allowed to react with trifluoroacetic acid to generate a macroinitiator, which finally induced the cationic graft polymerization of isobutyl vinyl ether (IBVE). This procedure yielded a graft copolymer with a polymethacrylate backbone and poly(IBVE) side chains containing a small amount of the IBVE homopolymer. (B) The anionic polymerization of **4** was carried out in THF without quenching, to produce a solution of macroinitiator. Then, in contrast to the first method, the polymer was not separated and toluene and IBVE were introduced into the system. Further, the cationic graft polymerization of IBVE was induced by adding an activator ($EtAlCl_2$). The THF, which was used as solvent in the anionic polymerization of **4**, acted as a Lewis base in the cationic polymerization step. This procedure yielded a pure graft copolymer with controlled molecular weight and narrow molecular weight distribution (M_w/M_n = 1.15–1.17), consisting of a polymethacrylate backbone and poly(IBVE) side chains.

1.8.1 INTRODUCTION

In recent years, great attention was paid to graft copolymers, because of their unique molecular architecture, particular morphology and increased number of applications.[1–5] They have been widely used for the preparation of compatibilizers for polymer blends,[6–11] membranes for separation of gases or liquids,[12–14] hydrogels,[15] drug deliverers,[16] thermoplastic elastomers,[17] etc. A number of methods have been employed for their synthesis, such as the macromonomer method,[18–20] radiation-induced polymerization,[21–23] ring opening olefin methathesis polymerization,[24] polycondensation reaction[25] and iniferter-induced polymerization.[26] However, the living polymerization technique is undoubtedly most suitable for the preparation of well-defined graft copolymers, in which both the backbone and the side chains possess designed molecular weights and narrow molecular weight distributions and the position, the number of side chains and the composition of the graft copolymer can be controlled. Se and co-workers[27] prepared a well-defined block–graft copolymer using the living anionic polymerization method. A block copolymer consisting of polystyrene and poly((4-vinylphenyl)dimethylvinylsilane) and living polyisoprene were first prepared separately, and their coupling reaction generated a graft copolymer. In this method, the unreacted polyisoprene had to be removed by repeated dissolution and precipitation. Recently, we have prepared a graft copolymer consisting of a poly methacrylate backbone and poly(alkyl vinyl ether) side chains,[28] by using the anionic polymerization of 1-(iso-butoxy)ethyl methacrylate followed by the cationic polymerization of alkyl vinyl ether. In that method, the poly(1-(isobutoxy) ethyl methacrylate) had to be isolated from its tetrahydrofuran (THF) solution and purified carefully before it was used as macroinitiator for the cationic polymerization of alkyl vinyl ether. In the present paper, an improved continuous method is reported for the synthesis of a similar graft copolymer with controlled molecular weight and narrow molecular weight distribution, in which the above lengthy purification is no longer needed.

2-(Vinyloxy)ethyl methacrylate (VEMA, **1** in Scheme 1.8.1) is a bifunctional monomer possessing an anionically as well as a cationically polymerizable C=C double bond. The C=C double bond located in a position of the carbonyl is expected to undergo anionic polymerization and a functional polymer (**2**, in Scheme 1.8.1; abbreviated as PMA), having a cationically polymerizable C=C double bond in each side chain, to be formed. On the other hand, the C=C double bond of the ester group of VEMA is expected to undergo cationic polymerization to generate another functional polymer (**3**, in Scheme 1.8.1; abbreviated as PVE) with a reactive methacryloyl group in each repeating unit. Obviously, two different functional polymers can be obtained from the same monomer, by its selective anionic or cationic polymerization.

A graft copolymer was prepared by two different routes. In route A (Scheme 1.8.2), PMA (**2**) was allowed to react with trifluoroacetic acid to generate a dormant macroinitiator (**5**, in Scheme 1.8.2). In route B, the vinyl of the ester group of VEMA was selectively reacted with acetic acid to produce a new monomer, 2-[1-acetoxy-ethoxy] ethyl methacrylate (**4**, in Scheme 1.8.1) which generated, via anionic polymerization, the dormant macroinitiator (**6**, in Scheme 1.8.2) with a molecular structure similar to that of **5**. The side chains of both **5** and **6** were activated by Lewis acids[29–31] to generate partly dissociated carbocations, which induced the cationic polymerization of vinyl ether. This yielded a graft copolymer (**7**, in Scheme 1.8.2) with a polymethacrylate backbone and poly(alkyl vinyl ether) side chains.

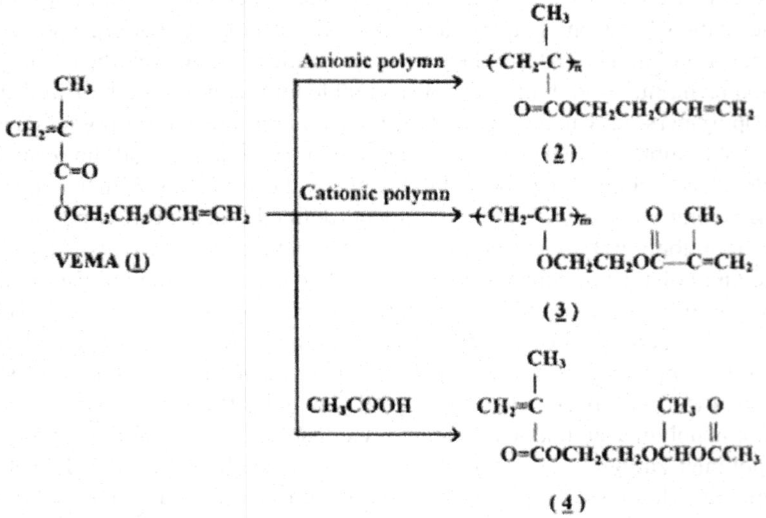

VEMA (1): $CH_2{=}C(CH_3){-}C({=}O){-}OCH_2CH_2OCH{=}CH_2$

Anionic polymn → (2): $-\!\!\left[CH_2{-}C(CH_3)\right]\!\!-_n$ with side chain $O{=}COCH_2CH_2OCH{=}CH_2$

Cationic polymn → (3): $-\!\!\left[CH_2{-}CH\right]\!\!-_m$ with side chain $OCH_2CH_2OC({=}O){-}C(CH_3){=}CH_2$

CH_3COOH → (4): $CH_2{=}C(CH_3){-}C({=}O)OCH_2CH_2OCH(CH_3)OC({=}O)CH_3$

SCHEME 1.8.1

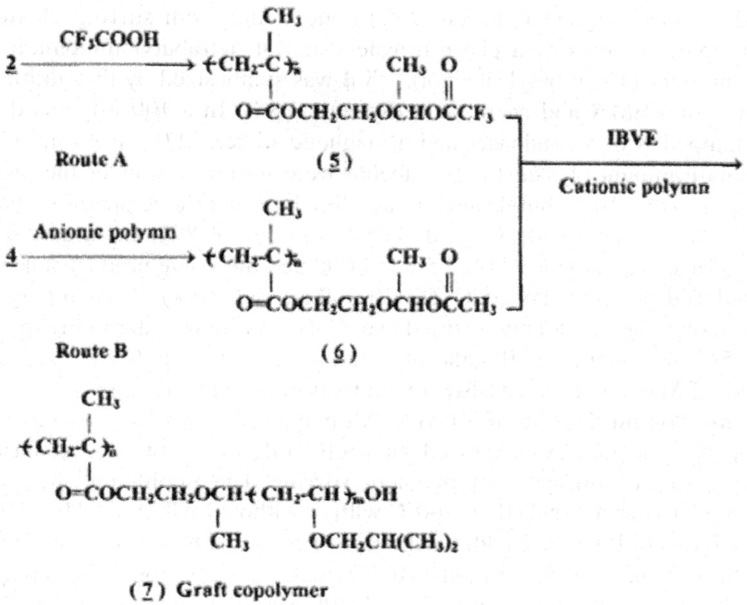

Route A: 2 —(CF_3COOH)→ $-(CH_2-C(CH_3))_n-$ with side chain $O{=}COCH_2CH_2OCH(CH_3)OCCF_3$ (with $C{=}O$) (5)

Route B: 4 —(Anionic polymn)→ $-(CH_2-C(CH_3))_n-$ with side chain $O{=}COCH_2CH_2OCH(CH_3)OCCH_3$ (with $C{=}O$) (6)

(5) or (6) —(IBVE, Cationic polymn)→

$-(CH_2-C(CH_3))_n-$ with side chain $O{=}COCH_2CH_2OCH(CH_3)-(CH_2-CH(OCH_2CH(CH_3)_2))_m-OH$

(7) Graft copolymer

SCHEME 1.8.2

1.8.2 EXPERIMENTAL SECTION

Materials. Tetrahydrofuran (THF) was dried with CaH_2 under reflux for more than 24 h, distilled, purified in the presence of naphthalene sodium, and finally distilled from a solution of (1,1-diphenylhexyl)lithium (DPHL) just before use. Toluene was washed with concentrated sulfuric acid and then with water, dried over $MgSO_4$, and distilled twice over CaH_2 before use. Et_2O was dried over CaH_2 and distilled in the presence of $LiAlH_4$. Hexane was first dried and distilled over CaH_2 and then distilled from a solution of *n*-BuLi. Isobutyl vinyl ether (IBVE; Aldrich, 99%) was washed with 10% aqueous sodium hydroxide solution and then with water, dried overnight with potassium hydroxide, and distilled twice over CaH_2 prior to polymerization.[29,31] 1,1-Diphenylethylene (DPE, Aldrich, 97%) was doubly distilled over CaH_2 and then distilled in the presence of triphenylmethyllithium under reduced pressure.[32] Lithium chloride (Aldrich, 99.99%) was dried at 120°C for 24 h and dissolved in THF.[33] *n*-BuLi (Aldrich, 1.6 M solution in hexane), $ZnCl_2$ (Aldrich, 1.0 M solution in diethyl ether), and $EtAlCl_2$ (Aldrich, 1.8 M solution in toluene) were diluted with purified hexane, diethyl ether and toluene, respectively. 1-Isobutoxyethyl trifluoroacetate (IETA) and 1-isobutoxyethyl acetate (IEA) were prepared by the addition reactions between IBVE and trifluoroacetic acid (Aldrich, 99%) or acetic acid (Aldrich, 99.8%), respectively.[29–31]

Syntheses of 2-(vinyloxy)ethyl Methacrylate (VEMA, 1 in Scheme 1.8.1) and 2-[1-Acetoxyethoxy]ethyl Methacrylate (4, in Scheme 1.8.1). VEMA was prepared through the reaction between 2-chloroethyl vinyl ether (Aldrich, 99%)

and sodium methacrylate (Aldrich, 99%) under reflux with stirring, in the presence of small amounts of a phase transfer catalyst (tetrabutylammonium iodide) and an inhibitor (4-*tert*-butylcatechol).[34–36] **4** was synthesized by the addition reaction between VEMA and acetic acid (Scheme 1.8.1). In a 100 mL round-bottom flask equipped with a condenser and a magnetic stirrer, 21.0 g (0.14 mol) VEMA and a small amount of 4-*tert*-butylcatechol were introduced under the protection of nitrogen. After 4-*tert*-butylcatechol had dissolved and the temperature had been raised to 70°C, acetic acid (8.1 g, 0.14 mol; Aldrich, 99.8%) was dropwise added with a syringe. The reaction lasted 5 h at 70°C and the crude product was distilled under reduced pressure (bp: 65°C–66°C/1.0 Torr; yield 86%). Prior to polymerization, the monomer was doubly distilled over CaH_2. As shown later in Figures 1.8.2A and 1.8.5A, the chemical shifts and their intensities in the ^1H NMR spectra of the prepared VEMA and **4** are consistent with their molecular structures.

Anionic Polymerization of VEMA. All polymerizations, anionic, cationic and graft copolymerizations, were carried out in a 100 mL round-bottom glass flask under an overpressure of nitrogen with magnetic stirring. The anionic polymerization of VEMA was performed in THF at −60°C with or without LiCl. After THF, DPE, and a THF solution of LiCl were added with dry syringes, the flask was cooled to −45°C and *n*-BuLi (in hexane) was introduced. The deep red color of DPHL appeared at once, and the reaction between *n*-BuLi and DPE was allowed to continue for 15 min. Then, the system was cooled to −60°C and the polymerization reaction was induced by the addition of a prechilled THF solution of VEMA to the above DPHL solution. After a selected time, the reaction was quenched with a small amount of methanol and the polymer solution was poured into a methanol–water mixture to precipitate the polymer. The polymer was reprecipitated by pouring its THF solution into the above mixture and finally dried in vacuum overnight.

Cationic Polymerization of VEMA. IETA, IEA and **4** were separately used as initiators for the cationic polymerization of VEMA, which was induced by adding, with a dry syringe, a solution of $ZnCl_2$ in Et_2O or of $EtAlCl_2$ in toluene to a toluene solution of VEMA and an initiator at 0°C or 25°C (see Table 1.8.2). The polymerization was terminated with methanol containing a small amount of ammonia. The quenched reaction mixture was washed with a 10% aqueous sodium thiosulfate solution (for $ZnCl_2$) or with a dilute hydrochloric acid (for $EtAlCl_2$) and then with water, evaporated to dryness under reduced pressure, and vacuum-dried to obtain the polymer.

Addition of Trifluoroacetic Acid to the Vinyls of the Side Chains of PMA and the Preparation of a Graft Copolymer by a Discontinuous Method. PMA prepared by the anionic polymerization of VEMA was freeze-dried from its benzene solution for 8 h and then vacuum-dried for more than 10 h. The purified toluene or a mixture of toluene and CCl_4 was introduced into the flask that contained the dried PMA. After the polymer had dissolved and the solution had been cooled to 0°C, trifluoroacetic acid was added and the addition reaction was allowed to last for 1.5 h. Subsequently, toluene and IBVE were introduced and the graft copolymerization was induced by the addition of a Et_2O solution of $ZnCl_2$. The termination of the reaction and the purification of the obtained copolymer were carried out as described above for the cationic polymerization of VEMA.

Anionic Polymerization of 4 and the Preparation of a Graft Copolymer by a Continuous Method. The anionic polymerization of **4** was carried out as for VEMA. However, poly(**4**) was not separated from its THF solution, to which toluene and IBVE were added. After the temperature was raised to 25°C, a toluene solution of $EtAlCl_2$ was introduced to start the graft copolymerization. The termination reaction and the purification of the produced copolymer were carried out as described for the cationic polymerization of VEMA.

Measurements. ^1H NMR spectra were recorded in $CDCl_3$ or CD_3OD on a VXR-400 spectrometer. M_n and M_w/M_n of the polymer were determined by gel permeation chromatography (GPC) on the basis of a polystyrene calibration curve. The GPC measurements were carried out using THF as solvent, at 30°C, with a 1.0 mL/min flow rate and a 1.0 cm/min chart speed. Three polystyrene gel columns (Waters, 7.8 × 300 mm; two Linear, Part No. 10681, and one HR 4E, Part No. 44240) were used, which were connected to a Waters 515 precision pump.

1.8.3 RESULTS AND DISCUSSION

VEMA was polymerized by either anionic or cationic polymerization to produce soluble polymers.[37,38] However, its living cationic polymerization was achieved only when the HI/I_2 initiating system was employed. Aoshima et al.[36] carried out the cationic polymerization of VEMA with HI/I_2 as initiator, in toluene, at −15°C to −40°C and obtained a polymer with controlled molecular weight and narrow molecular weight distribution. In the present paper, both the anionic and cationic polymerization of VEMA were carried out under different conditions than those previously used.

Anionic Polymerization of VEMA. LiCl has a positive effect on the anionic polymerization of the acrylic monomers,[39,40] because a μ-type complex[41] is formed between LiCl and the propagating site, which, by prevention of side reactions, markedly narrows the molecular weight distribution. In the present paper, this polymerization technique is employed in order to prepare a functional polymer (PMA) of VEMA, with controlled molecular weight and narrow molecular weight distribution.

The initiator, DPHL, was synthesized via the reaction of *n*-BuLi with DPE in the ratio $[DPE]/[n\text{-}BuLi]_0 = 1.2$ (Table 1.8.1), at −45°C, for 15 min. The anionic polymerization was carried out in THF at −60°C in the presence ($[LiCl]/[n\text{-}BuLi]_0 = 2$) or absence of LiCl. As shown in Table 1.8.1, VEMA can be anionically polymerized, with or without LiCl addition. However, the addition of LiCl controls the anionic polymerization of VEMA. The determined molecular weight of PMA is in good agreement with that calculated by assuming complete monomer conversion and that each initiator molecule generates a polymer chain (PMA-4 to PMA-8, in Table 1.8.1). The molecular weight distribution is narrow, nearly monodisperse. Figure 1.8.1 compares the GPC curves. Obviously, the sharp peak (curve B) of PMA-7 prepared in the presence of LiCl is comparable to that of polystyrene standard (curve C, $M_n = 10{,}000$, $M_w/M_n = 1.06$) and the molecular weight distribution of PMA-7 ($M_n = 9210$, $M_w/M_n = 1.07$) is much narrower than that of PMA-2 (curve A, $M_n = 3390$, $M_w/M_n = 1.24$), which was obtained in the absence of LiCl, even though the molecular weight of PMA-7 is larger. These results indicate that the anionic

TABLE 1.8.1
Anionic Polymerization of VEMA[a]

No.	$[n\text{-BuLi}]_0$ (mM)	[DPE] (mM)	[LiCl] (mM)	$[M]_0$ (M)	$10^{-3} M_n$ calcd	$10^{-3} M_n$ obsd[b]	M_w/M_n[b]
PMA-1	33.3	40.0	66.7	0.43	2.25	2.61	1.10
PMA-2	22.2	26.6	0	0.43	3.26	3.39	1.24
PMA-3	16.7	20.0	0	0.43	4.26	4.65	1.23
PMA-4	16.7	20.0	33.3	0.43	4.26	4.49	1.08
PMA-5	13.5	16.2	27.0	0.52	6.24	6.72	1.06
PMA-6	14.5	17.4	29.0	0.65	7.23	7.42	1.10
PMA-7	10.0	12.0	20.0	0.52	8.35	9.21	1.07
PMA-8	5.1	6.2	11.0	0.65	20.2	23.2	1.12

[a] Polymerization was carried out in THF at −60°C for 1 h. The monomer conversion was 100% in each case.
[b] Determined by GPC.

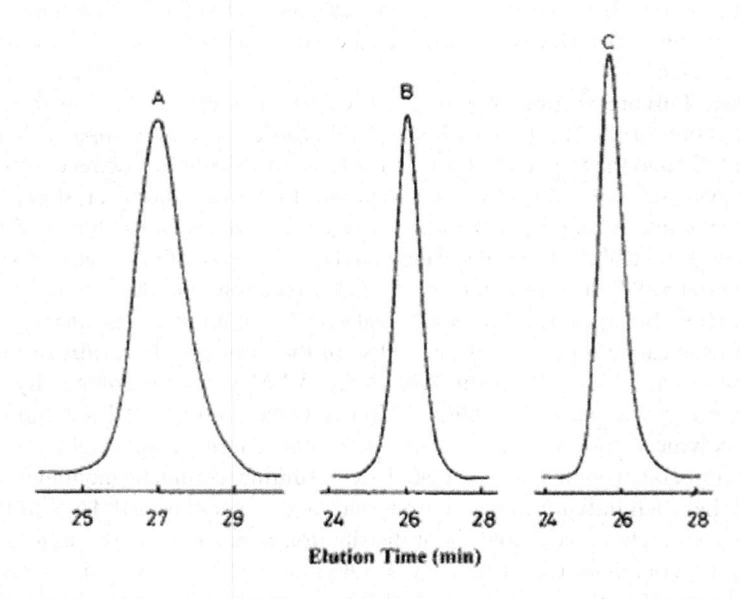

FIGURE 1.8.1 GPC traces of PMAs prepared in the absence (A, PMA-2 in Table 1.8.1, $M_n = 3390$, $M_w/M_n = 1.24$) or presence (B, PMA-7 in Table 1.8.1, $M_n = 9210$, $M_w/M_n = 1.07$) of LiCl and a standard polystyrene (C; Pressure Chemical Company, $M_n = 10{,}000$, $M_w/M_n = 1.06$. Lot No. 70111).

polymerization of VEMA in the presence of LiCl can proceed smoothly in a living manner without chain transfer or terminating reactions.

Parts A and B of Figure 1.8.2 depict the ^{1}H NMR spectra of the monomer (VEMA) and its anionically obtained polymer (PMA), respectively. After polymerization, the peaks **a** and **b** corresponding to the α-methyl and C=C double bond of the methacryloyl group, shifted to 0.71.3 (**a′**) and 1.7–2.2 ppm (**b′**), respectively. However, the other absorptions including the C=C double bond of vinyloxy group (**e** and **f**) did not change. In addition, PMA is soluble in THF, benzene, toluene, chloroform, ethyl acetate, etc. Evidently, no cross-linking reaction occurred during the anionic polymerization of VEMA, and the PMA thus obtained is a functional polymer, with a vinyloxy group in each side chain.

Cationic Polymerization of VEMA. The living cationic polymerization of a number of vinyl monomers, such as alkyl vinyl ether, isobutylene, and styrene-type monomers, was already achieved.[42–45] Higashimura et al.[29–31] demonstrated that

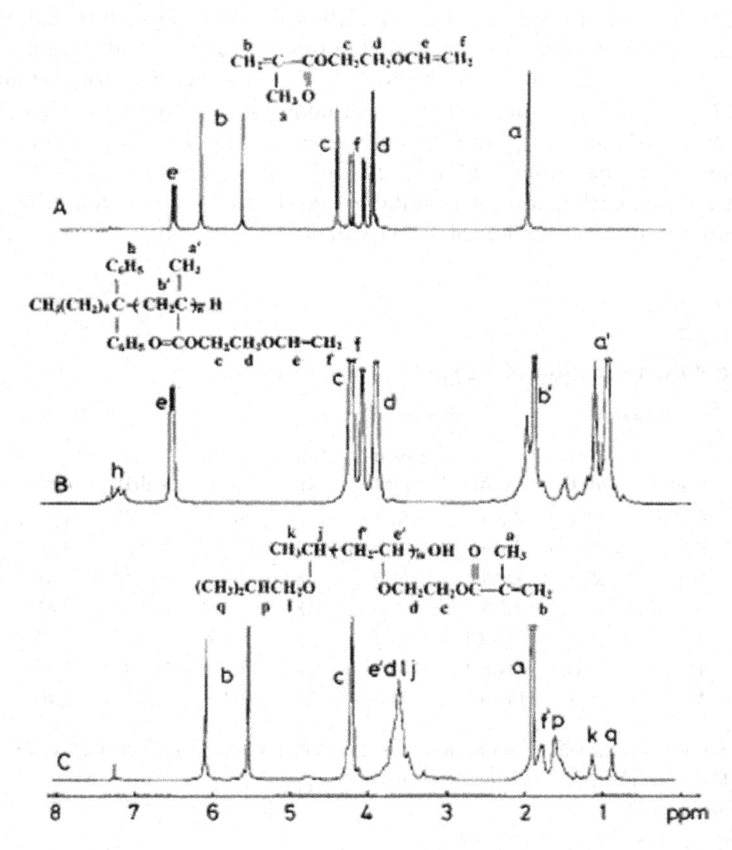

FIGURE 1.8.2 ^{1}H NMR spectra of VEMA (A), PMA (B, PMA-7 in Table 1.8.1) prepared by the anionic polymerization of VEMA and PVE (C, PVE–2 in Table 1.8.2) prepared by the cationic polymerization of VEMA.

the adducts of acetic acid derivatives with alkyl vinyl ethers can induce the living cationic polymerizations of vinyl ethers in the presence of an activator, such as $ZnCl_2$ or $EtAlCl_2$. In the present paper, three similar adducts, 1-isobutoxyethyl trifluoroacetate (IETA), 1-isobutoxyethyl acetate (IEA) and 2-[1-acetoxyethoxy] ethyl methacrylate (**4**, in Scheme 1.8.1), were employed as initiators for the cationic polymerization of VEMA. The polymerization was carried out in toluene at 0°C (for PVE–1 to PVE–4, in Table 1.8.2) or 25°C (for PVE–5 to PVE–7, in Table 1.8.2), and the reaction was started by adding the activator ($ZnCl_2$ in Et_2O or $EtAlCl_2$ in toluene) to the toluene solution of VEMA and initiator. As shown in Table 1.8.2, VEMA can be cationically polymerized regardless of the initiator and activator used. However, the molecular weight distributions of the produced polymers are different. When IEA/$ZnCl_2$ initiating system was used, the molecular weight distribution of the polymer was broad (Figure 1.8.3A, PVE–4 in Table 1.8.2, M_n = 3320, M_w/M_n = 1.47). This happened because the Lewis acid employed, $ZnCl_2$, is too weak and the nucleophilicity of the counteranion ($^-OCOCH_3...ZnCl_2$) too strong. However, when a very strong Lewis acid, $EtAlCl_2$, was used, the molecular weight distribution was also broad (Figure 1.8.3B, PVE–5 in Table 1.8.2, M_n = 3350, M_w/M_n = 1.30), because the counteranion ($^-OCOCH_3...EtAlCl_2$) is too weakly nucleophilic. In contrast, when a weak Lewis base, THF, was added to the above system, a narrowly disperse polymer was obtained (Figure 1.8.3C, PVE–7 in Table 1.8.2, M_n = 3460, M_w/M_n = 1.11). In this case, the carbocation was suitably stabilized by the added base.[30,31] The above cationic polymerization of VEMA demonstrates again that it is important

TABLE 1.8.2
Cationic Polymerization of VEMA[a]

	Initiator		Activator				$10^{-3}M_n$		
No.	Name	Amt (mM)	Name	Amt (mM)	Time (h)	$[M]_0$ (M)	calcd	obsd[b]	M_w/M_n[b]
PVE–1	IETA[c]	33.0	$ZnCl_2$	16.7	3	0.40	1.89	1.81	1.29
PVE–2	IETA[c]	16.0	$ZnCl_2$	8.3	4	0.40	3.90	3.29	1.30
PVE–3	IEA[d]	25.0	$ZnCl_2$	13.0	8	0.40	2.50	2.41	1.38
PVE–4	IEA[d]	16.0	$ZnCl_2$	7.0	8	0.40	3.90	3.32	1.47
PVE–5	**4**[e]	15.0	$EtAlCl_2$	45.0	3	0.32	3.33	3.35	1.30
PVE–6[f]	**4**[e]	33.0	$EtAlCl_2$	120	24	0.40	1.89	1.62	1.13
PVE–7[f]	**4**[e]	15.0	$EtAlCl_2$	90.0	24	0.32	3.33	3.46	1.11

[a] Polymerization was carried out in toluene at 0°C for PVE–1 to PVE–4 or 25°C for PVE–5 to PVE–7. The yields of polymers were quantitative in all cases.
[b] Determined by GPC.
[c] IETA: 1-(isobutoxy)ethyl trifluoroacetate.
[d] IEA: 1-(isobutoxy)ethyl acetate.
[e] **4**: 2-(1-acetoxyethoxy)ethyl methacrylate (see Scheme 1.8.1).
[f] Adding THF as Lewis base, [THF] = 1.5 M.

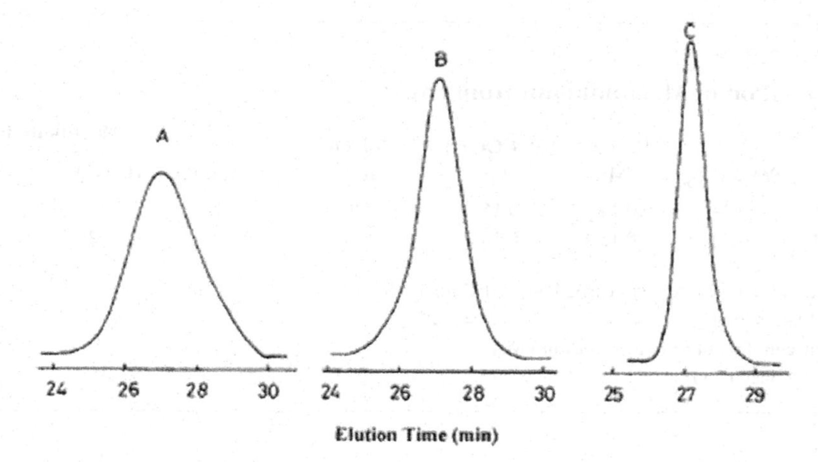

FIGURE 1.8.3 GPC traces of PVEs: (A) PVE–4 (in Table 1.8.2, $M_n = 3320$, $M_w/M_n = 1.47$) prepared by using IEA and $ZnCl_2$ as initiator and activator, respectively; (B) PVE–5 (in Table 1.8.2, $M_n = 3350$, $M_w/M_n = 1.30$) prepared by using **4** and $EtAlCl_2$ as initiator and activator in the absence of a Lewis base; (C) PVE–7 (in Table 1.8.2, $M_n = 3460$, $M_w/M_n = 1.11$) prepared by using **4**, $EtAlCl_2$, and THF as initiator, activator, and Lewis base, respectively.

to regulate the nucleophilic interaction between the growing carbocation and the counteranion ($^-OCOR...MX_n$).

As depicted in Figure 1.8.2C, the ^1H NMR spectrum of the cationically obtained polymer (PVE) of VEMA is free of peaks **e** and **f** corresponding to the C=C double bond of vinyloxy group in the monomer (Figure 1.8.2A), but retains the absorptions (**a** and **b**) of methacryloyl groups.

PVE can be dissolved in THF, benzene, chloroform, etc. Thus, VEMA was selectively polymerized via the cationic polymerization of its vinyloxy group without any cross-linking during polymerization. The PVE thus obtained is a functional polymer, with a methacryloyl group in each repeating unit.

Consequently, two different functional polymers could be prepared from a unique monomer, VEMA, via anionic and cationic polymerization, respectively.

Synthesis of a Macroinitiator from PMA and the Graft Copolymerization via a Discontinuous Method. The preparation of a graft copolymer was first tried using a discontinuous route (route A in Scheme 1.8.2). PMA prepared by the anionic polymerization of VEMA was freeze-dried from its benzene solution and then vacuum-dried for more than 10 h. The addition reaction between the vinyloxy group of the dried PMA and trifluoroacetic acid was carried out in toluene at 0°C for the ratio $[CF_3COOH]/[C{=}C] = 1.1$, for 1.5 h. After the addition reaction, the molecular weight of the polymer became larger (Table 1.8.3), the GPC curve of PMA disappeared (Figure 1.8.4A) and a new peak (Figure 1.8.4B) emerged in the high molecular weight region. This means that the addition reaction generated a new polymer (**5** in Scheme 1.8.2). **5** is a dormant macroinitiator for the cationic polymerization of alkyl vinyl ether, because each of its side chains has a structure

TABLE 1.8.3
Preparation of Macroinitiator from PMA[a]

No.	PMA-1[b] (g)	[C=C][c] (M)	[CF_3COOH] (M)	Toluene (mL)	CCl_4 (mL)	Macroinitiator $10^{-3}M_n$[d]	Macroinitiator M_w/M_n[d]
MI–1	0.44	0.14	0.16	20	0	3.82	1.13
MI–2	0.50	0.16	0.18	15	5	3.92	1.15

[a] The addition reaction was carried out at 0°C for 1.5 h.
[b] PMA-1 in Table 1.8.1: $M_n = 2610$, $M_w/M_n = 1.10$.
[c] Concentration of repeating units in PMA.
[d] Determined by GPC.

similar to that of IETA, which was used as initiator for the cationic polymerizations of IBVE[31] and VEMA described above.

The cationic graft polymerization of IBVE to the macroinitiator **5** was carried out in toluene at 0°C (Table 1.8.4). IBVE and $ZnCl_2$ (in Et_2O) were introduced into a toluene solution of **5** and the polymerization reaction was allowed to last for 3 h. As shown in Figure 1.8.4, the peak B of the macroinitiator (MI-2 in Table 1.8.3) disappeared completely after the graft copolymerization and two new peaks (C-a and C-b) emerged. Peak C-a corresponds to the produced graft copolymer, which has a much larger molecular weight (GP-2 in Table 1.8.4, $M_n = 9740$) than its macroinitiator (MI-2 in Table 1.8.3, $M_n = 3920$). On the other hand, Peak C-b

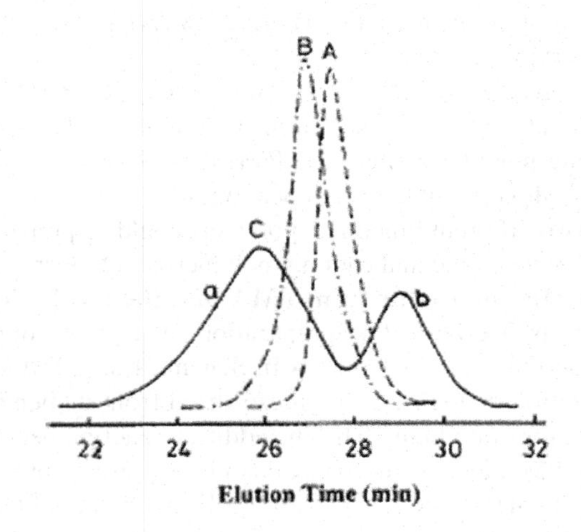

FIGURE 1.8.4 GPC traces for preparation of graft copolymers by a discontinuous method: (A) PMA (PMA-1 in Table 1.8.1, $M_n = 2610$, $M_w/M_n = 1.10$); (B) macroinitiator (MI-2 in Table 1.8.3, $M_n = 3920$, $M_w/M_n = 1.15$) prepared by the addition of trifluoroacetic acid to PMA-1; (C), after polymerization of IBVE (see GP-2 in Table 1.8.4), (a) a graft copolymer ($M_n = 9740$, $M_w/M_n = 1.37$), (b) a homopolymer of IBVE.

TABLE 1.8.4
Synthesis of Graft Copolymer by a Discontinuous Method[a]

No.	Macroinitiator (mM)	$[I]_0$[b]	$[IBVE]_0$ (M)	$[ZnCl_2]$ (mM)	Graft Copolymer[c] $10^{-3}M_n$[d]	M_w/M_n[d]
GP-1	MI-1[e]	56.4	0.28	28.0	9.32	1.39
GP-2	MI-2[e]	64.2	0.32	32.0	9.74	1.37

[a] Graft copolymerization was carried out in toluene at 0°C for 3h. Monomer conversion was 100% in each case.
[b] Concentration of repeating units in macroinitiator.
[c] Data corresponding to the graft copolymer, not including the homopolymer (see Figure 1.8.4).
[d] Determined by GPC.
[e] See Table 1.8.3

with a low molecular weight (M_n = 510) was caused by an excess of trifluoroacetic acid. In the preparation of the macroinitiator, an excess of trifluoroacetic acid must be added to ensure a complete addition of trifluoroacetic acid to the vinyloxy group. Consequently, a graft copolymer, with a polymethacrylate backbone and poly(IBVE) side chains can be prepared by this discontinuous method. However, the homopolymer of IBVE had to be removed in order to obtain the pure graft copolymer. In addition, the graft copolymer had a broad molecular weight distribution (M_w/M_n = 1.37–1.39, Table 1.8.4) and PMA used for the preparation of the macroinitiator had to be purified carefully. To overcome these shortcomings, another continuous method was tried.

Anionic Polymerization of 4 and the Preparation of a Graft Copolymer by a Continuous Method. 4 (in Scheme 1.8.1) can be used not only as an initiator for the cationic polymerization of VEMA (as noted above), but can also undergo anionic polymerization, because of the presence of the electron-deficient C=C double bond in the methacryloyl group; the poly(**4**) thus obtained is expected to be a dormant macroinitiator for the cationic polymerization of IBVE (**6**, route B in Scheme 1.8.2). As shown in Table 1.8.5, the anionic polymerization of **4** was carried out under conditions similar to those used for the anionic polymerization of VEMA. In the presence of LiCl, the anionic polymerization of **4** proceeded smoothly, without side reactions. The molecular weight of the macroinitiator poly(**4**) (determined by injecting a small amount of the THF polymerization solution into the GPC) could be controlled and the molecular weight distribution was narrow. However, when we tried to separate the poly(**4**), after quenching the polymerization system with a methanol–water mixture, another polymer, poly(2-hydroxyethyl methacrylate), precipitated slowly in about 3 h, because of the elimination of 1-acetoxyethyl group ($^{+}CH(CH_3)OCOCH_3$) of the side chains of poly(**4**). Parts A and B of Figure 1.8.5 depict the ^1H NMR spectra of **4** (Figure 1.8.5A) and of the precipitated polymer (Figure 1.8.5B). The absorptions (**e**, **f**, and **g**) due to the 1-acetoxyethyl groups disappeared completely, and were replaced by a new peak (i) in the spectrum of

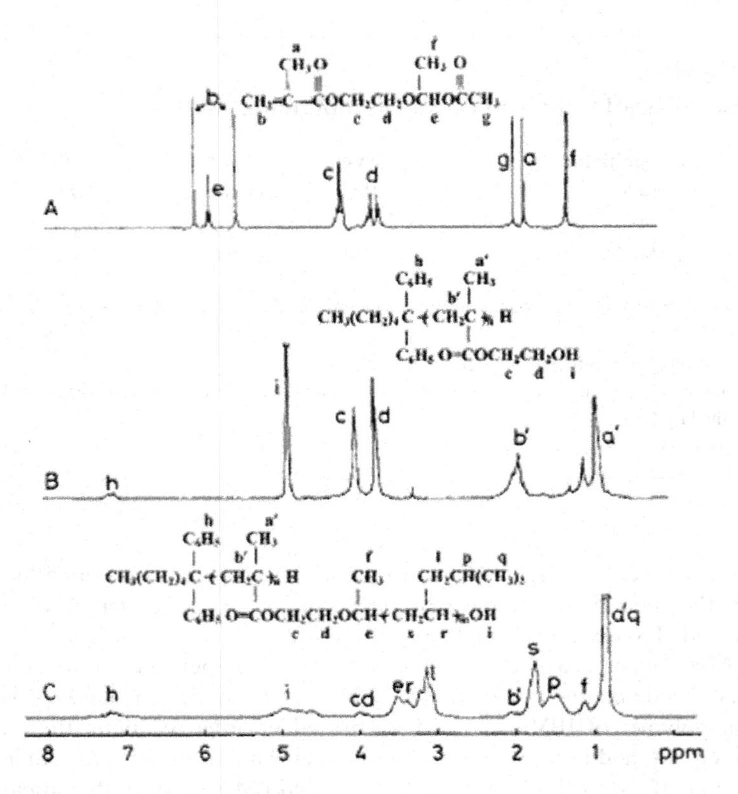

FIGURE 1.8.5 ^{1}H NMR spectra of **4** (A, dissolved in $CDCl_3$), poly(2-hydroxyethyl methacrylate) (B, dissolved in CD_3OD) obtained from MI-3 (Table 1.8.5) after elimination of its side chains by quenching with a methanol–water mixture, and graft copolymer (C, GP-3 in Table 1.8.6, dissolved in $CDCl_3$).

the precipitated polymer, which can be assigned to a hydroxyl group. These results indicate that, using the above procedure, poly(2-hydroxyethyl methacrylate) with a narrow molecular weight distribution can be prepared, but not the macroinitiator (**6** in Scheme 1.8.2) which can be further employed for graft copolymerization. However, fortunately, the graft copolymer could be prepared by a continuous method without the separation of the macroinitiator. The THF solution of the macroinitiator (MI-3 to MI-5 in Table 1.8.5) was not quenched; instead purified toluene and IBVE were added to the solution. The system remained transparent and no polymer precipitated. After the system was brought to room temperature (25°C), an activator $EtAlCl_2$ (in toluene) was added and the cationic graft polymerization of IBVE was allowed to proceed for 4 h (Table 1.8.6). THF, which was used as solvent in the preparation of the macroinitiator, acted as a Lewis base which stabilized the carbocation in the cationic polymerization step. As shown in Figure 1.8.6, the GPC peak (A) of the macroinitiator (MI-5 in Table 1.8.5, M_n = 5050, M_w/M_n = 1.06) disappeared entirely after the graft copolymerization. A new single peak (B) due to the graft copolymer (GP-5 in Table 1.8.6, M_n = 26,800, M_w/M_n = 1.12) emerged, and no peak belonging

TABLE 1.8.5
Preparation of Macroinitiator by the Anionic Polymerization of 4[a]

No.	$[n\text{-BuLi}]_0$ (mM)	[DPE] (mM)	[LiCl] (mM)	$[M]_0$ (M)	$10^{-3}M_n$ calcd	$10^{-3}M_n$ obsd[b]	M_w/M_n[b]
MI-3	26.4	31.7	64.0	0.37	3.27	4.04	1.09
MI-4	17.6	21.1	23.0	0.37	4.78	4.60	1.17
MI-5	15.0	18.0	30.0	0.34	5.11	5.05	1.06
MI-6	9.6	11.5	24.0	0.34	7.89	8.80	1.20

[a] Polymerization was carried out in THF at –60°C for 1 h. The monomer conversion was 100% in each case.
[b] Determined by GPC.

to the homopolymer remained. As shown in Table 1.8.6, the graft copolymers (GP-3, GP-4, and GP-5) have controllable molecular weights and narrow molecular weight distributions (M_w/M_n = 1.15–1.17). These results indicate that all of the side chains of the macroinitiator were consumed in initiating the cationic polymerization of IBVE and that the side chains have an almost equal length.

In the ^{1}H NMR spectrum of the graft copolymer (Figure 1.8.5C, GP-3 in Table 1.8.6), besides the absorptions (**l**, **p**, **q**, **r**, **s**) of the poly(IBVE) side chains, the peaks (**b′**, **c**, **d**, **f**, **h**) corresponding to the main chain can be still observed. Both the GPC and ^{1}H NMR measurements indicate that a pure graft copolymer having controlled molecular weight and narrow molecular weight distribution, with a polymethacrylate backbone and poly(IBVE) side chains, was prepared by the continuous polymerization method.

TABLE 1.8.6
Synthesis of Graft Copolymer by a Continuous Method[a]

No.	Macroinitiator	$[I]_0$[b] (mM)	[IBVE] (M)	$[EtAlCl_2$ (M)	[THF] (M)	Graft Copolymer $10^{-4}M_n$ obsd	Graft Copolymer $10^{-4}M_n$ obsd[c]	Graft Copolymer M_w/M_n[c]
GP-3	MI-3[d]	92.6	0.62	0.11	2.0	1.66	1.65	1.17
GP-4	MI-4[d]	77.2	0.64	0.12	2.0	2.12	1.76	1.15
GP-5	MI-5[d]	66.1	0.66	0.15	1.8	2.73	2.68	1.16

[a] Graft copolymerization was carried out in toluene at 25°C for 4 h. Monomer conversion was 100% in each case.
[b] Concentration of repeating units in macroinitiator.
[c] Determined by GPC.
[d] See Table 1.8.5.

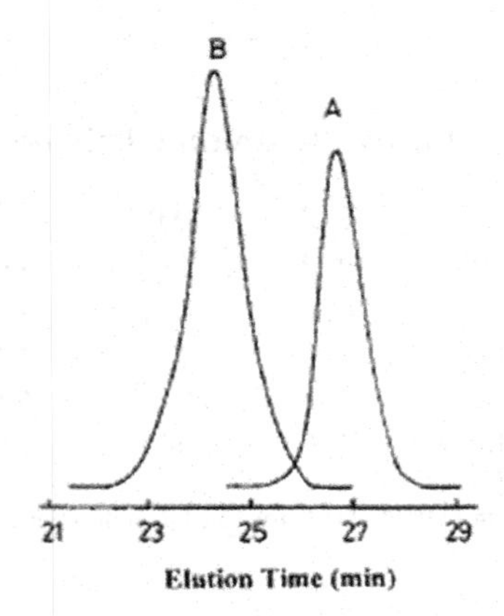

FIGURE 1.8.6 GPC traces for the preparation of a graft copolymer by the continuous method: (A) macroinitiator (MI-5 in Table 1.8.5, M_n = 5050, M_w/M_n = 1.06) prepared by the anionic polymerization of **4**; (B) graft copolymer (GP-5 in Table 1.8.6, M_n = 26800, M_w/M_n = 1.16).

1.8.4 CONCLUSION

2-Vinyloxyethyl methacrylate (VEMA) can be subjected to anionic or cationic polymerization to generate two kinds of functional polymers. Under the anionic polymerization conditions, the C=C double bond of the methacryloyl group of VEMA could be polymerized in THF, using DPHL as initiator, at −60°C, in the presence of LiCl. A functional polymer, with a cationically polymerizable C=C double bond in each side chain, was thus obtained. The determined molecular weight of the polymer was in good agreement with that calculated assuming that each initiator molecule generates one polymer chain and its distribution was narrow (M_w/M_n = 1.06–1.12). On the other hand, the C=C double bond of the vinyloxy group of VEMA can undergo cationic polymerization. A polymer with controlled molecular weight and narrow molecular weight distribution (M_w/M_n = 1.11–1.13) was prepared using 2-[1-(acetoxy)ethoxy]ethyl methacrylate (**4**) /$EtAlCl_2$ as initiator in the presence of THF, a weak Lewis base. The polymer thus obtained had a reactive methacryloyl group in each repeating unit.

The vinyloxy groups of the side chains of the anionically prepared polymer of VEMA could react with trifluoroacetic acid to generate a macroinitiator, which induced the cationic graft polymerization of isobutyl vinyl ether (IBVE) in toluene in the presence of $ZnCl_2$. In this manner, a graft copolymer with a polymethacrylate backbone and poly(IBVE) side chains could be prepared. However, this graft copolymer contained a small amount of the IBVE homopolymer, because of the excess trifluoroacetic acid which had to be employed.

The anionic polymerization of **4**, which proceeds under conditions similar to those for the anionic polymerization of VEMA, produced a THF solution of a macroinitiator. After toluene and IBVE were introduced into the system, the addition of an activator ($EtAlCl_2$) induced the cationic graft polymerization of IBVE to the macroinitiator. The THF, which was used as solvent in the anionic polymerization of **4**, acted as a Lewis base in the cationic polymerization step. By this continuous method, a pure graft copolymer with controlled molecular weight and narrow molecular weight distribution (M_w/M_n = 1.15–1.17), consisting of a polymethacrylate backbone and poly(IBVE) side chains, could be prepared.

REFERENCES

1. Sawamoto, M. *Int. J. Polym. Mater.* 1991, *15*, 197.
2. Lubnin, A. V.; Kennedy, J. P. *J. Macromol. Sci., Pure Appl. Chem.* 1994, *A31*, 1943.
3. Hasegawa, H.; Hashimoto, T.; Hyde, S. T. *Polymer* 1996, *37*, 3825.
4. Miyata, T.; Takagi, T.; Uragami, T. *Macromolecules* 1996, *29*, 7787.
5. Se, K.; Watanabe, O.; Isono, Y.; Fujimoto, T. *Makromol. Chem., Macromol. Symp.* 1989, *25*, 249.
6. Tang, S. C.; Hu, C. P.; Ying, S. K. *Polym. J.* 1990, *22*, 70.
7. Lohse, D. J.; Datta, S.; Kresge, E. N. *Macromolecules* 1991, *24*, 561.
8. Kobayashi, T.; Sato, M.; Takeno, N.; Mukaida, K. *Eur. Polym. J.* 1993, *29*, 1625.
9. Feng, H. Q.; Tian, J.; Ye, C. H. *J. Appl. Polym. Sci.* 1996, *61*, 2265.
10. Amorim, M. C. V.; Oliveira, C. M. F.; Tavares, M. I. B. *J. Appl. Polym. Sci.* 1996, *61*, 2245.
11. Sek, D.; Kaczmarczyk, B. *Polymer* 1997, *38*, 2925.
12. Nagase, Y.; Naruse, A.; Matsui, K. *Polymer* 1990, *31*, 121.
13. Nagase, Y.; Mori, S.; Egawa, M.; Matsui, K. *Makromol. Chem., Macromol. Chem. Phys.* 1990, *191*, 2413.
14. Miyata, T.; Takagi, T.; Kadato, T.; Uragami, T. *Macromol. Chem. Phys.* 1995, *196*, 1211.
15. ualeh, A. J.; Steiner, C. A. *Macromolecules* 1991, *24*, 112.
16. Muramatsu, N.; Yoshida, Y.; Kondo, T. *Chem. Pharm. Bull.* 1990, *38*, 3175.
17. Eisenbach, C. D.; Heinemann. T. *Macromol. Chem. Phys.* 1995, *196*, 2669.
18. Hashimoto, K.; Shinoda, H.; Okada, M.; Sumitomo, H. *Polym. J.* 1990, *22*, 312.
19. Geetha, B.; Mandal, A. B.; Ramasami, T. *Macromolecules 26*, 4083.
20. Oshea, M. S.; George, G. A. *Polymer* 1994, *35*, 4190.
21. Hegazy, E. S. A.; Dessouki, A. M.; Elsawy, N. M.; Elghaffar, M. A. A. *J. Polym. Sci., Polym. Chem.* 1993, *31*, 527.
22. Osman, M. B. S.; Hegazy, E. A.; Mostafa, A. E. K. B.; Abdelmaksoud, A. M. *Polym. Int.* 1995, *36*, 47.
23. Yang, J. M.; Hsiue, G. H. *J. Biomed. Mater. Res.* 1996, *31*, 281.
24. Grutke, S.; Hurley, J. H.; Risse, W. *Macromol. Chem. Phys. 195*, 2875.
25. Eisenbach, C. D.; Heinemann, T. *Macromolecules* 1995, *28*, 2133.
26. Yamashita, K.; Ito, K.; Tsuboi, H.; Takahama, S.; Tsuda, K. *J. Appl. Polym. Sci.* 1990, *40*, 1445.
27. Se, K.; Yamazaki, H.; Shibamoto, T.; Takano, A.; Fujimoto, T. *Macromolecules* 1997, *30*, 1570.
28. Ruckenstein, E.; Zhang, H. M. *Macromolecules* 1997, *30*, 6852.
29. Aoshima, S.; Higashimura, T. *Macromolecules* 1989, *22*, 1009.
30. Kishimoto, Y.; Aoshima, S.; Higashimura, T. *Macromolecules* 1989, *22*, 3877.
31. Kamigaito, M.; Sawamoto, M.; Higashimura, T. *Macromolecules* 1991, *24*, 3988.
32. Zhang, H. M.; Ruckenstein, E. *J. Polym. Sci., Polym. Chem.* 1997, *35*, 2901.
33. Varshney, S. K.; Hautekeer, J. P.; Fayt, R.; Jerome, R.; Teyssie, Ph. *Macromolecules* 1990, *23*, 2618.
34. Kato, K.; Ichijo, T.; Ishii, K.; Hasegawa, M. *J. Polym. Sci., Part A-1* 1971, *9*, 2109.
35. Watanabe, S.; Kato, M. *J. Polym. Sci, Polym. Chem.* 1984, *22*, 280.
36. Aoshima, S.; Hasegawa, O.; Higashimura, T. *Polym. Bull.* 1985, *13*, 229.

37. Haas, H. C.; Simon, M. S. *J. Polym. Sci.* 1955, *17*, 421.
38. Lal, J.; Devlin, E. F.; Trick, G. S. *J. Polym. Sci.* 1960, *44*, 523.
39. Fayt, R.; Forte, R.; Jacobs, C.; Jerome, R.; Ouhadi, T.; Teyssie, Ph.; Varshney, S. K. *Macromolecules* 1987, *20*, 1442.
40. Varshney, S. K.; Jerome, R.; Bayard, P.; Jacobs, C.; Fayt, R.; Teyssie, Ph. *Macromolecules* 1992, *25*, 4457.
41. Kunkel, D.; Muller, A. H. E.; Janata, M.; Lochman, L. *Makromol. Chem., Macromol. Symp.* 1992, *60*, 315.
42. Sawamoto, M. *Prog. Polym. Sci.* 1991, *16*, 111.
43. Majoros, I.; Nagy, A.; Kennedy, J. P. *Adv. Polym. Sci.* 1994, *112*, 1.
44. Fodor, ZS.; Faust, R. J. Macromol. *Sci.; Pure Appl. Chem.,* A32, 575.
45. Higashimura, T.; Ishihama, Y.; Sawamoto, M. *Macromolecules* 1993, *26*, 744.

1.9 Monomer [1-(Isobutoxy)ethyl Methacrylate] That Can Undergo Anionic Polymerization and Can Also Be an Initiator for the Cationic Polymerization of Vinyl Ethers: *Preparation of Combilke Polymers**

Eli Ruckenstein and Hongmin Zhang

Chemical Engineering Department, State University of New York at Buffalo, Amherst, New York 14260

ABSTRACT 1-(Isobutoxy)ethyl methacrylate (BOEMA) was prepared by the reaction between isobutyl vinyl ether and methacrylic acid. It can be used both as an initiator for the cationic polymerization of vinyl ether and as a monomer that can undergo anionic polymerization. Using zinc chloride as activator, BOEMA can induce the cationic polymerization of isobutyl vinyl ether or ethyl vinyl ether in toluene at 0°C to generate an end-functional polymer. On the other hand, BOEMA can be polymerized itself by using (1,1-diphenylhexyl)lithium as an initiator, in tetrahydrofuran, at −60°C, in the presence or absence of lithium chloride. The poly(BOEMA) obtained possesses a very narrow molecular

* *Macromolecules* 1997, 30, 6852–6855.

weight distribution ($\bar{M}_w/\bar{M}_n = 1.04-1.09$). It can also be employed as a macroinitiator for the cationic polymerization of isobutyl vinyl ether or ethyl vinyl ether, by using zinc chloride as activator in toluene at 0°C. Thus a graft copolymer consisting of polymethacrylate as the main chain and poly(isobutyl vinyl ether) or poly(ethyl vinyl ether) as side chains can be prepared.

1.9.1 INTRODUCTION

The molecular design of polymers that possess well-defined structures, selected molecular compositions, and particular properties is becoming an increasingly important route to high performance materials.[1–4] However, the number of monomers suitable for a certain kind of polymerization is limited. For instance, an alkyl methacrylate can be polymerized by radical or anionic polymerization but not by cationic polymerization. Similarly, vinyl ethers can undergo cationic polymerization but not radical or anionic polymerization. It is, therefore, necessary to combine two polymerization methods to prepare block or graft copolymers of the above two kinds of monomer.

Major progress was achieved recently in the living cationic polymerization of vinyl monomers.[5–8] Higashimura and his co-workers[9,10] discovered numerous initiator systems which induced the living cationic polymerization of vinyl ethers. For instance, they employed a series of acetic acid derivatives (RCOOH, where R can be CF_3, CCl_3, $CHCl_2$, CH_2Cl or CH_3), which, combined with zinc chloride, generated suitable nucleophilic counteranions which, by stabilizing the growing carbocation, allowed the isobutyl vinyl ether (IBVE) to undergo controlled cationic polymerization.[11] These substituted acetic acids react with IBVE to form esters. For instance, the addition reaction between IBVE and acetic acid generates an adduct, 1-isobutoxyethyl acetate (BOEA). BOEA cannot, by itself, induce the polymerization of IBVE,[12] but its ester linkage can be activated with zinc chloride to generate a partly dissociated carbocation, which can initiate the polymerization of a vinyl ether. The growing carbocation is stabilized by the nucleophilic counteranion ($^{-}OCOCH_3\cdots ZnCl_2$).

In the present paper, a similar initiator, 1-(isobutoxy) ethyl methacrylate (BOEMA, **1** in Scheme 1.9.1), which is an adduct of IBVE with methacrylic acid (MAA) is prepared. Because its molecular structure is similar to that of BOEA, BOEMA is expected to be an initiator for the cationic polymerization of vinyl ether in the presence of the activator $ZnCl_2$ (Scheme 1.9.1, left). On the other hand, because of the presence of the electron-deficient C=C double bond in the methacrylate group, BOEMA is also an anionically polymerizable monomer, which can be also polymerized using an anionic initiator like (1,1-diphenylhexyl)lithium (DPHL). Every side chain of the poly(BOEMA) (**5**, Scheme 1.9.1) has a molecular structure similar to that of BOEA. Consequently, poly(BOEMA) constitutes a dormant macroinitiator for the cationic polymerization of any vinyl ether, which can be activated with $ZnCl_2$ to generate **6** in Scheme 1.9.1. The cationic polymerization of IBVE or ethyl vinyl ether (EVE) leads to a graft copolymer (**7**, Scheme 1.9.1) consisting of polymethacrylate as the main chain and poly(IBVE) or poly(EVE) as side chains.

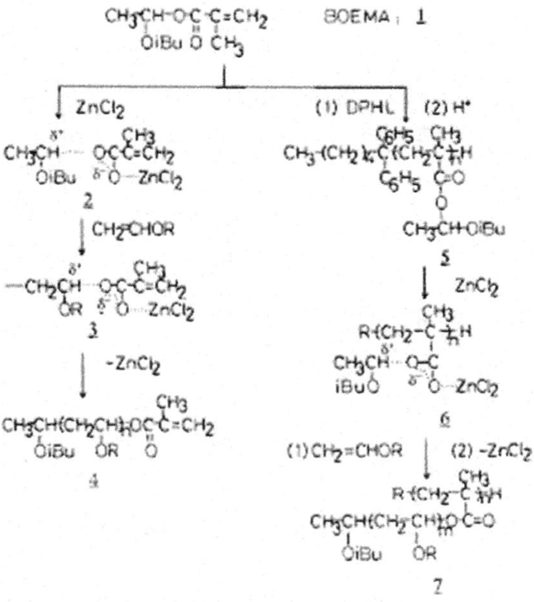

SCHEME 1.9.1

1.9.2 EXPERIMENTAL SECTION

Materials. Tetrahydrofuran (THF, solvent for the anionic polymerization of BOEMA) was stirred with CaH_2 under reflux for more than 24 h, distilled, purified in the presence of naphthalene sodium, and finally distilled from a solution of DPHL just before use.[13] Toluene, which is used as solvent for the cationic polymerization and for the grafting of poly(IBVE) or poly(EVE) on poly(BOEMA) via cationic polymerization, was washed with concentrated sulfuric acid and then with water, dried over $MgSO_4$, and distilled twice over calcium hydride before use. Et_2O was dried over CaH_2 and distilled in the presence of $LiAlH_4$. Hexane was first dried and distilled over CaH_2 and then distilled from a solution of *n*-BuLi. IBVE or EVE (Aldrich, 99%; bp 82°C and 33°C, respectively) was washed with 10% aqueous sodium hydroxide solution and then water, dried overnight over potassium hydroxide, and distilled twice over CaH_2 prior to polymerization.[11,12] 1,1-Diphenylethylene (DPE; Aldrich, 97%) was doubly distilled from CaH_2 and then distilled in the presence of triphenylmethyllithium under reduced pressure (bp 75°C/1.0 Torr).[14] Lithium chloride (Aldrich, 99.99%) was dried at 120°C for 24 h and dissolved in THF.[15] $ZnCl_2$ (Aldrich, 1.0 M solution in diethyl ether) and *n*-BuLi (Aldrich, 1.6 M solution in hexane) were diluted with purified Et_2O and hexane, respectively.

Synthesis of 1-(Isobutoxy)ethyl Methacrylate (BOEMA). BOEMA was prepared by the reaction of IBVE with methacrylic acid (MAA).[11] MAA (66 mL, 0.77 mol) was dropwise added to IBVE (122 mL, 0.93 mol) with magnetic stirring

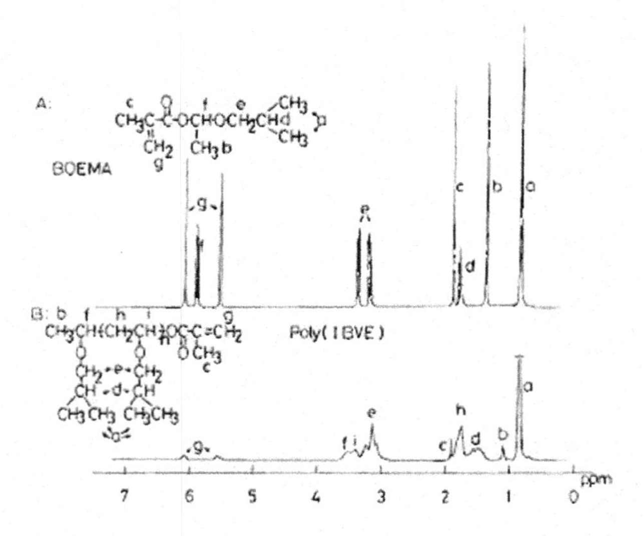

FIGURE 1.9.1 ^{1}H NMR spectra of BOEMA (A) and poly(IBVE) (B; no. 2 in Table 1.9.1).

at 55°C and kept at this temperature for 8 h. The crude product was distilled twice from calcium hydride under reduced pressure (32°C/0.2 Torr) and pure BOEMA was obtained as a sweet smelling, colorless oil, with an yield of 82% based on MAA. The purity (>99%) was determined by gas chromatography measurements. As shown in Figure 1.9.1, the chemical shifts and their intensities in the ^{1}H NMR spectrum of the prepared BOEMA are consistent with its molecular structure.

Polymerizations. Polymerization was carried out in a 100 mL round-bottom glass flask under an overpressure of nitrogen with magnetic stirring. The cationic polymerization of IBVE or EVE was initiated by adding, via a dry syringe, an Et_2O solution of $ZnCl_2$ to a toluene solution of the monomer and BOEMA at 0°C. The polymerization was terminated with methanol (ca. 2 mL) containing a small amount of ammonia. The quenched reaction mixture was washed with 10% aqueous sodium thiosulfate solution and then with water, evaporated to dryness under reduced pressure, and vacuum dried (40°C, 1.0 Torr) to obtain the product polymer.

The anionic polymerization of BOEMA was carried out in THF at −60°C in the presence or absence of LiCl. After THF, DPE and a THF solution of LiCl were added with dry syringes, the flask was cooled to −60°C and *n*-BuLi (in hexane) was added. The deep red color of DPHL appeared immediately, and the reaction between *n*-BuLi and DPE was allowed to continue for about 15 min. The polymerization reaction was induced by the addition of a prechilled THF solution of BOEMA to the above DPHL solution. After a certain time (40 or 60 min), the reaction was quenched with a small amount of methanol and the solution was poured into a cooled mixture of ethanol and water (−20°C) to precipitate the polymer. The polymer was reprecipitated from its THF solution with the above mixture and dried in vacuum (40°C, 1.0 Torr) overnight.

Graft copolymerization was carried out under conditions similar to those for cationic polymerization described above. Poly(BOEMA) was freeze-dried from its benzene solution for 8 h, then vacuum dried (50°C, 1.0 Torr) for more than 10 h. The purified toluene and monomer (IBVE or EVE) were introduced into the flask that contained the dried poly(BOEMA). After the polymer was completely dissolved and cooled to 0°C, $ZnCl_2$ (in Et_2O) was added to induce the graft copolymerization reaction. The termination of the reaction and the purification of the obtained copolymer were carried out as described for the cationic polymerization.

Measurements. ^{1}H NMR spectra were recorded in $CDCl_3$ on a VXR-400 spectrometer. The $\bar{M}_n$ and $\bar{M}_w/\bar{M}_n$ of the polymer were determined by gel permeation chromatography (GPC) on the basis of a polystyrene calibration curve. The GPC measurements were carried out with THF as solvent, at 30°C, using two polystyrene gel columns (Waters, Linear, 7.8 × 300 mm, Part No. 10681) connected to a Waters 515 precision pump.

1.9.3 RESULTS AND DISCUSSION

1. **Cationic Polymerization of IBVE or EVE.** Aoshima and co-workers[16] prepared a poly(vinyl ether) macromer having a methacrylate group at one end. The adduct [$CH_2{=}C(CH_3)COOCH_2CH_2OCHICH_3$] obtained via the reaction of 2-(vinyloxy)ethyl methacrylate with hydrogen iodide was employed as initiator to induce the living cationic polymerization of EVE in the presence of a small amount of iodine. Obviously, the methacrylate group is located at the initial end of the polymer. In contrast, in the present paper, the methacrylate group constitutes a counteranion which is moving to the terminus of the polymer as the chain propagating reaction proceeds (Scheme 1.9.1, left).

   The cationic polymerization of IBVE or EVE was carried out in toluene at 0°C. When $ZnCl_2$ was added to the mixture of toluene, monomer, and BOEMA, the polymerization system became either transparent (no. 3, Table 1.9.1), or yellow (no. 2) or red (no. 1), depending on the initial concentrations of BOEMA and $ZnCl_2$. The monomer conversion can reach 100%

TABLE 1.9.1
Cationic Polymerization of IBVE or EVE with BOEMA as Initiator[a]

				$\bar{M}_n$			
No.	$[BOEMA]_0$ (mM)	$[M]_0$ (M)	$[ZnCl_2]$ (mM)	calcd	GPC[b]	NMR[c]	$\bar{M}_w/\bar{M}_n$[b]
1	110	IBVE, 1.08	20.0	1190	1760	1200	2.70
2	53.6	IBVE, 1.12	10.4	2280	3180	2430	2.81
3	13.6	IBVE, 1.14	5.27	8530	8670		2.60
4	16.7	EVE, 1.54	13.8	3580	5860		2.96

[a] The polymerization was carried out in toluene at 0°C; for nos. 1, 2 and 3, the polymerization time was 6 h and the monomer conversion = 100%; for no. 4, the polymerization time was 3 h and the monomer conversion = 51%.

[b] Determined from the GPC measurements.

[c] Calculated from the ^{1}H NMR spectra of poly(IBVE).

in 6 h for IBVE, and the number average molecular weight, determined either by GPC or ^{1}H NMR, increased as the ratio $[M]_0/[BOEMA]_0$ became larger. However, the molecular weight distribution of the polymer obtained is broad $\bar{M}_w/\bar{M}_n = 2.60 - 2.96$ for both monomers. This may be a result of the strong nucleophilicity of the counteranion $[CH_2{=}C(CH_3)COO^-]$[11].

Figure 1.9.1 depicts the ^{1}H NMR spectra of BOEMA and poly(IBVE). After the polymerization of IBVE, some of the peaks of the moiety originating from BOEMA can be still observed. The peaks b and f are shifted, the peaks c and g did not change, and the other peaks are overlapped with those due to poly(IBVE). In addition, the molecular weight calculated from the ^{1}H NMR spectra is close to the theoretical value calculated assuming that one polymer chain is generated by each BOEMA molecule present.

2. **Anionic Polymerization of BOEMA.** Teyssie and co-workers[15,17,18] suggested that the addition of a ligand, such as LiCl, can be effective in the living anionic polymerization of *tert*-butyl acrylate and methyl methacrylate because a μ-type complex[19] is formed between LiCl and the growing site, which impedes side reactions from occurring at the propagating site, thus leading to monodisperse polymers. In this paper, we employ this polymerization technique for the anionic polymerization of BOEMA, to prepare nearly monodisperse poly(BOEMA).

The initiator, DPHL, was synthesized by the reaction of *n*-BuLi with DPE at a ratio $[DPE]/[n\text{-BuLi}]_0 = 1.2$ (Table 1.9.2). The anionic polymerization was carried out in THF at −60°C for $[LiCl]/[n\text{-BuLi}]_0$ ratios of 0, 0.5, 1.5, 2.0, and 3.0 (no. 5 to no. 9, Table 1.9.2). As shown in Table 1.9.2, poly(BOEMA) can be obtained, regardless if LiCl is added or not. However, the molecular weight distribution of the obtained polymer is narrower for larger $[LiCl]/[n\text{-BuLi}]_0$ ratios, and nearly monodisperse poly(BOEMA) can be obtained

TABLE 1.9.2
Anionic Polymerization of BOEMA[a]

					$\bar{M}_n$			
No.	$[n\text{-BuLi}]_0$ (mM)	[DPE] (mM)	[LiCl] (mM)	$[M]_0$ (M)	calcd	GPC[b]	NMR[c]	$\bar{M}_w/\bar{M}_n$[b]
5	50.4	60.5	0	0.61	2500	2800		1.17
6	152	182	64	0.56	920	900	930	1.13
7	69.4	83.3	117	0.51	1600	1800	1590	1.07
8	50.4	60.5	101	0.61	2490	2630	2650	1.04
9	42.9	57.1	150	0.63	2970	3200		1.05
10	20.8	25.0	42.0	0.61	5690	7900		1.09
11	12.3	14.8	25.0	0.61	8050	13200		1.07

[a] The polymerization was carried out in THF at −60°C for 40 (nos. 5–9) and 60 (nos. 10 and 11) min, respectively. The monomer conversion was 100%.

[b] See Table 1.9.1.

[c] Calculated from the ^{1}H NMR spectra of poly(BOEMA)s.

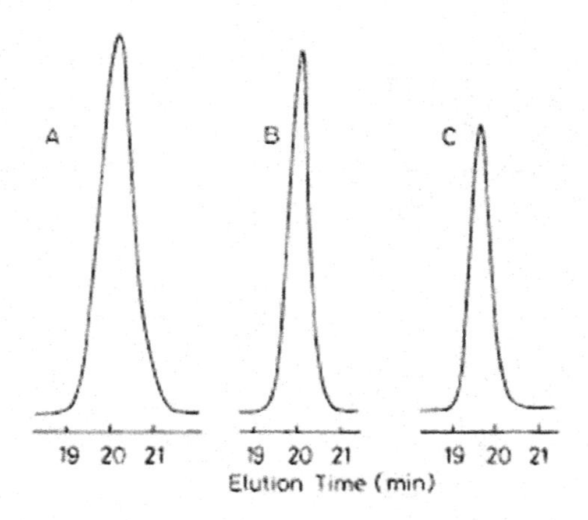

FIGURE 1.9.2 GPC traces of poly(BOEMA)s prepared in the absence (A; no. 5 in Table 1.9.2) or presence (B; no. 8 in Table 1.9.2) of LiCl and standard polystyrene (C; Pressure Chemical Company, $M_n = 4000$, $M_w/M_n = 1.06$, Lot No. 61110).

when $[LiCl]/[n\text{-}BuLi]_0 > 2$ (no. 8 and no. 9, Table 1.9.2). Polymerizations no. 5 and no. 8 of Table 1.9.2 differ only in the values of the ratio $[LiCl]/[n\text{-}BuLi]_0$. As shown in Figure 1.9.2, the GPC curve B of poly(BOEMA) prepared in the presence of LiCl ($[LiCl]/[n\text{-}BuLi]_0 = 2$, no. 8) is much narrower than that for no. 5 obtained in the absence of LiCl (curve A) and as narrow as that of the polystyrene standard (C).

The calculated number-average molecular weight of poly(BOEMA) is comparable to those determined by GPC or ^{1}H NMR when the molecular weight is small (nos. 6–9). When the concentration of *n*-BuLi is low (nos. 10 and 11), the difference between the calculated and determined values is large. The initiator efficiency decreases to 72% and 61% for no. 10 and no. 11, respectively, but the molecular weight distribution remains narrow ($\bar{M}_w/\bar{M}_n = 1.09$ and 1.07). These results indicate that the initiator is involved in an undetermined side reaction during the initiation stage. Alkyl acrylates or methacrylates can generally be purified prior to polymerization with a strong Lewis acid, like trialkylaluminum. BOEMA cannot be, however, purified with this reagent, because the presence of a strong Lewis acid results in its decomposition. The search for a suitable reagent for the purification of BOEMA is ongoing.

The ^{1}H NMR spectrum of poly(BOEMA) (no. 7, Table 1.9.2) is depicted in Figure 1.9.3A. Compared to the spectrum of BOEMA (Figure 1.9.1A), the absorptions (g) of the C=C double bond have disappeared, the peak (c) corresponding to the α-methyl group is shifted from 1.89 ppm to between 0.68 and 1.21 ppm. New peaks, (g′ and h), originating from the methylene group of the main chain and the phenyl group of DPHL emerge. In contrast, the absorptions (a, b, d, e, f) of the ester group did not change after polymerization. The chemical shift and the intensity of the peaks are consistent with the molecular structure of poly(BOEMA), and the number-average

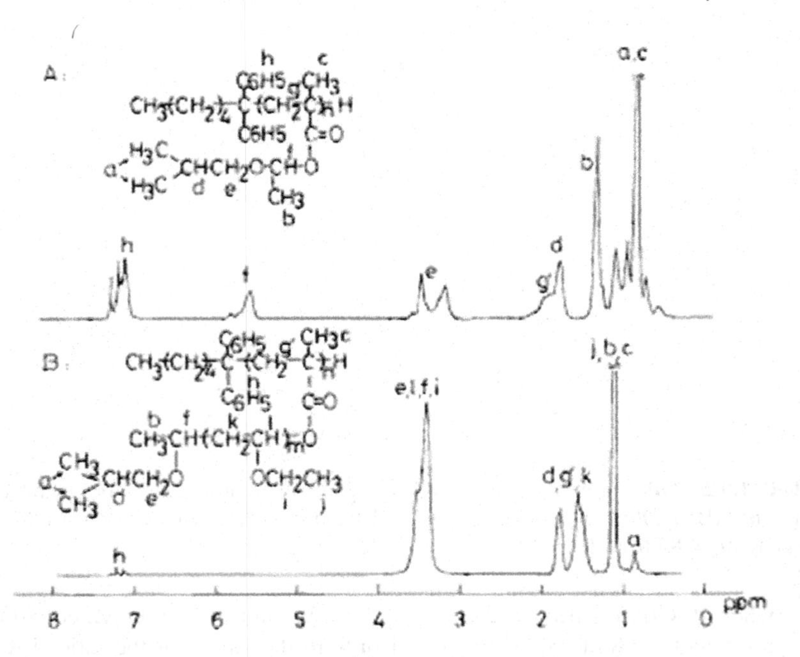

FIGURE 1.9.3 ^{1}H NMR spectra of poly(BOEMA) (A; no. 7 in Table 1.9.2) and poly(BOEMA-*g*-EVE) (B; for the preparation conditions, see Figure 1.9.4 B and the text).

molecular weight calculated from the ^{1}H NMR spectrum is in good agreement with the theoretical value, calculated assuming that one polymer chain is formed from each DPHL molecule present (Table 1.9.2).

3. **Grafting of Poly(IBVE) or Poly(EVE) to Poly-(BOEMA).** The grafting of poly(IBVE) or poly(EVE) to poly(BOEMA) was carried out in toluene at 0°C. When the prechilled monomer was added to the toluene solution of poly(BOEMA) in the absence of $ZnCl_2$, no reaction took place and the poly(BOEMA) could be recovered without any change, indicating that poly(BOEMA) is in its stable dormant state and cannot alone induce the cationic polymerization of vinyl ether. In contrast, when $ZnCl_2$ was added to the toluene solution of poly(BOEMA) and monomer (IBVE or EVE), the graft copolymerization could proceed. Two examples are presented in Figure 1.9.4. The freeze-dried poly(BOEMA) (no. 8 in Table 1.9.2, 0.5 g) was dissolved in 34 mL of toluene and after EVE (4.0 g) was added, the system was cooled to 0°C. When $ZnCl_2$ (1.20 mL, 1.0 M in Et_2O) was introduced, the system acquired a yellow color which changed gradually to red. The polymerization reaction lasted 4 h. As shown in Figure 1.9.4, the peak (A) of poly(BOEMA) used as macroinitiator disappeared completely and a broad peak (B) corresponding to the graft copolymer emerged. This result indicates that all of the poly(BOEMA) was consumed to initiate the graft polymerization of EVE. The molecular weight determined by GPC on the basis of a polystyrene calibration curve is 1.9×10^4, which is smaller than

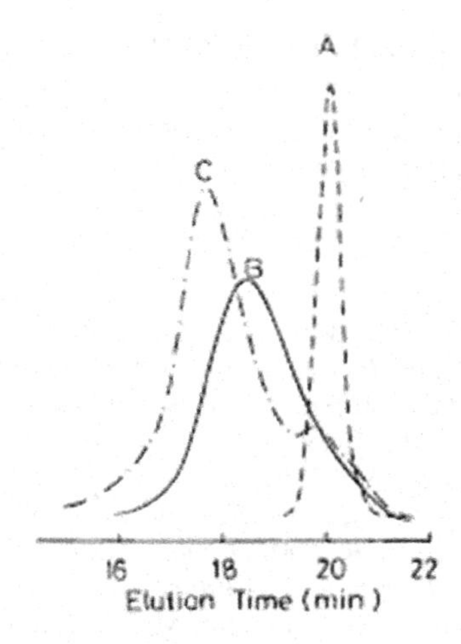

FIGURE 1.9.4 GPC traces: (A) poly(BOEMA) (no. 8 in Table 1.9.2, M_n = 2630, M_w/M_n = 1.04; (B) poly(BOEMA-*g*-EVE) prepared by the cationic polymerization of EVE in toluene at 0°C from poly(BOEMA) (no. 8, 0.50 g) in the presence of $ZnCl_2$ (concentration of the repeating units of poly(BOEMA) $[BOEMA]_0$ = 60 mM, $[ZnCl_2]$ = 30 mM, $[EVE]_0$ = 1.59 M; polymerization time, 4 h; monomer conversion = 100%); (C) poly(BOEMA-*g*-EVE) prepared under the same conditions as B, except $[ZnCl_2]$ = 15 mM.

the theoretical value (2.3×10^4). The molecular weight distribution is broad because the nucleophilicity of the large anion is strong, and as a result, the rate of initiation is small compared to that of propagation. When the amount of $ZnCl_2$ was reduced to 0.60 mL (1.0 M, in Et_2O) while the other conditions were not changed, no yellow or red color appeared in the system and a small amount of poly(BOEMA) remained unreacted (Figure 1.9.4, peak C).

The ^1H NMR spectrum of the graft copolymer is presented in Figure 1.9.3B. After graft copolymerization of EVE, the peak (f) in the spectrum of poly(BOEMA) disappeared. However, the absorption (a) of the methyls of the ester groups remained unchanged and can be observed at 0.85 ppm. The other peaks overlapped with those of poly(EVE) of the side chains. Both the GPC and ^1H NMR measurements indicate that a pure graft copolymer can be obtained by the method of Scheme 1.9.1.

1.9.4 CONCLUSION

1-Isobutoxyethyl methacrylate (BOEMA) can be used as both an initiator for the cationic polymerizations of vinyl ethers and a monomer that can undergo anionic polymerization. In the presence of an activator, $ZnCl_2$, in toluene at 0°C, BOEMA can induce the cationic polymerization of isobutyl vinyl ether or ethyl vinyl ether to generate an end-functional polymer, which possesses a methacrylate group at one end. On the other hand, BOEMA itself can be polymerized with the anionic initiator (1,1-diphenylhexyl)lithium. When the polymerization was carried out in the presence of LiCl, in THF, at −60°C, the poly(BOEMA) obtained had a very narrow molecular weight distribution ($\bar{M}_w/\bar{M}_n = 1.04-1.09$). Further, the poly(BOEMA) was used in the presence of $ZnCl_2$ as a macroinitiator for the cationic graft copolymerization of isobutyl vinyl ether or ethyl vinyl ether.

REFERENCES

1. Lubnin, A. V.; Kennedy, J. P. *J. Macromol. Sci., Pure Appl. Chem.* **1994**, *31*, 1943.
2. Zaschke, B.; Kennedy, J. P. *Macromolecules* **1995**, *28*, 4426.
3. Faust, R. Makromol. Chem., *Macromol. Symp.* **1994**, *85*, 295.
4. Kanaoka, S.; Sawamoto, M.; Higashimura, T. *Macromolecules* **1992**, *25*, 6414.
5. Sawamoto, M. *Prog. Polym. Sci.* **1991**, *16*, 111.
6. Majoros, I.; Nagy, A.; Kennedy, J. P. *Adv. Polym. Sci.* **1994**, *112*, 1.
7. Gyor, M.; Wang, H. C.; Faust, R. *J. Macromol. Sci., Pure Appl. Chem.* **1992**, *A29*, 639.
8. Fodor, Zs.; Faust, R. *J. Macromol. Sci., Pure Appl. Chem.* **1995**, *A32*, 575.
9. Higashimura, T.; Aoshima, S.; Sawamoto, M. *Makromol. Chem., Macromol. Symp.* **1988**, *13/14*, 457.
10. Sawamoto, M.; Higashimura, T. *Makromol. Chem., Macromol, Symp.* **1990**, *31*, 131.
11. Kamigaito, M.; Sawamoto, M.; Higashimura, T. *Macromolecules* **1991**, *24*, 3988.
12. Aoshima, S.; Higashimura, T. *Macromolecules* **1989**, *22*, 1009.
13. Fujimoto, T.; Zhang, H. M.; Kazama, T.; Isono, Y.; Hasegawa, H.; Hashimoto, T. *Polymer* **1992**, *33*, 2208.
14. Zhang, H. M.; Ishikawa, H.; Ohata, M.; Kazama, T.; Isono, Y.; Fujimoto, T. *Polymer* **1992**, *33*, 828.
15. Varshney, S. K.; Hautekeer, J. P.; Fayt, R.; Jerome, R.; Teyssie, Ph. *Macromolecules* **1990**, *23*, 2618.
16. Aoshima, S.; Ebara, K.; Higashimura, T. *Polym. Bull.* **1985**, *14*, 425.
17. Fayt, R.; Forte, R.; Jacobs, C.; Jerome, R.; Ouhadi, T.; Teyssie, Ph.; Varshney, S. K. *Macromolecules* **1987**, *20*, 1442.
18. Varshney, S. K.; Jerome, R.; Bayard, P.; Jacobs, C.; Fayt, R.; Teyssie, Ph. *Macromolecules* **1992**, *25*, 4457.
19. Kunkel, D.; Muller, A. H. E.; Janata, M.; Lochmann, L. *Makromol. Chem. Macromol. Symp.* **1992**, *60*, 315.

1.10 Block–Graft and Star-Shaped Copolymers by Continuous Transformation from Anionic to Cationic Polymerization*

Eli Ruckenstein and Hongmin Zhang

Chemical Engineering Department, State University of New York at Buffalo, Amherst, New York 14260

ABSTRACT The anionic block copolymerization of methyl methacrylate (MMA) and 2-(1-acetoxyethoxy)ethyl methacrylate (**2**) was first carried out by the sequential polymerization of MMA and **2**, using (1,1-diphenylhexyl)lithium (DPHL) as the initiator, in the presence of LiCl ($[LiCl]/[DPHL]_0 = 4$), in tetrahydrofuran (THF), at −60°C. The block copolymer obtained [poly(MMA-*b*-**2**)] possesses a controlled molecular weight, designed composition, and narrow molecular weight distribution ($M_w/M_n = 1.05–1.13$). Then, toluene, isobutyl vinyl ether (IBVE), and $EtAlCl_2$ were added to the THF solution of poly(MMA-*b*-**2**) obtained during the previous step. Every side chain of the poly(**2**) segment of poly(MMA-*b*-**2**) could be activated by a Lewis acid, $EtAlCl_2$, to initiate the living cationic polymerization of isobutyl vinyl ether at room temperature. The THF, which was used as the solvent in the preparation of poly(MMA-*b*-**2**), acted as a weak Lewis base which stabilizes the propagating carbocation in the cationic polymerization step of IBVE. In this manner, a block–graft copolymer consisting of poly(MMA-*b*-**2**) as the block backbone and poly(IBVE) as the side chains attached to the poly(**2**) segment was prepared. This copolymer has a designed graft number, controlled molecular weights of the backbone and side chains, and a narrow molecular weight distribution ($M_w/M_n = 1.11–1.16$). In a similar way, a star-shaped copolymer consisting of one poly(MMA) arm and several poly(IBVE) arms could be prepared, by restricting the polymerization degree

* *Macromolecules* 1998, 31, 2977–2982.

of **2** to a small value during the preparation step of poly(MMA-*b*-**2**). This star-shaped copolymer also possesses a designed arm number, controlled molecular weight, and narrow molecular weight distribution (M_w/M_n = 1.10–1.20).

1.10.1 INTRODUCTION

The advances in living polymerization have made the design and preparation of multiple-composition copolymers, such as block and graft copolymers, possible.[1–3] Numerous block copolymers could be prepared by the sequential monomer addition technique. However, it is more difficult to prepare graft copolymers than block copolymers. Although a number of graft copolymers have been obtained,[4–12] it was difficult to control the molecular weights of the backbone and side chains, the positions of the side chains, and their number. Even using a living polymerization technique, the graft copolymer was generally synthesized by a discontinuous route,[13–15] in which the precursor polymer had to be separated from the polymerization solution and purified carefully. Recently, we reported a continuous preparation method of a novel graft copolymer consisting of a polymethacrylate backbone and poly(isobutyl vinyl ether) [poly(IBVE)] side chains, by combined anionic and cationic polymerizations.[16] The scope of the present paper is to extend the latter method to more complex molecular architectures. Figure 1.10.1a illustrates schematically the block–graft copolymer which was prepared. It consists of an AB block backbone and C side chains attached to the B segment. This copolymer has the following characteristics. (1) The molecular weights of the backbone and side chains, hence the total molecular weight of the block–graft copolymer, can be controlled. (2) The C side chains are located only in one part of the backbone, and their number can be selected. (3) Both the backbone and the side chains possess narrow molecular weight distributions.

The synthesis method of the above block–graft copolymer is based on compounds, such as 2-(1-acetoxy-ethoxy)ethyl methacrylate (**2**, in Scheme 1.10.1), which are both monomers and dormant initiators for living polymerization. Because of the presence of the electron-deficient C=C double bond in the methacryloyl group, **2** can be subjected to anionic polymerization. On the other hand, in the presence of a

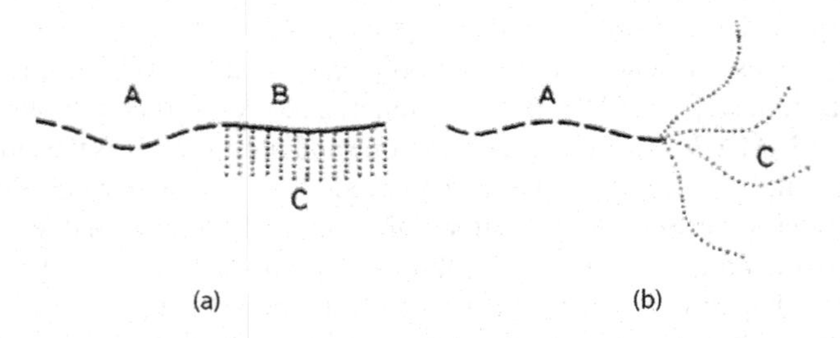

FIGURE 1.10.1 Schematic representation of the molecular architecture of the model copolymers prepared in the present paper. (a) Block–graft copolymer consisting of an AB block backbone and C side chains attached to the B segment. (b) Star-shaped copolymer consisting of only one A arm and several C arms.

Lewis acid, such as $ZnCl_2$ or $EtAlCl_2$, the linkage between the acetoxy and the ethoxy group can be activated to generate a partly dissociated carbocation, which can induce the cationic polymerization of IBVE.[17–19] In a basic environment, this linkage is stable enough to allow the anionic polymerization of **2** to proceed smoothly. As shown in Scheme 1.10.1, the block–graft copolymer was prepared by a continuous three-step route. In the first step, the living poly(methyl methacrylate) [poly(MMA)] was prepared by the anionic polymerization of MMA. In the second step, the block copolymer of MMA and **2** [poly(MMA-*b*-**2**)] was obtained by the continuous addition of **2** to the above living poly(MMA) solution. In the last step, IBVE and toluene were continuously (without the separation of the block copolymer from the THF solution) introduced into the above system containing poly(MMA-*b*-**2**) and the graft copolymerization of IBVE was induced by adding the activator, $EtAlCl_2$. Every side chain in the poly(**2**) segment became an initiating site for the cationic polymerization of IBVE. THF, which was used as the solvent in the first and second steps of the preparation of poly(MMA-*b*-**2**), acted as a Lewis base in the last step, thus stabilizing the propagating carbocation. By this continuous transformation from anionic to cationic polymerization, a block–graft copolymer containing poly(MMA-*b*-**2**) as the backbone and poly(IBVE) as the side chains attached to the poly(**2**) segment was obtained.

Since the above three steps take place in a living manner, the polymerization degree in every step can be controlled by the amount of added monomer. When the polymerization degrees of MMA and IBVE (*n* and *r*, see **4** in Scheme 1.10.1) are large and that of *2* (*m*) is small, the poly(**2**) segment is very short and a star-shaped copolymer can be obtained, which contains one poly(MMA) arm and several poly(IBVE) arms (b, in Figure 1.10.1). For instance, if the polymerization degree of **2** is restricted to

$$\text{-}\!\!\left(CH_2\text{-}\underset{COOCH_3}{\overset{CH_3}{C}}\right)_{n-1}\!CH_2\underset{COOCH_3}{\overset{CH_3}{C}}{}^{-}Li^{+} \xrightarrow[\text{Anionic block copolymerization in THF}]{m\ CH_2{=}\overset{CH_3}{C}\text{—}\overset{O}{\overset{\|}{C}}OCH_2CH_2OCHO\overset{CH_3\ O}{\overset{\|}{C}}CH_3\ (2)}$$

(1) Living poly(MMA)

$$\text{-}\!\!\left(CH_2\text{-}\underset{\underset{OCH_3}{C=O}}{\overset{CH_3}{C}}\right)_{n}\!\!\left(CH_2\underset{\underset{OCH_2CH_2O\overset{CH_3}{C}HO\overset{O}{\overset{\|}{C}}CH_3}{C=O}}{\overset{CH_3}{C}}\right)_{m} \xrightarrow[\text{Cationic graft copolymerization}]{\text{IBVE, Toluene, }EtAlCl_2}$$

(3) Poly(MMA-b-2)

$$\left(CH_2\text{-}\underset{\underset{OCH_3}{C=O}}{\overset{CH_3}{C}}\right)_{n}\!\!\left(CH_2\underset{\underset{OCH_2CH_2O\overset{CH_3}{C}H\left(CH_2\text{-}\overset{OCH_2CH(CH_3)_2}{CH}\right)_r}{C=O}}{\overset{CH_3}{C}}\right)_{m}$$

(4) Block-graft or star-shaped copolymer

SCHEME 1.10.1

four ($m = 4$), the star-shaped copolymer will have five arms ($1 + m$), one poly(MMA) arm, and four poly(IBVE) arms. Of course, as for the block–graft copolymers, the arm number and the molecular weights of both the poly(MMA) arm and the poly(IBVE) arms are controllable and the star-shaped copolymer possesses a narrow molecular weight distribution.

1.10.2 EXPERIMENTAL SECTION

Materials. Tetrahydrofuran (THF) was dried with CaH_2 under reflux for more than 24 h, distilled, purified in the presence of naphthalene sodium, and finally distilled from a solution of (1,1-diphenylhexyl)lithium (DPHL) just before use. Toluene was washed with concentrated sulfuric acid and then with water, dried over $MgSO_4$, and distilled twice over CaH_2 before use. Hexane was first dried and distilled over CaH_2 and then distilled from a solution of *n*-BuLi. Methyl methacrylate (MMA) and isobutyl vinyl ether (IBVE; Aldrich, 99%) were washed with 10% aqueous sodium hydroxide solution and then with water, dried overnight with $MgSO_4$ and potassium hydroxide, respectively, and distilled twice over CaH_2 prior to polymerization.[17,19] 1,1-Diphenylethylene (DPE; Aldrich, 97%) was doubly distilled over CaH_2 and then distilled in the presence of (triphenylmethyl)lithium under reduced pressure.[20] Lithium chloride (Aldrich, 99.99%) was dried at 120°C for 24 h and dissolved in THF.[21] *n*-BuLi (Aldrich, 1.6 M solution in hexane) and $EtAlCl_2$ (Aldrich, 1.8 M solution in toluene) were diluted with purified hexane and toluene, respectively.

Syntheses of 2-(Vinyloxy)ethyl Methacrylate (VEMA) and 2-(1-Acetoxyethoxy) ethyl Methacrylate (2, in Scheme 1.10.1). VEMA was prepared through the reaction between 2-chloroethyl vinyl ether (Aldrich, 99%) and sodium methacrylate (Aldrich, 99%) under reflux with stirring, in the presence of small amounts of a phase-transfer catalyst (tetrabutylammonium iodide) and an inhibitor (4-*tert*-butylcatechol).[22] **2** was synthesized by the addition reaction between VEMA and acetic acid. In a 100 mL round-bottom flask equipped with a condenser and a magnetic stirrer, 21.0 g (0.14 mol) of VEMA and a small amount of 4-*tert*-butylcatechol were introduced under the protection of nitrogen. After 4-*tert*-butylcatechol dissolved and the temperature was raised to 70°C, acetic acid (8.1 g, 0.14 mol; Aldrich, 99.8%) was added dropwise with a syringe. The reaction was allowed to last 5 h at 70°C, and the crude product was distilled under reduced pressure (bp 65°C–66°C/1.0 Torr; yield 86%). Prior to polymerization, the monomer was doubly distilled over CaH_2. As shown later in Figure 1.10.5A, the chemical shifts and their intensities in the ^1H NMR spectrum of the prepared **2** are consistent with its molecular structure.

Preparation of the Block Macroinitiator by the Anionic Block Copolymerization of MMA and 2. All polymerizations, anionic block and cationic graft copolymerizations, were carried out in a 100-mL round-bottom glass flask under an overpressure of nitrogen with magnetic stirring. The anionic block copolymerization of MMA and **2** was performed in THF at −60°C in the presence of LiCl ($[LiCl]/[DPHL]_0 = 4$). After THF, DPE, and a THF solution of LiCl were added with dry syringes, the flask was cooled to −40°C and *n*-BuLi (in hexane) was introduced. The deep red color of DPHL appeared at once, and the reaction between *n*-BuLi and DPE was allowed to continue for 15 min. Then, the system was cooled to −60°C and the polymerization reaction was induced by the addition of a prechilled THF solution of MMA. After 30 min, the prechilled **2** was sequentially added and the block copolymerization was allowed to last 90 min. Without polymer separation, this THF solution was directly used in the next step for the preparation of block–graft or star-shaped copolymer.

Synthesis of Block–graft or Star-Shaped Copolymer by the Cationic Polymerization of IBVE to the Block Macroinitiator. After the anionic block copolymerization of MMA and **2**, purified toluene and IBVE were continuously added to the THF solution of poly(MMA-*b*-**2**). After the temperature was raised to 25°C, a toluene solution of $EtAlCl_2$ was introduced to start the graft copolymerization and the reaction was allowed to last 4 h. Then, the polymerization was terminated with an aqueous solution of ammonia. The quenched reaction mixture was washed with dilute hydrochloric acid and then with water, evaporated to dryness under reduced pressure, and vacuum-dried to obtain the polymer

Measurements. ^1H NMR spectra were recorded in $CDCl_3$ or CD_3OD on a VXR-400 spectrometer. M_n and M_w/M_n of the polymer were determined by gel permeation chromatography (GPC) on the basis of a polystyrene calibration curve. The GPC measurements were carried out using THF as the solvent, at 30°C, with a 1.0 mL/min flow rate and a 1.0 cm/min chart speed. Three polystyrene gel columns (Waters, 7.8 × 300 mm; two Linear, Part No. 10681, and one HR 4E, Part No. 44240) were used, which were connected to a Waters 515 precision pump.

1.10.3 RESULTS AND DISCUSSION

Preparation of the Block Macroinitiator by the Anionic Block Copolymerization of MMA and 2. The anionic polymerization technique is most suitable for the synthesis of block copolymers, since it provides a controlled molecular weight and composition, narrow molecular weight distribution, and well-defined chain structure.[23] The backbone of the block–graft copolymer prepared in this paper is an AB-type block copolymer consisting of poly(MMA) and poly(**2**) segments, which was prepared by the sequential anionic polymerization of MMA and **2**.

LiCl is often used as an additive in the anionic polymerization of the acrylic monomers,[24–26] because the formation of a μ-type complex between LiCl and the propagating site prevents the occurrence of side reactions. A polymer with a well-defined molecular architecture and narrow molecular weight distribution can thus be obtained. Both MMA and **2** used for the preparation of the backbone of the block–graft copolymer belong to the acrylic monomers. LiCl is, therefore, employed to prepare a well-defined block macroinitiator.

The initiator, DPHL, was prepared via the reaction of *n*-BuLi with DPE in the ratio $[DPE]/[n\text{-BuLi}]_0 = 1.2$ (Table 1.10.1), at −40°C, for 15 min. The anionic block copolymerization was carried out in THF at −60°C in the presence of LiCl ($[LiCl]/[n\text{-BuLi}]_0 = 4$) by the sequential polymerization of MMA and **2**. As shown in Table 1.10.1, the calculated number-average molecular weight of the block copolymer is close to the determined value and the molecular weight distribution is very narrow ($M_w/M_n = 1.05–1.13$). Figure 1.10.2 shows that the GPC peak (a) of the living poly(MMA) precursor ($M_n = 2640$, $M_w/M_n = 1.09$) prepared in the first step

TABLE 1.10.1
Preparation of the Block Macroinitiator (BMI) by the Anionic Block Copolymerization of MMA and 2[a]

No.	$[n\text{-BuLi}]_0$ (mM)	$[\text{MMA}]_0$ (M)	$[2]_0$ (M)	$10^{-3}M_n$ calcd	$10^{-3}M_n$ obsd[b]	M_w/M_n[b]	n/m[c]
BMI-1	28	0.67	0.31	5.04	4.86	1.07	23.1/10.7
BMI-2	19	0.67	0.31	7.39	7.77	1.05	37.7/17.4
BMI-3	14	0.78	0.36	11.3	12.5	1.13	61.3/28.4
BMI-4	26	0.60	0.093	3.27	3.21	1.07	22.3/3.4
BMI-5	17	0.67	0.083	5.23	5.46	1.06	41.1/5.1

[a] The initiator DPHL was first prepared by the reaction of n-BuLi with DPE ($[\text{DPE}]/[n\text{-BuLi}]_0 = 1.2$), at −40°C, for 15 min. The block copolymerization was carried out in THF at −60°C in the presence of LiCl ($[\text{LiCl}]/[\text{DPHL}]_0 = 4$). The polymerization times were 30 and 90 min for MMA and **2**, respectively. The yields of the block copolymers were quantitative in all cases.

[b] Determined by GPC.

[c] n and m: polymerization degrees of MMA (molecular weight: 100) and **2** (molecular weight: 216) in the block copolymer, respectively. The calculation method is explained in ref 31.

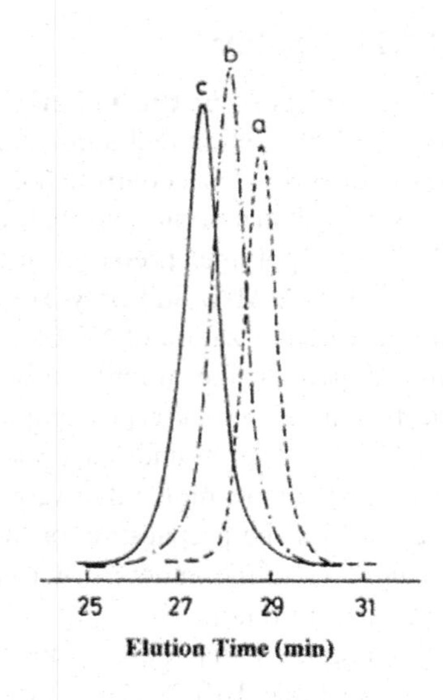

FIGURE 1.10.2 GPC traces for the continuous three-step synthesis of a block–graft copolymer. (a) Living poly(MMA) ($M_n = 2640$, $M_w/M_n = 1.09$) prepared in the first step by the anionic polymerization of MMA. (b) Block macroinitiator (poly(MMA-b-**2**), BMI-1 in Table 1.10.1, $M_n = 4860$, $M_w/M_n = 1.07$) prepared in the second step by the anionic block copolymerization of **2** starting with the above living poly(MMA). (c) Block–graft copolymer (BGP-1 in Table 1.10.2, $M_n = 6500$, $M_w/M_n = 1.12$) prepared in the last step by the cationic graft copolymerization of IBVE to the above block macroinitiator.

shifted toward a higher molecular weight (b) after the second-step polymerization of **2**, because of the formation of the diblock copolymer, poly(MMA-*b*-**2**) (BMI-1 in Table 1.10.1, M_n = 4860). In addition, no peak for the poly(MMA) precursor is present and the molecular weight distribution of the block copolymer is narrow (M_w/M_n = 1.07). This indicates that all the living sites of poly(MMA) were consumed in initiating the second-step polymerization of **2** and that the transformation of the anionic living site from poly(MMA) to poly(**2**) was faster than the propagation rate of **2**. For BMI-1, BMI-2, and BMI-3 (Table 1.10.1), the molecular weight of the poly(MMA) segment is almost equal to that of the poly(**2**) segment. However, the total molecular weight of the block copolymer increases as the ratio ($[MMA]_0$ + $[\mathbf{2}]_0$)/$[n\text{-BuLi}]_0$ becomes larger. Therefore, the polymerization degrees (*m*) of **2** are different. These block copolymers will be used as macroinitiators in the syntheses of the block–graft copolymers with various numbers of side chains. Compared to the above block copolymers, the polymerization degrees of **2** in BMI-4 and BMI-5 (Table 1.10.1) are small (*m* = 3.4 and 5.1, respectively). These block copolymers will be used as macroinitiators for the syntheses of the star-shaped copolymers.

Synthesis of the Block–Graft Copolymer by Cationic Graft Copolymerization of IBVE to the Block Macroinitiator. The living cationic polymerization of a number of vinyl monomers, such as alkyl vinyl ether, isobutylene, and styrene-type monomers, was carried out by a number of researchers.[27–30] Higashimura et al.[17–19] found that 1-(isobutoxy)ethyl acetate (IBEA) can be used as an initiator for the cationic polymerization of alkyl vinyl ether in the presence of a Lewis acid. The linkage between the acetoxy group and ethoxy group can be activated by the Lewis acid to generate a partly dissociated carbocation, which induces the cationic polymerization of alkyl vinyl ether. However, when a very strong Lewis acid, $EtAlCl_2$, was used, the cationic polymerization could not be controlled. In the latter case, the molecular weight distribution of the obtained polymer was broad, because the counteranion ($^{-}OCOCH_3 \cdots EtAlCl_2$) possesses a too weak nucleophilicity. However, when a weak Lewis base, such as THF, 1,4-dioxane, or ethyl acetate, was introduced into the above system, a living cationic polymerization could be achieved.[17,18] In this case, the propagating site was suitably stabilized by the added Lewis base.

Every side chain of the poly(**2**) segment in the poly(MMA-*b*-**2**) prepared above has a molecular structure (**3**, in Scheme 1.10.1) similar to that of the initiator, IBEA. In the presence of a Lewis acid, $EtAlCl_2$, every side chain of the poly(**2**) segment will become an initiating site for the cationic polymerization of IBVE. In addition, the THF, which was used as the solvent in the first and second polymerization steps for the preparation of the block macroinitiator, can act as a Lewis base in the last polymerization step for the synthesis of the block–graft copolymer. These conditions provide the possibility of preparing a block–graft copolymer by a continuous synthetic route.

After the anionic block copolymerization of MMA and **2**, the prepared block macroinitiator, poly(MMA-*b*-**2**), was not separated from its THF solution, to which purified toluene and IBVE were added. After the system acquired the room temperature (25°C), $EtAlCl_2$ (in toluene) was added to induce the cationic graft copolymerization of IBVE and the reaction was allowed to last 4 h (Table 1.10.2). As shown in Figure 1.10.2, the GPC peak (b) of the block macroinitiator (BMI-1 in Table 1.10.1,

TABLE 1.10.2
Synthesis of the Block–Graft Copolymer (BGP) by Cationic Graft Copolymerization of IBVE to the Block Macroinitiator[a]

No.	Macroinitiator[b]	$[I]_0$[c] (mM)	$[IBVE]_0$ (M)	N[d]	M_k[e]	$10^{-4}M_{n(graft)}$ calcd[f]	$10^{-4}M_{n(graft)}$ obsd[g]	M_w/M_n[g]
BGP-1	BMI-1	147	0.32	10.7	218	0.67	0.65	1.12
BGP-2	BMI-2	84	0.28	17.4	333	1.28	1.30	1.12
BGP-3[h]	BMI-2	81	0.54	17.4	667	1.86	1.81	1.11
BGP-4	BMI-2	84	0.84	17.4	1000	2.44	2.11	1.14
BGP-5	BMI-3	138	0.60	28.4	435	2.37	1.88	1.16

[a] The graft copolymerization was carried out using $EtAlCl_2$ as the activator ($[EtAlCl_2]$ = 0.20 M) and THF as a weak Lewis base ([THF] = 2.6 M), in toluene, at 25°C, for 4 h. The conversion was 100% in each case.

[b] See Table 1.10.1.

[c] Concentration of the repeating unit of **2** in the block macroinitiator.

[d] The number of side chains (N) is equal to the polymerization degree (m) of **2** in the block macroinitiator.

[e] Calculated molecular weight of a poly(IBVE) side chain, $M_k = M\text{IBVE}([\text{IBVE}]_0/[\text{I}]_0)$.

[f] The calculation method is explained in ref 31.

[g] Determined by GPC.

[h] Two times monomer addition experiment. After the cationic polymerization of IBVE in BGP-2 had proceeded for 2 h, an equal amount of fresh IBVE was added and the polymerization was allowed for an additional 2 h.

M_n = 4860, M_w/M_n = 1.07) disappears entirely after the graft copolymerization. A new peak (c) due to the block–graft copolymer (BGP-1 in Table 1.10.2) emerges and no peak belonging to the block macroinitiator remains. This block–graft copolymer possesses a narrow molecular weight distribution (M_w/M_n = 1.12); its graft number is 10.7 (the polymerization degree of **2**: m = 10.7; see Table 1.10.1 and ref. 31), and the molecular weight of each poly(IBVE) side chain is 218 (Table 1.10.2). The determined total molecular weight of the block–graft copolymer (M_n = 6500) is in good agreement with the calculated value ($M_{n(calcd)}$ = 6700). The above results indicate that all side chains of the poly(**2**) segment in BMI-1 were consumed in initiating the cationic polymerization of IBVE and that the side chains have an almost equal length.

The possibility of molecular control of the side chains was investigated by performing the following parallel experiments. A block macroinitiator (BMI-2) was first prepared. Its THF solution was equally divided between two flasks. The graft copolymerization in the two flasks was carried out under the same conditions but with different amounts of IBVE. In the preparation of BGP-2 and BGP-4 (Table 1.10.2), the initial concentrations of IBVE were 0.28 and 0.84 M and the designed molecular weights of the side chains were 333 and 1000, respectively. As shown in Figure 1.10.3, the GPC peaks of BGP-2 (B) and BGP-4 (C) after graft copolymerization appear at

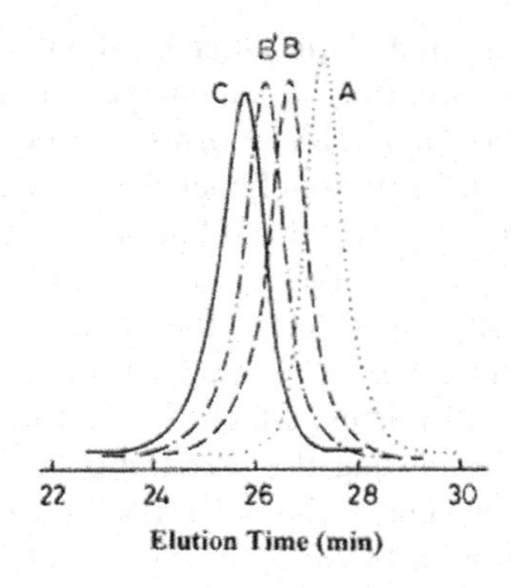

FIGURE 1.10.3 GPC traces for the molecular control of the block–graft copolymers. (A) Block macroinitiator (poly(MMA-*b*-**2**), BMI-2 in Table 1.10.1, M_n = 7770, M_w/M_n = 1.05) prepared by the anionic block copolymerization of MMA and **2**. (B) Block–graft copolymer (BGP-2 in Table 1.10.2, $M_n = 1.30 \times 10^4$, M_w/M_n = 1.12) prepared by the cationic graft copolymerization of IBVE to the above block macroinitiator (BMI-2) using a ratio of $[IBVE]_0/[I]_0$ = 3.33. (B′) BGP-3 (Table 1.10.2, $M_n = 1.81 \times 10^4$, M_w/M_n = 1.11) prepared by adding the same amount of IBVE to the above complete polymerization system of B. (C) BGP-4 (Table 1.10.2, $M_n = 2.11 \times 10^4$, M_w/M_n = 1.14). The amount of IBVE was 3 times as large as that of B ($[IBVE]_0/[I]_0$ = 10).

higher molecular weights than that of the block macroinitiator (A, M_n = 7770, M_w/M_n = 1.05) and BGP-2 and BGP-4 have almost the same narrow molecular weight distribution (M_w/M_n = 1.12 and 1.14, respectively). However, the molecular weight of BGP-4 ($M_n = 2.11 \times 10^4$) is obviously larger than that of BGP-2 ($M_n = 1.30 \times 10^4$).

The living characteristic of the cationic graft copolymerization of IBVE to the block macroinitiator was confirmed by a two times monomer addition experiment. After the cationic polymerization of IBVE during the preparation of BGP-2 has proceeded for 2 h, an equal amount of fresh IBVE was added and the polymerization was allowed to last an additional 2 h. As shown in Figure 1.10.3, a new peak (B′; BGP-3 in Table 1.10.2, $M_n = 1.81 \times 10^4$) between peaks B (BGP-2) and C (BGP-4) appears. However, its molecular weight distribution (M_w/M_n = 1.11) is as narrow as those of BGP-2 and BGP-4. BGP-2, BGP-3, and BGP-4 possess the same number of side chains, because the same block macroinitiator was employed (BMI-2; average polymerization degree of **2**, m = 17.4). However, their side-chain lengths are different, resulting in different total molecular weights. As shown in Table 1.10.2, as the side chain becomes longer (from BGP-2 to BGP-4), the determined molecular weight of the block–graft copolymer increases. They are close to the calculated values when the side chains are short and/or the graft number is small (BGP-1 to BGP-3). However, as expected, they are smaller than the calculated values when the side chains are longer (BGP-4) and/or the graft number is large (BGP-5). All the above results indicate that every side chain of the poly(**2**) segment of the block macroinitiator is a living site for the cationic graft copolymerization of IBVE, that the polymerization proceeds in a living manner, and that the molecular weight of each side chain can be controlled as well as that of the block backbone.

Synthesis of the Star-Shaped Copolymer by the Cationic Polymerization of IBVE to the Block Macroinitiator. A star-shaped copolymer consisting of one poly(MMA) arm and several poly(IBVE) arms can be prepared in a way similar to that used for the block–graft copolymer. However, the block macroinitiator should possess a long chain of poly(MMA) but only a few repeating units of **2** at one of its ends. The side chains of these repeating units of **2** are the living sites for the cationic polymerization of IBVE in the presence of $EtAlCl_2$. In this manner, a star-shaped copolymer can be generated. The length of the poly(MMA) arm and the number of poly(IBVE) arms, which is equal to the number of repeating units of **2**, can be controlled during the preparation of the block macroinitiator. In addition, the molecular weight of the poly(IBVE) arms can also be controlled. For instance, in the preparation of SP-2 and SP-4 (Table 1.10.3), the same block macroinitiator (BMI-5 of Table 1.10.1) was employed so that their arm number is the same ($N = 1 + m = 6.1$). However, the amount of IBVE used for SP-4 was 3 times larger than that used for SP-2. As shown in Figure 1.10.4, the molecular weight of SP-4 (peak C, $M_n = 2.82 \times 10^4$, $M_w/M_n = 1.16$) is larger than that of SP-2 (peak B, $M_n = 1.25 \times 10^4$, $M_w/M_n = 1.10$). Furthermore, as in the preparation of the block–graft copolymer, when the same amount of fresh IBVE was added to the completely polymerized SP-2, the propagating reaction took again place and a star-shaped copolymer with a larger molecular weight than that of SP-2 was obtained (SP-3 in Table 1.10.3, peak B′ in Figure 1.10.4, $M_n = 1.78 \times 10^4$, $M_w/M_n = 1.20$). As shown in Table 1.10.3, although the arm number and the molecular weight of the poly(MMA) arm ($M_n = 4110$) in SP-2, SP-3, and SP-4 are the same, the total molecular weights of the star-shaped copolymers are different, because their side-chain lengths are different. The molecular weight of the poly(IBVE) arm is smaller

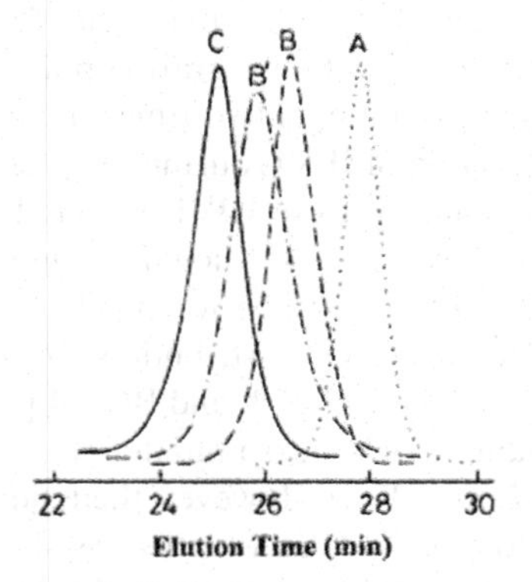

FIGURE 1.10.4 GPC traces for the molecular control of star-shaped copolymers. (A) Block macroinitiator (poly(MMA-*b*-**2**), BMI-5 in Table 1.10.1, $M_n = 5460$, $M_w/M_n = 1.06$) prepared by the anionic block copolymerization of MMA and **2**. (B) Star-shaped copolymer (SP-2 in Table 1.10.3, $M_n = 1.25 \times 10^4$, $M_w/M_n = 1.10$) prepared by the cationic polymerization of IBVE to the above block macroinitiator (BMI-5), using a ratio of $[IBVE]_0/[I]_0 = 14.2$. (B′) SP-3 (Table 1.10.3, $M_n = 1.78 \times 10^4$, $M_w/M_n = 1.20$) prepared by adding the same amount of IBVE to the above complete polymerization system of B. (C) SP-4 (Table 1.10.3, $M_n = 2.82 \times 10^4$, $M_w/M_n = 1.16$). The amount of IBVE was about 3 times as large as that of B ($[IBVE]_0/[I]_0 = 46.4$).

TABLE 1.10.3
Synthesis of the Star-Shaped Copolymer (SP) by the Cationic Polymerization of IBVE to the Block Macroinitiator[a]

No.	Macroinitiator[b]	$[I]_0$[c] (mM)	$[IBVE]_0$ (M)	N[d]	M_k[e]	$10^{-4}M_{n\ (star)}$ calcd[f]	$10^{-4}M_{n\ (star)}$ obsd[g]	M_w/M_n[g]
SP-1	BMI-4	34	0.74	4.4	2180	1.05	1.00	1.19
SP-2	BMI-5	24	0.34	6.1	1420	1.25	1.25	1.10
SP-3[h]	BMI-5	23	0.65	6.1	2830	1.97	1.78	1.20
SP-4	BMI-5	22	1.02	6.1	4640	2.89	2.82	1.16

[a] The cationic polymerization of IBVE was carried out using $EtAlCl_2$ as the activator ($[EtAlCl_2]$=0.20 M) and THF as a weak Lewis base ([THF]=2.8 M), in toluene, at 25°C, for 4 h. The conversion was 100% in each case.

[b] See Table 1.10.1.

[c] Concentration of the repeating unit of **2** in the block macroinitiator.

[d] Total arm number of star-shaped copolymer $N = m + 1$, where m is the polymerization degree of **2** in the block macroinitiator.

[e] Calculated molecular weight of a poly(IBVE) arm, $M_k = M_{IBVE}([IBVE]_0/[I]_0)$.

[f] The calculation method is explained in ref. 31.

[g] Determined by GPC.

[h] Two times monomer addition experiment. After the cationic polymerization of IBVE in SP-2 had proceeded for 2.5 h, an equal amount of fresh IBVE was added and the polymerization was allowed for an additional 2 h.

than that of the poly(MMA) arm in SP-2 and SP-3 but is larger in SP-4. As the molecular weight of the poly(IBVE) arm becomes larger (from SP-2 to SP-4), the total molecular weight of the star-shaped copolymer increases. These molecular weights are near the calculated values but, as expected, smaller. Therefore, the cationic polymerization of IBVE in the preparation of the star-shaped copolymers also proceeds in a living manner, and the molecular weight of the poly(IBVE) arm can be controlled as well as that of the block backbone.

^{1}H NMR Spectra of the Block–Graft and Star-Shaped Copolymers. Figure 1.10.5 depicts the ^{1}H NMR spectra of **2** (A), the block–graft copolymer (B, BGP-1), and the star-shaped copolymer (C, SP-1). In the ^{1}H NMR spectrum (B) of BGP-1, the components are in equal amounts and the absorption characteristics of each of the compounds can be easily detected. Indeed, peaks **c**, **e**, and **p** correspond, respectively, to the methyl in the ester group of the poly(MMA) segment, the methylene in the ester group of poly(**2**) segment, and the two methyls in the poly(IBVE) side chains. Compared to the above block–graft copolymer, in the ^{1}H NMR spectrum of the star-shaped copolymer (C), the absorptions corresponding to the poly(**2**) segment can hardly be observed, because the weight ratio of the three components is MMA:**2**:IBVE = 3:1:10. This copolymer can be considered to be a star-shaped copolymer consisting of just two components, the poly(MMA) arm and the poly(IBVE) arms.

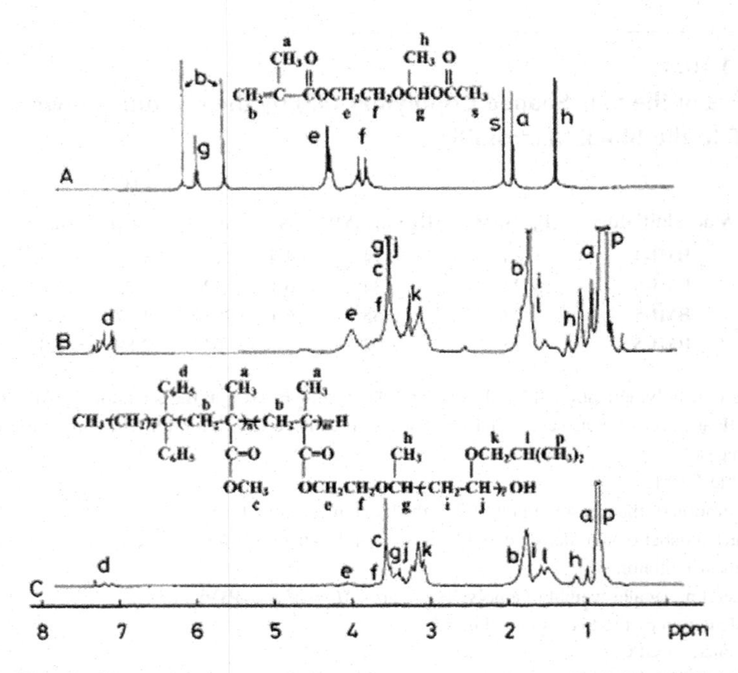

FIGURE 1.10.5 ^{1}H NMR spectra of **2** (A), block–graft copolymer (B, BGP-1 in Table 1.10.2, see the molecular structure in C), and star-shaped copolymer (C, SP-1 in Table 1.10.3).

1.10.4 CONCLUSION

The diblock copolymer of MMA and 2-(1-acetoxy-ethoxy)ethyl methacrylate (**2**) [poly(MMA-*b*-**2**)] can be prepared by the sequential anionic block copolymerization of MMA and **2**, in the presence of LiCl, using (1,1-diphenylhexyl)lithium as the initiator, in THF, at −60°C. This block copolymer has a controllable molecular weight and composition and a narrow molecular weight distribution (M_w/M_n = 1.05–1.13).

A block–graft copolymer consisting of a poly(MMA-*b*-**2**) block backbone and poly(isobutyl vinyl ether) [poly(IBVE)] side chains attached to the poly(**2**) segment can be prepared by a continuous transformation from anionic to cationic polymerization. After the anionic block copolymerization of MMA and **2** in THF, toluene, IBVE, and $EtAlCl_2$ were introduced into the THF solution of poly(MMA-*b*-**2**) at room temperature. Under these conditions, every side chain of the poly(**2**) segment in poly(MMA-*b*-**2**) can be activated by $EtAlCl_2$ to initiate the living cationic polymerization of IBVE. In this manner, a block–graft copolymer with designed graft number, controlled molecular weights of both the backbone and side chains, and narrow molecular weight distribution (M_w/M_n = 1.11–1.16) can be obtained.

Using a similar procedure, a star-shaped copolymer consisting of one poly(MMA) arm and several poly(IBVE) arms can be prepared, by restricting the polymerization degree of **2** to a small value. This star-shaped copolymer also possesses a designed arm number, controllable molecular weight, and narrow molecular weight distribution (M_w/M_n = 1.10–1.20).

REFERENCES

1. Rempp, P.; Franta, E.; Herz, J. *Anionic Polymerization*; McGrath, J. E., Ed.; ACS Symposium Series 166; America Chemical Society: Washington, DC, 1981.
2. Sawamoto, M. *Int. J. Polym. Mater.* **1991**, *15*, 197.
3. Lubnin, A. V.; Kennedy, J. P. *J. Macromol. Sci, Pure Appl. Chem.* **1994**, *A31*, 1943.
4. Hashimoto, K.; Shinoda, H.; Okada, M.; Sumitomo, H. *Polym. J.* **1990**, 22, 312.
5. Geetha, B.; Mandal, A. B.; Ramasami, T. *Macromolecules 26*, 4083.
6. Oshea, M. S.; George, G. A. *Polymer* **1994**, *35*, 4190.
7. Hegazy, E. S. A.; Dessouki, A. M.; Elsawy, N. M.; Elghaffar, M. A. A. *J. Polym. Sci., Polym. Chem.* **1993**, *31*, 527.
8. Osman, M. B. S.; Hegazy, E. A.; Mostafa, A. E. K. B.; Abdelmaksoud, A. M. *Polym. Int.* **1995**, *36*, 47.
9. Yang, J. M.; Hsiue, G. H. *J. Biomed. Mater. Res.* **1996**, *31*, 281.
10. Grutke, S.; Hurley, J. H.; Risse, W. *Macromol. Chem. Phys. 195*, 2875.
11. Eisenbach, C. D.; Heinemann, T. *Macromolecules* **1995**, *28*, 2133.
12. Yamashita, K.; Ito, K.; Tsuboi, H.; Takahama, S.; Tsuda, K. *J. Appl. Polym. Sci.* **1990**, *40*, 1445.
13. Ruckenstein, E.; Zhang, H. M. *Macromolecules* **1997**, *30*, 6852.
14. Se, K.; Watanabe, O.; Isono, Y.; Fujimoto, T. *Makromol. Chem., Macromol. Symp.* **1989**, *25*, 249.
15. Se, K.; Yamazaki, H.; Shibamoto, T.; Takano, A.; Fujimoto, T. *Macromolecules* **1997**, *30*, 1570.
16. Zhang, H. M.; Ruckenstein, E. *Macromolecules* **1998**, *31*, 746.
17. Aoshima, S.; Higashimura, T. *Macromolecules* **1989**, 22, 1009.
18. Kishimoto, Y.; Aoshima, S.; Higashimura, T. *Macromolecules* **1989**, *22*, 3877.
19. Kamigaito, M.; Sawamoto, M.; Higashimura, T. *Macromolecules* **1991**, *24*, 3988.
20. Zhang, H. M.; Ruckenstein, E. *J. Polym. Sci., Polym. Chem.* **1997**, *35*, 2901.
21. Varshney, S. K.; Hautekeer, J. P.; Fayt, R.; Jerome, R.; Teyssie, Ph. *Macromolecules* **1990**, *23*, 2618.
22. Aoshima, S.; Hasegawa, O.; Higashimura, T. *Polym. Bull.* **1985**, *13*, 229.
23. Norhay, A.; McGrath, J. E. *Block Copolymers*, Academic Press: New York, 1977.
24. Fayt, R.; Forte, R.; Jacobs, C.; Jerome, R.; Ouhadi, T.; Teyssie, Ph.; Varshney, S. K. *Macromolecules* **1987**, *20*, 1442.
25. Varshney, S. K.; Jerome, R.; Bayard, P.; Jacobs, C.; Fayt, R.; Teyssie, Ph. *Macromolecules* **1992**, *25*, 4457.
26. Kunkel, D.; Muller, A. H. E.; Janata, M.; Lochman, L. *Makromol. Chem., Macromol. Symp.* **1992**, *60*, 315.
27. Sawamoto, M. *Prog. Polym. Sci.* **1991**, *16*, 111.
28. Majoros, I.; Nagy, A.; Kennedy, J. P. *Adv. Polym. Sci.* **1994**, *112*, 1.
29. *Fodor, Zs.; Faust, R. J. Macromol. Sci.; Pure Appl. Chem.* **1995**, *A32*, 575.
30. Higashimura, T.; Ishihama, Y.; Sawamoto, M. *Macromolecules* **1993**, *26*, 744.

31. (a) The polymerization degrees of MMA (n) and **2** (m) were calculated as follows: $n = [(M_{n(block)} - M_{initiator}) W_{MMA}/(W_2 + W_{MMA})]/M_{MMA}$ and $m = [(M_{n(block)} - M_{initiator}) W_2/(W_2 + W_{MMA})]/M_2$, where $M_{n(block)}$ is the molecular weight of the block copolymer, poly(MMA-*b*-**2**), determined by GPC; $M_{initiator}$ is the molecular weight of the initiator moiety (1,1-diphenyl-hexyl); W_{MMA} and W_2 are the weights of the monomers, MMA and **2**, respectively; and M_{MMA} and M_2 are the molecular weights of the two monomers, MMA and **2**, respectively. (b) The molecular weight (M_k) of each poly(IBVE) side chain (or arm) was calculated as follows: $M_k = M_{IBVE}([IBVE]_0/[I]_0)$, where M_{IBVE} is the monomer molecular weight of IBVE, $[IBVE]_0$ is the initial concentration of IBVE, and $[I]_0$ is the concentration of the repeating unit of **2** in the block macroinitiator, poly(MMA-*b*-**2**). (c) The total molecular weight of the block–graft or star-shaped copolymer was calculated as follows: $M_n = M_{n(block)} + mM_k - 59m + 17m = M_{n(block)} + m(M_k - 42)$. After the cationic graft copolymerization of IBVE, the acetoxy group ($^-OCOCH_3$, molecular weight = 59) at each end of the side chain (or arm end) was replaced by a hydroxyl group (OH^-, molecular weight = 17).

2 Degradable and De-Cross-Linkable Polymers

CONTENTS

Degradable and de-cross-linkable polymers are an important category of polymeric materials that can be converted to small molecules or low-MW polymeric species after fulfilling the designed applications. Because conventional non-degradable polymers have generated severe environmental issues, the development of degradable and de-cross-linkable polymers represents a significant direction of polymer research. According to structure-property relationship, each of these polymers possesses cleavable functional groups that are responsible for their specific degradable and de-cross-linkable properties. Therefore, it is critical to control the cleavable chemical structures of these polymers through innovative synthetic approaches. This chapter describes the synthesis and characterization of four types of degradable and de-cross-linkable polymers. Each of them has a different set of cleavable functionalities, and is associated with a different preparation method.

The first type of polymer, poly(L-lactide-*co*-trimethylene carbonate), possess in-chain ester and carbonate groups and is prepared by copolymerization approach (Section 2.1). Aliphatic polyesters with hydrolysable in-chain ester groups are an essential class of degradable polymers. Although poly(L-lactide) is one of the most commonly used types of aliphatic polyesters, chemical modification of poly(L-lactide) by introducing carbonate in-chain groups can tune its mechanical and degradable properties and may expand its applicability.

The second type of polymer, polyacetal, has in-chain acetal groups and is synthesized by self-polyaddition of hydroxyalkyl vinyl ethers (Section 2.2). Because acetal groups are acid-sensitive, polyacetal can undergo quick cleavage under acidic conditions. It should be noted that linear polyacetal can also be prepared by polyaddition of divinyloxyl compounds with diol to novel degradable polymers. Moreover, de-cross-linkable acetal-containing polymer network can be obtained by addition reactions of two precursor polymers with hydroxyl groups and vinyloxyl groups, respectively.

The third type of polymer gains degradable and de-cross-linkable properties through dicyclopentadiene (DCPD) groups, which can undergo thermally reversible cleavage to form cyclopentadiene groups upon heating via the reversible Diels-Alder reaction (Section 2.3). The DCPD groups are introduced to linear oligomers or cross-linked polymer networks by the reactions of potassium dicyclopentadienedicarboxylate with dihalides or halide-containing polymers.

The fourth type of polymer, as quaternary ammonium-containing crosslinked networks, has thermally reversible de-cross-linking properties based on the dequaternization of quaternary ammoniums (Section 2.4). These networks are prepared by crosslinking reactions of chlorine or tertiary amine-containing polymers with ditertiary amines and dihalide compounds, respectively.

2.1 Molten Ring-Open Copolymerization of L-Lactide and Cyclic Trimethylene Carbonate*

Eli Ruckenstein and Yumin Yuan

Department of Chemical Engineering, State University of New York at Buffalo, Clifford C. Furnus Hall, P.O. Box 604200, Buffalo, New York 14260, USA

ABSTRACT The biodegradable poly(L-lactide-*co*-trimethylene carbonate) [poly(LLA-*co*-TMC)] was synthesized by molten ring-open copolymerization and the structure and properties of the copolymer were investigated by gel permeation chromatography (GPC), NMR, FTIR, differential scanning calorimeter (DSC), and Instron testing. The copolymer with a higher L-lactide (LLA) content had a higher initial molecular weight (M_n), but a lower final M_n. In contrast, the copolymer with a higher trimethylene carbonate (TMC) content had a lower initial M_n, but its M_n increased with increasing reaction time and TMC content. The glass transition temperature (T_g), the melting temperature (T_m) and the crystallinity of the copolymer decreased with increasing TMC content. The elongation of the copolymer significantly increased with increasing TMC content while the toughness passed through a maximum.

2.1.1 INTRODUCTION

The modification of the properties of the brittle biodegradable polyesters such as poly (hydroxyal kanonate), polylactide (PLA), and polyglycolide, has been intensively investigated.[1–5] Usually a biodegradable or biocompatible rubber was introduced to toughen the brittle aliphatic polyester, so that the modified material remained biodegradable or biocompatible. The modifications could be achieved by both physical blending and chemical copolymerization. Owing to the high crystallinity of the aliphatic polyesters, few miscible blends with a biodegradable or biocompatible rubber have been reported. Poly(3-hydroxybutyrate)/poly(ethylene oxide) is one of them.[6] In contrast, the modification via copolymerization provided a number of

* *Journal of Applied Polymer Science*, 1998, 69, 1429–1434.

advantages because the architecture of the biomaterial can be tailor made by random or sequence copolymerization (anionic or complex). Thus, a random, block or star copolymer can be synthesized. The copolymer properties such as glass transition temperature, melting temperature, crystallinity, toughness, and biodegradability can be easily controlled by varying the monomer composition in the feed, such as to meet different needs. Generally, the elongation and toughness of the brittle polyester can be significantly improved by incorporating rubber moieties in the copolymer.[7–10]

The aliphatic polycarbonate, such as poly(trimethylene carbonate) (PTMC) rubber, can be hydrolyzed both *in vivo* and *in vitro*, and hence, possesses potential applications in biomedical and environmental areas.[11–13] The introduction of carbonate linkages into a polyester constitutes an additional route for the improvement of the performance of polyesters. Because the carbonate linkage is more hydrophobic than the ester one, the copolymer is expected to be more stable to hydrolysis *in vitro* and, therefore, to have a much longer shelf life than PLA. However, because the hydrolysis of carbonate is faster *in vivo*,[14] the copolymer can be used as a bioabsorbable material. Indeed, poly(glycolide-*co*-TMC) was used as a flexible, strong, and absorbable monofilament suture.[4] Recently, Cai and Zhu[15] reported the synthesis of an amorphous poly(*D*, L-lactide-*co*-trimethylene carbonate) and evaluated its biodegradability both *in vivo* and *in vitro*.

In this article, a semicrystalline biodegradable copolymer poly(L-lactide-*co*-trimethylene carbonate) [poly(LLA-*co*-TMC)] was prepared by the molten ring-open copolymerization. The effects of the trimethylene carbonate (TMC) content on the molecular weight, thermal, and mechanical properties of the copolymer were investigated.

2.1.2 EXPERIMENTAL SECTION

2.1.2.1 Materials

L-Lactide was purified by recrystallization from dry ethyl acetate and dried in a vacuum oven at room temperature for 24 h. The catalyst stannous octoate [$Sn(Oct)_2$, Sigma, St. Louis, MO] was used as received. All other chemicals were purchased from Aldrich (Milwaukee, WI) and were purified by the standard methods.

2.1.2.2 Monomer Synthesis

The monomer 1,3-dioxan-2-one or cyclic trimethylene carbonate (TMC) was synthesized as follows[16]: 20 g (0.263 mol) 1,3-propanol, and 57.1 g (0.526 mol) ethyl chloroformate were dissolved in 300 mL tetrahydrofuran (THF) at 0°C. Subsequently, 56.2 g (0.552 mol) triethylamine (TEA) was added dropwise to the flask within 30 min, and the mixture was stirred for 2 additional h at 0°C. The precipitated TEA·HCl salt was removed by filtration and the excess THF distilled under reduced pressure. The residue was crystallized and recrystallized from the mixture of THF-ether two to six times, and the obtained white TMC crystal was dried under vacuum at room temperature before use. The structure and purity of TMC were confirmed by ^{1}H-NMR (2.10 ppm, m, 2H, CCH_2C; 4.44 ppm, t, 4H, CCH_2O).

2.1.2.3 Polymerization

The molten ring-open polymerization was carried out in a flask equipped with a magnetic stirrer bar, using $Sn(Oct)_2$, as catalyst.[17] Selected amounts of monomers (total 3 g) and catalyst (for details see Tables 2.1.1 and 2.1.2) were introduced into a flask inside a glove box filled with nitrogen and the reactants were degassed in vacuum. Then the flask was filled with nitrogen again, sealed, immersed into an oil

TABLE 2.1.1
Effect of Reaction Time on M_n and M_w/M_n of Poly(LLA-*co*-TMC)[a]

Monomer Feed LLA/TMC mol/mol	Time h	Molecular Weight (M_n) g/mol	M_w/M_n
90/10	12	61,800	1.29
90/10	24	44,000	1.63
90/10	48	65,400	1.49
90/10	96	20,000	3.33
70/30	12	26,200	2.32
70/30	24	22,690	1.93
70/30	48	27,430	2.43
70/30	91	41,500	1.95

[a] The molar ratio of monomer to catalyst is 500/1.

TABLE 2.1.2
Effect of Monomer Composition on the M_n and M_w/M_n[a]

LLA/TMC mol/ mol (in the feed)	Time h	TMC Content mol % (calc.) (in copolymer)	TMC wt% Calc.	TMC Content wt% (found) (by ^{1}H-NMR)	M_n g/mol	M_w/M_n
90/10	24	5.3	7.3	—	44,000	1.63
70/30	24	11.1	15.0	—	22,690	1.93
0/100	24	100	100	100	8,850	1.76[c]
100/0	96	0	0	—	[b]	—
90/10	96	5.3	7.3	7.8	20,000	3.33
80/20	95	11.1	15.0	14.2	37,840	1.54
70/30	91	17.7	23.3	23.7	41,500	1.95
70/30	94	17.7	23.3	18.1	9,740[d]	1.85
60/40	96	25.0	32.1	—	48,000	1.86
50/50	95	33.3	41.5	—	21,000	1.78

[a] The molar ratio of monomer to catalyst is 500/1.
[b] PLLA is not soluble in THF.
[c] The molar ratio of monomer to catalyst is about 100/1.
[d] TMC was recrystallized twice, in all the other experiments the TMC was recrystallized six times.

bath at 110°C, and kept under magnetic stirring until the viscosity became very high. After 12–96 h of reaction, the polymer was removed from the flask by its dissolution in chloroform, purified by precipitation in hexane, washed with hexane two to three times and, finally, dried under reduced pressure.

2.1.2.4 Characterization

2.1.2.4.1 NMR Analysis

The ^{1}H-NMR analysis of the poly(LLA-*co*-TMC) samples was carried out with a GEMINI-300 NMR spectrometer using $CDCl_3$ as the solvent. The copolymer composition was calculated using the ^{1}H-NMR spectrum.

2.1.2.4.2 FTIR Analysis

The FTIR analysis was performed with a Perkin-Elmer 1760X FTIR instrument. The samples were dissolved in chloroform and cast onto NaCl plates. The scanning number was 32 and the resolution 2 cm^{-1}.

2.1.2.4.3 Molecular Weight Determination

The molecular weight data were obtained by gel permeation chromatography (GPC, Waters) at 30°C, using THF as eluent at a flow rate of 1 mL/min. Polystyrene standards were used to calibrate the molecular weight.

2.1.2.4.4 Thermal Analysis

The thermal behavior of the polymers was investigated using a DuPont 910 differential scanning calorimeter (DSC), under a nitrogen atmosphere. The samples were scanned from room temperature to 180°C, at a heating rate of 10°C/min, hold at 180°C for 1 min, and then quenched into liquid nitrogen. The samples were rescanned from −10°C or −50°C to 180°C (depending on the T_g of copolymer or homopolymer) at a heating rate of 10°C/min. The data were collected during the second scanning. The glass transition temperature (T_g) was taken as the midpoint of the heat capacity change; the cool crystallization temperature and the melting point were provided by the exo- and endo-thermal peaks, respectively.

2.1.2.4.5 Mechanical Properties Determination

Thin films of PLLA and copolymers were prepared by hot pressing at a temperature above their melting points and by cooling down quickly to about 5°C–10°C. The obtained films (0.10–0.15-mm thickness) were cut into dumbbell-shaped forms, as indicated by ASTM D.638-58T. The tensile testing was performed at room temperature, using an Instron Universal Testing Instrument (Model 1000), with an elongation rate of 10 mm/min. The yield and tensile strengths were calculated on the basis on the initial cross-section area of the specimen. The toughness (MJ/m^3), defined as the energy needed to break a sample of unit area and unit length, is given by the area under the stress-strain curve.

2.1.3 RESULTS AND DISCUSSION

The ring-open copolymerization of L-lactide and trimethylene carbonate was carried out in the molten state at 110°C for 0.5–4 days. The effect of the reaction time on the number-average molecular weight (M_n) of the copolymers obtained is presented in Table 2.1.1. Three general trends can be noted: (a) the M_n of poly(LLA-*co*-7.3 wt% TMC) (high LLA content) is higher than that of poly(LLA-*co*-23 wt% TMC) (high TMC content) for the first 48 h; (b) the M_n of poly(LLA-*co*-7.3 wt% TMC) becomes, however, lower than that of poly(LLA-*co*-23 wt%) after 4 days of reaction; (c) the M_n does not change appreciably between 12 and 48 h of reaction. This happens because: (1) the reaction rate of LLA is faster than that of cyclic TMC; (2) the concentration of LLA is higher in LLA-7.3 wt% TMC than that in LLA-23 wt% TMC; and (3) the depolymerization reaction due to thermal degradation becomes significant only after a sufficient long time of heating at 110°C and the degradation affects more strongly the polymers with a higher content of LLA than those with a lower content. For these reasons, an optimum reaction time for a maximum M_n could be identified. The time is longer for higher TMC content; it is less than 12 h for the poly(LLA-*co*-7.3 wt% TMC) and greater than 48 h for the poly(LLA-*co*-23 wt% TMC). The degradation is associated with an increase in the ratio M_w/M_n; hence, an increase in the broadness of the molecular weight distribution.

The effect of the monomer ratio on M_n of the copolymer is presented in Table 2.1.2. The L-lactide content in the copolymer was calculated based on the ^{1}H-NMR spectra, and was found to be close to the monomer feed composition. Several factors affect, for a given catalyst concentration, the molecular weight of copolymer: (a) the purities of the reactants; and (b) the feed composition. The purity of TMC had a critical effect on the M_n of the copolymer. For instance, the TMC recrystallized twice from a THF-ether mixture possessed some impurities, detected by NMR, and a low molecular weight (less than 10^4) copolymer was obtained. In contrast, when the TMC was recrystallized six times, much less impurities were detected by NMR and a copolymer with a higher molecular weight (4×10^4) was obtained. This happens because impurities, such as traces of water and various functional groups, can initiate polymerization, and the larger the number of the initiation centers, the lower the M_n. For molar ratios LLA/TMC greater than 1.5 and a reaction time of 4 days, one can see that M_n increases with increasing TMC content. This happens because less depolymerization takes place. However, for a molar ratio of 1 and a reaction time of 4 days, M_n is smaller than for a molar ratio of 1.5, because of the slower reaction rate of TMC. For a reaction time of 24 h (Table 2.1.2), M_n increases with increasing LLA/TMC molar ratio, because the reaction of LLA is more rapid than that of TMC and less depolymerization occurs.

Figure 2.1.1 presents the ^{1}H-NMR spectrum of a copolymer containing about 80 mol % (77 wt%) L-lactide. It was reported that some carbon dioxide is eliminated during the cation ring-open polymerization of TMC and, hence, that some ether linkages are formed in the polycarbonate.[18] However, no elimination of carbon dioxide was detected in our experiments, because no peak at 3.4 ppm, which can be assigned to —CH_2—O—CH_2—, is present.

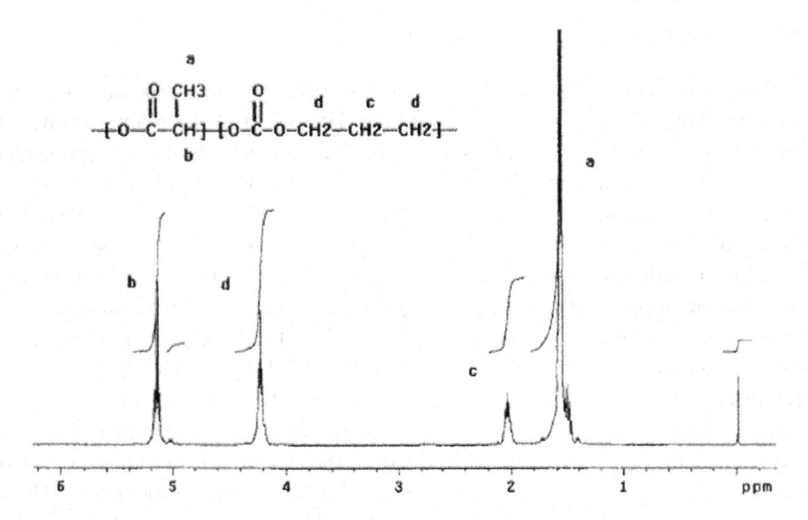

FIGURE 2.1.1 ^{1}H NMR spectrum of poly(LLA-*co*-23 wt% TMC).

Figure 2.1.2 depicts the FTIR spectra of pure PLLA (a), poly(L-lactide-*co*-41 wt% TMC) (b), and pure PTMC (c). The peaks at 1757 and 1185 cm^{-1} in (a) can be assigned to the C=O stretching and the —C—C—O asymmetric stretching frequencies of PLLA, respectively, and the peaks at 1745 and 1247 cm^{-1} in (c) to the C=O stretching and O—C—O asymmetric stretching frequencies of PTMC, respectively. The copolymer (b) has a C=O stretching frequency at 1752 cm^{-1}, located between those of PLLA and PTMC, but retains the stretching frequencies

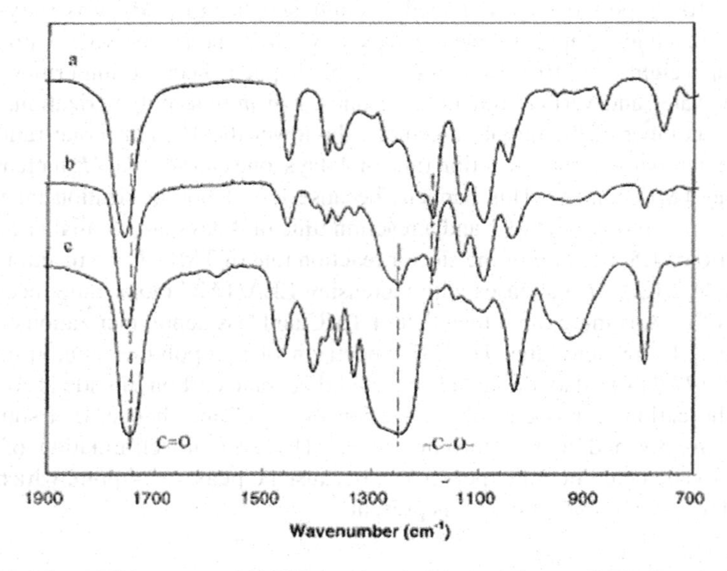

FIGURE 2.1.2 FTIR spectra of (a) PLLA, (b) poly(LLA-*co*-41.5 wt% TMC) and (c) PTMC.

at 1185 and 1248 cm^{-1} of the C—C—O and O—C—O asymmetric stretchings of the ester and the carbonate moieties, respectively.

The DSC traces of copolymers are presented in Figure 2.1.3. The T_g and T_m of the copolymers decrease with increasing TMC content. For higher TMC contents, the heat of fusion of the PLLA crystals becomes smaller, indicating that the extent of crystallinity decreases.

The dependence of T_g on composition is in good agreement with the Fox equation:

$$1/T_g = w_1/T_{g1} + w_2/T_{g2}$$

where T_{g1} and T_{g2}, w_1 and w_2 are the T_gs and the weight fractions of the individual homopolymers, respectively. The results are plotted in Figures 2.1.3 and 2.1.4.

The mechanical properties of poly(LLA-*co*-TMC) are listed in Table 2.1.3. For the pure PLLA, the elongation at break is very low, about 6%. With increasing TMC content in the copolymer, the elongation at break increases. At 15 wt% TMC, the elongation becomes about 15%, while when the TMC content increases to 32 wt%, the copolymer acquires a rubber behavior and the elongation increases significantly, up to 375%. This high elongation at break is associated with a T_g of 22°C, which is lower than the testing temperature (room temperature). The copolymer with 41.5 wt% TMC has a T_g of about 17°C, but being too soft no data could be collected regarding the mechanical properties, because no suitable film could be prepared. As expected, the yield and tensile strengths as well as the Young modulus decrease with increasing TMC content. The toughness increases with increasing TMC content as long as the latter is not too high. For too high values (greater than about 40 wt%), the material acquires a low toughness due to its low strength.

The copolymer properties can be adjusted by varying the monomer feed composition. The toughness of the copolymer can be tremendously increased by increasing the TMC content to 32 wt% (Table 2.1.3). The T_g can be also controlled.

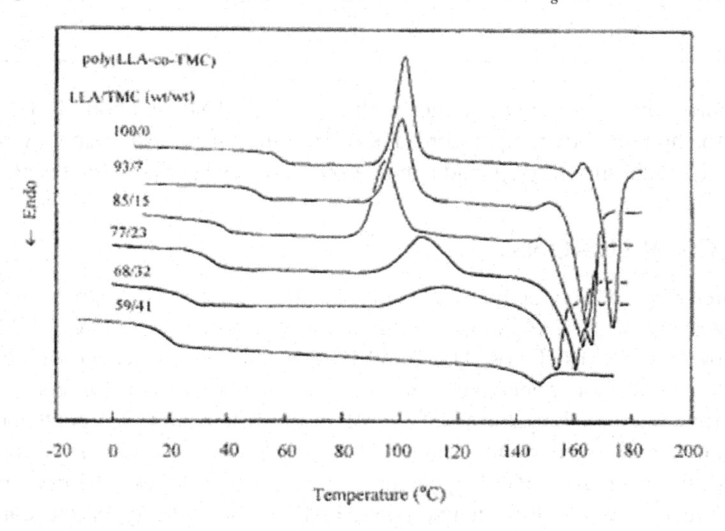

FIGURE 2.1.3 DSC traces of poly(LLA-*co*-TMC) with TMC content 0–41.5 wt%.

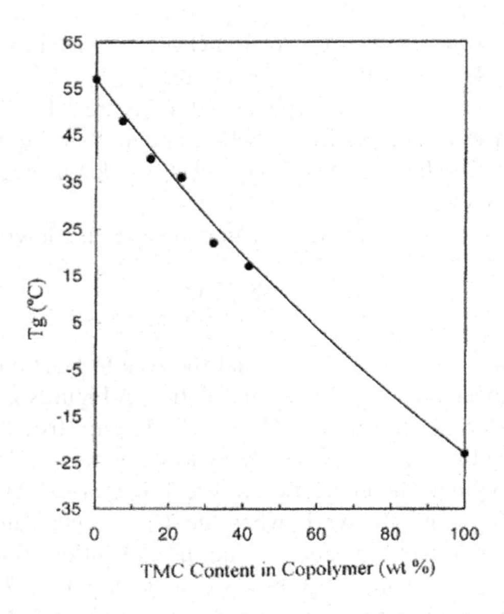

FIGURE 2.1.4 The dependence of T_g of the copolymer on the TMC content (•): experimental points; (—): Fox equation.

TABLE 2.1.3
The Mechanical Properties of Poly(LLA-*co*-TMC)

TMC Content wt%	Yield Strength MPa	Tensile Strength MPa	Young Modulus MPa	Elongation %	Toughness MJ/m³
0 (PLLA)	61	54	1400	6	2.5
15	62	55	1480	15	7
32	33	39	650	375	105

For instance, the T_g of the copolymer with 23 wt% of TMC is about 36°C, which is close to the human body temperature. This copolymer possesses great toughness and good biodegradability *in vivo*, and can be used as a bioabsorbable material.

2.1.4 CONCLUSIONS

The molten ring open copolymerization of L-lactide (LLA) and trimethylene carbonate (TMC) was investigated and the structure and properties of poly(LLA-*co*-TMC) were studied by GPC, NMR, FTIR, DSC, and Instron testing. The reaction time and the monomer composition affect the molecular weight (M_n) of copolymers. When the reaction time is not too long, the M_n of the copolymer with a high LLA content is greater than that of the copolymer with a high TMC content. However, the M_n of copolymers increased with increasing PTMC content for long reaction times. The glass transition temperature (T_g), the melting temperature, and the crystallinity of the copolymers decrease with increasing TMC content due to the amorphous nature of PTMC.

REFERENCES

1. E. A. Dawes, Ed., *Novel Biodegradable Microbial Polymers*, Kluwer Academic Publishers, Dordrecht, the Netherlands, 1990.
2. H. Brandl, R. A. Gross, R. W. Lenz, and R. C. Fuller, *Advances in Biochemical Engineering/Biotechnology*, Vol. 41, T. K. Ghose and A. Fiechter, Eds., Springer, Berlin, Germany, 1990, p. 77.
3. G. T. Rodeheaver, T. A. Powell, J. G. Thacker, and R. F. Edlich, *Am. J. Surg.*, **154**, 544, (1987).
4. R. Katz, D. Mukherjee, A. L. Kaganov, and X. Gordons, *Surg. Gynecol. Obstet.*, **161**, 213, (1985).
5. S. Matsumura, K. Tsukada, and K. Toshima, *Macromolecules,* **39**, 3122, (1997).
6. M. Avella, E. Martuscelli, and P. Greco, *Polymer*, **32**, 1647, (1991).
7. J. A. Nijenhuis, E. Colstee, D. W. Grijpma, and A. J. Pennings, *Polymer,* **37**, 5849, (1996).
8. Y. Yuan and E. Ruckenstein, *Polym. Bull.*, **40**, 485, (1998).
9. D. W. Grijpma, R. D. A. van Hofslot, H. Super, J. A. Nijenhuis, and A. J. Pennings, *Polym. Eng. Sci.*, **34**, 1674, (1994).
10. J. M. Vion, R. Jerome, P. Teyssie, M. Aubin, and R. E. Prud'homme, *Macromolecules*, **19**, 1828, (1986).
11. A.-C. Albertsson and M. Eklund, *J. Polym. Sci., Part A: Polym. Chem.*, **32**, 265, (1994).
12. X. Chen, S. P. McCarthy, and R. A. Gross, *Macromolecules*, **30**, 3470, (1997).
13. Y. Hori, Y. Gonda, Y. Takahashi, and T. Hagiwara, *Macromolecules*, **29**, 804, (1996).
14. K. J. Zhu, R. W. Hendren, K. Jensen, and C. G. Pitt, *Macromolecules*, **24**, 1736, (1991).
15. J. Cai and K. J. Zhu, *Polym. Int.*, **42**, 373, (1997).
16. T. Ariga, T. Takata, and T. Endo, *J. Polym. Sci., Part A: Polym. Chem.*, **31**, 581, (1993).
17. J. A. Nijenhuis, D. W. Grijpma, and A. J. Pennings, *Macromolecules*, **25**, 6419, (1992).
18. H. R. Kricheldorf and B. Weegen-Schulz, *Macromolecules*, **26**, 5991, (1993).

2.2 Self-Polyaddition of Hydroxyalkyl Vinyl Ethers*

Hongmin Zhang and Eli Ruckenstein

Chemical Engineering Department, State University of New York at Buffalo, Amherst, New York 14260

ABSTRACT With tetrahydrofuran as a solvent and pyridium *p*-toluenesulfonate as a catalyst, the hydroxyalkyl vinyl ethers 2-hydroxyethyl vinyl ether (2E), 4-hydroxybutyl vinyl ether (4B), and 6-hydroxyhexyl vinyl ether (6H) underwent step-growth self-polyaddition, generating polymers with an acetal main-chain structure. The molecular weight of the resulting polymers increased gradually during the initial polymerization period at room temperature. However, decomposition occurred after about 22–24 h, and the presence of a large amount of catalyst accelerated the latter process. The three monomers exhibited different polymerization capabilities. In contrast to the smooth polymerization of 6H, cyclization side reactions usually took place during the polymerizations of 4B and 2E, which resulted in low polymer yields and low molecular weights because of the formation of unreactive small cyclic acetals. In the self-polyaddition of 4B, this side reaction was greatly restricted at high concentrations of the monomer. Higher temperatures (60°C–70°C) remarkably accelerated the self-polyaddition process to produce polymers with high molecular weights. However, the polymerizations at high temperatures had to be terminated within about 2 h to avoid the severe decomposition of the polymers. Copolymers were also obtained via the copolyaddition of any two of the monomers. The easiness of the incorporation of the monomers into the copolymers was in the sequence 6H > 4B > 2E. Poly(6H), poly(4B), poly(2E), and the copolymers possessed different hydrophilicities and were stable in basic, neutral, and even weak acidic media but exhibited degradation in the presence of a strong acid.

* *Journal of Polymer Science: Part A: Polymer Chemistry* 2000, 38, 3751–3760.

2.2.1 INTRODUCTION

Vinyloxyl compounds can undergo an addition reaction either with a carboxyl group[1] in the absence of a catalyst or with a phenol or alcohol[2] in the presence of an acidic catalyst. This reaction has been extended to bifunctional compounds to prepare polymers. For instance, the polyaddition between a vinyloxyl compound, such as ethylene glycol divinyl ether, and a dicarboxylic acid, such as adipic acid, generated a polymer with a hemi acetal ester main-chain structure[3] that was reported to be thermally degradable. When a bivinyloxyl compound was reacted with a diphenol or a diol, an acetal polymer was obtained[4] that was degradable by acids.

The cationically polymerizable monomers can be mainly classified into two categories, isobutene[5] and alkyl vinyl ethers.[6] Higashimura et al.[7] developed various initiator systems and investigated the living cationic polymerizations of a number of vinyl ethers, including the functional ones. For instance, the hydroxyl group of the bifunctional monomer 2-hydroxyethyl vinyl ether (2E) was protected with a silyl,[8(a)] benzoyl,[8(b)] or acetyl group.[8(c)] After the cationic polymerization, the protecting group was eliminated by hydrolysis, and a well-defined water-soluble polyalcohol was obtained. This polymerization technique was also applied to the syntheses of multicomponent amphiphilic copolymers,[9] such as block and star-shaped copolymers. During these reactions, a cationic chain polymerization took place in which the C=C double bond of the monomer participated alone in the reaction, resulting in a polymer with a carbon–carbon main-chain structure (Route 1 in Scheme 2.2.1). In contrast to this chain reaction, this article is concerned with the step-growth polymerization of hydroxyalkyl vinyl ethers (HAVE in Scheme 2.2.1). The 2-hydroxyethyl vinyl ether (2E), 4-hydroxybutyl vinyl ether (4B), and 6-hydroxyhexyl vinyl ether (6H) contain both a vinyloxyl and a hydroxyl group. In the absence of a catalyst, no reaction can take place between these two groups, and for this reason, these monomers are stable under normal conditions. However, in the presence of a suitable catalyst, such as pyridium *p*-toluenesulfonate (PTS), the addition reaction between the vinyloxyl group of one molecule and the hydroxyl group of another one can occur to form a dimer. The further polyaddition between the monomer and oligomers generates a polymer. As shown in route 2 in Scheme 2.2.1, the resulting polymer possesses a polyacetal main chain instead of a carbon–carbon backbone. For this reason, the polymers thus obtained might be degradable thermally or in the presence of an acid.

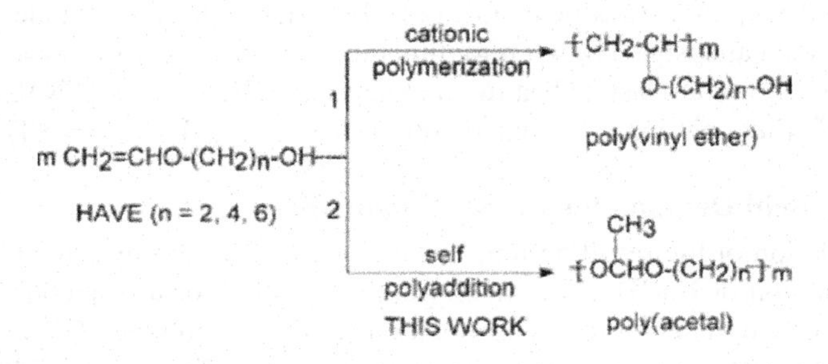

SCHEME 2.2.1

2.2.2 EXPERIMENTAL

2.2.2.1 MATERIALS

Tetrahydrofuran (THF) was dried with CaH_2 under reflux for more than 24 h and was distilled just before use. 2E (Aldrich, 97%), 4B (Aldrich, 99%), and 6H (Aldrich, 98%) were dried with magnesium sulfate and distilled under reduced pressure prior to polymerization. Ethyl vinyl ether (EVE) was dried with CaH_2 and distilled before use. PTS (Aldrich, 98%)[10] and sodium ethoxide (Aldrich, 21 wt% solution in ethyl alcohol) were used as received.

2.2.2.2 POLYMERIZATION

Both the self-polyaddition of 2E, 4B, and 6H and the copolyaddition of two monomers were carried out in well-dried round-bottom flasks, under the protection of nitrogen, with magnetic stirring. With nitrogen purging, a certain amount of monomer (2E, 4B, 6H, or two of them) was added, and this was followed by the addition of purified THF with a dry syringe. After the flask was kept at a selected temperature, the polymerization reaction was started by the addition of a trace amount of PTS as the catalyst. If necessary, the polymerization process was monitored by gel permeation chromatography (GPC); for this purpose, samples (ca. 0.1 mL) were taken out with dry syringes under the protection of nitrogen. The polyaddition was terminated by the introduction of an excess amount of EVE into the polymerization system. About 1 h later, an ethanol solution of sodium ethoxide was dropwise added until a pH of 10 was reached. Then, the system was evaporated to dryness, and the polymer was vacuum-dried overnight at 55°C to obtain the yield. For additional purification, the polymer was precipitated by its THF solution being poured into water [for poly(4B) or poly(6H)] or hexane [for poly(2E)]. The polymer thus obtained was washed with the precipitant and vacuum-dried for more than 24 h at 55°C.

2.2.2.3 IN SITU INVESTIGATION OF THE POLYMERIZATION SYSTEMS OF 2E, 4B, AND 6H

To identify the side reactions during polyaddition, *in situ* ^{1}H NMR measurements were carried out. For instance, 0.2 g of 4B was transferred into a well-dried NMR tube with a dry syringe, to which 1 mL of CaH_2-dried THF-d_8 was added. After the addition of the catalyst (PTS, 4.6 mg), this tube was purged with nitrogen, and the reaction was allowed to last 22 h at room temperature. Before the ^{1}H NMR measurement, the THF-d_8 solution was diluted with $CDCl_3$ ($V_{CDCl_3}/V_{THF-d_8} = 3/1$).

2.2.2.3.1 Acid Degradation of the (Co)polymers

The degradation of the resulting (co)polymers was examined by the addition of an aqueous solution of NaOH, pure water, pure acetic acid, or an aqueous solution of hydrochloric acid to one of four THF solutions of the (co)polymer. GPC was used to verify whether degradation occurred. For instance, 0.6 g of poly(6H) [Table 2.2.2, 13 (shown later)] was dissolved in 12 mL of THF. This THF solution was equally divided into four flasks, to which 0.2 mL of hydrochloric acid (1 N), 0.2 mL of acetic acid (99.8%), 0.2 mL of distilled water, or 0.2 mL of an aqueous solution of NaOH (5 N) was added. After being stirred for 12 h, each of the systems was neutralized until the pH reached 7. After filtration, the filtrate was subjected to GPC measurements.

2.2.2.3.2 Measurements

The ^{1}H NMR spectra were recorded in $CDCl_3$ or in a mixture of $CDCl_3$ and THF-d_8 (3/1 v/v) on an Inova-500 spectrometer. The number-average molecular weight (M_n) and the molecular weight distribution [weight-average molecular weight/number-average molecular weight (M_w/M_n)] of the (co)polymers were determined by GPC with a polystyrene calibration curve. The GPC measurements were carried out with THF as a solvent at 30°C with a 1.0 mL/min flow rate and a 1.0 cm/min chart speed. Three polystyrene gel columns (Waters, 7.8 × 300 mm; one HR 5E, part no. 44228, one linear, part no. 10681 and one HR 4E, part no. 44240) were used; they were connected to a Waters 515 precision pump.

2.2.3 RESULTS AND DISCUSSION

2.2.3.1 Polymerization Mechanism

Three hydroxyalkyl vinyl ethers, 2E, 4B, and 6H, were subjected to polymerization. They possessed the same vinyloxyl and hydroxyl groups but different spacers, which separated the vinyloxyl group from the hydroxyl one. This allowed us to investigate the effect of the molecular architecture on the polymerization capability and properties of the resulting polymers. However, as mentioned previously, because two polymerization routes (Scheme 2.2.1) were possible for every monomer, it was important to confirm the polymerization mechanism.

The polymerization was carried out in THF, in the presence of PTS as a catalyst, under the protection of nitrogen. Regardless of the changes in the polymerization conditions, such as the amount of catalyst, monomer concentration, and polymerization temperature, polymers with the same molecular structure were obtained for a certain monomer. For example, Figure 2.2.1A–C depicts the ^{1}H NMR spectra of poly(2E), poly(4B), and poly(6H), respectively. These spectra clearly indicate that the polymerization of each monomer proceeded by a self-polyaddition mechanism and the corresponding acetal polymer was obtained. Indeed, in the spectrum A of poly(2E), the characteristic peak (b) of the acetal methine proton [—O**CH**(CH_3)O—] is present at 4.80 ppm, and its integral intensity is exactly one-third of the intensity of the methyl protons [—OCH(**CH_3**)O—, peak a]. In addition, the methylene protons (CH_2CH_2) provide two pairs of multiplets, which also constitute a distinctive characteristic of the acetal polymers.[3,4(b)] If the polymerization had proceeded via a chain-reaction mechanism, a polymer with a carbon–carbon main-chain structure should have been obtained. Then, the ^{1}H NMR spectrum of the polymer of 2E should have been quite different from spectrum A (Figure 2.2.1), and only two peaks (1.50 ppm, CH_2 in the main chain; 3.45 ppm, CH in the main chain and —OCH_2CH_2O—in the side chain) should have appeared.[8(a)] It is obvious from Figure 2.2.1 that the latter polymerization mechanism did not occur.

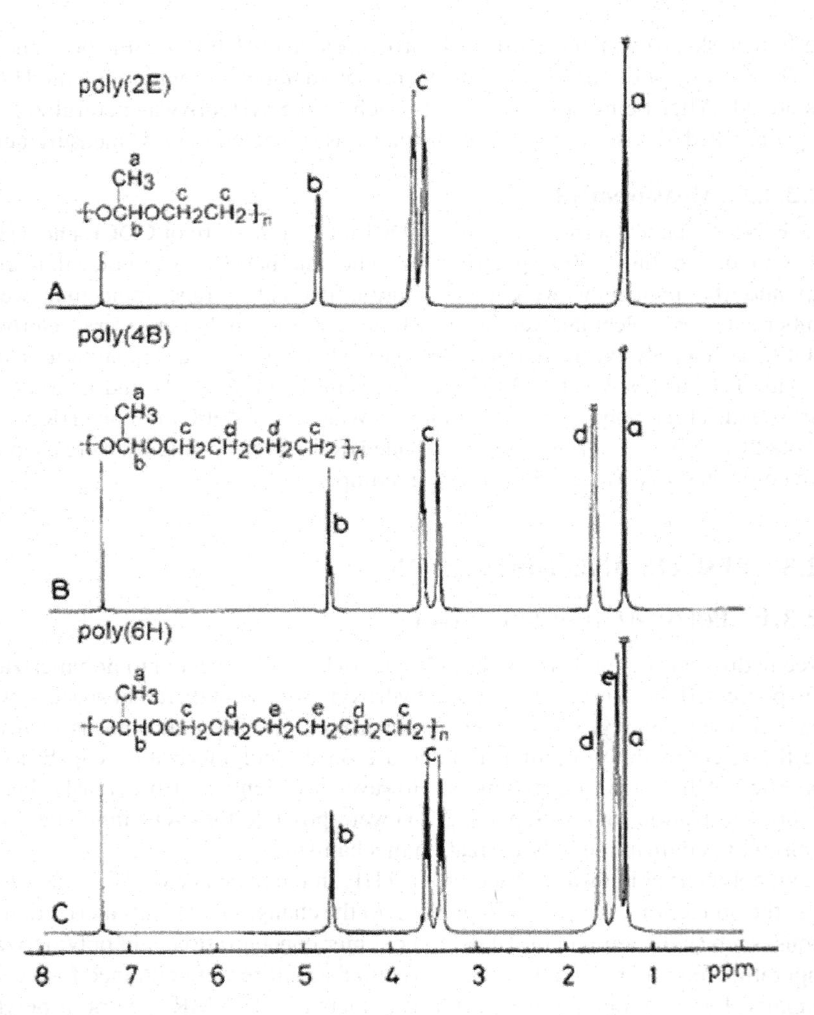

FIGURE 2.2.1 ^{1}H NMR spectra (in $CDCl_3$) of (A) poly(2E) (Table 2.2.2, 11), (B) poly(4B) (Table 2.2.2, 16), and (C) poly(6H) (Table 2.2.2, 12).

2.2.3.2 Self-Polyaddition of 2E, 4B, and 6H

The reaction time, the amount of catalyst, the structure of the monomer, the monomer concentration, and the temperature affected the self-polyaddition process.

2.2.3.2.1 Reaction Time

We investigated the effect of the reaction time by following the polymerization process by GPC measurements. Figure 2.2.2 presents the GPC chromatograms of poly(6H)s obtained for different polymerization times. As shown in Figure 2.2.2A, the addition product exhibited a multiple peak distribution after 1.5 h, implying that dimers, tetramers, and small oligomers were present. Although the monomers were consumed

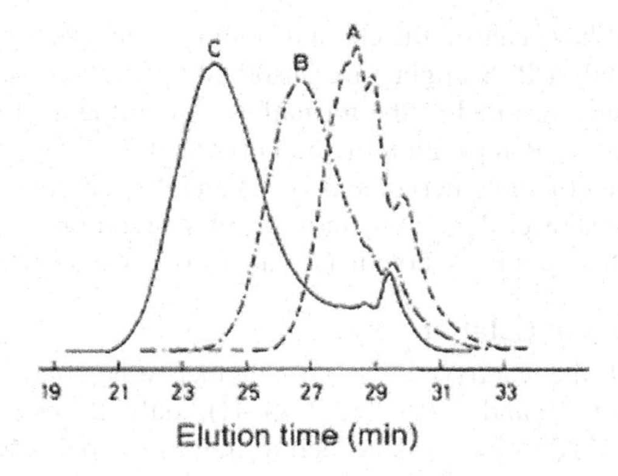

FIGURE 2.2.2 Polymer formation during the step-growth self-polyaddition of 6H, as monitored by GPC: (A) 1.5 h, M_w = 1360, M_w/M_n = 1.24; (B) 3.5 h, M_w = 3940, M_w/M_n = 1.71; and (C) 23.5 h, M_w = 24,600, M_w/M_n = 3.52. The polymerization of 6H (2.56 g) was carried out in THF (10 mL), in the presence of PTS (0.016 g) as a catalyst, at room temperature (23°C).

rapidly, the molecular weight of the product was low. As the polymerization proceeded, the GPC chromatogram acquired a single peak [Figure 2.2.2B], the molecular weight increased gradually, and the molecular weight distribution became broader [Figure 2.2.2B, C]. These are typical characteristics of a step-growth polymerization process.[11]

Generally speaking, for the traditional step-growth polymerization, the longer the polymerization time is, the higher the molecular weight of the resulting polymer is. However, when the self-polyaddition of 2E, 4B, or 6H was carried out at room temperature, a maximum of the molecular weight appeared after about 22–24 h (Table 2.2.1, 1, M_w = 24,000), and for longer reaction times, the molecular weight decreased gradually. As shown in Table 2.2.1 (1), the molecular weight of poly(6H) became much smaller (M_w = 10,200) after 75 h, indicating that decomposition took place. Because of the acetal repeating units of the resulting polymer, the presence of

TABLE 2.2.1
Effect of the Reaction Time on the Self-Polyaddition of 6H[a]

Number 1	Time (h)	1.5	3.5	6.5	23	48.5	52	75
6H (2.52 g; 17.5 mmol)	M_w[b]	1,340	3,850	10,900	24,000	17,300	14,500	10,200
PTS (0.020 g; 0.080 mmol)	M_w/M_n[b]	1.21	1.67	2.37	3.48	2.83	2.63	2.16
Number 2	Time (h)	1.0	3.0	6.0	22	47.5	50.5	75
6H (2.58 g; 17.5 mmol)	M_w[b]	1,260	3,660	10,700	22,300	15,800	11,400	6,650
PTS (0.056 g; 0.22 mmol)	M_w/M_n[b]	1.19	1.65	2.35	3.14	2.71	2.62	2.04

[a] The polymerization was carried out in THF (V_{THF} = 10 mL) at 23°C.
[b] Determined by GPC.

a strong acid might have caused the chain decomposition. After the polymerization lasted 20 h, the catalyst PTS might have absorbed a trace amount of water, and its hydrolysis might have resulted in the formation of a strong acid, *p*-toluenesulfonic acid, that caused the decomposition of the polyacetal. Similar decomposition was also observed during the polymerizations of 4B and 2E. Therefore, the polymerization time should be shorter than 24 h, and the polymerization should be terminated with an alkali, such as sodium ethoxide (see the Experimental section).

2.2.3.2.2 Amount of Catalyst

In the absence of the catalyst PTS, no polymer was obtained. As shown in Table 2.2.1 (1 and 2) and Table 2.2.2 (3–11), only a trace amount of PTS ($mol_{monomer}/mol_{PTS}$ = 200–400) was needed to induce the polyaddition. In contrast, a larger amount of catalyst had a negative effect. The amount of catalyst used for 2 (mol_{6H}/mol_{PTS} = 80, Table 2.2.1) was larger than that for 1 (mol_{6H}/mol_{PTS} = 200). Almost the same molecular weights were obtained within 22–24 h. However, the presence of a larger amount of catalyst accelerated the polymer decomposition at later stages of polymerization (Table 2.2.1, 2). Similar observations were made for the other two monomers (Table 2.2.2, 3–11). Therefore, the molar ratio between the monomer and the catalyst should lie in the range from 200 to 400.

2.2.3.2.3 Molecular Structure of the Monomer

Although 2E, 4B, and 6H had similar molecular architectures, the polymerization capabilities were quite different because of their different spacer lengths. As shown in Table 2.2.2 (3–11), under similar conditions, the polymerization of 6H proceeded more smoothly than those of 2E and 4B. Comparing 6 with 7 and 8, one can observe that in contrast to the high polymer yield (92%) and large molecular weight (M_w = 17,500) of poly(6H), the polymer yields for 4B and 2E were 61 and 57% and the molecular weights were 5720 and 4850, respectively. These differences were obviously caused by their different molecular structures.

To investigate what happened during the polymerization processes of 2E and 4B, *in situ* ^1H NMR measurements of the three polymerization systems were carried out. The polymerization was performed with THF-d_8 as a solvent, in a NMR tube, under conditions similar to those employed for 6–8 in Table 2.2.2 (see the Experimental section). After 22 h, the polymerization systems were diluted with $CDCl_3$ ($V_{CDCl_3}/V_{THF\text{-}d_8}$ = 3/1) and subjected to ^1H NMR measurements. As shown in Figure 2.2.3A–C, the absorptions (6.45 ppm, =CH; 3.95 and 4.20 ppm, CH_2=) corresponding to the CH_2=CH—O— groups of 2E, 4B, or 6H were no longer present, indicating that the monomer was consumed in each case. The spectrum [Figure 2.2.3C] of the 6H polymerization system is almost the same as the spectrum of Figure 2.2.1C and is consistent with the acetal structure of poly(6H). Consequently, the monomer 6H was completely consumed to form its acetal polymer without any side reaction. However, the spectra of the 2E and 4B systems [Figure 2.2.3A, B] are different from those of poly(2E) and poly(4B) [Figure 2.2.1A, B]. Indeed, two types of acetal methine protons [peaks b and b' in Figure 2.2.3A] and two types of methylene protons (peaks c and c')

TABLE 2.2.2
Self-Polyaddition of the Hydroxyalkyl Vinyl Ethers[a]

Number	M/g (mmol)[b]	PTS/g (mmol)[c]	V_{THF} (mL)	V_{THF} (mL/g)[d]	Temperature (°C)	Time (h)	Yield (%)	M_w[e]	M_w/M_n[e]
3	6H/2.41 (16.7)	0.012 (0.05)	12	4.9	23	24	86	20,600	3.27
4	4B/2.54 (21.9)	0.012 (0.05)	12	4.7	23	24	53	9,590	2.35
5	2E/2.58 (29.2)	0.012 (0.05)	12	4.7	23	24	55	3,670	1.63
6	6H/2.44 (16.9)	0.053 (0.21)	12	4.9	23	24	92	17,500	2.88
7	4B/2.34 (20.2)	0.057 (0.23)	12	5.1	23	24	61	5,720	2.08
8	2E/2.28 (25.8)	0.058 (0.23)	12	5.3	23	24	57	4,850	1.99
9	6H/2.41 (16.7)	0.108 (0.43)	12	5.0	23	24	100	11,800	2.61
10	4B/2.13 (18.3)	0.119 (0.47)	12	5.6	23	24	69	4,420	1.98
11	2E/2.56 (29.0)	0.112 (0.45)	12	4.7	23	24	55	5,420	2.15
12	6H/2.14 (14.9)	0.010 (0.04)	6	2.8	23	23	91	26,800	3.26
13	6H/2.35 (16.3)	0.010 (0.04)	2	0.85	23	23	94	13,100	2.27
14	6H/2.45 (17.0)	0.010 (0.04)	0	0	23	23	93	4,820	1.71
15	4B/2.28 (19.6)	0.010 (0.04)	6	2.6	23	23.5	73	13,600	2.54
16	4B/2.53 (21.8)	0.010 (0.04)	2	0.79	23	23	81	20,000	2.54
17	4B/2.59 (22.3)	0.010 (0.04)	0	0	23	21	83	5,320	1.75
18	2E/2.18 (24.8)	0.010 (0.04)	6	2.8	23	24	60	650	1.32
19	2E/2.59 (29.4)	0.010 (0.04)	2	0.77	23	24	Trace		
20	2E/2.06 (23.3)	0.010 (0.04)	25	12	23	22	40	1,120	1.39
21	6H/2.26 (15.7)	0.010 (0.04)	6	2.7	70	2	93	62,400	4.76
22	4B/2.55 (21.9)	0.010 (0.04)	2	0.78	70	2	64	26,500	2.93
23	2E/2.05 (23.3)	0.010 (0.04)	10	4.9	70	2	43	3,590	1.98
24	6H/2.54 (17.6)	0.052 (0.21)	12	4.7	60	22	93	36,600	3.68
25	4B/2.46 (21.2)	0.057 (0.22)	12	4.9	60	24	Trace		
26	2E/2.52 (28.6)	0.055 (0.22)	12	4.8	60	24	Trace		

[a] The polymerization was carried out in THF, under the protection of nitrogen, with magnetic stirring.
[b] M = monomer.
[c] Used as a catalyst.
[d] THF volume (mL) for 1.0 g of monomer.
[e] Determined by GPC.

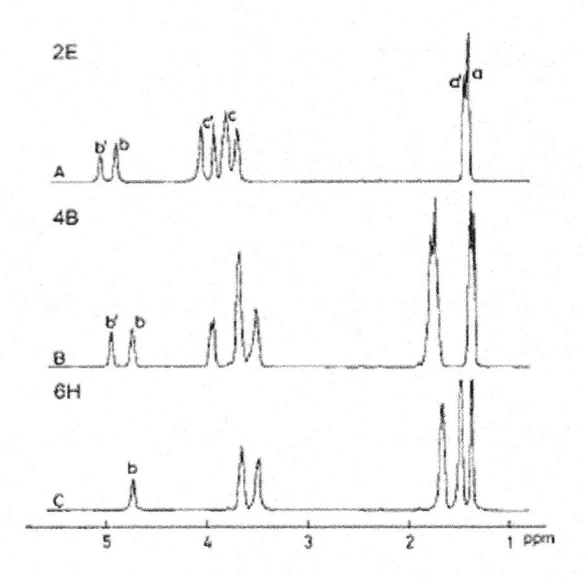

FIGURE 2.2.3 In situ ^{1}H NMR investigation of the polymerization systems of (A) 2E, (B) 4B, and (C) 6H. The solvent was a mixture of $CDCl_3$ and THF-d_8 (3/1 v/v).

appear in the spectrum of the 2E system, indicating that two kinds of acetal products were present in its polymerization system. One of them could be attributed to the acetal polymer of 2E (peaks a, b, and c). The other acetal product (peaks a', b', and c') could be easily removed completely by simple vacuum drying. Therefore, the unknown product was an acetal compound with a low molecular weight. As shown in eqs 1–3 (Scheme 2.2.2), while the polymerization of 2E or 4B was proceeding, a cyclization side reaction has occurred that produced small cyclic molecules with the acetal structure. The competition between the polymerization and the formation of these unreactive acetal byproducts resulted in low polymer yields and low molecular weights.

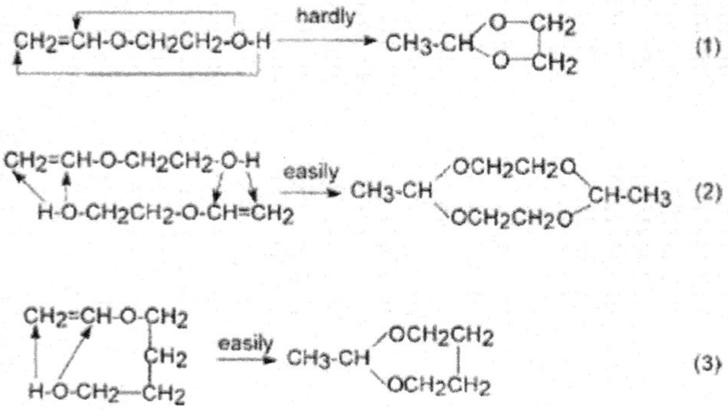

CH2=CH-O-CH2CH2-O-H —hardly→ CH3-CH(O-CH2)(O-CH2) (1)

CH2=CH-O-CH2CH2-O-H + H-O-CH2CH2-O-CH=CH2 —easily→ CH3-CH(OCH2CH2O)(OCH2CH2O)CH-CH3 (2)

CH2=CH-O-CH2-CH2-CH2-O-H —easily→ CH3-CH(OCH2CH2)(OCH2CH2) (3)

SCHEME 2.2.2

2.2.3.2.4 Monomer Concentration

The amount of solvent (THF) had different effects on the polymerizations of 2E, 4B, and 6H. For 6H, as shown in Table 2.2.2 (3 and 12–14), when the amount of THF was reduced from 4.9 to 2.8 mL for 1.0 g of monomer, both the polymer yield (86%–91%) and the molecular weight (20,600–26,800) increased. However, the further reduction of THF caused a great decrease in the molecular weight (13 and 14). In contrast, a more concentrated system enhanced the polymerization of 4B. As shown in Table 2.2.2 (4 and 15–17), when the amount of THF was reduced from 4.7 (4) to 0.79 mL for 1.0 g of 4B, the polymer yield increased from 53% to 81% and the molecular weight increased from 9590 to 20,000. This means that the cyclization side reaction was restricted in more concentrated systems. It is likely that the side reaction was an inner, less stable molecular cyclization (eq 3).

In contrast to 4B, the reduction of the amount of THF had the opposite effect on the polymerization process of 2E. As shown in Table 2.2.2 (5 and 18–20), almost no polymer was obtained when the amount of THF was decreased to 0.77 mL for 1.0 g of 2E (19). This means that the cyclization side reaction was enhanced in a concentrated system. It is likely that the side reaction was a more stable bimolecular cyclization (eq 2) that led to a 10-membered cyclic acetal, instead of an inner molecular reaction (eq 1).

2.2.3.2.5 Reaction Temperature

A higher temperature (60°C–70°C) greatly enhanced the self-polyaddition processes of 6H and 4B, but there was no positive effect for 2E. For instance, when the self-polyaddition of 6H lasted 2h at 70°C, poly(6H) with the high M_w of 62,400 was obtained at a high yield (93%; Table 2.2.2, 21). However, the yield and molecular weight of poly(2E) obtained under similar conditions (23) were even lower than those obtained at room temperature (5). In addition, a long reaction time at higher temperatures had a negative effect on both the polymer yield and molecular weight. For instance, when the self-polyaddition of 4B (25) or 2E (26) lasted 24 h at 60°C, almost no polymer was obtained. Under similar conditions (60°C, 22 h; see 24), poly(6H) was obtained, but its molecular weight was much smaller (M_w = 36,600) than that obtained in 2 h. This was obviously caused by the polymer decomposition. Therefore, the high temperature enhanced not only the self-polyaddition of the monomer but also the decomposition of the polymer, and the decompositions of poly(4B) and poly(2E) were much faster than that of poly(6H). Consequently, for the self-polyaddition at higher temperatures (60°C–70°C), the reaction time should be about 2 h.

2.2.3.3 Copolyaddition of Two Hydroxyalkyl Vinyl Ethers

If the copolyaddition between two hydroxyalkyl vinyl ethers proceeds smoothly, the properties of the resulting polymer, including the hydrophilicity and glass-transition temperature, are expected to be different from those of their homopolymers. As shown in Table 2.2.3 (27–29), both the polymer yield and molecular weight were dependent on the binary combinations of the three monomers, in the sequence 6H/4B > 6H/2E > 4B/2E. The effect of the monomer concentration was also investigated (Table 2.2.3, 30–35). For 6H/4B (30 and 31), a more concentrated

TABLE 2.2.3
Copolyaddition of Two Hydroxylalkyl Vinyl Ethers[a]

Number	M_1/g (mmol)[b]	M_2/g (mmol)[b]	V_{THF} (mL)	PTS/g (mmol)[c]	Temperature (°C)	Time (h)	Yield (%)	M_w[d]	M_w/M_n[d]
27	6H/1.15 (7.95)	4B/1.20 (10.3)	12	0.061 (0.24)	23	26	81	13,000	2.67
28	6H/1.05 (7.27)	2E/1.18 (13.4)	12	0.061 (0.24)	23	26	78	9,680	2.47
29	4B/1.40 (12.1)	2E/1.30 (14.8)	12	0.061 (0.24)	23	26	66	7,750	2.24
30	6H/1.06 (7.38)	4B/1.11 (9.58)	10	0.010 (0.04)	23	22	82	16,500	3.03
31	6H/1.04 (7.21)	4B/1.06 (9.12)	4	0.010 (0.04)	23	22.5	88	25,300	2.99
32	6H/1.15 (7.95)	2E/1.26 (14.3)	10	0.010 (0.04)	23	23	81	8,320	2.22
33	6H/1.16 (8.04)	2E/1.08 (12.3)	4	0.010 (0.04)	23	23	89	8,310	2.05
34	4B/1.18 (10.1)	2E/1.35 (15.4)	10	0.010 (0.04)	23	25	70	7,310	2.21
35	4B/1.15 (9.88)	2E/1.29 (14.6)	4	0.010 (0.04)	23	25	79	6,140	2.11
36	6H/1.33 (9.19)	4B/1.08 (9.31)	4	0.012 (0.05)	65	2	84	45,700	3.92
37	6H/1.38 (9.59)	2E/1.19 (13.4)	10	0.012 (0.05)	65	2	76	23,600	3.85
38	4B/1.58 (13.6)	2E/1.21 (13.7)	4	0.012 (0.05)	65	2	70	17,300	2.82
39	6H/2.51 (17.4)	4B/0.53 (4.52)	6	0.012 (0.05)	65	2	87	47,000	4.20
40	6H/1.17 (8.12)	4B/1.13 (9.68)	12	0.058 (0.23)	60	24	59	1,950	1.39
41	6H/1.61 (11.2)	4B/0.50 (4.30)	4	0.012 (0.05)	23	24	90	24,700	3.10
42	6H/0.54 (3.77)	4B/1.54 (13.2)	4	0.012 (0.05)	23	24	84	18,400	2.75
43	6H/2.15 (14.9)	2E/0.36 (4.04)	6	0.012 (0.05)	23	24	90	28,700	3.45
44	4B/2.22 (19.1)	2E/0.58 (6.54)	4	0.012 (0.05)	23	24	79	15,800	2.58

[a] The polymerization was carried out in THF, under the protection of nitrogen, with magnetic stirring.
[b] M = monomer.
[c] Used as a catalyst.
[d] Determined by GPC.

system provided a higher polymer yield and a greater molecular weight. However, when 2E participated in the copolymerization (Table 2.2.3, 32–35), the change in the monomer concentration had no obvious effect on the copolymerizations of 6H/2E and 4B/2E. As for homopolymerization, the higher temperatures enhanced the copolymerization process (Table 2.2.3, 36–39). Compared to the copolymers (Table 2.2.3, 27–35) prepared at room temperature for 22–26 h, copolymers with higher molecular weights were obtained at 65°C in 2 h (36–39). However, a longer reaction time at a higher temperature (24 h at 60°C; 40) also resulted in a severe decomposition of the copolymer. Therefore, the copolymerization at higher temperatures should be terminated within a relatively short time (ca. 2 h).

The composition constitutes the most important factor because the properties of the copolymers greatly depended on the molar ratios of the two components. As shown in Figure 2.2.4, regardless of the combinations of the monomers

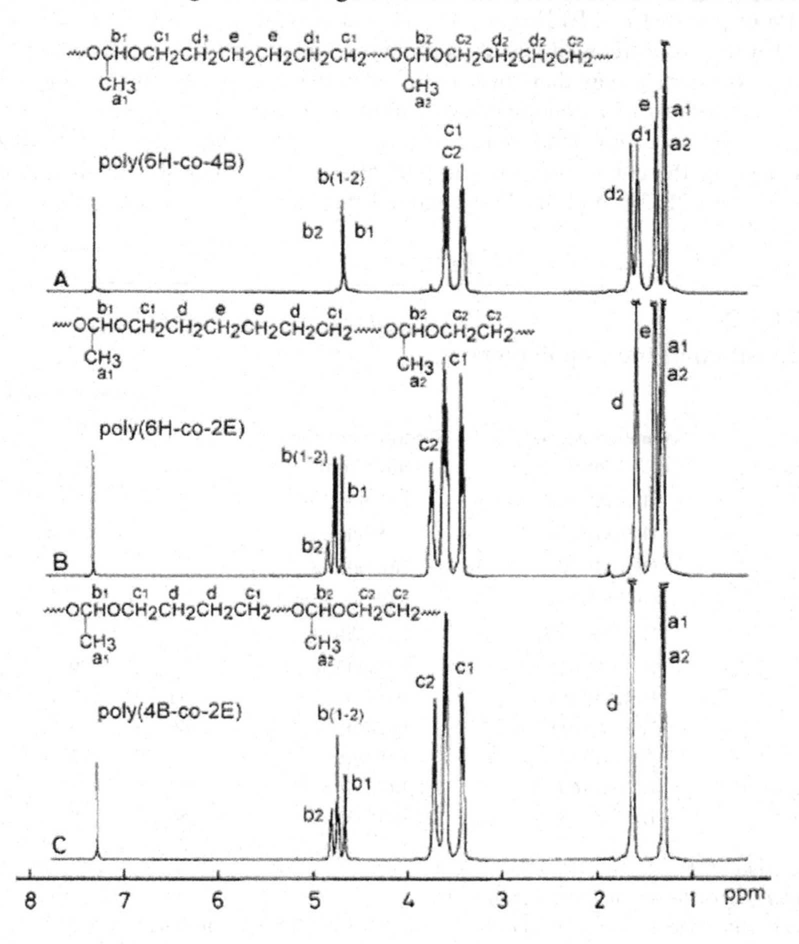

FIGURE 2.2.4 ^{1}H NMR spectra (in $CDCl_3$) of the copolymers: (A) poly(6H-*co*-4B) (Tables 2.2.3 and 2.2.4, 36), (B) poly(6H-*co*-2E) (Tables 2.2.3 and 2.2.4, 37), and (C) poly(4B-*co*-2E) (Tables 2.2.3 and 2.2.4, 38).

(6H/4B, 6H/2E, or 4B/2E), both components were incorporated into the copolymers in all cases. As demonstrated by the ^1H NMR spectrum of poly(6H-*co*-4B) [Figure 2.2.4A and Tables 2.2.3 and 2.2.4, 36), in addition to the absorptions (a_1, b_1, c_1, d_1, and e) of the 6H units, the peaks (a_2, b_2, c_2, and d_2) corresponding to the 4B units were also present. Furthermore, from the intensity ratio, the molar ratio of the two components were calculated. As summarized in Table 2.2.4, when 6H participated in the copolymerization with 4B or 2E, its molar content in the copolymer was always higher than its molar content in the monomer mixture. For the copolyaddition of 4B and 2E, the molar content of 4B was always higher in the copolymer than in the monomer mixture. Therefore, the reactivity of the three monomers had the sequence 6H > 4B > 2E, and the higher the reactivity was, the greater the molar content incorporated into the copolymer was.

Comparing the ^1H NMR spectrum of poly(6H-*co*-4B) [Figure 2.2.4A] with the spectra of poly(6H-*co*-2E) [Figure 2.2.4B] and poly(4B-*co*-2E) [Figure 2.2.4C], we note that the absorptions of acetal methine protons in poly(6H-*co*-4B) emerged in a different way (one group) than those of the other two copolymers [three groups: b_1, b_2, and $b_{(1-2)}$]. Because two monomers took part in the reaction, three types of connections between the two components were possible. As an example, Figure 2.2.5 schematically presents the connecting ways in poly(4B-*co*-2E). Obviously, the three connections, 4B-4B, 2E-2E, and 4B-2E, generated three kinds of acetal methine protons,

TABLE 2.2.4
Composition of the Copolymer[a]

Number	Monomer M_1/M_2 (mol %/mol %)[b]	Copolymer M_1/M_2 (mol %/mol %)[c]	Methine Content (mol %)[d]		
			b_1	b_2	$b_{(1-2)}$
31HB	H(44)/B(56)	H(50)/B(50)			
36HB	H(50)/B(50)	H(56)/B(44)			
39HB	H(80)/B(20)	H(82)/B(18)			
41HB	H(72)/B(28)	H(75)/B(25)			
42HB	H(22)/B(78)	H(29)/B(71)			
33HE	H(40)/E(60)	H(44)/E(56)	22	30	48
37HE	H(42)/E(58)	H(57)/E(43)	34	17	49
43HE	H(79)/E(21)	H(84)/E(16)	66	3	31
38BE	B(50)/E(50)	B(53)/E(47)	29	22	49
34BE	B(40)/E(60)	B(42)/E(58)	17	37	46
44BE	B(75)/E(25)	B(76)/E(24)	56	8	36

[a] See Table 2.2.3.
[b] Molar ratio of the two monomers employed.
[c] Molar ratio of the two monomer units in the resulting copolymer (determined by ^1H NMR).
[d] Molar ratios of the three types of methine protons (see Figure 2.2.5 and the text) in the main chain of the copolymer, as determined by ^1H NMR.

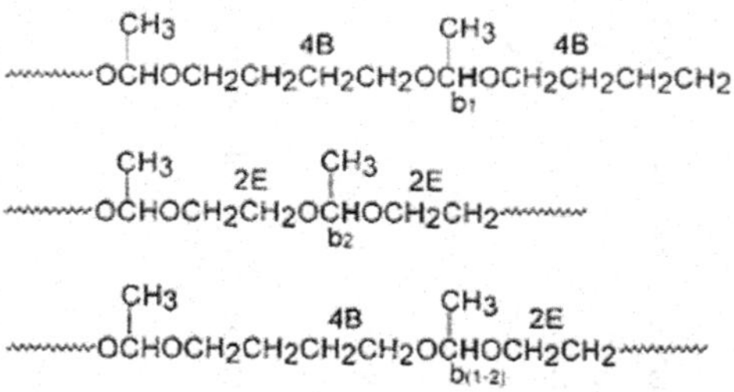

CH3 4B CH3 4B
~OCHOCH2CH2CH2CH2OCHOCH2CH2CH2CH2~
b1

CH3 2E CH3 2E
~OCHOCH2CH2OCHOCH2CH2~
b2

CH3 4B CH3 2E
~OCHOCH2CH2CH2CH2OCHOCH2CH2~
b(1-2)

FIGURE 2.2.5 Three types of acetal methine protons in the copolymer of 4B and 2E. Protons b_1, b_2, and $b_{(1-2)}$ were generated via the connections 4B-4B, 2E-2E, and 4B-2E, respectively.

namely, b_1, b_2, and $b_{(1-2)}$, respectively. From their intensity ratios, we calculated their molar contents (Table 2.2.4), which also represented the molar contents of the three types of segments in the copolymer. Three kinds of acetal methine protons were also present in the copolymer of 6H and 4B. However, because their chemical shifts are the same (4.66 ppm; see Figure 2.2.1), they did not separate into three groups.

2.2.3.4 Solubility of the (Co)polymers

Because of the acetal main-chain structure, the (co)polymers exhibited excellent solubilities. As shown in Table 2.2.5, they were soluble in benzene, chloroform, dioxane, THF, ethyl acetate, and acetone. As expected, the different spacers of the three monomers affected the hydrophilicities of their corresponding polymers. For instance, poly(6H) with a long spacer (6—CH_2—) was soluble in typical nonpolar solvents, such as hexane, but was insoluble in methanol and water. In contrast, poly(2E) with a short spacer (2—CH_2—) was soluble in methanol and even in water but insoluble in hexane. Poly(4B) dissolved slowly in alcohol and was partially soluble in both hexane and water. In addition, most of the copolymers were insoluble in water or hexane but soluble in any of the other solvents mentioned.

2.2.3.5 Stability and Acid Degradation of the (Co)polymers

The pH dependence of the stability and degradation of the (co)polymers was investigated by the treatment of their THF solutions with a strong alkali, water, a weak acid, or a strong acid. When they were treated with sodium hydroxide (pH 14), water (pH 7), or acetic acid (pH 4) overnight, no polymer degradation occurred, indicating that the (co)polymers were stable in basic, neutral, and even weak acidic media. For this reason, the (co)polymers were purified by pouring their THF solutions into water to precipitate the (co)polymer. However, in the presence of a strong acid (pH 1), such as HCl, the (co)polymers exhibited degradation. For instance, 3 mL of a 5 wt% THF solution of poly(6H) (Table 2.2.2, 21) was treated with 0.2 mL of hydrochloric acid (1.0 M) overnight. As shown in Figure 2.2.6B, the complete degradation of this polymer was confirmed by GPC. The other (co)polymers exhibited similar acid degradations.

TABLE 2.2.5
Solubilities of the Homopolyacetals and Copolyacetals[a]

Number[b]	Composition	Hexane	Bz	$CHCl_3$	DMF[c]	Dioxane	THF	EA[d]	Acetone	EtOH	MeOH	H_2O
12	Poly(6H)	S	S	S	I	D	S	S	S	P	I	I
25	Poly(4B)	P	S	S	D	D	S	S	S	D	D	P
11	Poly(2E)	I	S	S	S	S	S	S	S	S	S	S
31	HB (50/50)	P	S	S	P	D	S	S	S	P	D	I
33	HE (53/47)	I	S	S	S	S	S	S	S	S	S	I
38	BE (53/47)	I	S	S	S	D	S	S	S	S	S	I
44	BE (76/24)	I	S	S	D	D	S	S	S	S	S	I

[a] The experiment was carried out at room temperature. The amounts of (co)polymer and solvent were 0.03 g and 1 mL, respectively. S = soluble; P = partially soluble; D = dissolved slowly; I = insoluble.
[b] See Tables 2.2.2 and 2.2.3.
[c] *N,N*-Dimethylformamide.
[d] Ethyl acetate.

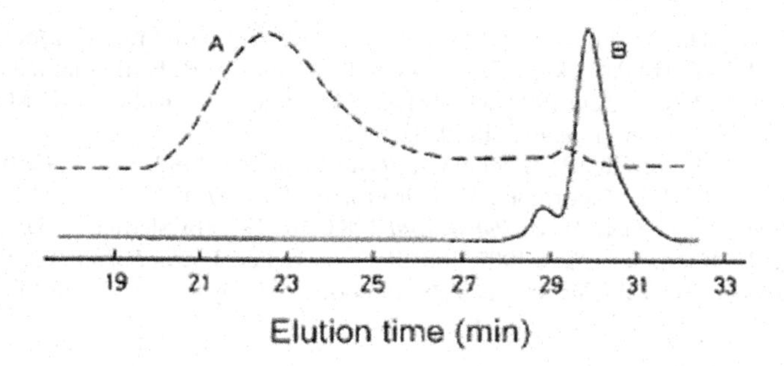

FIGURE 2.2.6 Acid degradation of poly(6H) as demonstrated by GPC traces of (A) poly(6H) (Table 2.2.2, 21; M_w = 62,400, M_w/M_n = 4.76) and (B) its HCl-treated product.

2.2.4 CONCLUSION

2E, 4B, 6H were polymerized in THF in the presence of PTS as a catalyst. The polymerization proceeded by a step-growth self-polyaddition mechanism and not by a chain reaction. A hydroxyl group of one molecule reacted with a vinyloxyl group of another molecule to generate a polymer with an acetal main-chain structure and not a carbon–carbon backbone. Under selected conditions, the copolymerization of two monomers, 6H/4B, 6H/2E, or 4B/2E, also proceeded smoothly, generating the corresponding copolymers with acetal chain structure.

REFERENCES

1. (a) Aoshima, S.; Higashimura, T. *Macromolecules* 1989, 22, 1009; (b) Kamigaito, M.; Sawamoto, M.; Higashimura, T. *Macromolecules* 1991, 24, 3988; (c) Zhang, H. M.; Ruckenstein, E. *Polym Bull* 1997, 39, 399; (d) Zhang, H. M.; Ruckenstein, E. *Macromolecules* 1998, 31, 7575; (e) Nakane, Y.; Ishidoya, M.; Endo, T. *J Polym Sci Part A: Polym Chem* 1999, 37, 609.
2. (a) Fieser, L. C.; Fieser, M. *Reagents for Organic Synthesis*; Wiley: New York, 1967; p. 256; (b) Larhed, M.; Hallberg, A. *J Org Chem* 1997, 62, 7858.
3. Otsuka, H.; Endo, T. *Macromolecules* 1999, 32, 9059.
4. (a) Heller, J.; Penhale, D. W. H.; Helwing, R. F. *J Polym Sci Polym Lett Ed* 1980, 18, 293; (b) Ruckenstein, E.; Zhang, H. M. *J Polym Sci Part A: Polym Chem* 2000, 38, 1848.
5. (a) Majoros, I.; Nagy, A.; Kennedy, J. P. *Adv Polym Sci* 1994, 112, 1; (b) Jacob, S.; Kennedy, J. P. *Adv Polym Sci* 1999, 146, 1; (c) Charleux, B.; Faust, R. *Adv Polym Sci* 1999, 142, 1.
6. (a) Sawamoto, M. *Prog Polym Sci* 1991, 16, 111; (b) *Cationic Polymerization: Mechanism, Synthesis, and Applications*; Matyjaszewski, K., Ed.; Marcel Dekker: New York, 1996.
7. (a) Miyamoto, M.; Sawamoto, M.; Higashimura, T. *Macromolecules* 1984, 17, 265; (b) Kojima, M.; Sawamoto, M.; Higashimura, T. *Macromolecules* 1989, 22, 1552; (c) Higashimura, T.; Kishimoto, Y.; Aoshima, S. *Polym Bull* 1987, 18, 111.
8. (a) Higashimura, T.; Ebara, K.; Aoshima, S. *J Polym Sci Part A: Polym Chem* 1989, 27, 2937; (b) Higashimura, T.; Aoshima, S.; Sawamoto, M. *Makromol Chem Macromol Symp* 1986, 3, 99; (c) Aoshima, S.; Nakamura, N.; Uesugi, M.; Sawamoto, M.; Higashimura, T. *Macromolecules* 1985, 18, 2097.

9. (a) Fukui, H.; Yoshihashi, S.; Sawamoto, M.; Higashimura, T. *Macromolecules* 1996, 29, 1862; (b) Patrickios, C. S.; Forder, C.; Armes, S. P.; Billingham, N. C. *J Polym Sci Part A: Polym Chem* 1997, 35, 1181; (c) Matsumoto, K.; Kubota, M.; Matsuoka, H.; Yamaoka, H. *Macromolecules* 1999, 32, 7122.
10. (a) Scrimin, P.; Tecilla, P.; Tonellata, U. *J Am Chem Soc* 1992, 114, 5086; (b) Freeman, F.; Kim, D. S. H. L.; Rodriguez, E. *J Org Chem* 1992, 57, 1722.
11. (a) Percec, V.; Auman, B. C. *Polym Bull* 1983, 10, 385; (b) Shaffer, T. D.; Antolin, K.; Percec, V. *Makromol Chem* 1987, 188, 1033; (c) Heller, J. *Adv Polym Sci* 1993, 107, 41; (d) Kunisada, H.; Yuki, Y.; Kondo, S.; Nishimori, Y.; Masuyama, A. *Polym J* 1991, 23, 1455.

2.3 Thermally Reversible Linking of Halide-Containing Polymers by Potassium Dicyclopentadienedicarboxylate*

Xiaonong Chen and Eli Ruckenstein

Chemical Engineering Department, State University of New York at Buffalo, Amherst, New York 14260

ABSTRACT A novel crosslinker for thermally reversible covalent (TRC) linking of halide-containing polymers is suggested. Chlorine-containing polymers such as chloro-methylstyrene copolymers, chlorinated polypropylene, polyvinylchloride, chlorinated polyisoprene, and polyepichlorohydrin were crosslinked with potassium dicyclopentadienedicarboxylate (KDCPDCA). The crosslinker was prepared by reacting potassium ethoxide with dicyclopentadienedicarboxylic acid. Because of the low solubility of KDCPDCA in organic solvents, a phase transfer catalyst, benzyltrimethyl-ammonium bromide, was employed for the crosslinking reaction. The crosslinking reaction occurred at a higher rate in a polar solvent, such as dimethylformamide, than in a nonpolar one, such as toluene, and was affected by the nature of the chlorine-containing polymer. Some of the polymers crosslinked even at room temperature. The chain-extending reaction between KDCPDCA and a α,ω-dihalide compound such as α,α'-dichloro-*p*-xylene, 1,4-dichlorobutane, or 1,4-dibromobutane also was carried out to obtain linear oligomers. The IR spectra indicated that the crosslinking and chain-extending reactions were based on the esterification between the halide—carbon bonds of the polymer and the COOK groups of KDCPDCA. The flowability at 195°C and solubility on heating in a dichlorobenzen-maleic compound mixture of the crosslinked polymers indicated that the TRC crosslinking occurred via the

* *Journal of Polymer Science: Part A: Polymer Chemistry*, 1999, 37, 4390–4401.

reversible Diels–Alder cyclopentadiene/dicyclopentadiene conversion as long as the polymer was thermally stable and did not contain olefinic C=C bonds. The TRC linking also was confirmed by the rapid decrease of the specific viscosity of the obtained linear oligomers on heating.

2.3.1 INTRODUCTION

Thermally reversible covalent (TRC) crosslinked polymers are expected to be similar to thermosetting polymers in their physical properties and solvent resistance and can be remolded by thermal processing technology. Consequently, TRC crosslinking technology can be used for manufacturing recyclable materials and improving the mechanical properties of traditional thermoplastic materials. Additionally, linear polymers with TRC linkages in their backbone are expected to exhibit lower viscosities on heating than typical linear polymers and to be processed at lower shear rates

TRC crosslinkers were reviewed by Engle and Wagener.[1] In the past decades, TRC linkages were developed that were generated via the reversible Diels–Alder cycloaddition of a diene, such as furan, to a dienophile, such as maleimide,[2–4] and by the reversible Diels–Alder cycloadditive dimerization of cyclopentadiene (CPD) or of its derivatives.[5–13]

The first report on TRC linkages based on the thermally reversible dimerization reaction of CPD was made in the early 1970s.[5] Thereafter, a series of CPD derivatives were synthesized and employed as cyclopentadienylation agents for different polymers or as crosslinkers in the polymerization of vinyl monomers. Cyclopentadienyl sodium (CPDNa) or cyclopentadienyl lithium (CPDLi) was used for the cyclopentadienylation of chlorine-containing polymers such as polyvinylchloride (PVC), polychloroprene, and polyepichlorohydrin (PECH) to yield polymers with pendant CPD groups.[5,6,10] Cyclopentadienylations of chlorinated isobutylene-isoprene rubber and chlorinated ethylene-propylene rubber were carried out with dimethylcyclopentadienylaluminum (Me_2AlCPD) by Kennedy and Castner.[7,8] Thermally reversible chain-extensible α,ω-di(3-cyclopentadienyl propyldimethylsilyl)polyiso-butylene also was prepared by Kennedy & Carlson[9] by reacting CPDNa and BrC_3Si—PIB—SiC_3Br. TRC crosslinked polysphazenes were obtained by Salamone et al.[11] via the substitution reaction of chlorine in poly(dichlorophosphazene) by cyclopentadienyl-ethoxide group or its dimer. Addition polymerizable monomers containing CPD or dicyclopentadiene (DCPD) groups, such as allyl-CPD,[5,12] cyclopentadienylethanol acrylate,[12] and vinyl ester of dicyclopentadienedicarboxylic acid,[5] were employed for copolymerization with vinyl monomers to generate crosslinkable or crosslinked vinyl copolymers. Recently, diglycide dicyclopentadienedicarboxylate was synthesized and copolymerized with epoxide monomers by ring-opening polymerization to produce a novel TRC crosslinked thermoplastic polyether.[13]

The polymers crosslinked via the dimerization of CPD were the focus of the previous thermally reversible Diels–Alder crosslinkings because CPD easily could be obtained in large amounts as a by-product of the thermal operations in the petrochemical and coal-chemical industries. Another possible reason is that CPD-containing compounds with different functional groups could be obtained without difficulty because of the high reactivity of CPD.

In this article, we report a novel TRC crosslinking of halide-containing polymers that employs potassium dicyclopentadienedicarboxylate (KDCPDCA) as a crosslinker, with the crosslinkage being generated through the esterification between COOK and a halide—carbon group of the polymer chain. Compared to Me_2AlCPD, the new crosslinker, KDCPDCA, can be prepared easily and safely with a high yield and is chemically stable even when exposed to air and moisture. In addition, KDCPDCA is a weak base compared to CPDNa or CPDLi; consequently, side reactions (e.g., dehydrochlorination) that can induce an irreversible crosslinked structure can be retarded during the crosslinking reaction.

2.3.2 EXPERIMENTAL

2.3.2.1 Materials

Styrene (St; Aldrich, 99%), butyl acrylate (BA; Fluka, >99%), and chloromethyl-styrene (CMS; Aldrich, 90%) were distilled under reduced pressure before use, and cyclohexane (Aldrich; 99+%) was purified by distillation. The solvents and following chemicals were used as supplied: CPDNa (Aldrich; 2.0M solution in tetrahydrofuran [THF]), 4-*tert*-butylcatehol (TBC; Aldrich, 97%), KOH (Baker; 87.5%), potassium ethoxide (Aldrich; 95%), lauroyl peroxide (LPO; Aldrich; 97%), potassium persulfate (Aldrich; 99.99%), dodecyl sulfate sodium salt (Aldrich; 98%), benzyltrimethylammonium bromide ($BzMe_3NBr$; Aldrich; 97%), 1,4-dichlorobutane (DCB; Aldrich; 99%), 1,4-dibromobutane (DBB; Aldrich; 99%), α,α'-dichloro-*p*-xylene (DCX; Aldrich; 98%), maleic anhydride (Aldrich; 99%), and maleimide (Aldrich; 99%).

The following commercial polymers were used as received: PVC (Aldrich; inherent viscosity = 0.51; M_w = 71,600 by gel permeation chromatography [GPC]), chlorinated polypropylene (CPP; Aldrich; 32 wt% chlorine; M_w = 150,000), chlorinated polyisoprene (CPIP, Aldrich; 65 wt% chlorine; M_w = 59,000 by GPC), PECH (Aldrich; M_w = 700,000), polychloromethylstyrene (PCMS; Aldrich; 60/40 mixture of 3- and 4-isomers; M_w = 55,000).

2.3.2.2 General Procedures

^1H NMR spectra were recorded in $CDCl_3$ (D 99.8%, 0.05 v % tetramethylsilicone) or D_2O (D 99.9%, 0.75% 3-(trimethylsilyl) propionic-2,2,3,3-d_4 acid, sodium salt) on a VXR-400 spectrometer. The Fourier transform infrared spectra were recorded on a Perkin–Elmer 1760-X spectrometer. GPC measurements were carried out using THF as the solvent at 30°C with a 1.0 mL/min flow rate and 1.0 cm/min chart speed. Three polystyrene gel columns (Waters; 7.8 mm × 300 mm; one HR5E, Part No. 44228, one Linear, Part No. 10681, and one HR4E, Part No. 44240) were used, which were connected to a Waters 515 precision pump.

2.3.2.3 Preparation Methods

2.3.2.3.1 Dicyclopentadienedicarboxylic Acid (DCPDCA)

The modified Peter method[14] was used in the preparation of DCPDCA. To 2 kg of dry ice (excess), an 800 mL CPDNa solution in THF (1.6 mole CPDNa) was added with stirring. The mixture was allowed to react overnight, after which it was mixed with distilled water. The aqueous solution was neutralized with 5 wt% HCl acid to generate a powder that was separated from the aqueous phase by filtration. After being washed with distilled water 5 times, the obtained filter cake was dried under reduced pressure at a temperature smaller than 60°C. 133 g of a light-yellow powder (yield = 76%) thus was obtained with a melting point of 208°C.[14] It had a purity, determined by neutralization titration, of 98.9%.

2.3.2.3.2 Potassium Salt of Dicyclopentadienedicarboxylic Acid (KDCPDCA)

To a solution of DCPDCA (56.5 g, 0.254 mol) in 270 mL of anhydrous ethanol, a solution of potassium ethoxide (42.5 g, 0.480 mol) in 400 mL of anhydrous THF/ether (v/v 1 : 1) was added slowly with stirring and cooling (ice/water bath). The clear solvent was removed after an overnight reaction to yield a light-yellow slurry that after filtration was washed with anhydrous ether. Then, the product was dried under reduced pressure at a temperature lower than 60°C to yield 69.2 g of a light-yellow powder (yield = 97%).

2.3.2.3.3 Potassium Salt of Maleic Acid (KMA)

5.6 g of maleic anhydride (0.0565 mol) was dissolved in 20 mL of THF containing 1.1 mL of distilled water. Twelve hours later, a KOH (7.2 g, 0.112 mol) solution in 80 mL of ethanol was added with stirring under cooling (ice/water bath). This was followed by the treatment employed in the KDCPDCA preparation and yielded 10.4 g of a light-yellow powder (yield = 96%) with a melting point higher than 275°C, the temperature at which it decomposed.

2.3.2.3.4 Solution (s) Polymerization (Copolymer s-PSt-CMS-BA)

19.5 g of St, 18.7 g of BA, and 3.2 g of CMS were polymerized in 100 mL of cyclohexane at 65°C for 13 h with 0.87 g of LPO used as the initiator. The polymer was precipitated with methanol containing 2 wt% TBC, and the product was vacuum dried to yield 18.6 g of the copolymer (yield = 44.9%).

2.3.2.3.5 Emulsion (e) Polymerization (Copolymers e-PSt-CMS-BA and e-PSt-CMS)

The polymerizations were carried out under the conditions listed in Table 2.3.1. The polymer was separated from the emulsion by adding methanol (containing 2 wt% TBC), kept in water overnight, and washed twice with water before being vacuum dried.

2.3.2.3.6 Crosslinking Reaction

The reaction was carried out with magnetic stirring under a N_2 atmosphere in a solution of a chlorine-containing polymer by adding KDCPDCA powder, $BzMe_3NBr$ as

TABLE 2.3.1
Emulsion Copolymerization of CMS[a]

Sample	Monomers (mol)	Water (mL)	Potassium Persulfate (g)	Polymerization Time (h)	Yield (%)
e-PSt-CMS-BA	St/CMS/BA 0.090/0.011/0.082	100	0.31	7.0	95.0
e-PSt-CMS	St/CMS 0.287/0.017	150	0.45	9.5	92.7

[a] 1.0 wt% (based on water) sodium dodecyl sulfate was used as the emulsifier; the polymerization temperature was 45°C.

the catalyst, and 0.3 wt% (based on the solvent) TBC as the stabilizer of the C=C double bonds of the crosslinker. The time duration to the moment when the magnetic stirring bar could no longer move was considered the gelation time. The gel was cut into small particles, and the solvent was extracted from the gel with methanol. The obtained polymer particles then were submerged in water for 12 h, washed three times with fresh water and twice with methanol, and then dried under reduced pressure at a temperature lower than 60°C.

2.3.2.3.7 Chain-Extending Reaction of Dihalide Compounds by Dipotassium Dicarboxylate

A α,ω-dihalide compound was allowed to react with KDCPDCA or KMA just as in the crosslinking reaction mentioned previously. The product was treated with water and methanol before being dried to remove the unreacted reagents.

2.3.2.4 Measurements

2.3.2.4.1 Gel Content of the Crosslinked Polymer (Gel wt%)

0.1 to 0.2 g of the crosslinked polymer was wrapped well in a medium-fast filter paper (Whatman, Φ 90 mm) and subjected for 10 h to extraction with refluxed THF in a Soxhlet extractor. The paper package then was dried under reduced pressure at a temperature lower than 60°C until a constant weight was reached. The gel content was calculated from the weight ratio of the weight increase of the filter paper, and the original weight of the polymer tested.

2.3.2.4.2 Insoluble Polymer Percentage of the Crosslinked Polymer After Its Treatment with a Maleic Compound

0.1 to 0.2 g of the crosslinked polymer was introduced into 10 mL of dichlorobenzen containing about 3 wt% maleimide or maleic anhydride and about 1 wt% TBC and was heated for 5 h in a 190°C oil bath. The mixture then was filtered through a medium-fast filter paper, wrapped well in such paper, and subjected to Soxhlet extraction as discussed in the previous paragraph.

2.3.2.4.3 Specific Viscosity of Linear Polymer (η_{sp})

A dilute solution of the polymer (<1 wt%) in a solvent mixture containing 68.5 wt% benzophenone, 30.3 wt% dichlorobenzen, and 1.2 wt% TBC was employed. The flow-out times of the solvent mixture (t_0) and polymer solution (t) were measured with a Cannon-Fenske viscometer tube (Kimax Size 100). η_{sp} was calculated as $[t/t_0] - 1$. The testing temperature was controlled by an oil bath and was lowered during the testing from 200 or 210°C to 80°C.

2.3.2.4.4 Thermocompression of the Crosslinked Polymer

A polymer tablet with a thickness of about 1.5 to 2 mm was made first from small particles (ca. 0.5 mm) of the crosslinked polymer by using a macro-micro KBr die. Then, the obtained tablet was placed between two plate molds and pressed for about 10 min with a Carver Laboratory Press (Model C) at a temperature of about 195°C under a pressure of 700 psi or less. Subsequently, the tablet and molds were pressed again in another Carver Press at room temperature under a pressure of 1300 psi or less to produce a polymer film with a thickness of about 0.3 mm. The obtained film then was cut into small pieces and layered to a total thickness of about 1.5 to 2 mm, and the cycling experiment was repeated several times.

2.3.3 RESULTS AND DISCUSSION

2.3.3.1 Characterization of KDCPDCA

The melting-point test showed that the obtained KDCPDCA did not melt even at a temperature as high as 275°C, the temperature at which it decomposed. Furthermore, KDCPDCA dissolved in water but was insoluble in common organic solvents such as THF, dimethylformamide (DMF), ethyl ether, methanol, and toluene.

The structure of DCPDCA, confirmed by UV[14] and ^{1}H NMR spectra[13] indicated that each carboxyl group was conjugated with a double bond at the position of Carbon A or B:

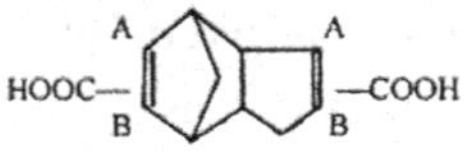

A A
HOOC— —COOH
B B

Figure 2.3.1 presents the ^{1}H NMR spectrum of the obtained KDCPDCA. Seven peaks of H atoms connected with unsaturated carbons can be seen in the range 5.9 to 6.7 ppm, and the ratio of the total peak integral of olefinic H to the total peak integral (1.0–3.6 ppm) of H atoms connected to saturated carbons was found to be 0.246 : 1 (0.250 : 1 calculated from the DCPDCA structure). The obtained KDCPDCA had the same structure as DCPDCA and consisted of a mixture of several isomeric dimers of potassium cyclopentadiene-carboxylate. No carboxylic acid (COOH) peak of DCPDCA (at about 12 ppm[13]) was found in the spectrum. This indicates that the excess DCPDCA was removed during the after-treatment of the system, and all the remaining COOH was converted completely to COOK.

Figure 2.3.2 presents the IR spectrum of the obtained KDCPDCA. The peaks at 1542 $cm^{-1}$and 1407 cm^{-1} can be assigned to COOK (1540–1590 cm^{-1} in the

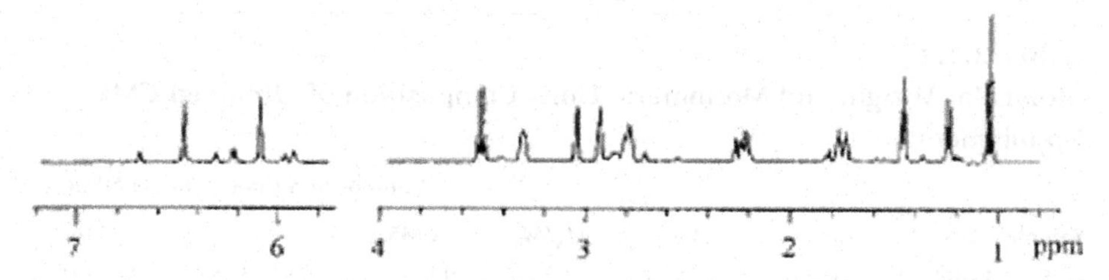

FIGURE 2.3.1 ^1H NMR spectrum of prepared KDCPDCA recorded in a D_2O solution.

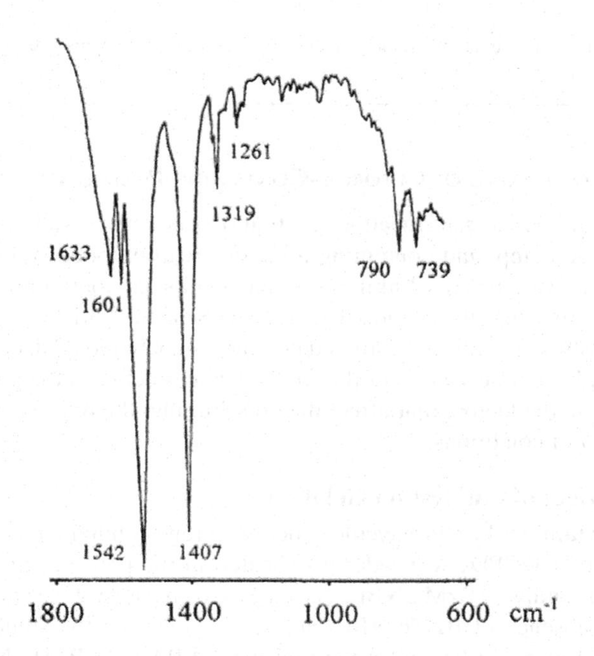

FIGURE 2.3.2 IR spectrum of prepared KDCPDCA (KBr tablet).

literature [15]) and COO— (1390–1410 cm^{-1} in the literature [17]), respectively. The peaks at 1633 cm^{-1} and 1601 cm^{-1} can be assigned to the two C=C double bonds conjugated with the carboxyl groups (1631 cm^{-1} and 1599 cm^{-1} were reported for diester dicyclopentadienedicarboxylic acid [13]). The 1319, 1261, 790, and 739 cm^{-1} peaks can be assigned to the structure of the CPD dimer (1340, 1250, 775, and 730 cm^{-1} were reported in the spectrum of DCPD[18]). Consequently, these results and those reported in literature agree.

2.3.3.2 Characterization of CMS Copolymers

The molecular weights (by GPC) and chlorine contents (by ^1H NMR with $CDCl_3$ as the solvent) of the obtained CMS copolymers are listed in Table 2.3.2.

TABLE 2.3.2
Molecular Weight and Monomeric Units Composition of Obtained CMS Copolymers[a]

				Composition (mol %; by ^{1}H NMR)		
Sample	M_w	M_n	M_w/M_n	**CMS**	**St**	**BA**
s-PSt-CMS-BA	51,300	27,900	1.86	9.6 (5.9)	51.2 (52.8)	39.2 (41.2)
e-PSt-CMS-BA	749,000	151,000	4.95	5.8 (6.0)	51.4 (49.2)	42.8 (44.8)
e-PSt-CMS	672,000	190.000	3.53	4.7 (5.6)	95.3 (94.4)	

[a] The numbers in parentheses are the values calculated from the initial amounts of the monomers used the copolymerization.

2.3.3.3 Crosslinking of Chlorine-Containing Polymers

It is well known that an esterification reaction occurs between a potassium carboxylate and halide compound containing a halide—carbon bond when the halide is sufficiently reactive. The crosslinking reaction of chlorine-containing polymers with KDCPDCA is also an esterification reaction (see Scheme 2.3.1).

The decrosslinking of the crosslinked polymer is not based on deesterification but on the reversible dedimerization of the DCPD bridges between the polymer chains.

To determine the factors that affect the crosslinking, the reaction was conducted under a variety of conditions.

2.3.3.3.1 Effect of Catalyst on Gelation

A solution containing the styrene/chloromethylstyrene copolymer (*e*-PSt-CMS) and a suspension of KDCPDCA powder was heated in the presence and absence of a phase transfer catalyst ($BzMe_3NBr$). As indicated in Table 2.3.3, no gel was generated in the absence of $BzMe_3NBr$ (Nos. 1–3). However, the solution gelated in only 25 min after being heated in the presence of 0.1 wt% $BzMe_3NBr$ in the solvent. The gelation was the result of the catalyzed esterification between KDCPDCA and the pendant CH_2—Cl groups of the polymer chain. A phase transfer catalyst was necessary for the esterification because of the low solubility of KDCPDCA in organic solvents.

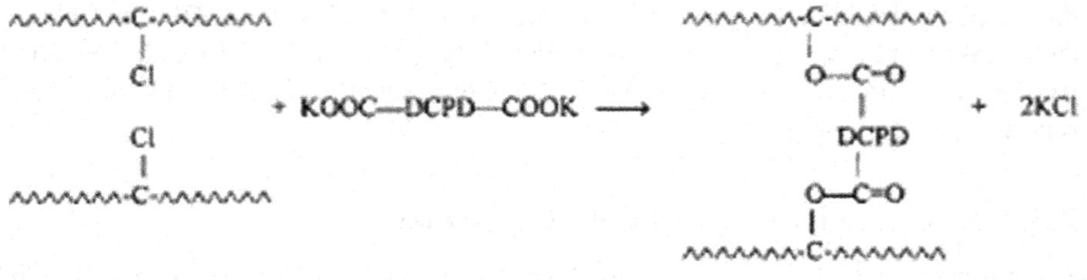

~~~C~~~ (Cl) + Cl (~~~C~~~) + KOOC—DCPD—COOK ⟶ ~~~C~~~ –O–C=O–DCPD–O–C=O– ~~~C~~~ + 2KCl

SCHEME 2.3.1
~~~

TABLE 2.3.3
Gelation Time of a Suspension of KDCPDCA in a e-PSt-CMS Solution

Sample Number	Solvent[a]	$BzMe_3NBr$ (wt% based on solvent)	$COOK/CH_2Cl$ (mole ratio)	Temperature (°C)	Gelation Time	Reaction Time (h)
1	CHO	0	0	145	No gelation	4
2	CHO	0	1/0.8	100	No gelation	4
3	CHO/DMF	0	1/0.8	100	No gelation	2
4	CHO/DMF	0.1	1/0.8	100	25 min	2

[a] CHO = cyclohexanone; DMF = dimethylformamide. CHO/DMF was 3/2 (v/v) in Nos. 3 and 4. The concentration of *e*-PSt-CMS was 6.9 wt% in Nos. 1 and 2, 4.5 wt% in Nos. 3 and 4.

2.3.3.3.2 Effect of Solvent on Gelation

The crosslinking reaction of *e*-PSt-CMS was carried out in toluene, cyclohexanone, or DMF. As seen in Table 2.3.4, the rate of gelation increased in the following sequence: toluene < cyclohexanone < DMF. For comparison, *e*-PSt-CMS also was crosslinked irreversibly with KMA, which possesses carboxyl groups conjugated with the C=C double bond similar to KDCPDCA. The results indicate that the crosslinking reaction was more efficient in DMF than in toluene.

2.3.3.3.3 Crosslinking Reaction of Different Chlorine-Containing Polymers and the Temperature Effect on Reaction

Chloromethylstyrene Containing Polymers. As shown in Tables 2.3.4 and 2.3.5, the CMS containing polymers crosslinked rapidly at 90°C; because of its lower molecular weight, the *s*-PSt-CMS-BA (No. 11) needed additional time to achieve gelation, when compared to *e*-PSt-CMS-BA. The crosslinking reaction took place for the PCMS (No. 14) even at room temperature (24 ± 1°C). This result indicates a high crosslinking reactivity of the chloromethyl group of the CMS monomeric unit.

TABLE 2.3.4
Effect of Different Solvents on Crosslinking of e-PSt-CMS[a]

Sample Number	Crosslinker	Solvent	Temperature (°C)	Gelation Time	Reaction Time (h)	Yield (%)	Gel Content (wt%)
5	KDCPDCA	Toluene	110	8.5 h	13.5	97.6	96.0
6	KDCPDCA	CHO	100	2 h	3	98.6	89.6
7	KDCPDCA	DMF	90	11 min	2	94.3	100.0
8	KMA	Toluene	110	No gelation[b]	16	99.0	48.8
9	KMA	DMF	100	50 min	2	95.9	100.0

[a] A 7 wt% *e*-PSt-CMS solution was used; the mole ratio of $CH_2Cl/COOK$ was 0.8/1; and 0.05 wt% (based on solvent) $BzMe_3NBr$ was added as a catalyst.

[b] The viscosity increased during the reaction, but no gelation occurred.

TABLE 2.3.5
Crosslinking Reaction of CMS Polymers with KDCPDCA[a]

Sample Number	Polymer[b]	CH_2Cl/COOK (mole ratio)	Temperature (°C)	Gelation Time	Reaction Time (h)	Yield (%)	Gel Content (wt%)
10	*e*-PSt-CMS-BA	0.9/1	90	10 min	2	97.2	100.0
11	*s*-PSt-CMS-BA	1.6/1	90	25 min	2	96.9	100.0
12	*s*-PSt-CMS-BA	1.6/1[c]	90	4h	7	93.4	98.7
13	PCMS	10/1	90	33 min	2	95.5	96.7
14	PCMS	10/1	Room temperature	60 h	113	95.3	87.5

[a] 0.1 wt% (based on solvent) $BzMe_3NBr$ was used as a catalyst.
[b] 9 wt% solution in DMF.
[c] KMA replaced KDCPDCA.

Crosslinking Reaction of Commercial Chlorine-Containing Polymers. PECH, PVC, CPP, and CPIP were reacted with KDCPDCA under a variety of conditions. As shown in Table 2.3.6, all these chlorine-containing polymers could be crosslinked. However, their reactivities were very different. CPP exhibited the highest reactivity. Its solution gelated at a relatively high rate even at room temperature. PECH could be gelated at room temperature but very slowly. PVC was less reactive than CPP and PECH because no gel was formed at room temperature. Despite its high chlorine content (65 wt%), CPIP could be crosslinked only at high temperatures (>100°C). The reaction results for PVC clearly indicate that DMF is a better solvent than cyclohexanone (CHO) for this kind of crosslinking. The experimental results regarding the crosslinking reaction allow one to conclude that the crosslinking reactivity of a chlorine-containing polymer is affected by the nature of the polymer.

2.3.3.3.4 IR Analysis of Crosslinked Polymers

IR spectra of the chlorine-containing polymers were recorded before and after crosslinking. Figure 2.3.3 presents the spectra of *e*-PSt-CMS (b) and crosslinked *e*-PSt-CMS (a; No. 7 in Table 2.3.4) and shows that a new strong absorption peak (1726 cm^{-1}) is present in the spectrum of the crosslinked product. This peak can be attributed to the ester carboxyl group(C=O, 1710–1730 cm^{-1}) connected to a C=C double bond[15] and is due to the formation of an ester between KDCPDCA and a chloromethyl group during the crosslinking reaction. Figure 2.3.4 presents the spectra of PECH (b) and its crosslinked product (a; No. 15 in Table 2.3.6). The peak at 1713 cm^{-1} can be attributed to an ester carboxyl group connected to a C=C double bond and is due to the esterification crosslinking reaction. The two peaks at 1634 cm^{-1} and 1603 cm^{-1}, which are present in the spectrum of the crosslinked PECH, can be assigned to the two C=C bonds of the —DCPD—. The peak at 1553 cm^{-1} can be assigned to the carboxyl group of COOK and is due to the unreacted COOK of the crosslinker. The spectra of the other crosslinked polymers provided similar results.

TABLE 2.3.6
Crosslinking Reaction of Different Chlorine-Containing Polymers with KDCPDCA[a]

Sample Number	Polymer[b]	Solvent	Cl/COOK (mole ratio)	Temperature (°C)	Gelation Time	Reaction Time (h)	Yield (%)	Gel Content (wt%)
15	PECH	DMF	10/1	90	20 min	2	99.3	89.4
16	PECH	DMF	10/1	RT[d]	380 h	380		
17	PVC	CHO	10/1	130	0.5 h	1	93.9	94.7
18	PVC	CHO	10/1	100	4h	4.5	90.3	66.1
19	PVC	DMF	10/1	RT[d]	No gelation	380		
20	PVC	DMF	10/1	90	0.5 h	2	95.7	97.6
21	CPP	DMF/toluene[c]	10/1	90	13 min	2	100.0	68.9
22	CPP	DMF/toluene[c]	10/1	RT[d]	12 h	113	98.1	92.4
23	CPIP	DMF	16/1	100	No gelation	8		
24	CPIP	DMF	16/1	130	3h	3	82.0	81.7

[a] 0.2 wt% (based on solvent) $BzMe_3NBr$ was used as a catalyst.
[b] 9 wt% solution was used in the reaction.
[c] DMF/toluene 5 = 1/2 (v/v).
[d] RT (room temperature) = 24 ± 1°C.

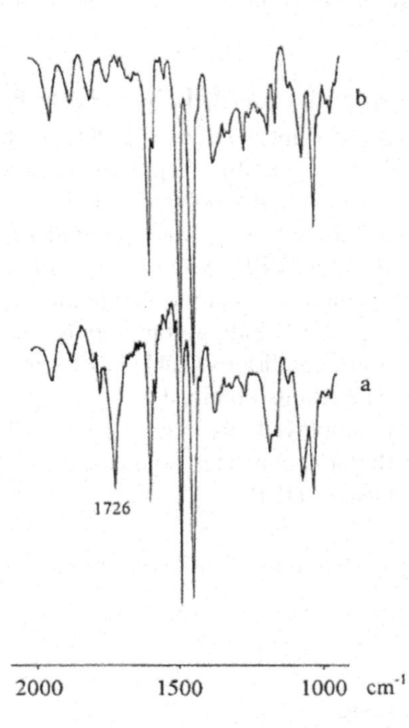

FIGURE 2.3.3 IR spectra of *e*-PSt-CMS: (a) crosslinked and on a KBr tablet and (b) before crosslinking and on a film.

b
a
1603
1713
1634
1553
1800
1400
1000
600 cm⁻¹

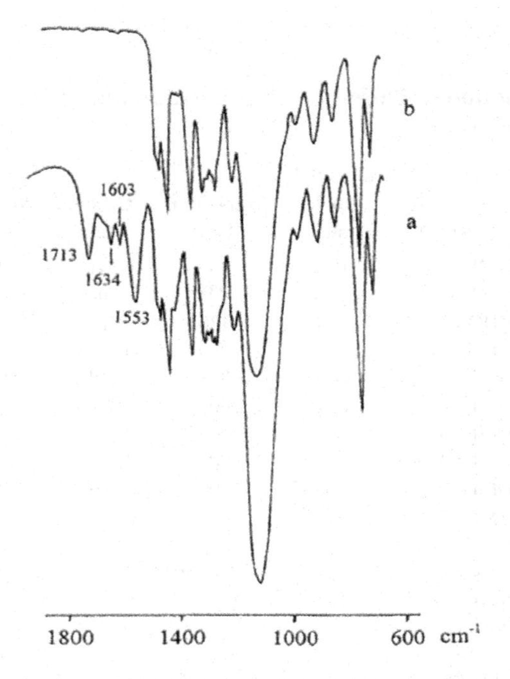

FIGURE 2.3.4 IR spectra of PECH: (a) crosslinked and on a KBr tablet and (b) before crosslinking and on a film.

2.3.3.3.5 Chain Extension of a α,ω-Dihalide Linear Polymer by KDCPDCA

The reversible dimerization/dedimerization of CPD can be employed to prepare thermally reversible chain-extended linear polymers in which a CPD group is located at each of the two ends of the polymer chain. A CPD-ended linear polyisobutylene that exhibited TRC bonding was prepared by Kennedy and Carlson[9] by reacting CPDNa and BrC_3Si—PIB—SiC_3Br. Our aim was to prepare TRC linear polymers by extending the α,ω-dihalide compounds by using KDCPDCA as an extender (Scheme 2.3.2). α,ω-Dihalide compounds were allowed to react with KDCPDCA under a variety of conditions (Table 2.3.7). For comparison, KMA also was used as an extender. The results obtained are listed in Table 2.3.7. The reaction was affected by the solvent employed; the polar solvents (CHO and DMF) exhibited much higher yields than the nonpolar ones (toluene). Also, the reactivity decreased in the sequence: DCX > DBB > DCB.

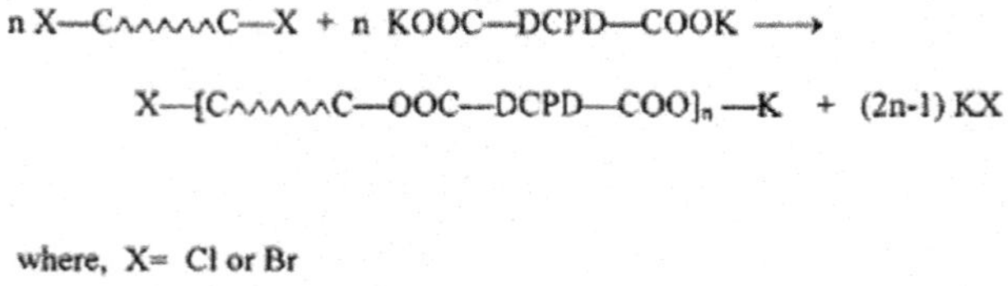

n X—C∧∧∧∧∧C—X + n KOOC—DCPD—COOK ⟶
X—[C∧∧∧∧∧C—OOC—DCPD—COO]n—K + (2n-1) KX
where, X= Cl or Br

SCHEME 2.3.2

TABLE 2.3.7
Reaction of Dihalide Compounds with Dipotassium Salts[a]

Sample Number	Dihalide Compound	Dipotassium Salt	Solvent	Reagents/Solvent (g/100 g)	Temperature (°C)	Reaction Time (h)	Yield (%)
25	DCX	KDCPDCA	Toluene	24.3	110	14	61.5
26	DCX	KDCPDCA	DMF	23.9	90	5	80.6
27	DCX	KMA	Toluene	27.0	110	8.5	64.6
28	DBB	KDCPDCA	Toluene	43.9	110	11	9.0
29	DBB	KDCPDCA	CHO	32.3	125	10	38.1
30	DBB	KDCPDCA	DMF	24.1	90	5	27.1
31	DCB	KDCPDCA	Toluene	36.0	110	11	0.49
32	DCB	KDCPDCA	CHO	20.5	135	9.5	14.8
33	DCB	KDCPDCA	DMF	19.8	90	5	16.8

[a] The mole ratio of dihalide compound/dipotassium salt was 1/1; 1 mol % $BzMe_3NBr$ (based on the mole number of COOK) was used as a catalyst.

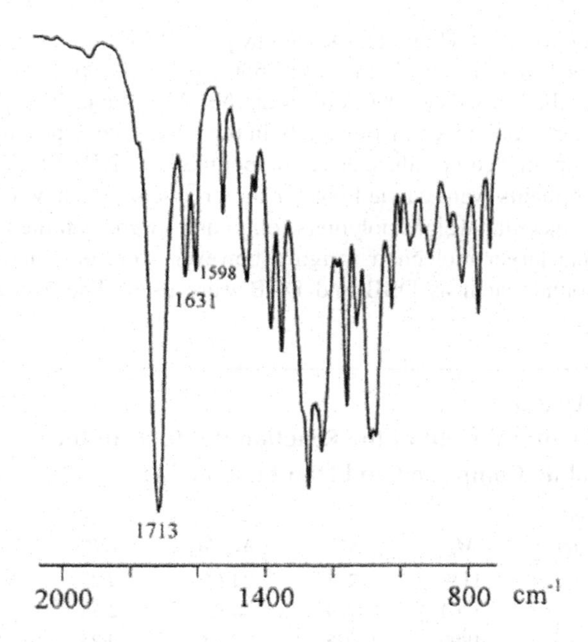

FIGURE 2.3.5 IR spectrum of the reaction product of DCX/KDCPDCA (KBr tablet).

IR spectra were recorded for the reaction products. Two of the spectra are shown in Figures 2.3.5 and 2.3.6. The absorption peaks at 1713 cm^{-1} (ester carboxyl), 1631 cm^{-1}, and 1598 cm^{-1} (—DCPD—structure) are present in the spectrum (Figure 2.3.5) of the reaction product between DCX and KDCPDCA (No. 26 in Table 2.3.7). The fact that there is no COOK absorption peak indicates that all the COOK groups of KDCPDCA

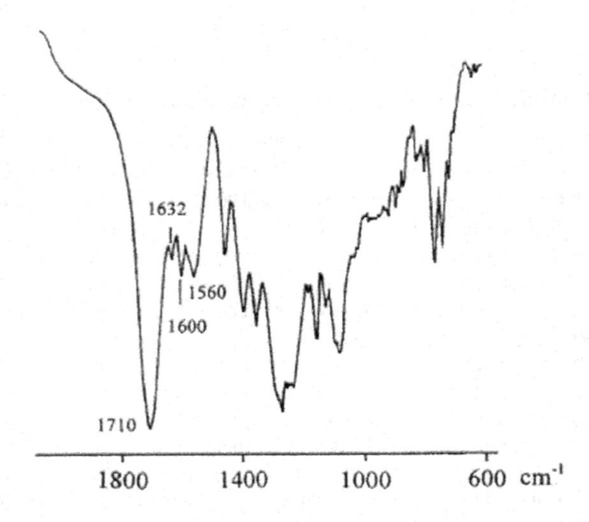

FIGURE 2.3.6 IR spectrum of the reaction product of DBB/KDCPDCA (KBr tablet).

were converted into esters. However, the COOK peak (1560 cm^{-1}) is present, together with the peaks at 1710 cm^{-1}, 1632 cm^{-1}, and 1600 cm^{-1} in the spectrum (Figure 2.3.6) of the product of the DBB/KDCPDCA reaction (No. 29 in Table 2.3.7). The esterification reaction clearly took place incompletely in the latter case. One can conclude that the dihalide compound can produce ester compounds with KDCPDCA.

GPC measurements were carried out for the products. As shown in Table 2.3.8, relatively low molecular weight polymers (oligomers) were obtained in our experiments; somewhat larger molecular weights, however, were obtained when flexible dihalide compounds such as DBB and DCB were used. The average number of

TABLE 2.3.8
Molecular Weight of the Reaction Products of the Dihalide Compound and Dipotassium Salt

Sample Number	M_w	M_n	M_w/M_n	DP[a]	N[b]
25	1139	733	1.55	2.27	3.55
26	1464	734	1.99	2.28	3.55
27	1990	928	2.14	4.25	7.51
29	6167	2332	2.64	8.50	16.00
32	1961	1054	1.86	3.84	6.68

[a] *DP* is the degree of polymerization that is equal to the average value of n in Scheme 2.3.2.

[b] N is the average repeat number of esterifications completed between the dihalide compound and dipotassium salt and is given by the expression $N = 2n-1$.

esterifications per extended chain, N, was found to be as large as 3.5 ~ 16. Therefore, one can conclude that KDCPDCA has enough esterification reactivity for the extension of dihalide ended linear polymers.

2.3.3.4 Thermally Reversible Behavior of the Linkage Based on DCPD Bridges

2.3.3.4.1 Thermocompression of Crosslinked Polymers

Several chlorine-containing polymers that were crosslinked with KDCPDCA were compressed at about 195°C to determine whether the crosslinked CMS homopolymer, CMS-St copolymer, and CMS-St-BA copolymer would exhibit good flowability. Indeed, from these crosslinked polymers, continuous, transparent thin films of about 0.3 mm in thickness with smooth surfaces were obtained via thermocompression. The polymer film subsequently was cut into small pieces, layered to a thickness of about 1.5 to 2 mm, placed in the mold, and subjected again to compression. The cycle was repeated three or four times. A larger number of cycles could not be conducted because of the irreversible changes produced on repeated heating at high temperatures. Similar results were obtained through the compression cycles of the crosslinked CPP. However, noncontinuous, nonsmoothed surfaces and brittle products were obtained when the crosslinked CPIP was compressed under the same conditions; this occurred because of the irreversible Diels–Alder reaction between CPD and the C=C double bonds of the CPIP chains (Scheme 2.3.3). During the compression cycles of the crosslinked PVC or PECH, an irritating odor was generated, and the formed film acquired a dark color. Obviously, the degradation of the PVC or PECH samples due to their dehydrochlorination at a high temperature was responsible for this behavior.

Consequently, the thermocompression experiments indicate that TRC crosslinking behavior depends not only on the reversible conversion of CPD/DCPD but also on the nature and thermostability of the polymer to be crosslinked. In other words, TRC crosslinked polymers can be obtained only from those polymers that do not contain olefinic C=C double bonds and are stable at about 195 ~ 220°C because the latter temperature is required for the dedimerization of the DCPDCA ester.[16]

2.3.3.4.2 Solubility of Crosslinked Polymers

Kennedy and Salamone[8,11] reported that the crosslinked polymers based on DCPD linkages did not dissolve in any solvent even on heating because the monomer-dimer equilibrium ensured that some crosslinkages always remained present. However, these polymers could be dissolved on heating in a solvent containing maleic anhydride because the CPD groups dissociated from DCPD were trapped by the maleic anhydride molecules, which is a stronger dienophile than CPD, so the Diels–Alder dimerization of CPD was prevented.

Our solubility experiments (Table 2.3.9) provided results similar to those of Kennedy and Salamone despite the higher temperature required for the dissociation of the crosslinkages. As shown in Table 2.3.9, the prepared polymers had high gel contents and, except for PECH, could dissolve at least partially in maleic anhydride or

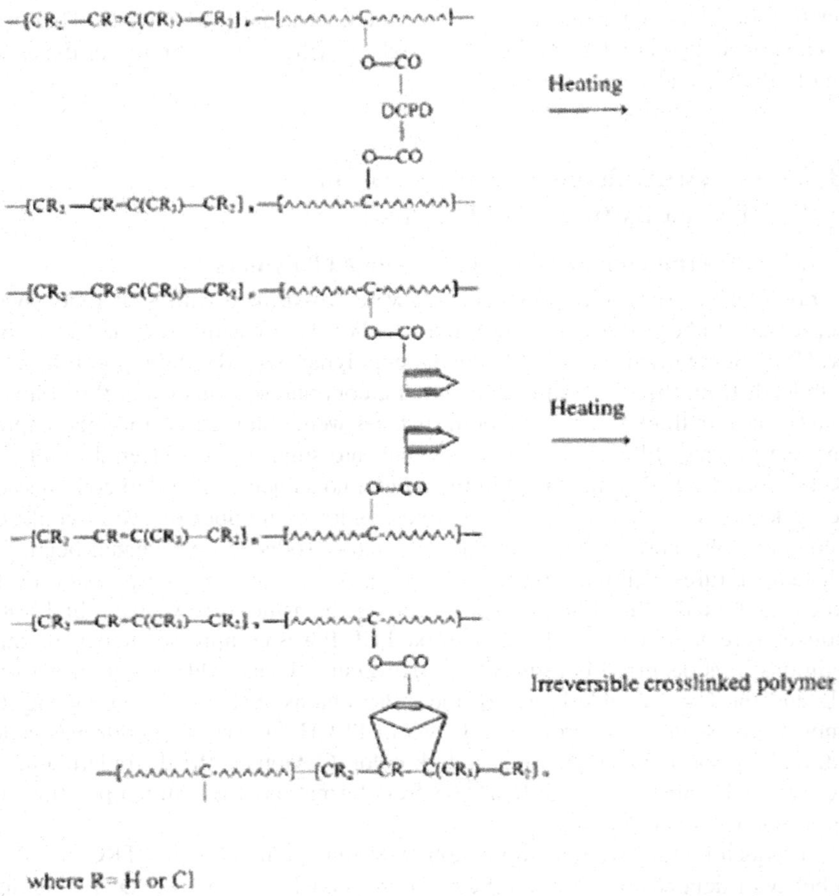

—[CR₂—CR=C(CR₃)—CR₂]ₙ—[∧∧∧∧∧∧C∧∧∧∧∧∧]—
O—CO
DCPD
O—CO
—[CR₂—CR=C(CR₃)—CR₂]ₙ—[∧∧∧∧∧∧C∧∧∧∧∧∧]—

Heating →

—[CR₂—CR=C(CR₃)—CR₂]ₙ—[∧∧∧∧∧∧C∧∧∧∧∧∧]—
O—CO
O—CO
—[CR₂—CR=C(CR₃)—CR₂]ₙ—[∧∧∧∧∧∧C∧∧∧∧∧∧]—

Heating →

—[CR₂—CR=C(CR₃)—CR₂]ₙ—[∧∧∧∧∧∧C∧∧∧∧∧∧]—
O—CO
—[∧∧∧∧∧∧C∧∧∧∧∧∧]—[CR₂—CR—C(CR₃)—CR₂]ₙ

Irreversible crosslinked polymer

where R= H or Cl

SCHEME 2.3.3

maleimide containing dichlorobenzen on heating at 195°C. The highly insoluble 74.7 weight percent in the PStCMS sample crosslinked with KMA was expected because KMA is an irreversible crosslinker. The PECH crosslinked with KDCPDCA could not dissolve in either maleic anhydride or maleimide containing dichlorobenzen on heating, probably because of the degradation of the polymer with the formation of double bonds that reacted irreversibly with the CPD moieties.

2.3.3.4.3 Relationship Between the Specific Viscosity and Temperature of the Linear Oligomer Solutions

The specific viscosities (η_{sp}) of the linear oligomers obtained from the reactions of KDCPDCA with DCX (No. 25 in Table 2.3.8) or DCB (No. 32 in Table 2.3.8) and KMA with DCX (No. 27 in Table 2.3.8) were determined at various temperatures. One can see from the curves in Figure 2.3.7 that the viscosity of the products of DCX and KMA decreased slowly with increasing temperature (Curve C). Similar

TABLE 2.3.9
Solubility of Crosslinked Polymers Based on DCPD Linking Bridges

Crosslinked Polymer	Insoluble % (Refluxed 10 h in THF)	Insoluble % (Treated with a Maleic Compound before Being Refluxed in THF)
e-PSt-CMS, No. 4	84.5	3.8[b]
e-PSt-CMS, No. 6	89.6	0[c]
e-PSt-CMS, No. 7	100.0	21.6[b]
e-PSt-CMS, No. 9[a]	100.0	74.7[b]
PECH, No. 15	89.4	>100[d]
PVC, No. 17	94.7	0[c]
PVC, No. 18	66.1	0[b]
CPP, No. 22	92.4	0[b]

[a] This was a product crosslinked by KMA.
[b] Heated 5 h in maleimide-containing dichlorobenzen at 195°C.
[c] Heated 5 h in maleic anhydride-containing dichlorobenzen at 195°C.
[d] A weight increase was found when the crosslinked product was treated for 5 h at 195°C with either maleimide-containing or maleic anhydride-containing dichlorobenzen.

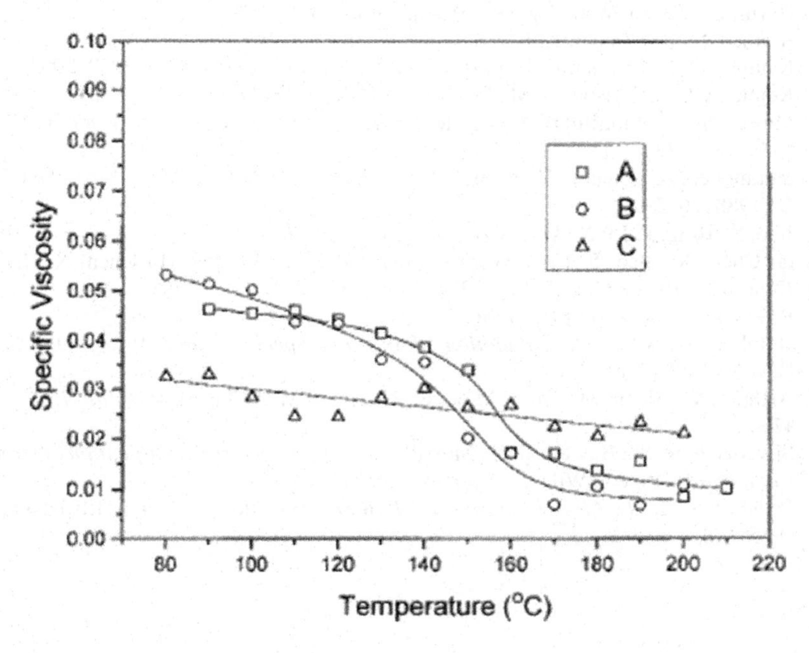

FIGURE 2.3.7 Relationship between the specific viscosity and temperature of the prepared linear oligomers: (A) linear oligomer from KDCPDCA/DCX (No. 25 in Table 2.3.7), (B) linear oligomer from KDCPDCA/DCB (No. 32 in Table 2.3.7), and (C) linear oligomer from KMA/DCX (No. 27 in Table 2.3.7).

slow viscosity changes occurred in the low and high temperature ranges in the solutions of the products of DCX/KDCPDCA and DCB/KDCPDCA. However, the viscosities of the latter two oligomer solutions decreased sharply at a temperature of about 150 ~ 170°C. The high reduction of viscosity is believed to be the result of the dissociation of the —DCPD— moieties in the backbone of the linear oligomer. The viscosity measurements provided a simple proof of the thermally reversible behavior of the linkages achieved via the esterification between chlorine-carbon and KDCPDCA.

2.3.4 CONCLUSION

A new crosslinker, potassium dicyclopentadienedicarboxylate, was employed to prepare (1) thermally reversible covalently crosslinked chlorine-containing polymers and (2) thermally reversible chain-extended α,ω-dihalide compounds.

REFERENCES

1. Engle, L. P.; Wagener, K. B. *J Macromol Sci Rev Macromol Chem Phys* 1993, C33 (3), 239.
2. Craven, J. M. U.S. Patent 3,435,003, 1969.
3. Chujo, Y.; Sada, K.; Saegusa, T. *Macromolecules* 1990, 23, 2636.
4. Canary, S. A.; Stevens, M. P. *J Polym Sci Polym Chem Ed* 1992, 30, 1755.
5. Takeshita, Y.; Uoi, M.; Hirai, Y.; Uchiyama, M. German Patent 2,164,022, 1972.
6. Harumi, A.; Shyuzo, K. Japanese Patent 48-36293, 1973.
7. Kennedy, J. P.; Castner, K. F. U.S. Patent 4,138,441, 1979.
8. Kennedy, J. P.; Castner, K. F. *J Polym Sci Polym Chem Ed* 1979, 17, 2039, 2055.
9. Kennedy, J. P.; Carlson, G. M. *J Polym Sci Polym Chem Ed* 1983, 21, 3551.
10. Masatoshi, M.; Fumihiko, A.; Toshio, U.; Yasuhiro, I.; Kuniharu, N. *Makromol Chem* 1985, 186, 473.
11. Salamone, J. C.; Chung, Y.; Clough, S. B.; Watterson, A. C. *J Polym Sci Polym Chem Ed* 1988, 26, 2923.
12. Jiao, S.; Bing, J.; Li, X. *Acta Polym Sinica* 1994, 6, 702.
13. (a) Chen, X.; Jiao, S. *Chin Synth Rubber Ind* 1996, 19, 159; (b) Chen, X.; Jiao, S. *Chin Synth Rubber Ind* 1998, 21, 336.
14. Peters, D. *J Chem Soc* 1959, 1761.
15. Dolphin, D.; Wick, A. *Tabulation of Infrared Spectral Data*; Wiley-Interscience: New York, 1977; pp. 331–332.
16. Arthurs, M.; Sloan, M.; Drew, M. G. B.; Nelson, S.M. *J Chem Soc Dalton Trans* 1975, 1794.
17. Silverstein, R. M.; Bassler, G. C.; Morrill, T. C. *Spectrometric Identification of Organic Compounds*, 4th ed.; Wiley: New York, 1981; p. 167.
18. Pouchert, C. J. *The Aldrich Library of FT-IR Spectra*; Aldrich Chem: Milwaukee, WI, 1985; Vol. 1, p. 53B.

2.4 Covalent Cross-Linking of Polymers through Ionene Formation and Their Thermal De-Cross-Linking*

Eli Ruckenstein and Xiaonong Chen

Chemical Engineering Department, State University of New York at Buffalo, Amherst, New York 14260

ABSTRACT Quaternization and dequaternization of tertiary amines were employed to generate thermally reversible covalent ionene networks. Chlorine or tertiary amine containing polymers were cross-linked with ditertiary amines and dihalide compounds, respectively, to generate ionene networks. The cross-linking reactivity of the dihalide decreased with decreasing carbonium ion character of its alkyl, and the reactivity of the ditertiary amine was dependent on the steric effect of its alkyl groups and the flexibility of the alkyl connecting the two nitrogen atoms of its molecule. IR and NMR tests, reactive solubility experiments and conductivity, differential scanning calorimetry, and flowability (at 215°C) determinations have been carried out to investigate the thermal reversibility of cross-linked polymers. The thermal de-cross-linking/re-cross-linking was markedly affected by the nature of the tertiary amine and halide that took part in quaternization. Poor reversibility was found when the Cl-containing polymer was cross-linked with a diamine whose nitrogen atoms belonged to a saturated ring, such as dipiperidinomethane and 1,4-dimethylpiperazine. The de-cross-linking and re-cross-linking were rapid when either one or both bridging alkyls of the nitrogen atoms possessed a higher carbonium ion character than the two nonbridging alkyls. The (co)-polymers obtained through the (co)-polymerization of chloromethylstyrene, 2-(dimethylamino)ethyl acrylate,

* *Macromolecules* 2000, 33, 8992–9001.

or vinylpyridine could be effectively quaternized with selected cross-linkers under moderate conditions to generate networks that exhibited thermal reversibility.

2.4.1 INTRODUCTION

Ionenes are polymers that contain quaternary amines in their backbones and can be obtained through the Menschutkin reaction between a dihalide and a ditertiary amine. Numerous papers regarding the preparation, the properties, and the applications of linear ionenes have been reported since the systematic research conducted by Rembaum's group.[1] As well-known, the reversible Menschutkin reaction of a quaternary ammonium occurs at elevated temperatures.[2] In fact, the dequaternization of ionenes at temperatures around 200°C was frequently regarded as a negative feature (decomposition) of ionene polymers.[3] However, very limited results were reported about the thermally reversible dequaternization at high temperatures and requaternization at low temperatures of ionenes. In the past 3 decades, a number of papers[3b–d,4] became available regarding the elastomeric ionene polymers that contained ionene segments and soft blocks in their backbone. These polymers were considered physically cross-linked elastomers because of ion pair interactions. There are also reports dealing with the cross-linking of polymers via ionene covalent bridges.[5] One of them was concerned with a rubber that contained about two ionene bridges per polymer chain and could de-cross-link under mechanical shear and re-cross-link under resting conditions, due to reversible dequaternization.[5d] While the possibility of a thermally reversible cross-linking via ionene bridges was mentioned by Engle and Wagener,[6] no experimental investigation was yet carried out. Only the thermally reversible chain extension of poly(tetramethylene oxide) (PTMO) was studied by Leir et al.[7]

Numerous attempts have been made in the past 3 decades to develop thermally reversible covalent (TRC) cross-linking of polymers, because such cross-linked polymers are expected to be similar to the thermosetting polymers as concerns their physical properties and solvent resistance and can be remolded by thermal processing technologies.[6,8] Several thermally reversible systems, reviewed in ref 6, have been proposed. It should be, however, emphasized that the reactions do not fully reverse; nevertheless, they reverse enough for the de-cross-linking to provide thermoplasticity to the cross-linked polymers.

In the present paper the quaternization/dequaternization of quaternary ammoniums was applied to polymer cross-linking through ionene covalent bridges. These experiments were attempted because (1) no catalyst and no third component are required for the quaternization/dequaternization equilibrium which is controlled only by temperature; (2) the polymer network can be generated at a high rate under moderate conditions, because of the facile formation of the ionene moieties; (3) the dequaternization temperatures of most ionenes are around 200 °C,[3,9] the temperature employed in the thermal processing of many traditional polymers; and (4) the electrostatic resistance of the cross-linked polymers can be improved by the ammonium salt ionic pairs, which provide a higher conductivity than the hydrocarbon chains.

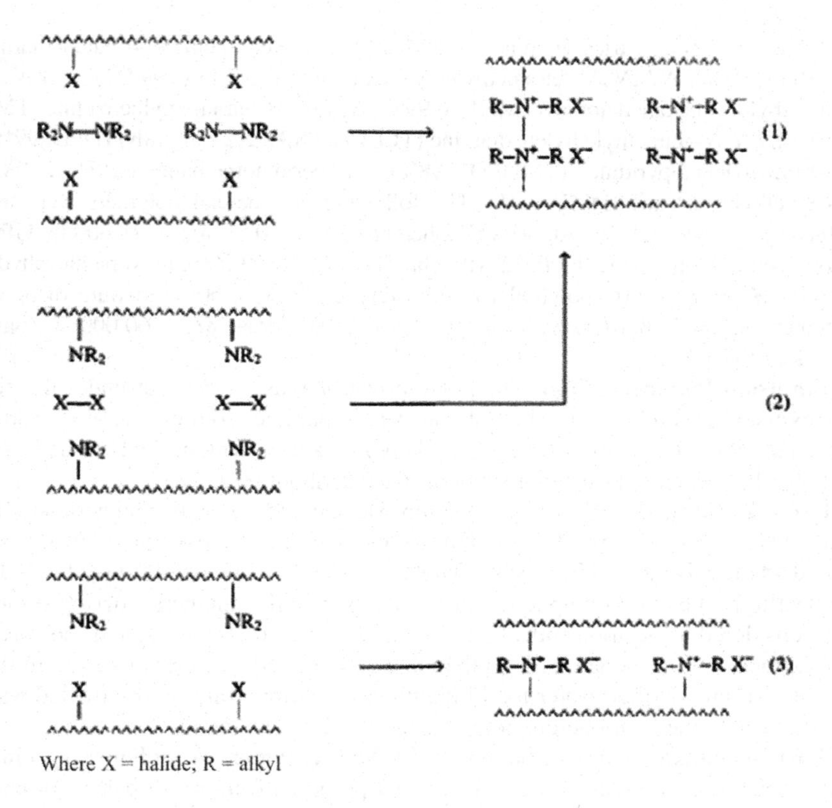

SCHEME 2.4.1

For the above reasons, the cross-linking reactions between (1) Cl-containing polymers and bifunctional tertiary amines, (2) polymers containing pendant tertiary amino groups and dihalide compounds, and (3) Cl-containing polymers and polymers containing pendant tertiary amino groups were carried out (Scheme 2.4.1). The thermal reversibility of these cross-linked polymers was investigated through IR and NMR tests, reactive solubility, differential scanning calorimetry (DSC), and conductivity and flowability measurements on heating.

2.4.2 EXPERIMENTAL SECTION

Materials. All materials were purchased from Aldrich except when noted otherwise. Styrene (St, 99%), butyl acrylate (BA, Fluka, >99%), chloromethylstyrene (CMS, 90%), and 2-(dimethylamino)ethyl acrylate (DMAEA, 98%) were distilled under reduced pressure before use. The solvents and the following chemicals were used as supplied: potassium persulfate (KPS, 99.99%), dodecyl sulfate sodium salt (98%), 4-*tert*-butylcatehol (TBC, 97%), 1,3-dichloroacetone (DCA, 95$^+$%), 1,4-dichlorobutane (DCB, 99%), 1,4-dibromobutane (DBB, 99%), α,α′-dichloro-*p*-xylene (DCX, 98%), α,α′-dibromo-*p*-xylene (DBX, 97%), benzyl bromide (BzB, 98%), benzyl chloride (BzC, Mallinckrodt, Analytical Reagent, boiling range 177.1°C – 179.3°C), *N*,*N*-dimethylbenzylamine

(DMBA, 99+%), triethylamine (99%), *N,N,N′,N′*-tetramethyl-1,4-butanediamine (TMBDA, 98%), *N,N,N′,N′*-tetramethylethylenediamine (TMEDA, 99.5%), *N,N,N′,N′*-tetramethyl-1,6-hexanediamine (TMHDA, 99%), *N,N,N′,N′*-tetramethylbenzidine (TMB, 97%), *N,N,N′,N′*-tetraethylethylenediamine (TEEDA, 98%), 2,2′-dipyridyl (DPD, 99+%), 4,4′-trimethylenedipyridine (TMDPD, 98%), dipiperidinomethane (DPDM, 98%), 1,4-dimethylpiperazine (DMP, 98%). The following commercial polymers were used as received: poly(vinyl chloride) (PVC, inherent viscosity 0.51, M_w = 71 600 by GPC), chlorinated polypropylene (CPP, 32 wt% chlorine, M_w = 150 000), polyepichlorohydrin (PECH, M_w = 700 000), poly(chloromethylstyrene) (PCMS, 60/40 mixture of 3- and 4-isomers, M_w = 55 000), poly(4-vinylpyridine) (PVP, average M_w = 60 000, T_g (onset, annealed) 137°C).

Emulsion Polymerization. The polymerization was carried out under the conditions listed in Table 2.4.1. The polymer was separated from emulsion by adding methanol (containing 2 wt% TBC), then kept in water overnight, and washed twice with distilled water before being vacuum-dried at about 60°C.

Cross-Linking Reaction. The reaction was carried out with magnetic stirring under a N_2 atmosphere by adding to a polymer solution the cross-linker and 0.1 wt% (based on the solvent) of TBC (as inhibitor to avoid the formation of free radicals during heating). The duration up to the moment at which the magnetic stirring stopped was considered as gelation time. The gel was cut into small pieces, and the solvent was extracted from the gel with methanol. The obtained polymer pieces were then submerged into distilled water for 12 h and washed three times with distilled water and twice with methanol before being vacuum-dried at about 60°C.

Thermocompression of the Cross-Linked Polymer. A polymer specimen with a thickness of about 1.5–2 mm was first prepared from small pieces (of about 0.5 mm size) of cross-linked polymer, using a macro–micro KBr die. Then the obtained specimen was placed between two plates and pressed for about 10 min with a Carver Laboratory Press (model C) at about 215°C under a pressure of about 700 psi. Subsequently, the polymer specimen was pressed again at room temperature (about 15 min) in another Carver Press under a pressure of about 1200 psi, to produce a polymer film with a thickness of about 0.2 mm. The obtained film was cut into small pieces which were piled up in four or five layers before being again subjected to compression. The two plates were coated with Silicone Fluid SF 96/50 (Thomas Scientific) to facilitate the release after compression.

Measurements. ^{1}H NMR spectra were recorded in $CDCl_3$ (99.8 atom % D, 0.05 vol % tetramethylsilicone) or in a mixture of $CDCl_3$ and THF-d_8 (99.5 atom % D) on a VXR-400 spectrometer. The cross-linked polymer was ground as a fine powder before being introduced into the NMR tube. The FT-IR spectra were recorded on a Perkin-Elmer 1760-X spectrometer. Gel permeation chromatography (GPC, Waters) measurements were carried out using THF as solvent, at 30°C. The gel content of the cross-linked polymer (denoted gel %) was determined by wrapping 0.1 g of cross-linked polymer in a filter paper (Whatman, Φ90 mm) and subjecting it for 10 h to extraction with refluxed THF in a Soxhlet equipment. The package was then vacuum-dried at a temperature of about 60°C, until a constant weight was reached. The gel % was calculated from the weight increase above that of the filter paper and the original weight of the polymer tested. The insoluble percentage of cross-linked

TABLE 2.4.1
Emulsion Polymerization[a]

Polymer	Monomer Ratio (wt/wt)	Monomer/ H_2O (wt/wt)	KPS/Monomer (wt/wt)	Temp (°C)	Time (h)	Yield (%)	*Mw*(by GPC)		Composition (mol % by ^1H NMR)		
							$10^{-5}\ M_n$	$10^{-5}\ M_w$	BA	St	Functional Unit
PBSC	BA/St/CMS 7.3/8.1/1.0	21.6/100	1.4/100	45	7	95	1.51	7.49	42.8	51.4	5.8 (CMS)
PBN	BA/DMAEA 10.5/1.0	19.2/100	1.6/100	45	8	95	0.42	1.43	94.5		5.5 (DMAEA)
PSN	St/DMAEA 18.9/1.0	26.1/100	0.5/100	40	10	96	0.46	2.02		95.9	4.1 (DMAEA)

[a] 1 wt% (based on water) dodecyl sulfate sodium salt was used as the emulsifier. BA = butyl acrylate; St = styrene; CMS = chloromethylstyrene; DMAEA = 2-(dimethylamino) ethyl acrylate.

polymer after its treatment with a monofunctional halide compound or tertiary amine (denoted gel′ %) was obtained as follows: 0.1 g of cross-linked polymer was introduced into 5 g of a solvent containing a monofunctional compound and 0.3 wt% (based on solvent) TBC and heated in an oil bath at various temperatures. The mixture was then filtered through a medium fast filter paper, well wrapped up with such paper, and subjected to the same extraction and drying procedures employed in the determination of the gel % described above. The following method was employed in the conductivity measurements. Two stainless steel plates (5 mm × 20 mm), each connected to one of the electrodes of a conductivity cell (cell constant, $K = 1.10$, at 25°C), were placed parallel to each other at a distance δ of 0.3 cm into a small test tube (volume ca. 2 mL). A mixture of a polymer solution and a cross-linker was then added to the test tube that was plugged with a rubber stopper and further sealed by winding with a Teflon tape. The conductivity cell was connected to a Jenway 4010 conductivity meter to detect the conductivity during gelation and during heating (at a rate of about 8 °C/min) and cooling (at a rate of about –3°C/min) cycle of the polymer gels. Differential scanning calorimetry (DSC) was carried out on a Perkin-Elmer 7 series thermal analysis system under a helium atmosphere at a rate of 10°C/min.

2.4.3 RESULTS AND DISCUSSION

Preparation of Reactive Copolymers. In our experiments, chloromethylstyrene (CMS) and 2-(dimethylamino)ethyl acrylate (DMAEA) were employed as functional monomers; they were copolymerized with styrene (St) and/or butyl acrylate (BA), to obtain the rubberlike BA-St-CMS copolymer (PBSC in Table 2.4.1), the BA-DMAEA copolymer (PBN in Table 2.4.1), and the plasticlike St-DMAEA copolymer (PSN in Table 2.4.1).

Cross-Linking of Chlorine-Containing Polymers. Nine bifunctional tertiary amines were employed as cross-linkers for PCMS and PBSC. Table 2.4.2 shows that the CMS (co)polymers could be cross-linked with all the diamines employed. No gelation was, however, observed when the PCMS solution was heated at a relatively high temperature (100°C) for a long time, after a monofunctional tertiary amine was added (C2 and C3 in Table 2.4.2), or heated without adding the above amine at a higher temperature (140°C, C1). Figure 2.4.1 presents the IR spectra of the PCMS (b) and its cross-linked product with tetramethyl-1,6-hexanediamine (TMHDA) (a). The peaks at 1670, 1213, and 1159 cm^{-1} can be assigned to the quaternary salt (1640, 1218, and 1154 cm^{-1} were reported in the literature[10]). These results clearly indicate that the gelation was caused by the formation of covalent ionene bridges between the polymer chains and not by the ion pair interactions between the quaternary salt moieties. The cross-linking reactivity was found to depend on the structure of the diamine employed. As shown in Table 2.4.2, tetraethylethylenediamine (TEEDA) provided much lower cross-linking reactivity than tetramethylethylenediamine (TMEDA). This was most likely caused by the stronger steric effect of ethyl than methyl groups. The steric effect can also explain the low gelation rate when dipiperidinomethane (DPDM) was used as cross-linker. When 2,2′-dipyridyl (DPD) was employed, the gelation time became twice as large than for 4,4′-trimethylenedipyridine (TMDPD), which contains a flexible trimethylene moiety between its two N atoms. The gelation rate increased with increasing number of methylene groups between the two N atoms

TABLE 2.4.2
Cross-Linking and De-Cross-Linking of Polymers Containing CMS Monomeric Unit

No.	Polymer[a]	Cross-Linker[b]	Amino/Halide (mol/mol)	Temp (°C)	Gelation Time	Reaction Time (h)	Yield (%)	Gel %[c]	Gel′ %[d]	Flowability at 215°C (Y/N?)
C1	PCMS	No	0	140	no gelation	4				
C2	PCMS	Et_3N	0.2/1	100	no gelation	22				
C3	PCMS	$BzMe_2N$	0.2/1	100	no gelation	22				
C4	PCMS	TMHDA	0.1/1	65	7 min	2	95	100	19	Y
C5	PBSC	DMP	2/1	65	8 h	9	100	100	88	N
C6	PBSC	DPDM	2/1	65	26 h	28	95	100	48	N
C7	PBSC	DPDM	2/1	100	4 h	4	100	100		
C8	PBSC	DPD	2/1	65	24 h	25	97	100	6	Y
C9	PBSC	TMDPD	2/1	65	12 h	13	100	100		Y
C10	PBSC	TMB	2/1	65	32 h	32	96	100	5	Y
C11	PBSC	TMB	2/1	100	4 h	4	100	99		
C12	PBSC	TEEDA	2/1	65	24 h	25	95	100	91	N
C13	PBSC	TEEDA	2/1	100	4 h	4	100	100		
C14	PBSC	TMEDA	2/1	65	3 h	4	99	100	12	Y
C15	PBSC	TMBDA	2/1	65	1 h, 25 min	2	97	100	10	Y
C16	PBSC	TMHDA	2/1	65	50 min	2	98	100	8	Y

[a] PCMS = poly(chloromethylstyrene), 9.5 wt% solution in dimethylformamide (DMF); PBSC = butyl acrylate–styrene–chloromethylstyrene copolymer of Table 2.4.1, 6 wt% solution in DMF.

[b] TMHDA = tetramethyl-1,6-hexanediamine; DMP = 1,4-dimethylpiperazine; DPDM = dipiperidinomethane; DPD = 2,2′-dipyridyl; TMDPD = 4,4′-trimethylenedipyridine; TMB = tetramethylbenzidine; TEEDA = tetraethylethylenediamine; TMEDA = tetramethylethylenediamine; TMBDA = tetramethyl-1,4-butanediamine.

[c] The gel in the cross-linked product.

[d] The gel after the cross-linked product together with a mixture of benzyl bromide/dichlorobenzene (1/6 by weight) was heated 4 h in a 190°C oil bath.

FIGURE 2.4.1 IR spectra of PCMS: (a) cross-linked with TMHDA (C4 in Table 2.4.2, KBr tablet); (b) before cross-linking (film).

of the tetramethyldiamine molecule (no. C14–16 in Table 2.4.2). For tetramethylbenzidine (TMB) as cross-linker, which contains a rigid biphenyl group between its two N atoms, the gelation time was very long at 65°C.

To examine the cross-linkability of Cl-containing polymers with low reactive C—Cl sites, polyepichlorohydrin (PECH, containing primary Cl), poly(vinyl chloride) (PVC, containing secondary Cl), and chlorinated polypropylene (CPP, containing tertiary Cl) were reacted with TMHDA or poly(4-vinylpyridine) (PVP) (Table 2.4.3). No gelation of the PVC or PECH solutions was observed when they were heated with TMHDA, but partly cross-linked PECH and PVC could be obtained when PVP was used as a multifunctional cross-linker. A high gel % was obtained when CPP was allowed to react with either TMHDA or PVP. The weak carbonium ion character of the secondary and primary C—Cl can be considered responsible for the poor cross-linkabilities of PVC and PECH, since the Menschutkin reaction involves a nucleophilic substitution between an alkyl halide and a tertiary amine.[2]

Cross-Linking of Polymers Containing Tertiary Amine Pendants. Five α, *ω*-dihalides were employed as cross-linkers for the polymers containing tertiary amine pendant groups, such as BA-DMAEA copolymer (PBN in Table 2.4.1), St-DMAEA copolymer (PSN in Table 2.4.1), and PVP. As shown in Table 2.4.4, the gelation times of PSN and PVP were 20 (N8) and 13 min (N15), respectively, at room temperature (ca. 24°C), for α,α′-dibromo-*p*-xylene (DBX) as cross-linker, indicating a high cross-linking reactivity of this dihalide. No gelation was reached, however, when PSN and PVP were reacted with a monohalide, such as benzyl bromide (BzB,

TABLE 2.4.3
Cross-Linking and De-Cross-Linking of Chlorine-Containing Polymers

No.	Polymer[a]	Cross-Linker[b]	Temp (°C)	Gelation Time (h)	Reaction Time (h)	Yield (%)	Gel %[c]	Gel′ %[d]	Flowability at 215°C (Y/N?)
CL1	CPP	TMHDA	65	no gelation	25				
CL2	CPP	TMHDA	100	12[e]	12	85	98	0	Y
CL3	CPP	PVP	100	14	15	91	92	0	Y
CL4	PECH	TMHDA	65	no gelation	25				
CL5	PECH	TMHDA	100	no gelation	22				
CL6	PECH	PVP	100	14	15	100	78	0	
CL7	PVC	TMHDA	65	no gelation	25				
CL8	PVC	PVP	100	28	28	100	56	0	

[a] CPP = chlorinated polypropylene, 10 wt% solution in the mixed solvent of dimethylformamide (DMF) and toluene (1/2 by volume); PECH = poly(epichlorohydrin), 9 wt% solution in DMF; PVC = poly(vinyl chloride), 9 wt% solution in DMF.

[b] TMHDA = tetramethyl-1,6-hexanediamine; PVP = poly(4-vinylpyridine), 10 wt% solution in DMF; amino/halide (mol/mol) was 0.1/1.

[c] The gel in the cross-linked product.

[d] The gel after the cross-linked product together with a mixture of benzyl chloride/dichlorobenzene (1/6 by weight) was heated 5 h in a 190°C oil bath.

[e] No gelation was observed during the reaction and the obtained mixture gelated when it was cooled to room temperature after reaction.

TABLE 2.4.4
Cross-Linking and De-Cross-Linking of Polymers Containing Tertiary Amino Groups

No.	Polymer[a]	Cross-Linker[b]	Halide/Amino (mol/mol)	Temp (°C)	Gelation Time	Reaction Time (h)	Yield (%)	Gel %[c]	Gel′ %	Flowability at 215°C (Y/N?)
N1	PBN	DBX	2/1	65	1 h, 40 min	5	93	100	0[d]	Y
N2	PBN	DCX	2/1	65	10.5 h	16	100	100	0[d]	Y
N3	PBN	DCA	2/1	65	10 h	16	95	100	0[d]	Y
N4	PBN	DBB	2/1	65	no gelation	21				
N5	PBN	DBB	2/1	100	no gelation	19				
N6	PBN	DCB	2/1	65	no gelation	21				
N7	PBN	DCB	2/1	100	no gelation	19				
N8	PSN	DBX	2/1.8	RT[f]	20 min	30	100	96	0[e]	Y
N9	PSN	DCX	2/1.8	60	3.5 h	5	100	98	0[e]	Y
N10	PSN	DCA	2/1.8	60	2 h, 40 min	5	100	87	16[d]	Y
N11	PSN	DBB	2/1.8	60	14 h	17	98	96	0[e]	Y
N12	PSN	DCB	2/1.8	60	no gelation	17				
N13	PSN	DCB	2/1.8	100	no gelation	9				
N14	PSN	BzB	2.4/1	100	no gelation	26				
N15	PVP	DBX	0.1/1	RT[f]	13 min	29	99	100	6[d]	Y
N16	PVP	DCX	0.1/1	60	1 h, 30 min	5	100	99	19[d]	Y
N17	PVP	DCA	0.1/1	60	12 min	5	99	100	28[d]	Y
N18	PVP	DBB	0.1/1	60	1 h, 35 min	5	100	99	86[d]	N

(Continued)

TABLE 2.4.4 (*Continued*)
Cross-Linking and De-Cross-Linking of Polymers Containing Tertiary Amino Groups

No.	Polymer[a]	Cross-Linker[b]	Halide/Amino (mol/mol)	Temp (°C)	Gelation Time	Reaction Time (h)	Yield (%)	Gel %[c]	Gel′ %	Flowability at 215°C (Y/N?)
N19	PVP	DCB	0.1/1	60	no gelation	17				
N20	PVP	DCB	0.1/1	100	no gelation	9				
N21	PVP	BzC	0.2/1	100	no gelation	26				
N22	PVP	PCMS	1/1	RT[f]	4h	96	100	100		

[a] PBN = butyl acrylate–dimethylaminoethyl acrylate copolymer of Table 2.4.1, 7 wt% solution in dimethylformamide (DMF); PSN = styrene–(dimethylamino)ethyl acrylate copolymer of Table 2.4.1, 7.5 wt% solution in DMF; PVP = poly(4-vinylpyridine), 10 wt% solution in DMF.

[b] DBX = α,α′-dibromo-*p*-xylene; DCX = α,α′-dichloro-*p*-xylene; DCA = 1,3-dichloroacetone; DBB = 1,4-dibromobutane; DCB = 1,4-dichlorobutane; BzB = benzyl bromide; BzC = benzyl chloride; PCMS = poly(chloromethylstyrene), 9.5 wt% solution in DMF.

[c] The gel in the cross-linked product.

[d] The gel after the cross-linked product together with a mixture of dimethylbenzylamine/dichlorobenzene (1/10 by weight) was heated 4 h in a 190°C oil bath.

[e] The gel after the cross-linked product together with a mixture of BzB/benzophenone (1/4 by weight) was heated 2 h in a 210°C oil bath.

[f] RT = room temperature, 24 ± 1°C.

N14) and benzyl chloride (BzC, N21), respectively, at large halide/amine mole ratios and a high temperature (100°C) for as long as 26 h. These results clearly indicate that the formation of covalent ionene bridges and not the ion pair interactions was responsible for the cross-linking of these polymers. No gelation was, however, observed for 1,4-dichlorobutane (DCB) as cross-linker, indicating a low reactivity of the primary chlorine of the DCB. As expected, the cross-linking reactivity decreased in the carbonium ion character sequence DBX > DCA > DCX > DBB > DCB.

Thermal De-Cross-Linking/Re-Cross-Linking of the Networks Generated with Polymers Containing Tertiary Amine Pendants. The cross-linked polymers were insoluble even on heating in pure dichlorobenzene or benzophenone. However, they could be dissolved at least partially after heated in the same solvent containing dimethylbenzylamine (DMBA) or benzyl bromide (BzB) (Table 2.4.4). As shown in Scheme 2.4.2, when the polymer was heated in the pure solvent, the thermodynamic equilibrium between dequaternization and quaternization allowed some of the ionene bridges to remain, and this kept the polymer insoluble. The addition of BzB or DMBA to the solvent displaced, however, the equilibrium in the direction of dequaternization, resulting in an enhanced dissociation of the polymer.

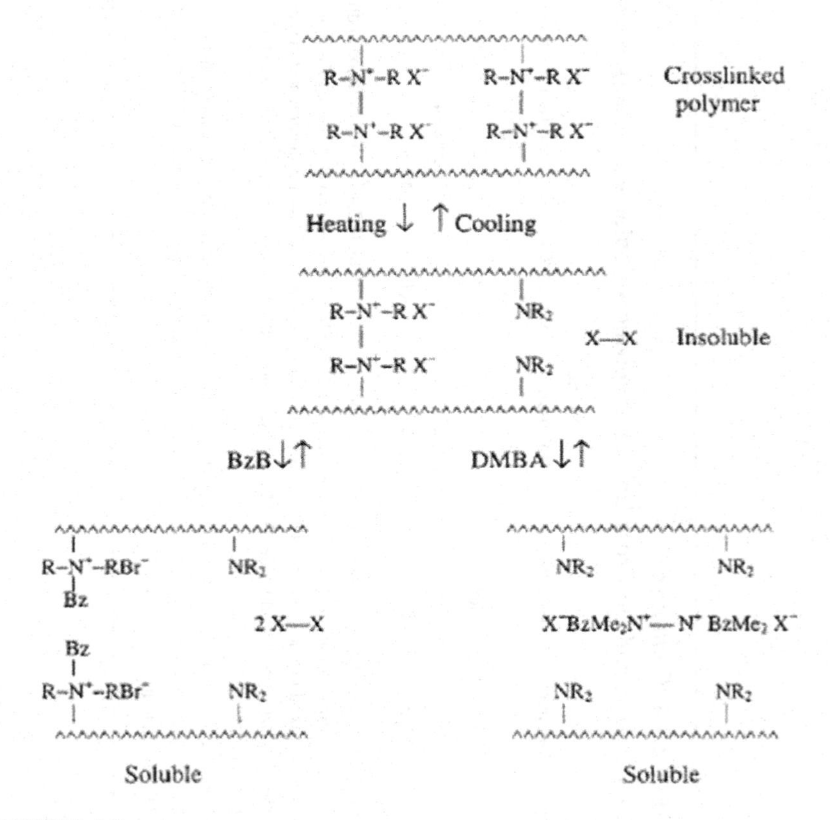

SCHEME 2.4.2

The flowability, which constitutes an effective indication of de-cross-linking on heating, was investigated by thermocompression (heating/cooling compression cycle). A cross-linked polymer was regarded as thermally flowable when (1) a continuous, transparent thin film of about 0.2 mm in thickness could be obtained after each of the successive thermocompression cycles and (2) the film did not dissolve or crack into small gel pieces when it was submerged into dimethylformamide (DMF) immediately after each compression cycle. As shown in Table 2.4.4, all cross-linked polymers except PVP cross-linked with 1,4-dibromobutane (DBB, N18) exhibited good flowability on heating during compression. No notable change in the flowability was observed even after five cycles when the PSN cross-linked with 1,3-dichloroacetone (DCA, N10) or DBB (N11) was subjected to five successive thermocompressions. The film thus obtained was found to merely swell when introduced in DMF immediately after the fifth cycle. However, the PVP film, cross-linked with DCA (N17), could dissolve in DMF after the fifth cycle, indicating that no re-cross-linking occurred during the cooling stage of the fifth cycle. Most likely the DCA, generated via dequaternization, escaped from the system because the heating temperature (215°C) was much higher than its boiling point (173°C). The thermocompression and swelling experiments indicated that the de-cross-linking during heating and re-cross-linking during cooling were rapid, the former lasting about 10 min and the latter about 15 min.

To investigate the effect of the structure of the reactants on the reversible conversion of the ionene network, it is useful to examine the mechanism of dissociation of the ionene bridges. Generally, several competitive reactions may occur during dequaternization of an ammonium salt because an alkyl halide can be generated from the halide anion X^- and any one of the four alkyl groups (in the present case two bridging and two nonbridging alkyls) connected with each of the nitrogen atoms. Of course, the competitive reactions that result in the generation of nonbridging alkyl halides could not induce the dissociation of the network. Two factors ensure that the dequaternization will lead to the de-cross-linking of the polymer. First, the tension that exists along the linking bridges, especially when the network is subjected to shear stress, enhances the likelihood of dequaternization via the dissociation of the bridges. Second, the reverse Menschutkin reaction as well as its forward reaction involve the formation of a carbonium ion in the transition state. The alkyl that generates the strongest carbonium ion in the transition state will dissociate prior to the other three alkyls, resulting in the corresponding alkyl halide and tertiary amine. Indeed, only the alkyl that possessed the higher carbonium ion strength dissociated during the dequaternization of ammonium salts.[2b,c] Consequently, for the dissociation of the ionene network to occur during the dequaternization of the cross-linked PSN and PBN, the bridging alkyls should possess greater carbonium ion strength than the nonbridging ones. The cross-linked DMAEA containing polymers can dissociate in two ways, with the formation of either —OCH_2CH_2—halide (and the corresponding Me_2N—) or —OCH_2CH_2—NMe_2 (and the corresponding halide of the cross-linker). In contrast, the de-cross-linking of the cross-linked PVP can occur only through the formation of the stable aromatic pyridine group and the corresponding halide cross-linker. However, the carbonium ion strength of butylene is lower than that of benzyl, and this constitutes most likely the reason for the high gel′ % (86%) and poor flowability of the DBB cross-linked PVP (N18 in Table 2.4.4).

A side reaction that might occur is the Hoffmann degradation. Indeed, traces of the Hoffmann degradation products were detected even when a diamine and a dihalide were polymerized at room temperature.[11] In addition, the mass spectrometry of the pyrolysis of tetraalkylammonium bromides indicated that the intensity of $[HBr]^+$ ions (one of the degradation products) was 0.2–2.6% (mostly <1%) of the total ion current (ion source temperature 210–230°C).[12] The Hoffmann degradation might be one of the side reactions during the dequaternization of the cross-linked polymers that possess βH in their ammonium moieties. Such a degradation and the subsequent irreversible changes of the generated C=C bonds might result in irreversible cross-linking.

The IR and ^{1}H NMR spectra of DCA and DBB cross-linked PSN (N10 and N11 in Table 2.4.4) were recorded before and after the thermocompression cycles. Figure 2.4.2 presents the IR spectra of the non-cross-linked PSN (a) and DBB

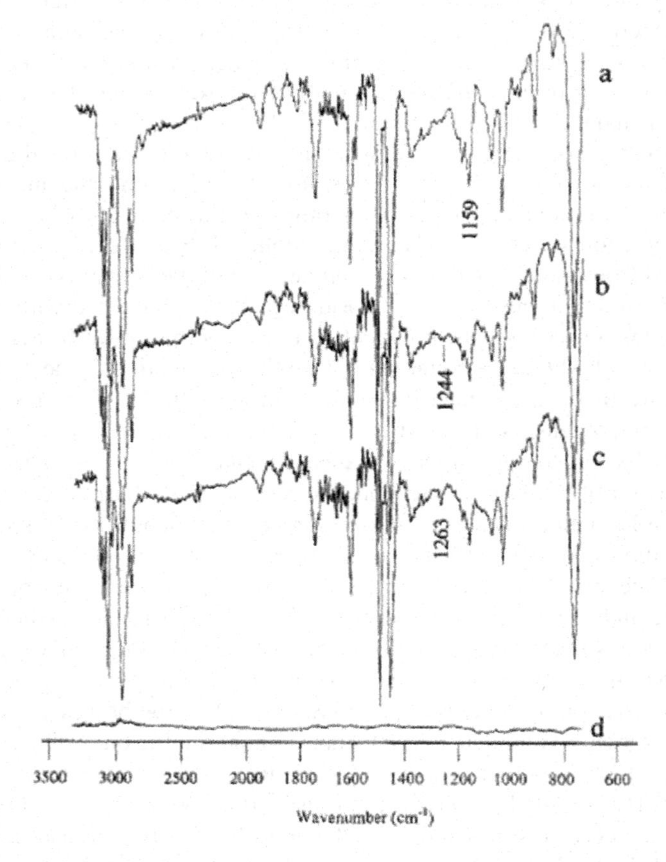

FIGURE 2.4.2 IR spectra of PSN (KBr tablet): (a) without cross-linking (PSN in Table 2.4.1); (b) cross-linked with DBB (N11 in Table 2.4.4); (c) polymer N11 in Table 2.4.4 after five thermocompression cycles; (d) differential spectrum of the polymer N11 (c–b).

cross-linked ones, before (b) and after five compression cycles (c). The peak at 1159 cm^{-1} can be assigned to $-NMe_2$ of the DMAEA functional monomeric unit (1164.5 cm^{-1} in the literature[13]), the peak at 1244 cm^{-1} to the quaternary ammonium moieties (1244.2 cm^{-1} in the literature[14]) of the generated ionene bridges, and the peak at 1263 cm^{-1} to the unreacted —CH_2Br groups (1260.6 cm^{-1} in the literature[15]). Spectra b and c and their differential (Figure 2.4.2d) indicate that almost no detectable change in the polymer chain structure occurred after the polymer was subjected to five compression cycles. The —CH_2Br peak (1263 cm^{-1}) became somewhat stronger, indicating that some of the —CH_2Br groups generated via dequaternization on heating did not requaternize during the cooling stage of the cycles. A similar conclusion was obtained from the spectra recorded for DCA cross-linked PSN, before (Figure 2.4.3a) and after five compression cycles (Figure 2.4.3b) and their differential (Figure 2.4.3c). In the latter case, the peak at 1269 cm^{-1} can be assigned to the unreacted —CH_2Cl groups of DCA (1260 cm^{-1} in the literature[16]).

The ^1H NMR spectra of the cross-linked PSN were recorded by dispersing and swelling polymer powders in a solvent mixture of $CDCl_3$ and THF-d_8 (about 7/3 by volume). Figure 2.4.4 presents the ^1H NMR spectra of the non-cross-linked PSN

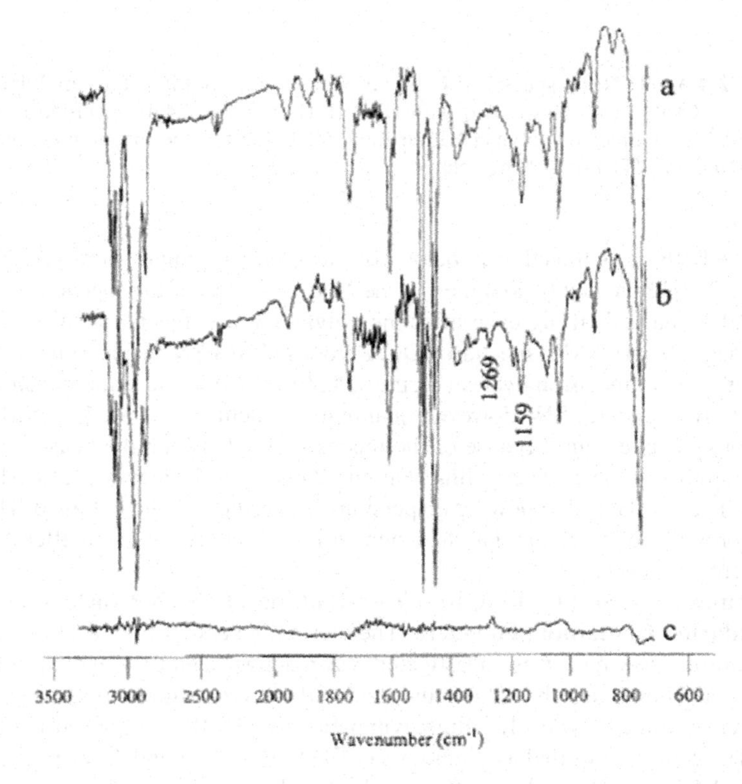

FIGURE 2.4.3 IR spectra of DCA cross-linked PSN (N10 in Table 2.4.4, KBr tablet): (a) before any thermocompression; (b) after five thermocompression cycles; (c) differential spectrum (a–b).

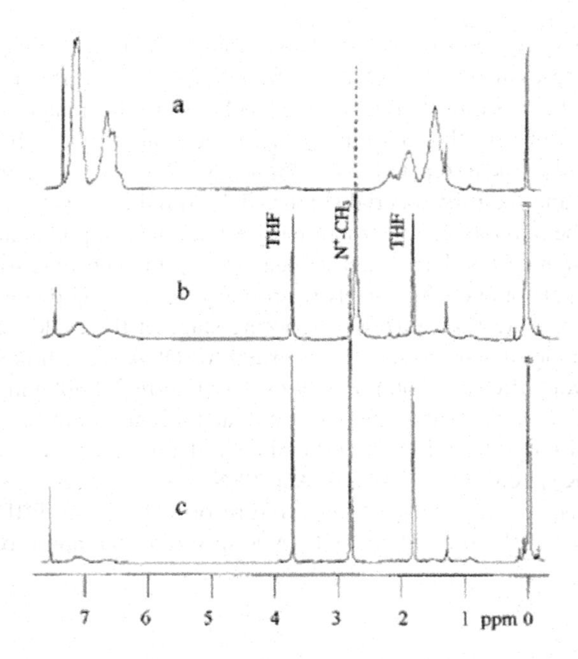

FIGURE 2.4.4 ^1H NMR spectra of PSN: (a) without cross-linking (PSN in Table 2.4.1), recorded in $CDCl_3$; (b) cross-linked with DBB (N11 in Table 2.4.4), recorded in $CDCl_3$/THF-d_8 (7/3 by volume); (c) polymer N11 in Table 2.4.4 after five thermocompression cycles, recorded in $CDCl_3$/THF-d_8 (7/3 by volume).

(a) and the DBB cross-linked ones, before (b) and after five compression cycles (c). The peak at 2.7–2.8 ppm can be assigned to the N^+CH_3 groups of the ionene bridges.[17,18] This peak remained strong even after the polymer was subjected to five compression cycles. The ^1H NMR spectra of DCA cross-linked PSN before (Figure 2.4.5a) and after five compression cycles (Figure 2.4.5b) provided the same conclusion as the DBB cross-linked PSN. However, quantitative calculations based on NMR data could not be carried out, because of the inconsistency between the peak intensities and the number of the corresponding H atoms. This inconsistency was caused by the heterogeneity of the tested powder dispersions. Nevertheless, the obtained ^1H NMR spectra provide an indication that the ionene bridges were regenerated after the thermocompression cycles.

Thermal De-Cross-Linking/Re-Cross-Linking of the Networks Generated with Chlorine-Containing Polymers. The thermal reversibility of the cross-linked Cl-containing polymers was investigated via reactive solubility and thermocompression. As shown in Tables 2.4.2 and 2.4.3, all the cross-linked polymers, except PBSC cross-linked with 1,4-dimethylpiperazine (DMP), dipiperidinomethane (DPDM), or tetraethylethylenediamine (TEEDA) (C5, C6, and C12, respectively, in Table 2.4.2), could be almost completely dissolved in dichlorobenzene containing BzB or BzC and exhibited flowability on heating. Insoluble, continuous, and transparent thin films were obtained from PBSC cross-linked with 2,2′-dipyridyl

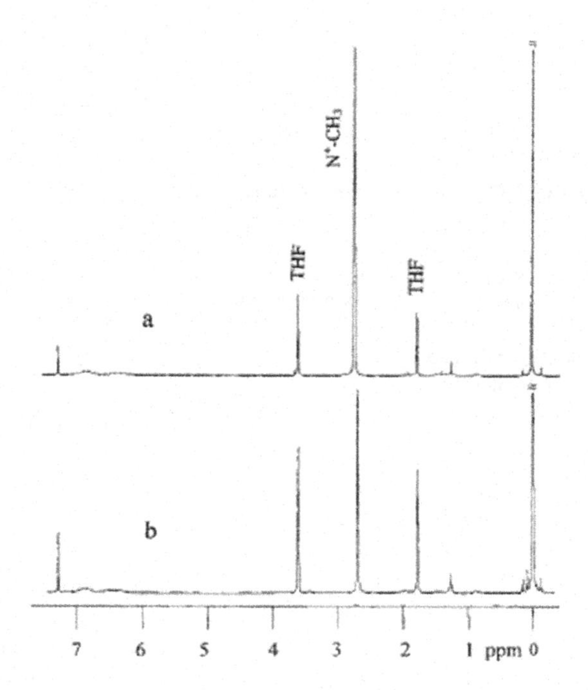

FIGURE 2.4.5 ^{1}H NMR spectra of DCA cross-linked PSN (N10 in Table 2.4.4), recorded in $CDCl_3$/THF-d_8 (7/3 by volume): (a) before any thermocompression; (b) after five thermocompression cycles.

(DPD), 4,4′-trimethylenedipyridine (TMDPD), or tetramethyl-1,4-butanediamine (TMBDA), even when they were subjected to five thermocompression cycles.

With DMP, DPDM, and TEEDA as cross-linkers, high insoluble fractions (gel′ %) were found after reactive solubility experiments were carried out with the cross-linked PBSC. No continuous films could be obtained by the thermocompression of the three cross-linked polymers. Furthermore, the compressed samples cracked into small gel particles when they were submerged into DMF. The poor reversibilities of the above three cross-linked polymers were probably due to a variety of reasons. Because of the tension in the ring, the reverse Menschutkin reaction might have occurred in a ring-opening way that did not dissociate the network generated by DMP (Scheme 2.4.3) or DPDM (Scheme 2.4.4). When the TEEDA cross-linked polymer was subjected to heating, a competitive reaction that generated ethyl chloride from either of the two nonbridging ethyl groups might have resulted in an incomplete dissociation of the network, since the nonbridging ethyl groups of TEEDA possess a carbonium ion strength comparable to that of the bridging-ethylene group. Additionally, the Hoffmann degradation might have occurred since there are at least two alkyl groups that possess βH in each of the amine moieties of DMP, DPDM, and TEEDA.

Conductivity Investigations of the Polymer System during Cross-Linking and De-Cross-Linking. When the conductivity of the PBSC/PBN mixture in benzophenone (CM4 in Table 2.4.5) was determined at 100°C, it increased from 6/δ μS/cm

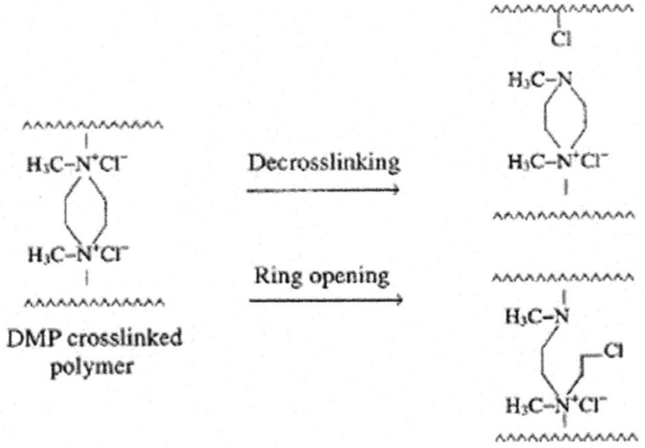

SCHEME 2.4.3

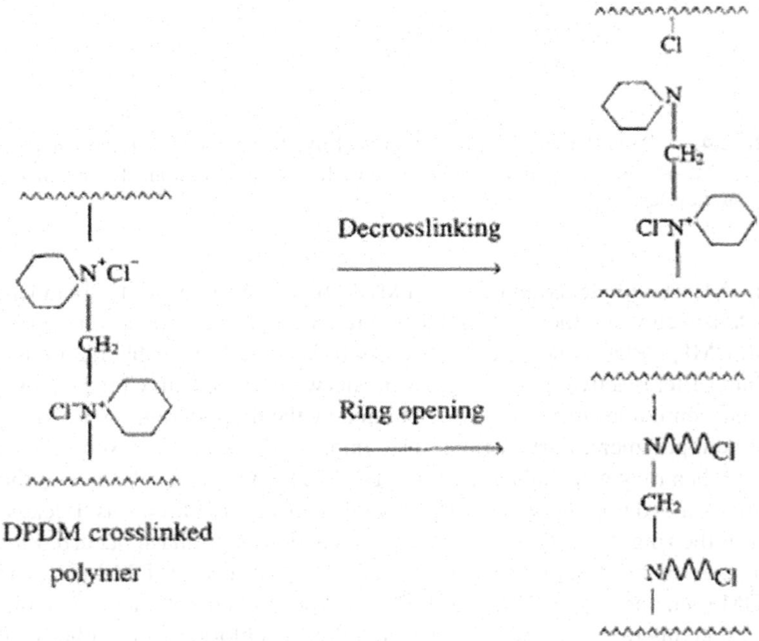

SCHEME 2.4.4

(where δ ca. 0.3 cm is the thickness of the solution layer) to 13/δ μS/cm during the first hour, to 24/δ μS/cm after an additional 3 h, after which it remained constant. A different time dependence was recorded when a PCMS solution contained a high concentration of PVP (CM2 in Table 2.4.5) or TMHDA (CM3 in Table 2.4.5). Indeed, Figure 2.4.6 shows that the conductivity of the PCMS solution remained almost unchanged in the

TABLE 2.4.5
Cross-Linking Systems for Conductivity Measurements

No.	Cl-Containing Polymer[a]	Cross-Linker or Amino-Containing Polymer[b]	Amino/Chlorine (mol/mol)	Solvent[c]	Polymer Concn (wt%)	Heating Time at 100°C (h)
CM1	PCMS	no	0/1	1-chloro-2-nitrobenzene	14	1
CM2[d]	PCMS	PVP	1/1	1-chloro-2-nitrobenzene	10	1
CM3[d]	PCMS	TMHDA	1/1	1-chloro-2-nitrobenzene	14	4
CM4[d]	PBSC	PBN	1/1	benzophenone	11	5

[a] PCMS = poly(chloromethylstyrene); PBSC = butyl acrylate–styrene–chloromethylstyrene copolymer of Table 2.4.1.
[b] PVP = poly(4-vinylpyridine); TMHDA = tetramethyl-1,6-hexanediamine; PBN = butyl acrylate–(dimethylamino)ethyl acrylate copolymer of Table 2.4.1.
[c] Containing 0.3 wt% TBC.
[d] Gelation was observed after the heating.

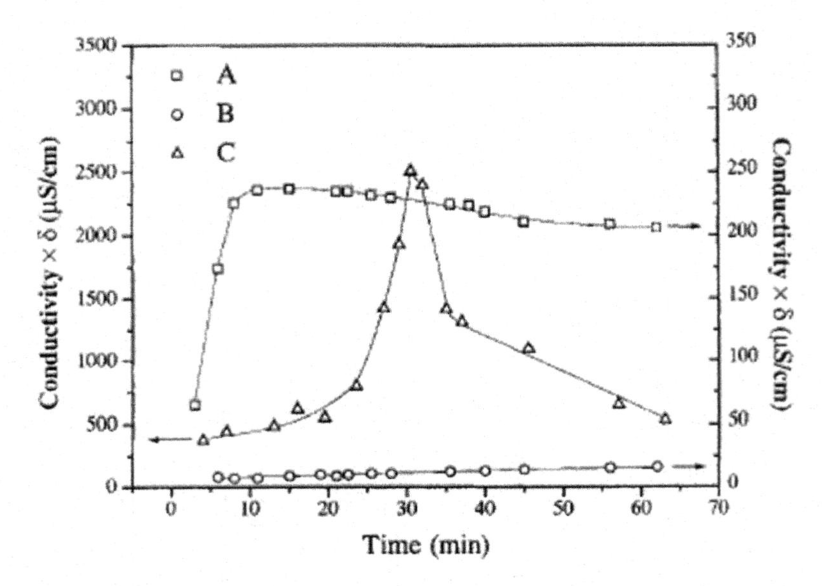

FIGURE 2.4.6 Conductivity of polymer mixtures recorded at 100°C: (A) PCMS/PVP solution (CM2 in Table 2.4.5); (B) PCMS solution (CM1 in Table 2.4.5); (C) PCMS/TMHDA solution (CM3 in Table 2.4.5). δ (ca. 0.3 cm) is the thickness of the solution layer.

absence of a cross-linker (CM1 in Table 2.4.5, curve B). However, the conductivity of the solution of PCMS and PVP (curve A) increased sharply at the beginning, due to the formation of ionene ion pairs, and decreased after a maximum was attained. A maximum in the conductivity was also attained (curve C) when the cross-linker TMHDA was introduced into the PCMS solution. The decrease in the conductivity might be the result of the formation of a network with an increasing cross-link density which caused an increasing resistance to the motion of the generated ions. These conductivity results provided additional evidence that the cross-linking of the polymer was caused by the quaternization between the halide and the tertiary amine.

The conductivity of the cross-linked polymer mixtures (polymer gels, Table 2.4.5) was determined at various temperatures. Figure 2.4.7 presents the temperature dependence of the conductivity recorded during a cycle of heating (curve B) and cooling (curve C) of a PCMS/PVP gel. The conductivity increased with increasing temperature from 90°C to 130°C and 170°C to 205°C but remained almost unchanged in the temperature range 130°C–170°C by decreasing little with rising temperature. A low increase of the conductivity was recorded when a PCMS solution free of cross-linker was heated under the same conditions (curve A). Obviously, the range of temperatures in which the conductivity remained almost constant was a result of the competition between the reduction of the number of ion pairs due to the dequaternization of the ionene bridges with increasing temperature and the increase of the conductivity with increasing temperature. The conductivity returned to its original level when after the heating of PCMS/PVP gel the system was cooled to 90°C. In other words, the dequaternized ionene moieties were regenerated rapidly

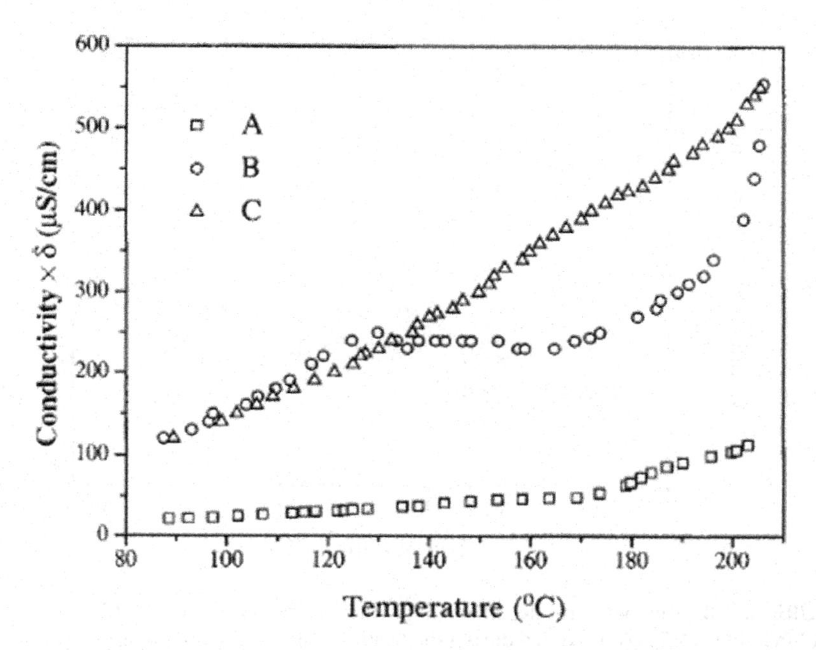

FIGURE 2.4.7 Temperature dependence of the conductivity of (A) PCMS solution (CM1 in Table 2.4.5); (B) PCMS/PVP gel (CM2 in Table 2.4.5), recorded during the heating stage; and (C) the same gel as (B), recorded during the cooling stage. δ (ca. 0.3 cm) is the thickness of the gel layer.

during the cooling period. A similar conductivity–temperature relationship was observed when the PCMS/TMHDA gel was subjected to the heating and cooling cycle (Figure 2.4.8). In this case, the conductivity remained almost constant during heating between 123°C and 130°C and between 165°C and 170°C.

An interesting result was observed when the PBSC/PBN gel, which possessed a lower cross-link density in its network than PCMS/PVP and PCMS/TMHDA, was subjected to the heating/cooling cycles. As shown in Figure 2.4.9, the conductivity increased from 80°C to 160°C and decreased with a further rise in temperature (curve A). This probably occurred because at a temperature greater than 160°C the increase in conductivity due to the temperature was less than the decrease in conductivity caused by dequaternization. During cooling (curve B), the conductivity increased when the temperature was lowered from 213°C to 165°C and acquired exactly the starting level at the end of the cycle. After the polymer gel was cooled to room temperature and kept for about 24 h at the latter temperature, it was subjected to a second heating/cooling cycle. As during the first cycle, the conductivity decreased after 165°C (curve C) during the heating stage. The two heating/cooling cycles provided similar behaviors regarding the change in conductivity. There were, however, large differences at the high temperatures, which indicated that the process was not completely reversible. The conductivity measurements provided evidence that the dequaternization occurred through dissociation of the covalent ionene

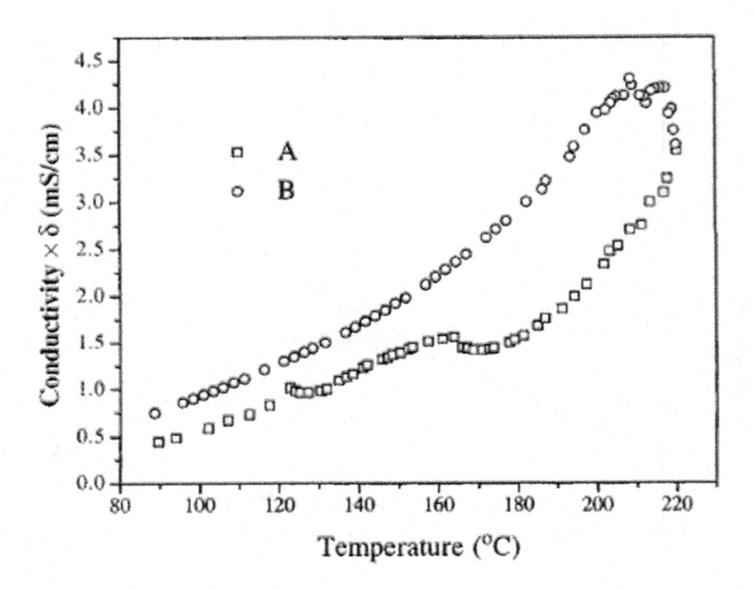

FIGURE 2.4.8 Temperature dependence of the conductivity of PCMS/TMHDA gel (CM3 in Table 2.4.5): (A) recorded during the heating stage; (B) recorded during the cooling stage. δ (ca. 0.3 cm) is the thickness of the gel layer.

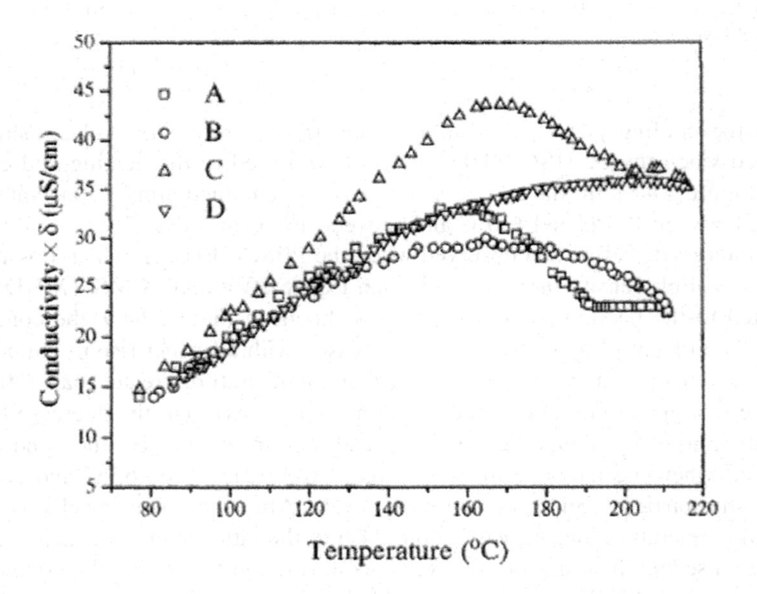

FIGURE 2.4.9 Temperature dependence of the conductivity of PBSC/PBN gel (CM4 in Table 2.4.5): (A) recorded during the heating stage of the first cycle; (B) recorded during the cooling stage of the first cycle; (C) recorded during the heating stage of the second cycle; (D) recorded during the cooling stage of the second cycle. δ (ca. 0.3 cm) is the thickness of the gel layer.

bridges, almost free of CH_3Cl formation (through the dequaternization of the non-bridging methyl groups). Indeed, otherwise, the conductivity could not have returned to its original level, because CH3Cl having a low boiling point (–24.2°C) would have escaped from the liquid phase and requaternization could not have occurred.

DSC Analysis. The thermal analysis of dequaternization is expected to provide some information about the reversibility of the ionene network. Therefore, the cross-linked PCMS/PVP system (N22 in Table 2.4.4) which contains a large number of ion pairs was subjected to the DSC test, after its structure was confirmed by the IR spectrum. Compared to the spectra of PVP (Figure 2.4.10b) and PCMS (Figure 2.4.1b), the peaks at 994, 1069, 1220, and 1416 cm^{-1}, which can be assigned to PVP,[19] and the peak at 1267 cm^{-1}, which is the characteristic absorption peak of chloromethylene of PCMS,[20] are relatively weak in the spectrum of the cross-linked PCMS/PVP (Figure 2.4.10a). The strong peaks at 1152, 1465, and 1637 cm^{-1}, which can be assigned to the quaternary salt of pyridine,[21] indicate that a covalent network of PCMS/PVP was generated by the quaternization between the chloromethylene group of PCMS and the pyridine ring of PVP.

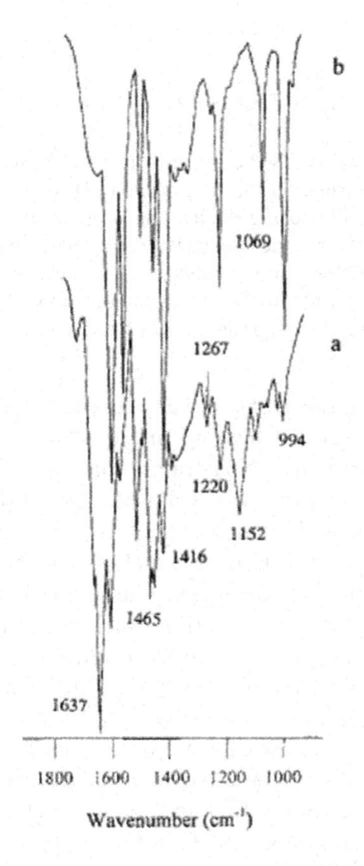

FIGURE 2.4.10 IR spectra of PVP: (a) cross-linked with PCMS (N22 in Table 2.4.4, KBr tablet); (b) before cross-linking (KBr tablet).

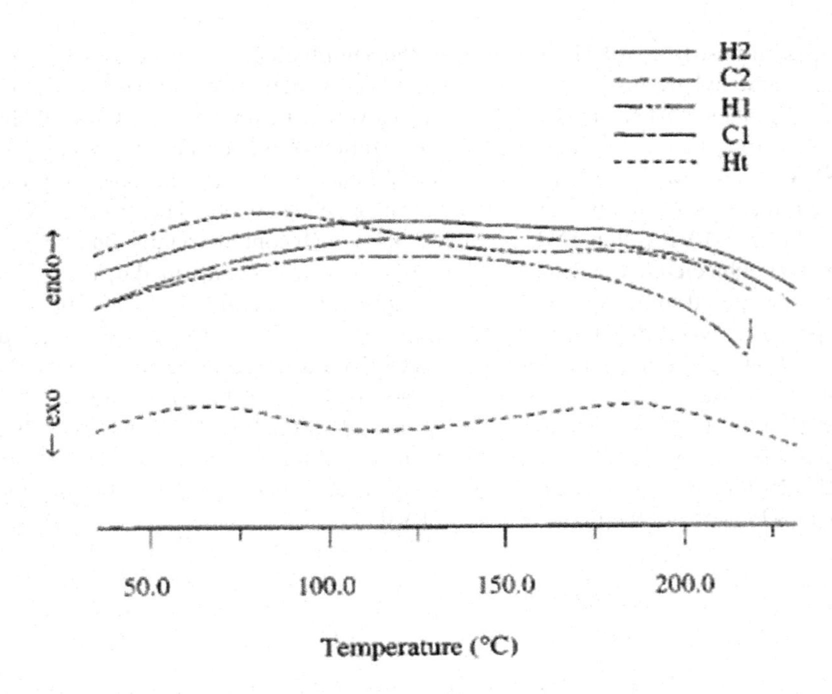

FIGURE 2.4.11 DSC curves of PVP cross-linked with PCMS (N22 in Table 2.4.4): (H1) recorded during the heating stage of the first cycle; (C1) recorded during the cooling stage of the first cycle; (H2) recorded during the heating stage of the second cycle; (C2) recorded during the cooling stage of the second cycle; (Ht) recorded during the heating stage after the polymer was heated at 220°C for 7 min, cooled to 110°C and kept at 110°C for 2 h, and finally cooled to room temperature. All DSC tests were carried out under a helium atmosphere, at a heating rate of 10°C/min and a cooling rate of –10°C/min.

Figure 2.4.11 presents the DSC curves recorded during a first heating (H1) and cooling (C1) and a second heating and cooling (H2 and C2) (carried out immediately after the first one). Nonstraight curves were recorded for both the heating and cooling stages of the two cycles because of the change of the specific heat of quaternary salts with temperature.[9] An anomalous thermal transition peak can be seen at 145°C–150°C in curve H1. As reported by Burns et al.,[9] the decomposition temperature of a quaternary ammonium salt is characterized by a broad and symmetrical thermal transition. Such broad thermal transition peaks were also observed for the ionene polyelectrolytes.[3a,22,23] The peak observed in Figure 2.4.11 is located in the range (130°C–170°C) in which the conductivity of the PCMS/PVP gel remained almost constant during heating. Consequently, this peak can be attributed to the thermal transition during the dequaternization of the PCMS/PVP ionene network. No thermal transition can be noted in the cooling curves (C1 and C2) of the first and second cycle, and only a small change in curvature can be noted around 150°C in the heating curve (H2) of the second cycle. The above curves indicated that after the first dequaternization the polymer remained almost in the same state during the first cooling, the second heating, and the second cooling. This occurred because both polymers have high glass transition temperatures, and hence their segments have low mobilities. However, after the treatment described below, a thermal transition similar to that observed during the first heating could be detected. The polymer was wrapped in an aluminum foil and heated at 220°C for 7 min, was cooled in 10 min to 110°C (a temperature close to the glass transition temperatures of PCMS and PVP) at which it was kept for 2 h in order to stimulate the thermal motion of the segments of PCMS and PVP chains, and finally was cooled to room temperature. As shown by curve Ht in Figure 2.4.11, a thermal transition peak which started at about 70°C and ended at about 190°C was present in a somewhat different temperature range than for curve H1, indicating that the ionene network was regenerated during the cooling program employed.

2.4.4 CONCLUSION

The results obtained from the cross-linking reactions, IR and ^{1}H NMR tests, reactive solubility experiments, thermocompression cycles, conductivity measurements, and DSC indicated that covalent networks could be obtained via the formation of ionene bridges between polymer chains and that such networks exhibited thermally reversible cross-linking due to quaternization on cooling and dequaternization on heating. The cross-linking reactivity and thermal de-cross-linking/re-cross-linking behavior were markedly affected by the structures of the two kinds of functional groups involved in the quaternization, the tertiary amine, and halide–carbon groups. Reversibility can be achieved if (among the alkyl groups connected to the N atoms) the bridging alkyl groups provide an enough high carbonium ion strength compared to those of the nonbridging alkyl groups and if no or few βH are present in the ammonium salt moieties to initiate the Hoffmann degradation. The polymers containing a functional monomeric unit, such as chloromethylstyrene, 2-(dimethylamino)-ethyl acrylate, and vinylpyridine, could be quaternized effectively with selected cross-linkers under moderate conditions to generate ionene networks that exhibited thermal reversibility.

REFERENCES

1. Rembaum, A.; Baumgartner, W.; Eisenberg, A. *J. Polym. Sci, Polym. Lett.* **1968**, *6*, 159. (b) Somoano, R.; Yen, S. P. S.; Rembaum, A. *J. Polym. Sci., Polym. Lett.* **1970**, *8*, 467. (c) Noguchi, H.; Rembaum, A. *Macromolecules* **1972**, *5*, 253. (d) Rembaum, A.; Noguchi, H. *Macromolecules* **1972**, *5*, 261. (e) Hadek, V.; Noguchi, H.; Rembaum, A. *Macromolecules* **1971**, *4*, 494.
2. (a) Leffek, K. T.; Matheson, A. F. *Can. J. Chem.* **1972**, *50*, 986. (b) Ko, E. C. F.; Leffek, K. T. *Can. J. Chem.* **1972**, *50*, 1297. (c) Ko, E. C. F.; Leffek, K. T. *Can.J. Chem.* **1970**, *48*, 1865. (d) Burns, J. T.; Leffek, K. T. *Can. J. Chem.* **1969**, *47*, 3725. (e) Leffek, K. T.; Tsao, F. H. C. *Can. J. Chem.* **1968**, *46*, 1215. (f) Gordon, J. E. *J. Org. Chem.* **1965**, *30*, 2760.
3. (a) Prasad, B. B.; Easo, S.; Kumar, A. *Polym. J.* **1994**, *26*(3), 251. (b) Yano, S.; Tadano, K.; Jerome, R. *Macromolecules* **1991**, *24*, 6439. (c) Klun, T. P.; Wendling, L. A.; Van Bogart, J. W. C.; Robbins, A. F. *J. Polym. Sci., Polym. Chem. Ed.* **1987**, *25*, 87. (d) Grassl, B.; Galin, J. C. *Macromolecules* **1995**, *28*, 7035.

4. (a) Dieterich, D.; Keberle, W.; Witt, H. *Angew. Chem, Int. Ed. Engl.* **1970**, *9*(1), 40. (b) Watanabe, M.; Takizawa, Y.; Shinohara, I. *Polymer* **1983**, *24*, 491. (c) Loveday, D.; Wilkes, G. L.; Bheda, M. C.; Shen, Y. X.; Gibson, H. W. *J. Macromol. Sci., Pure Appl. Chem.* **1995**, *A32*(1), 1. (d) Ikeda, Y.; Murakami, T.; Yuguchi, Y.; Kajiwara, K. *Macromolecules* **1998**, *31*, 1246.
5. (a) Dieterich, D.; Bayer, O.; Peter, J.; Muller, E. *DBP* 1156977, 1962. (b) Wieden, H.; Rellensmann, W.; Dieterich, D.; Nischk, G. E. Brit. Pat. 1150634, 1965. (c) Dolezal, T.; Edwards, D. C.; Wunder, R. H. *Rubber World* **1968**, *158*, 46. (d) Buckler, E. J.; Briggs, G. J.; Dunn, J. R.; Lasis, E.; Wei, Y. K. *Rubb. Chem. Technol.* **1978**, *51*, 872. (e) Rutkowska, M. *J. Appl. Polym. Sci.* **1986**, *31*, 1469.
6. (a) Engle, L. P.; Wagener, K. B. *J. Macromol. Sci., Rev. Macromol. Chem. Phys.* **1993**, *C33* (3), 239. (b) Wagener, K. B.; Engle, L. P. *J. Polym. Sci., Polym. Chem. Ed.* **1993**, *31*, 865.
7. (a) Leir, C. M.; Stark, J. E. *J. Appl. Polym. Sci.* **1989**, *38*, 1535. (b) Feng, D.; Venkateshwaran, L. N.; Wilkes, G. L.; Leir, C. M.; Stark, J. E. *J. Appl. Polym. Sci.* **1989**, *38*, 1549.
8. (a) Craven, J. M. U. S. Pat. 3435003, 1969. (b) Takeshita, Y.; Uoi, M.; Hirai, Y.; Uchiyama, M. Ger. Pat. 2164022, 1972. (c) Harumi, A.; Shyuzo, K. Jpn. Pat. 48-36293, 1973. (d) Kennedy, J. P.; Castner, K. F. *J. Polym. Sci., Polym. Chem. Ed.* **1979**, *17*, 2055. (e) Salamone, J. C.; Chung, Y.; Clough, S. B.; Watterson, A. C. *J. Polym. Sci., Polym. Chem. Ed.* **1988**, *26*, 2923. (f) Chujo, Y.; Sada, K.; Saegusa, T. *Macromolecules*, **1990**, *23*, 2636. (g) Canary, S. A.; Stevens, M. P. *J. Polym. Sci., Polym. Chem. Ed.* **1992**, *30*, 1755. (h) Jiao, S. K.; Bing, J. L.; Li, X. Y. *Acta Polym. Sin.* **1994**, *6*, 702. (i) Chen, X. N.; Ruckenstein, E. *J. Polym. Sci., Polym. Chem. Ed.* **1999**, *37*, 4390. (j) Ruckenstein, E.; Chen, X. N. *J. Polym. Sci., Polym. Chem. Ed.* **2000**, *38*, 818. (k) Chen, X. N.; Jiao, S. K. *Acta Polym. Sin.* **1999**, *5*, 564. (l) Fawcett, A. H.; McGonigle, E.; Hohn, M.; Russell, E. *Polym. Prepr.* **1999**, *40*(1), 232.
9. Burns, J. A.; Verrall, R. E. *Thermochim. Acta* **1974**, *9*, 277.
10. Pouchert, C. J. *The Aldrich Library of FT-IR Spectra*; Aldrich Chem. Co.: Milwaukee, WI, 1985; Vol. 1, pp. 1320B and 1322A.
11. Wang, J.; Meyer, W. H.; Wegner, G. *Macromol. Chem. Phys.* **1994**, *185*, 1777.
12. Verrall, R. E.; Burns, J. A. *Can. J. Chem.* **1974**, *52*, 3438.
13. (a) Silverstein, R. M.; Bassler, G. C.; Morrill, T. C. *Spectrometric Identification of Organic Compounds*, 4th ed.; John Wiley & Sons: New York, 1981; p. 168. (b) Pouchert, C. J. *The Aldrich Library of FT-IR Spectra*; Aldrich Chem. Co.: Milwaukee, WI, 1985; Vol. 1, p. 675C.
14. Pouchert, C. J. *The Aldrich Library of FT-IR Spectra*; Aldrich Chem. Co.: Milwaukee, WI, 1985; Vol. 1, p. 1322C.
15. Pouchert, C. J. *The Aldrich Library of FT-IR Spectra*; Aldrich Chem. Co.: Milwaukee, WI, 1985; Vol. 1, p. 74D.
16. Pouchert, C. J. *The Aldrich Library of FT-IR Spectra*; Aldrich Chem. Co.: Milwaukee, WI, 1985; Vol. 1, p. 417A.
17. (a) Noguchi, H.; Rembaum, A. *Macromolecules* **1972**, *5*, 253. (b) Noguchi, H.; Rembaum, A. *J. Polym. Sci., Polym. Lett.* **1969**, *7*, 383. (c) Rembaum, A.; Singer, S.; Keyzer, H. *J. Polym. Sci., Polym. Lett.* **1969**, *7*, 395. (d) Chen, X. N.; Ruckenstein, E. *J. Polym. Sci., Polym. Chem. Ed.*, **2000**, *38*, 1662.
18. When the ^{1}H NMR spectrum of a linear ionene was recorded using its solution in D_2O, a normal peak of the N^+CH_3 was present at 3.1 ppm. However, when the solvent, D_2O, was replaced by the mixture of $CDCl_3$/THF-d_8, a nonsolvent for ionene, the chemical shift of the N^+CH_3 became 2.8 ppm and the intensity of this peak became abnormally strong.[17d]

19. Pouchert, C. J. *The Aldrich Library of FT-IR Spectra*; Aldrich Chem. Co.: Milwaukee, 1985; Vol. 2, pp. 1209 and 1223.
20. Hoffenberg, D. S. *Ind. Eng. Chem, Prod. Res. Dev.* **1964**, *3*(2), 113.
21. Pouchert, C. J. *The Aldrich Library of FT-IR Spectra*; Aldrich Chem. Co.: Milwaukee, 1985; Vol. 2, pp. 731 and 757.
22. Gordon, J. E. *J. Org. Chem.* **1965**, *30*, 2760.
23. The broad thermal transition is the result of the low starting temperature and high completing temperature of dequaternization. The dequaternization of ammonium salts can start at a temperature below 100°C.[22]

3 Semi- and Interpenetrating Polymer Network Pervaporation Membranes

CONTENTS

Pervaporation is an important energy-efficient technique for the separation of azeotropic mixtures, compounds with close boiling points, and mixtures consisting of heat-sensitive components. Pervaporation has critical industrial applications, especially for the dehydration of ethanol–water mixtures. Pervaporation membranes are essential for pervaporation separation, because they selectively allow the desired component(s) of the liquid mixture to transfer through it by vaporization. A variety of polymer membranes have been investigated for pervaporation separation. Polymer membranes based on a single polymeric component, such as hydrophilic poly(vinyl alcohol) (PVA) or hydrophobic polydimethylsiloxane (PDMS), may have significant pervaporation selectivity, but are often restricted by other required material properties. On the other hand, semi- and interpenetrating polymer network (IPN) membranes are highly promising materials for applications as pervaporation membranes because of their superb comprehensive properties. Their multiple polymeric components can be tuned to optimize selectivity and permeability of pervaporation separation, while their crosslinked network structures can provide significant mechanical properties and dimensional stability. This chapter describes systematic studies in the preparation and assessment of semi-IPN (SIPN) and IPN membranes.

Based on the high selectivity and permeability of poly(acrylic acid) (PAA) and the remarkable mechanical strength and film-forming capability of PVA, hydrophilic PAA-PVA semi- and interpenetrating polymer network pervaporation membranes are investigated to target the synergistic properties of the two base polymeric components (Section 3.1). As compared with IPN membranes with very low permeability, SIPN membranes can allow for significant permeation rate.

Hydrophilic IPN membranes consisting of PVC and polyacrylamide (PAAM), both unsupported and supported on polyethersulfone (PESF) ultrafiltration membranes, are also studied (Section 3.2). Relative to crosslinked PVA membranes, the PVA-PAAM IPN membranes possess greatly enhanced mechanical properties and selectivity for pervaporation of ethanol–water mixtures. The supported PVA-PAAM IPN membranes exhibit a higher permeability but a somewhat lower selectivity than the unsupported ones.

Hydrophobic IPN membranes consisting of PDMS and polystyrene (PS), supported on PESF ultrafiltration membranes, are investigated (Section 3.3). The PDMS-PS IPN membranes exhibit superior mechanical properties and film-forming capability as compared to the crosslinked PDMS membranes. With the change of composition of the ethanol–water mixtures, these IPN membranes also show an interesting inversion of selectivity.

3.1 Poly(acrylic acid)–Poly(vinyl alcohol) Semi- and Interpenetrating Polymer Network Pervaporation Membranes*

Eli Ruckenstein and Liang Liang

Department of Chemical Engineering, State University of New York at Buffalo, Buffalo, New York 14260

ABSTRACT A series of poly(acrylic acid) (PAA)–poly(vinyl alcohol) (PVA) semi-interpenetrating (SIPN) and interpenetrating (IPN) polymer network membranes were prepared by crosslinking PVA alone or by crosslinking both PVA and PAA. Glutaraldehyde and ethylene glycol were used as crosslinking agents for the PVA and PAA networks, respectively. The presence of PAA increases the permeability of the membranes while the presence of PVA improves their mechanical and film-forming properties. The mechanical properties of the membranes were investigated via tensile testing. These hydrophilic membranes are permselective to water from ethanol water mixtures and to ethanol from ethanol–benzene mixtures. The IPN membranes were employed for the former mixtures and the SIPN membranes for the latter, because the IPN ones provided too low permeation rates. The permeation rates and separation factors were determined as functions of the IPN or SIPN composition, feed composition, and temperature. For the azeotropic ethanol–water mixture (95 wt% ethanol), the separation factor and permeation rate at 50°C of the PAA-PVA IPN membrane, containing 50 wt% PAA, were 50 and 260 g/m^2 h, respectively. For the ethanol–benzene mixture, the PAA-PVA SIPN membranes had separation factors between 1.4 and 1200 and permeation rates between 6 and 550 g/m^2 h, respectively, depending on the feed composition and temperature.

* *Journal of Applied Polymer Science*, 1996, 62, 973–987.

3.1.1 INTRODUCTION

Pervaporation can be effectively used in the separation of azeotropic, close-boiling point, and heat sensitive mixtures. Many pervaporation membranes have been employed for the dehydration of water–organic mixtures. Typical membranes are poly(acrylic acid) (PAA), poly(vinyl alcohol) (PVA), polyacrylonitrile (PAN), and chitosan. Both the mechanical and separation properties of the membranes used in pervaporation can be improved by using graft, block, blend, and crosslinked polymers.[1–6] Few articles have been concerned with interpenetrating polymer network (IPN) membranes.[7–9] Compared to those prepared by graft, block, or blend polymerization, the IPN or semi-IPN (SIPN) membranes have higher stabilities at high temperatures and in various liquids.

The PAA membranes exhibit high water selectivity and permeation rate because of their carboxylic groups. Of course, the PAA membranes must be crosslinked to avoid their dissolution in water. However, the crosslinked PAA membrane is brittle and its film-forming capability is low. Compared to PAA, PVA has a lower hydrophilicity but higher mechanical strength and film-forming capability. For these reasons, we prepared PAA–PVA IPN membranes by the sequential IPN technique; and their pervaporation performance for ethanol–water mixtures was investigated. We also prepared PAA–PVA SIPN membranes, and their pervaporation performance was investigated for the ethanol–benzene mixture for which the boiling points of the two components are almost the same. The higher affinity of PAA for ethanol than benzene suggested that we use the above membrane for the separation of ethanol from the ethanol–benzene mixtures. In the latter case, the SIPN membrane was preferred to the IPN one, because the latter provided a permeation rate that was too low.

3.1.2 EXPERIMENTAL

3.1.2.1 MATERIALS

PVA (Aldrich, molecular weight 124,000), PAA (25 wt% in water, Aldrich, molecular weight 240,000), glutaraldehyde (GAL, 25 wt% in water, Aldrich), ethylene glycol (EG, 99%, Fisher), hydrochloric acid (37 wt% in water, Aldrich), benzene (99%, Aldrich), and ethanol (99%, Aldrich) were used as received. Water was deionized and double distilled.

3.1.2.2 PREPARATION OF PAA–PVA IPN MEMBRANES

To obtain the PAA–PVA IPN membranes, GAL and EG were used as crosslinking agents for the PVA and PAA networks, respectively. The compositions of the casting solutions are listed in Table 3.1.1. The solution was mixed with a magnetic stirrer for 10 min, cast on a glass plate, and allowed to evaporate at room temperature for 24 h. During this process, a crosslinked PVA network containing macromolecules of PAA was generated. Subsequently, the membranes were heated in an oven at 120°C for 3 h to generate the crosslinked PAA network. The membranes were dried in a vacuum oven at room temperature for 24 h. The thickness of PAA–PVA IPN membrane was in the range of 20–30 μm.

TABLE 3.1.1
Compositions of Casting Solutions Used to Prepare PAA–PVA IPN Membranes

Membrane No.	PAA/PVA (wt Ratio)	PAA (g)	PVA (g)	GAL (g)	EG (g) × 10^2	HCl (1*N*) (mL)	H_2O (mL)
I	70/30	0.58	0.25	0.025	1.2	0.125	8
II	50/50	0.25	0.25	0.025	0.5	0.125	3.5
III	30/70	0.11	0.25	0.025	0.2	0.125	2
IV	10/90	0.03	0.25	0.025	0.05	0.125	—

Note: For all membranes GAL = 0.5 wt% in the crosslinked PVA network, EG = 2.0 wt% in the crosslinked PAA network.

3.1.2.3 Preparation of PAA-Crosslinked PVA SIPN Membranes

The compositions of the casting solutions used to prepare the PAA–PVA SIPN membranes are listed in Table 3.1.2. The solution containing PAA, PVA, the crosslinking agent GAL, and the catalyst HC1 was first mixed using magnetic stirring for 10 min. Then, the solution was cast on a glass plate and allowed to evaporate at room temperature for 24 h. The transparent SIPN membranes thus obtained were dried for 24 h in a vacuum oven at room temperature. The thickness of the dry membrane was in the range 20–30 μm.

3.1.2.4 Tensile Testing

The tensile testing of the PAA–PVA SIPN and IPN samples was performed by preparing sheets of the size required by ASTM D638-58 T and using an Instron Universal Testing Instrument (model 1000) at room temperature. The extension speed of the instrument was 10 mm/min.

TABLE 3.1.2
Compositions of Casting Solutions Used to Prepare PAA–PVA SIPN Membranes

Membrane No.	PAA/PVA (wt Ratio)	PAA (g)	PVA (g)	GAL (g)	HCl (1*N*) (mL)	H_2O (mL)
I	80/20	1.0	0.25	0.025	0.125	14
II	70/30	0.58	0.25	0.025	0.125	8
III	60/40	0.37	0.25	0.025	0.125	6

Note: For all membranes GAL = 0.5 wt% in the crosslinked PVA network.

3.1.2.5 Pervaporation

The pervaporation experiments were performed using the apparatus described previously.[8] The membrane was located on a porous glass support. The temperature of the permeation cell was controlled using a thermostated water bath. The effective membrane area was 9.6 cm^2. The downstream pressure was kept at 3 ± 1 torr by a vacuum pump. The steady state can be achieved after running the apparatus for about 2 h.

The permeated sample was collected in a cold trap cooled with liquid nitrogen. The composition of the sample was determined using a gas chromatograph equipped with a Porapack Q column at the temperatures of 160°C and 215°C for the ethanol–water and ethanol–benzene mixtures, respectively, and a thermal conductivity detector. Helium was used as the carrier gas.

The permeation rate, J, can be calculated using the expression

$$J = \frac{Q}{At}$$

where Q is the total amount of permeate during the experimental time interval t at steady state and A is the effective membrane surface area.

The selectivity of the membrane can be characterized via the separation factor, $\alpha_{i/j}$, which is defined as

$$\alpha_{i/j} = \frac{(C_i/C_j)_{\text{permeate}}}{(C_i/C_j)_{\text{feed}}}$$

where $(C_i/C_j)_{\text{permeate}}$ is the weight ratio of components i and j in the permeate and $(C_i/C_j)_{\text{feed}}$ is their weight ratio in the feed. Component i is the preferentially pervaporated one.

3.1.3 RESULTS AND DISCUSSION

3.1.3.1 Tensile Strength and Elongation of PAA–PVA SIPN and IPN Membranes

Figure 3.1.1 shows that with increasing content of PAA in the PAA-crosslinked PVA SIPN membranes, the tensile strength increases and the elongation at break decreases. PAA is more brittle than PVA because of the rigidity caused by the strong interactions among the carboxyl groups. When the content of PAA in the SIPN was higher than 80 wt%, the membrane became so brittle that it could not be separated from the glass plate. Membranes containing about 60 wt% PAA are suitable from a mechanical point of view.

The tensile strength and elongation at break for the PAA–PVA IPN membranes are presented in Figure 3.1.2 as a function of the PAA content. One can see that with increasing PAA content both the tensile strength and the elongation at break decrease. The degree of crosslinking of the PAA–PVA IPN membranes is much higher than that of the PAA–PVA SIPN membranes, because both polymers are

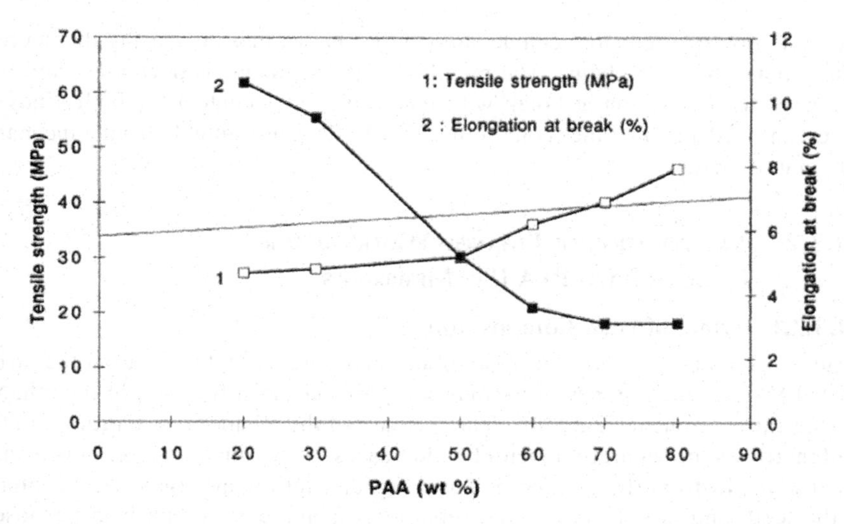

FIGURE 3.1.1 The tensile strength and elongation at break against the PAA content for the PAA–PVA SIPN membranes. For all membranes GAL = 0.5 wt% in the crosslinked PVA network.

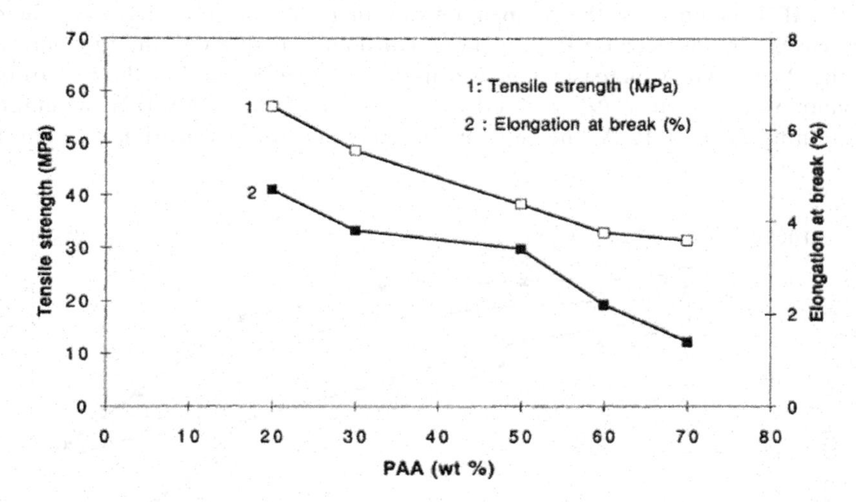

FIGURE 3.1.2 The tensile strength and elongation at break against the PAA content for the PAA–PVA IPN membranes. For all membranes GAL = 0.5 wt% in the crosslinked PVA network; EG = 2.0 wt% in the crosslinked PAA network.

crosslinked. In addition to the crosslinking in the individual networks, there are interactions among the chains, as well as some crosslinking between the two networks due to the reaction between COOH and OH. For the SIPN membranes, the tensile strength increases with increasing PAA content because of increasing interactions among the chains. For the IPN membranes, the tensile strength decreases with increasing PAA content, because the crosslinking of the latter by the EG molecules does not allow enough close contact among the chains, thus decreasing the interactions among the PAA chains. The mobility of the segments is decreased, and this decreases the elongation at break with increasing PAA content for both kinds of membranes. IPN membranes containing 50 wt% PAA are suitable from a mechanical point of view.

3.1.3.2 Pervaporation of Ethanol–Water Mixture Through PAA–PVA IPN Membranes

3.1.3.2.1 Effect of Feed Composition

Figures 3.1.3 and 3.1.4 present the total permeation rate and separation factor for PAA–PVA IPN membranes of various compositions, as a function of the ethanol concentration in the feed at 50°C. The increase of the ethanol concentration in the feed increases the separation factor but decreases the permeation rate. This occurs because the hydrophilic membrane has a higher swelling at larger water contents in the feed, and this decreases the separation factor. For comparison purposes, Figure 3.1.5 presents the ethanol concentration in the permeate as a function of its concentration in the feed, together with the vapor–liquid equilibrium curve at 50°C.[10] Figures 3.1.3 and 3.1.4 also show that with increasing PAA content in the IPN membranes, the permeation rate increases but the separation factor decreases. This is because PAA is more hydrophilic than PVA, and the increase of the PAA content increases the swelling of the membrane. For the azeotropic ethanol–water mixture (95 wt% ethanol) and for a PAA–PVA IPN membrane containing 50 wt% PAA, the separation factor and the permeation rate are 50

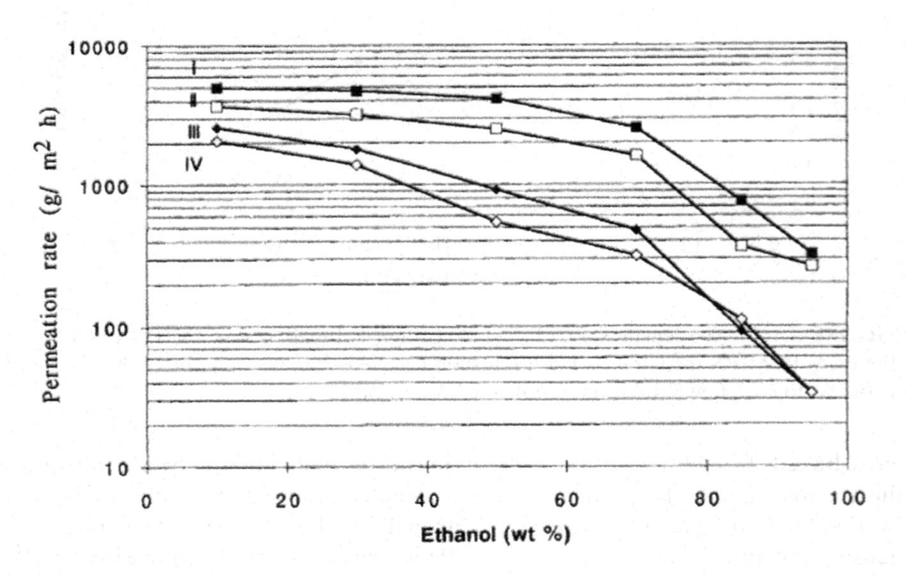

FIGURE 3.1.3 Effect of ethanol composition in the feed on the total permeation rate at 50°C for the pervaporation of ethanol–water mixtures. The membranes are those of Table 3.1.1.

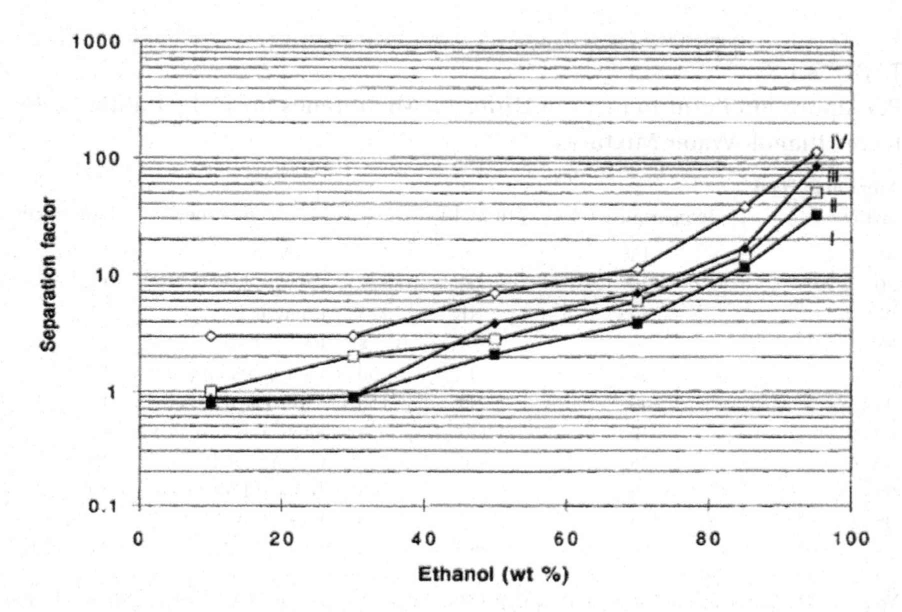

FIGURE 3.1.4 Effect of ethanol composition in the feed on the separation factor at 50°C for the pervaporation of ethanol–water mixtures. The membranes are those of Table 3.1.1.

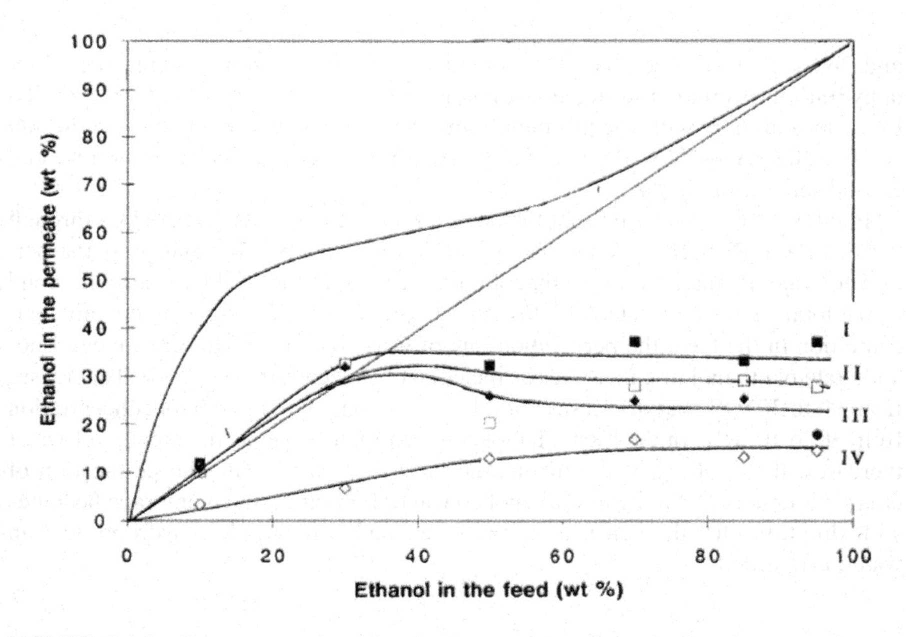

FIGURE 3.1.5 Permeate composition in the pervaporation of ethanol–water mixtures through various of PAA–PVA IPN membranes as a function of the feed composition at 50°C. The membranes are those of Table 3.1.1.

TABLE 3.1.3
Pervaporation Performances of Different Membranes for Dehydrating Water from Ethanol–Water Mixtures

Ethanol in Feed (wt%)	Temperature (°C)	J (g/m^2 h)	α	Membranes	References
96	15	5	350	PMA	11
80	70	20	250	PVA-*g*-PAA	12
90	40	31	123	Chitosan	13
88	50	9	314.8	PAA-*g*-PVA	14
80	30	500	30	PVA-*g*-(PS-*co*-MA)	15
90	40	~ 50	150	Crosslinked PVA	16
90	40	~ 200	~ 23	Photocrosslinked PVA	17
90	75	~ 130	~ 280	Crosslinked PVA	18
90	25	78	46	Nylon 6-PAA blend	19
95	75	60	4100	PVA–PAAM IPN	8
95	50	260	50	PAA–PVA IPN	This study

Note: PMA, poly(maleimide-*co*-acrylonitrile); PVA, poly(vinyl alcohol); PAA, poly(acrylic acid); PS, polystyrene; MA, maleic anhydride; PAAM, polyacrylamide; IPN, interpenetrating polymer network.

and 260 g m^2 h, respectively. For comparison purposes, some results regarding dehydration of ethanol–water mixtures at or near the azeotropic point obtained by various authors with various membranes are listed in Table 3.1.3.[8,11–19] One can see that the PAA–PVA IPN membrane has a relatively high permeation rate and a good separation factor.

Figures 3.1.6 and 3.1.7 present the permeation rates of water and ethanol through various PAA–PVA IPN membranes at 50°C, respectively. The change of the permeation rate of water with the ethanol concentration in the feed has the same trend as the total permeation rate from Figure 3.1.3; that is, with decreasing ethanol concentration in the feed, the permeation rate of water increases. However, the permeation rate of ethanol first increases as the ethanol concentration in the feed decreases from 90 to 30 wt%, but decreases with the further decrease of ethanol concentration from 30 to 10 wt%. In the high ethanol concentration range, as the amount of water increases, the swelling of the membrane increases; as a result, the permeation of ethanol is enhanced. In the low ethanol concentration range, the membrane becomes so hydrophilic that the permeation rate of ethanol, which is less hydrophilic than water, is retarded.

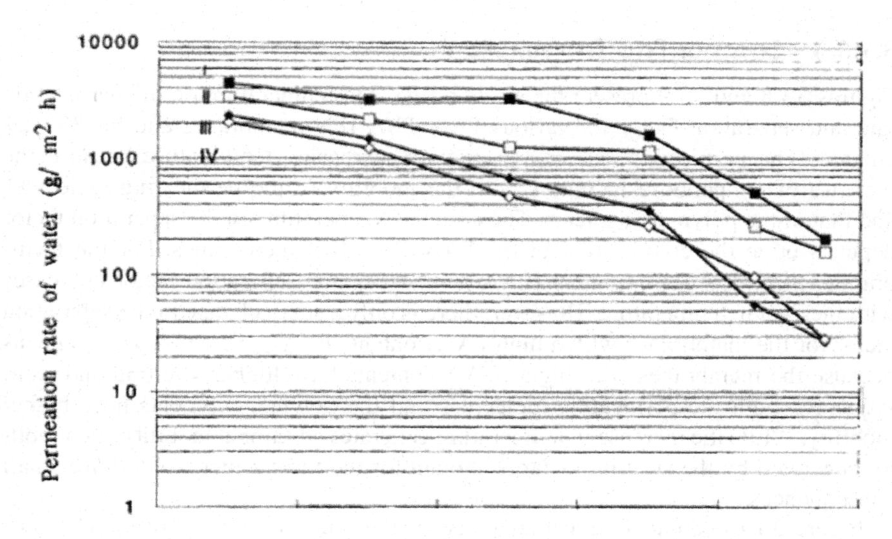

FIGURE 3.1.6 Effect of the ethanol composition in the feed on the permeation rate of water at 50°C for the pervaporation of ethanol–water mixtures. The membranes are those of Table 3.1.1.

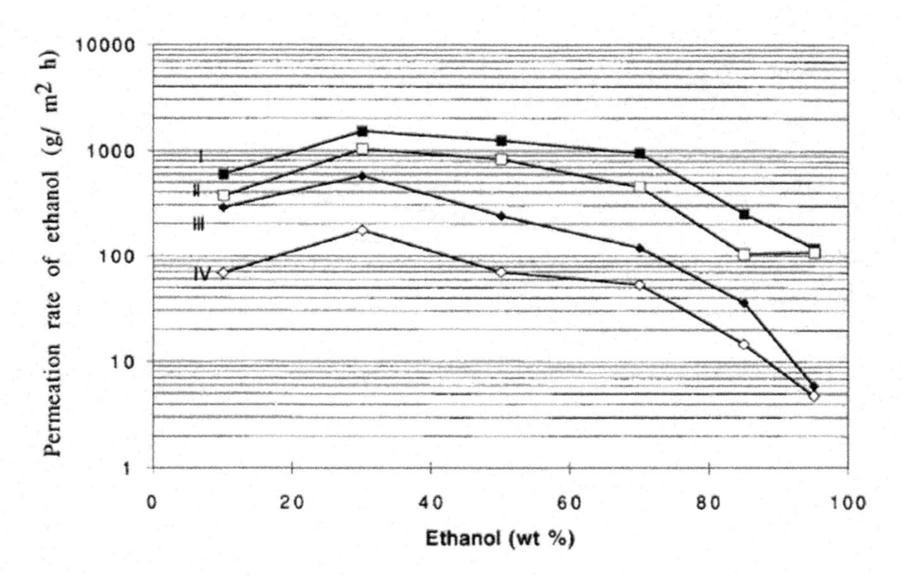

FIGURE 3.1.7 Effect of the ethanol composition in the feed on the permeation rate of ethanol at 50°C for the pervaporation of ethanol–water mixtures. The membranes are those of Table 3.1.1.

3.1.3.3 Effect of Feed Temperature

Figures 3.1.8 and 3.1.9 present the effect of the feed temperature on the permeation rate and separation factor for various PAA–PVA IPN membranes and for 85 wt% ethanol. The permeation rate increases with increasing temperature because the increase of temperature increases both the mobility of the permeating molecules and that of the polymer segments. The effect of temperature on the separation factor depends upon the composition of the PAA–PVA IPN membranes. For the membranes with a low PAA content (PAA = 10 wt%), the separation factor decreases with increasing temperature. However, there is only a small change in the separation factor for the membranes with a high PAA content (PAA ≥ 30 wt%). This happens because the membranes with higher PAA contents have higher swelling and, consequently, both the permeating molecules and the polymer segments have higher mobility. While the increase in temperature generates additional mobility, the mobility generated by the swelling is large enough for the former increase to have small consequences.

Figure 3.1.10 contains Arrhenius plots for the total permeation rates. The activation energies are listed in Table 3.1.4, which shows that the activation energies decrease with increasing PAA content in the membrane.

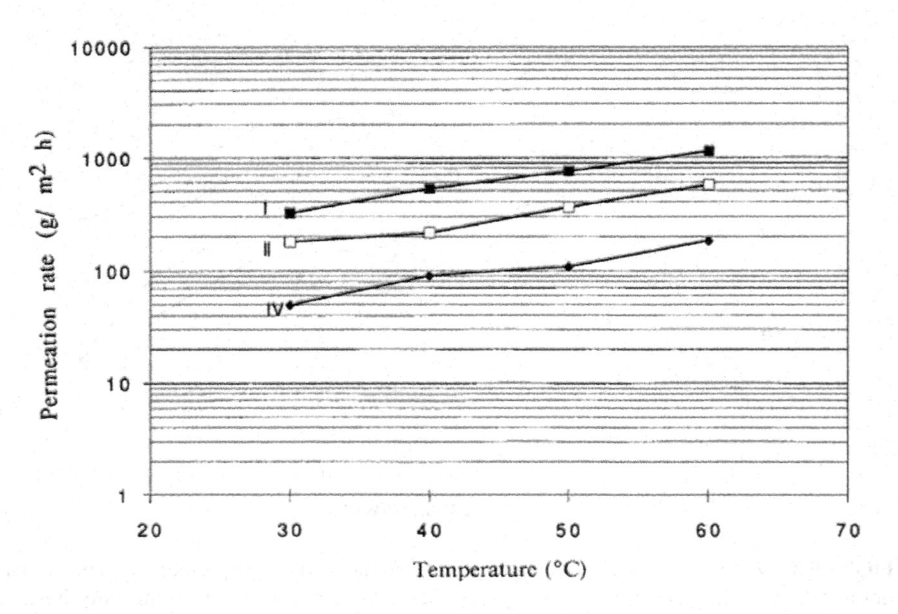

FIGURE 3.1.8 Dependence of the total permeation rate on the feed temperature for 85 wt% ethanol in ethanol–water mixtures and membranes I, II, and IV of Table 3.1.1.

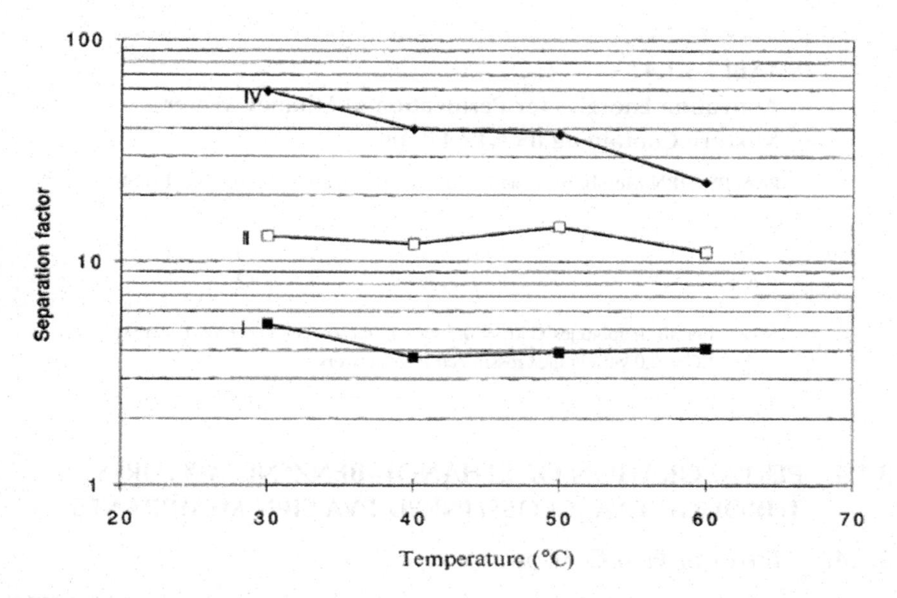

FIGURE 3.1.9 Dependence of the separation factor on the feed temperature for 85 wt% ethanol in ethanol–water mixtures and membranes I, II, and IV of Table 3.1.1.

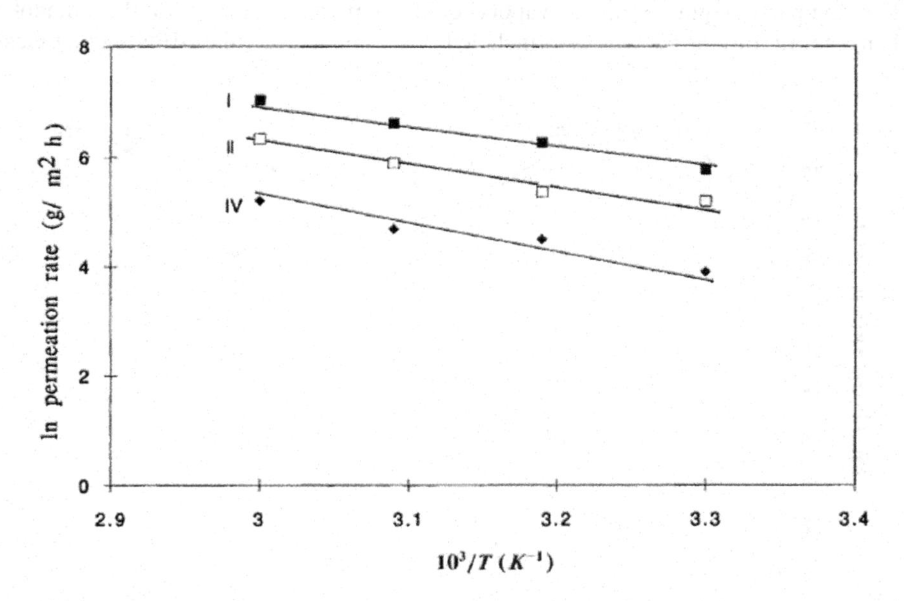

FIGURE 3.1.10 Arrhenius plots of the total permeation rate vs. temperature for 85 wt% ethanol in ethanol–water mixture and membranes I, II, and IV of Table 3.1.1.

TABLE 3.1.4
Activation Energies for Permeation of Ethanol–Water Mixture Containing 85 wt% Ethanol

PAA–PVA IPN Membranes (wt Ratio)	Activation Energy (kcal/mol)
70/30	7.3
50/50	7.4
10/90	9.3

Note: For all membranes GAL = 0.5 wt% in the crosslinked PVA network, EG = 2.0 wt% in the crosslinked PAA network.

3.1.4 PERVAPORATION OF ETHANOL–BENZENE MIXTURES THROUGH PAA–CROSSLINKED PVA SIPN MEMBRANES

3.1.4.1 Effect of Feed Composition

Figures 3.1.11 through 3.1.13 present the separation characteristics of PAA–PVA SIPN membranes for the ethanol-benzene mixture at 50°C. Figure 3.1.11 shows the relationship between the ethanol concentrations in the permeate and in the feed. For comparison purposes, the vapor–liquid equilibrium curve[20] of the ethanol-benzene mixture at 50°C is also included. One can see that the SIPN membranes

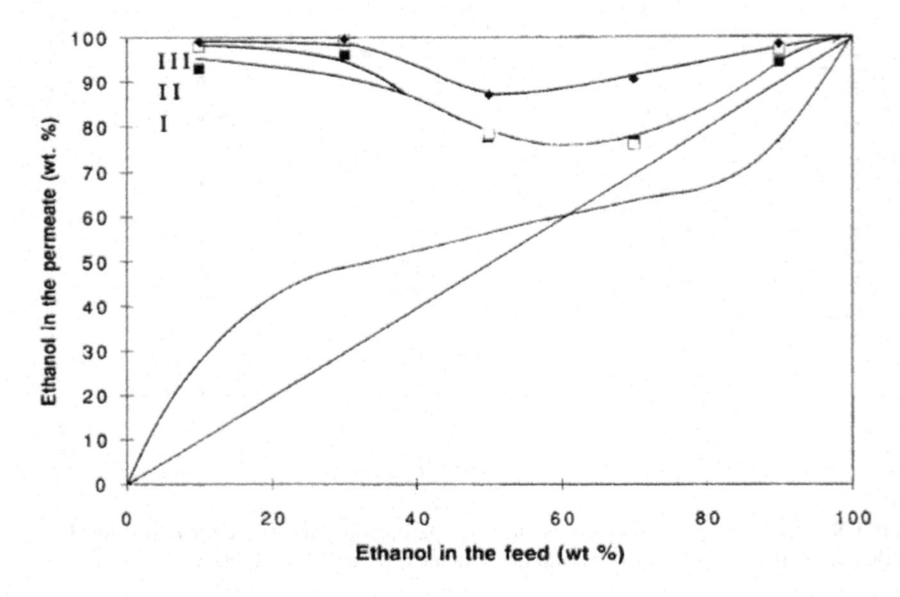

FIGURE 3.1.11 Permeate composition at 50°C in the pervaporation of ethanol–benzene mixtures through various PAA–PVA SIPN membranes, as a function of the feed composition. The membranes are those of Table 3.1.2.

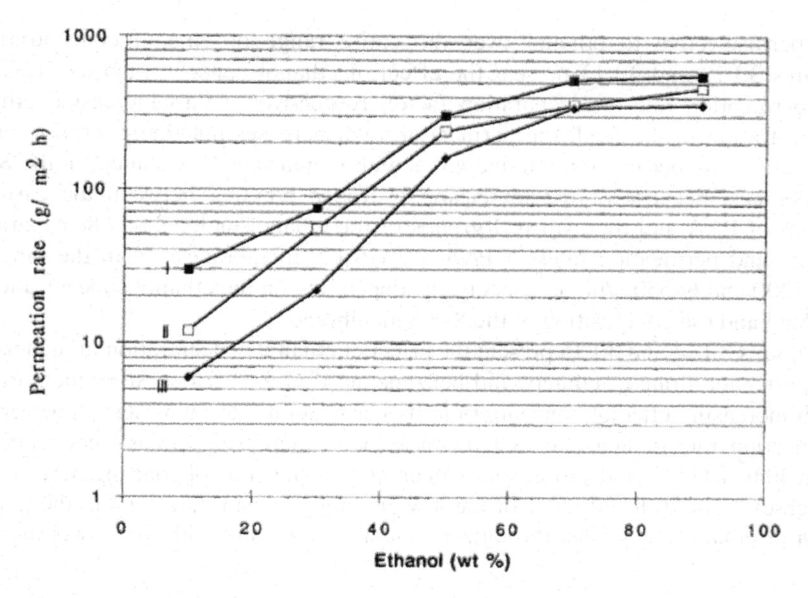

FIGURE 3.1.12 Effect of ethanol composition in the feed on the total permeation rate at 50°C for pervaporation of ethanol–benzene mixtures. The membranes are those of Table 3.1.2.

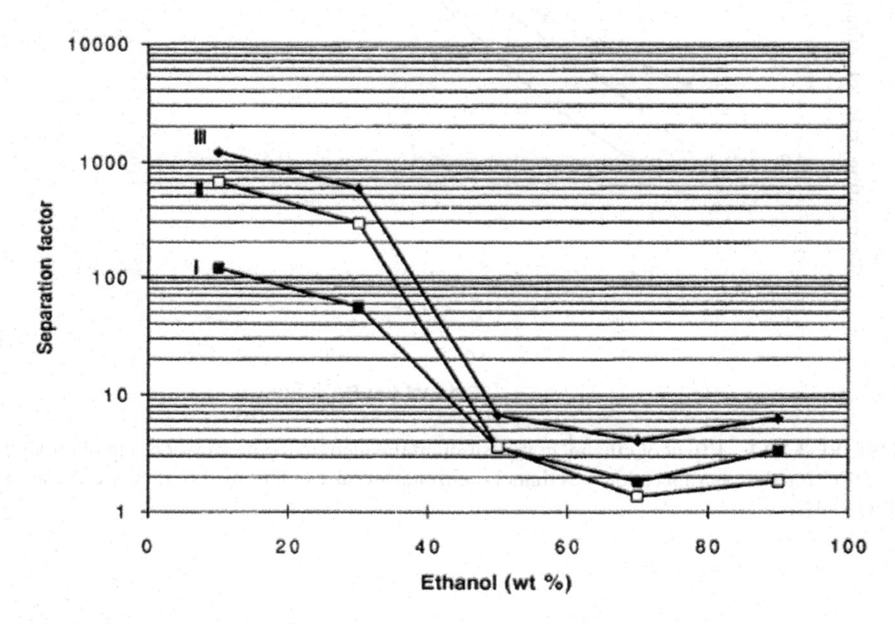

FIGURE 3.1.13 Effect of ethanol composition in the feed on the separation factor at 50°C for pervaporation of ethanol–benzene mixtures. The membranes are those of Table 3.1.2.

are permselective to ethanol over the entire range of ethanol concentrations. Figures 3.1.12 and 3.1.13 present the effect of ethanol concentration (at 50°C) on the permeation rate and separation factor, respectively. With increasing ethanol concentration in the feed, the permeation rate increases but the separation factor decreases. The permeation rate increases with increasing PAA content in the SIPN membranes, because of the increasing hydrogen bonding between the carboxyl groups of PAA and the hydroxyl groups of the ethanol molecules. The separation factors and permeation rates of PAA–PVA SIPN membranes are in the range of 1.4–1200 and 6–550 g/m^2 h, respectively, depending on the ethanol concentration in the feed and the composition of the SIPN membrane.

Figures 3.1.14 and 3.1.15 present the effect of ethanol concentration in the feed on the permeation rate of ethanol and benzene at 50°C for various SIPN membranes. With increasing ethanol concentration, its permeation rate increases. However, the permeation rate of benzene first increases with increasing ethanol concentration from 10 to 70 wt%, and subsequently decreases as the ethanol concentration further increases from 70 to 90 wt%. In the low ethanol concentration region, the permeation of ethanol molecules through the membrane occurs with little swelling; and

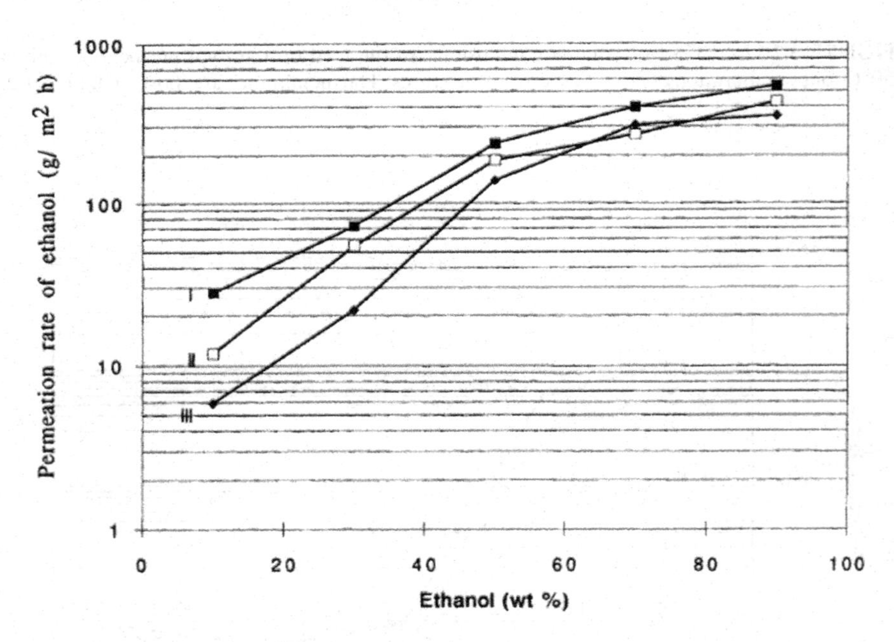

FIGURE 3.1.14 Effect of ethanol composition in the feed on the permeation rate of ethanol at 50°C for the pervaporation of ethanol–benzene mixtures. The membranes are those of Table 3.1.2.

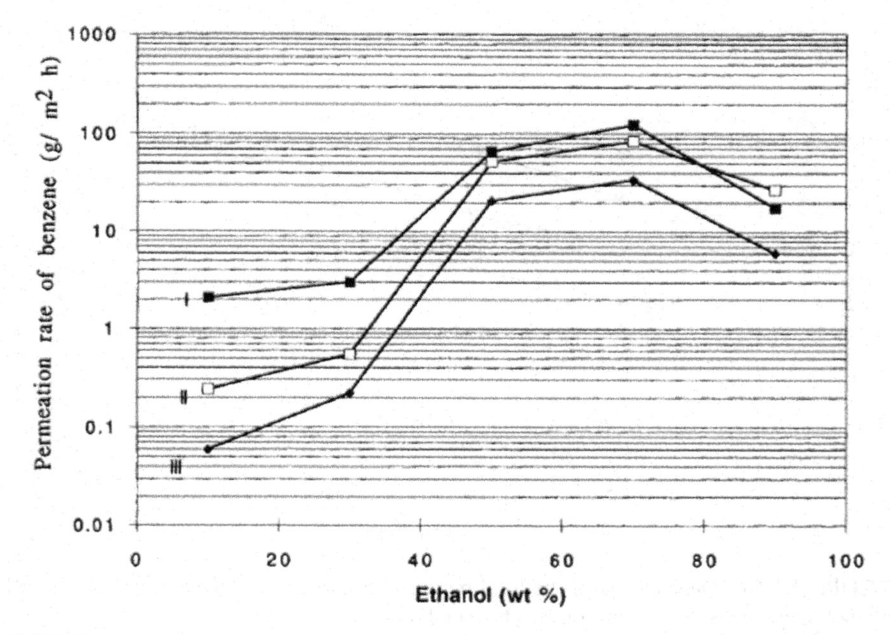

FIGURE 3.1.15 Effect of ethanol composition in the feed on the permeation rate of benzene at 50°C for the pervaporation of ethanol–benzene mixtures. The membranes are those of Table 3.1.2.

their adsorption on the hydrophilic sites of the membrane partially hydrophobize the membrane, thus enhancing the transfer of the benzene molecule. In the high ethanol concentration region, the swelling of the membrane increases and the membrane is increasingly hydrophilized. As a result, the permeation rate of benzene through the membrane decreases.

3.1.4.2 Effect of Feed Temperature

The dependencies of the permeation rate and separation factor on the feed temperature for 50 wt% ethanol in the feed are presented in Figures 3.1.16 and 3.1.17, respectively. With increasing feed temperature, the permeation rate increases and the separation factor decreases. Figure 3.1.18 shows that there is a linear relationship between the inverse of the permeation rate and the inverse of the absolute temperature. The activation energies calculated from Figure 3.1.18 are listed in Table 3.1.5. A comparison with the results of Table 3.1.4 shows that, as expected, the activation energies for the SIPN membranes are lower than those for the IPN membranes.

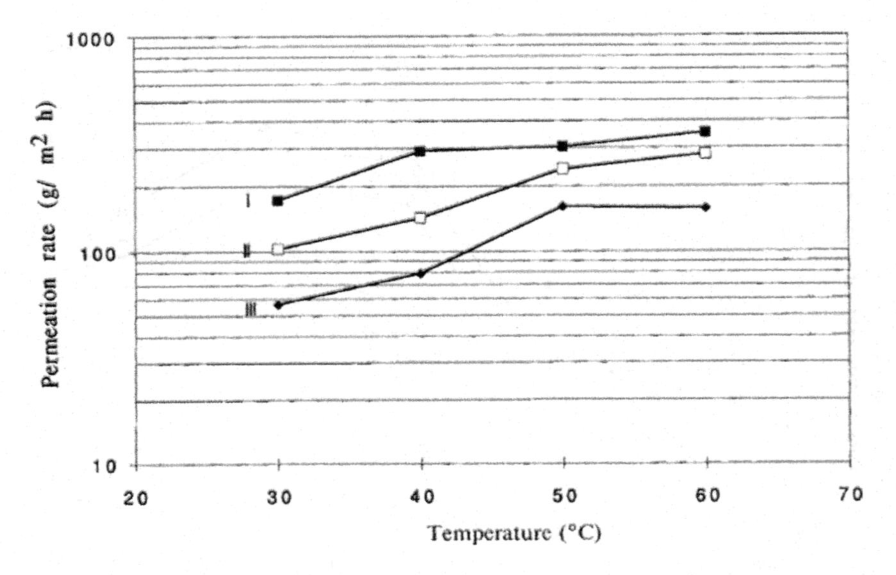

FIGURE 3.1.16 Dependence of the total permeation rate on the temperature for 50 wt% ethanol in ethanol–benzene mixtures. The membranes are those of Table 3.1.2.

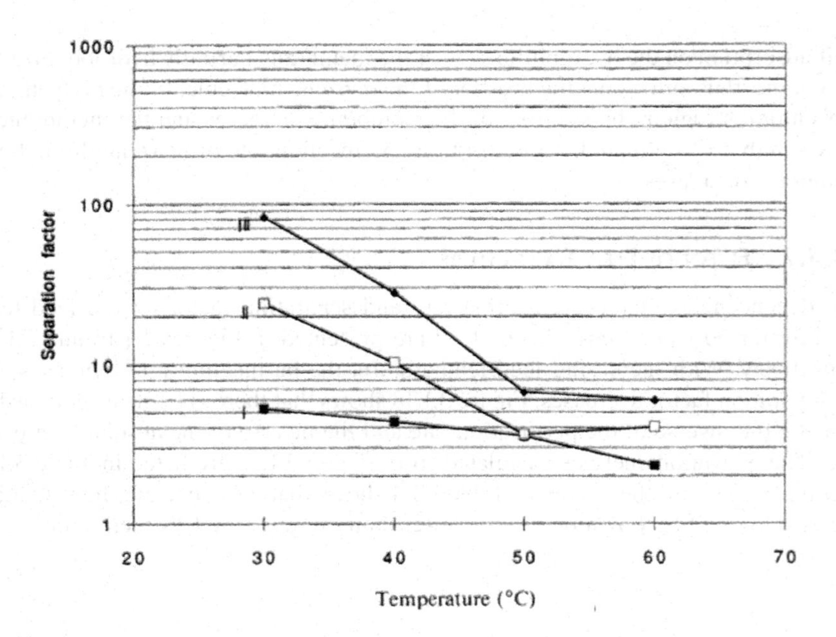

FIGURE 3.1.17 Dependence of the separation factor on the temperature for 50 wt% ethanol in ethanol–benzene mixtures. The membranes are those of Table 3.1.2.

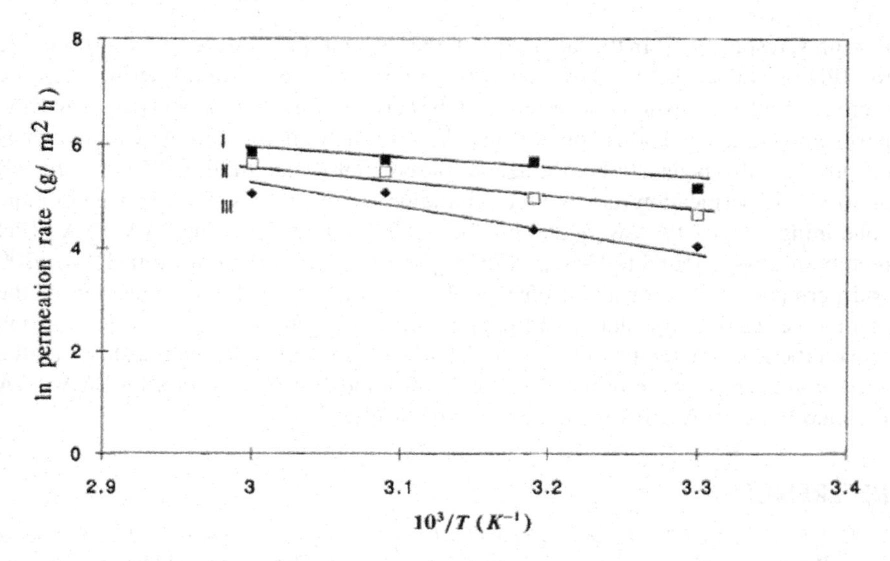

FIGURE 3.1.18 Arrhenius plots of total permeation rate vs. temperature for 50 wt% ethanol in ethanol–benzene mixtures. The membranes are those of Table 3.1.2.

TABLE 3.1.5
Activation Energies for Ethanol–Benzene Mixture Containing 50 wt% Ethanol

PAA–PVA SIPN Membranes (wt Ratio)	Activation Energies (kcal/mol)
80/20	4.6
70/30	6.6
60/40	7.5

Note: For all membranes GAL = 0.5 wt% in the crosslinked PVA network.

3.1.5 CONCLUSION

PAA–PVA IPN membranes as well as PAA-crosslinked PVA SIPN membranes were prepared. For the PAA–PVA IPN membranes, the tensile strength and elongation at break decrease with increasing PAA content; the tensile strength increases and the elongation at break decreases with increasing PAA content in the PAA-crosslinked PVA SIPN membranes. An explanation was proposed for the different behaviors of the two kinds of membranes. For application purposes the content of PAA should be about 50 and 60 wt% for the IPN and SIPN membranes, respectively.

The separation characteristics of IPN and SIPN membranes were investigated as a function of membrane composition, feed composition, and temperature, by dehydrating ethanol–water mixtures and separating ethanol from ethanol–benzene mixtures, respectively. In the latter case, the SIPN membranes were employed because the IPN ones provided too low permeation rates. While the overall permeation rate increases with increasing concentration of the permselective component in the feed, the permeation rate of the other component passes through a maximum. An explanation is proposed for this behavior. For the azeotropic ethanol–water mixture (95 wt% ethanol) at 50°C, the permeation rate and the separation factor of PAA–PVA IPN membrane containing 50 wt% PAA were 260 g/m^2 h and 50, respectively. The PAA–PVA SIPN membranes were found to have at 50°C separation factors ranging from 1.4 to 1200 and permeation rates of about 6560 g/m^2 h, depending on the composition of the membrane, feed composition, and temperature. The dependence of the permeation rate on the absolute temperature is consistent with an Arrhenius relationship, and the activation energies were in the range of 7.3–9.3 and 4.6–7.5 kcal/mol for PAA–PVA IPN and PAA–PVA SIPN membranes, respectively.

REFERENCES

1. J. L. Rapin, in *Proceedings of the Third International Conference on Pervaporation Processes in the Chemical Industry*, R. Bakish, Ed., Bakish Materials Corp., Englewood, NJ, 1988, p. 364.
2. H. C. Park, R. M. Meertens, M. H. V. Mulder, and C. A. Smolders, *J. Membr. Sci.*, **90**, 265 (1994).
3. Y. M. Lee and B. K. Oh, *J. Membr. Sci.*, **85**, 13 (1993).
4. H. C. Park, N. E. Ramaker, M. H. V. Mulder, and C. A. Smolders, *Sep. Sci. Technol.*, **30**, 419 (1995).
5. G. H. Koops, J. A. M. Nolten, M. H. V. Mulder, and C. A. Smolders, *J. Membr. Sci.*, **81**, 57 (1993).
6. E. Ruckenstein and F. Sun, *J. Membr. Sci.*, **81**, 191 (1993).
7. Y. K. Lee, T. M. Tak, D. S. Lee, and K. C. Kim, *J. Membr. Sci.*, **52**, 157 (1990).
8. L. Liang and E. Ruckenstein, *J. Membr. Sci.*, **106**, 167 (1995).
9. L. Liang and E. Ruckenstein, *J. Membr. Sci.*, **114**, 227 (1996).
10. J. Gmehling and U. Onken, *Vapor–Liquid Equilibrium Data Collection*, Vol. 1, Part 1, DECHEMA, Germany, 1977, p. 163.
11. M. Yoshikawa, H. Yokoi, K. Sanui, and N. Ogata, *J. Polym. Sci., Polym. Lett. Ed.*, **22**, 125 (1984).
12. M. Yoshikawa, N. Ogata, and I. Shimitza, *J. Membr. Sci.*, **26**, 107 (1986).
13. T. Uragami, M. Saito, and K. Takigama, *Makromol. Chem. Rapid Commun.*, **9**, 361 (1988).
14. T. Hirotsu, Jpn. Pat. 62-254807A, (1987).
15. C. Wan Yen and H. Chao Yi, *Angew. Makromol. Chem.*, **219**, 169 (1994).
16. Y. S. Kang, S. W. Lee, U. Y. Kim, and J. S. Shim, *J. Membr. Sci.*, **51**, 215 (1990).
17. T. Hirotsu, K. Ichimura, K. Mizoguchi, and E. Wakamura, *J. Appl. Polym. Sci.*, **36**, 1717 (1988).
18. R. T. M. Huang and C. K. Yeom, *J. Membr. Sci.*, **51**, 273 (1990).
19. F. Y. Xu and R. Y. M. Huang, *J. Appl. Polym. Sci.*, **36**, 1121 (1988).
20. J. Gmehling and U. Onken, *Vapor–Liquid Equilibrium Data Collection*, Vol. 1, Part 2a, DECHEMA, Germany, 1977, p. 409.

3.2 Pervaporation of Ethanol–Water Mixtures through Polyvinyl Alcohol–Polyacrylamide Interpenetrating Polymer Network Membranes Unsupported and Supported on Polyethersulfone Ultrafiltration Membranes: *A Comparison**

Eli Ruckenstein and Liang Liang
Department of Chemical Engineering, State University of New York at Buffalo, Buffalo, NY 14260, USA

ABSTRACT Polyvinyl alcohol–polyacrylamide interpenetrating polymer network (PVA-PAAM IPN) membranes, both unsupported and supported on polyethersulfone (PESF) ultrafiltration membranes, were prepared and investigated regarding the pervaporation of water–ethanol mixtures. Compared to the crosslinked polyvinyl alcohol membranes, the PVA-PAAM IPN membranes exhibit improved thermostability, mechanical properties and selectivity.

* *Journal of Membrane Science* 1996, 110, 99–107.

The PVA-PAAM IPN membranes were synthesized by the sequential IPN technique. The thicknesses of the unsupported membranes were in the range 30–40 μm, those of the supported ones were about 4 μm, and that of the PESF membrane was 90 μm. The selection of the compositions of the PVA-PAAM IPN layer of the supported membranes was made by examining the pervaporation performance of the unsupported ones. Furthermore, the effects of the feed composition, feed temperature and operating time on the performance of the PVA-PAAM IPN supported membrane was investigated. Depending on the feed composition and temperature, the supported membranes had separation factors between 30 and 28,300 and permeation rates between 30 and 3800 g/(m^2 h). High selectivities were obtained at high ethanol concentrations. For the azeotropic water–ethanol mixture (95 wt% ethanol), the supported membrane had at 60°C a separation factor of 13,000 and a permeation rate of 80 g/(m^2 h). Compared to the unsupported membranes, the supported ones had a higher permeability but a somewhat lower selectivity.

3.2.1 INTRODUCTION

Pervaporation is an energy efficient method of separating azeotropic mixtures, close-boiling point compounds, and mixtures consisting of heat-sensitive compounds. For this reason, in recent years, there has been increased interest in its use. Much research was carried out regarding the water–ethanol mixture [1]. Numerous membranes were prepared and used for the dehydration of ethanol–water mixtures by pervaporation [2–11]. Among them, the crosslinked polyvinyl alcohol (PVA) membranes had been frequently investigated because their hydroxy groups have through hydrogen bonding strong interactions with water [12]. The crosslinked PVA supported on a polyacrylonitrile (PAN) membrane, produced by Gesellshaft für Trenn-technik (GFT) mbH Germany, has been commercialized in 1982 [13]. Since then, numerous attempts were made to improve the separation capability of the PVA membranes. Two approaches were employed. In one of them, a more suitable crosslinking agent [14–16], such as, maleic acid, was used. In this case, the selectivity of the crosslinked PVA membrane for water was enhanced because not all the carboxylic groups of the crosslinker are involved in crosslinking and the remaining ones have strong interactions with water through hydrogen bonding. In the other one, the properties of the membrane were controlled by grafting a hydrophobic polymer to PVA, such as polystyrene or polymethyl methacrylate [17–19]. Polyacrylamide (PAAM) is a brittle material which cannot be used in pervaporation because of its poor film-forming capability. However, compared to PVA, PAAM has a higher thermostability [20], a higher swelling in water [21], and a higher selectivity for water from ethanol–water mixtures [22]. For this reason, polyvinyl alcohol–polyacrylamide interpenetrating polymer network (PVA-PAAM IPN) membranes were prepared via the sequential IPN technique and investigated regarding their performance in dehydrating, through pervaporation, ethanol–water mixtures [23]. The results showed that the mechanical properties and the selectivity of PVA-PAAM IPN membranes were greatly enhanced compared to those of the crosslinked PVA membranes. The emphasis in the present paper is on membranes containing a PVA-PAAM IPN layer supported on polyethersulfone (PESF) ultrafiltration membranes. We prefer to call them supported membranes and not composite membranes as they are usually called. The effects of the composition of the water–ethanol mixture, feed temperature and operating time on the pervaporation performance of the supported membranes are investigated. Results regarding the pervaporation through unsupported (free of PESF) membranes are also included, in order to select the most appropriate compositions for the supported layer and to compare them with the supported ones. Because in the supported case the thickness of the pervaporation membrane is much smaller, its permeation rate is expected to be higher than that of the unsupported one.

3.2.2 EXPERIMENTAL

3.2.2.1 Materials

The acrylamide (AAM, Aldrich) and potassium persulfate ($K_2S_2O_8$, Aldrich) were purified by recrystallization from methanol and water, respectively. Polyvinyl alcohol (PVA, Aldrich) of molecular weight 124 000, *N, N′*-methylenebisacrylamide (BisAAM, Aldrich), glutaraldehyde (GAL, 25 wt% in water), hydrochloric acid (37 wt% in water, Aldrich); sodium hydrogen carbonate (99.7%, Aldrich) and ethanol (99%, Aldrich) were used without purification. Water was deionized and double distilled. Polyethersulfone ultrafiltration membranes (Gelman Scientific Co.), with a protein molecular weight cut-off of 10 000 and thicknesses of about 90 μm, were employed.

3.2.2.2 PVA-PAAM IPN Unsupported Membrane

The unsupported membrane was prepared as in the previous paper [23]. In summary, the following procedure was employed. PVA, AAM, BisAAM and potassium persulfate were dissolved in water. The reaction was carried out in a one-necked 100 mL round bottomed flask, with magnetic stirring, in a N_2 atmosphere. The mixture was heated at about 70°C for 8 h to generate a PAAM crosslinked network, and cooled to room temperature. Then, GAL and hydrochloric acid were dissolved in the solution and the mixture was cast on a flat glass plate and dried at ambient temperature. This was followed by the neutralization of the residual acid with a dilute solution of sodium hydrogen carbonate, by washing with water and finally by drying. The thickness of the dry membrane, measured with a micrometer with an accuracy of ±10 μm, was in the range of 30–40 μm (which represents an average of 6 measurements in different points of the membrane).

3.2.2.3 PVA-PAAM IPN Supported Membrane

An aqueous mixture containing PVA and the crosslinked PAAM network was cast on the surface of a PESF ultrafiltration membrane. The crosslinking agent GAL and catalyst HC1 were added to the solution prior to casting. As a result, a PVA-PAAM IPN layer was deposited on the PESF membrane. The supported membrane was dried at room temperature for 24 h. Three layers were deposited one after another in order to ensure that the supported membrane was pinhole-free. The thickness of the dry top layer, measured by scanning electron microscopy (SEM), was about 4 μm, while the thickness of the dry support was about 90 μm. The presence or absence of pinholes was determined via pervaporation experiments at the azeotropic composition. In this case, in the absence of pinholes, the separation factor a should be very large. The membranes with low values of α (<50) have been disregarded.

3.2.2.4 Scanning Electron Microscopy (SEM)

The PVA-PAAM IPN supported membranes were fractured in liquid nitrogen, sputtered with carbon and their cross-sections investigated with SEM (Hitachi S-800).

3.2.2.5 Pervaporation

The pervaporation apparatus was described previously [23]. The membrane was placed on a porous glass support. The effective area of the membrane was 9.6 cm^2. The feed solution was maintained at a constant temperature by recirculation through a heat exchanger placed in a thermostated water bath. The permeated vapor was collected in a liquid nitrogen trap. The composition of the permeate was analyzed chromatographically. The separation factor of the membrane was calculated using the expression

$$\alpha = \frac{\left(Y_{H_2O}/Y_{EtOH}\right)}{\left(X_{H_2O}/X_{EtOH}\right)}$$

where Y_{H_2O}/Y_{EtOH} is the weight ratio of water to ethanol in the permeate and X_{H_2O}/X_{EtOH} is the weight ratio of water to ethanol in the feed mixture.

3.2.3 RESULTS AND DISCUSSION

3.2.3.1 Suitable Compositions for the PVA-PAAM IPN Supported Membrane

The only difference between the unsupported and supported membranes is the thickness of the separating layer of PVA-PAAM IPN. It is, therefore, reasonable to select for the composition of the deposited layer one of the efficient compositions of the unsupported membrane. For this reason, the unsupported membrane was investigated in some detail.

Figure 3.2.1 presents the effect of GAL content in the PVA network on the total permeation rate J [g/(m^2 h)] and separation factor α for the PVA-PAAM IPN unsupported membrane that contained 1.0 wt% BisAAM in the PAAM network and 16 wt% PAAM in the PVA-PAAM IPN. The pervaporation runs were carried out at 75°C, for an ethanol concentration of 85 wt%. One can see that the separation factor increases with increasing GAL content in the PVA network, and the permeation rate decreases. This occurs because an increase in the extent of crosslinking decreases the mobility of the polymer chains. The resulting membrane has a more compact

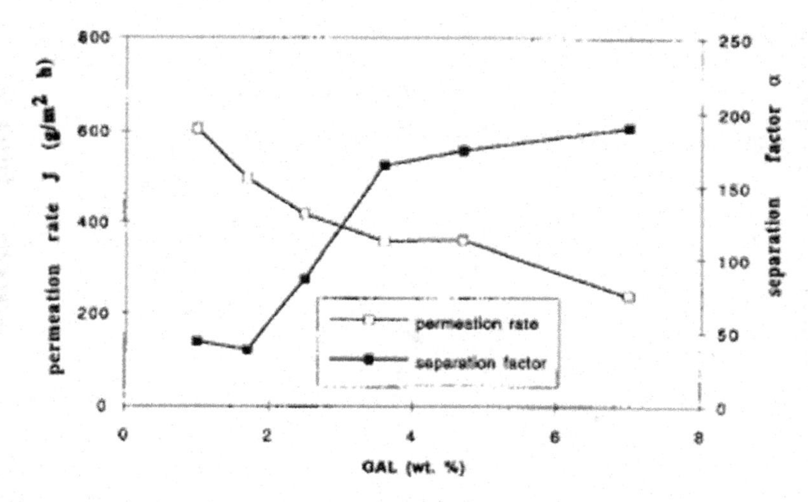

FIGURE 3.2.1 Effect of GAL content in the PVA network on the pervaporation performance at 75°C, for 85 wt% ethanol and PVA-PAAM IPN unsupported membrane. BisAAM = 1.0 wt% in the PAAM network and PAAM = 16 wt% in the PVA-PAAM IPN.

network structure and its swelling decreases [23]. As a result, the solubilities and diffusivities of the components of the liquid mixture decrease. Consequently, the permeation rate through the membrane decreases. Since less water is sorbed and the presence of water stimulates the diffusion of the ethanol molecules, the separation factor increases. A reasonable permeation rate of the membrane is obtained when the content of GAL in the PVA network is about 1.0 wt%. Hence, 1.0 wt% GAL in the PVA network constitutes a suitable concentration for a PVA-PAAM IPN layer. It is important to note that a PVA-PAAM IPN membrane containing 1.0 wt% GAL in the PVA network has also good mechanical and swelling properties [23].

The effect of BisAAM content in the PAAM network on the permeation rate and separation factor for the PVA-PAAM IPN unsupported membrane that contained 1.0 wt% GAL in the PVA network and 16 wt% PAAM in the PVA-PAAM IPN was investigated for the pervaporation of a mixture containing 85 wt% ethanol, at 75°C. Figure 3.2.2 shows that with increasing BisAAM content in the PAAM network, the separation factor increases and the permeation rate decreases. The increase of crosslinking in both the PVA and PAAM networks increases the separation factor. The permeation rate changes more steeply in the range 1–3 wt% BisAAM in the PAAM network. We selected 2 and 4 wt% BisAAM in the PAAM network to prepare the PVA-PAAM IPN supported membranes because for these values the permeation rate and the separation factor have acceptable values.

Figure 3.2.3 shows the effect of the PAAM content in the PVA-PAAM IPN unsupported membrane on the permeation rate and separation factor for 1.0 wt% GAL in the PVA network and 4.0 wt% BisAAM in the PAAM network. The pervaporation experiments were carried out at 75°C for a 85 wt% ethanol mixture. With increasing PAAM content, both the permeation rate and separation factor increase. This occurs because compared to PVA, PAAM sorbs a larger amount of water and has a higher selectivity for water from water–ethanol mixtures. It is clear that to improve

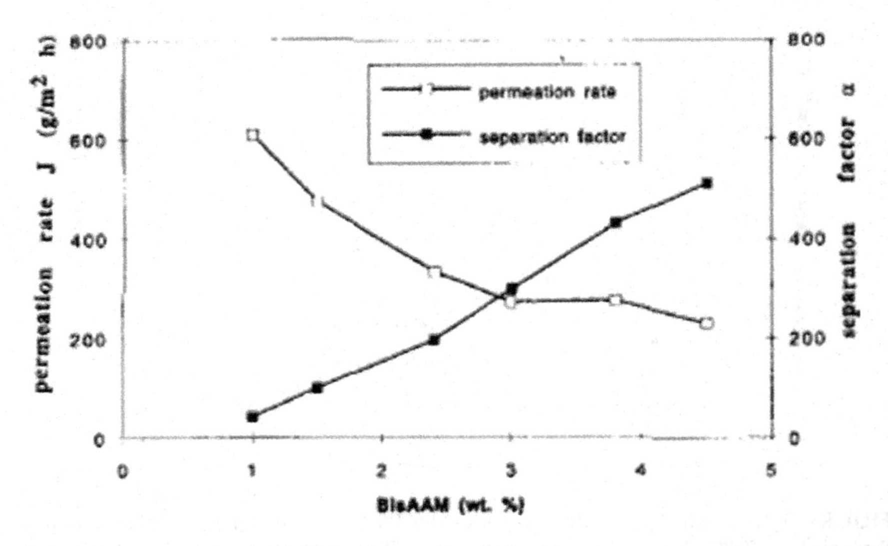

FIGURE 3.2.2 Effect of BisAAM content in the PAAM network on pervaporation performance at 75°C, for 85 wt% ethanol and PVA-PAAM IPN unsupported membrane. GAL = 1.0 wt% in the PVA network and PAAM = 16 wt% in the PVA-PAAM IPN.

the separation capability of the membrane, the content of PAAM in the PVA-PAAM IPN should be increased as much as possible. However, the PVA-PAAM IPN unsupported membranes become too brittle for PAAM contents greater than 16 wt%.

From the above discussion, one can conclude that a PVA-PAAM IPN supported layer should contain: 1.0 wt% GAL in the PVA network, 2 or 4 wt% BisAAM in the

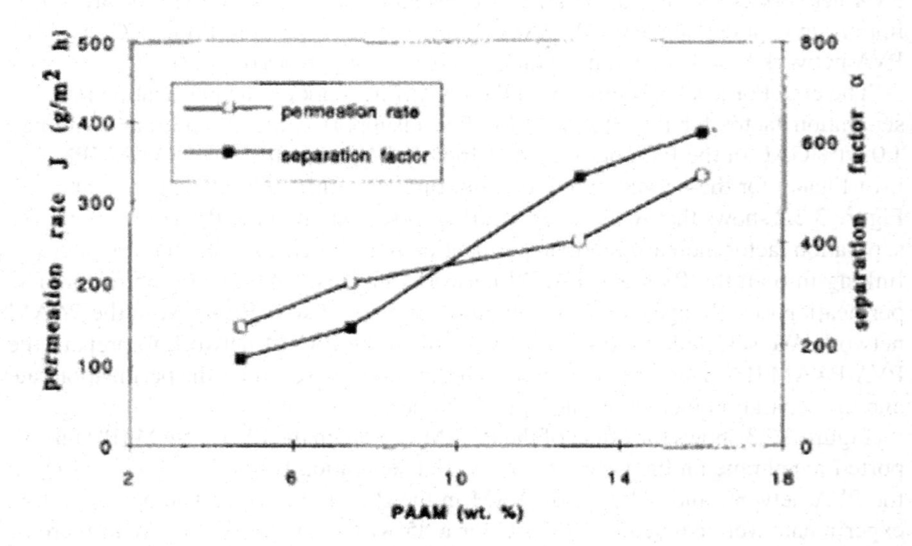

FIGURE 3.2.3 Effect of PAAM content in PVA-PAAM IPN membrane on the pervaporation performance at 75°C, for 85 wt% ethanol and PVA-PAAM IPN unsupported membrane. GAL = 1.0 wt% in the PVA network and BisAAM = 4.0 wt% in the PAAM network.

TABLE 3.2.1
The PVA-PAAM IPN Supported Membranes Employed[a]

Membrane	PAAM (wt%) in the PVA-PAAM IPN Layer	GAL (wt.%) in the PVA Network	BisAAM (wt%) in the PAAM Network	Thickness of the PVA-PAAM IPN Layer (μm)
SHM-1	16	1.0	2.0	4.3
SHM-2	16	1.0	4.0	4.0

[a] The thickness of polyethersulfone ultrafiltration membrane used as support was about 90 μm.

PAAM network and 16 wt% PAAM. Table 3.2.1 lists the compositions and characteristics of the PVA-PAAM IPN supported layers used in this study.

3.2.3.2 Characteristics of the PVA-PAAM IPN Supported Membranes

SEM micrographs of the PVA-PAAM IPN supported membranes are presented in Figure 3.2.4. The morphology of the PESF layer has a typical finger-structure [24]. One can see that there are no pores larger than about 100 Å (which is the resolution of SEM) in the PVA-PAAM IPN membranes.

3.2.3.3 Effect of the Feed Composition

The separation characteristics of the PVA-PAAM IPN supported membranes for water–ethanol mixtures at 50°C are presented in Figures 3.2.5 through 3.2.7. Figure 3.2.5 shows that the PVA-PAAM IPN supported membranes are permselective

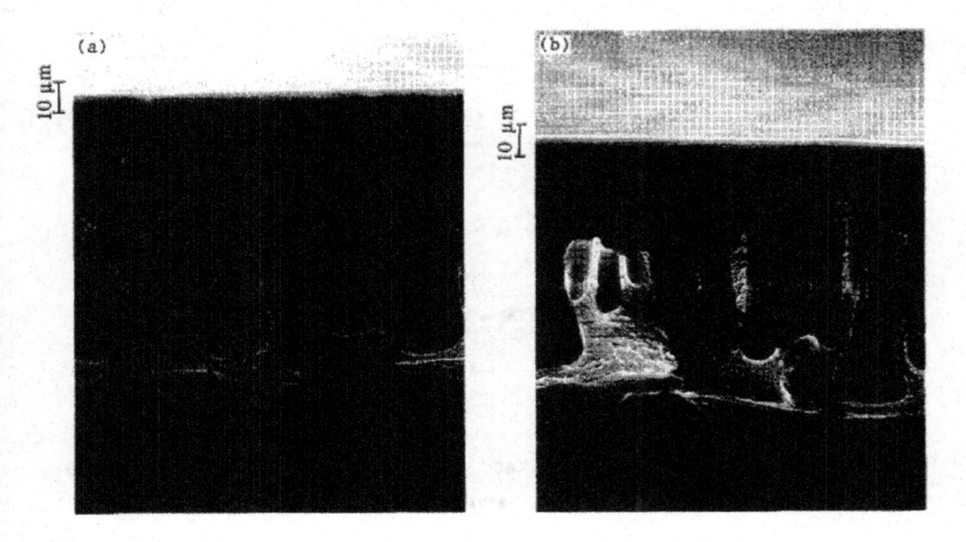

FIGURE 3.2.4 SEM micrographs of PVA-PAAM IPN supported membranes. (a) SHM-1 membrane (Table 3.2.1), (b) SHM-2 membrane (Table 3.2.1).

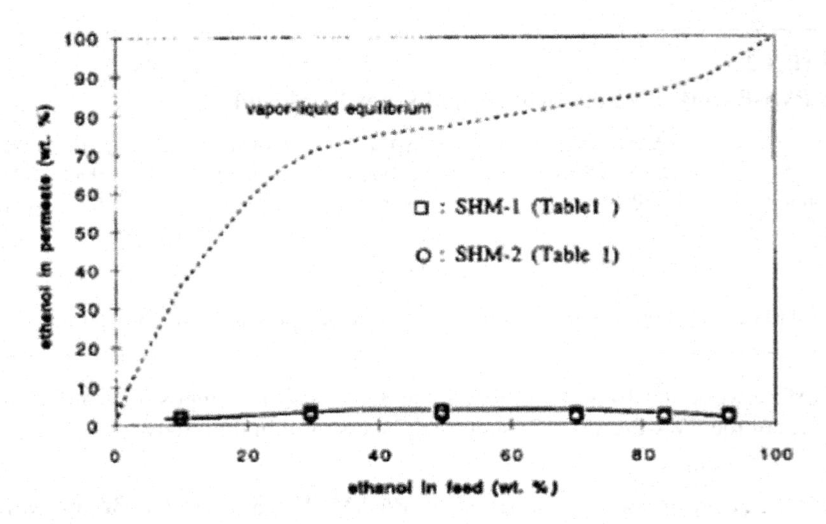

FIGURE 3.2.5 Effect of feed composition on the permeate composition at 50°C for supported membranes.

to water over the entire range of ethanol concentrations. Figures 3.2.6 and 3.2.7 show the effect of ethanol concentration, at 50°C, on the permeation rate and separation factor, respectively. As the ethanol concentration in the feed increases, the separation factor increases, but the permeation rate decreases. Figures 3.2.6 and 3.2.7 also show that with increasing content of BisAAM in the PAAM network, the separation factor increases tremendously at high ethanol concentrations.

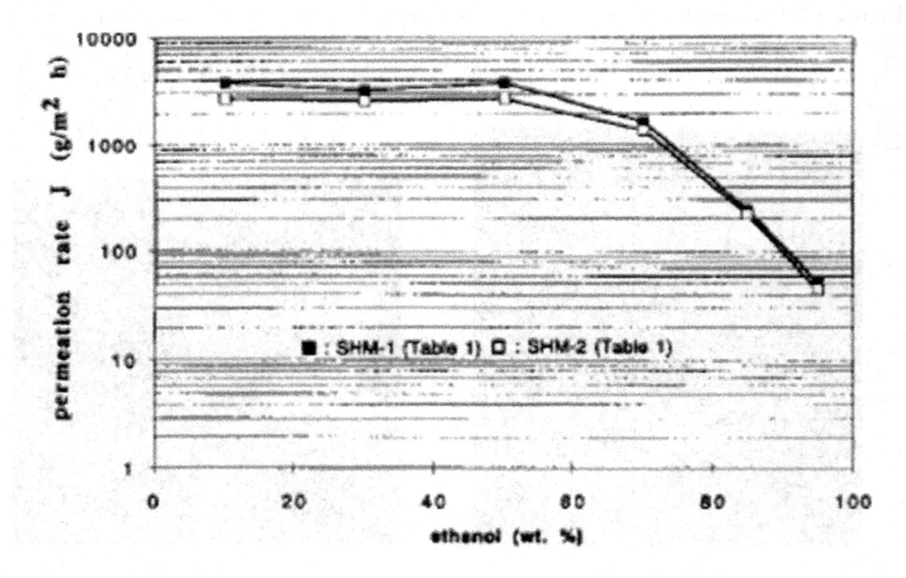

FIGURE 3.2.6 Effect of feed composition on the permeation rate at 50°C for supported membranes.

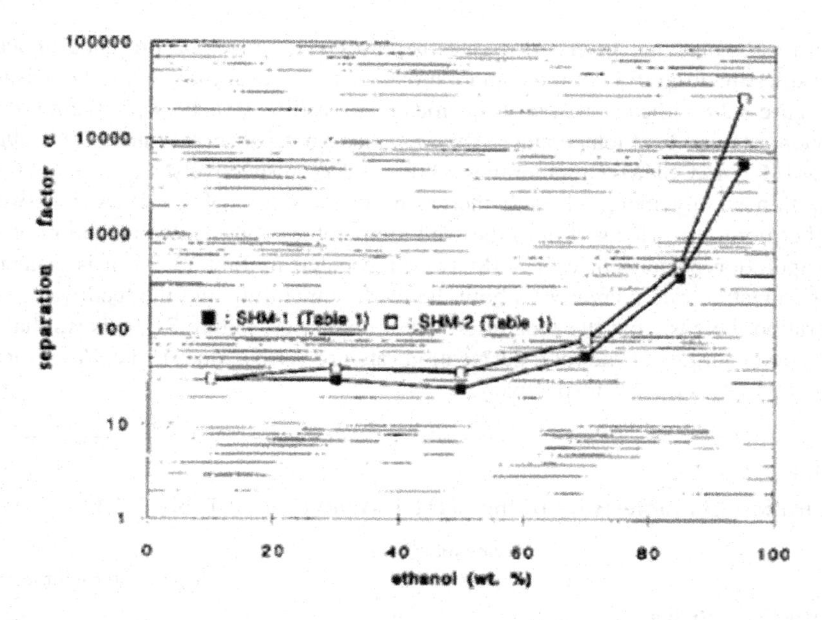

FIGURE 3.2.7 Effect of feed composition on the separation factor at 50°C for supported membranes.

3.2.3.4 Effect of the Feed Temperature

The dependence of the separation factor and the permeation rate on the feed temperature for 95 wt% ethanol (the azeotropic composition) is presented in Table 3.2.2. It should be noted that in some cases the ethanol concentration in the permeate is so low that it could not be detected chromatographically. For both SHM-1 and SHM-2 membranes (Table 3.2.1) the permeation rate increases with increasing feed temperature, but the separation factor decreases. This is because with increasing feed temperature, the frequency and amplitude of the polymer chain motions become larger.

TABLE 3.2.2
Pervaporation Performances of PVA-PAAM IPN Supported Membranes for 95 wt% Ethanol at Various Temperatures

Temperature (°C)	SHM-1 J [g/(m² h)]	α[a]	SHM-2 J [g/(m² h)]	α
30	32	very large	28	very large
40	45	very large	32	very large
50	53	5900	45	28300
60	110	4800	84	13000

[a] Very large indicates that the concentration of ethanol in the permeate could not be detected chromatographically.

As a result, the total permeation rate increases and, since the presence of water intensifies the migration of the ethanol molecules, the separation factor decreases.

Figure 3.2.8 indicates that an Arrhenius relationship exists between the total permeation rate and feed temperature. The activation energies are 9.9 and 12.0 kcal/mol for SHM-1 and SHM-2 membranes (Table 3.2.1), respectively. As expected, the activation energy increases with increasing crosslinking of the PAAM network. Furthermore, the effects of both the feed temperature and feed composition on the pervaporation performance of SHM-1 are presented in Table 3.2.3. It is clear that the permeation rate increases as both the feed temperature and the feed water concentration increase, but the separation factor decreases. Figure 3.2.9 shows that the activation energies of the total permeation rate are 7.4, 8.5 and 9.9 kcal/mol for 70, 85 and 95 wt% ethanol, respectively.

TABLE 3.2.3
Separation Characteristics of the SHM-1 Membrane of Table 3.2.1

Ethanol Concentration in the Feed (wt%)	Permeation Rate J [g/(m² h)]			Separation Factor α		
	70	**85**	**95**	**70**	**85**	**95**
30°C	538	103	32	421	7484	very large[a]
40°C	983	206	45	293	628	very large[a]
50°C	1674	367	53	274	383	5900
60°C	2496	768	110	51	431	4800

[a] The concentration of ethanol in the permeate could not be detected chromatographically.

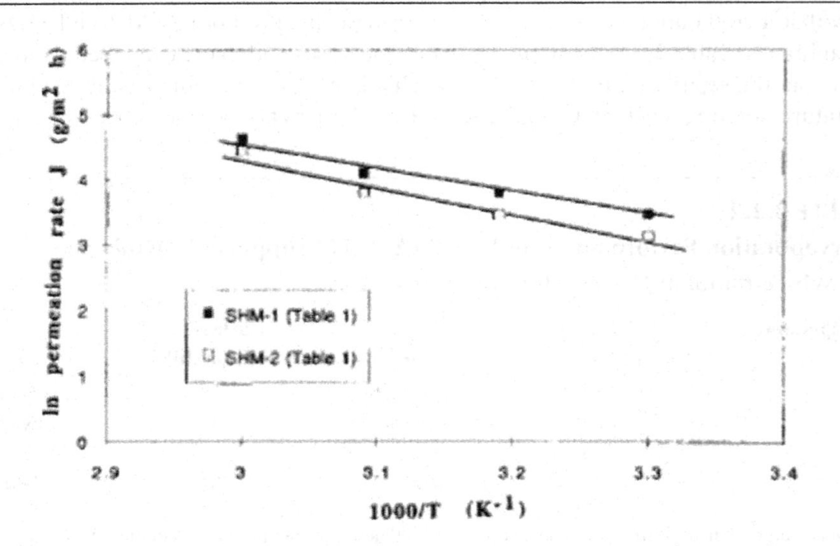

FIGURE 3.2.8 Arrhenius plots for the total permeation rate through supported membranes vs. temperature, for 95 wt% ethanol.

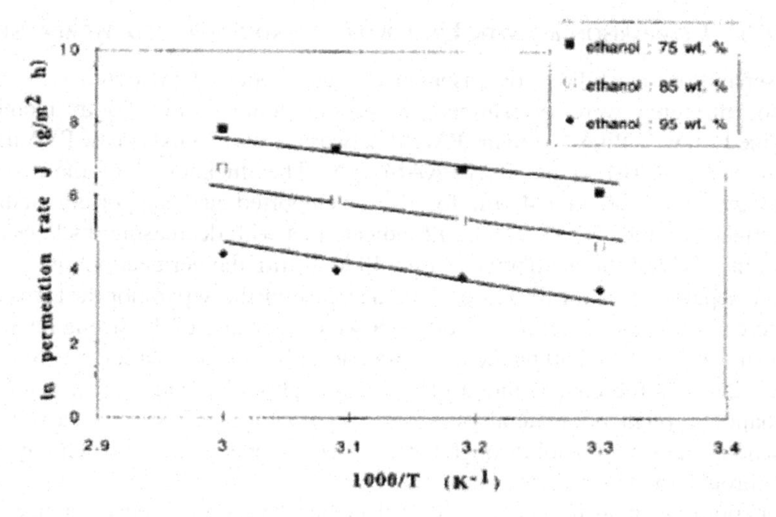

FIGURE 3.2.9 Arrhenius plots for the total permeation rate vs. temperature for the supported membrane SHM-1 of Table 3.2.1.

3.2.3.5 Effect of the Operating Time

The relationship between the separation performances and operating time was investigated for the membrane SHM-1 (Table 3.2.1) and the results are plotted in Figure 3.2.10. One can see that the membrane has a stable separation performance, i.e., the permeation rate and separation factor remain almost unchanged during 6 days of pervaporation. This is due to the stable crosslinked structure of the PVA-PAAM IPN supported layer of the membrane.

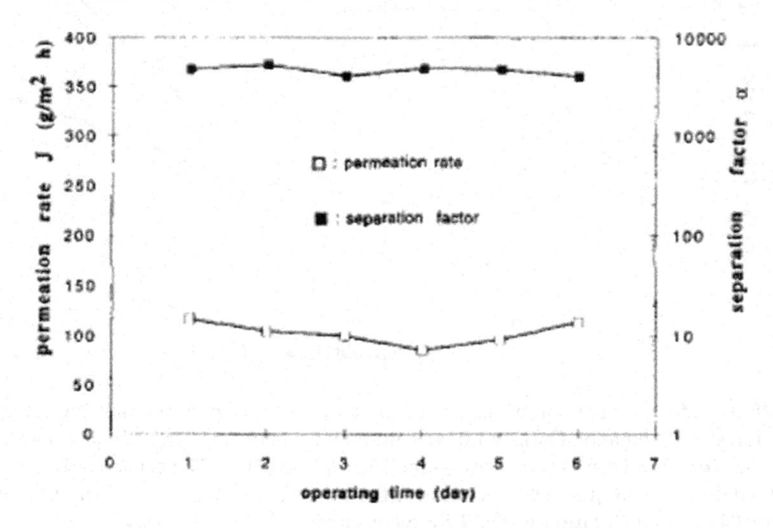

FIGURE 3.2.10 Pervaporation performance of PVA-PAAM IPN supported membrane as a function of operating time for membrane SHM-1 of Table 3.2.1, 95 wt% ethanol and 60°C.

3.2.3.6 Comparison between Unsupported and Supported Membranes

The separation capabilities of unsupported and supported membranes for water–ethanol mixtures were determined at various temperatures. Both membranes contained 4.0 wt% BisAAM in the PAAM network, 1.0 wt% GAL in the PVA network and 16 wt% PAAM in the PVA-PAAM IPN. The thicknesses of the separating membranes were 40 and 4 μm for the unsupported and supported membrane, respectively. Figures 3.2.11 and 3.2.12 indicate that with decreasing thickness of the separating layer of the membranes (from 40 to 4 μm), the permeation rate increases and the separation factor decreases. The decrease of the separation factor is due to the increase of the permeation rate of water which, because of the strong interactions between water and ethanol molecules, stimulates also the permeation rate of ethanol. Table 3.2.4 lists the concentration of ethanol in permeate and the permeation rate of ethanol through both membranes at various temperatures. It is clear that the permeation rates of ethanol through the supported membranes are much higher than those through the unsupported ones.

For comparison purposes, Table 3.2.5 lists the pervaporation performances of different supported membranes used by various groups for dehydrating a water–ethanol mixture with a composition near the azeotropic one.

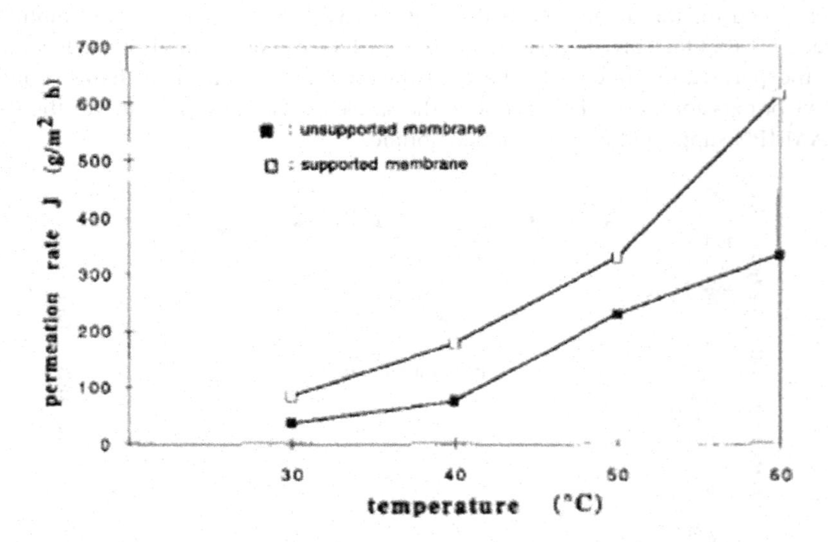

FIGURE 3.2.11 Comparison of the permeation rates of unsupported and supported membranes for 85 wt% ethanol. GAL = 1.0 wt% in the PVA network, BisAAM = 4.0 wt% in the PAAM network and PAAM = 16 wt% in the PVA-PAAM IPN. The thickness of the unsupported membrane = 40 μm. The thickness of the supported membrane = 4 μm for the PVA-PAAM IPN layer and 90 μm for the PESF membrane.

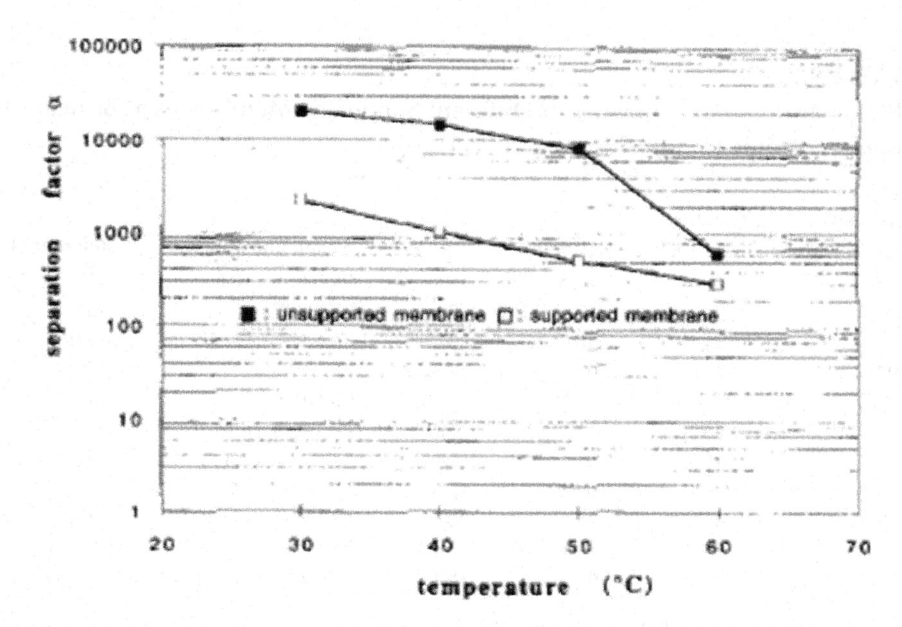

FIGURE 3.2.12 Comparison of the separation factors of unsupported and supported membranes for 85 wt% ethanol. GAL = 1.0 wt% in the PVA network, BisAAM = 4.0 wt% in the PAAM network and PAAM = 16 wt% in the PVA-PAAM IPN. The thickness of the unsupported membrane = 40 μm. The thickness of the supported membrane = 4 μm for the PVA-PAAM IPN layer and 90 μm for the PESF membrane.

TABLE 3.2.4
Comparison between Unsupported and Supported Membranes[a]

	Unsupported Membrane			Supported Membrane		
Temperature (°C)	Total Permeation Rate J [g/(m² h)]	Permeation Rate of Ethanol [g/(m² h)]	Ethanol (wt%) in Permeate	Total Permeation Rate J [g/(m² h)]	Permeation Rate of Ethanol [g/(m² h)]	Ethanol (wt%) in Permeate
30	37	0.01	0.03	84	0.21	0.25
40	75	0.03	0.04	178	0.38	0.55
50	230	0.14	0.06	300	2.43	1.10
60	334	3.07	0.92	612	11.60	1.90

[a] For both unsupported and supported membranes, GAL = 1.0 wt% in the PVA network, BisAAM = 4.0 wt% in the PAAM network and PAAM = 16 wt% in the PVA-PAAM IPN. The thickness of unsupported membrane = 40 μm. The thickness of supported membrane = 4 μm for the PVA-PAAM IPN layer and 90 μm for the polyethersulfone ultrafiltration membrane. The concentration of ethanol in the feed was 85 wt%.

TABLE 3.2.5
Pervaporation Performances of Different Supported Membranes at or near the Azeotropic Composition

Ethanol in Feed (wt%)	Temperature (°C)	J [kg/(m² h)]	α	Membrane Material[a]	References
95	60	0.15	1500	PVA-PSSA/PAN	[25]
95.5	78	4.0	450	PVA/PSF	[26]
95	50	~0.04	30,000	PAA/PAN	[27]
95	60	0.02	4800	PVA/PAN	[28]
95	50	~0.10	10,800	PAA/PSF	[29]
95	60	1.63	3510	PAA-polyion/PESF	[30]
95	70	0.2	500	PVA-chitosan/PSF	[31]
90	25	0.42	13.5	PVAc/Nylon-4	[32]
88.5	40	29.7	0.14	PAA-*co*-PAAM/PP	[33]
95	60	0.08	13,000	PVA-PAAM IPN/PESF	this study
95	60	0.11	4800	PVA-PAAM IPN/PESF	this study

[a] PVA, polyvinyl alcohol; PSF, polysulfone; PAA, polyacrylic acid; PAN, polyacrylonitrile; PESF, polyethersulfone; PSSA, polystyrene-sulfonic acid; PVAc, polyvinyl acetate; PP, polypropylene; PAAM, polyacrylamide; IPN, interpenetrating polymer network.

3.2.4 CONCLUSION

The PVA-PAAM IPN supported membranes were prepared by depositing an aqueous solution of PVA and GAL, which also contained a crosslinked PAAM network, on the surface of a PESF ultrafiltration membrane. The reaction between PVA and GAL generated the effective separating layer of PVA-PAAM IPN supported on the PESF membrane. Most of the experiments have been carried out with PVA-PAAM IPN layers containing 1.0 wt% GAL in the PVA network, 2.0 wt% BisAAM in the PAAM network and 16 wt% PAAM in the PVA-PAAM IPN. The separation factors and permeation rates of the PVA-PAAM IPN supported membranes are in the range 30–28,300 and 30–3800 g/(m² h), respectively, depending upon the ethanol concentration in the feed, feed temperature and crosslinking extent of the PVA-PAAM IPN layer. The dependence of the total permeation rate on the temperature is consistent with an Arrhenius relationship, with an activation energy in the range of 7.4–12.0 kcal/mol. The supported membrane had stable pervaporation performance with time, the permeability and selectivity remaining almost unchanged during 6 days.

REFERENCES

1. S. Zhang and E. Drioli, Pervaporation membranes, *Sep. Sci. Technol.*, 30(1) (1995)1.
2. R.Y.M. Huang and C.K. Yeom, Pervaporation separation of aqueous mixtures using crosslinked poly(vinyl alcohol) (PVA) II. Permeation of ethanol–water mixtures, *J. Membrane Sci.*, 51 (1990) 273.

3. Y.S. Kang, S.W. Lee, U.Y. Kim and J.S. Shim, Pervaporation of water–ethanol mixtures through crosslinked and surface-modified poly(vinyl alcohol) membrane, *J. Membrane Sci.*, 51 (1990) 215.
4. Y. Wei and R.Y.M. Huang, Pervaporation with latex membranes. I. Latex membrane preparation and characterization, *J. Membrane Sci.*, 82 (1993) 27.
5. E. Nagy, O. Borali and J. Stelmaszek, Pervaporation of alcohol-water mixtures on cellulose hydrate membrane, *J. Membrane Sci.*, 16 (1983) 79.
6. T. Uragami, T. Matsuda, H. Okuna and T. Miyata, Structure of chemically modified chitosan membranes and their characteristics of permeation and separation of aqueous ethanol solution, *J. Membrane Sci.*, 88 (1991) 243.
7. I. Terada, M. Nakamura and M. Nakao, Water/ethanol permeation properties through poly(hydroxymethylene) and poly(hydroxymethylene-*co*-fluorodefin) membrane by pervaporation method, *Desalination*, 70 (1988) 455.
8. E. Ruckenstein and H.H. Chen, Preparation of a water-permselective composite membrane by the concentrated emulsion method: Its swelling and permselectivity characteristics, *J. Appl. Polym. Sci.*, 42 (1991) 2434.
9. E. Ruckenstein and J.S. Park, The separation of water–ethanol mixtures by pervaporation through hydrophilic-hydrophobic composite membranes, *J. Appl. Polym. Sci.*, 40 (1990) 213.
10. H. Uramoto and N. Kamabata, Separation of alcohol-water mixture by pervaporation through a reinforced polyvinylpyridine membrane, *J. Appl. Polym. Sci.*, 50 (1993) 115.
11. M. Yoshikama, H. Hara, M. Tanigaki, M. Guiver and T. Matsuura, Modified polysulfone membrane II. Pervaporation of aqueous ethanol solution through modified polysulfone membranes bearing various hydroxyl groups, *Polym. J.*, 24 (1992) 1049.
12. T.M. Aminabhavi, R.S. Khinnavar, S.R. Harogoppad, U.S. Aithal, Q.T. Nauyen and K.C. Hansen, Pervaporation separation of organic-aqueous and organic-organic binary mixtures, *J. Macromol. Chem. Phys.*, 34 (1994) 139.
13. H.E.A. Bruschke, Japanese Pat., Kokai, 59-109204 (1984).
14. H.E.A. Bruschke, German Pat., DE 3, 220, 570 Al (1983).
15. R.Y.M. Huang and J.W. Rhim, Separation characteristics of pervaporation membrane: separation processes using modified poly(vinyl alcohol) membranes, *Polym. Int.*, 30 (1993) 123.
16. C.K. Yeom and R.Y.M. Huang, Pervaporation separation of aqueous mixtures using crosslinked poly(vinyl alcohol) I. Characterization of the reaction between PVA and amic acid, *Angew. Makromol. Chem.*, 184 (1991) 27.
17. W.Y. Chiang and C.M. Hu, Separation of liquid mixtures by using polymer membranes I. Water-alcohol separation by pervaporation through PVA-*g*-MMA/MA membrane, *J. Appl. Polym. Sci.*, 43 (1991) 2005.
18. W.Y. Chiang and C.C. Huang, Separation of liquid mixtures by using polymer membranes IV. Water–alcohol separation by pervaporation through modified acrylonitrile grafted polyvinyl alcohol copolymer (PVA-*g*-AN) membranes, *J. Appl. Polym. Sci.*, 48 (1993) 199.
19. W.Y. Chiang and C.Y. Hsia, Separation of liquid mixtures by using polymer membranes V. Water–ethanol separation by pervaporation through poly(vinyl alcohol)-*graft*-poly (styrene-*co*-maleic anhydride) membranes, *Angew. Makromol. Chem.*, 219 (1994) 169.
20. J. Brandrup and E.H. Immergut, *Polymer Handbook*, 3rd ed., Wiley, New York, 1989, p. VI/217, VI/221.
21. H. Kaur and P.R. Chatterji, Interpenetrating hydrogel networks. 2. Swelling and mechanical properties of the gelatin-polyacrylamide interpenetrating networks, *Macromolecules*, 23 (1990) 4848.
22. E. Ruckenstein and L. Liang, Polyacrylamide-reactive styrene/ unsaturated polyester microgel composites, *J. Appl. Polym. Sci.*, 57 (1995) 605.

23. L. Liang and E. Ruckenstein, Polyvinyl alcohol-polyacrylamide interpenetrating polymer network membranes and their pervaporation characteristics for ethanol–water mixtures, *J. Membrane Sci.*, 106 (1995) 167.
24. H.K. Lonsdale, The growth of membrane technology, *J. Membrane Sci.*, 10 (1982) 81.
25. S. Takegami, H. Yamada and S. Tsujii, Dehydration of water/ethanol mixtures by pervaporation using modified poly(vinyl alcohol) membranes, *Polym. J.*, 24 (1992) 1239.
26. H.A. Ballweg, A.E.H. Bruschke, H.W. Schneider and F.G. Twel, *Paper presented at Fuel Alcohol Conference*, New Zealand, 1982.
27. H. Ohya, M. Shibata, Y. Negishi, Q.H. Guo and H.S. Choi, The effect of molecular weight cut off of PAN ultrafiltration support layer on separation of water–ethanol mixtures through pervaporation with PAA-PAN composite membrane, *J. Membrane Sci.*, 90 (1994) 91.
28. H. Ohya, K. Matswuoto, Y. Negishi, T. Hino and H.S. Choi, The separation of water and ethanol by pervaporation with PVA-PAN composite membranes, *J. Membrane Sci.*, 68 (1992) 141.
29. H.S. Choi, T. Hino, M. Shibata, Y. Negishi and H. Ohya, The characteristics of a PAA-PSf composite membrane for separation of water–ethanol mixtures through pervaporation, *J. Membrane Sci.*, 72 (1992) 259.
30. H. Karakane, M. Tsuyumoto, Y. Maeda and Z. Honda, Separation of water–ethanol by pervaporation through polyion complex composite membrane, *J. Appl. Polym. Sci.*, 42 (1991) 3229.
31. L.G. Wu, C.L. Zhu and M. Liu, Study of a new pervaporation membrane Part 1. Preparation and characteristics of the new membrane, *J. Membrane Sci.*, 90 (1994) 159.
32. K.R. Lee, R.Y. Chew and J.Y. Lai, Plasma deposition of vinyl acetate onto Nylon-4 membrane for pervaporation and evapomeation separation of aqueous alcohol mixtures, *J. Membrane Sci.*, 75 (1992) 171.
33. T. Hirotsu and H. Nakajima, Water–ethanol permseparation by pervaporation through the plasma graft copolymeric membrane of acrylic acid and acrylamide, *J. Appl. Polym. Sci.*, 36 (1988) 177.

3.3 Pervaporation of Ethanol–Water Mixtures through Polydimethylsiloxane-Polystyrene Interpenetrating Polymer Network Supported Membranes*

Liang Liang and Eli Ruckenstein

Department of Chemical Engineering, State University of New York at Buffalo, Buffalo, NY 14260, USA

ABSTRACT The pervaporation performances of polydimethylsiloxane-polystyrene interpenetrating polymer network (PDMS-PS IPN) layers supported on polyethersulfone (PESF) ultrafiltration membranes for ethanol–water mixtures were investigated. The PDMS-PS IPN layers were prepared via the sequential IPN technique. The mechanical properties and the film-forming capability of the PDMS-PS IPN are superior to those of the crosslinked PDMS. An inversion of the selectivity of PDMS-PS IPN layers supported on PESF membranes was observed; at low ethanol concentrations in the feed, the membrane was more selective for ethanol, while at high concentrations it was more selective for water. An explanation is suggested for this inversion. For a mixture containing 10 wt% ethanol, which is of interest in biotechnology, the membrane was ethanol selective, and the permeation rate and separation factor of the PDMS-PS IPN supported membrane at 60°C, with a crosslinked PS content of 40 wt%, were 160 g/m^2 h and 5.5, respectively.

* *Journal of Membrane Science* 1996, 114, 227–234.

3.3.1 INTRODUCTION

Two kinds of pervaporation membranes were used for the separation of water–ethanol mixtures, namely, the water-permselective and ethanol-permselective membranes. While numerous investigations have been carried out regarding the water-permselective membranes (some of them are [1–7]), much fewer have been performed concerning the ethanol-permselective ones [8]. Silicon-containing polymers, particularly polydimethylsiloxane (PDMS), have been the most employed for the selective pervaporation of organics from organic-water mixtures [9]. However, PDMS has poor mechanical and film-forming properties [10], and it is therefore difficult to prepare ultrathin membranes from this polymer. In order to overcome these disadvantages, the PDMS membranes were improved by the block or graft copolymerization of PDMS with other polymers, by the blending of PDMS with zeolites and by the plasma polymerization of PDMS [11–14].

The properties of polymers can be controlled by the formation of interpenetrating polymer networks (IPN) [15]. As is well known, polystyrene (PS) is more hydrophobic than PDMS and has a higher tensile strength. Consequently, one can expect that the mechanical, film-forming and selectivity characteristics of the PDMS membranes can be improved by replacing them with the PDMS-PS IPN membranes. In the present paper, we report the preparation of PDMS-PS IPN layers supported on polyethersulfone ultrafiltration membranes and their separation characteristics in the selective pervaporation of ethanol from ethanol–water mixtures.

3.3.2 EXPERIMENTAL

3.3.2.1 Materials

The inhibitor *tert*-butylcatechol was removed from styrene (S, Aldrich, 99%) and divinylbenzene (DVB, technical, 80%, Aldrich) by passing them through a packed column (Aldrich). The initiator, azobisisobutyronitrile (AIBN, Aldrich) was recrystallized from methanol. α,ω-Dihydroxypolydimethylsiloxane (PDMS, Scientific Polymers Inc.) with a weight average molecular weight of 36,000, tetraethylorthosilicate (TAOS, 99.9%, Aldrich), dibutyltin dilaurate (Aldrich, 95%), toluene (Aldrich, 99%) and ethanol (Aldrich, 99%) were used without further purification. Water was distilled and deionized. Polyethersulfone (PESF) ultrafiltration membranes, with a protein molecular weight cut-off of 10000 (Gelman Scientific Co.), were used as supports.

3.3.2.2 Preparation of Crosslinked PDMS Supported Membranes

PDMS (0.5 g), the crosslinking agent TAOS (0.1 g) and the catalyst for crosslinking dibutyltin dilaurate (0.013 g) were dissolved in 2.0 mL toluene. After the solution became sufficiently viscous (about 20 min), it was cast on the surface of a PESF membrane. The system containing some crosslinked PDMS, formed during about 2 h at ambient temperature, was introduced into an oven at 60°C for 4 h to complete the crosslinking. When the film of crosslinked PDMS was generated on glass, it was impossible to detach the dried film without completely damaging it.

3.3.2.3 Preparation of PDMS-PS IPN Supported Membranes

The compositions of the casting solutions used to prepare PDMS-PS IPN layers are listed in Table 3.3.1. After it became sufficiently viscous (about 20 min), the mixture was cast on the surface of a PESF membrane. In order to prevent the evaporation of styrene and DVB, the system was covered with a glass dish. The crosslinked PDMS network, which was first generated at room temperature during about 2 h, was swollen with the styrene and DVB absorbed from the solution. The system was introduced into an oven at 60°C for completing the crosslinking of PDMS and for the polymerization of styrene and its crosslinking by DVB. This process lasted about 12 h. All the supported membranes were dried in a vacuum oven at ambient temperature for 24 h before they were used in pervaporation. When the PDMS-PS IPN film was generated on glass, it was easy to detach the film without any damage. Compared to PDMS, it is clear that the film forming capability is much higher for the PDMS-PS IPN.

3.3.2.4 Tensile Measurements

The tensile properties of the crosslinked PDMS and PDMS-PS IPN unsupported membranes were determined by preparing sheets of the size required by ASTM D. 638–58T. The thicknesses of the unsupported dry membranes, measured using a micrometer, which had an accuracy of ±10 μm, were in the range of 200–300 μm. An Instron universal testing instrument (model 1000) was used, at room temperature, to perform the tensile measurements. The elongation speed of the instrument was 10 mm/min.

3.3.2.5 FT-IR Spectra

The infrared absorption spectra of different polymers were recorded on a Mattson Alpha Centauri FT-IR instrument.

TABLE 3.3.1
Composition of the Casting Solutions Used to Prepare PDMS-PS IPN Supported Membranes[a]

		PDMS Network		PS Network			
Membrane	PS (wt%) in IPN	PDMS (g)	TAOS (g)	S (g)	DVB (g)	Toluene (mL)	Thickness of the Top Layer of the Membrane (μm)
I	0	0.5	0.1	0	0	2.4	20
II	17	0.5	0.1	0.1	0.01	2.4	17
III	23	0.5	0.1	0.15	0.015	2.5	17
IV	30	0.5	0.1	0.2	0.02	2.7	19
V	40	0.5	0.1	0.33	0.033	3.2	15

[a] In all membranes: TAOS = 16 wt% in the PS network; DVB = 9 wt% in the PS network.

3.3.2.6 Scanning Electron Microscopy (SEM)

The PDMS-PS IPN supported membranes were fractured in liquid nitrogen. The fractured section was coated with evaporated carbon. The cross-section of the supported membrane was investigated using a Hitachi S-800 SEM. The thickness of the top separating layers, determined via SEM, was in the range 15–20 μm.

3.3.2.7 Pervaporation Experiments

The pervaporation apparatus was presented schematically in a previous paper [16]. The effective area of the membrane was 9.6 cm^2 and the downstream pressure was 3 ± 1 torr. After about 2 h, a steady state was achieved. The vapor was collected by condensation in a cold trap cooled with liquid nitrogen. The permeated sample was analyzed with a gas chromatograph equipped with a Porapak Q column heated at 160°C and with a thermal conductivity detector. Helium was used as the carrier gas.

The permeation rate (J) at steady state was calculated using the expression:

$$J = \frac{Q}{At}$$

where Q is the total amount of permeate at steady state during the experimental time interval t, and A is the effective membrane surface area.

The permselectivity of the membrane was expressed via the separation factor (α) defined as

$$\alpha = \frac{\left(Y_{\mathrm{EtOH}}/Y_{\mathrm{H_2O}}\right)}{\left(X_{\mathrm{EtOH}}/X_{\mathrm{H_2O}}\right)}$$

where $Y_{\mathrm{EtOH}}/Y_{\mathrm{H_2O}}$ is the weight ratio of ethanol to water in the permeate and $X_{\mathrm{EtOH}}/X_{\mathrm{H_2O}}$ is the weight ratio of ethanol to water in the feed mixture.

3.3.3 RESULTS AND DISCUSSION

3.3.3.1 FT-IR Spectra of Different Polymers

Figure 3.3.1 shows the FT-IR spectra of different polymers. In Figure 3.3.1a, the absorption peaks at 2963 and 2856 cm^{-1} can be assigned to the symmetric and asymmetric stretching vibrations of the CH_3 groups of the PDMS, respectively. The absorption peaks due to the Si-CH_3 bonds in the PDMS are located at 800 and 1258 cm^{-1} [17]. In Figure 3.3.1b, the absorption peak at 1413 cm^{-1} can be assigned to the aromatic skeletal vibration of PS and the peak at 700 cm^{-1} to the aromatic C-H out-of-plane deformation vibration [18]. The main absorption peaks of PDMS and PS can be also observed in the FT-IR spectrum of PDMS-PS IPN (Figure 3.3.1c). This confirms the IPN structure of the system.

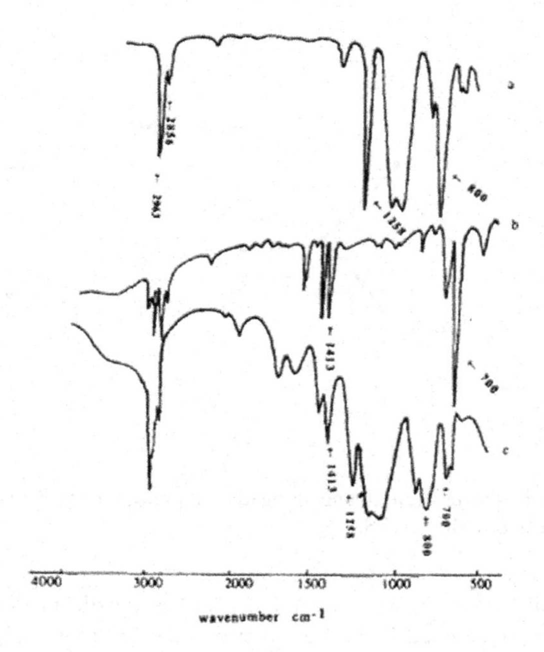

FIGURE 3.3.1 FT-IR spectra of different polymers. (a) α,ω-Dihydroxy PDMS. (b) Crosslinked PS (DVB = 9 wt% in the PS network), (c) PDMS-PS IPN. (In all figures: TAOS = 16 wt% in the PDMS network; DVB = 9 wt% in the PS network.)

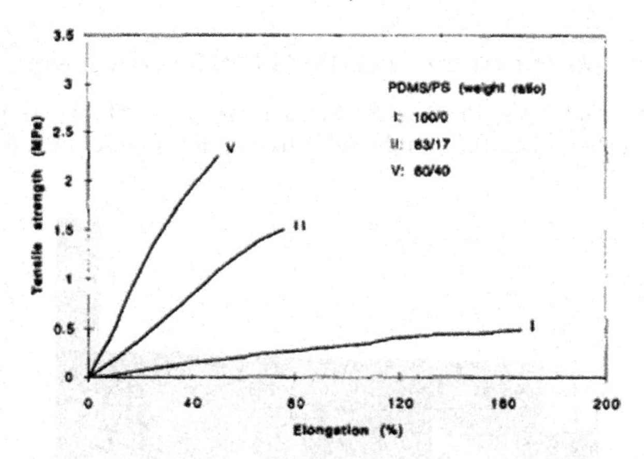

FIGURE 3.3.2 Stress-strain curves for the membranes I, II and V of Table 3.3.1.

3.3.3.2 Mechanical Behavior

Figure 3.3.2 presents the stress-strain curves of the crosslinked PDMS and of two PDMS-PS IPN unsupported membranes. The stress-strain curve of the crosslinked PDMS has a typical elastomeric behavior [19], that is, the elongation at break is high (160%) and the tensile strength is low (0.5 MPa). Compared to PDMS, the stress-strain curves of the PDMS-PS IPN membranes exhibit a smaller elongation at break

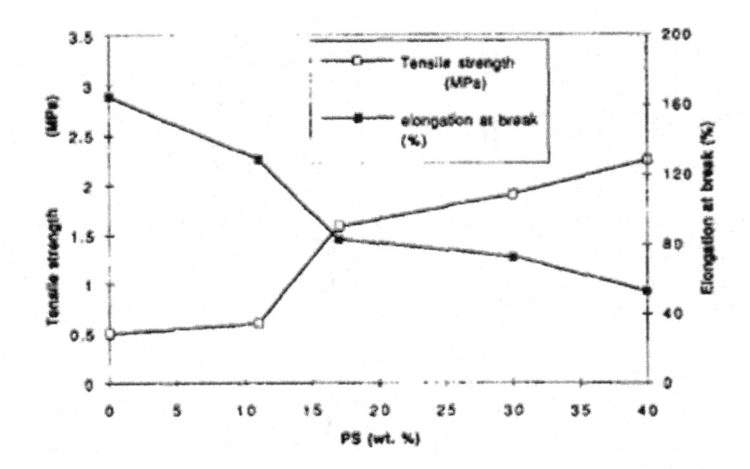

FIGURE 3.3.3 The tensile strength and elongation at break for PDMS-PS IPN membranes against the content of crosslinked PS.

but a larger tensile strength. The dependence of the tensile properties of the IPN membrane on the crosslinked PS content is presented in Figure 3.3.3. With increasing crosslinked PS content, the elongation at break decreases but the tensile strength increases. It is clear that the poor mechanical properties of crosslinked PDMS membrane are improved by the PDMS-PS IPN membranes.

3.3.3.3 SEM Micrographs of PDMS-PS IPN Supported Membrane

A SEM micrograph of a PDMS-PS IPN supported membrane is presented in Figure 3.3.4. It shows that the membrane is free of pores larger than the resolution

FIGURE 3.3.4 SEM micrograph of PDMS-PS IPN supported membrane. PDMS/PS = 60/40 (weight ratio). The thickness of the supported film is 15 μm.

of SEM (about 100 Å) and that the thickness of the top layer of the supported membrane is about 15 μm.

3.3.4 PERVAPORATION OF THE ETHANOL–WATER MIXTURES

3.3.4.1 Effect of Crosslinked PS Content in the IPN Supported Membrane

Figure 3.3.5 shows the effect of the weight percent of crosslinked PS in the IPN membrane on the permeation rate and separation factor for 16 wt% TAOS in the PDMS network and 9 wt% DVB in the PS network. The pervaporation experiments were carried out at 60°C for an ethanol–water mixture containing 10 wt% EtOH. As the content of crosslinked PS increases, the permeation rate decreases. Because the hydrophobicity of the membrane increases with increasing crosslinked PS content, the separation factor increases.

3.3.4.2 Effect of the Feed Temperature

The effect of the temperature on the permeation rate and separation factor for an ethanol–water mixture containing 10 wt% EtOH is presented in Figures 3.3.6 and 3.3.7, respectively. With increasing temperature, both the permeation rate and the separation factor increase. The permeation rate increases monotonically with increasing temperature because the mobility of the permeating molecules is enhanced both by the temperature and by the higher mobility of the

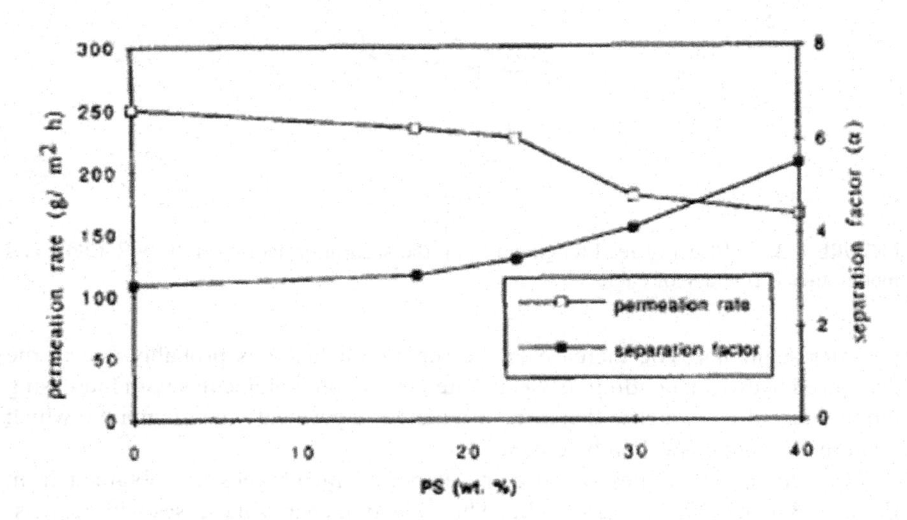

FIGURE 3.3.5 Permeation rate and separation factor against the crosslinked PS content. EtOH, 10 wt%; temperature, 60°C.

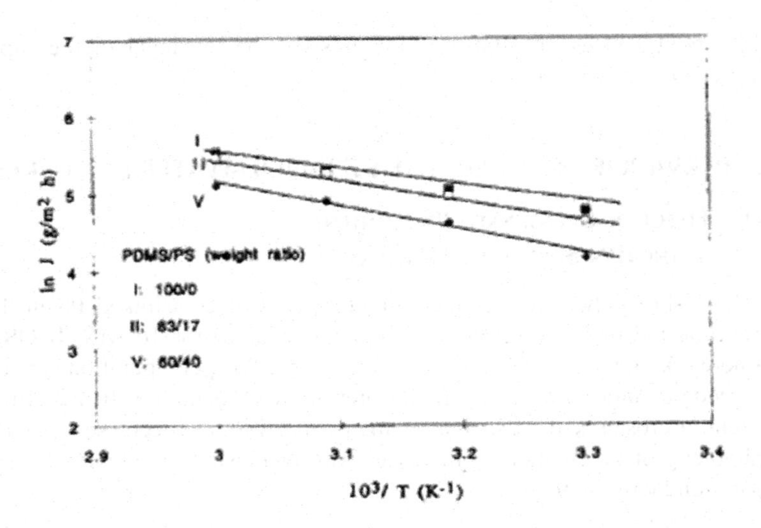

FIGURE 3.3.6 Arrhenius plots of permeation rate vs. temperature for 10 wt% EtOH and membranes I, II and V of Table 3.3.1.

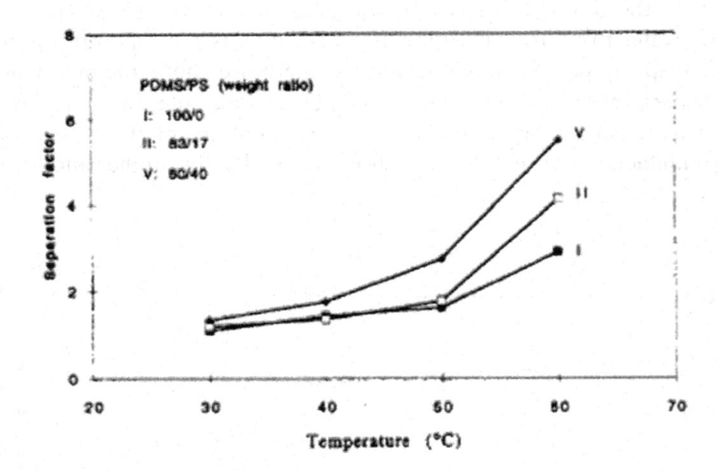

FIGURE 3.3.7 Effect of feed temperature on the separation factor for 10 wt% EtOH and membranes I, II and V of Table 3.3.1.

polymer segments. The increase of the separation factor is probably due to the decreased hydrogen bonding between water and alcohol molecules with increasing temperature. Less water is therefore stimulated to permeate the membrane which is more selectively swollen by ethanol.

The activation energies of different supported membranes are obtained from Figure 3.3.6 and listed in Table 3.3.2. The activation energy increases with increasing crosslinked PS content in the IPN, because for a higher crosslinking the thermal mobility of the polymer segments is lower.

TABLE 3.3.2
The Activation Energies for the Permeation Rate of a Water–Ethanol Mixture Containing 10 wt% Ethanol

Membrane No.[a]	Activation Energy (kcal/mol)
I	4.6
II	5.5
V	7.2

[a] The membranes are those of Table 3.3.1.

3.3.4.3 Effect of Feed Composition

Figure 3.3.8 presents the permeation rate at 60°C for different PDMS-PS IPN supported membranes as a function of the concentration in the feed. For comparison, the permeation rate for the crosslinked PDMS supported membrane is also included. The figure shows that the permeation rate increases with increasing ethanol concentration in the feed. This occurs because the swelling of the membranes increases with increasing ethanol concentration [20]. Furthermore, one can see that the permeation rate increases with increasing crosslinked PDMS content in the IPN.

Figure 3.3.9 presents the effect of ethanol concentration in the feed on the separation factor. The separation factor increases with decreasing ethanol concentration in

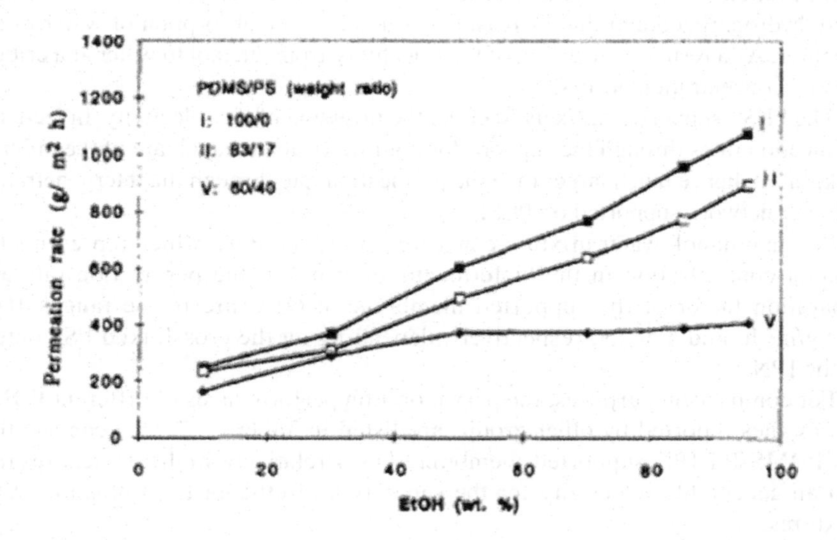

FIGURE 3.3.8 Effect of feed composition on the permeation rate for pervaporation at 60°C and membranes I, II and V of Table 3.3.1.

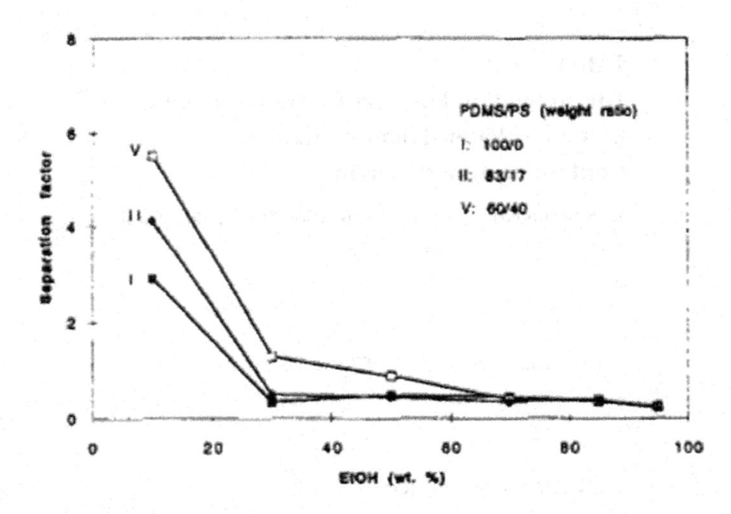

FIGURE 3.3.9 Effect of feed composition on the separation factor for pervaporation at 60°C and membranes I, II and V of Table 3.3.1.

the feed. However, the supported membranes preferentially permeate ethanol only in the low ethanol concentration range. In the high ethanol concentration range, the membranes permeate preferentially water. This inversion was also found for other supported membranes and mixtures [21,22]. A possible explanation is as follows. The top layer is hydrophobic, and, therefore, its swelling increases as the ethanol concentration increases. Because of the OH groups of ethanol, the layer is increasingly hydrophilized and this increasingly stimulates the absorption of water in the top layer. As a result an inversion of the selectivity from ethanol to water at a critical ethanol concentration occurs.

The PESF support is unlikely to effect the inversion of the selectivity. Indeed, the permeation rates through the support, for both water and ethanol, are of the order of 15 kg/m^2 h, hence much larger than the permeation rate through the interpenetrating polymer network supported on PESF.

For an ethanol–water mixture containing 10 wt% EtOH, which represents the ethanol concentration in the biofermentation process, the permeation rate and separation factor of the supported membrane at 60°C are in the ranges 160–250 g/m^2 h, and 5.5–2.9, respectively, depending on the crosslinked PS content of the IPN.

For comparison purposes, the pervaporation performances of different PDMS membranes reported by other groups are listed in Table 3.3.3. One can see that the PDMS-PS IPN supported membrane has a relatively high permeation rate and an acceptable selectivity for the separation of ethanol from ethanol–water mixtures.

TABLE 3.3.3
Pervaporation Performances of Different PDMS Supported and Unsupported Membranes

Membrane[a]	Ethanol Concentration in the Feed (wt%)	Temperature (°C)	J (g/m² h)	α	References
PDMS-PS graft copolymer supported on a PESF microfiltration membrane	10	60	130	6.2	[10]
PDMS-PSF block copolymer	10	25	27	6.2	[11]
PDMS-PPP graft copolymer	7.0	30	19	40.0	[20]
PDMS	10	30	20	5.0	[23]
PDMS	11.9	25	14	7.1	[24]
PDMS	8	30	25	10.8	[25]
PDMS-vinyl substituted copolymer	40	25	33	6.7	[26]
PDMS-PI graft copolymer	6.6	48	32	6.6	[27]
PDMS-PS IPN	10	60	160	5.5	this study

[a] PDMS, polydimethylsiloxane; PESF, polyethersulfone; PSF, polysulfone; PI, polyimide; PPP, poly(l-phenyl-1-propyne); PS, polystyrene; IPN, interpenetrating polymer network.

3.3.5 CONCLUSION

Polydimethylsiloxane–polystyrene interpenetrating polymer network supported membranes were prepared by casting solutions which contained α,ω-dihydroxy polydimethylsiloxane, tetraethylorthosilicate, styrene and divinylbenzene, on the surface of polyethersulfone ultrafiltration membranes. Dibutyltin dilaurate and azobisisobutyronitrile were used as catalyst for the reaction between α, ω-dihydroxypolydimethylsiloxane and tetraethylorthosilicate and initiator for the crosslinked polymerization of styrene and divinylbenzene, respectively. A crosslinked PDMS network was first formed at room temperature, followed by the formation of a crosslinked PS network. The tensile strength of the PDMS-PS IPN composite increases with increasing crosslinked PS content, but the elongation at break decreases. Regarding the pervaporation of ethanol–water mixtures, the permeation rate of PDMS-PS IPN supported membrane decreases and the separation factor increases with increasing crosslinked PS content. Both the permeation rate and the separation factor increase with increasing feed temperature. The activation energies of the supported membranes are in the range 4.57–7.20 kcal/mol. The composite membrane is either selective to ethanol or water, depending on the concentration of ethanol in the feed. At low concentrations, it is selective for ethanol. An explanation is suggested for this inversion. The permeation rate and separation factor are at 60°C in the range 160–250 g/m^2 h and 5.5–2.9 for an ethanol–water mixture containing 10 wt% EtOH, respectively, depending upon the content of crosslinked PS in the IPN membrane.

REFERENCES

1. Y. Maeda and M. Kai, in R.Y.M. Huang (Ed.), Recent progress in pervaporation membranes for water–ethanol separation, *Pervaporation Membrane Separation Process*, Elsevier, Amsterdam, the Netherlands, 1991, p. 391.
2. H.E.A. Bruschke, *Jpn. Pat. Kokai*, 59–109204(1984).
3. H. Karakane, M. Tsuyumote, Y. Maeda and Z. Honda, Separation of water–ethanol by pervaporation through polyion complex composite membrane, *J. Appl. Polym. Sci.*, 42 (1991) 3229.
4. E. Ruckenstein and H.H. Chen, Preparation of a water-permselective composite membrane by the concentrated emulsion method: Its swelling and permselectivity characteristics, *J. Appl. Polym. Sci.*, 42 (1991) 2434.
5. K.R. Lee, R.Y. Chen and J.Y. Lai, Plasma deposition of vinyl acetate onto Nylon-4 membrane for pervaporation and evapomeation separation of aqueous alcohol mixtures, *J. Membrane Sci.*, 75 (1992) 171.
6. Y. Wei and R.Y.M. Huang, Pervaporation with latex membranes. I. Latex membrane preparation and characterization, *J. Membrane Sci.*, 82 (1993) 27.
7. H. Yamagishita, C. Maejima, D. Kitanaoto and T. Nakane, Preparation of asymmetric polyimide membranes for water/ethanol separation in pervaporation by the phase inversion process, *J. Membrane Sci.*, 86 (1994) 231.
8. T. Masuda, M. Takatsuka, B.Z. Tang and T. Higashimura, Pervaporation of organic liquid-water mixtures through substituted polyacetylene membranes, *J. Membrane Sci.*, 4(1990) 69.
9. S. Zhang and E. Drioli, Pervaporation membrane, *Sep. Sci. Technol.*, 90 (1995) 1.
10. S. Takegemi, H. Yamada and S. Tsujii, Pervaporation of ethanol/water mixture using novel hydrophobic membranes containing polydimethylsiloxane, *J. Membrane Sci.*, 75 (1992) 93.
11. K. Okamoto, A. Butsuen, S. Tsuru, S. Nishioka, K. Tanaka, H. Kita and S. Asakawa, Pervaporation of water–ethanol mixtures through polydimethylsiloxane block-copolymer membranes, *Polym. J.*, 19(1987) 734.
12. E. Akiyama, Y. Takamura and Y. Nagaze, Synthesis of soluble polyimide/polydimethylsiloxane graft copolymer and application to separation membranes, *Makromol. Chem.*, 193 (1992) 1509.
13. M.D. Jia, K.V. Peinemann and R.D. Behling, Pervaporation and characterization of thin-film zeolite-PDMS composite membranes, *J. Membrane Sci.*, 73 (1992) 119.
14. T. Kashiwagi, K. Okabe and K. Okita, Separation of ethanol from ethanol water mixtures by plasma–polymerized membranes from silicone compounds, *J. Membrane Sci.*, 36 (1988) 353.
15. L.H. Sperling, *Interpenetrating Polymer Networks and Related Materials*, Plenum, New York, 1981.
16. L. Liang and E. Ruckenstein, Polyvinyl alcohol-polyacrylamide interpenetrating polymer network membranes and their pervaporation characteristics for ethanol–water mixtures, *J. Membrane. Sci.*, 106(1995) 167.
17. F.F. Bentley, L.D. Smithson and A.L. Rozek, *Infrared Spectra and Characteristics Frequencies –700–300 cm^{-1}*, Interscience, Wiley, New York, 1968.

18. C.D. Craver (Ed.), *The Coblentz Society Desk Book of Infrared Spectra*, 2nd edn., The Coblentz Society, Kirkwood, MO, 1982.
19. L.H. Sperling, *Introduction to Physical Polymer Science*, 2nd edn., John Wiley & Sons, New York, 1992, p. 508.
20. Y. Nagase, S. Mori and K. Matsui, Chemical modification of poly(substituted-acetylene) IV. Pervaporation of organic liquid-water mixture through poly(l-phenyl-1-propyne)/polydimethylsiloxane graft copolymer membrane, *J. Appl. Polym. Sci.*, 37 (1989) 1259.
21. W. Gudernatsch, T. Menzel and H. Strethmann, Influence of composite membrane structure on pervaporation, *J. Membrane Sci.*, 61 (1991) 19.
22. J. Bai, A.F. Fouda, T. Matsuura and J.D. Hazlett, A study on the pervaporation and performance of polydimethylsiloxane-coated polyetherimide membrane in pervaporation, *J. Appl. Polym. Sci.*, 48 (1993) 499.
23. M. de Carmo Conclaves, G. de Souza Sorrentino Marquez and F. Galembeck, Pervaporation and dialysis of water–ethanol solution by using silicone rubber-membrane, *Sep. Sci. Technol.*, 18 (1983) 893.
24. Z. Changliu, L. Moe and X. We, Separation of ethanol–water mixtures by pervaporation-membrane separation process, *Desalination*, 62 (1987) 299.
25. K. Ishihara and K. Matsui, Pervaporation of ethanol–water mixture through composite membranes composed of styrene-fluoroalkyl acrylate graft copolymers and cross-linked polydimethylsiloxane membrane, *J. Appl. Polym. Sci.*, 34 (1987) 437.
26. M. Tanigaki, M. Yoskikawa and W. Egachi, Selective separation of alcohol from aqueous solution through polymer membrane, in R. Bakish (Ed.), *Proceedings of the 2nd. International Conference Pervaporation Processes*, San Antonio, Englewood, NJ, 1987, p. 126.
27. T. Kashiwagi, K. Okabe and K. Okita, Separation of ethanol from ethanol/water mixtures by plasma-polymerized membranes from silicone compounds, *J. Membrane Sci.*, 36 (1988) 353.

4 Soluble Conducting Polymers

CONTENTS

As the low-weight alternative of metals for conductivity-based applications, intrinsically conducting polymers with conjugated chain structures are critically important materials in modern society. They have been used not only in electronics to replace metals but also in a wide range of new applications, such as printing electronic circuits, transparent displays, solar cells, and radar-absorptive coatings on stealth aircraft. A broad variety of conducting polymers have been synthesized by chemical or electrochemical synthesis. However, a major technical barrier that restricts conducting polymers from broad large-scale applications is their poor processability. Because typically they are not thermoplastics and have low solubility due to their rigid conjugated main-chain structures, processing of conducting polymers is very difficult and depends heavily on dispersion techniques which are not often capable of efficiently producing structured materials. Therefore, it is critically important to design, synthesize and study soluble conducting polymers. This chapter collects six papers on soluble conducting polymers. Based on structural features, these soluble conducting polymers can be classified into three types, each corresponding to a specific preparation approach.

The first type is polyanilines co-doped with multiple acids, which are prepared by copolymerization of salts of aniline with different acids (Section 4.1). Such co-doped polyanilines are soluble in organic solvents, and the solubility enhancement can be attributed to the favorable entropy factor in dissolving the complexes of polyanilines with acids relative to merely polyanilines. They also demonstrate high conductivity in solid form (up to 14 S/cm).

The second type is poly(ethylene oxide) (PEO)-grafted polyanilines and polydiphenylamines (Sections 4.2–4.3). These PEO-grafted conducting polymers are water-soluble due to the presence of hydrophilic PEO grafts. The PEO-grafted polyanilines are prepared by copolymerization of aniline and PEO-modified aniline. The PEO-grafted polydiphenylamines are synthesized by the grafting-onto reaction of tosyloyl-functionalized PEO with amine-functionalized polydiphenylamine, and these copolymers show interesting fluorescence behavior in water.

The third type is grafted copolymers with non-conjugated backbones and conducting grafts (Sections 4.4–4.6). They are obtained through the synthesis of linear or hyperbranched aniline-functionalized vinyl polymers via radical copolymerization, followed by the formation of conducting polymer grafts from the backbone aniline functionalities. The grafts are prepared from oxidation polymerization of anilines or sodium diphenylamine sulfonate. With the presence of ionic moieties, these grafted copolymers are water-soluble.

It should be noted that although the soluble conducting polymers demonstrated in these articles are based on polyanilines or polydiphenylamines, the corresponding synthetic strategies can be applied to the preparation of soluble conducting polymers with other types of structural units.

4.1 Soluble Polyaniline Co-doped with Dodecyl Benzene Sulfonic Acid and Hydrochloric Acid*

Wusheng Yin and Eli Ruckenstein
Department of Chemical Engineering, State University of New York at Buffalo, Buffalo, NY 14260, USA

ABSTRACT A chemical oxidative polymerization of aniline dodecyl benzenesulfonic acid (ANIDBSA) and aniline hydrochloric acid (ANIHCl) was performed in an aqueous solution because the former monomer stimulates the solubility and the latter the conductive structure of the synthesized polymer. A co-doped polyaniline (PANI) was thus obtained, which is soluble in common organic solvents such as chloroform, and exhibits, without any additional doping, a higher conductivity than the insoluble HCl-doped PANI compressed pellet, and much higher conductivity than that prepared from pure ANIDBSA. The PANI doped with DBSA and HCl was characterized using FTIR, EDX and UV spectroscopies. At an ANIHCl/ANIDBSA molar ratio in the feed of 3:7 and no additional HCl, the polymer exhibited a maximum conductivity of 7.9 S/cm and a maximum yield of 30.8%. If additional HCl was introduced, the conductivity could reach a value as high as 14 S/cm.

4.1.1 INTRODUCTION

Polyaniline (PANI) was first synthesized in 1862 [1] and has been extensively studied as a conducting polymer since the 1980s [2,3]. It can be obtained in several forms with different structures of the repeat unit, depending on the synthesis conditions. The main forms are presented in Scheme 4.1.1. The emeraldine base constitutes a form of PANI which after protonation provides the highest electrical conductivity [4]. Being soluble in *N*-methyl-2-pyrrolidinone (NMP) the emeraldine base can be used for the processing of films and fibers [5]. However, it is difficult to dope the emeraldine product because of the slow diffusion of the protons in the solid. Consequently, finding a procedure to prepare a soluble PANI has remained an interesting goal.

* *Synthetic Metals* 2000, 108, 39–46.

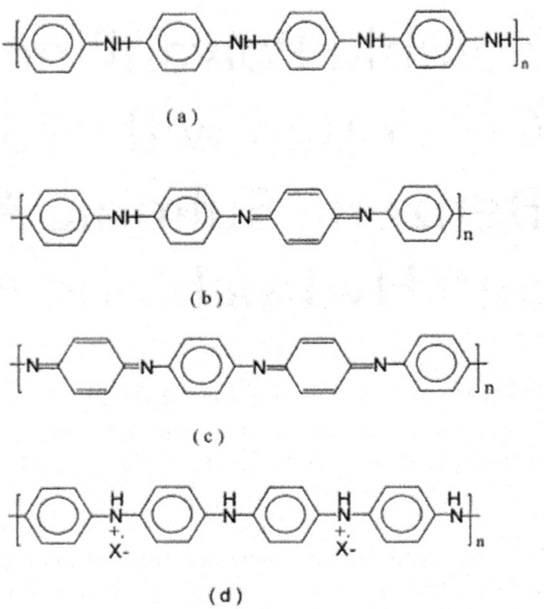

(a)

(b)

(c)

(d)

SCHEME 4.1.1 Four forms of PANI: (a) leucomeraldine base, (b) emeraldine base, (c) pernigraniline base, and (d) metallic emeraldine salt.

Recently, several methods have been developed to improve the processibility of PANI by increasing its solubility in various solvents. (1) By postprocessing, which involves (a) neutralization of the emeraldine salt to form the emeraldine base and (b) the protonation of the latter base with a second protonic acid. Cao et al. [6,7] reported that the PANI doped with dodecyl benzenesulfonic acid (DBSA) or camphor sulfonic acid became soluble in various organic solvents such as chloroform and xylene. This procedure could be used to fabricate a flexible light-emitting diode (LED) [8]. (2) By grafting copolymerization of aniline (ANI) onto a modified polymeric surfactant [9] or by using a polymeric acid such as polystyrene sulfonic acid as template for the oxidative polymerization [10]. These approaches significantly improved the processibility of PANI. (3) By introducing some substituents such as alkyl groups [11] and sulfonic acid groups [12] onto the backbone of PANI. These groups allowed the undoped and doped forms of PANI to become soluble in some common organic solvents and even water. In addition, a soluble doped PANI was prepared via an emulsion polymerization method [13].

The typical polymerization of ANI in a HCl aqueous solution results in an unprocessible PANI, whose conductivity is in the range from very small to 10 S/cm. A PANI with improved processibility was synthesized via the polymerization in an aqueous solution of DBSA aniline salt [14]. However, its conductivity was very low (10^{-1} to 10^{-2} S/cm), and the molecular weight of the PANI obtained was also low because of the high polymerization temperature [15].

In the present paper, the chemical oxidative polymerization of aniline dodecyl benzenesulfonic acid (ANIDBSA) and aniline hydrochloride acid salts was investigated. This pathway was selected, because one expects the ANIDBSA salt moieties of the polymer to enhance its solubility, hence the processibility, while the aniline hydrochloride acid salt moieties to provide a good conductive structure for the resulting polymer. Indeed, the obtained polymer could be prepared in a single step, could be dissolved in common organic solvents, and exhibited a larger conductivity than the compressed pellet prepared from HCl-doped PANI and a much higher conductivity than that prepared from ANIDBSA alone.

4.1.2 EXPERIMENTAL SECTION

4.1.2.1 Chemicals

Aniline (Aldrich, 99%) was distilled under reduced pressure prior to use. Hydrochloric acid (Aldrich, 37%), DBSA (Fischer, 98%), ammonium persulfate (APS, Aldrich, 99%), methanol (Aldrich, 99%) and chloroform were used as received. Deionized water was employed.

4.1.2.2 Polymer (PANI Doped with DBSA and HCl) Preparation

To prepare a polymer based on ANIDBSA and aniline hydrochloric acid (ANIHCl), ANI (1.86 g) was added with stirring to a flask (located in an ice bath) containing distilled water (80 g), DBSA and HCl in a selected molar ratio (in most cases 7:3) with a molar ratio of aniline/(DBSA + HCl) = 1. APS (1.2 g) was dissolved into 20 g deionized water and then added slowly to the ANI solution with stirring for 1 h at a temperature controlled between $0 \pm 1°C$ by an ice bath; this was followed by stirring for 1.5 h at room temperature. The polymer was precipitated with methanol, washed with distilled water several times and then dried in vacuum (10 mm Hg) at about 45°C for more than 12 h. The product was weighed for the determination of the yield.

4.1.2.3 Measurement of the Electrical Conductivity

The polymer was ground and then pressed under 6.5×10^5 psi at 180°C for 1–2 min. The obtained sheet was cut in a rectangular shape, and the electrical conductivity of the films was measured at room temperature by the standard four-point method.

4.1.2.4 PANI Yield

Polymer (PANI doped with DBSA and HCl), 0.1 g, was weighed and added to 40 mL of 0.1 M aqueous NH_4OH solution to remove DBSA and HCl. After stirring for more than 24 h, the mixture was filtered and then the precipitate was dried in vacuum (10 mm Hg) at about 45°C for more than 12 h. The obtained powder was weighed and the yield calculated. The yield was also calculated on the basis of the nitrogen content determined by a standard elemental analysis.

4.1.2.5 Fourier Transform Infrared (FTIR) Analysis

The powders of the product and KBr were ground together into a fine powder and pressed into a pellet. The analysis was performed with a Perkin-Elmer (Model 1760X) instrument.

4.1.2.6 UV–VIS Analysis

The product was ground into a fine powder, dissolved in chloroform, and the spectrum obtained using a Shimadzu UV-210A spectrophotometer.

4.1.2.7 Energy Dispersive Spectroscopy (EDX) Analysis

The EDX analysis was performed with a PGT/TMIX field emission microscopy equipment.

4.1.3 RESULTS AND DISCUSSION

The polymerization of ANIDBSA and ANIHCl was carried out in an aqueous solution. After a small amount of APS was slowly dropped into the solution, the slightly white tint of the reaction mixture changed to green and finally acquired a dark green coloration. The resulting polymer could dissolve in chloroform. Table 4.1.1 lists the main absorption peaks of the PANI base and of the resulting doped polymers. The absorption peak of the C=C stretching of the quinoid ring is shifted from 1590 cm^{-1} for the undoped PANI to 1562 cm^{-1} for the polymer DR46 (the two numbers represent the numerator and denominator of the molar ratio ANIHCl/ANIDBSA) or 1566 cm^{-1} for the polymers DR37 and DR55, while that due to the $C_{aromatic}$-N stretching vibration is shifted from 1313 cm^{-1} for PANI to 1303 cm^{-1} for the polymer DR46 or 1307 cm^{-1} for the polymers DR37 and DR55. In addition, there is a characteristic broad band extended from 4000 to 3000 cm^{-1}, which is due to the free charge carrier absorption of the doped polymer [16]. These absorptions of the polymer are in complete agreement with those of the PANI doped with a protonic acid (1562, 1302 cm^{-1}) [17].

TABLE 4.1.1
FTIR Peaks of the As-Synthesized Polymers and PANI Base

Code[a]	C=C (quinoid ring) (cm^{-1})	$C_{aromatic}$-N (cm^{-1})
PANI	1590	1313
DR37	1566	1307
DR46	1562	1303
DR55	1566	1307

[a] ANIHCl/ANIDBSA (molar ratio): DR37 (3:7), DR46 (4:6), and DR55 (5:5). Other polymerization conditions are described in Section 4.1.2.

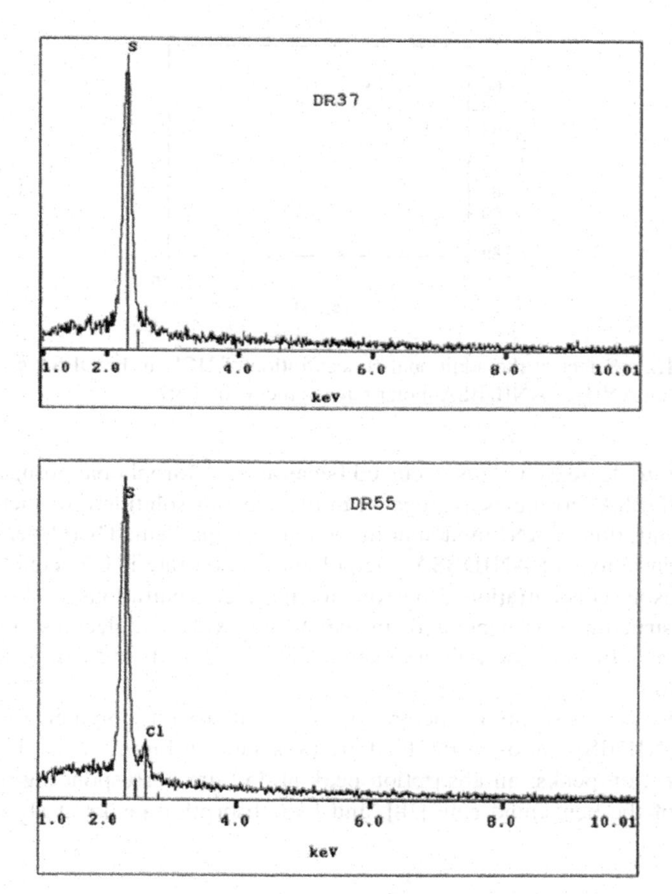

FIGURE 4.1.1 EDX spectra of DR37 and DR55 of Table 4.1.1.

The EDX spectra of polymers DR37 and DR55 (Figure 4.1.1) reveal that (i) only the sulfur peak is present in the polymer DR37, even though the Cl/S atomic ratio in the feed was as high as 30:70, and (ii) both the chlorine and sulfur peaks are present in the polymer DR55 with a Cl/S atomic ratio of 16:84, which is, however, much lower than the ratio (50:50) in the feed. Consequently the polymer was mainly doped with DBSA, and the amount of HCl in the polymer increased slowly with increasing ANIHCl/ANIDBSA molar ratio. If for a given ratio of ANIHCl/ANIDBSA additional HCl acid was introduced in the solution, the amount of DBSA in the polymer decreased with increasing HCl concentration. For a ratio of 3:7, the weight percentage of DBSA in the polymer determined via elemental analysis, is plotted in Figure 4.1.2 as a function of the concentration of HCl in solution.

Visual observations have shown that the color of the system changed with increasing rate from white to green, and further to dark green when the ANIHCl/ANIDBSA molar ratio increased at a constant concentration of ANI (equal to the sum of molar concentration of DBSA and HCl), passed through a maximum rate of change, after

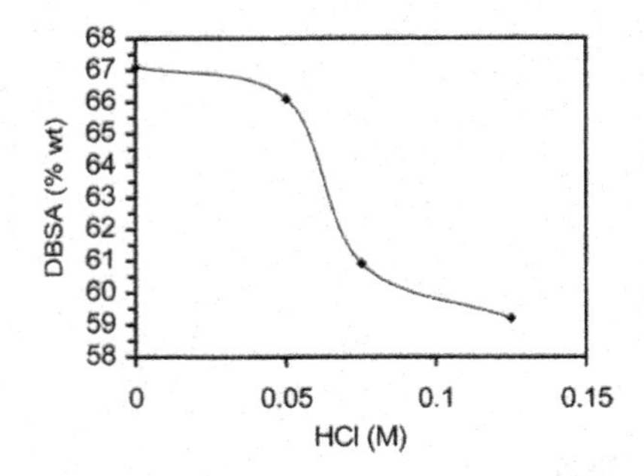

FIGURE 4.1.2 Effect of the additional concentration of HCl on the DBSA content in the polymer for an ANIHCl/ANIDBSA molar ratio in the feed of 3:7.

which the rate decreased. This occurred because the hydrophobic groups ($C_{12}H_{25}$–) of the ANIDBSA molecules aggregated in the aqueous solution thus increasing the local concentration of ANI monomer in the reaction medium. Therefore, for not too large concentrations of ANIDBSA, the polymerization rate is expected to increase with increasing concentration. However, for large concentrations of ANIDBSA, a "gel-like" structure was generated, in which the oxidant molecules could hardly penetrate, and the rate of polymerization decreased with increasing ANIDBSA concentration.

The UV–VIS absorption spectra of the synthesized polymer for different ANIHCl/ANIDBSA ratios in the feed are presented in Figure 4.1.3. They exhibit three absorption peaks: an absorption peak at 350 nm corresponding to the π-π* transition of the benzenoid ring [18], and two absorption peaks at about 430 and

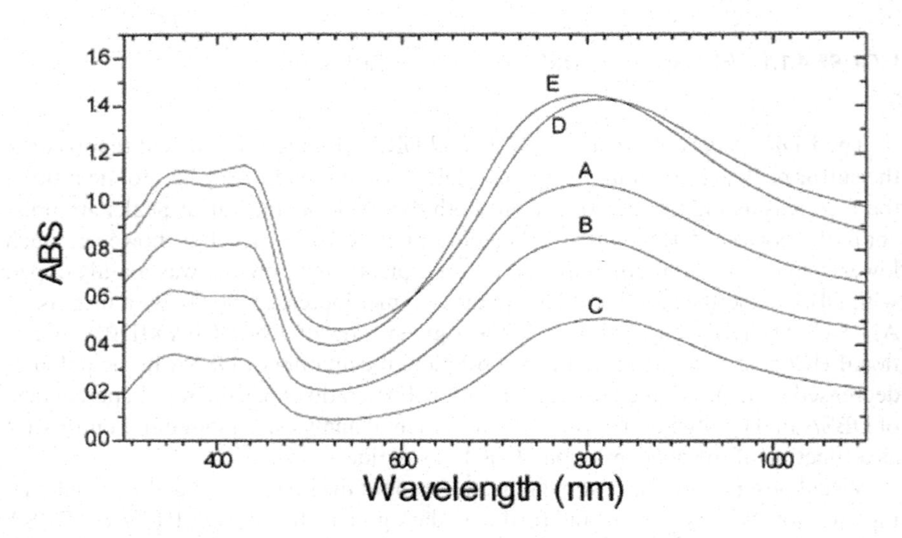

FIGURE 4.1.3 UV spectra of the polymers prepared for various ANIHCl/ANIDBSA molar ratios: 0:10 (A), 1:9 (B), 2:8 (C), 3:7 (D), and 1.7:1 (E).

800 nm, which can be assigned to the polaron band transitions. These three peaks constitute a typical emeraldine salt spectrum. Figure 4.1.3 shows that by increasing the ANIHCl/ANIDBSA molar ratio from 0 to 3:7, the polaron band exhibits a bathochromic shift from 800 to 825 nm, and with a further increase of the ratio to 1:1.7, the polaron band is shifted back to 800 nm. This indicates that the polarons in the polymer prepared using a molar ratio of about 3:7 are more delocalized than for any other ratio. This optimum might occur because of an optimum in the effective conjugation length as the molar ratio of ANIHCl/ANIDBSA increases. Indeed, for moderate ANIHCl amounts, the effective conjugation length is expected to increase as the ANIHCl/ANIDBSA molar ratio increases, because ANIHCl favors the formation of a conductive structure and this leads to the red shift of the polaron band. However, at high ANIHCl/ANIDBSA molar ratios, a blue shift of the polaron band take places because the effective conjugation length decreases as the ANIHCl/ANIDBSA molar ratio increases. This happens because for sufficiently large amounts of ANIHCl, the polymer becomes less compatible with the solvent (chloroform) and the molecules become increasingly coiled to decrease their contact with the solvent [19].

The effect of ANIHCl/ANIDBSA molar ratio in the feed on the conductivity and yield of the resulting polymer is presented in Figure 4.1.4 for the case in which the number of moles of ANI is equal to the number of moles of HCl and DBSA. With increasing ANIHCl/ANIDBSA molar ratio, both the conductivity and yield first increase and then decrease, passing through a maximum conductivity of 7.9 S/cm and a maximum yield of 30.8% at a molar ratio of about 3:7. The conductivity of the polymer is comparable to that of the insoluble HCl doped PANI compressed pellet, but much larger than that prepared from ANIDBSA alone [14]. Most importantly, the polymer is soluble in some common organic solvents. The maxima in the conductivity and yield can be explained by taking into account the aggregation of ANIDBSA induced by its hydrophobic hydrocarbon chains. A high local concentration of

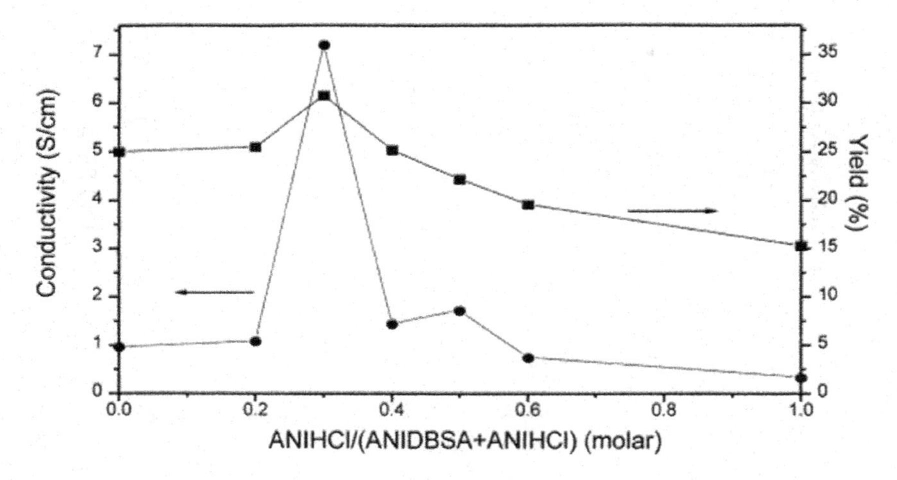

FIGURE 4.1.4 Effect of ANIHCl/ANIDBSA molar ratio on the conductivity and yield of the polymer. (Reaction time: 1.5 h at 0°C followed by 2.5 h at room temperature.)

ANI is thus generated, which tends to increase the polymerization rate. However, at low ANIHCl/ANIDBSA molar ratios, the ANIDBSA aggregates generate a "gel-like" structure and the polymerization rate of ANIHCl and ANIDBSA and the molecular weight of the polymer are affected by the slow diffusion of the oxidant in the gel. For large ANICl/ANIDBSA molar ratio, the extent of aggregation is smaller, but the diffusion of the oxidant is gradually intensified. Because of the two opposite effects, the conductivity and yield exhibit maximal values for an ANI-HCl/ANIDBSA molar ratio of about 3:7.

The effect of an additional HCl amount in the solution on the conductivity and yield of the resulting polymer is presented in Figure 4.1.5 for an ANIHCl/ANIDBSA molar ratio of 3:7. With increasing concentration from 0 to 0.125 M, the conductivity increases from 0.7 to 13.9 S/cm, and the yield passes through a maximum of 24.2% at 0.025 M HCl. The UV–VIS spectra of these polymers are given in Figure 4.1.6, which shows that the polaron band is shifted to shorter wavelengths (blue shift) with increasing additional HCl. Such a blue shift occurs because the segments of PANI doped with HCl tend to decrease the contact with solvent (chloroform) by coiling the chains. The conductivity increases with increasing concentration of additional HCl because of the good conductive structure it generates.

Figure 4.1.7 presents the effect of the concentration of the oxidant, APS, on the conductivity and yield of polymer for an ANIHCl/ANIDBSA molar ratio of 3:7. When $(NH_4)_2S_2O_8$ increases from 1.0 to 1.4 g, the conductivity and yield of the polymer increase from 1.5 to 10.1 S/cm, and 19% to 28.3%, respectively. With a further increase in the amount of $(NH_4)_2S_2O_8$ to 1.6 g, the conductivity and yield decrease to 1.9 S/cm and 27.2%, respectively. However, the polaron bands of these polymers

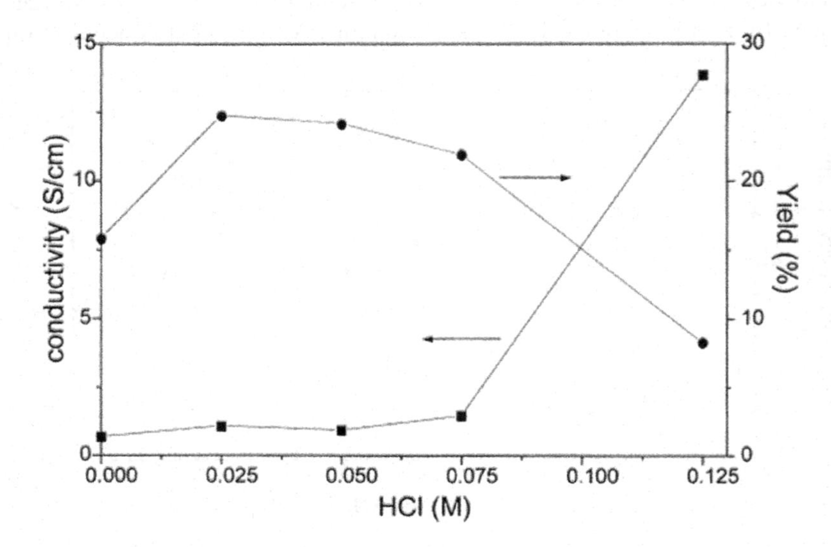

FIGURE 4.1.5 Effect of the additional HCl on the conductivity and yield of the polymer for an ANIHCl/ANIDBSA molar ratio in the feed of 3:7.

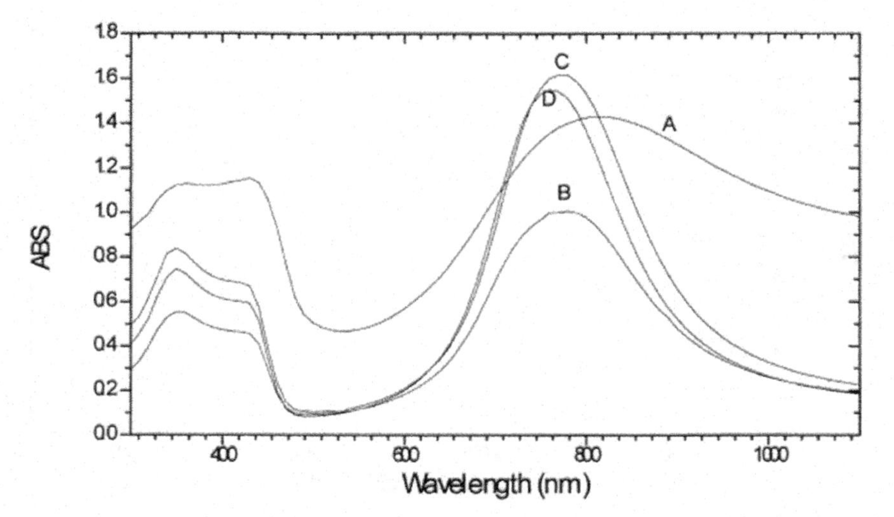

FIGURE 4.1.6 UV spectra of the polymer prepared for various additional HCl concentrations and an ANIHCl/ANIDBSA molar ratio in the feed of 3:7 — 0 M (A), 0.05 M (B), 0.075 M (C), and 0.125 M (D).

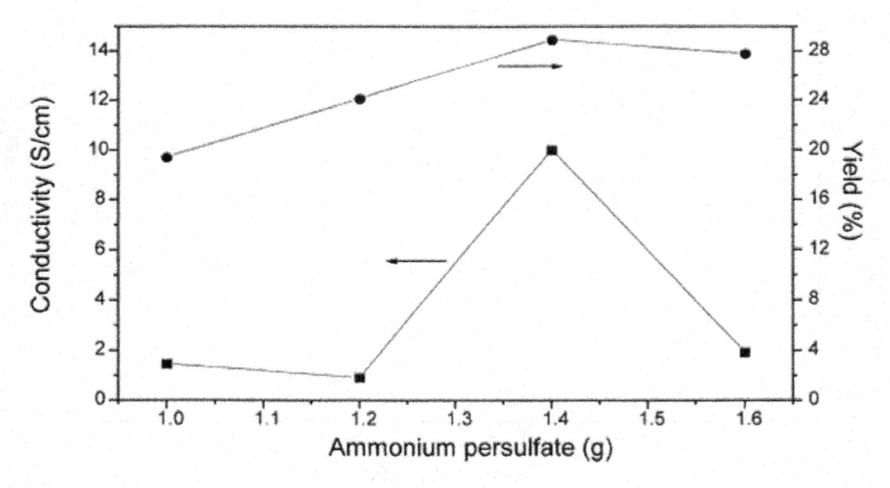

FIGURE 4.1.7 Effect of the amount of APS on the conductivity and yield of the polymer for an ANIHCl/ANIDBSA molar ratio in the feed of 3:7.

remained unchanged with the change in the oxidant concentration (Figure 4.1.8). This occurs because the HCl/DBSA ratio in the polymer does not change with increasing amount of oxidant. The decreases of conductivity and yield for large amounts of oxidant are probably caused by side reactions.

As shown in Figure 4.1.9, the conductivity and yield of the polymer are strongly dependent on the reaction temperature. When the reaction temperature of the first

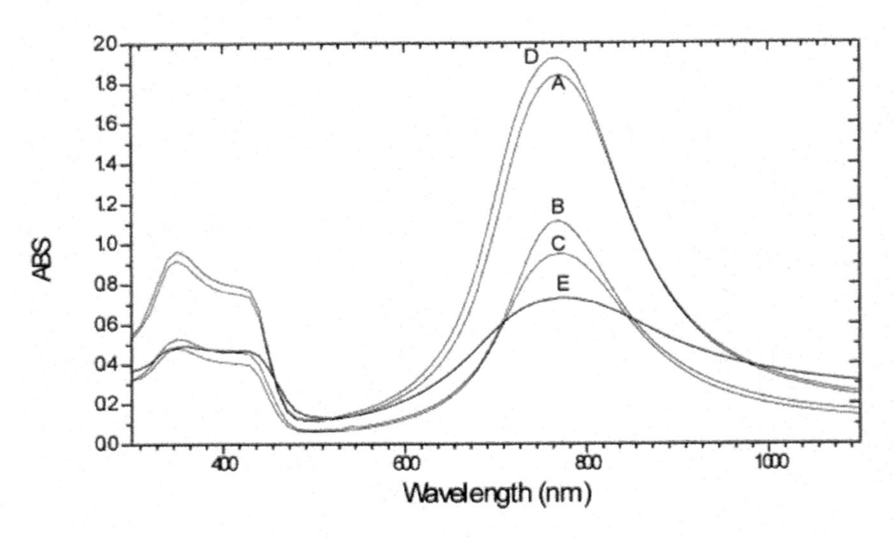

FIGURE 4.1.8 UV spectra of the polymers prepared in 0.05 M HCl solution, and for an ANIHCl/ANIDBSA molar ratio in the feed of 3:7 and for various amounts of APS: 0.8 g (A), 1 g (B), 1.2 g (C), 1.4 g (D), and 1.6 g (E).

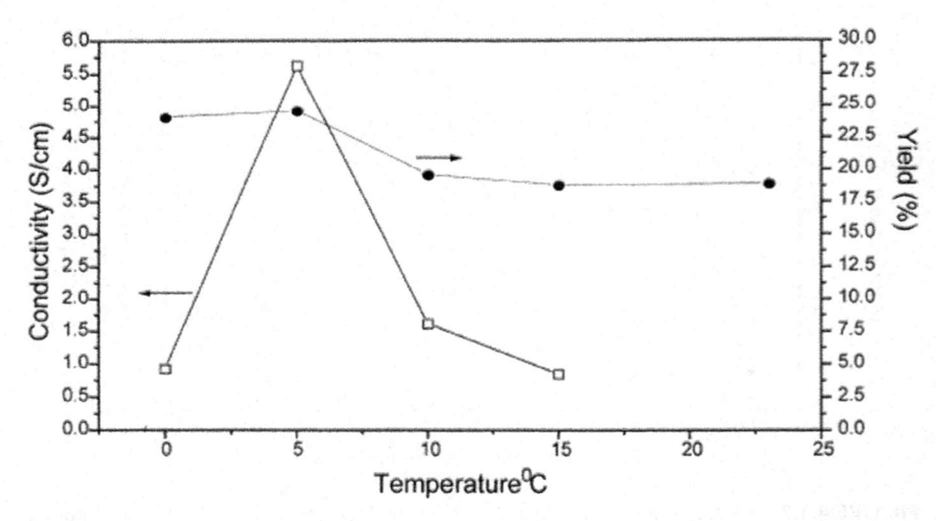

FIGURE 4.1.9 Effect of reaction temperature on the conductivity and yield of the polymer for an ANIHCl/ANIDBSA molar ratio in the feed of 3:7 and additional HCl concentration of 0.05 M.

stage is increased from 0°C to 5°C, the conductivity increases from 0.9 to 5.6 S/cm, while the yield remains almost unchanged. With a further increase to 15°C, both the conductivity and yield decrease to 0.8 S/cm and 18.4%, respectively. This occurs because side reactions, such as the over-oxidative reaction and the reverse hydrolysis reactions, are stimulated by higher temperatures. The polaron band of the polymer

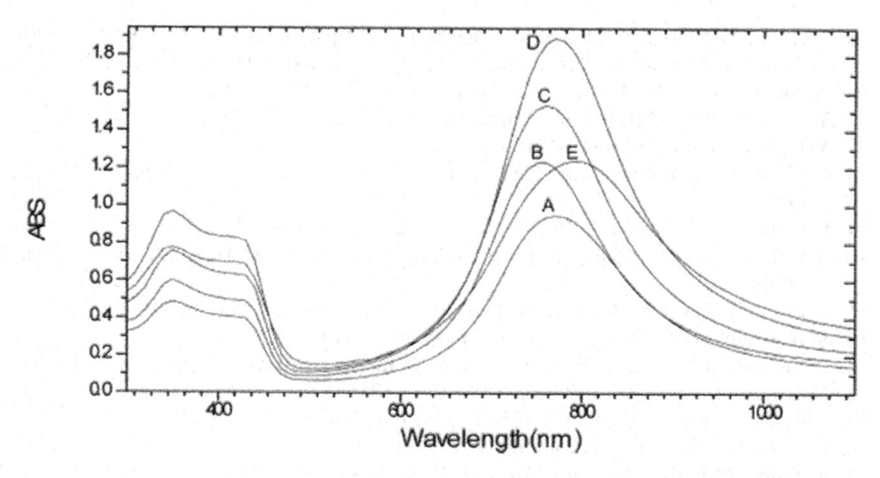

FIGURE 4.1.10 UV spectra of the polymers prepared in 0.05 M HCl solution for an ANIHCl/ANIDBSA molar ratio in the feed of 3:7 and for various temperatures: 0°C (A), 5°C (B), 10°C (C), 15°C (D), and 23°C (RT) (E).

is shifted toward a shorter wavelength (blue shift) (Figure 4.1.10) when the reaction temperature increases from 0°C to 5°C, but is shifted back toward a longer wavelength (red shift) with a further increase in temperature to 23°C (room temperature). This behaviour is caused by (i) the increased dissociation of the ANIDBSA aggregates with increasing temperature, which can shift the polaron band toward a longer wavelength; and (ii) the pernigraniline structure is formed at sufficiently high temperatures due to the side reactions, and this shifts the polaron band toward a longer wavelength.

4.1.4 CONCLUSION

A soluble polymer (PANI doped with DBSA and HCl) could be obtained directly by the in situ doping polymerization, whose conductivity reached a value as large as 13.9 S/cm. The conductivity and yield of the polymer were strongly dependent on the polymerization conditions, such as ANIHCl/ANIDBSA ratio, additional HCl solution, oxidant amount and reaction temperature. At an ANIHCl/ANIDBSA molar ratio of 3:7 in the feed, the polymer exhibited its maximum conductivity of 7.9 S/cm and maximum yield of 30.8%. When additional HCl was introduced in the feed, the conductivity could be increased to 13.9 S/cm. The absorption peaks at 430 and at about 800 nm, which are characteristic for emeraldine were clearly identified in the obtained polymers.

REFERENCES

1. H.L. Letheby, *J. Chem. Soc.* 15 (1862) 161.
2. A.F. Diaz, J.A. Logan, *J. Electroanal. Chem.* 111 (1980) 111.
3. E.M. Genes, A. Boyl, M. Lapkowski, C. Trintavis, *Synth. Met.* 36 (1990) 139.

4. A.G. MacDiamid, J.C. Chiang, A.F. Richter, A.J. Epstein, *Synth. Met.* 18 (1987) 285.
5. M. Abe, A. Ohtani, Y. Umemoto, S. Akizuki, M. Ezoe, H. Higuchi, K. Nakmoto, A. Okuno, Y. Noda, *J. Chem. Soc., Chem. Commun.* 22 (1989) 1736.
6. Y. Cao, P. Smith, A.J. Heeger, *Synth. Met.* 48 (1992) 91.
7. Y. Cao, P. Smith, *Polymer* 34 (1993) 3139.
8. G. Gustafsson, Y. Cao, G.M. Treac, F. Klavetter, N. Colaneri, A.J. Heeger, *Nature* 357 (1992) 477.
9. R.F. Bay, S.P. Armes, C.J. Pickett, K.S. Ryder, *Polymer* 32 (1991) 2456.
10. L.A. Samuelson, A. Anagnostopoulos, K.S. Alva, J. Kumar, S.K. Tripathy, *Macromolecules* 31 (1998) 4376.
11. J.-W. Chevalier, J.-Y. Bergeron, L.H. Dao, *Macromolecules* 25 (1992) 3325.
12. S.-A. Chen, G.-W. Huang, *J. Am. Chem. Soc.* 116 (1994) 7938.
13. P.J. Kinlen, J.L. Ding, C.R. Graham, E.E. Remsen, *Macromolecules* 31 (1998) 1735.
14. N. Kuramoto, A. Tomita, *Polymer* 38 (1997) 3055.
15. P.N. Adam, A.P. Monkman, *Synth. Met.* 87 (1997) 165.
16. C.-T. Kue, C.-H. Chen, *Synth. Met.* 99 (1999) 163.
17. Y. Wang, M.F. Rubner, *Synth. Met.* 47 (1992) 255.
18. F.L. Liu, F. Wudl, M. Nowak, A.J. Heeger, *J. Am. Chem. Soc.* 108 (1986) 8311.
19. W. Zhang, Y. Min, A.G. MacDiarmid, M. Angelopoulos, Y.-H. Liao, A.J. Epstein, *Synth. Met.* 84 (1997) 109.

4.2 Copolymers of Aniline and 3-Aminophenol Derivatives with Oligo(oxyethylene) Side Chains as Novel Water-Soluble Conducting Polymers*

Fengjun Hua and Eli Ruckenstein

Department of Chemical and Biological Engineering, State University of New York at Buffalo, Buffalo, New York 14260

ABSTRACT A novel water-soluble copolymer of aniline and aminophenol (AP) grafted with oligo(oxyethylene) (PEO) side chains (AP-*g*-PEO) was synthesized. The AP-*g*-PEO macromononer was prepared starting from an *N*-protected 3-aminophenol (protection group: *tert*-butoxycarbonyl) followed by the substitution of tosylated oligo(oxyethylene) on the hydroxide moiety of the AP and deprotection. Finally, AP-*g*-PEO was copolymerized with aniline at various feed mole ratios (AP-*g*-PEO/aniline). The copolymers and various intermediates were characterized by FTIR, MS, NMR, GPC, UV–vis, and chemical elemental analysis. The increase of the aniline content and the decrease of the PEO side chain length generated lower oligo(oxyethylene) grafted concentrations and solubilities in water but longer conjugation lengths and higher conductivities. Four AP-*g*-PEOs were prepared from four oligo(oxyethylene) methyl ethers with M_n = 164, 350, 750, and 2000. The poly((AP-*g*-PEO-750)-*co*-aniline) at a feed mole ratio of 3/1 was water-soluble and possessed a relatively high conductivity (0.12 S/cm). AP-*g*-PEOs were found to have low reactivities and to generate low homopolymerization degrees because of the torsional effects of the PEO

* *Macromolecules* 2004, 37, 6104–6112.

side chains on the backbone of the copolymer. A possible copolymerization mechanism was suggested. Furthermore, the copolymers with high PEO content had lower oxidation and doping levels of their backbones, which were confirmed by X-ray photoelectron spectroscopy.

4.2.1 INTRODUCTION

Polyaniline has been extensively studied in the past 20 years, as one of the important conducting polymers, because of its thermal stability and high conductivity.[1–4] However, this polymer was found to have low solubilities in the common solvents and poor processability. Numerous investigations have been carried out for the improvement of the solubility and processability by introducing various side chains such as alkyl,[5] alkoxy,[6] benzyl,[7] and aryl[8] substituents either at the aromatic ring or at nitrogen. The grafted side chains in these aniline derivatives increased the solubility of the PANI backbones. Water-soluble poly(aniline derivatives) could be also prepared by the direct polymerization of water-soluble aniline derivative monomers containing a sulfonic acid moiety at the phenyl ring or nitrogen.[9–13] The withdrawing effect of the sulfonic group caused, however, difficulties in homopolymerization and also decreased the conductivity. The copolymerizations of sulfonated aniline and aniline have been successfully carried out.[14–18] Another pathway was the modification of the backbone of polyaniline through grafting water-soluble side chains at either the ring or nitrogen. A typical example is the directly sulfonated polyaniline (SPAN) reported by Epstein et al.[19] A decrease of the conductivity because of the withdrawing effect was observed.[19–21] Yamaguchi et al. prepared PANI grafted with hydrophilic poly(ethylene oxide) (PEO) through a "graft from" process, in which a ring-opening polymerization of an epoxide occurred on the active anionic nitrogens of the PANI backbone.[22] Wang et al. grafted PEO chains to the active anionic nitrogens of PANI via a direct attachment under very strict experimental conditions. A lower grafting degree, below 30%, was, however, achieved through this "graft onto" process.[23] Less attention was paid to the grafting of PEO side chains to the phenyl ring, even though the modification that occurs at nitrogen was expected to produce a higher decrease of the conductivity than that which occurs at a phenyl ring. Park et al.[24] suggested a synthesis strategy in which poly(ethyleneoxy)-3-aminobenzoate was prepared by reacting 3-aminobenzoic acid with trityl chloride, followed by the chlorination and esterification of the benzoic acid moiety. After deprotection in acidic conditions, the resulted macromonomer could be copolymerized with aniline. The presence of benzoic acid or of its ester decreased the reactivity of the macromonomer because of their electron-withdrawing effect, resulting in a low-grafted copolymer (<35%). The aniline derivatives with a short-chain alkoxyl grafted on the phenyl ring, such as 2- or 3-methoxyaniline or 2- or 3-ethoxyaniline, were easily homopolymerized to generate conducting polymers with conductivities as high as that of polyaniline because the alkoxyl is an electron donor.[25] The homopolymerization of 2- or 3-aminophenol in acidic conditions has been also re-ported.[26] The low conductivity achieved, below 10^{-8} S/cm, might have been caused by the inductive effect of the hydroxide moiety of the AP.

In previous papers, we reported the preparation of a water-soluble polydiphenylamine (PDPA) with polyethylene oxide) (PEO) grafted onto the nitrogens of the PDPA backbone.[27] The grafting was achieved by substituting the tosyloyl end group of PEO-tosylate with an amine of the PDPA under mild conditions. The resulted PDPA-*g*-PEO with a comb-shaped architecture became soluble in organic solvents, such as chloroform or tetrahydrofuran, and even became water-soluble for PEO side chains of molecular weight $M_n = 2000$. The PEO-grafted polyaniline can be employed as battery material and biomaterial.[28–31]

In the present paper, an N-protected aniline derivative was first prepared by reacting 3-aminophenol (AP) with tert-butoxycarbonyl (*t*-Boc) as a temporary protecting group.[32] The protection was followed by the grafting of an oligo(oxyethylene) side chain at the 3-position by substituting the tosyloyl end group of tosylate oligo(oxyethylene) with the hydroxide moiety of the AP under mild conditions[31,33] to generate AP-Boc-*g*-PEO. Further, the AP-Boc-*g*-PEO was deprotected with trifluoroacetic acid (TFA) to obtain an aniline macromonomer possessing an oligo(oxyethylene) side chain at the 3-position (AP-*g*-PEO), which was (co) polymerized via chemical oxidation in the presence of 1.2 N HCl aqueous solution (see Scheme 4.2.1). The molecular structures of the amino-protected oligomers and copolymers were characterized by FTIR, MS, NMR, GPC, and elemental chemical analysis. Furthermore, the oxidation levels of these copolymers upon doping or undoping were examined by X-ray photoelectron spectroscopy (XPS).

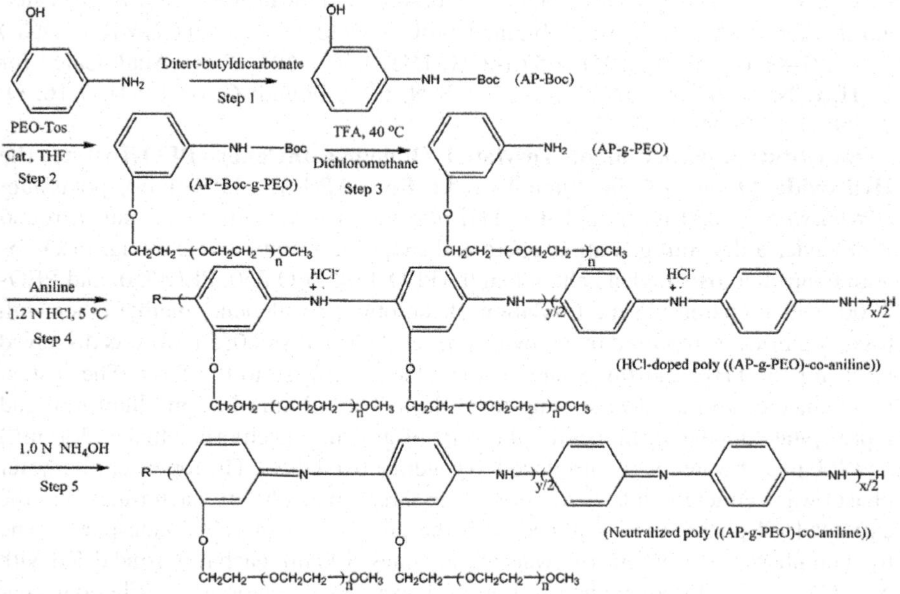

SCHEME 4.2.1 Synthesis pathway of copolymers of AP-*g*-PEO and aniline.

4.2.2 EXPERIMENTAL SECTION

Materials. 3-Aminophenol (99%), aniline (99%), di-*tert*-butyldicarbonate (97%), hydrochloric acid (37 wt%), ammonium persulfate (98 wt%), trifluoroacetic acid (TFA, 99%), triethylamine (99 wt%), *p*-toluenesulfonyl chloride (99+ wt%), and potassium *tert*-butylate (>95 wt%) were purchased from Aldrich and used without further purification. Oligo(oxyethylene) methyl ethers with different molecular weights, such as tri(ethylene glycol) monomethyl ether (PEO-150, $M_n = 164$) and poly(ethylene oxide) methyl ethers (PEO-350 ($M_n = 350$), PEO-750 ($M_n = 750$), and PEO-2000 ($M_n = 2000$)), were also purchased from Aldrich and used without further purification. The solvents, dimethylformamide (DMF), diethyl ether, and tetrahydrofuran (THF) from Aldrich and dichloro-methane and chloroform from Fisher, were of HPLC purity.

Amino Protecting Reaction. 3-Aminophenol (0.1 mol), di-*tert*-butyldicarbonate (0.1 mol), and 40 mL of DMF were introduced into a round flask under a nitrogen atmosphere. The system was subjected to intensive magnetic stirring for a few minutes and then cooled to 0°C in an ice-water bath. A triethylamine (0.105 mol) solution in DMF (40 mL) was added dropwise within 1 h. The reaction lasted an additional 12 h at 0°C. Further, 300 mL of chloroform and 100 mL of water were introduced to generate two layers. The pH of the top layer (the water phase) was carefully adjusted, using a 0.1 N HCl aqueous solution, at pH = 7, and extracted with chloroform three times (30 mL each time). The combined chloroform phases were first washed with a saturated $NaHCO_3$ aqueous solution and then washed with water until neutral. The chloroform solution was dried with Na_2SO_4, and the solvent was evaporated on a rotary evaporator. A light brown solid, N-protected aminophenol (AP-Boc), was obtained with a yield of 89%. ^1H NMR ($CDCl_3$) $\delta = 7.01–6.10$ (1,3-Ph, 4 H), 4.57 (–OH, 1H), 3.66 ($-NH_2$, 2 H). Anal. Calcd for $C_{11}H_{15}O_3N$: C, 63.15; H, 7.18; O, 22.97; N, 6.70. Found: C, 63.12; H, 7.16; O, 22.99; N, 6.73. MS (*m/z*): 209.1.

Substitution Reaction of Tosylated Oligo(oxyethylene) (PEO-Tos) to the Hydroxide Moiety of the Phenol of AP-Boc. AP-Boc (0.050 mol), potassium *tert-butylate* (0.050 mol), and dry THF (250 mL) were introduced into a round flask under a dry nitrogen atmosphere and subjected to intensive stirring. PEO-Tos with controlled oxyethylene chain length (PEO-150, PEO-350, PEO-750, and PEO-2000) was prepared via the tosylation of the oligo(oxyethylene) methyl ether with tosyol chloride as reported in a previous paper.[27] PEO-Tos (0.055 mol) was dissolved in 50 mL of THF, and this solution was added dropwise to the flask. The system was subjected to intensive stirring for 3 days. The mixture became light red, and a precipitate appeared, indicating the start of grafting. Dichloromethane (300 mL) and 100 mL of water were introduced to generate two layers. The top layer (the water phase) was extracted with dichloromethane three times (30 mL each time), and the extracted oil phase was combined with the previous dichloromethane phase. The total oil phase was washed with water three times (100 mL each time) and dried with Na_2SO_4. The dichloromethane was removed on a rotary evaporator. Further purification was carried out via a chromatographic method. A light brown liquid, namely AP-Boc-*g*-PEO, was thus obtained with a yield of about 70%.

Deprotecting Reaction of AP-Boc-*g*-PEO and Copolymerization with Aniline. AP-Boc-*g*-PEO (0.030 mol), TFA (0.060 mol), and dichloromethane (40 mL) were introduced into a round flask equipped with a condensor, subjected to intensive stirring, and its temperature raised to 40°C. The reaction lasted 2 h at 40°C, after which the solvent and the excess of TFA were removed on a rotary evaporator under high vacuum at a temperature <30°C. A brown liquid, AP-*g*-PEO, was obtained with a yield of 98%.

AP-*g*-PEO and aniline, with feed mole ratios of AP-*g*-PEO/ aniline of 10/1, 6/1, 3/1, 1/1, or 1/2, and a total concentration of 1.2 N were dissolved in 45 mL of 1.2 N HCl aqueous solution with intensive stirring, and the system was cooled to 5°C. Ammonium persulfate (0.010 mol) dissolved in 1.2 N HCl aqueous solution (15 mL) was added dropwise into the flask. The color of the mixture changed slowly from brown to green, indicating the beginning of copolymerization. After 18 h, a green precipitate was separated using a supercentrifuge (Dupont, RC-5) at 3500 rpm for 30 min. The green particles of the copolymers were washed with water several times until no red supernatant was identified. Further purifications of the HCl-doped copolymers were carried out through dissolution–precipitation in diethyl ether. The yields were 32, 57, 62, 67, and 75% for feed mole ratios of 10/1, 6/1, 3/1, 1/1, and 1/2, respectively. The dedoping was carried out through the neutralization of the above HCl-doped copolymers with a 1.0 N NH_4OH aqueous solution. A blue solid was obtained with a yield above 90% in all cases.

The homopolymerization of AP-*g*-PEOs was also carried out for comparison purposes. The yield was in all cases below 37%.

Characterization. Proton (^1H) NMR, MS, UV-vis absorption, and FTIR measurements have been carried out on a 400 MHz INOVA-400, Mass Spectroscope (PHI 7200 model), a Thermo Spectronic Genesys-6, and a Perkin-Elmer FTIR 1760, respectively. The NMR solutions were prepared by dissolving the polymer in deuterated water or deuterated chloroform (5 g/L). Gel permeation chromatography (GPC, Waters) was used to evaluate the molecular weights of the homopolymers and copolymers on the basis of a polystyrene calibration curve. The GPC was equipped with three 30 cm long columns filled with a Waters Styragel, a Waters HPLC 515 pump, and a Waters 410 RI detector. The GPC measurements have been carried out using DMF as eluent at 60°C, at a flow rate of 1.0 mL/min and 1.0 cm/min chart speed.

The chemical elemental analysis of the polymer samples was carried out on a Perkin-Elmer model 2400 C, H, N analyzer. The chlorine and sulfur contents were determined by the oxygen flask method. The room temperature conductivities of the compressed pellets of the various copolymers were determined using the conventional four-point method. The surface elemental analysis of compressed disks was carried out on a VG ESCA/SIMSLAB MK II with a Mg Kα radiation source (1253.6 eV). Each copolymer sample was first dissolved in DMF, and subsequently DMF was removed using a rotary evaporator. The solid sample was dried under high vacuum at room temperature for at least 7 days under a nitrogen atmosphere and ground into a powder which was compressed into a disk (of thickness around 1 mm) between two Teflon films.

4.2.3 RESULTS AND DISCUSSION

Synthesis and Characterization. The copolymers of aminophenol with oligo(oxyethylene) side chains at the 3-position of the phenyl ring (AP-*g*-PEO) and aniline were prepared via the successive four steps of Scheme 4.2.1. The *t*-Boc-*N*-protected aminophenol (AP-Boc) was grafted via substitution with tosylated oligo(oxyethylene) (PEO-Tos-150, PEO-Tos-350, PEO-Tos-750, or PEO-Tos-2000 prepared from PEO-150 (M_n = 164), PEO-350 (M_n = 350), PEO-750 (M_n = 750), and PEO-2000 (M_n = 2000), respectively) to generate the corresponding oligomer (AP-Boc-*g*-PEO). Figure 4.2.1 presents the FTIR spectra of PEO-750-Tos (spectrum a), AP-Boc (spectrum b), and AP-Boc-*g*-PEO-750 (spectrum c); the characteristic peak of the hydroxide moiety at 3300 cm^{-1} disappeared in AP-Boc-*g*-PEO-750 because of the formation of an ether bond after substitution, and the ether bond peaks (C–O–C) of the PEO side chains and of the phenoxyl (=C–O–C) appeared around 1108 cm^{-1} as a multipeak. The carbonyl vibration peak at 1734 cm^{-1}, which can be assigned to *t*-Boc, is still present in the two compounds, indicating that the *t*-Boc was not eliminated during substitution. Figure 4.2.2 displays a typical MS spectrum of AP-Boc-*g*-PEO-750 (ESI MS), where the values of the main molecular ionic peaks in the MS spectra reflect the following formula:

$$M_n = 23(Na^+) + 31(-OCH_3) + 208\,(\,(-O)C_6H_3\text{-}NH\overset{O}{\overset{\|}{C}}\text{-}O\text{-}C(CH_3)_3\,) + 44\,(-CH_2CH_2O-)n$$

where *n* is the number of ethylene oxides in the oligo(oxyethylene) side chains. Furthermore, in the ^1H NMR spectra ($CDCl_3$, Figure 4.2.3A), the signals in the range 8.0–6.0 ppm of the phenyl rings of AP-Boc or AP-Boc-*g*-PEO-750 were shifted downfield compared with those of 3-aminophenol because of the deshielding effect caused by the tert-butoxycarbonyl. The characteristic signal at 1.44 ppm, which can be assigned to the protons of methane of *t*-Boc, remained almost the same in the two cases (spectra b and c in Figure 4.2.3A), indicating that the protecting groups were not lost during substitution. In Figure 4.2.3B of the expanded NMR spectra of AP-Boc and AP-Boc-*g*-PEO-750, the protons of the phenyl ring could be assigned to the chemical formula inserted. The signals at 4.18, 3.65, and 3.41 ppm can be assigned to the protons of phenoxymethylene, the oxyethylene, and oxymethyl of the PEO side chains, respectively. The NMR spectrum of AP-*g*-PEO-750 obtained after deprotection (spectrum d of Figure 4.2.3A) shows that the signals at 7.10 ppm (peak i) and 1.40 ppm (peak j) present in AP-Boc-*g*-PEO-750 (which can be assigned to the protons of the amide (–CONH–) and methyl ($-CH_3$) of the *t*-Boc, respectively) disappeared. The signals due to the protons of the phenyl ring were shifted upfield because the deshielding effect caused by the *tert*-butoxycarbonyl was no longer present. The main peaks in the MS spectrum after deprotection were found to differ from those of AP-Boc-*g*-PEO-750 (see Figure 4.2.2) by 108 Da, corresponding exactly to the molecular weight of the *t*-Boc moiety.

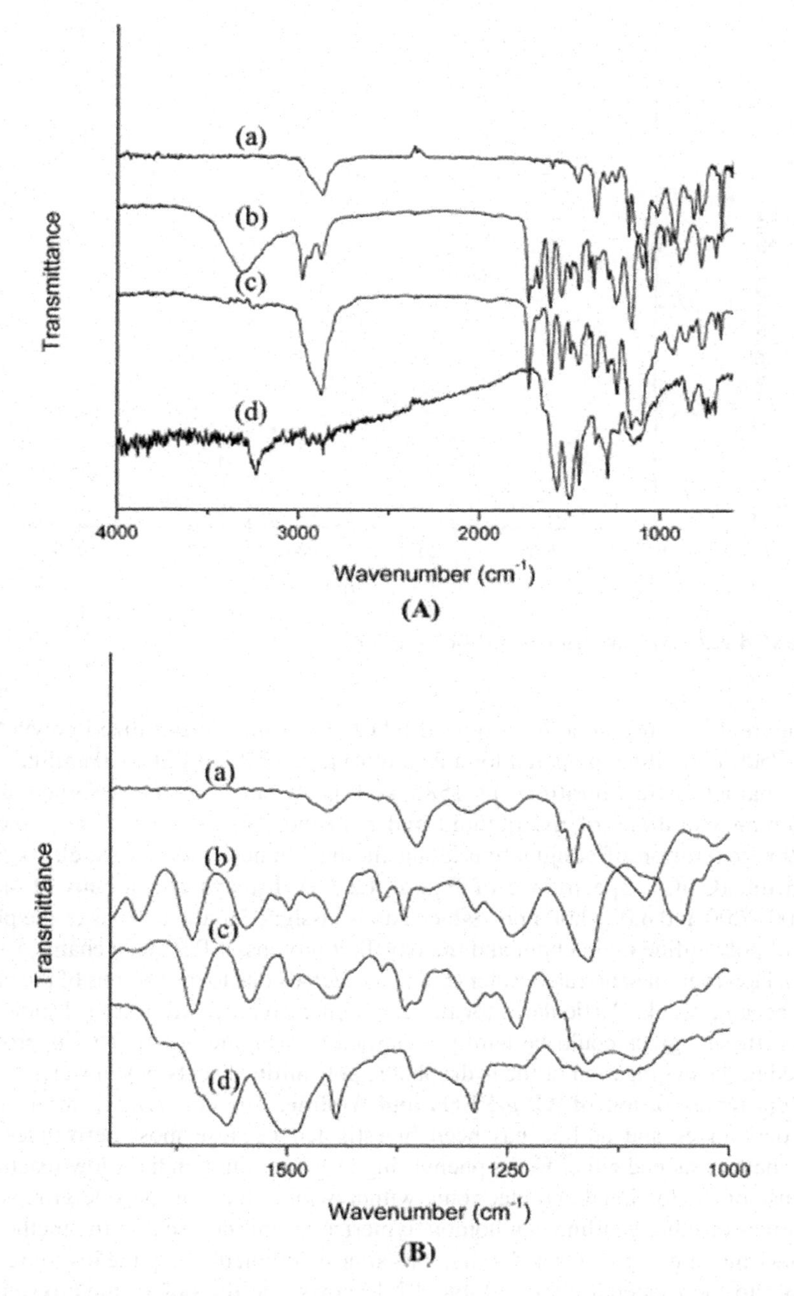

FIGURE 4.2.1 FTIR spectra of (a) PEO-750-Tos, (b) AP-Boc, (c) AP-Boc-*g*-PEO-750, and (d) poly9(AP-*g*-PEO-750)-*co*-aniline) (3/1).

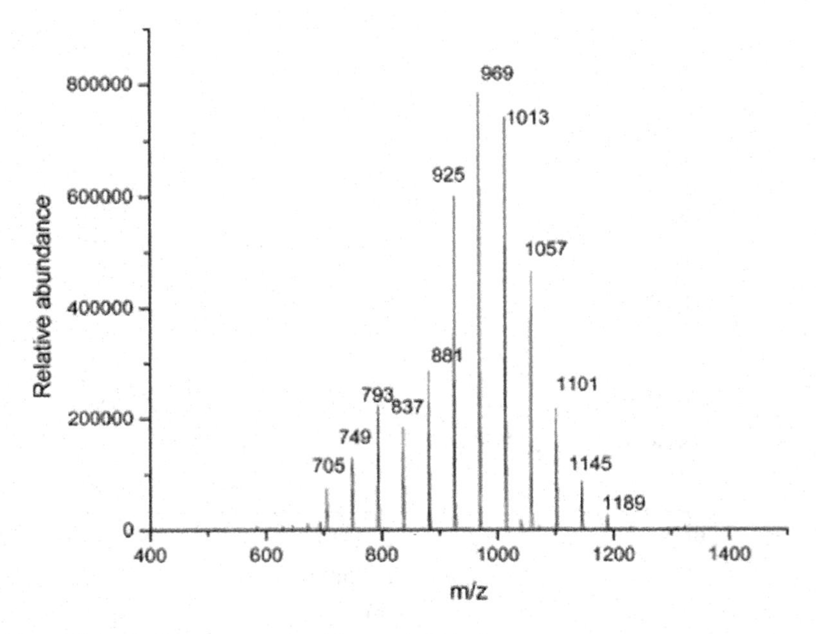

FIGURE 4.2.2 MS spectrum of AP-Boc-*g*-PEO-750.

Figure 4.2.1c also contains a typical FTIR spectrum of neutralized poly9(AP-*g*-PEO-7500-*co*-aniline) prepared for a feed mole ratio of AP-*g*-PEO-750/aniline = 3/1. The characteristic vibrations at 1585 and 1490 cm^{-1} can be assigned to the stretching vibrations of the quinoid and benzenoid rings, respectively, indicating the generation of conjugation along the backbone (oxidated state). Its NMR spectrum ($CDCl_3$, spectrum e of Figure 4.2.3A) displays two groups of signals at 8.00–7.00 and 4.00–3.00 ppm, which can be assigned to the protons of the phenyl ring of polyaniline main chain and the oxyalkyl protons of PEO side chains, respectively. However, in deuterated water (D_2O), the signals due to the protons of the phenyl ring became weak. Particularly for the copolymer prepared from a feed mole ratio of 1/1, these signals could be hardly identified (see Figure 4.2.4). This is probably related to the aggregation of the hydrophobic polyaniline backbones in water.[27]

Copolymerization of AP-*g*-PEOs and Aniline. The copolymerization of aniline derivatives and aniline has been investigated because most derivatives with side chains attached either to the phenyl ring or to the nitrogen have low reactivities because of the torsional and electronic withdrawing effects of the side groups. The methoxy- or ethoxyaniline can homopolymerize as aniline because the methoxy or ethyoxy moieties are electron donors, thus almost complementing the loss of reactivity due to the torsional effects. Table 4.2.1 shows that the poly(2-methoxyaniline) (poly(*o*-anisidine)) and polyaniline exhibit comparable polymerization degrees (153.0 and 169.5, respectively). The UV–vis spectra of PANI and poly(*o*-anisidine)

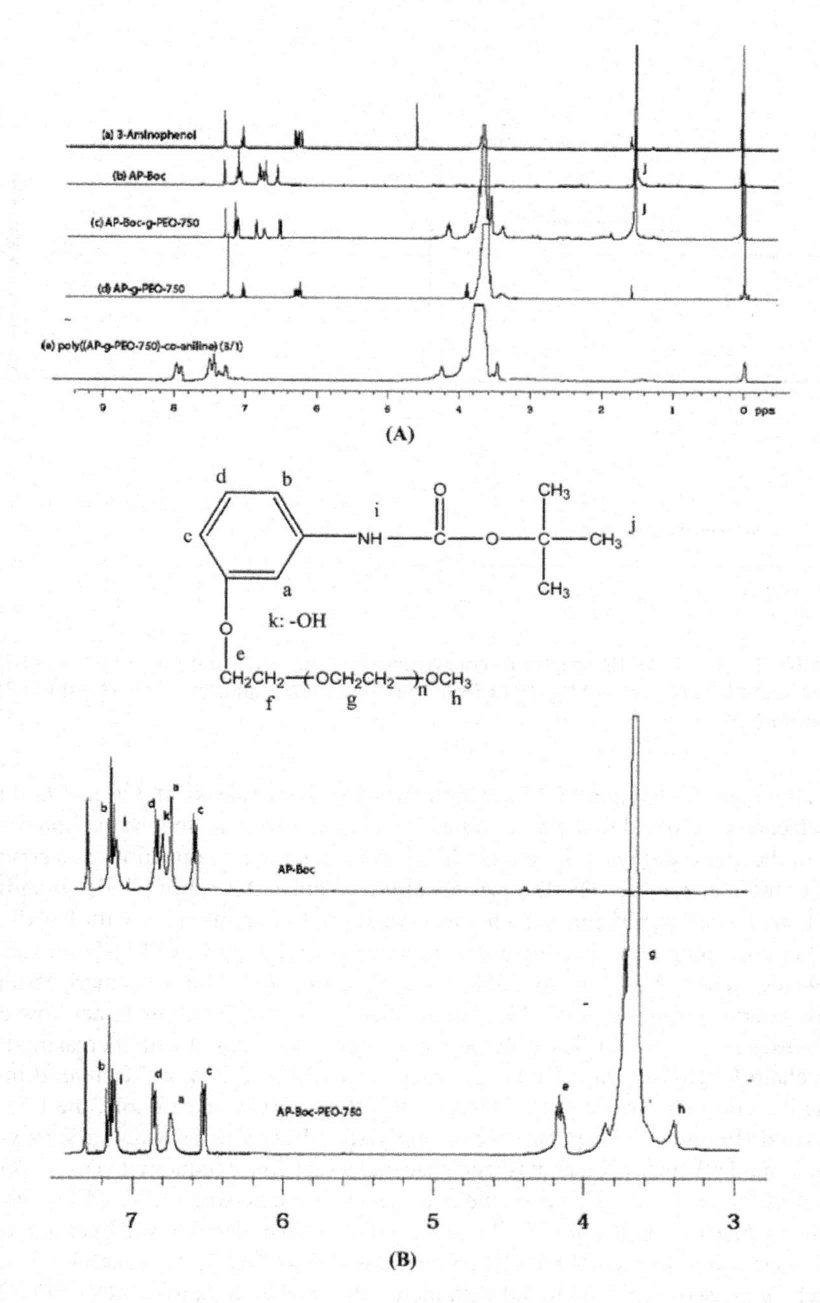

FIGURE 4.2.3 (A) ^1H NMR spectra of (a) 3-aminophenol, (b) AP-Boc, (c) AP-Boc-*g*-PEO-750, (d) AP-*g*-PEO-750, and (e) poly((AP-*g*-PEO)-*co*-aniline) (3/1) in $CDCl_3$, 5 g/L. (B) Two expanded ^1H NMR spectra of (b) AP-Boc and (c) AP-Boc-*g*-PEO-750 in $CDCl_3$, 5 g/L.

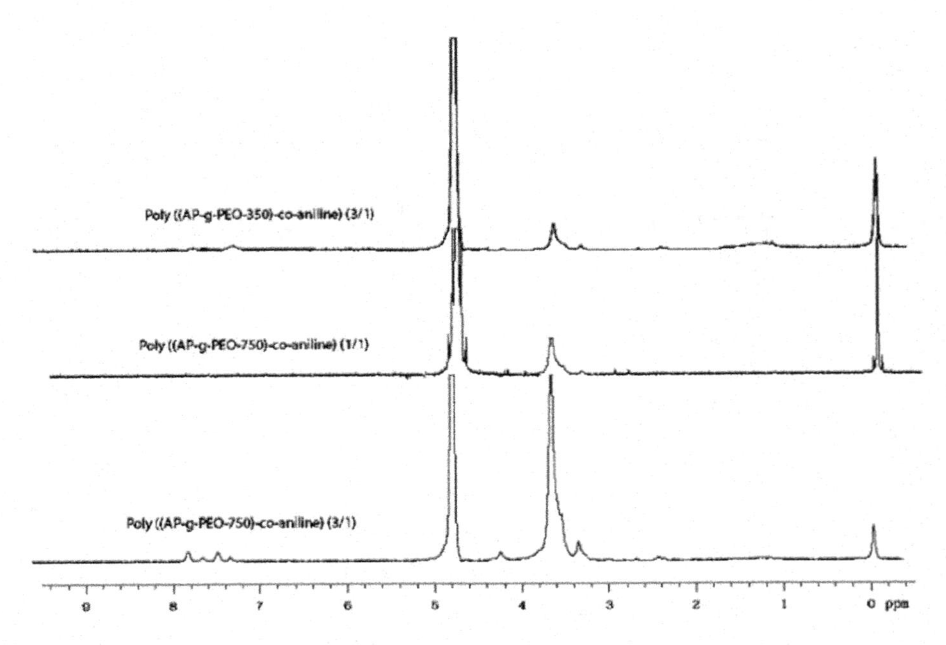

FIGURE 4.2.4 ^{1}H NMR spectra of copolymers in D_2O, 5 g/L: (a) poly((AP-*g*-PEO-750)-*co*-aniline) (3/1), (b) poly((AP-*g*-PEO-750)-*co*-aniline) (1/1), and (c) poly((AP-*g*-PEO-350)-*co*-aniline) (3/1).

(spectra a and b of Figure 4.2.5) exhibit two absorbance peaks at 320 and 640 nm, which can be assigned to the π–π* transition of the benzenoid and the exciton transition of the quinoid rings, respectively. They show that their conjugations and conductivities are comparable. The UV–vis spectrum of poly((AP-*g*-PEO-750)-*co*-aniline) (3/1) has a peak at 605 nm, which was blue-shifted in comparison with PANI, and a lower area integral, indicating a lower conjugation. Four AP-*g*-PEOs with various PEO side chains ($M_{n,\ PEO}$ = 164, 350, 750, and 2000) were homopolymerized under acidic conditions and produced low molecular weight homopolymers and low conductivities (<10^{-3} S/cm). The polymerization degree decreased with increasing PEO side chain length, indicating that the reactivity of AP-*g*-PEOs has decreased in the same direction (see Table 4.2.1). Using AP-*g*-PEO-750 as an example, the UV–vis spectrum (Figure 4.2.5, spectrum e) of polyAP-*g*-PEO-750 provided a very weak peak around 600 nm in the neutralized state and a very low conductivity (<10^{-7} S/cm) in the HCl-doped state. However, the copolymerization between AP-*g*-PEO-750 and aniline with a feed mole ratio of 3/1 could produce high molecular weight copolymers (polymerization degree = 98.6). The mole ratio AP-*g*-PEO-750/aniline in the copolymers increased from 0.44 to 4.0 with increasing feed mole ratio of AP-*g*-PEO-750/aniline from 1/2 to 10/1, and the conductivity decreased in the same direction. In Figure 4.2.6, the blue shift compared to PANI increased with increasing feed mole ratio; i.e., the peak at 640 nm of the polyaniline was shifted to 595 nm for poly((AP-*g*-PEO-750)-*co*-aniline) at a feed mole ratio = 6/1, along with a decrease in the area integral, indicating a decrease of conjugation. Furthermore, the GPC traces of these

TABLE 4.2.1
Physical Parameters of the Copolymers of AP-g-PEOs and Aniline

Name	HCl Concn, Temp	AP-g-PEO/ Aniline,[a] Feed Mole Ratio	M_n (polymerization degree)[b]	AP-g-PEO/ Aniline in Copolymer (y/x)[c]	PEO wt Fraction in Copolymer[d]	Conductivity, S/cm[e]
polyaniline	1.2 N, 5°C		15760 (169.5)			4.57
poly(*o*-anisidine)	1.2 N, 5°C		18970 (153.0)			3.53
poly(AP-*g*-PEO-150)	3.0 N, 5°C		9550 (36.4)			2.34×10^{-3}
poly(AP-*g*-PEO-350)	3.0 N, 5°C		11350 (24.0)			5.43×10^{-5}
poly(AP-*g*-PEO-750)	3.0 N,5°C		5675 (6.50)			$<10^{-7}$
poly(AP-*g*-PEO-2000)	3.0 N, 5°C		12060 (5.7)			$<10^{-7}$
poly((AP-*g*-PEO-750)-*co*-aniline)	1.2 N, 5°C	10/1	9650 (13.7)	4.0	0.850	7.4×10^{-5}
poly((AP-*g*-PEO-750)-*co*-aniline)	1.2 N, 5°C	6/1	20460 (30.4)	3.2	0.845	4.8×10^{-4}
poly((AP-*g*-PEO-750)-*co*-aniline)	1.2 N, 5°C	3/1	56450 (98.6)	1.7	0.821	0.12
poly((AP-*g*-PEO-750)-*co*-aniline)	1.2 N, 5°C	1/1	41690 (102.7)	0.7	0.757	0.30
poly((AP-*g*-PEO-750)-*co*-aniline)	1.2 N, 5°C	1/2	40630 (125.1)	0.44	0.702	0.83
poly((AP-*g*-PEO-150)-*co*-aniline)	1.2 N, 5°C	3/1	27590 (126.9)	2.4	0.504	0.67
poly((AP-*g*-PEO-350)-*co*-aniline)	1.2 N, 5°C	3/1	37480 (112.1)	2.0	0.693	0.44
poly((AP-*g*-PEO-2000)-*co*-aniline)	3.0 N, 5°C	3/1	26540 (17.6)	0.4	0.858	$<10^{-7}$

[a] AP-*g*-PEOs denote aminophenol derivatives with various oligo(oxyethylene) side chains at the 3-position of 3-aminophenol.

[b] Number average molecular weights of the neutralized polymers were determined by GPC on the basis of a PS calibration curve; using DMF as eluent at a rate of 1 mL/min and at 60°C. The polymerization degree is expressed as $DP_n = (M_{n,\,GPC} \times (1 + (y/x)))/((y/x) \times M_{n,\,AP\text{-}g\text{-}PEO} + M_{n,\,aniline} - 2)$.

[c] The mole ratio of AP-*g*-PEO/aniline in copolymer, y/x, was determined from Table 4.2.3 via chemical elemental analysis, i.e., $y/x = (C/N - 6)/(7 - C/N + (M_n - 31)/22))$, where $M_{n,\,AP\text{-}g\text{-}PEO}$ is the molecular weight of the oligo(oxyethylene).

[d] PEO weight fraction was calculated from footnote *c* using the expression $(M_{n,\,AP\text{-}g\text{-}PEO} \times (y/x))/((M_{n,\,AP\text{-}g\text{-}PEO} + 109) \times (y/x) + 91)$.

[e] Conductivities of polymeric compressed pellets determined via the four-point method. Each polymer was dried under vacuum for at least 7 days before the measurement and ground into a powder which was further compressed into a disk.

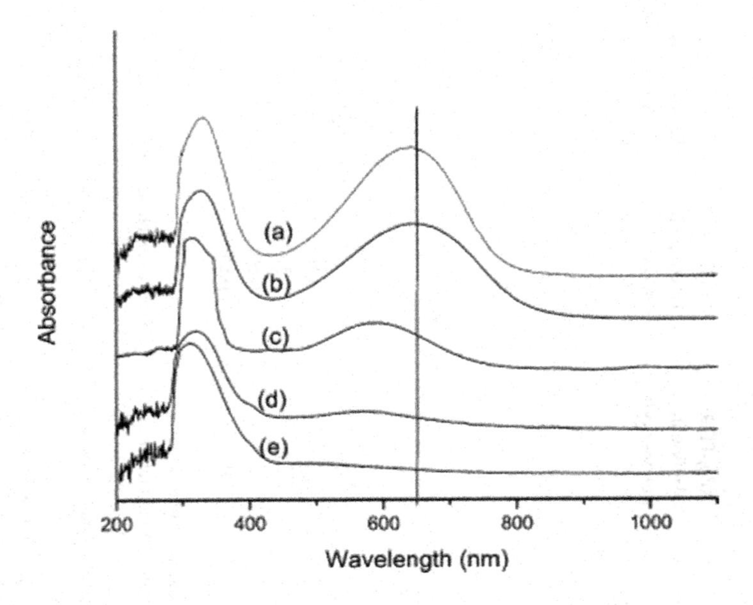

FIGURE 4.2.5 UV–vis spectra of (a) neutralized PANI, (b) neutralized poly(*o*-anisidine), (c) neutralized poly((AP-*g*-PEO-750)-*co*-aniline) (3/1), (d) neutralized poly(2-aminophenol), and (e) neutralized poly(AP-*g*-PEO-750) (all in DMF).

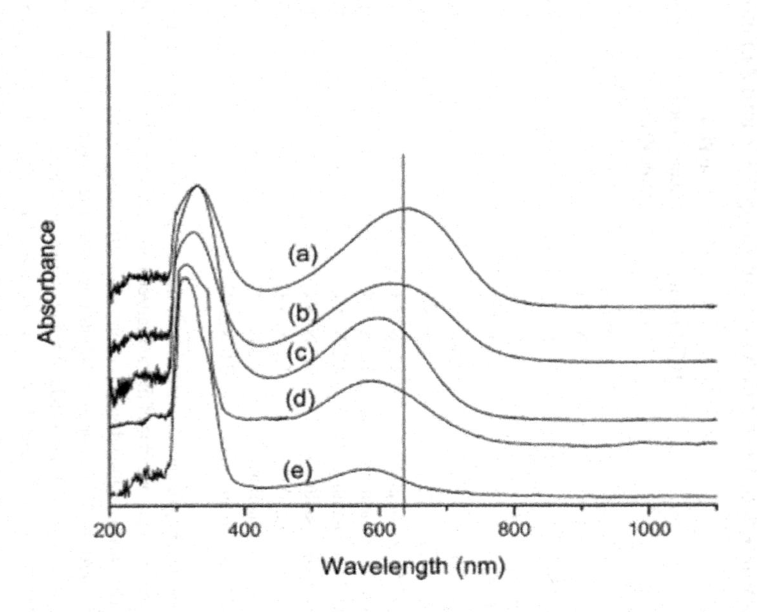

FIGURE 4.2.6 UV–vis spectra of (a) neutralized PANI, (b) neutralized poly((AP-*g*-PEO-750)-*co*-aniline) (1/2), (c) neutralized poly((AP-*g*-PEO-750)-*co*-aniline) (1/1), (d) neutralized poly((AP-*g*-PEO-750)-*co*-aniline) (3/1), and (e) neutralized poly((AP-*g*-PEO-750)-*co*-aniline) (6/1) (all in DMF).

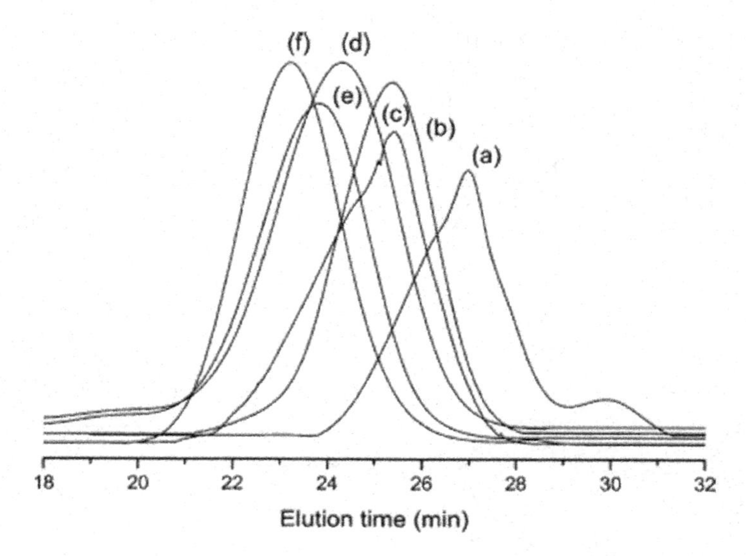

FIGURE 4.2.7 GPC traces of neutralized polymers: (a) poly(AP-*g*-PEO-750), (b) polyaniline, (c) poly((AP-*g*-PEO-750)-*co*-aniline) (6/1), (d) poly((AP-*g*-PEO-750)-*co*-aniline) (1/1), (e) poly((AP-*g*-PEO-350)-*co*-aniline) (3/1), and (f) poly((AP-*g*-PEO-750)-*co*-aniline (3/1).

copolymers (Figure 4.2.7) demonstrated that their molecular weight varied with the feed mole ratio. (The number-average molecular weights are listed in Table 4.2.1.) The polymerization degrees of the copolymers prepared at the feed mole ratios of 1/2, 1/1, 3/1, 6/1, and 10/1 were 125.1, 102.7, 98.6, 30.4, and 13.7, respectively. As the feed mole ratio increased, the backbone length became shorter and possessed a lower conjugation. The grafted PEO-750 side chain of AP-*g*-PEO-750 provided a steric barrier which weakened the polaron formation under oxidation conditions, and as a result, the reactivity decreased. The higher diffusion barriers generated by the longer side chains regarding the propagation of the radical species decreased the rate of growth of the chains. Indeed, for the same feed mole ratio, the polymerization degree decreased as the oligo(oxyethylene) side chain length increased (see Table 4.2.1) because of the decreased reactivity of AP-*g*-PEO with longer side chains. A possible copolymerization mechanism via chemical oxidation under acidic conditions is suggested in Scheme 4.2.2. The copolymers resulted from AP-*g*-PEOs with different side chains and the same feed mole ratio exhibited a blue shift in the UV–vis spectra, i.e., the peak at 625 nm of poly((AP-*g*-PEO-150)-*co*-aniline) (3/1) becoming 595 nm for poly((AP-*g*-PEO-2000)-*co*-aniline) (3/1) (see Figure 4.2.8).

Solubility. The solubility of the copolymers was dramatically improved by the grafting with oligo(oxy-ethylene) (Table 4.2.2). The copolymers prepared from AP-*g*-PEO-750 and aniline in the range of feed mole ratios 1/1 to 10/1 can dissolve in most commercial organic polar solvents. The poly((AP-*g*-PEO-750)-*co*-aniline) (3/1) exhibited a high water solubility and a relatively high conductivity (0.12 S/cm).

R_1 = oligo(oxyethylene) and R_2 = R_1 or H

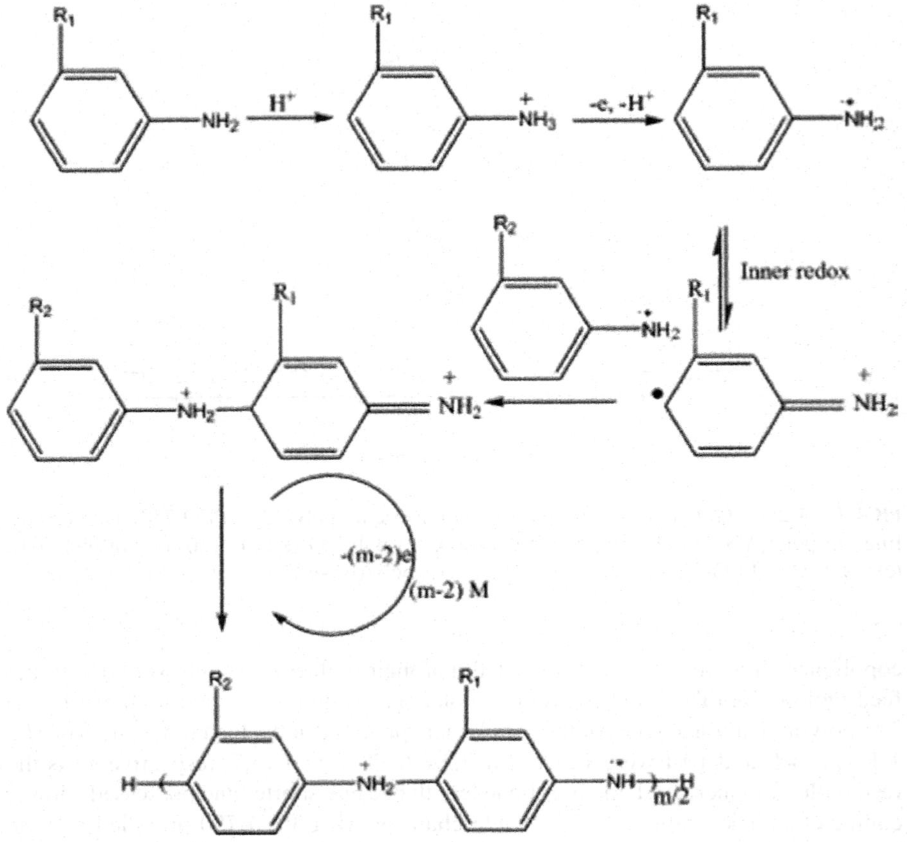

SCHEME 4.2.2 A possible copolymerization mechanism.

Surface Elemental Analysis of the Copolymers. The copolymers were also examined using X-ray photoelectron spectroscopy (XPS). In Table 4.2.3, the surface and bulk elemental stoichiometries of the copolymers prepared from various feed mole ratios are listed. The carbon and oxygen contents obtained by XPS were higher than those provided by the chemical elemental analysis, mostly because of the contamination of the surface (the adsorbed carbon dioxide), which occurs in any XPS analysis.[34] Figure 4.2.9 presents the wide-scale XPS spectra of the HCl-doped and neutralized polymers. They exhibited binding energies around 284, 398, and 533 eV, which can be assigned to C 1s, N 1s, and O 1s, respectively. The peaks due to S 2p and Cl 2p at 167 and 199 eV, respectively, could be hardly identified in the neutralized polymers, indicating that the chloride ions and the tosyol groups have been completely removed. Of course, the chloride peak appeared after HCl doping.

It should be noted that the C, N, and O cores are present in several energetic states which involve different components upon doping or dedoping (see Table 4.2.4). The

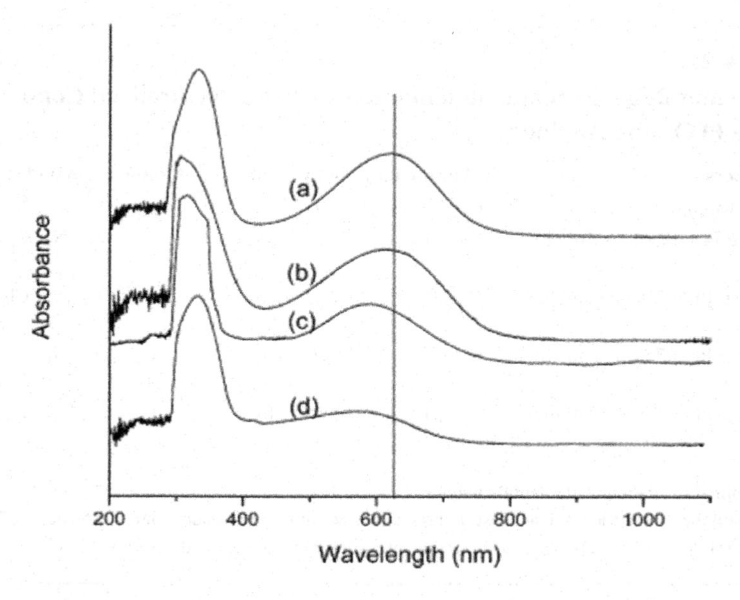

FIGURE 4.2.8 UV–vis spectra of (a) neutralized poly((AP-*g*-PEO-150)-*co*-aniline) (3/1), (b) neutralized poly((AP-*g*-PEO-350)-*co*-aniline) (3/1), (c) neutralized poly((AP-*g*-PEO-750)-*co*-aniline) (3/1), and (d) neutralized poly((AP-*g*-PEO-2000)-*co*-aniline) (3/1) (all in DMF).

TABLE 4.2.2
Solubility of the Various Copolymers in Neutralized State[a]

Neutralized Polymers	NMP	THF	Chloroform	DMF	Water
Polyaniline	S	I	I	S	I
poly(*o*-anisidine)	S	PS	PS	S	I
poly(AP-*g*-PEO-150)	S	PS	PS	S	SS
poly (AP-*g*-PEO-350)	S	S	S	S	PS
poly(AP-*g*-PEO-750)	S	S	S	S	S
poly(AP-*g*-PEO-2000)	S	S	S	S	S
poly((AP-*g*-PEO-750)-*co*-aniline) (10/1)	S	S	S	S	S
poly((AP-*g*-PEO-750)-*co*-aniline) (6/1)	S	S	S	S	S
poly((AP-*g*-PEO-750)-*co*-aniline) (3/1)	S	S	S	S	S
poly((AP-*g*-PEO-750)-*co*-aniline) (1/1)	S	S	PS	S	PS
poly((AP-*g*-PEO-750)-*co*-aniline) (1/2)	S	PS	SS	S	SS
poly((AP-*g*-PEO-150)-*co*-aniline) (3/1)	S	PS	PS	S	SS
poly((AP-*g*-PEO-350)-*co*-aniline) (3/1)	S	S	S	S	PS
poly((AP-*g*-PEO-2000)-*co*-aniline) (3/1)	S	S	S	S	S

[a] Keywords: S = soluble up to 1.0 g/L, stable for 1 month; I = insoluble; SS = slightly soluble, precipitation after 1 day, with a colored supernatant; PS = partially soluble, no precipitation within 1 week, but precipitation by centrifugation at 7000 rpm for 10 min.

TABLE 4.2.3
Surface and Bulk Elemental Stoichiometries of the Neutralized Copolymers of AP-*g*-PEOs and Aniline

Copolymers	Atomic Ratio in the Bulk[a]	Atomic Ratio on the Surface[b]
poly(*o*-anisidine)	$C_{7.38}H_{9.03}O_{1.05}N_{100}Cl_{0.01}S_{0.00}$	$C_{13.13}O_{2.42}N_{1.00}Cl_{0.00}S_{0.00}$
poly(AP-*g*-PEO-750)-*co*-aniline) (6/1)	$C_{31.70}H_{59.87}O_{13.20}N_{1.00}Cl_{0.00}S_{0.00}$	$C_{38.70}O_{19.02}N_{1.00}C_{10.01}S_{0.03}$
poly((AP-*g*-PEO-750)-*co*-aniline) (3/1)	$C_{27.20}H_{49.87}O_{10.91}N_{1.00}Cl_{0.00}S_{0.00}$	$C_{34.57}O_{16.10}N_{1.00}Cl_{0.00}S_{0.00}$
poly((AP-*g*-PEO-750)-*co*-aniline) (1/1)	$C_{19.86}H_{31.29}O_{7.10}N_{1.00}Cl_{0.00}S_{0.01}$	$C_{25.20}O_{10.61}N_{1.00}C_{10.01}S_{0.03}$
poly((AP-*g*-PEO-750)-*co*-aniline) (1/2)	$C_{16.29}H_{24.54}O_{5.20}N_{1.00}Cl_{0.01}S_{0.00}$	$C_{20.70}O_{7.90}N_{1.00}C_{10.01}S_{0.01}$

[a] Determined by chemical elemental analysis.

[b] Each polymer was dried for at least 7 days under vacuum and ground into a powder, followed by compression as a disk. The disk surface concentrations were determined by XPS.

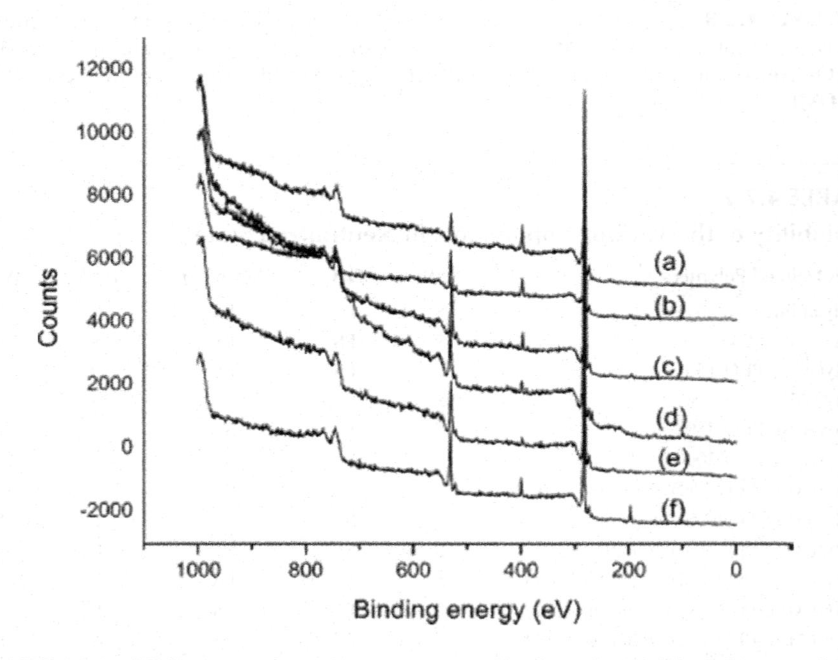

FIGURE 4.2.9 XPS spectra of copolymers: (a) neutralized poly(*o*-anisidine), (b) neutralized poly((AP-*g*-PEO-750)-*co*-aniline) (3/1), (c) neutralized poly((AP-*g*-PEO-750)-*co*-aniline) (1/1), (d) neutralized poly((AP-*g*-PEO-350)-*co*-aniline (1/2), (e) neutralized poly(AP-*g*-PEO-750), and (f) HCl-doped poly((AP-*g*-PEO-750)-*co*-aniline) (3/1).

TABLE 4.2.4
Element Analysis of a Typical HCl-Doped Copolymer Poly((AP-*g*-PEO-750)-*co*-aniline) (3/1) by XPS

Element (Orbital)	Possible Components	Binding Energy (eV)	FWHM (eV)[a]	Component Conc, %
N 1s	=N–	398.20	1.09	2.3
	–NH–	399.41	1.86	53.3
	$-N_{+}-$	400.83	2.68	36.7
	$-N_{+}H-$	402.31	2.95	7.7
Cl 2p	Cl^-	199.00	1.5	100

[a] FWHM denotes the full width at half-maximum of the fitted peak.

oxidation and doping levels of the copolymers can be determined by analyzing the N 1s core levels.[35] In the XPS data analysis, the Shirley background was subtracted before the curve fitting.[36] The AugerScan Demo software was employed for curve fitting. The experimental spectra were fit into components of Gaussian–Lorentzian line shape. The component compositions were determined from the ratios among the area integrals of the fitted peaks. Figure 4.2.10 presents the core XPS spectra of N 1s of HCl-doped poly((AP-*g*-PEO-750)-*co*-aniline) (3/1) and HCl-doped poly(*o*-anisidine). The N 1s core has four fitted components for the core levels of 398.20, 399.41, 400.83, and 402.31 eV of Figure 4.2.10a, listed in Table 4.2.4. They can be assigned to the undoped imine and amine, doped imine, and positively charged amine, with the component concentrations 2.3, 53.3, 36.7, and 7.7%, respectively.[23,35,37] The doping level attained was 44.4%, which is much lower than the 62.4% of the HCl-doped poly(*o*-anisidine). (Its four N 1s core components have the compositions of 2.8, 34.8, 53.6, and 8.8% in Figure 4.2.10b.) The total oxidation level of 39.0% of the HCl-doped poly((AP-*g*-PEO-750)-*co*-aniline) (3/1) is much lower than the 56.4% of HCl-doped poly(o-anisidine). This indicates that in the copolymer the chain conformation produced a lower conjugation because of the higher steric barrier of the PEO side chains. Thus, the doping level was depressed, and the electron delocalization was weakened along the backbone. In this case, a lower conductivity was expected, which is consistent with the UV–vis results presented above. In a series of parallel XPS experiments, a decrease of the doping level was observed with increasing graft side chain length and feed mole ratio AP-*g*-PEO/aniline.

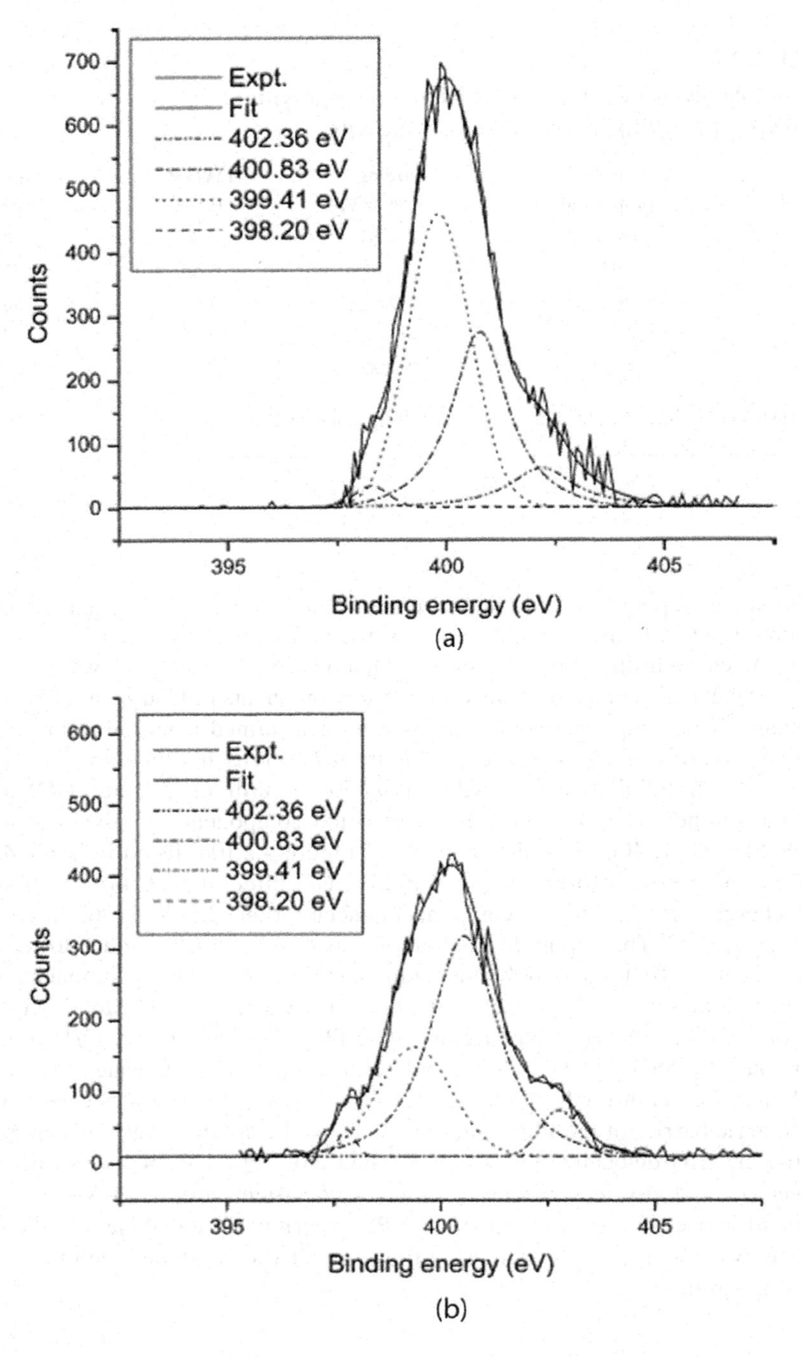

FIGURE 4.2.10 XPS N 1s core level spectra of (a) HCl-doped poly((AP-*g*-PEO-750)-*co*-aniline) (3/1) and (b) HCl-doped poly(*o*-anisidine).

4.2.4 CONCLUSIONS

A simple synthetic strategy to prepare various soluble and processable copolymers of aniline and a macromonomer, aminophenol, grafted with various oligo(oxyethylene) side chains at the 3-position (AP-*g*-PEOs), was developed. The copolymerization and copolymer compositions were changed using various feed mole ratios AP-*g*-PEO/aniline and oligo(oxyethylene) side chain lengths. By increasing the aniline content of the feed mole ratio or decreasing the side chain length of PEO, the copolymerization produced longer chain copolymers with higher conjugation and conductivity. The copolymer of poly((AP-*g*-PEO-750-*co*-aniline) with a feed mole ratio of 3/1 was water-soluble and exhibited a relatively high conductivity (0.12 S/cm).

ACKNOWLEDGMENT

We are thankful to Dr. J. R. Errington for allowing us to use his equipment. Dr. W. N. Richard from Surface and Thin Films Lab of the University of Buffalo carried out the XPS measurements.

REFERENCES AND NOTES

1. MacDiarmid, A. G.; Chiang, J. C.; Richter, A. F.; Epstein, A. J. *Synth. Met.* **1987**, *18,* 285.
2. Oyama, N.; Ohaka, T. *Synth. Met.* **1987**, *18,* 375.
3. McManus, P. M.; Cushman, R. J.; Yang, S. C. *J. Phys. Chem.* **1987**, *91,* 744.
4. Yue, J.; Epstein, A. J. *J. Am. Chem. Soc.* **1990**, *112,* 2800.
5. Leclerc, M.; Guay, J.; Dao, L. H. *J. Electroanal. Chem.* **1988**, *251,* 21. Watanabe, A.; Mori, K.; Iwabuchi, A.; Iwasaki, Y.; Nakamura, Y. *Macromolecules* **1989**, 22, 3521. Wei, Y.; Jang, G.; Chan, C.; Hueh, K.; Harihara, R.; Patel, S. A.; Whitecar, C. K. *J. Phys. Chem.* **1990**, *94,* 7716. Zheng, W.; Levon, K.; Laakso, J.; Osterholm, J. E. *Macromolecules* **1994**, *27,* 7754.
6. Prevost, V.; Petit, A.; Pla, F. *Synth. Met:.* **1999**, *104,* 79. Guo, R.; Barisci, J. N.; Innis, P. C.; Too, C. O.; Wallace, G. G.; Zhou, D. *Synth. Met.* **2000**, *114, 267.*
7. Chevalier, J. W.; Bergeron, J. Y.; Dao, L. H. *Polym. Commun.* **1989**, *30,* 308.
8. Gua, J.; Leclerc, M.; Dao, H. L. *J. Electroanal. Chem.* **1988**, *251,* 31. Guay, J.; Dao, H. L. *Polym. Commun.* **1989**, *30,* 149. Guay, J.; Paynter, R.; Dao, L. H. *Macromolecules* **1990**, *23,* 3598.
9. DeArmitt, C.; Armes, S.; Winter, J.; Uribe, F. A.; Gottesfeld, S.; Mombourquette, C. *Polymer* **1993**, *34,* 158.
10. Shimizu, S.; Saitoh, T.; Yuasa, M.; Yano, K.; Maruyama, T.; Watanae, K. *Synth. Met.* **1997**, *85,* 1337.
11. Uzawa, M.; Zhou, D.; Innis, P.; Wallace, G. G.; Shimizu, S.; Maeda, S. *Synth. Met.* **2000**, *114,* 287.
12. Moon, H.; Park, J. *Solid State Ionics* **1999**, *120,* 1.
13. Chan, H.; Neuendorf, J.; Ng, S.; Wong, P. M. L.; Young, D. J. *Chem. Commun.* **1998**, 1327.
14. Ohno, N.; Wang, H. J.; Yan, H.; Toshima, N. *Polym. J.* **2001**, *33,* 165.
15. Roy, B. C.; Gupta, M. D.; Bhowmik, L.; Ray, J. K. *Synth. Met:.* **1999**, *100,* 233.
16. Roy, B. C.; Gupta, M. D.; Bhowmik, L.; Ray, J. K. *Synth. Met.* **2002**, *130,* 27.
17. Planes, G. A.; Morales, G. M.; Miras, M. C.; Barbero, C. *Synth. Met.* **1998**, *97,* 223.

18. Yin, W.; Ruckenstein, E. *Synth. Met.* **2000**, *108,* 39. Yin, W.; Ruckenstein, E. *Macromolecules* **2000**, 33, 1129. Hua, F.; Ruckenstein, E. *J. Polym. Sci, Part A* **2004**, *42,* 1429.
19. Yue, J.; Wang, Z. H.; Cromack, K. R.; Epstein, A. J.; MacDiarmid, A. G. *J. Am. Chem. Soc.* **1991**, *113,* 2665. Yue, J.; Gordon, G.; Epstein, A. J. *Polymer* **1992**, 33, 4409. Wei, X. L.; Wang, Y. Z.; Long, S. M.; Bobeczko, C.; Epstein, A. J. *J. Am. Chem. Soc.* **1996**, *118,* 2545.
20. Wei, Y.; Focke, W. W.; Wnek, G. E.; Ray, A.; MacDiarmid, A. G. *J. Phys. Chem.* **1989**, *93,* 495.
21. Ito, S.; Murata, K.; Teshima, S.; Aizawa, R.; Asako, Y.; Takahashi, K.; Hoffman, B. M. *Synth. Met.* **1998**, *96,* 161.
22. Yamaguchi, I.; Yasuda, T.; Yamamoto, T. *J. Polym. Sci., Part A: Polym. Chem.* **2001**, *39,* 3137. Moon, D. K.; Ezuka, M.; Maruyama, T.; Osakada, K.; Yamamoto, T. *Macromolecules* **1993**, *26,* 364.
23. Wang, P.; Tan, K. L.; Zhang, F.; Kang, E. T.; Neoh, K. G. *Chem. Mater.* **2001**, *13,* 581.
24. Moon, H. S.; Park, J. K. *Macromolecules* **1998**, *31,* 6461.
25. Epstein, A. J.; MacDiarmid, A. G. Mol. *Cryst. Liq. Cryst.* **1988**, *160,* 165. Mav, I.; Zigon, M. *Polym. Int.* **2002**, *51,* 1072. Paterno, L. G.; Mattoso, L. H. C. *J. Appl. Polym. Sci.* **2002**, *83,* 1309. Raposo, M.; Oliveira, O. N. *Langmuir* **2002**, *18,* 6866.
26. Rivas, B. L.; Sanchez, C. O.; Bernede, J. C.; Mollinie, P. *Polym. Bull. (Berlin)* **2002**, *49,* 257.
27. Hua, F.; Ruckenstein, E. *Macromolecules* **2003**, *36,* 9971. Hua, F.; Ruckenstein, E. *Langmuir* **2004**, *20,* 3954.
28. Yamada, K.; Ito, A.; Iwamoto, N.; Haraguchi, T.; Kajiyama, T. *Polym. J.* **2000**, *32,* 222.
29. Lim, V. W. L.; Kang, E. T.; Neoh, K. G. *Synth. Met.* **2001**, *119,* 261.
30. Armand, M. *Solid State Ionics* **1994**, *69,* 309. Novak, P.; Mueller, K.; Santhanam, K. S. V.; Hass, O. *Chem. Rev.* **1997**, *97,* 207.
31. Lauter, U.; Meyer, W. H.; Wegner, G. *Macromolecules* **1997**, *30,* 2092. Bruce, P. G. *Solid State Electrochemistry* VCH Publishers: New York, 1994.
32. Brady, S. F.; Hirschmann, R.; Veber, D. F.: *J. Org. Chem.* **1977**, *42,* 143. Roos, E. C.; Bernabe, P.; Hiemstra, H.; Speckamp, W. N.; Kaptein, B.; Boesten, W. H. *J. Org. Chem.* **1995**, *60,* 1733. Debenham, J. S.; Fraser-Reid, B. *J. Org. Chem.* **1996**, *61,* 432. Gordon, K. H.; Balasubramanian, S. *Org Lett.* **2001**, 3, 53. Reddy, P. G.; Pratap, T. V.; Kumar, G. D. K.; Mohanty, S. K.; Baskaran, S. ***Eu.*** *J. Org Chem.* **2002**, 3740. Ramesh, C.; Mahender, G.; Ravindranath, N.; Das, B. *Tetrahedron* **2003**, *59,* 1049.
33. Ouchi, M.; Inoue, Y.; Nagamune, S.; Nagamura, S.; Wada, K.; Hakushi, T. *Bull. Chem. Jpn.* **1990**, *63,* 1260.
34. Kumar, S. N.; Gaillard, F.; Bouyssoux, G.; Sartre, A. *Synth. Met.* **1990**, *36,* 111.
35. Wei, X. L.; Fahlman, M.; Epstein, A. J. *Macromolecules* **1999**, *32,* 3114.
36. Shirley, D. A. *Phys. Rev. B* **1972**, *123,* 4709. Briggs, D., Seah, M. P., Eds. *Practical Surface Analysis;* John Wiley: New York, 1990; Vol. 1.
37. Kimartin, P. A.; Wright, G. A. *Synth. Met.* **1997**, *88,* 163.

4.3 Water-Soluble Conducting Poly(ethylene oxide)-Grafted Polydiphenylamine Synthesis through a "Graft Onto" Process*

Fengjun Hua and Eli Ruckenstein

Department of Chemical Engineering, State University of New York at Buffalo, Buffalo, New York 14260

ABSTRACT A synthesis method to prepare a fully grafted polydiphenylamine (PDPA) with poly(ethylene oxide) (PEO) through a "graft onto" process is reported. Three graft copolymers, PDPA-*g*-PEO-350, PDPA-*g*-PEO-750, and PDPA-*g*-PEO-2000, were obtained by substituting the tosyloyl end group of tosylate PEO with an amine-functionalized PDPA. The latter polymer was prepared via an acid-mediated chemical polymerization of diphenylamine. These graft copolymers became water-soluble when the grafted PEO molecular weight (M_n) was above 750. Furthermore, the copolymers can be oxidized, in aqueous acidic solutions in the presence of air, to produce PDPA in a conducting state, containing, depending on the degree of oxidation, the *N,N′*-diphenylbenzidine radical cation ($DPB^{\bullet +}$) or the *N,N′*-phenylbenzidine dication (DPB^{2+}). The conductivities of the HCl-doped graft copolymers were found to be in the range 10^{-1}–10^{-4} S/cm, dependent on the grafted PEO side-chain length. As expected, the PDPA-*g*-PEO copolymers with longer PEO side chains possessed lower conductivities. Furthermore, for the HCl-doped PDPA-*g*-PEO-2000, the crystallization of the PEO-2000 side chains appears to be responsible for the low conductivity. Its conductivity could be raised 10-fold upon heating at 55°C, above the PEO-2000 melting point (50°C).

* *Macromolecules* 2003, 36, 9971–9978.

4.3.1 INTRODUCTION

In the past decades, polyaniline has attracted increasing attention, owing to its high electronic conductivity and environmental stability.[1–4] This conductive polymer can be easily prepared by electrochemical or chemical oxidation of aniline in aqueous acidic media. However, its poor processability and solubility in organic compounds because of the stiffness of its backbone and the hydrogen-bonding interaction of the inter-amino moieties of adjacent chains have stimulated attempts for their improvement. Numerous investigations have been carried out regarding its processability and solubility by introducing various side chains into the rigid PANI backbone. The PANI combined with alkyl,[5] alkoxy,[6] benzyl,[7] and aryl[8] substituents on the aromatic ring or nitrogen has been successfully prepared. The grafted side chains increased the solubility of PANI by increasing the entropy of dissolution and decreasing the interactions between the main chains.[9] Sometimes, the densely grafted PANI with side chains looks like a comb or a brush.[10–12] The microstructure of the brushed PANI is expected to provide some unique physical properties and special molecular conformations.[11]

Poly(ethylene oxide) (PEO) has been frequently employed, owing to its unique physical and biochemical properties.[13,14] The incorporation of PEO into a PANI backbone can endow the modified PANI with a number of properties, including higher water solubility, more facile processability, and biocompatibility.[15,16] Recently, it was found a most useful application in secondary lithium polymer batteries.[17,18] PEO side chains were grafted on PANI, and this allowed lithium salts to dissolve in the polymer matrix.[19] In general, the modification by grafting could be carried out using three synthetic routes: "graft onto" (coupling attachment of the side chains to the main chain), "graft from" (direct grafting polymerization of the monomer to form side chains from active points on the backbone), and "graft through" (homopolymerization of macromonomers).[20–22] Moon suggested a typical grafting through route by the oxidative homopolymerization of an aniline substituted with PEO.[20] Yamaguchi successfully prepared PANI grafted with PEO through a "graft from" process, in which a ring-opening polymerization of an epoxide proceeded on active anionic nitrogens of the PANI backbone.[21] Wang et al.[22] grafted PEO chains to the active anionic nitrogens of PANI via direct attachment under very strict experimental conditions. A lower grafting degree, below 30%, was, however, reached through this "graft onto" process.

Polydiphenylamine (PDPA) was prepared mostly by electrochemical polymerization of diphenylamine, which because of the phenyl substituent on nitrogen, occurred via a 4,4′ C–C phenyl–phenyl coupling mechanism. This mechanism differs from the preferential head-to-tail polymerization of aniline.[8,23–27] Thus, the resulting structure is intermediate between that of PANI and that of poly(*p*-phenylene) and allows a more facile dissolution of its dedoped form in organic compounds, such as tetrahydrofuran or chloroform, etc. In contrast, the dedoped form of PANI can dissolve only in some special solvents, such as *N*-methyl-2-pyrrolidinone (NMP). Recently, Wen et al. synthesized PDPA by a chemical method in a 3 N methanesulfonic acid solution.[28]

In a previous paper we reported the synthesis of polysulfonic diphenylamine salt and of its water-soluble PEO block copolymer.[29] In the present paper, we suggest a simple synthesis strategy for polydiphenylamine-*graft*-PEO (PDPA-*g*-PEO), which is also expected to be water-soluble. First, diphenylamine was polymerized in a highly concentrated HCl aqueous solution (3 N) using ammonium persulfate as initiator. PDPA was neutralized with a 1 N ammonium aqueous solution, followed by reduction with hydrazine to produce a secondary amine-functionalized PDPA. Second, tosylate-ended PEO chains were grafted onto the nitrogens of the PDPA backbone through a "graft onto" process via a nucleophilic substitution mechanism. The chemical structures of various prepolymers and graft copolymers with various PEO chain lengths were characterized by FTIR, ^{1}H or ^{13}C NMR, UV–vis, and GPC, and their electronic conductivities were determined.

4.3.2 EXPERIMENTAL SECTION

Materials. Diphenylamine (99%), hydrochloric acid (37 wt%), ammonium persulfate (98 wt%), hydrazine (98 wt%), *p*-toluenesulfonyl chloride (99+ wt%), and potassium *tert*-butylate (>95 wt%) were purchased from Aldrich and used without further purification. Poly(ethylene oxide) methyl ethers, PEO-350 (M_n = 350), PEO-750 (M_n = 750), and PEO-2000 (M_n = 2000) were also purchased from Aldrich and used without further purification. The solvents, such as acetone, diethyl ether, and tetrahydrofuran (THF) from Aldrich and chloroform from Fisher, were of HPLC purity.

Preparation of Polydiphenylamine. Polydiphenylamine (PDPA) was prepared by the chemical oxidation method in a 3 N aqueous HCl solution using ammonium persulfate as initiator. The synthesis method is sketched in step 1 of Scheme 4.3.1. Diphenylamine (6.0 g, 0.0354 mol), 120 mL of 3 N HCl solution, and 150 mL of ethanol were introduced into a 500 mL round-bottom flask. The system was subjected to intensive magnetic stirring for a few minutes and then cooled to 5°C in an ice–water bath. Ammonium persulfate (3.01 g, 0.0132 mol) was dissolved in 15 mL of 3 N HCl solution and added dropwise into the flask. The color of the reaction solution changed from transparent to green, indicating the start of polymerization. The system was subjected to intensive stirring for 16 h. A green precipitate, PDPA, was separated using a supercentrifuge (Dupont, RC-5) at 3500 rpm for 30 min. The PDPA particles were washed with ethanol/water mixtures (3/1, v/v) a few times (30 mL each time), until no green supernatant was identified, and finally dried under vacuum at room temperature for at least 24 h. The yield was 35%.

Neutralization and Reduction of PDPA. To prepare an amine-functionalized PDPA with secondary amine moieties on the PDPA backbone, the HCl-doped PDPA was neutralized after polymerization with a 1 N ammonium aqueous solution, followed by reduction with hydrazine as shown in steps 2 and 3 of Scheme 4.3.1. HCl-doped PDPA (1.8 g), 5 mL of ethanol, and 20 mL of 1 N NH_4OH solution were introduced into a 50 mL round-bottom flask and mixed with intensive stirring. The neutralization has taken place at room temperature for 18 h. The neutralized PDPA was separated by centrifugation at 3500 rpm for 30 min. The precipitate was

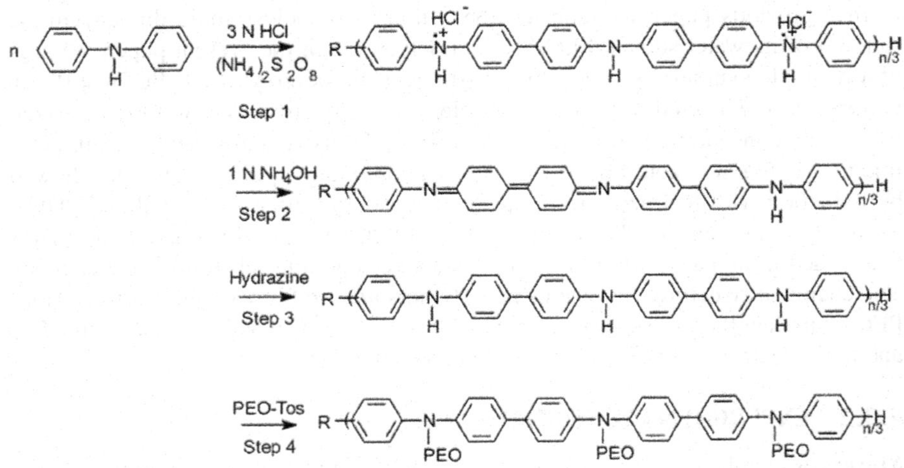

SCHEME 4.3.1 Synthesis strategy of polydiphenylamine grafted with poly(ethylene oxide) PDPA-*g*-PEO through a "graft onto" process.

washed a few times with ethanol/water mixtures (3/1) until no colored supernatant was observed. The dark purple PDPA particles were dried under vacuum at room temperature for at least 24 h.

The reduction was carried out in an inert nitrogen atmosphere at room temperature. The dried and neutralized PDPA (1.5 g), hydrazine (2.5 g), and 25 mL of ethanol were introduced into a 50 mL flask and mixed with intensive stirring. Nitrogen was bubbled through the mixture for half hour to remove the dissolved oxygen. The color of the reaction system changed from dark purple to light red. The reduction reaction was continued for 36 h. After reaction, 2.5 mL of water was added to the reaction system to precipitate the polymer. The reduced PDPA was separated by centrifugation at 3500 rpm for 30 min and washed with ethanol/water (5/1, v/v), followed by drying under vacuum at room temperature for at least 24 h. The polymer was protected under a nitrogen atmosphere before its further use. The yields for neutralization and reduction were 90% and 70%, respectively.

Graft Reaction of PEO and PDPA through a Graft onto Process. Hydroxyl-ended PEO was first converted to tosylate-functionalized PEO as described in refs 29–31. The tosylates can be substituted by secondary amines on a reduced PDPA backbone via a nucleophilic substitution mechanism.[29] PEO chains fully grafted the PDPA backbone when the amount of PEO-Tos was 3 times that of the amine moieties of the reduced PDPA. All operations were carried out under a nitrogen atmosphere. The reduced PDPA (0.52 g, 0.00307 mol of amine moieties), PEO-Tos-350 (3.23 g, 0.00928 mol), potassium *tert*-butylate (0.9 g), and 20 mL of dry THF were introduced into a 100 mL round-bottom flask and mixed with intensive stirring for a few minutes. The system was kept at room temperature for 6 days. Then, 20 mL of dry THF was added to produce precipitation. The precipitate was removed by centrifugation (3500 rpm, 30 min). The remaining THF solution was concentrated using a rotary evaporater. The concentrated THF solution was added to a 3.5-fold volume of diethyl ether to precipitate red particles. The thus-obtained PDPA-*g*-PEO was further purified through dissolution–precipitation in THF/diethyl ether (1/3.5, v/v) and dried under vacuum at room temperature for at least 24 h. PDPA-*g*-PEO-350 was obtained from PEO-Tos-350, PDPA-*g*-PEO-750 from PEO-Tos-750, and PDPA-*g*-PEO-2000 from PEO-Tos-2000. The yields in all three cases were about 65%.

Oxidation of the Graft Copolymers. The graft copolymers in the reduced state were oxidized in air to produce oxidized states.[32,33] PDPA-*g*-PEO-350 (0.16 g), PDPA-*g*-PEO-750 (0.22 g), and PDPA-*g*-PEO-2000 (0.84 g) were ground into fine powders and each dispersed into 10 mL of 3 N HCl solutions in 50 mL centrifuge tubes. Compressed air was bubbled through the solutions in order to oxidize the graft copolymers. The bubbling at room temperature lasted for at least 72 h. The color changed from red to green. A small sample was taken out periodically and its UV–vis spectrum determined. When the UV–vis spectrum displayed no change, the water was removed using a rotary evaporator under vacuum. The HCl-doped green graft copolymers were further dried under vacuum at room temperature for at least 24 h.

Characterization. Proton (^{1}H) and carbon (^{13}C) NMR, UV–vis absorption, and FTIR measurements were carried out on a 400 MHz InOvA-400, a Thermo Spectronic Genesys-6, and a Perkin-Elmer-FTIR 1760, respectively. The NMR solutions were prepared by dissolving the polymer in deuterated water or deuterated chloroform. Gel permeation chromatography (GPC, Waters) was used to obtain the molecular weights and the polydispersity indexes of the prepolymers and the graft copolymers on the basis of a polystyrene calibration curve. The GPC was equipped with three 30 cm long columns filled with Waters Styragel, a Waters HPLC 515 pump, and a Waters 410 RI detector. The GPC measurements were carried out using THF (containing 0.5 wt% triethylamine)[34] as eluent at 37°C, with a flow rate of 1.0 mL/min and 1.0 cm/min chart speed. The elemental analysis of the polymer samples was carried out on a Perkin-Elmer model 2400 C, H, N, analyzer. The chlorine and sulfur contents were determined by the oxygen flask method. The room temperature conductivities of the compressed pellets of various prepolymers and graft copolymers were determined using the conventional four-point probe method. The temperature dependence of the conductivity was determined using compressed pellets in a homemade heating stage possessing a temperature controller. Every determination was recorded after the pellet film was heated and stabilized for at least 5 min at the selected temperature. Differential scanning calorimetery (DSC, Perkin-Elmer, DSC-7) was carried out using a heating rate of 15 K/min in the range from −100°C to 200°C. The XRD of the polymer particles was recorded on a SIEMENS, D500, X-ray diffractometer, operated at 30 MA and 40 KVP MAX.

4.3.3 RESULTS AND DISCUSSION

Preparation of PDPA-*g*-PEO Copolymers. The synthesis strategy of graft copolymers is schematically presented in Scheme 4.3.1. The grafting process was carried out through a "graft onto" process, i.e., the tosylate-functionalized PEO reacted with the amine moieties on the PDPA backbone. The fully reduced PDPA was used in this process in order to avoid side reactions such as the cross-linking.[35] A high excess of PEO-Tos compared to that of the secondary amine moieties of PDPA ([tosylate]/[NH] = 3/1) ensured the generation of a fully grafted PDPA. The tosylate PEO was obtained via the esterification of HO-ended PEO with tosylol chloride, as described in ref. 31.

Figure 4.3.1 presents the FTIR spectra of PEO-2000, PEO-2000-Tos, neutralized PDPA, reduced PDPA, and neutralized PDPA-*g*-PEO-2000. The neutralized PDPA possesses the following characteristic bands: 3420, 1595, 1494, and 754 cm^{-1}, which can be assigned to –NH stretching vibration, stretching vibrations of quinoid ring, phenyl hydrogen, and C–H plane vibration of the para-substituted aromatic ring, respectively. The reduced PDPA has a very weak absorption at 1595 cm^{-1}, indicating that most quinoid rings disappeared because of reduction and acquired the structure presented in step 3 of Scheme 4.3.1. The spectrum of the neutralized PDPA-*g*-PEO-2000 displays a combination of the bands of neutralized PDPA and PEO-2000. The characteristic bands at 1109, 1595, and 1494 cm^{-1} can be assigned to the C–O–C stretching vibration of oxyethylene of the PEO side chain and the quinoid ring and phenyl group of the PDPA backbone, respectively. Furthermore, the characteristic –NH stretching vibration of PDPA is no longer present, owing to its replacement by the C–N connection between PEO and the PDPA backbone. The absence of the peak in the graft copolymer due to –NH at 3420 cm^{-1} demonstrates that a full substitution with PEO on the PDPA backbone has taken place. In addition, the characteristic absorption at 1182 cm^{-1}, which can be assigned to the stretching vibrations of C–O in S–O–C of PEO-Tos-2000, disappeared because the tosylate has been removed through the substitution reaction (step 4 of Scheme 4.3.1).

Figure 4.3.2 presents the UV–vis spectra in THF of HCl-doped PDPA, neutralized PDPA, reduced PDPA, PDPA-*g*-PEO-2000, and HCl-doped PDPA-*g*-PEO-2000. For HCl-doped PDPA, the two absorption bands at 320 and 630 nm can be assigned to the π–π^* transition of the benzenoid ring and polaron structure, respectively.[6–8] After neutralization, the band at 630 nm experienced a blue shift to 520 nm, which can be assigned to an exciton transition, consistent with the FTIR results. However, in the spectrum of reduced PDPA, the band at 630 nm fully disappeared whereas the band at 320 nm due to the benzenoid ring remained. Before oxidative doping, the PDPA-*g*-PEO-2000 exhibited a spectrum similar to that of the reduced PDPA, indicating that in reduced state the PDPA and the graft copolymer are nonconductive. After oxidation, the spectrum of HCl-doped PDPA-*g*-PEO-2000 exhibited a band at 610 nm due to the generation of *N, N*-diphenylbenzidine radical cation ($DPB^{\bullet+}$), which constitutes the polaron form of the PDPA unit.[6] This issue will be discussed in detail later.

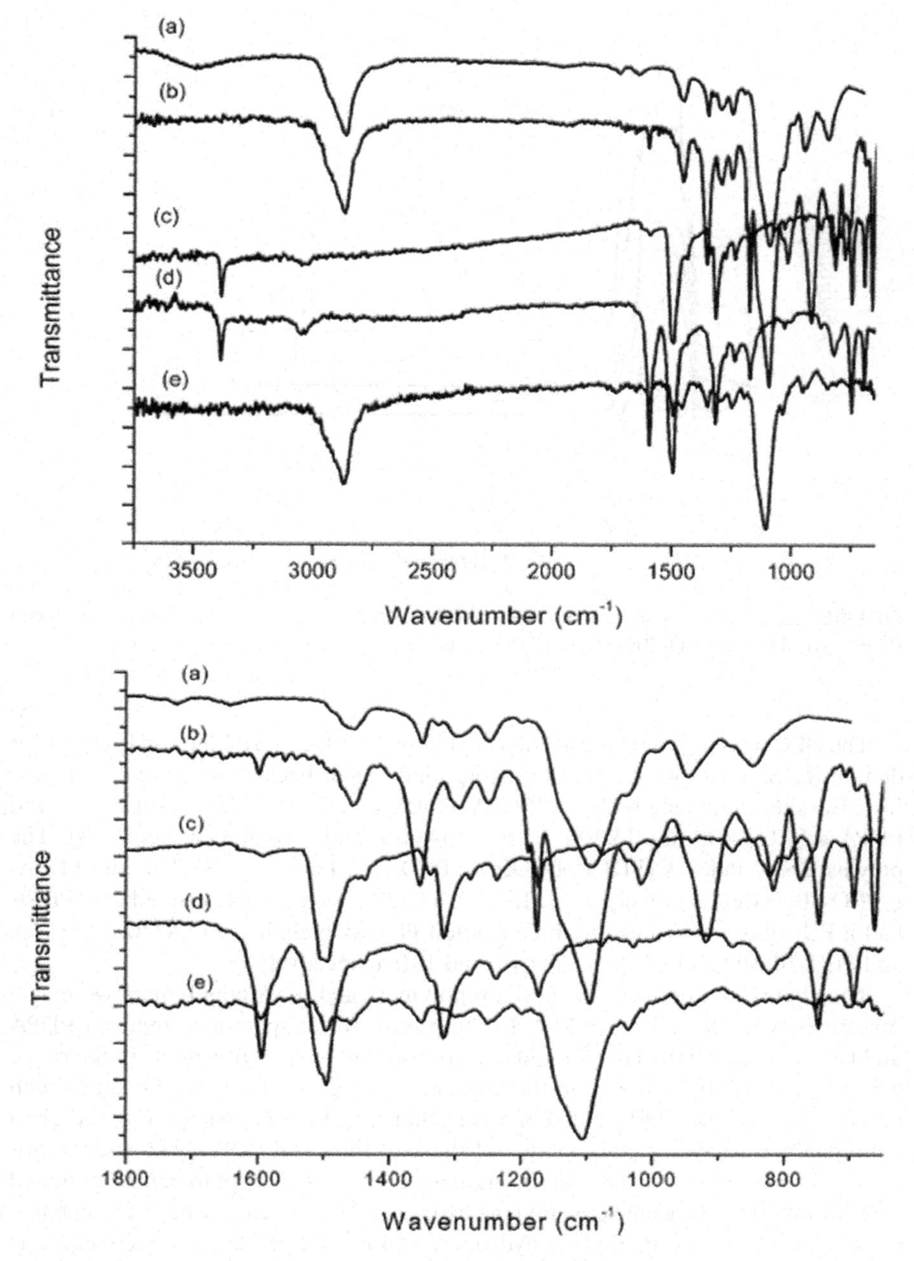

FIGURE 4.3.1 FTIR spectra of (a) PEO-2000, (b) PEO-Tos-2000, (c) reduced PDPA, (d) neutralized PDPA, and (e) neutralized PDPA-*g*-PEO-2000.

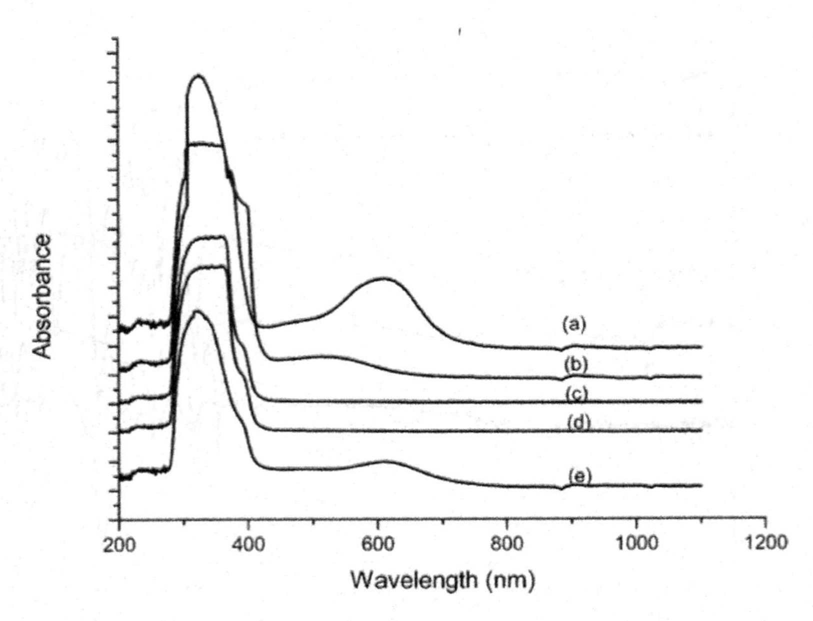

FIGURE 4.3.2 UV–vis spectra of (a) HCl-doped PDPA, (b) neutralized PDPA, (c) reduced PDPA, (d) PDPA-*g*-PEO-2000, and (e) HCl-doped PDPA-*g*-PEO-2000.

The GPC traces of the three grafted copolymers and reduced PDPA in Figure 4.3.3 demonstrated, as expected, that the molecular weight became larger with increasing PEO side chain length (M_ns of PDPA, PDPA-*g*-PEO-350, PDPA-*g*-PEO-750, and PDPA-*g*-PEO-2000 are 13,400, 35,600, 64,000, and 1,76,500, respectively). The polydispersity indexes (PDIs) of PDPA-*g*-PEO-350, PDPA-*g*-PEO-750, and PDPA-*g*-PEO-2000 were relatively large, 1.9, 2.3, and 2.6, respectively. The reduced PDPA had a PDI of 3.0, whereas the three grafted PEO side chains (PEO-350, PEO-750, and PEO-2000) had PDIs of 1.12, 1.10, and 1.10, respectively.

The chemical structures of these prepolymers and graft copolymers were also investigated by ^{1}H and ^{13}C NMR. In the proton NMR spectra of reduced PDPA and HCl-doped PDPA in deuterated chloroform, the wide multisignals in the range 6.8–7.8 ppm can be assigned to the aromatic hydrogens of the double-substituted phenyl rings of the PDPA backbone (see Figure 4.3.4). The spectra of the graft copolymers displayed a combination of those of PEO and PDPA. The wide multisignals in the range 6.8–7.8 and the signals at 3.60 and 3.37 ppm can be assigned to the aromatic hydrogens, oxyethylene hydrogens of ethylene oxide (EO) repeated units, and the end oxymethylene hydrogens of PEO side chains, respectively. The signal at 4.05 ppm, which can be assigned to the oxymethylene hydrogens connected with the tosylate, disappeared after substitution, being replaced by a new signal at 3.50 ppm, which can be assigned to the methylene hydrogens connected with the nitrogens of the PDPA backbone, indicating the connections between PEO side chains and PDPA backbone. Furthermore, the typical ^{13}C NMR spectra of

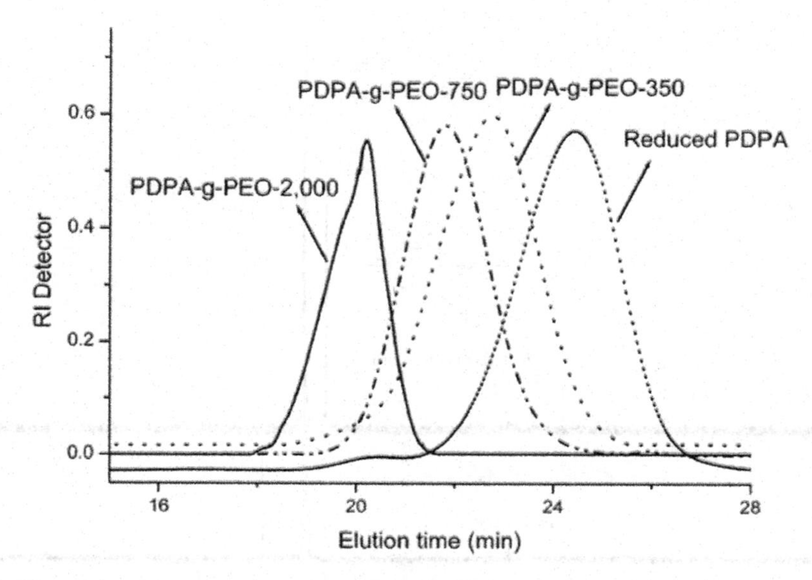

FIGURE 4.3.3 GPC traces of the reduced PDPA and various graft copolymers.

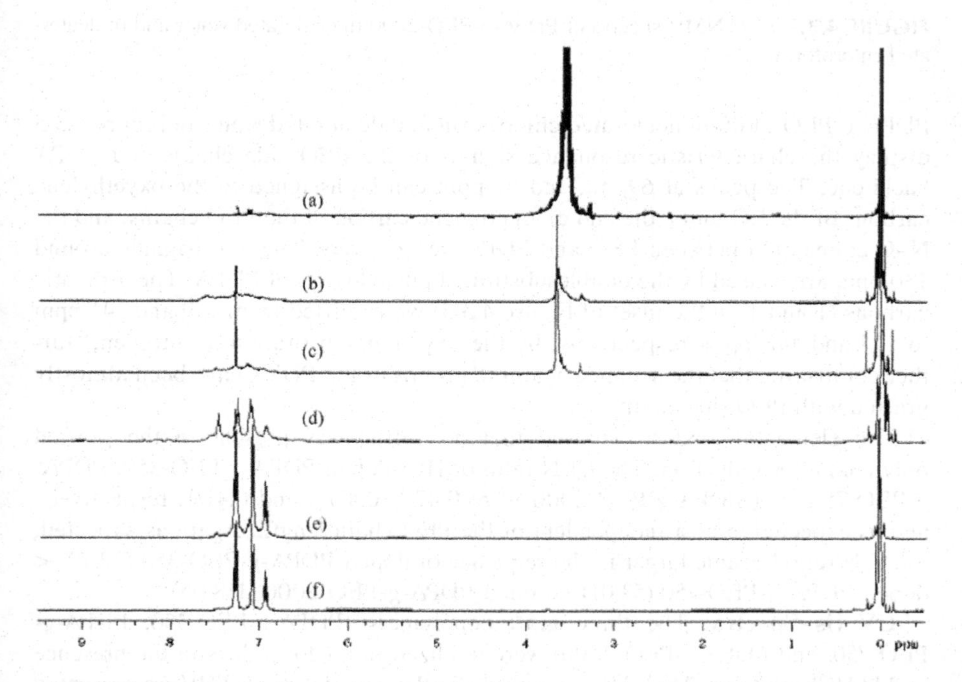

FIGURE 4.3.4 ^{1}H NMR spectra of DPA monomer, (a) PDPA-*g*-PEO-2000, (b) PDPA-*g*-PEO-750, (c) PDPA-*g*-PEO-350, (d) reduced PDPA, (e) HCl-doped PDPA, and (f) DPA monomer.

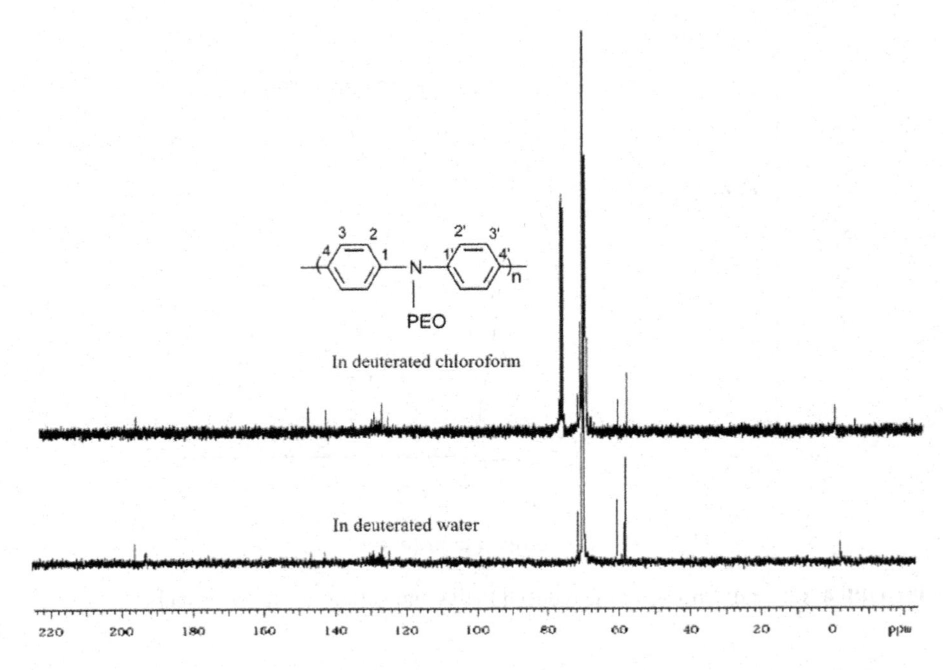

FIGURE 4.3.5 ^{13}C NMR spectra of PDPA-*g*-PEO-2000 in deuterated water and in deuterated chloroform.

PDPA-*g*-PEO-2000 in deuterated chloroform and deuterated water in Figure 4.3.5 display the characteristic resonance signals of the PEO side chains and PDPA backbone. The peaks at 69, 58, and 61 ppm can be assigned to the oxyethylene carbons of the EO units, the end oxymethylene carbon of the PEO chains, and the N–C connection between PEO and PDPA, respectively. The multisignals around 130 ppm are caused by the doublesubstituted phenyl rings of PDPA. The aromatic carbons (1 and 1′ in the inset of Figure 4.3.5) were shifted from 139 and 141 ppm to 142 and 148 ppm, respectively, by the oxyalkyl substitution at nitrogen,[5] further confirming that the secondary amine groups of the PDPA have been almostly grafted with PEO side chains.

The elemental analysis showed that no sulfur was present in the grafted polymers after reduction. The Cl/N ratio of HCl-doped PDPA-*g*-PEO-350, PDPA-*g*-PEO-750, and PDPA-*g*-PEO-2000 were 0.47/1, 0.43/1, and 0.41/1, respectively, and as expected almost independent of the PEO chain length. Again as expected, the C/N ratio became larger in the sequence of doped-PDPA-*g*-PEO-350 (32.12) < doped-PDPA-*g*-PEO-750 (53.01) < doped-PDPA-*g*-PEO-2000 (118.05).

UV–vis Spectra. The three graft copolymers (PDPA-*g*-PEO-350, PDPA-*g*-PEO-750, and PDPA-*g*-PEO-2000) were oxidized in air for 3 days in the presence of 3 N HCl solutions. Their UV–vis spectra in the doped state in THF are presented in Figure 4.3.6. The two absorptions at about 320 and 620 nm can be assigned to the π–π^* transition of the benzenoid rings and to the polaron form of the PDPA

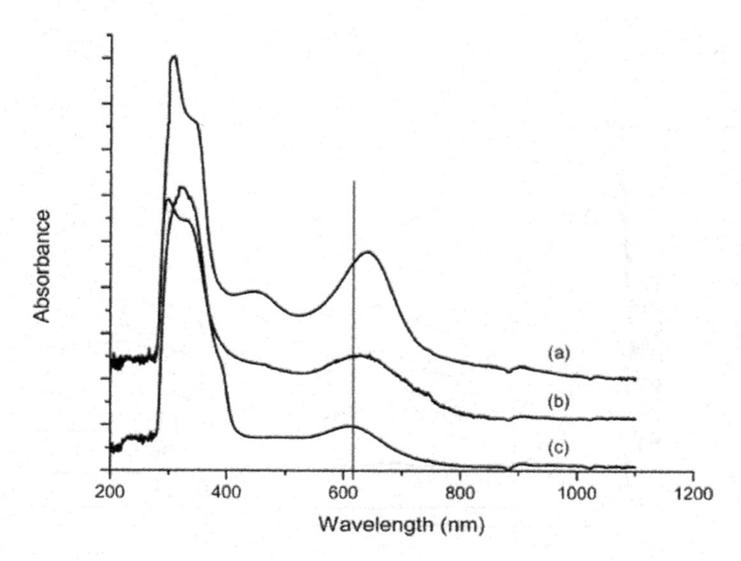

FIGURE 4.3.6 UV–vis spectra of the graft copolymers with various PEO chain lengths in THF: (a) HCl-doped PDPA-*g*-PEO-350, (b) HCl-doped PDPA-*g*-PEO-750, and (c) HCl-doped PDPA-*g*-PEO-2000.

unit, as also observed for HCl-doped PDPA in a THF solution mentioned before. The latter absorption peak exhibited a blue shift, which increased with increasing side-chain length. This indicates that the planar conformation of the PDPA backbone became increasingly nonplanar with longer side chains because the longer PEO side chains caused greater torsional twists, thus decreasing the effective conjugation of the PDPA backbone.

As mentioned in the first section, the reduced graft copolymers are in a nonconductive state, and they should be oxidized to become conductive. They have been oxidized with air, as described in Experimental Section, and the oxidation process was followed by UV–vis. A series of UV–vis spectra of a PDPA-*g*-PEO-2000 graft copolymer are presented in Figure 4.3.7a. An absorption peak at about 610 nm was generated gradually through the bubbling of air for 8 h. This peak can be assigned to the polaron *N*, *N′*-diphenylbenzidine radical cation ($DPB^{\bullet+}$), induced in the benzenoid ring because of the removal by oxidation of one electron belonging to the nitrogen of a DPA unit. The peak intensity increased with increasing oxidation time because of increasing number of polarons. After 3 days, the UV–vis spectrum did no longer change. A similar behavior occurred when the oxidation process was carried out in the presence of 80 wt% aqueous acetic acid solution. However, if the latter system was subjected to oxidation for 1 week, the peak at about 610 nm was weakened and a new absorption peak above 700 nm appeared, which can be assigned to the bipolaron diphenylbenzidine dication (DPB^{2+} (see Figure 4.3.7b)). After 1 month of oxidation, only the peak above 700 nm remained, indicating that the polaron form of the PDPA unit was converted to the higher oxidative state of the bipolaron. However, the conductivities of the above two

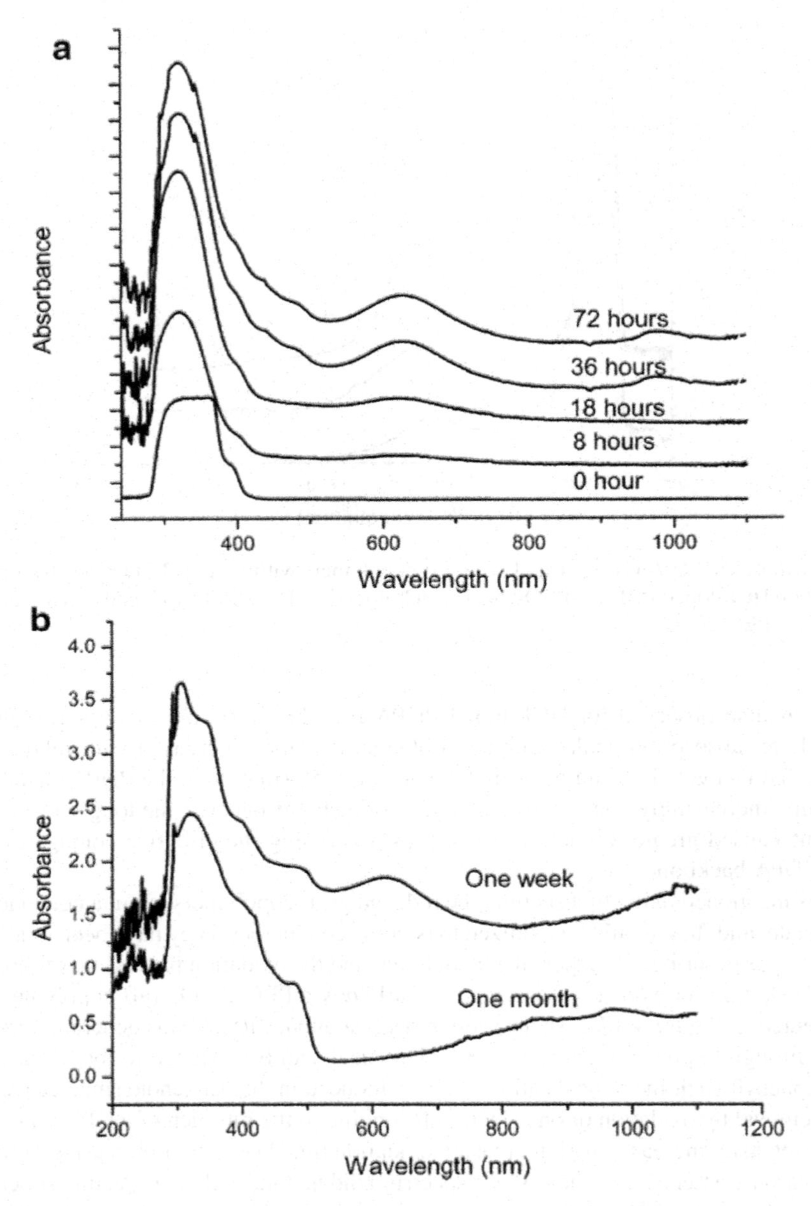

FIGURE 4.3.7 UV–vis spectra of the PDPA-*g*-PEO-2000 against the oxidation time in air in the presence of 3 N HCl solution (a) or 80% aqueous acetic acid solution (b).

overoxidized systems were not improved as happened with PANI.[32,33,37] The mechanism for the oxidative doping described above is detailed in Scheme 4.3.2.

Solubility. The PDPA grafted with amphiphilic PEO side chains is expected to increase the water solubility of PDPA. The three grafted copolymers were introduced into water to reach concentrations of 1.0 g/L. The HCl-doped PDPA-*g*-PEO-2000

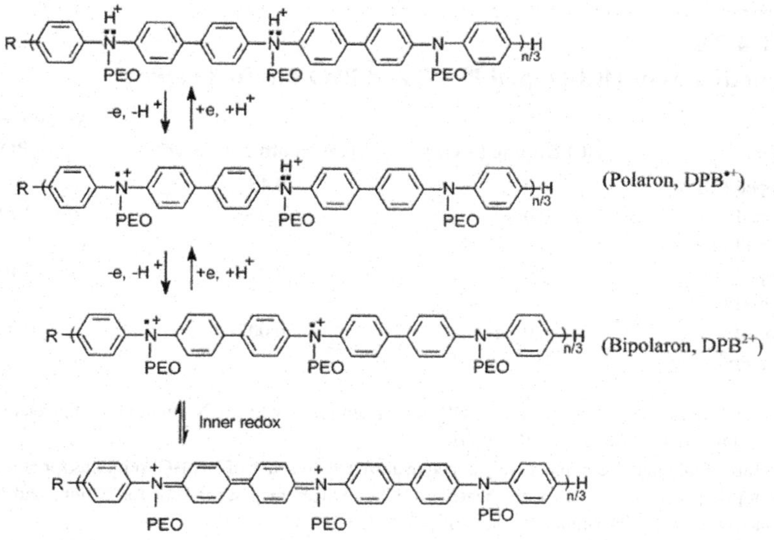

SCHEME 4.3.2 Reduction by hydrazine and oxidation in air in the presence of an aqueous acidic solution.

water solution remained stable under ambient conditions for several months. However, the HCl-doped PDPA-*g*-PEO-350 precipitated from solution after 1 day with a colored supernatant which exhibited absorptions in UV–vis as in Figure 4.3.6. The HCl-doped PDPA-*g*-PEO-750 could be precipitated under centrifugation at 7000 rpm for 10 min after 1 week. The solubility of the copolymers in the organic solvents, THF, chloroform, dichloromethane, ethanol, and NMP was also determined, and the results are listed in Table 4.3.1.

TABLE 4.3.1
Solubility of PDPAs and Graft Copolymers in Various Solvents[a]

	Solvents					
Polymer	NMP	THF	Chloroform	Ethanol	Dichloromethane	Water
HCl-doped PDPA	S	S	S	SS	PS	I
Neutralized PDPA	S	S	S	PS	S	I
PDPA-*g*-PEO-350[b]	S	S	S	S	S	SS
PDPA-*g*-PEO-750[b]	S	S	S	S	S	PS
PDPA-*g*-PEO-2000[b]	S	S	S	S	S	S[c]

[a] Keywords: S = soluble up to 1.0 g/L, stable for 1 month; I = insoluble; SS = slightly soluble, precipitation after 1 day, with a colored supernatant; PS = partially soluble, not precipitation within 1 week, but precipitation by centrifugation at 7000 rpm for 10 min after 1 week.

[b] The graft copolymers were oxidatively doped in air in the presence of a 3 N HCl solution for at least 72 h.

[c] The dissolution process was carried out upon heating to 40°C with intensive stirring for at least 2 min.

TABLE 4.3.2
Conductivities of HCl-Doped PDPA and PEO-Grafted PDPAs

Polymers	HCl Doping (S/cm)	Heat Treatment[c] (S/cm)	Volume Fraction of PDPA[d]
HCl-doped PDPA	0.37[a]	0.45	1.0
HCl-doped PDPA-*g*-PEO-350	0.042[b]	0.056	0.29
HCl-doped PDPA-*g*-PEO-750	0.007[b]	0.010	0.16
HCl-doped PDPA-*g*-PEO-2000	0.00056[b]	0.0051	0.065

[a] After HCl-mediated polymerization, PDPA was neutralized with a 1 N ammonium aqueous solution, followed by doping with a 3 N HCl solution.

[b] The graft copolymers were oxidatively doped in air in the presence of 3 N HCl solutions for at least 72 h.

[c] HCl-doped polymer samples were heated to 55°C, where they were kept for 5 min, and then the conductivities were determined by the four-point method.

[d] Based on the experimental density of PDPA film (d_1 = 1.33 g/cm^3) and the density of PEO specified in the Aldrich catalog (d_2 = 1.10 g/cm^3).

Conductivity. The HCl-doped polymers were ground into powders and compressed into films. The electronic conductivity of the film was determined by the four-probe method. The HCl-doped PDPA possessed the high conductivity of 0.37 S/cm, and could attain 0.45 S/cm upon heating at 55°C, as a result of the thermal active effect.[35] The conductivity decreased with increasing PEO side-chain length. The reduction in conductivity can be explained by the volumetric "dilute effect"[36] of the fully grafted PEO chains which increases with increasing side-chain length (see Table 4.3.2). In addition, longer PEO side chains induced larger nonplanar conformations.

Particularly, HCl-doped PDPA-*g*-PEO-2000 had a very low conductivity, around 5.6×10^{-4} S/cm. However, upon heating above 55°C, it increased up to 5.1×10^{-3} S/cm, a 10-fold increase. This increase might have been caused by the melting of the crystalline PEO-2000 side chains.

Figure 4.3.8 presents the DSC profiles of diphenylamine monomer, PEO-2000, PDPA in various states, and the three graft copolymers. DPA and PEO-2000 have a similar melting point, around 50°C. The neutralized PDPA and the reduced PDPA did not have melting transitions in the scanning range −100°C to 200°C, indicating that PDPA has an amorphous structure. Only the neutralized PDPA-*g*-PEO-2000 possessed a melting point, around 50°C, which can be assigned to the crystalline structure of the ordered PEO-2000 side chain. The PEO crystallization in PDPA-*g*-PEO-2000 was confirmed by the XRD results presented in Figure 4.3.9. The XRD profile of the neutralized PDPA-*g*-PEO-2000 exhibited two characteristic peaks for 2θ of 19 and 23, near to those of PEO-2000. Therefore, the nonconductive PEO crystals provided resistance to the transport of charges among the PDPA domains. Upon heating above the PEO crystal melting point, the molten PEO segments could better contact among themselves, thus enhancing the contacts among the PDPA backbones.

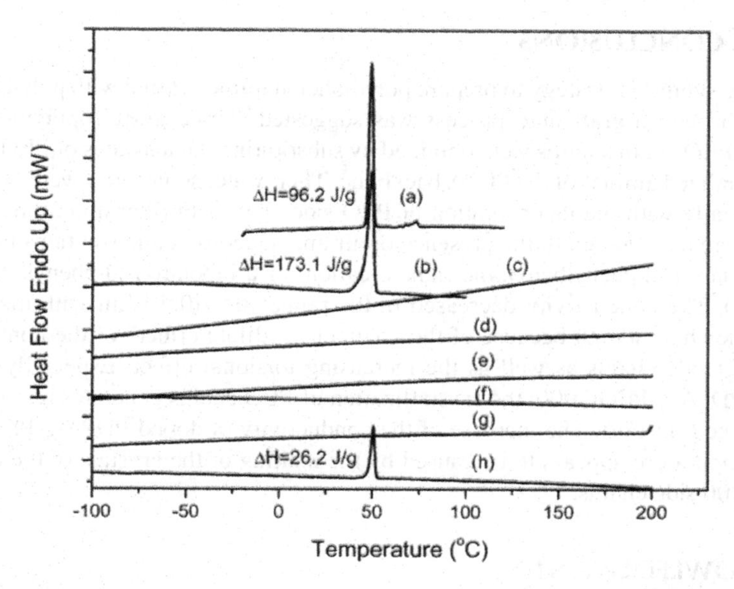

FIGURE 4.3.8 DSC profiles of (a) DPA monomer, (b) PEO-2000, (c) HCl-doped PDPA, (d) neutralized PDPA, (e) reduced PDPA, (f) neutralized PDPA-*g*-PEO-350, (g) neutralized PDPA-*g*-PEO-750, and (h) neutralized PDPA-*g*-PEO-2000.

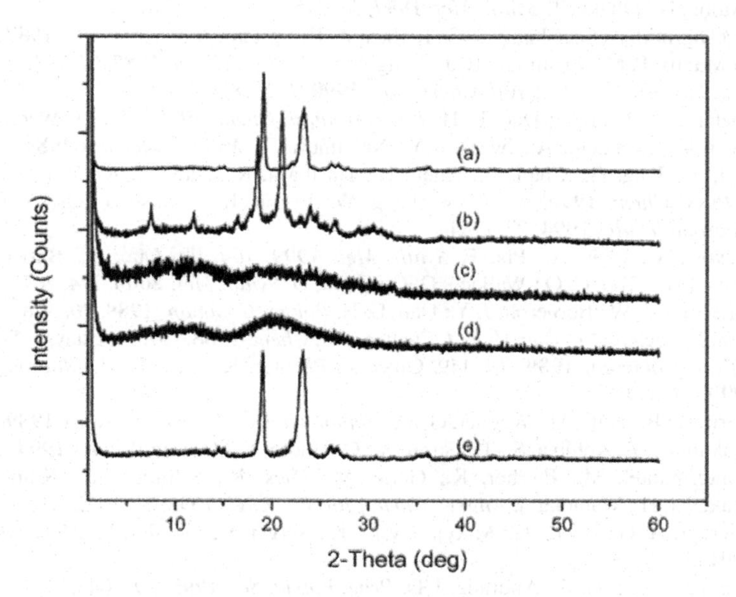

FIGURE 4.3.9 XRD profiles of (a) PEO-2000, (b) DPA monomer, (c) reduced PDPA, (d) PDPA-*g*-PEO-750, and (e) PDPA-*g*-PEO-2000.

4.3.4 CONCLUSIONS

A simple synthesis strategy to prepare polydiphenylamine grafted with poly(ethylene oxide) through a graft onto process was suggested. Three graft copolymers with various PEO chain lengths were obtained by substituting the tosylates of PEO-Tos on functionalized amines of the PDPA backbone. Their water solubilities were improved dramatically with the incorporation of PEO side chains. Furthermore, these copolymers can be oxidized in the presence of air and aqueous acidic solutions to generate a polaron (diphenylbenzidine radical cation) or a bipolaron (diphenylbenzidine dication). The conductivity decreased in the range 10^{-1}–10^{-4} S/cm with increasing PEO side chain length because of the volumetric "dilute effect" of the nonconductive PEO side chains as well as the increasing torsional effect. Especially for the doped PDPA-*g*-PEO-2000, the crystallization of PEO chains generates resistance to the charge transport. The increase of the conductivity of doped PDPA-*g*-PEO-2000 with temperature appears to be caused by the melting of the crystals of the ordered PEO-2000 side chains.

ACKNOWLEDGMENT

We are thankful to Drs. W. Anderson, H. Chopra, and J. R. Errington for allowing us to use their equipment.

REFERENCES

1. Oyama, N.; Ohaka, T. *Synth. Met.* **1987**, *18*, 375.
2. MacDiarmid, A. G.; Chiang, J. C.; Richter, A. F.; Epstein, A. J. *Synth. Met.* **1987**, *18*, 285.
3. McManus, P. M.; Cushman, R. J.; Yang, S. C. *J. Phys. Chem.* **1987**, *91*, 744.
4. Yue, J.; Epstein, A. J. *J. Am. Chem. Soc.* **1990**, *112*, 2800.
5. Leclerc, M.; Guay, J.; Dao, L. H. *J. Electroanal. Chem.* **1988**, *251*, 21. Watanabe, A.; Mori, K.; Iwabuchi, A.; Iwasaki, Y.; Nakamura, Y. *Macromolecules* **1989**, *22*, 3521. Wei, Y.; Jang, G.; Chan, C.; Hueh, K.; Harihara, R.; Patel, S. A.; Whitecar, C. K. *J. Phys. Chem.* **1990**, *94*, 7716. Zheng, W.; Levon, K.; Laakso, J.; Osterholm, J. E. *Macromolecules* **1994**, *27*, 7754.
6. Prevost, V.; Petit, A.; Pla, F. *Synth. Met.* **1999**, *104*, 79. Guo, R.; Barisci, J. N.; Innis, P. C.; Too, C. O.; Wallace, G. G.; Zhou, D. *Synth. Met.* **2000**, *114*, 267.
7. Chevalier, J. W.; Bergeron, J. Y.; Dao, L. H. *Polym. Commun.* **1989**, *30*, 308.
8. Gua, J.; Leclerc, M.; Dao, H. L. *J. Electroanal. Chem.* **1988**, *251*, 31. Guay, J.; Dao, H. L. *Polym. Commun.* **1989**, *30*, 149. Guay, J.; Paynter, R.; Dao, L. H. *Macromolecules* **1990**, *23*, 3598.
9. Stern, R.; Ballauff, M.; Wegner, G. *Macromol. Chem., Macromol. Symp.* **1989**, *23*, 373.
10. Tsukahara, Y.; Kohjiya, S.; Tsutsumi, K.; Okamoto, Y. *Macromolecules* **1994**, *27*, 1662.
11. Wintermantel, M.; Fischer, K.; Gerle, M.; Ries, R.; Schmidt, M.; Kajiwara, K.; Urakawa, H.; Wataoka, I. *Angew. Chem., Int. Ed. Engl.* **1995**, *34*, 1472.
12. Boerner, H. G.; Beers, K.; Matyjasewski, K.; Sergei, S.; Moeller, M. *Macromolecules* **2001**, *34*, 4375.
13. Lee, J. H.; Lee, H. B.; Andrade, J. D. *Prog. Polym. Sci.* **1995**, *20*, 1043.
14. Chen, Y. J.; Kang, E. T.; Neoh, K. G.; Wang, P.; Tan, K. L. *Synth. Met.* **2000**, *110*, 47.
15. Yamada, K.; Ito, A.; Iwamoto, N.; Haraguchi, T.; Kajiyama, T. *Polym. J.* **2000**, *32*, 222.

16. Lim, V. W. L.; Kang, E. T.; Neoh, K. G. *Synth. Met.* **2001**, *119*, 261.
17. Armand, M. *Solid State Ionics* **1994**, *69*, 309. Novak, P.; Mueller, K.; Santhanam, K. S. V.; Hass, O. *Chem. Rev.* **1997**, *97*, 207.
18. Lauter, U.; Meyer, W. H.; Wegner, G. *Macromolecules* **1997**, *30*, 2092. Bruce, P. G. *Solid-State Electrochemistry,* VCH Publishers: New York, 1994.
19. Moon, H.; Park, J. *Solid State Ionics* **1999**, *120*, 1.
20. Moon, H.; Park, J. *Macromolecules* **1998**, *31*, 6461.
21. Yamaguchi, I.; Yasuda, T.; Yamamoto, T. *J. Polym. Sci., Part A: Polym. Chem.* **2001**, *39*, 3137.
22. Wang, P.; Tan, K. L.; Zhang, F.; Kang, E. T.; Neoh, K. G. *Chem. Mater.* **2001**, *13*, 581.
23. Guay, J.; Dao, L. H. *J. Electroanal. Chem.* **1989**, *274*, 135.
24. Athawale, A. A.; Deore, B. A.; Chabukswar, V. V. *Mater. Chem. Phys.* **1999**, *58*, 94.
25. Rajendran, V.; Gopalan, A.; Vasudevan, T.; Wen, T. *J. Electrochem. Soc.* **2000**, *147*, 3014.
26. Chung, C.; Wen, T.; Gopalan, A. *Electrochim. Acta* **2001**, *47*, 423.
27. Wu, M.; Wen, T.; Gopalan, A. *J. Electrochem. Soc.* **2001**, *148*, D65.
28. Wen, T.; Chen, J.; Gopalan, A. *Mater. Lett.* **2002**, *57*, 280.
29. Hua, F. J.; Ruckenstein, E. *J. Polym. Sci., Part A: Polym. Chem.* **2004**, 42, 2179.
30. Ouchi, M.; Inoue, Y.; Nagamune, S.; Nagamura, S.; Wada, K.; Hakushi, T. *Bull. Chem. Soc. Jpn.* **1990**, *63*, 1260.
31. 32.75 g of PEO-350 (0.0936 mol), 25 mL of THF, and 5.59 g (0.1398 mol) of NaOH dissolved in 30 mL of water were introduced in a 250 mL round-bottom flask, after which the flask was located in a salt–ice bath around –5°C. The first 5.5 g of TosCl dissolved in 30 mL of THF was added dropwise within 1 h, and another 12.75 g of TosCl dissolved in 25 mL of THF was added within another 1 h. The reaction was continued for 4 h at –5°C, after which the temperature was raised to the room temperature and kept at that temperature for other 18 h. After reaction, 100 mL of diethyl ether was added for extraction three times. The diethyl ether solution was dried by passing through anhydrous Na_2SO_4 (30 g) columns, after which the diethyl ether was removed using a rotary evaporator. The PEO-Tos-350 prepared from PEO-350 was a transparent liquid, the PEO-750-Tos from PEO-750 was a paste, and PEO-Tos-2000 from PEO-2000 was a waxlike solid. The yield was in all three cases 90 wt%. The FTIR of PEO-Tos was 1594, 1182, and 784 cm^{-1}, which can be assigned to the stretching vibrations of the phenyl hydrogens, and of C–O and S–O in S–O–C. The ^1H NMR of PEO-Tos was 4.05 ppm (2H), due to the oxymethylene connected to tosylate, and 7.70 ppm (2H) and 7.22 (2H) ppm, which can be assigned to the double-substituted phenyl hydrogens of the tosylate moiety. The values of the main molecular ionic peaks in the MS spectra reflect the following formula: $M_n = 23\ (Na^+) + 15\ (-CH_3) + 155\ (CH_3-C_6H_4-SO_2-) + 44n\ (-OCH_2CH_2-)$, where n is the number of ethylene oxides in the PEO.
32. Asturias, G. E.; MacDiarmid, A. G.; McCall, R. P.; Epstein, A. J. *Synth. Met.* **1989**, *29*, 157.
33. Moon, D. K.; Ezuka, M.; Maruyama, T.; Osakada, K.; Yamamoto, T. *Macromolecules* **1993**, *26*, 364.
34. Mav, I.; Zigon, M.; Sebenik, A.; Vohlidal, J. *J. Polym. Sci., Part A: Polym. Chem.* **2000**, *38*, 3390.
35. Roy, B. C.; Gupta, M. D.; Bhoumik, L.; Ray, J. K. *Synth. Met.* **2002**, *130*, 27.
36. Lecterc, M.; Daprano, G.; Zotti, G. *Synth. Met.* **1993**, *55*, 231.
37. Asturias, G. E.; MacDiarmid, A. G.; Maccall, R. P.; Epstein, A. J. *Synth. Met.* **1989**, *29*, E157.

4.4 Water-Soluble Self-Doped Conducting Polyaniline Copolymer*

Wusheng Yin and Eli Ruckenstein

Department of Chemical Engineering, State University of New York at Buffalo, Buffalo, New York 14260

Polyaniline (PANI) has emerged as an important conducting polymer because of its good environmental stability, low cost, and high conductivity upon doping with acid.[1,2] The methods proposed to synthesize water-soluble conducting polyaniline involved either the attachment of water-soluble functional groups to its benzene rings or the replacement of in the N–H groups of PANI by water-soluble functional groups. Yue and Epstein[3,4] have synthesized sulfonic acid ring-substituted PANI (SPAN) by reacting the emeraldine base with fuming acid and obtained a conductivity of 0.1–1 S/cm. However, the polymer obtained became water-soluble only after its conversion to the salt form after undoping with a basic aqueous solution. The replacement of the sulfonic acid group with phosphoric acid yielded another water-soluble self-doped ring-substituted PANI, which exhibited, however, the low conductivity of 10^{-3} S/cm.[5] Hany et al.[6] attempted to synthesize a self-doped PANI by reacting the emeraldine base with propane sultone or butane sultone. However, the product had both a very low solubility and low conductivity (~10^{-9} S/cm). Through a variant of the latter method, an N-substituted water-soluble self-doped PANI, poly-(aniline-*co*-propanesulfonic acid aniline) [PAPSAH], was synthesized by reacting the emeraldine base first with NaH and subsequently with propane sultone. PAPSAH could be cast into free-standing films directly from its aqueous solution and provided a conductivity of about 10^{-2} S/cm without external doping.[7] Another approach for the preparation of N-substituted water-soluble self-doped conductive PANI was the copolymerization of aniline with a suitable substituted aniline to generate copolymers. The copolymer thus obtained exhibited improved solubility in an aqueous NH_4OH solution[8]; however, its conductivity was low (10^{-3} S/cm). In conclusion, the above approaches synthesized water-soluble self-doped polyanilines by modifying the structure of the emeraldine base.

A new water-soluble self-doped copolymer, namely, poly(2-acrylamido-2-methyl-1-propanesulfonic acid-*co*-aniline) [PAMPANI], and its salt have been

* *Macromolecules* 2000, 33, 1129–1131.

synthesized by us. Both PAMPANI and its salt could be dissolved in water and cast into films from their aqueous solutions.

The synthesis of PAMPANI is described in Scheme 4.4.1. *N*-(4-Anilinophenyl) methacrylamide (APMA) could be synthesized via the catalytic aminolysis reaction. The preparation procedure can be described as follows: 18.4 g (0.1 mol) of *p*-aminodiphenylamine, 40 g (0.4 mol) of methyl methacrylate, 1.24 g (0.05 mol) of dibutyltin oxide (as catalyst), and 0.02 g (0.0001 mol) of phenothiazine (as retardant of polymerization) were introduced into a 100 mL round-bottom three-neck flask fitted with a magnetic stirrer, a 40 cm vacuum jacketed Vigreux column, and a distillation head with a water-cooled condenser. The mixture was heated at low reflux and the methanol/methyl acrylate codistillation product removed continuously for over 11 h. The Vigreux column was then removed and replaced with a distillation head, and the remaining methyl methacrylate was distilled, while concurrently adding 25 mL of xylene. With stirring, an additional 25 mL of xylene was added at 130°C and the mixture cooled to room temperature. The product was separated by filtration, washed with xylene, dried, and purified by recrystallization from ethanol. The AMPA obtained exhibited the following characteristics; mp 107 ± 1°C. ^1H NMR ($CDCl_3$, δ ppm relative to Me_4Si): 2.0

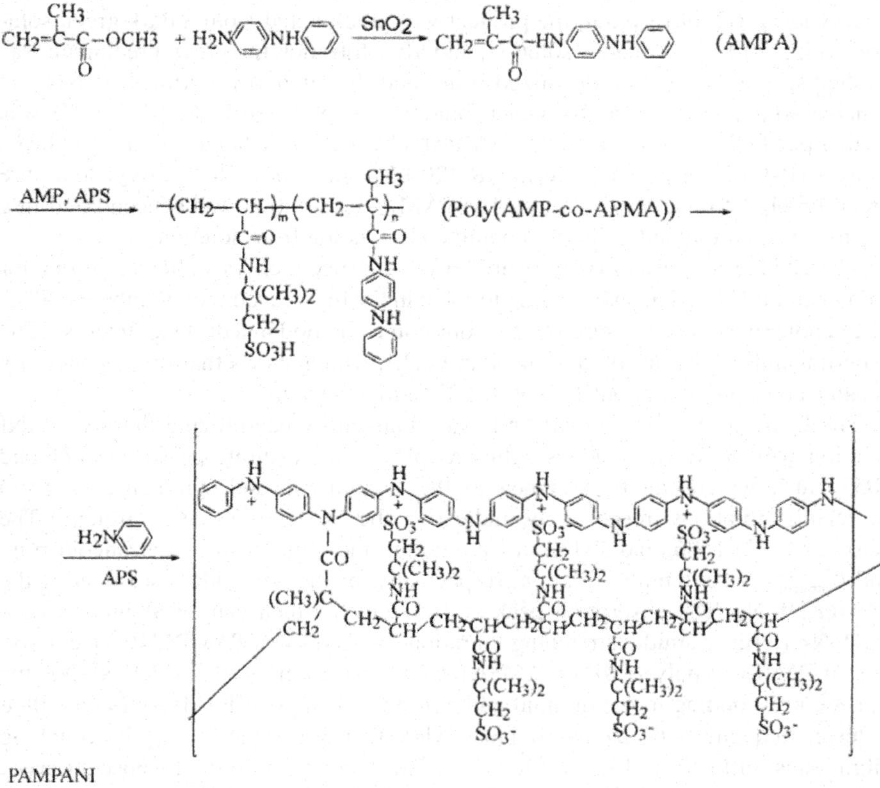

SnO_2 (AMPA)

AMP, APS (Poly(AMP-co-APMA))

H_2N–Ph, APS

PAMPANI

SCHEME 4.4.1

(s, 3H), 5.4 (s, 1H), 5.8 (s, 1H), 5.9 (s, 1H), 6.9–7.4 (m, 9H), 7.5 (s, 1H) MS (*m/e*): 252.2 (M^+), 183.2, 69. FTIR (cm^{-1}): 3353 (N–H), 3100–3059 (=CH), 1663 (C=O in amide), 1608 (C-NH def).

Poly(AMP-*co*-APMA) (AMP = 2-acrylamido-2-methyl-1-propanesulfonic acid) was prepared through a surfactant-free emulsion polymerization in water. At first, 4.5 g of AMP and 0.05 g of ammonium persulfate were dissolved in 40 g of distilled water present in a three-neck flask located in a water bath at 70°C under a nitrogen atmosphere, and the mixture was stirred for about 1 h. Then, 0.5 g of AMPA in 10 g of chloroform was slowly added dropwise and the reaction allowed to last for 4.5 h. This was followed by filtration, and the water of the filtrate was removed by evaporation to obtain a pale yellow poly(AMP-*co*-AMPA). FTIR (cm^{-1}): 3345–3245 (N–H); 1662 (CO–NH–alkyl); 1640 (CO–NH–aromatic); 1190 and 1054 (O=S=O, asymmetric and symmetric stretchings). ^1H NMR (D_2O, ppm): 1.3 (s, CH_3), 1.8–2.0 (m, CH_2), 3.0–3.4 (m, CH_2–SO_3 and CH–CO), and 7.0–7.4 (m, aromatic hydrogen). AIBN could not be used as initiator because a large amount of an insoluble precipitate (poly(AMPA)) was generated.

The graft copolymerization of aniline onto poly(AMP-*co*-AMPA) was carried out by the dropwise addition of 20 g of 2 wt% poly(AMP-*co*-AMPA) and 10 g of 8 wt% APS aqueous solutions to an aniline (0.8 g)/1 M HCl aqueous solution of pH < 2. After 4–6 h of reaction, the product was precipitated from a dark-green solution with a 1 M HCl aqueous solution, and after filtration the green precipitate was washed several times with deionized water and then undoped with a 1 M NH_4OH aqueous solution to yield a blue-violet solution. The polymer, PAMPANI–NH_4, was further purified and converted into PAMPANI by the ion exchange of NH_4^+ with H^+, using a H^+-type ion-exchange resin (IR 1200H resin from Aldrich), to yield a pure green PAMPANI aqueous solution. The PAMPANI solution was concentrated in a vacuum evaporator, and a film was obtained by casting from solution.

The conductivity was first measured after the specimen was allowed to dry for 24 h in air and later during its drying for 48 h in the high vacuum of 30 mbar at 40°C. The conductivity was measured by the four points method in a drybox under N_2. The experimental results are plotted in Figure 4.4.1, which shows that the conductivity became constant after about 12 h of drying in high vacuum.

The FTIR spectrum of PAMPANI–Na (obtained by neutralizing the PAMPANI solution with a NaOH aqueous solution (pH = 12)) exhibits peaks at 1171 and 1057 cm^{-1}, which can be assigned to the asymmetric and symmetric O=S=O stretching vibrations, respectively, indicating the existence of SO_3^- groups.[9] The peaks at 1592, 1506, and 1313 cm^{-1} are due to the quinoid ring, benzeneoid ring, and $C_{aromatic}$–N stretching vibration, respectively, of the emeraldine segment of the PAMPANI–Na. The absorption peak at 1651 cm^{-1}, which can be assigned to the –CO–N< (tertiary amide) stretching vibration, indicates that PAMPANI is a copolymer of PANI and poly(AMP-*co*-AMPA). Compared to those of PAMPANI–Na, the peaks corresponding to the quinoid and benzeneoid rings of PAMPANI are shifted to lower frequencies by 29 and 32 cm^{-1}, while that due to the $C_{aromatic}$–N stretching vibration is shifted from 1313 to 1303 cm^{-1}. These shifts are similar to those observed when emeraldine was doped with sulfonic acid[10,11] and reveal that PAMPANI is in a self-doped state.

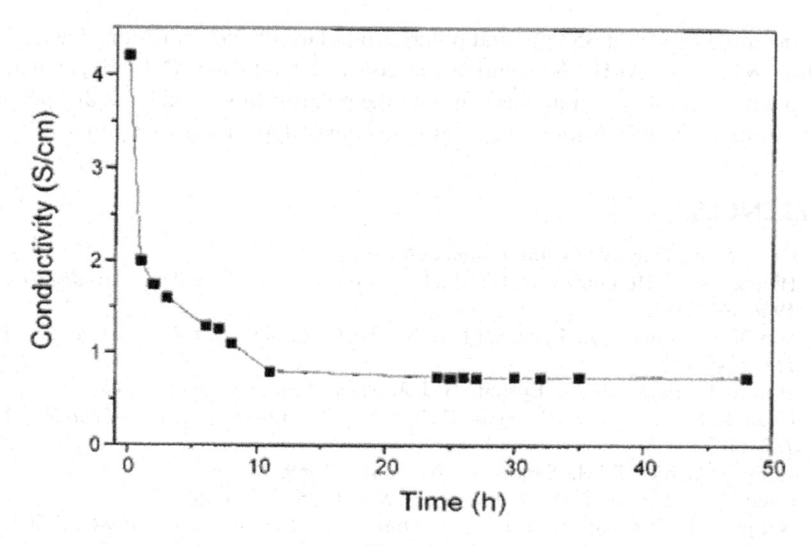

FIGURE 4.4.1 Conductivity vs the drying (in a vacuum) time of a thin film (thickness: 1 μm).

The ^{1}H NMR spectrum of PAMPANI (D_2O as solvent) exhibits multipeaks around 7.9, 8.5, and 8.7 ppm, which can be attributed to the hydrogens of the benzenic ring, a peak at 1.4 ppm, which can be assigned to the hydrogens of the methyl groups, broad peaks at about 2.0 ppm, which can be attributed to the hydrogens of $-CH_2-$, and peaks at about 3.2 ppm, which can be attributed to the hydrogens of $-CH_2-SO_3$ and –CH–CO[12] These results are in agreement with the FTIR results and with the structure proposed for the PAMPANI conducting polymer (Scheme 4.4.1).

The fact that PAMPANI is in a self-doped state is also confirmed by the electronic spectrum of its aqueous solution (Figure 4.4.2), which exhibits a π–π* transition of

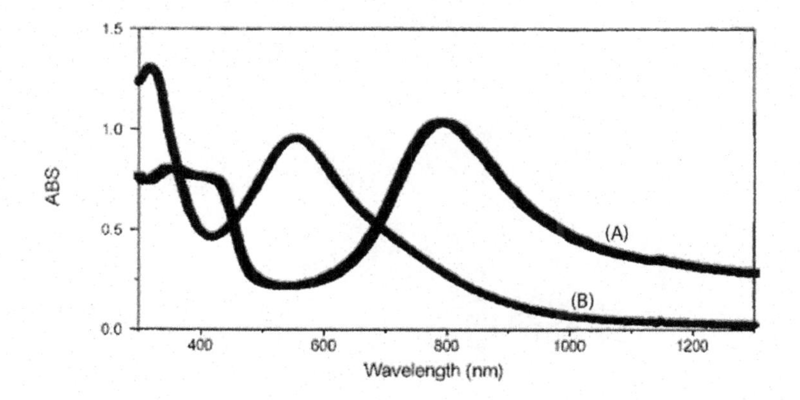

FIGURE 4.4.2 UV/vis spectra of PAMPANI: (A) aqueous solution, (B) aqueous solution after neutralization with 1 M NH_4OH.

the benzenoid rings[13] at 352 nm and polaron/bipolaron band transitions[14] at 477 and 791 nm. When the PAMPANI solution was converted to PAMPANI–NH_4 by neutralization with 1 M NH_4OH aqueous solution, the polaron/bipolaron band disappeared, and a strong exciton transition of the quinoid rings[15] appeared at 556 nm.

REFERENCES

1. Chen, S.-A.; Fang, W. G. *Macromolecules* **1991**, *24*, 1242.
2. Huang, W. S.; Humphrey, B. D.; MacDiarmid, A. G. J. *Chem. Soc, Faraday Trans.* 1 **1986**, *82*, 2385.
3. Wei, X.-L.; Wang, Y. Z.; Long, S. M.; Bobeczko, C.; Epstein, A. J. *J.Am. Chem. Soc.* **1996**, *118*, 2545.
4. Wei, X.-L.; Fahlman, M.; Epstein, A. J. *Macromolecules* **1999**, *32*, 3114.
5. Chan, H. S. O.; Ho, P. K. H.; Ng, S. C.; Tan, B. T. G.; Tan, K. L. *J.Am. Chem. Soc.* **1995**, *117*, 8517.
6. Hany, P.; Genies, E. M.; Santier, *C. Synth. Met.* **1989**, *31*, 369.
7. Chen, S.-A.; Hwang, G.-W. *J. Am. Chem. Soc.* **1995**, *117*, 10055.
8. Nguyen, M. T.; Kasai, P.; Miller, J. L.; Diaz, A. F. *Macromolecules* **1994**, *27*, 3625.
9. Avram, M. *Infrared Spectroscopy*; John Wiley & Sons: New York, 1972.
10. Yin, W.; Ruckenstein, E. *Synth. Met.* **2000**, *108*, 39.
11. Kue, C.-T.; Chen, C.-H. *Synth. Met.* **1999**, *99*, 163.
12. Pavia, D. L.; Lampman, G. M., Jr., G. S. K. *Introduction to Spectroscopy*; W.B. Saunders: London, UK, 1979.
13. Lu, F. L.; Wudle, F.; Nowak, M.; Heeger, A. J. J. *Am. Chem. Soc.* **1986**, *108*, 8, 8311.
14. Stafstrom, S.; Breadas, J. L.; Epstein, A. J.; Woo, H. S.; Tanner, D. B.; Huang, W. S.; MacDiarmid, A. G. *Phys. Rev. Lett.* **1987**, *59*, 1464.
15. Epstein, A. J.; Ginder, J. M.; Zuo, F.; Bigelow, R. W.; Woo, H.-S.; Tanner, D. B.; Richter, A. F.; Huang, W. S.; MacDiarmid, A. G. *Synth. Met.* **1987**, *18*, 303.

4.5 Preparation of Densely Grafted Poly(aniline-2-sulfonic acid-*co*-aniline)s as Novel Water-Soluble Conducting Copolymers*

Fengjun Hua and Eli Ruckenstein

Department of Chemical and Biochemical Engineering, State University of New York at Buffalo, Buffalo, New York 14260

ABSTRACT Various densely grafted polymers containing poly(aniline-2-sulfonic acid-*co*-aniline)s as side chains and polystyrene as the backbone were prepared. A styryl-substituted aniline macromonomer, 4-(4-vinylbenzoxyl)(*N*-*tert*-butoxycarbonyl)phenylamine (4-VBPA-^{t}BOC), was first prepared by the reaction of 4-aminophenol with the amino-protecting moiety di-*tert*-butoxyldicarbonate, and this was followed by substitution with 4-vinylbenzyl chloride. 4-VBPA-^{t}BOC thus obtained was homopolymerized with azobisisobutyronitrile as an initiator, and this was followed by deprotection with trifluoroacetic acid to generate poly[4-(4-vinylbenzoxyl)phenylamine] (PVBPA) with pendent amine moieties. Second, the copolymerization of aniline-2-sulfonic acid and aniline was carried out in the presence of PVBPA to generate densely grafted poly(aniline-2-sulfonic acid-*co*-aniline). Through the variation of the molar feed ratio of aniline-2-sulfonic acid to aniline, various densely grafted copolymers were generated with different aniline-2-sulfonic acid/aniline composition ratios along the side chains. The copolymers prepared with molar feed ratios greater than 1/2 were water-soluble and had conductivities comparable to those of the linear copolymers. Furthermore, these copolymers

* *Journal of Polymer Science: Part A: Polymer Chemistry*, 2005, 43, 1090–1099.

could self-dope in water through intermolecular or intramolecular interactions between the sulfonic acid moieties and imine nitrogens, and this generated large aggregates.

4.5.1 INTRODUCTION

High interest in densely grafted comblike (co)polymers or molecular brushes with an axis-symmetric distribution of the branching points along a backbone chain has arisen from the successful synthesis of visible macromolecules.[1] When the side chains are sufficiently long (number of repeating units > 10), the main chain acquires an extremely high stiffness because of the steric overcrowding of the side chains. As a result, single macromolecules can be visualized by transmission electron microscopy or atomic force microscopy (AFM).[2]

Polyaniline (PANI) is one of the important conducting polymers discovered in the past decades. To improve the solubility of PANI in organic compounds and its processability, many approaches have been employed, such as the copolymerization of aniline (ANI) with its derivatives[3] and suitable substitutions of the emeraldine base, either at the N sites or on the phenyl rings of the PANI backbone.[4] The substitutions on the PANI backbone can cause large torsional effects that reduce the conjugation length and hence the conductivity. Synthesis methods through modifications of the PANI backbone have also been employed to prepare water-soluble PANI. For example, through the addition of —SO_3H moieties on a PANI backbone through a direct reaction of PANI with fume sulfonic acid or a sulfur trioxide/triethyl phosphate complex, water-soluble sulfonated PANI was obtained.[5,6] However, lower conductivities were also obtained. In addition, 5- or 6-member-ring complexes were generated because the sulfonic acid moieties doped the adjacent amine nitrogen, generating positive charges localized on the nitrogens. Most water-soluble ANI monomers, such as aniline-2 (or -3)-sulfonic acid, could hardly be homopolymerized because of the withdrawing effect of the sulfonic acid moiety of the phenyl ring. These sulfonyl aniline derivatives could, however, be copolymerized with ANI or other ANI derivatives.[7] Another possibility for preparing a water-soluble PANI is to add poly(ethylene oxide) (PEO) to the nitrogen atoms, as reported by Wang et al.[8] and Yamaguchi et al.[9] Longer PEO chains grafted on the PANI backbone reduce, however, the conductivity because of torsional effects.

Our group developed synthesis strategies to improve both the water solubility and the processability of PANI.[10–12] In previous studies, PEO was incorporated into the PANI backbone by either the preparation of a diblock copolymer with poly(sulfonic diphenyl aniline) or by its grafting at the nitrogen atom or phenyl ring of reduced PANI.[11] In addition, a water-soluble copolymer of ANI and aminophenol (AP) grafted with oligomeric PEO side chains (AP-*g*-PEO) was synthesized.[12] The AP-*g*-PEO macromonomer was prepared from an N-protected 3-aminophenol (protection group = *tert*-butoxycarbonyl), and this was followed by the substitution of tosylated oligomeric PEO on the hydroxide moiety of AP and deprotection.

In this study, a synthesis strategy for preparing densely grafted polymers containing poly(aniline-2-sulfonic acid-*co*-aniline) [d-poly(SAN-*co*-ANI] as side chains was developed. As shown in Scheme 4.5.1, self-doped d-poly(SAN-*co*-ANI) could be obtained via seven steps. The intermediates, prepared prepolymers, and final copolymers were characterized with NMR, Fourier transform infrared (FTIR), mass spectrometry, and UV-vis spectroscopy. The conductivities of the various densely grafted copolymers obtained with various molar feed ratios were determined by the four-point method. The behavior of these copolymers in water was investigated with AFM and scanning electron microscopy (SEM).

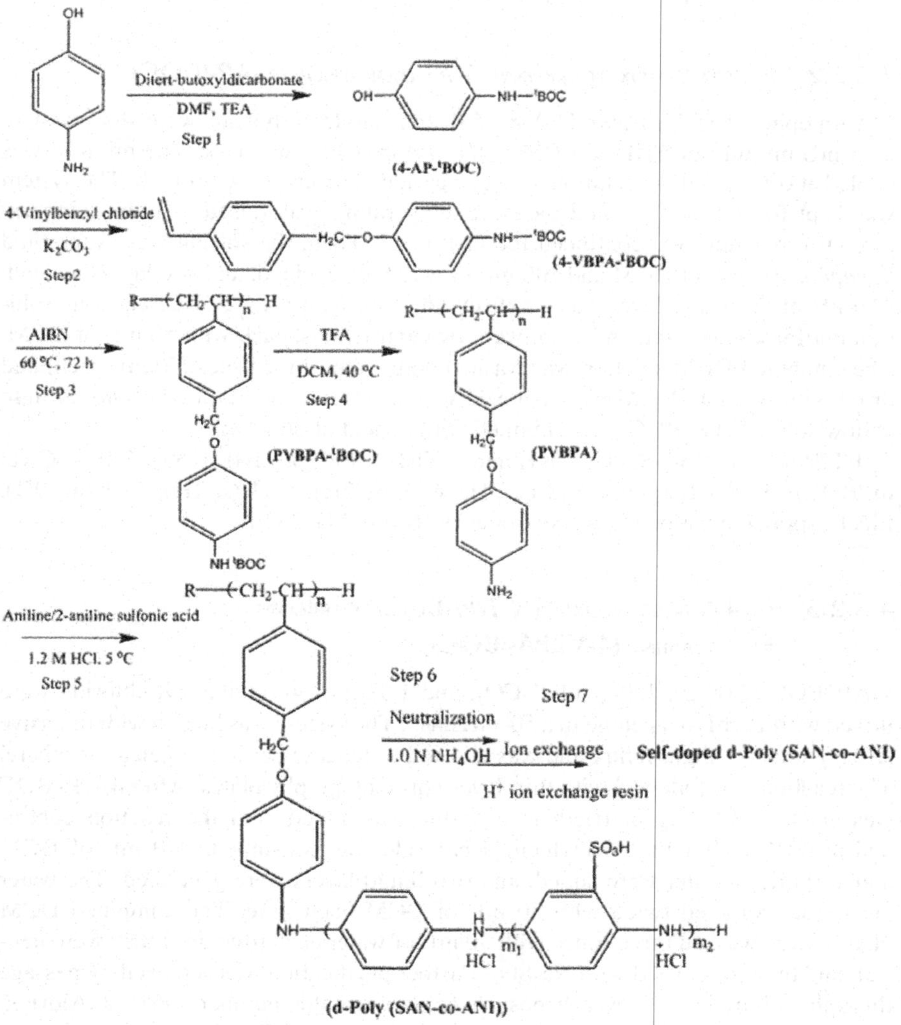

Ditert-butoxyldicarbonate
DMF, TEA
Step 1
(4-AP-^{t}BOC)
4-Vinylbenzyl chloride
K_2CO_3
Step2
(4-VBPA-^{t}BOC)
AIBN
60 °C, 72 h
Step 3
TFA
DCM, 40 °C
Step 4
(PVBPA-^{t}BOC)
(PVBPA)
Aniline/2-aniline sulfonic acid
1.2 M HCl, 5 °C
Step 5
Step 6
Neutralization
1.0 N NH_4OH
Step 7
Ion exchange
H^+ ion exchange resin
Self-doped d-Poly (SAN-co-ANI)
(d-Poly (SAN-co-ANI))

SCHEME 4.5.1 Synthesis strategy for d-poly(SAN-*co*-ANI) copolymers.

4.5.2 EXPERIMENTAL

4.5.2.1 MATERIALS

4-Aminophenol (>98 wt%), aniline-2-sulfonic acid (SAN; 95 wt%), ANI (99 wt%), di-*tert*-butyldicarbonate (97 wt%), hydrochloric acid (37 wt%), trifluoroacetic acid (TFA; 99 wt%), ammonium persulfate (98 wt%), and potassium carbonate (99 wt%) were purchased from Aldrich and used without further purification. The solvents dimethylformamide (DMF), tetrahydrofuran (THF), methyl ethyl ketone (MEK), and dichloromethane (DCM), purchased from Aldrich, were of high-performance-liquid-chromatography (HPLC) purity.

4.5.2.2 *N-TERT*-BUTOXYLCARBONYL 4-AMINOPHENOL (4-AP-^{t}BOC)

4-Aminophenol (5.4 g) and 11.0 g of di-*tert*-butyldicarbonate were dissolved in a 55-mL mixture of THF and DMF (1/1 v/v) in a 100-mL flask. The mixture was cooled at 0°C. Triethylene amine (5.1 g) was added dropwise within 1 h. The system was kept for 4 h at 0°C, and then the temperature was raised at room temperature. The reaction was continued for 12 h more. Then, the solvent was evaporated *in vacuo*, 300 mL of DCM and 100 mL of water were added, and two layers formed. The pH of the water layer was carefully adjusted with a 1.0 N HCl aqueous solution until it was neutral, and there were two extractions, each with 50 mL of DCM. The combined DCM solution was washed with water three times (50 mL each) and dried with Na_2SO_4. DCM was removed with a rotary evaporator *in vacuo*. A light yellow solid, 4-AP-^{t}BOC, was obtained with a yield of 86 wt%.

FTIR (cm^{-1}): 3550–3200 (—NH and —OH), 2900 (phenyl), 1750–1670 (—C=O in ^{t}BOC). ^{1}H NMR (ppm): 7.18 (s, 2H), 6.71 (s, 2H), 6.34 (s, 1H), 1.43 (m, 9H). Electrospray ionization mass spectrometry (Na^+ + M): 232.1.

4.5.2.3 4-(4-VINYLBENZOXYL)(*N-TERT*-BUTOXYCARBONYL)-PHENYLAMINE (4-VBPA-^{t}BOC)

4-AP-^{t}BOC (2.09 g), 1.38 g of K_2CO_3, and 1.52 g of 4-vinylbenzyl chloride were mixed with 15 mL of acetone in a 50-mL flask. The system was heated with intensive stirring under a slight reflux and was allowed to react under a nitrogen atmosphere. The reaction was followed with thin-layer-chromatography plates. After 4-AP-^{t}BOC was consumed, 1.0 g of triethylene amine was added, and the reaction continued at 60°C. After the 4-vinylbenzyl chloride was consumed, 300 mL of DCM and 100 mL of water were added, and two liquid layers were generated. The water layer was extracted twice with 50 mL of DCM each time. The combined DCM phases were washed three times with 50 mL of water each time until they were neutral, and they were dried with Na_2SO_4. Further purification was achieved by passage through an inhibitor remover disposable column (catalog number 30631–2, Aldrich) to eliminate the 4-AP-^{t}BOC excess (which inhibited the radical homopolymerization of 4-VBPA-^{t}BOC). DCM was removed with a rotary evaporator *in vacuo*. A light red solid, 4-VBPA-^{t}BOC, was obtained with a yield of 83 wt%.

FTIR (cm^{-1}): 3300–3150 (—NH), 2900 (phenyl), 1750–1670 (—C=O in ^{t}BOC), 957 (—CH= CH_2). ^{1}H NMR (ppm): 7.3–6.6 (m, 8H), 6.43 (m, 1H), 6.34 (s, 1H), 5.6 (m, H), 5.4 (s, H), 5.2 (s, 2H), 1.43 (m, 9H). Electrospray ionization mass spectrometry (Na^+ + M): 348.0.

4.5.2.4 Homopolymerization of 4-VBPA-^{t}BOC

4-VBPA-^{t}BOC (3.48 g) and 0.124 g of azobisisobutyronitrile (AIBN) were added to a 50-mL flask filled with 10 mL of MEK. The system was degassed three times through thawing–freezing cycles; then the temperature was raised to 60°C, and the reaction was allowed to continue for 72 h more under a dry nitrogen atmosphere. The increase with time of the viscosity indicated the start of homopolymerization. DCM (6 mL) was added, and the solution was precipitated in a mixture of methanol and water (5/1 v/v). Further purification was achieved by dissolution and precipitation in the aforementioned nonsolvent mixture. A light yellow precipitate, PVBPA-^{t}BOC, was obtained and dried *in vacuo* for at least 24 h at room temperature with a yield of 57 wt%.

FTIR (cm^{-1}): 3300–3150 (—NH), 2900 (phenyl), 1750–1670 (—C=O in ^{t}BOC). ^{1}H NMR (ppm): 7.6–6.2 (m, 8H), 5.15 (s, 2H), 2.56 (m, 1H), 2.21 (m, 2H), 1.43 (m, 9H).

4.5.2.5 General Procedure of Deprotection of PVBPA-^{t}BOC

PVBPA-^{t}BOC (2.4 g) was dissolved in 60 mL of DCM in a 100-mL flask. TFA (4.0 g) was added, and the system was mixed with intensive stirring. The system was sealed in a dry nitrogen atmosphere, and the temperature was raised with a slight reflux to about 60°C. The reaction lasted 6 h. Two layers were formed, and the DCM and excess of TFA were removed *in vacuo*. Triethylene amine (3.0 g) and 100 mL of DCM were added to precipitate the salt. DCM was washed with 50 mL of water until it was neutral and was dried with Na_2SO_4. After DCM was removed with a rotary evaporator, a red solid, poly[4-(4-vinylbenzoxyl)phenylamine] (PVBPA), was obtained with a yield of 90 wt%.

FTIR (cm^{-1}): 3350–3250 (—NH_2), 2900 (phenyl), 1610 (C—NH). ^{1}H NMR (ppm): 7.4–6.0 (m, 8H), 4.95 (s, 2H), 2.47 (s, 1H), 2.06 (m, 2H).

4.5.2.6 Synthesis of the Densely Grafted Copolymers

To a 100-mL flask, 0.348 g of PVBPA (containing 1 mmol of the amine moiety) dissolved in 3 mL of THF, 40 mL of a 1.2 N HCl aqueous solution, 10 mmol of ANI, and 10 mmol of SAN were added, and then the mixture was cooled to 5°C. Ammonium persulfate (5 mmol), dissolved in 10 mL of a 1.2 N HCl aqueous solution, was added dropwise within 1 h, and the reaction was continued for 16 h more. After that, the copolymer was separated by centrifugation at 3500 rpm for 30 min on a superspeed centrifuge (RC–5B, Dupont), and the precipitate was washed three times with water/ethanol (10/1 v/v) until a colorless supernatant was obtained. The precipitate was neutralized with a 1.0 N ammonium aqueous solution, and the neutralized copolymer was further purified by dialysis. The dialysis bag (Fisherbrand; the nominal molecular weight cutoff, 60,000, separated the linear and low-molecular-weight homopolymers or copolymers), containing a neutralized copolymer aqueous solution (the copolymer was dissolved in 50 mL of water), was kept in a 0.01 N ammonium aqueous solution for 7 days until no colored supernatant was detected. This was followed by dialysis in water for an additional 7 days for the solution to become neutral. A self-doped copolymer solution was prepared by the passage of the obtained solution through an H^+ ion-exchange column containing the IR 1200 H resin by the ion exchanging of NH_4^+ with H^+. Various copolymers with various SAN/ANI molar ratios were prepared by variations of the molar feed ratio of SAN to ANI.

4.5.2.7 Characterization

^{1}H NMR, ultraviolet–visible (UV–vis) absorption, and FTIR measurements were carried out on a 400–MHz Inova–400, a Thermo Spectronic Genesys–6, and a PerkinElmer FTIR 1760, respectively. The solutions for NMR were prepared by the dissolution of the polymer in deuterated DMF, deuterated chloroform, or deuterated water. Gel permeation chromatography (GPC; Waters) was used to evaluate the molecular weights of the polymers on the basis of a polystyrene calibration curve. The GPC instrument was equipped with three 30–cm–long columns filled with Waters Styragel, a Waters HPLC 515 pump, and a Waters 410 refractive-index detector. The GPC measurements were carried out with DMF as an eluent at 60°C at a flow rate of 1.0 mL/min and a chart speed of 1.0 cm/min. The elemental analysis of the polymer samples was carried out on a PerkinElmer model 2400 CHN analyzer. The chlorine and sulfur contents were determined by the oxygen flask method. The room-temperature conductivities of the compressed pellets of the densely grafted and doped copolymers were determined with the conventional four-point probe method. SEM was used to examine the surface morphology. The polymer samples were prepared through the spin coating of an aqueous polymer solution (0.4 g/L) on precleaned glass plates at 1000 rpm for 1 min, and this was followed by sputter coating with carbon (to prevent sample charging); the samples were examined with a Hitachi S–4000 field emission scanning electron microscope operated at 20 KeV. AFM (Quesant Scan Atomic Co.) was used to investigate the molecular conformation of the polymeric macromolecules and the structure of the aggregates. Aqueous polymer solutions were spin-coated for 1 min onto freshly cleaned silicon wafers at 1000 rpm and subsequently examined under AFM in the tapping mode.

4.5.3 RESULTS AND DISCUSSION

4.5.3.1 Synthesis and Characterization of the Intermediates, Prepolymers, and Densely Grafted Copolymers

As shown in Scheme 4.5.1, first, to protect the amine, a styryl–substituted ANI macromonomer, 4-VBPA-^{t}BOC, was prepared through the reaction of 4-aminophenol with di-*tert*-butoxyl- dicarbonate, followed by substitution with 4-vinylbenzyl chloride. The 4-VBPA-^{t}BOC was homopolymerized with AIBN as an initiator, and this was followed by deprotection with TFA to generate PVBPA with pendent amine moieties. Second, d-poly(SAN-*co*-ANI) was prepared through the copolymerization of PVBPA, SAN, and ANI.

Figures 4.5.1 and 4.5.2 present FTIR and ^{1}H NMR spectra of the intermediates (4-AP-^{t}BOC and 4-VBPA-^{t}BOC) and prepolymers (PVBPA-^{t}BOC and PVBPA). For 4-AP-^{t}BOC, the characteristic peaks at 3180 and 1690 cm^{-1} can be assigned to the stretching vibrations of the amine and hydrogen-associated carbonyl groups, respectively, indicating the presence of the amide moiety after amino protection (Figure 4.5.1a). After substitution with 4-vinylbenzyl chloride, the stretching vibration of the carbonyl group of the obtained 4-VBPA-^{t}BOC exhibited a slight blueshift to 1725 cm^{-1} because the hydroxyl group of 4-AP-^{t}BOC was capped and the hydrogen bonding between the hydroxide and carbonyl groups disappeared. Furthermore, for the prepolymer of PVBPA-^{t}BOC, this carbonyl stretching vibration around 1720 cm^{-1} was still present (Figure 4.5.1b and c), and this indicated that both the substitution and homopolymerization did not destroy the amino protection. However, this vibration absolutely disappeared after deprotection in the presence of TFA, as shown by Figure 4.5.1d for PVBPA, because of the removal of the butoxycarbonyl (BOC) group.

The ^{1}H NMR spectrum of the amino-protected AP (4-AP-^{t}BOC) exhibits four groups of peaks at 7.18, 6.71, 6.34, and 1.43 ppm [the integral area ratio is 2:2:1:9; Figure 4.5.2a], which can be assigned to the protons of the phenyl rings (4H), amine (1H), and methyl groups (9H), respectively. For 4-VBPA-^{t}BOC, the signals at 7.3–6.6, 6.34, and 5.20 ppm can be assigned to the phenyl, amine, and benzyl ether methylene moieties, and the signals at 6.43, 5.6, 5.4, and 1.43 ppm can be assigned to the double bond and methyl moieties; this confirms the presence of styryl and BOC moieties (see Figure 4.5.2b). For the prepolymer of PVBPA, the signal due to benzyl ether methylene was high-field-shifted from 5.20 to 4.95 ppm in comparison with that of PVBPA-^{t}BOC (Figure 4.5.2c and d) because of the deshielding effect caused by the removal of the BOC group. These two polymers (PVBPA-^{t}BOC and PVBPA) possessed weight-average molecular weights and polydispersities of 28,700 and 2.9 and 23,750 and 2.5, respectively, which were obtained by GPC on the basis of a polystyrene standard sample calibration curve, as shown in Figure 4.5.3.

In a typical FTIR spectrum of the self-doped copolymer with a molar feed ratio of SAN/ANI = 1/1 (Figure 4.5.1e), the peaks at 1020 and 1390 cm^{-1} can be assigned to the symmetric and asymmetric stretching vibrations of O=S=O, respectively, confirming the existence of the SO_3H moieties.[4,13] The peaks at 1495 and 1596 cm^{-1}, which can be assigned to the quinoid and benzoid rings, respectively, indicate that the copolymer was in an oxidative state. This result was also confirmed by the UV–vis spectra presented in Figure 4.5.4. For the neutralized copolymer, the absorption peak around 590 nm can be assigned to the exciton transition of the quinoid rings, which is redshifted to above 700 nm after self-doping (Figure 4.5.4a and b). The self-doped copolymer could be additionally doped when treated with 1.0 N HCl for 1 h (Figure 4.5.4c). The external doping occurred because the molar fraction of the SAN segments in the copolymer (ca. 0.40) could not provide enough protons and because the external doping by the small molecules of HCl was more favorable thermodynamically than the self-doping, which caused distortions of the chains.

Figure 4.5.5 presents the ^{1}H NMR spectra of densely grafted copolymers in deuterated water. For the neutralized copolymers prepared with a molar feed ratio of

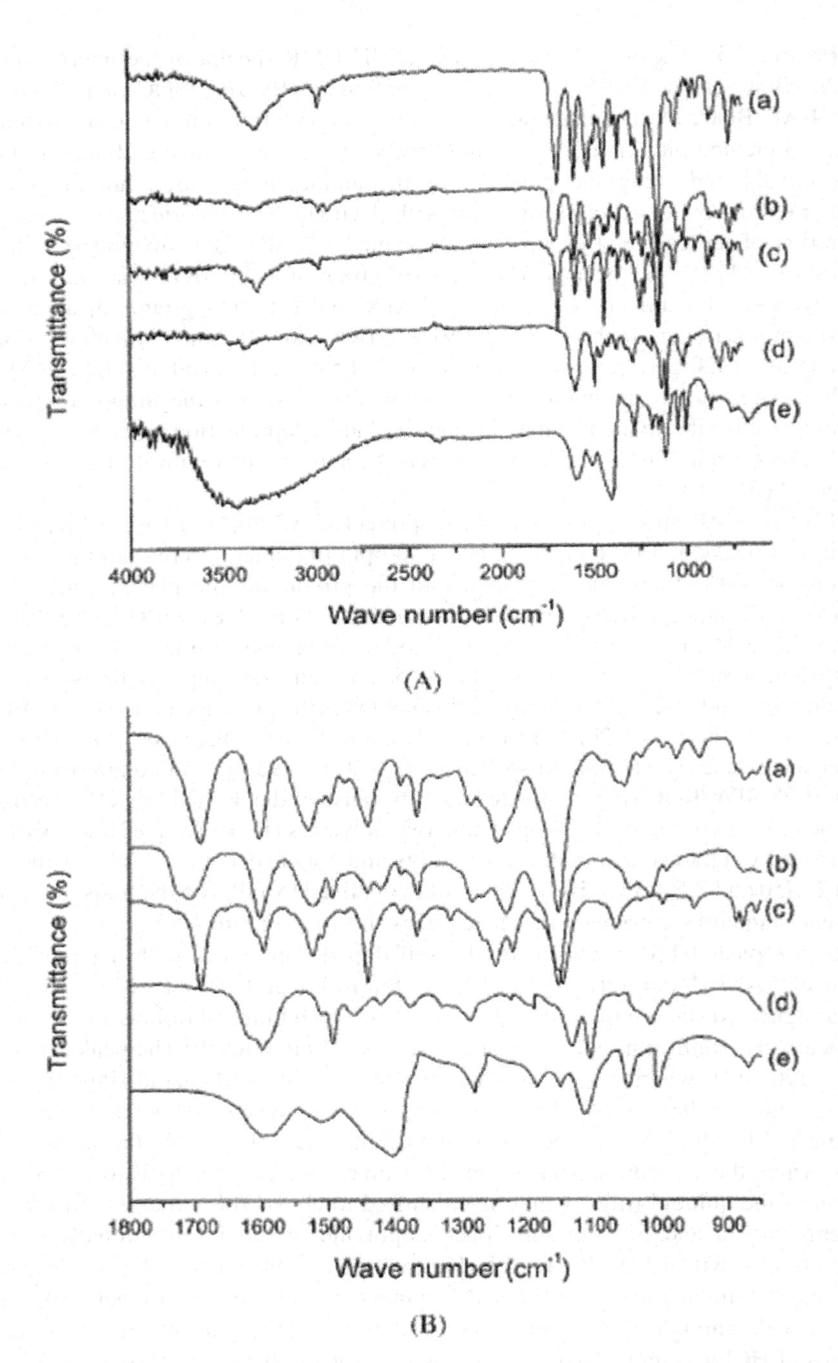

FIGURE 4.5.1 (A) Full–range and (B) partial–scale FTIR spectra of (a) 4-AP-^{t}BOC, (b) 4-VBPA-^{t}BOC, (c) PVBPA-^{t}BOC, (d) PVBPA, and (e) self-doped d-poly(SAN-*co*-ANI) (SAN/ANI = 1/1).

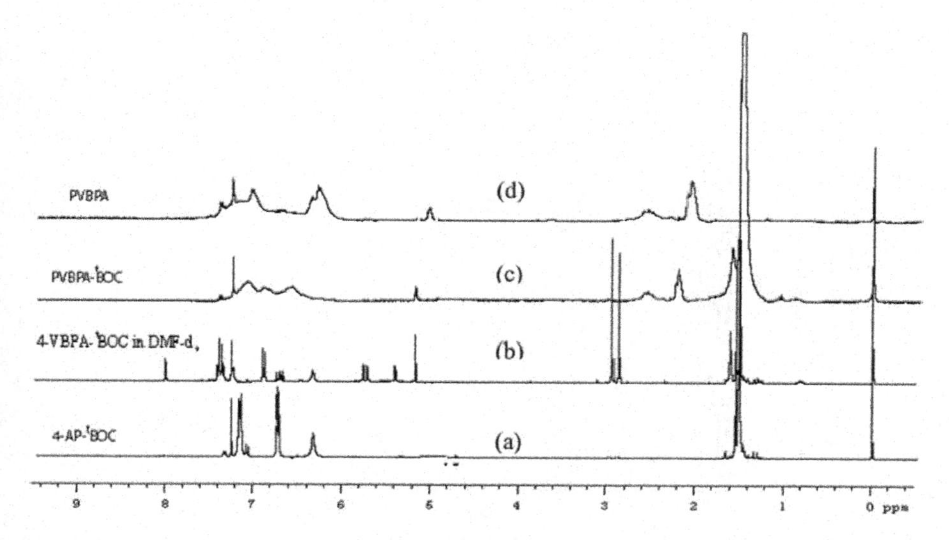

FIGURE 4.5.2 ^{1}H NMR spectra of the intermediates and prepolymers: (a) 4-AP-^{t}BOC in deuterated chloroform, (b) 4-VBPA-^{t}BOC in deuterated DMF, (c) PVBPA-^{t}BOC in deuterated chloroform, and (d) PVBPA in deuterated chloroform.

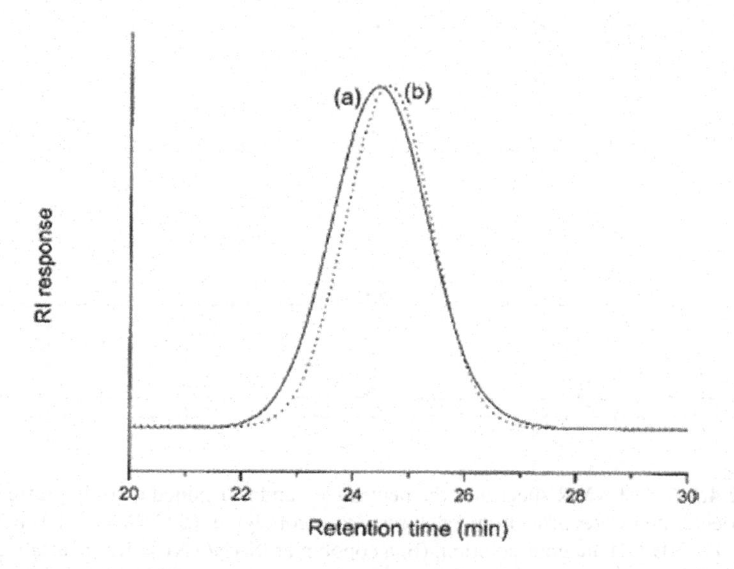

FIGURE 4.5.3 GPC traces of (a) PVBPA-^{t}BOC and (b) PVBPA with DMF as an eluent at 60°C.

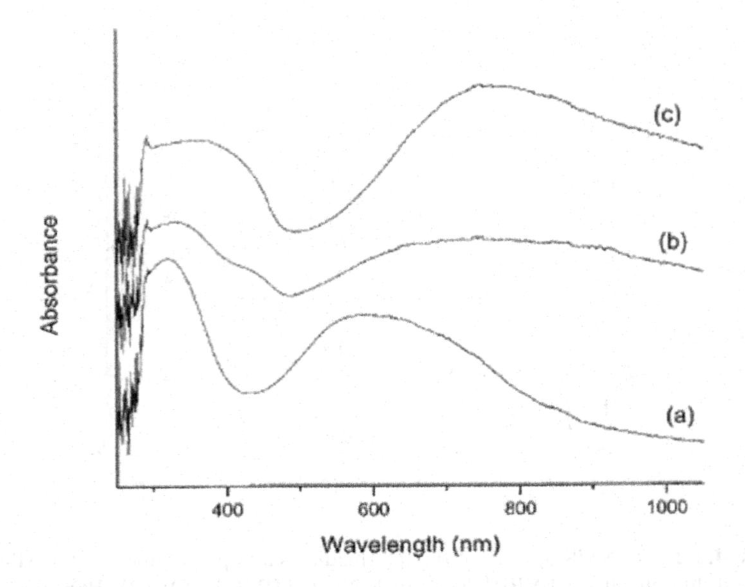

FIGURE 4.5.4 UV–vis spectra of densely grafted copolymer (SAN/ANI = 1/1) in aqueous solutions: (a) neutralized with a 1.0 N NH_4OH aqueous solution, (b) self-doped, and (c) externally doped with a 1.0 N HCl aqueous solution.

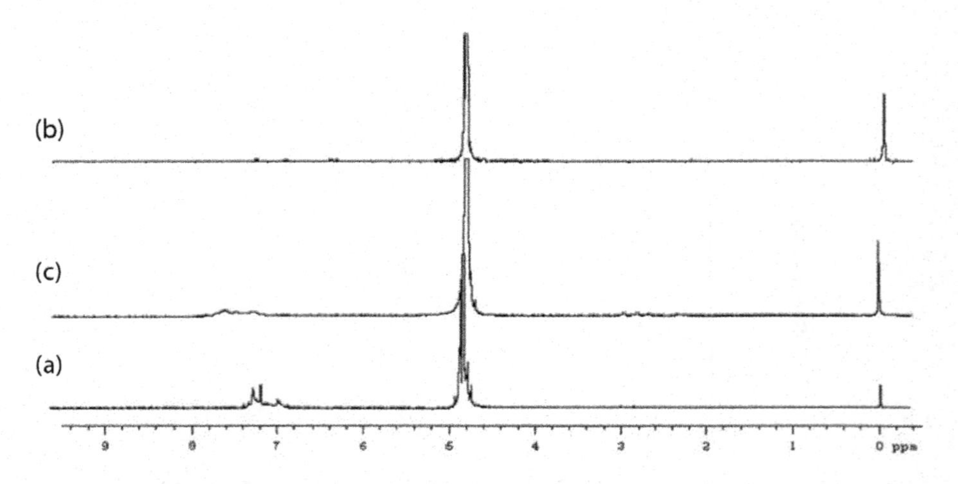

FIGURE 4.5.5 ^{1}H NMR spectra of the neutralized and self-doped densely grafted copolymers in deuterated water after 1 day of storage: (a) a copolymer (SAN/ANI = 1/1) neutralized with a 1.0 N NH_4OH aqueous solution, (b) a copolymer (SAN/ANI = 1/2) neutralized with a 1.0 N NH_4OH aqueous solution, and (c) a self-doped copolymer (SAN/ANI = 1/1).

SAN/ANI = 1/1, weak and broad signals above 6.8 ppm (Figure 4.5.5a), which can be assigned to the protons of the phenyl rings of poly(aniline-*co*-aniline-2-sulfonic salt) side chains, were observed after 1 day of storage. This indicates that in this case the densely grafted macromolecules aggregated because of the hydrophobic interactions due to the polystyrene backbones and ANI segments and a sufficient number of SAN segments exposed to water. Weaker signals in the range of 6.0–7.5 ppm (Figure 4.5.5b) were found for the neutralized copolymers prepared with the lower molar feed ratio of SAN/ANI = 1/2 because of the larger number of ANI segments incorporated into the side chains, which allowed for a larger fraction of the individual macromolecules to aggregate. After ion exchange, the wider signals for the former copolymer (SAN/ANI = 1/1) were shifted downfield above 7.2 ppm (Figure 4.5.5c) because the self-doping on the amine nitrogens by the SO_3H moieties generated a withdrawing effect on the phenyl rings induced by the positive charges localized on nitrogens.

4.5.3.2 Conductivities

Table 4.5.1 shows that the weight fraction of the SAN segments in the copolymer increased from 0.46 to 0.77 when the molar feed ratio (SAN/ANI) was increased from 1/2 to 3/1. However, the SAN weight fraction could not become larger than 0.80 even when the molar feed ratio was increased to 6/1. The reaction rate and yield decreased with an increasing molar feed ratio of SAN to ANI because of the increased amount of SAN, which did not polymerize easily with itself. The conductivity decreased with an increasing molar feed ratio because of the withdrawing effect of the sulfonic acid moieties. The densely grafted copolymers in the

TABLE 4.5.1
Physical Parameters of PANI and Densely Grafted Copolymers

	SAN/ANI (Molar Feed Ratio)	SAN/ANI in the Copolymer (m_2/m_1)[a]	SAN Weight Fraction in the Copolymer[b]	Conductivity (S/cm)[c]
PANI	–	–	–	4.57
l-Poly(SAN-*co*-ANI)	1/1	0.59	0.53	6.63×10^{-2}
d-Poly(SAN-*co*-ANI)	1/2	0.38	0.46	1.90×10^{-1}
d-Poly(SAN-*co*-ANI)	1/1	0.65	0.56	5.73×10^{-2}
d-Poly(SAN-*co*-ANI)	2/1	1.18	0.65	8.34×10^{-3}
d-Poly(SAN-*co*-ANI)	3/1	1.24	0.77	3.42×10^{-4}

[a] The molar ratio of SAN/ANI in the copolymer (m_2/m_1) was determined via chemical elemental analysis.
[b] The SAN weight fraction was estimated with the expression $M_{SAN}/[M_{SAN} + M_{ANI} \times (m_1/m_2)]$.
[c] The conductivities of the polymeric compressed pellets was determined via the four-point method. Each polymer was dried in vacuo for at least 7 days and ground into a powder, which was further compressed into a disk.

neutralized state of Table 4.5.1 exhibited higher water solubilities because of the larger number of sulfonic acid moieties. The conductivity of d-poly(SAN-*co*-ANI) (SAN/ANI = 1/1) was comparable to that of the linear copolymer l-poly(SAN-*co*-ANI) (SAN/ANI = 1/1). Because the nonconducting polystyrene backbone had a low composition ratio, about 3.5 wt%, in the entire densely grafted copolymer, its effect on the conductivity was small.

4.5.3.3 Morphology

The morphology of the self-doped copolymer was investigated with SEM and AFM. Figure 4.5.6 presents time-dependent SEM pictures of a self-doped copolymer prepared with a molar feed ratio of SAN/ANI = 1/1. After the neutralized copolymer was passed through an H^+ ion-exchange resin and kept for 1 h, spherical particles with a diameter of approximately 200 nm were detected [Figure 4.5.6a].

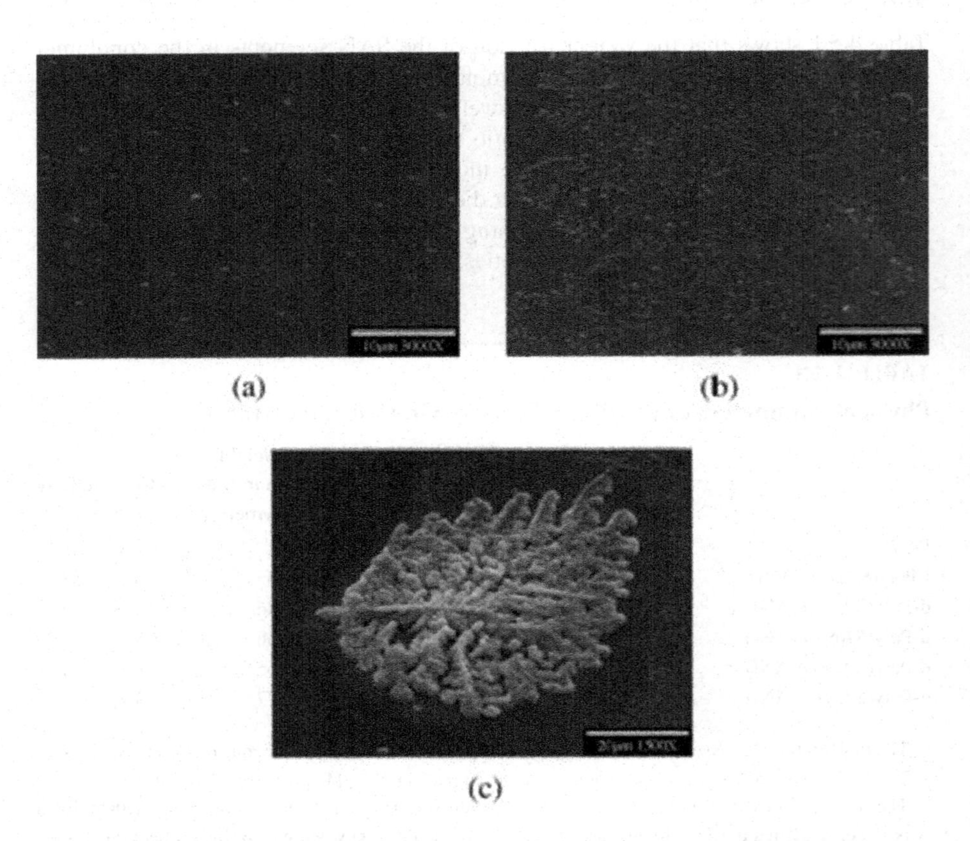

FIGURE 4.5.6 SEM pictures of a self-doped densely grafted copolymer (SAN/ANI = 1/1): (a) after 1 h, (b) after 3 days, and (c) after 7 days.

These particles combined into larger particles and finally formed numerous micrometer structures after 7 days [Figure 4.5.6c]. A similar change with time was observed with AFM, as shown in Figure 4.5.7. The stacking could be related to intermolecular and intramolecular doping. For a densely grafted copolymer, the SO_3H moieties of the SAN segments of the side chains could induce intrachain doping (see Scheme 4.5.2). This interaction caused torsion in the stiff polystyrene backbone and thus produced a spherical structure. The π–π interactions could also enhance the stacking. The d-poly(SAN-*co*-ANI) nanomacromolecules and their microstructures, induced by self-assembly, were expected to be useful for nanosize semiconductor polymer devices because their conductivities were in the range of semiconductors.

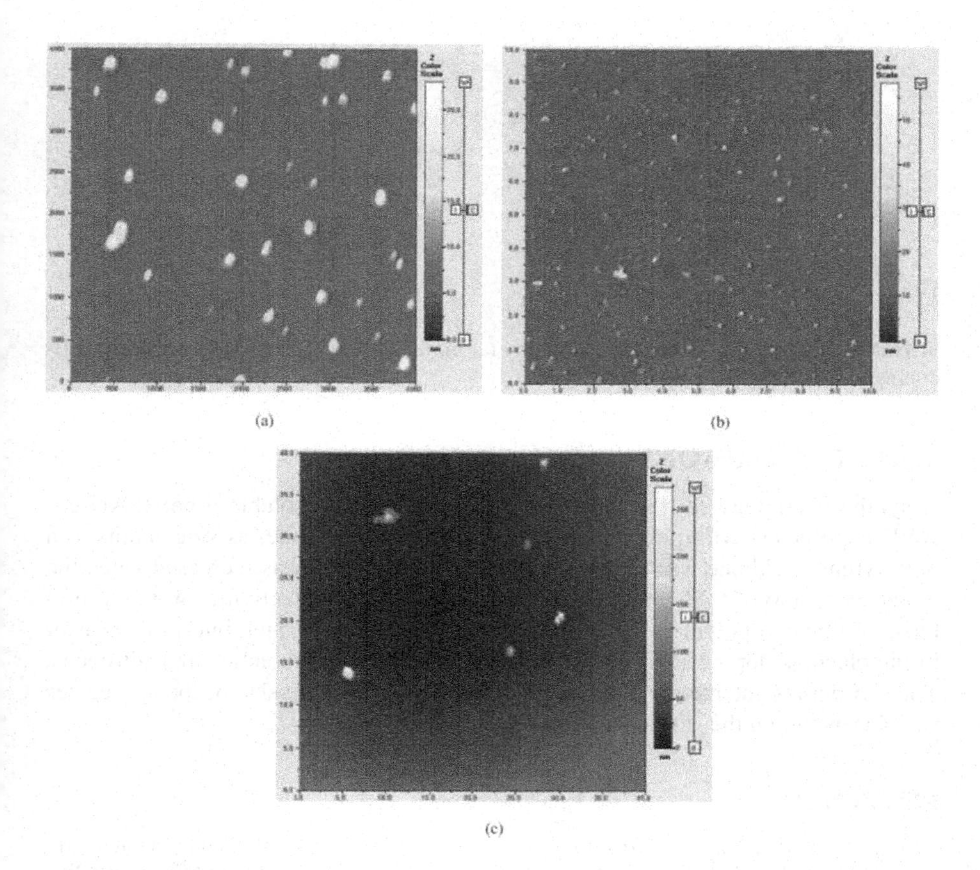

FIGURE 4.5.7 Tapping AFM images of self-doped d-poly(SAN-*co*-ANI) (SAN/ANI = 1/1): (a) after 1 h, (b) after 3 days, and (c) after 1 week.

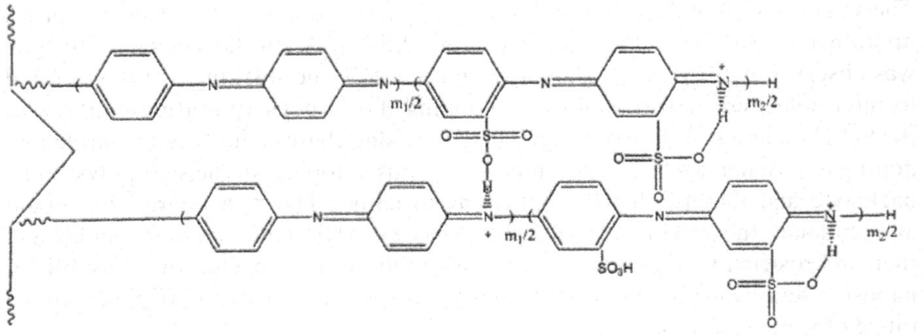

(a)

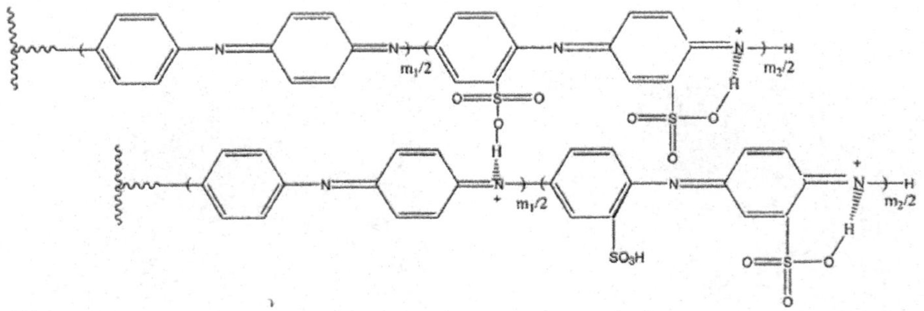

(b)

SCHEME 4.5.2 Self-doped aggregation by (a) intramolecular and (b) intermolecular self-doping of d-poly(SAN-*co*-ANI) copolymers.

4.5.4 CONCLUSIONS

A synthesis strategy for the preparation of various water-soluble d-poly(SAN-*co*-ANI) copolymers with poly(aniline-2-sulfonic acid-*co*-aniline) as side chains to a polystyrene backbone was developed. The water solubility was increased when the molar feed ratio of SAN to ANI was increased. Their conductivities were close to those of linear copolymers. They could dope themselves through intermolecular or intramolecular doping interactions between sulfonic acid moieties and nitrogens. The self-doping interaction caused aggregation in water. The size of the aggregates was dependent on the storage time.

REFERENCES

1. (a) Fredrickson, G. *Macromolecules* 1993, 26, 2825; (b) Wintermantel, M.; Schmidt, M.; Tsukahara, Y.; Kajiwara, K.; Kohjiya, S. *Macromol Rapid Commun* 1994, 15, 279; (c) Tsukahara, Y.; Kohjiya, S.; Tsutsumi, K.; Okamoto, Y. *Macromolecules* 1994, 27, 1662; (d) Wintermantel, M.; Fischer, K.; Gerle, M.; Ries, R.; Schmidt, M.; Kajiwara, K.; Urakawa, H.; Wataoka, I. *Angew Chem Int Ed Engl* 1995, 34, 1472;

(e) Wintermantel, M.; Gerle, M.; Fischer, K.; Schmidt, M.; Wataoka, I.; Urakawa, H.; Kajiwara, K.; Kohjiya, S.; Tsukahara, Y. *Macromolecules* 1996, 29, 978; (f) Dziezok, P.; Sheiko, S. S.; Fischer, K.; Schimdt, M.; Moeller, M. *Angew Chem Int Ed* 1997, 109.
2. (a) Djalali, R.; Hugenberg, N.; Fischer, K.; Schmidt, M. *Macromol Rapid Commun* 1999, 20, 444; (b) Gerle, M.; Fischer, K.; Roos, S.; Mueller, A. H. E.; Schimdt, M. *Macromolecules* 1999, 32, 2629.
3. (a) Kilmartin, A. P.; Wright, G. A. *Synth Met* 1997, 88, 153; (b) Kilmartin, A. P.; Wright, G. A. *Synth Met* 1997, 88, 163; (c) Kinlen, P. J.; Liu, J.; Ding, Y.; Graham, C. R.; Remsen, E. E. *Macromolecules* 1998, 31, 1735; (d) Roy, B. C.; Gupta, M. D.; Bhoumik, L.; Ray, J. K. *Synth Met* 2002, 130, 27.
4. (a) Cataldo, F.; Maltese, P. *Eur Polym J* 2002, 38, 1791; (b) Planes, G. A.; Morales, G. M.; Miras, M. C.; Barbero, C. *Synth Met* 1998, 97, 223; (c) DeArmitt, C.; Armes, S.; Winter, J.; Uribe, F. A.; Gottesfeld, S.; Mombourquette, C. *Polymer* 1993, 34, 158.
5. (a) Yue, J.; Epstein, A. J. *J Am Chem Soc* 1990, 112, 2800; (b) Yue, J.; Wang, Z. H.; Cromack, K. R.; Epstein, A. J.; MacDiarmid, A. G. *J Am Chem Soc* 1991, 113, 2665; (c) Yue, J.; Gordon, G.; Epstein, A. J. *Polymer* 1992, 33, 4409; (d) Wei, X. L.; Wang, Y. Z.; Long, S. M.; Bobeczko, C.; Epstein, A. J. *J Am Chem Soc* 1996, 118, 2545; (e) Wei, X. L.; Fahlman, M.; Epstein, A. J. *Macromolecules* 1999, 32, 3114.
6. Ito, S.; Murata, K.; Teshima, S.; Aizawa, R.; Asako, Y.; Takahashi, K.; Hoffman, B. M. *Synth Met* 1998, 96, 161.
7. (a) Kilmartin, A. P.; Wright, G. A. *Synth Met* 1997, 88, 153; 163; (b) Kinlen, P. J.; Liu, J.; Ding, Y.; Graham, C. R.; Remsen, E. E. *Macromolecules* 1998, 31, 1735; (c) Planes, G. A.; Morales, G. M.; Miras, M. C.; Barbero, C. *Synth Met* 1998, 97, 223; (d) Roy, B. C.; Gupta, M. D.; Bhowmik, L.; Ray, J. K. *Synth Met* 1999, 100, 233; (e) Chan, H. S. O.; Ho, P. K. H.; Ng, S. C.; Tan, B. T. G.; Tan, K. L. *J Am Chem Soc* 1995, 117, 8517; (f) Prevost, V.; Petit, A.; Pla, F. *Synth Met* 1999, 104, 79; (g) Guo, R.; Barisci, J. N.; Innis, P. C.; Too, C. O.; Wallace, G. G.; Zhou, D. *Synth Met* 2000, 114, 267.
8. Wang, P.; Tan, K. L.; Zhang, F.; Kang, E. T.; Neoh, K. G. *Chem Mater* 2001, 13, 581.
9. Yamaguchi, I.; Yasuda, T.; Yamamoto, T. *J Polym Sci Part A: Polym Chem* 2001, 39, 3137.
10. (a) Yin, W.; Ruckenstein, E. *Synth Met* 2000, 108, 39; (b) Yin, W.; Ruckenstein, E. *Macromolecules* 200, 33, 1129.
11. (a) Hua, F. J.; Ruckenstein, E. *Macromolecules* 2003, 36, 9971; (b) Hua, F. J.; Ruckenstein, E. *J Polym Sci Part A: Polym Chem* 2004, 42, 1429; (c) Hua, F. J.; Ruckenstein, E. *Langmuir* 2004, 20, 3954; (d) Hua, F. J.; Ruckenstein, E. *J Polym Sci Part A: Polym Chem* 2004, 17, 4756.
12. Hua, F. J.; Ruckenstein, E. *Macromolecules* 2004, 37, 6104.
13. Avram, M. *Infrared Spectroscopy*; Wiley: New York, 1972.

4.6 Hyperbranched Sulfonated Polydiphenylamine as a Novel Self-Doped Conducting Polymer and Its pH Response*

Fengjun Hua and Eli Ruckenstein

Department of Chemical and Biological Engineering, State University of New York at Buffalo, Buffalo, New York 14260

ABSTRACT A synthesis strategy for a hyperbranched sulfonated polydiphenylamine was developed. First, a hyperbranched polyvinylbenzoxylamine (H-PVBPA) was prepared by controlled atom transfer radical copolymerization of an amino-protected vinylbenzoxylamine (4-VBPA-^{t}BOC) and 4-chloromethyl styrene followed by deprotection. Second, H-PVBPA was reacted with sodium diphenylamine sulfonate under acidic conditions to generate hyperbranched sulfonated polydiphenylamine (H-PSDA). The H-PSDA exhibited conductivities of 3.7×10^{-2} and 1.2×10^{-2} S/cm in HCl-doped and self-doped states, respectively. The microstructure of H-PSDA in aqueous solutions was found to be sensitive to the pH. In the dedoped state (at pH $\geq$ 7), AFM identified nanoscale particles of uniform size around 40 nm. The self-doped H-PSDA aggregated at pH = 7 via intermolecular doping interaction, generating large and irregular particles (>200 nm). When the pH was decreased below 5, the aggregates could be dispersed as smaller particles of about 120 nm because of the replacement of the intermolecular self-doping by the external-doping interaction caused by the small molecules of HCl.

* *Macromolecules* 2005, 38, 888–898.

4.6.1 INTRODUCTION

In the past decade, dendrimers have been extensively studied as materials with novel physical properties.[1,2] They have a very compact structure and can be highly functionalized. They can be prepared by the step-growth polycondensation of AB_n monomers. However, the step-growth synthesis requires strict experimental conditions and even intermediate purifications. In contrast, the synthesis of hyperbranched polymers can provide dendritic molecules in a single, one-pot reaction. The self-condensation vinyl polymerization (SCVP) is a new synthesis method, which can be employed to prepare hyperbranched (co)polymers.[3] An initiator-monomer ("inimer") with the general structure AB*, in which A is a double bond and B* is a functional group, was employed. B* can be transformed to an active center which initiates the polymerization of the double bonds. This general approach was applied to various kinds of living polymerizations, i.e., ionic,[3] radical,[4–6] group transfer,[7] and even ring-opening polymerization,[8] because the B* group can be transformed into a cationic, radical, or anionic active center. In recent years, the SCVP synthesis concept was used to prepare hyperbranched polymers in a single, one-pot controlled radical polymerization. Three controlled (living) radical polymerizations, namely the stable free radical polymerization (SFRP),[9] the atom transfer radical polymerization (ATRP),[10] and the reversible addition–fragmentation chain transfer polymerization (RAFT),[11] have been developed. A radical SCVP of styrene was carried out in the presence of vinyl-functionalized 2,2,6,6-tetramethylpiperidinyloxy (TEMPO) by the SFRP method.[6,12] Matyjaszewski et al. developed an ATRP method, which allowed the controlled radical polymerization of some vinyl-functionalized inimers.[13] The 4-chloromethylstyrene (CMS) inimer was homopolymerized to hyperbranched polystyrene (H-PS) and also copolymerized with styrene by the ATRP method.[14] Yang et al. reported an inimer which could be copolymerized with styrene by the RAFT method.[15] Furthermore, Mueller et al.[16] synthesized the macroinimer acryoyl-functionalized poly(*tert*-butyl acrylate), which has a polymerizable double bond at one end and a bromine atom at the other end that can act as initiator in the ATRP method.

As an important conjugated polymer, polyaniline (PANI) was extensively studied in the past decades.[17–20] PANI was easily prepared from aniline via an acid-mediated radical mechanism. Its poor processability and solubility in common solvents were caused by the stiffness of its backbone and the inter-hydrogen bonding interaction with the amino moieties of the adjacent chains. To improve the solubility of PANI in organic compounds and its processability, many approaches have been employed, such as the copolymerization of aniline with its derivatives[21] and substitutions on the emeraldine base, either at the N sites or on the phenyl rings of the PANI backbone.[22] Synthesis methods through modifications of the PANI backbone were also employed to produce water-soluble polyaniline.[23,24]

In the present paper, a hyperbranched sulfonated polydiphenylamine was prepared. First, a styryl-substituted aniline macromonomer, namely 4-(4-vinylbenzoxyl) (*N-tert*-butoxycarbonyl) phenylamine (4-VBPA-^{t}BOC), was prepared by reacting 4-aminophenol with di-*tert*-butoxyl dicarbonate followed by the substitution with 4-vinylbenzyl chloride. The amino-protected 4-VBPA-^{t}BOC was subjected to atom radical copolymerization with CMS, which acted as an inimer. The subsequent deprotection with trifluoroacetic acid (TFA) produced hyperbranched poly(4-(4-vinylbenzoxyl) phenylamine) (H-PVBPA) containing pendent amine moieties. Second, the H-PVBPA was reacted with a water-soluble monomer, sodium diphenylamine sulfonate (SDAS) to generate hyperbranched sulfonated polydiphenylamine (H-PSDA). The conductivities of the final H-PSDA self-doped or HCl-doped were determined by the four-point method, and the electromagnetic shielding effectiveness (EMI SE) of the self-doped copolymer over a frequency range from 10 to 10^3 MHz was also measured. The behavior of the H-PSDA in aqueous solutions at various pH values was characterized by UV-vis, AFM, and SEM. Furthermore, the oxidation levels of these copolymers at various doping states were examined by X-ray photoelectron spectroscopy (XPS).

4.6.2 EXPERIMENTAL SECTION

Materials. 4-Aminophenol (98+ wt%), sodium diphenylamino-4-sulfonic, hydrochloric acid (37 wt%), aniline (99 wt%), di*tert*-butyldicarbonate (97 wt%), trifluoroacetic acid (TFA, 97 wt%), ammonium persulfate (98 wt%), 4-choromethyl styrene (97 wt%), *N, N,N′,N′,N″*-pentamethyldiethylenetriamine (PMDETA, 98 wt%), triethylamine (97 wt%), CuCl (99+ wt%), ethyl-2-chloropropionate (EPNCl, 97 wt%), and potassium carbonate (99 wt%) were purchased from Aldrich and used without further purification. The solvents, dimethylformamide (DMF), tetrahydrofuran (THF), and dichloromethane (DCM), purchased from Aldrich, were of HPLC purity. An H^+-type ion-exchange resin (H^+-type AMBERJET 1200 H resin) was purchased from Aldrich (Trademark of Rohm and Haas Co.).

N-tert-butoxylcarbonyl-4-aminophenol (4-AP-^{t}BOC) and 4-(4-vinylbenzoxyl) (*N-tert*-butoxycarbonyl) phenylamine) (4-VBPA-^{t}BOC) were prepared as described in refs 25 and 26, respectively.

ATRP of 4-VBPA-^{t}BOCA in the Presence of CMS. 4-VBPA-^{t}BOC (7.5 mmol), 0.25 mmol of CMS, 0.25 mmol of CuCl, and 0.375 mmol of PMDETA were introduced into a 50 mL flask containing 10 mL of THF. The system was degassed three times using thaw–freeze cycles, and then the temperature was raised to 60°C and the reaction allowed to proceed for 48 h more under a dry nitrogen atmosphere. The increase with time of the viscosity indicated the start of polymerization. Small samples were taken out periodically for NMR determinations. After reaction, 30 mL of THF was introduced to dilute the viscous paste in the flask. To remove the PMDETA and CuCl, the THF solution was passed through a neutral Al_2O_3 column. The light yellow solution obtained was first concentrated by removing the THF using a rotary evaporator and then subjected to precipitation in a mixture of methanol/water (5:1, v/v). Further purification was achieved by dissolution in THF followed by precipitation in the above nonsolvent mixture. A light yellow precipitate, hyperbranched PVBPA-^{t}BOC (H-PVBPA-^{t}BOC), was obtained and dried under vacuum for at least 24 h at room temperature with a yield of 46 wt%. FTIR (cm^{-1}): 3300–3150 (–NH), 2900 (phenyl), 1750–1670 (–C=O in ^{t}BOC). ^{1}H NMR (ppm): 7.4–6.7 (m, 8H), 6.34 (s, 1H), 5.05 (s, 2H), 2.75 (m, 1H), 1.90 (m, 2H), 1.53 (m, 9H).

A controlled linear PVBPA-^{t}BOC was prepared using the same procedure but replacing CMS with EPNCl.

General Procedure of Deprotection of H-PVBPA-^{t}BOC. H-PVBPA-^{t}BOC (2.4 g) was dissolved in 60 mL of DCM in a 100 mL flask. TFA (4.0 g) was added and the system subjected to intensive stirring. The system was sealed in a dry nitrogen atmosphere, the temperature was raised with a slight reflux to about 60°C, and the reaction was allowed to proceed for 6 h. Two layers were formed, and the DCM and the excess of TFA were removed under vacuum. Then, triethylamine (3.0 g) and 100 mL of DCM were introduced to precipitate the salt. The DCM was washed with 50 mL of water until neutral and dried with Na_2SO_4. After the DCM was removed using a rotary evaporator, a red solid, PVBPA, was obtained with a yield of 90 wt%. FTIR (cm^{-1}): 3350–3250 ($-NH_2$), 2900 (phenyl), 1610 (C–NH, def). ^1H NMR (ppm): 7.4–5.8 (m, 8H), 4.95 (s, 2H), 2.38 (s, 1H), 1.46 (m, 2H).

Copolymerization of Hyperbranched PVBPA and Sodium Diphenylamine Sulfonate (SDAS). H-PVBPA (0.133 g), 1.14 g of $(NH_4)_2S_2O_8$, and 2.7 g of SDAS were introduced into a 50 mL flask containing 30 mL of 1.2 N HCl. The system was subjected to intensive stirring and cooled to 5°C. After 8 h, the copolymer was separated by centrifugation at 3500 rpm for 30 min in a super-speed centrifuge (RC-5B, Dupont) and the precipitate was washed three times with water/ethanol (10:1, v/v) until a colorless supernatant was obtained. The precipitate was neutralized with a 1.0 N ammonium aqueous solution and the neutralized copolymer was further purified by dialysis. The neutralized solution of 0.01 g/L could be passed through the PTFE filter membrane with a pore size of 250 nm. The dialysis bag (Fisherbrand, Nominal MWCO 60,000, which separates the linear and low-molecular-weight homopolymers or copolymers), containing a neutralized copolymer aqueous solution (the copolymer was dissolved in 50 mL of water), was kept in a 0.01 N ammonium aqueous solution for 7 days until no colored supernatant was detected. This was followed by dialysis in water for additional 7 days for the solution to become neutral. A self-doping copolymer solution was prepared by passing the obtained solution through an H^+ ion-exchange column containing an IR 1200 H resin by ion exchanging NH_4^+ with H^+.[27]

Characterization. Proton (^1H) NMR, UV–vis absorption, and FTIR measurements were carried out on a 500 MHz INOVA-500, a Thermo Spectronic Genesys-6, and a Perkin- Elmer-FTIR 1760 spectrometer, respectively. The solutions for NMR were prepared by dissolving the polymer in deuterated DMF, deuterated chloroform, deuterated DMSO, or deuterated water. Gel permeation chromatography (GPC, Waters) was used to evaluate the molecular weights of the polymers on the basis of a polystyrene calibration curve. The GPC was equipped with three 30 cm long columns filled with Waters Styragel, a Waters HPLC 515 pump, and a Waters 410 RI detector. The GPC measurements were carried out at 60°C using DMF as eluent at a flow rate of 1.0 mL/min and 1.0 cm/min chart speed. The elemental analysis of the polymer samples was carried out on a Perkin-Elmer Model 2400 C, H, N analyzer. The chlorine and sulfur contents were determined by the oxygen flask method. The room-temperature conductivities of the compressed pellets of the self-doped and HCl-doped hyper-branched copolymers were determined using the conventional four-point probe method. The electromagnetic shielding effectiveness (EMI SE) was determined over a frequency range from 10 to 10^3 MHz using the coaxial cable method. The setup consisted of an Elgal (Israel) SET 19 A shielding effectiveness tester with its input and output connected to a HP 8510A network analyzer.

An HP APC-7 calibration kit was used to calibrate the system. The EMI SE values, expressed in dB, were calculated from the ratio of the incident to transmitted power of the electromagnetic wave using the following equation:

$$SE = 10\log[P_1/P_2] \quad (decibels,\ dB)$$

where P_1 and P_2 are the incident power and the transmitted power, respectively.

Scanning electron microscopy (SEM) was used to examine the surface morphology. The polymer samples were prepared through spin-coating polymer aqueous solutions of 0.1 g/L on freshly cleaned mica plates at 1000 rpm for 1 min followed by sputter-coating with carbon (to prevent sample-charging) and examined using a Hitachi S-4000 Field Emission Scanning Electron Microscope operated at 20 keV. The atomic force microscope (AFM, Quesant Scan Atomic Company) was used to investigate the molecular conformation of the macromolecules and the structure of the aggregates. Polymer aqueous solutions (0.1 g/L) at various pH values were spin-coated for 1 min onto freshly cleaned silicon wafers at 1000 rpm, and subsequently examined under AFM in the tapping mode. The XRD of the polymer particles was recorded on a SIEMENS, D5000, X-ray diffractometer with Cu Kα radiation of 1.5406 Å wavelength operated at a maximum of 30 MA and 40 KVP.

4.6.3 RESULTS AND DISCUSSION

Synthesis and Characterization. A hyperbranched sulfonated polydiphenylamine (**7**) was prepared via the six successive steps presented in Scheme 4.6.1. The vinyl-functionalized monomer, 4-VBPA-^{t}BOC (**3**), was first obtained via the amino-protected reaction of 4-aminophenol with ^{t}BOC to generate ^{t}BOC-protected aminophenol (4-AP-^{t}BOC, **2**). This reaction was followed by the substitution reaction of 4-chloromethyl styrene to the hydroxide moiety of 4-AP-^{t}BOC (**2**). Further, 4-VBPA-^{t}BOC (**3**) was subjected to ATRP copolymerization using CMS as inimer. The active site, the chloride moiety, initiated the polymerization of 4-VBPA-^{t}BOC, generating branching and thus forming a hyperbranched PVBPA-^{t}BOC (**4**). After deprotection with TFA, a hyperbranched PVBPA (**5**) was obtained by removing the ^{t}BOC moieties. Then, sodium diphenylamine sulfonate was allowed to react with the amine moieties of H-PVBPA to generate a hyperbranched poly(sodium diphenylamine sulfonate) (**6**). Finally, the ion-exchange reaction between **6** and the H^+ of an ion-exchange resin resulted in an acidified hyperbranched H-PSDA (**7**). Figure 4.6.1 presents the ^{1}H NMR spectra of compounds **2**, **3**, **4**, and **5**. After the ^{t}BOC protection, the proton signal of the amino moiety of **2** was shifted to a lower field, 6.34 ppm, because of the shielding due to the ^{t}BOC. For compound **3**, the proton signals which can be assigned to the vinyl moiety are located at 6.67, 5.6, and 5.2 ppm, respectively (deuterated DMF as solvent). The signals due to the amino and methyl moieties of ^{t}BOC are still present at 6.41 and 1.56 ppm, respectively, and the signal due to hydroxide moiety at 7.48 ppm disappeared, indicating that the styrene moiety was successfully attached to aminophenol. The vinyl-functionalized monomer (**3**) was copolymerized with CMS by the ATRP method (the polymerization kinetics will

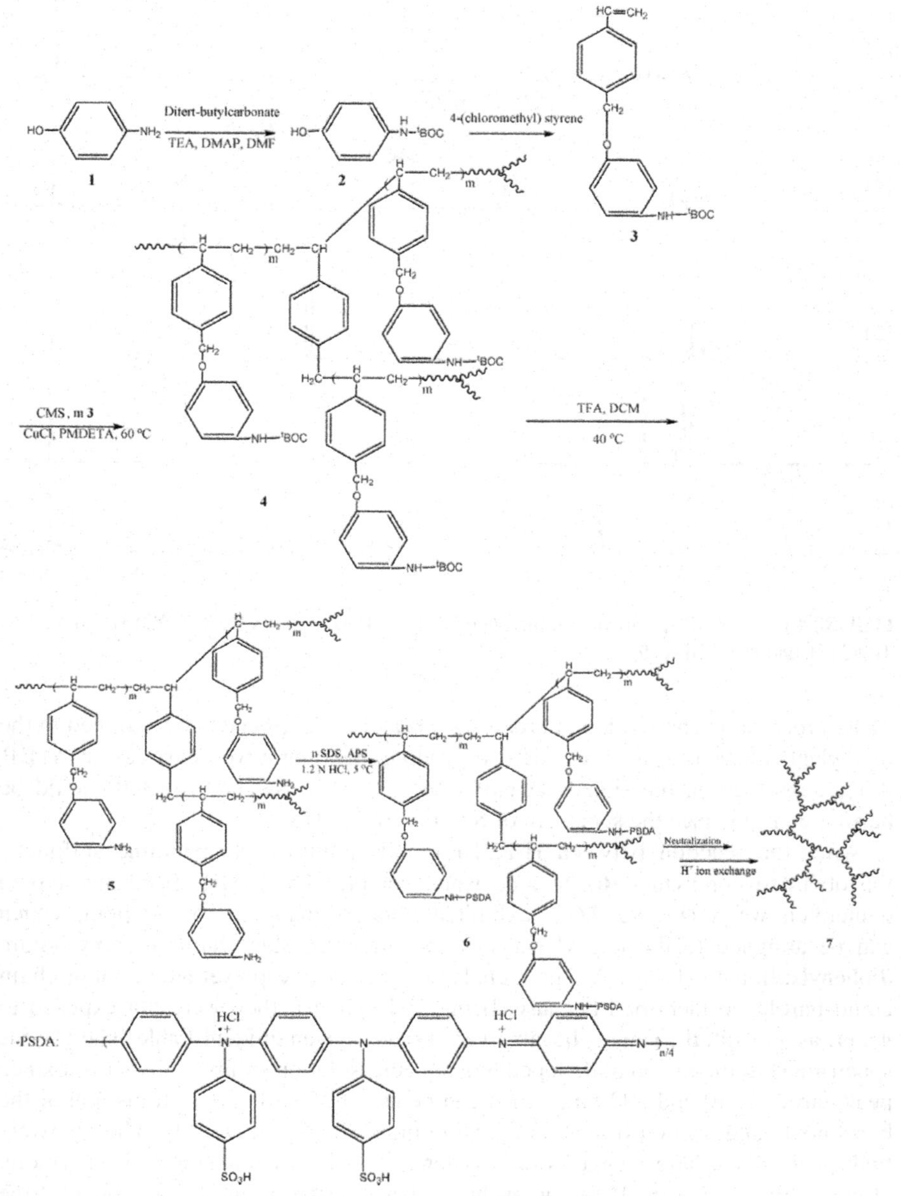

SCHEME 4.6.1 Synthesis strategy of hyperbranched PSDA.

be discussed later). The broad signals of the obtained hyperbranched PVBPA (**4**) at 7.4–6.7 ppm can be assigned to the substituted phenyl moieties of **4** and CMS, and the signals at 2.75 and 1.90 can be assigned to methylene and methane of the ethylene originated from the vinyl moiety (**3**). The other signals at 6.34 and 1.53 ppm can be assigned to ^{t}BOC, indicating that the polymerization did not destroy the

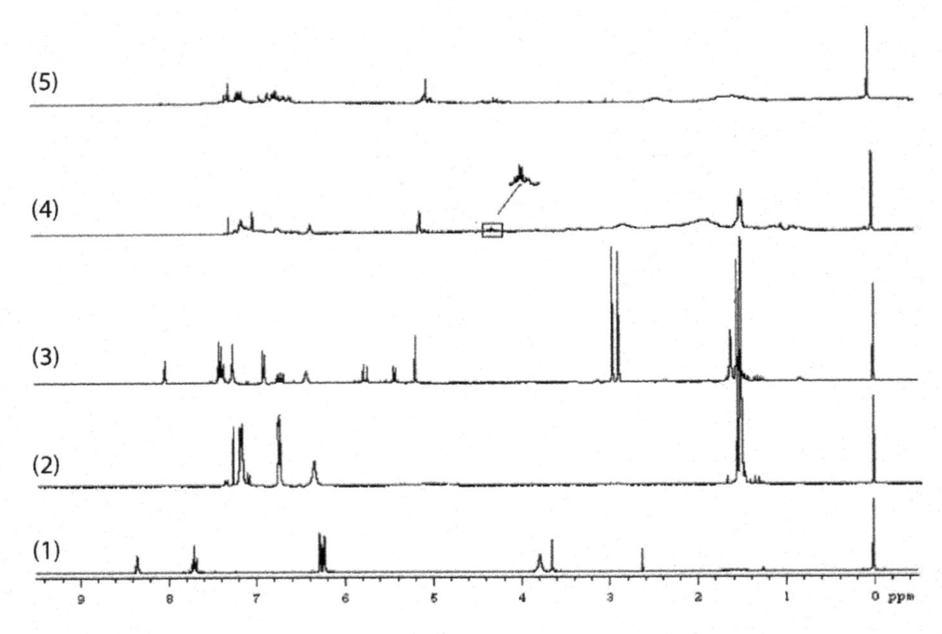

FIGURE 4.6.1 NMR spectra of 4-aminophenol (**1**), 4-AP-^{t}BOC (**2**), VBPA-^{t}BOC (**3**), H-PVBPA-^{t}BOC (**4**), and H-PVBPA (**5**).

^{t}BOC protection. The weak and broad signals at 4.2–4.7 ppm can be assigned to the methylene adjacent to the chloride end group, indicating a well-carried-out ATRP. After deprotection, the signals at 6.34 and 1.53 ppm assigned to ^{t}BOC could be hardly identified (see the spectrum of **5** in Figure 4.6.1).

After the reaction between H-PVBPA and sodium diphenylamine sulfonate, the obtained copolymer (**6**) became water soluble. The NMR spectrum of **6** in deuterated water (Figure 4.6.2) exhibited broad signals at 6.5–8.0 ppm, which can be assigned to the phenyl rings of the branched side chains of polysodium diphenylsulfonate (PSDAS). The signals assigned to the polyethylene main chain could hardly be identified because, being hydrophobic, they were not exposed to water; as a result, the signals became very broad or even unidentifiable. The UV-vis spectrum (Figure 4.6.3a) of dedoped **6** in aqueous solution exhibited two absorbance peaks around 310 and 590 nm, which can be assigned to the π–π^* transition of the benzenoids and exciton transition of the quinoid rings, respectively. The relatively high peak of the latter suggests the existence of a long conjugation. Furthermore, after acidification with H^+ by ion exchange between NH_4^+ and H^+, the peak of the self-doped copolymer was shifted above 800 nm (see Figure 4.6.3b) in aqueous solution, indicating the formation of polaron structures which contribute to the conductivity. It should be noted that the spectrum of the self-doped polymer (Figure 4.6.3b) is very similar to the spectrum of the doped oligoaniline in the semi-oxidized state, consistent with the chemical nature of the hyperbranched product.

Atom Transfer Radical Copolymerization between 4-VBPA-^{t}BOC and CMS. As noted above, 4-VBPA-^{t}BOC can be polymerized via the ATRP method.

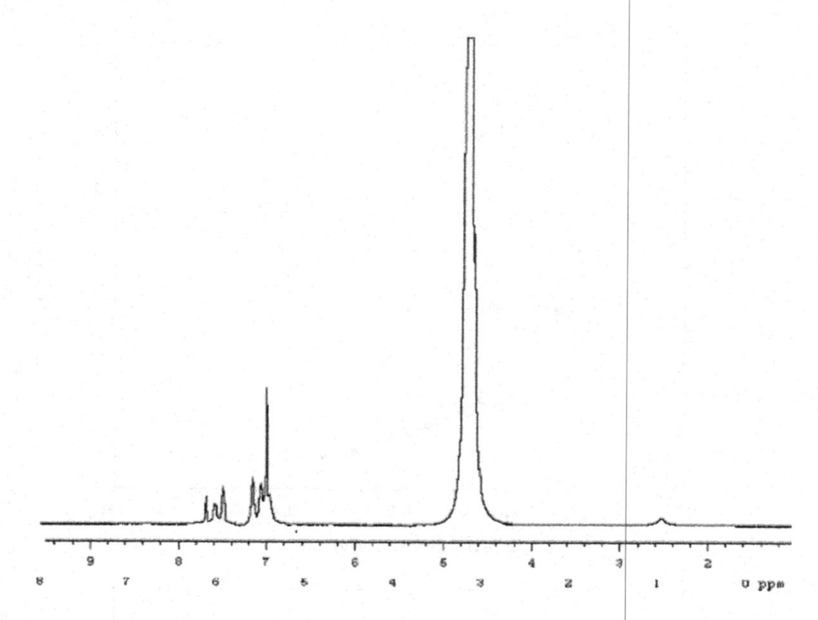

FIGURE 4.6.2 NMR spectra of self-doped H-PSDA (**6**) in deuterated water.

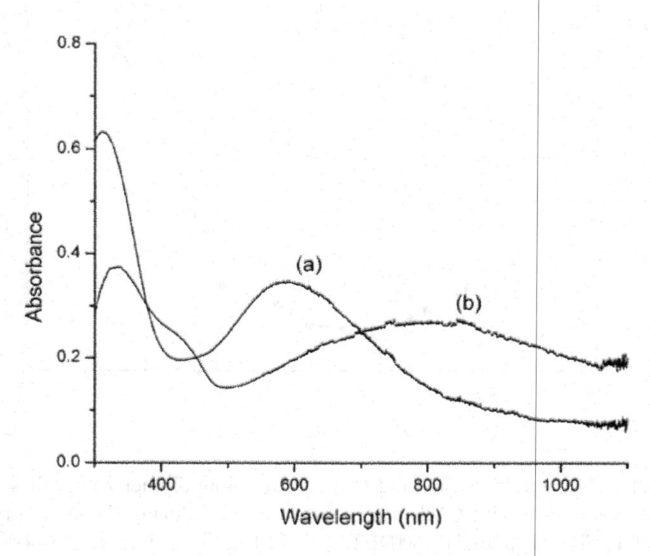

FIGURE 4.6.3 UV-vis spectra of (a) dedoped and (b) self-doped H-PSDA in aqueous solution.

The polymerization was followed by NMR and GPC. The signals which can be assigned to the vinyl moiety decreased with increasing reaction time. The conversion (α) of the double bonds was calculated using the expression $\alpha = (A^0_{5.6}/A^0_{6.34}\text{-}A^t_{5.6}/A^t_{6.34})/(A^0_{5.6}/A^0_{6.34})$, where $A^0_{5.6}$, $A^0_{6.34}$ and $A^t_{5.6}$, $A^t_{6.34}$ are the integral areas of the signals at 5.6 (vinyl proton) and 6.34 ppm (amino proton) at

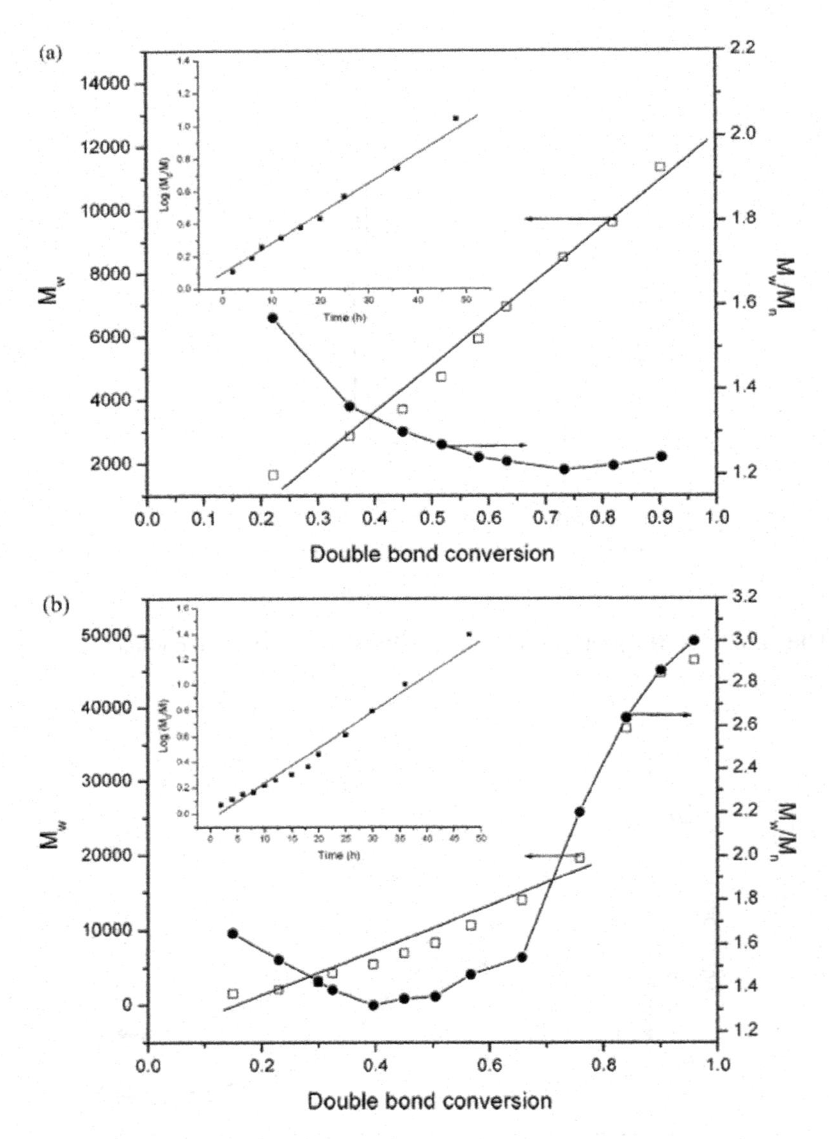

FIGURE 4.6.4 Dependence of M_w and PDI for (a) homopolymerization of 4-VBPA-^{t}BOC and (b) copolymerization with CMS on the reaction time during the ATRP processes. (a) [4-VBPA-^{t}BOC]/[EPNCl]/[CuCl]/[PMDETA] = 30:1:1:1.5, 60°C, and (b) [4-VBPA-^{t}BOC]/[CMS]/[CuCl]/[PMDETA] = 30:1:1:1.5, 60°C.

times $t = 0$ and t, respectively.[14–16] The kinetic results are plotted in Figure 4.6.4. In the absence of CMS, the weight-average molecular weight increased linearly with the double-bond conversion and the polydispersity of the molecular-weight distribution became as low as PDI = 1.2 (Figure 4.6.4a). Consequently, the homopolymerization of 4-VBPA-^{t}BOC proceeded in a living manner. However, in the presence of CMS, the molecular weight increased almost linearly with the conversion only for conversion smaller than 70%, consistent with a linear "living" polymerization process (see Figure 4.6.4b). It appears that the growing polymer chains are not significantly incorporated into each other until later during polymerization, when a significant deviation from the above behavior occurs. CMS has a double bond and a chloride atom. The latter atom can initiate the polymerization of 4-VBPA-^{t}BOC, and as a result, the growing PVBPA chains have a double bond at the chain end. This double bond can be incorporated into another growing macromonomer, generating a branching. The growing chain can either include additional monomer units or other chains of polystyrene. As a result, the molecular weight of the hyperbranched copolymer increased dramatically. The polydispersity of the hyper-branched copolymer became broad with a PDI up to 3.0 after 48 h (see Figure 4.6.4b). Figure 4.6.5 presents GPC traces of the intermediate copolymers as a function of time. The molecular weight increased dramatically in the late stages of copolymerization in the presence of CMS, becoming even 45,000 after 48 h, thus deviating from the homopolymerization linearity (see Figure 4.6.4a).

pH Response of H-PSDA. The UV spectra of the hyperbranched PSDA at various pH values are presented in Figure 4.6.6. In the dedoped state, the absorption peak around 590 nm was slightly blue-shifted with increasing pH. The sulfonic moieties no longer self-doped the nitrogens at high pH values. As already mentioned, in the self-doped state, the absorption occurred around 800 nm due to the polaron formation. The self-doping can be caused by intra- or intermolecular interactions between the sulfonic acid moieties and the nitrogens, as shown in Scheme 4.6.2.

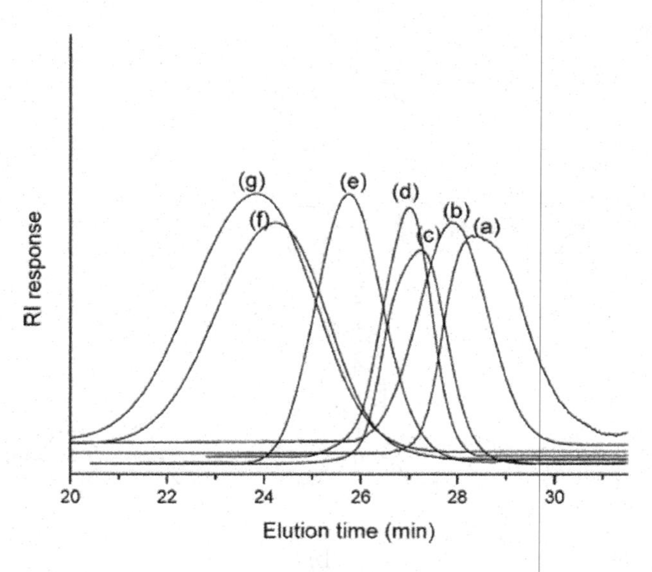

FIGURE 4.6.5 GPC traces of the prepolymers during the ATRP process of 4-VBPA-^{t}BOC in the presence of CMS as a function of the reaction time: (a) 2, (b) 6, (c) 10, (d) 12, (e) 18, (f) 25, and (g) 36 h. Reaction conditions are the same as those in Figure 4.6.4b.

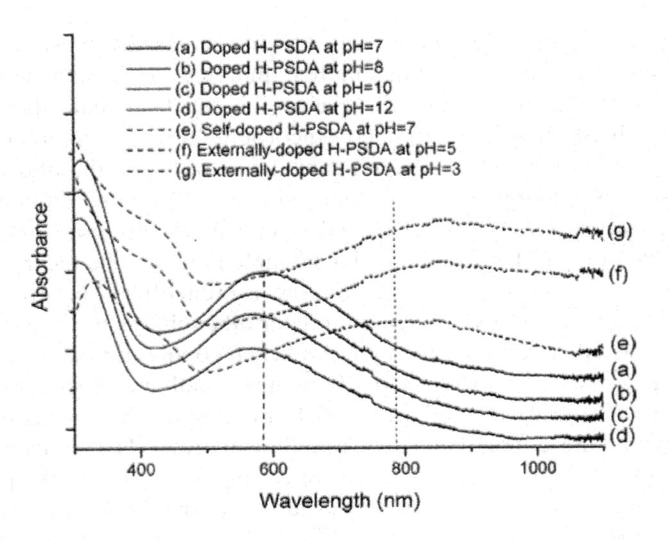

FIGURE 4.6.6 UV-vis spectra of the H-PSDA aqueous solutions at various pH values.

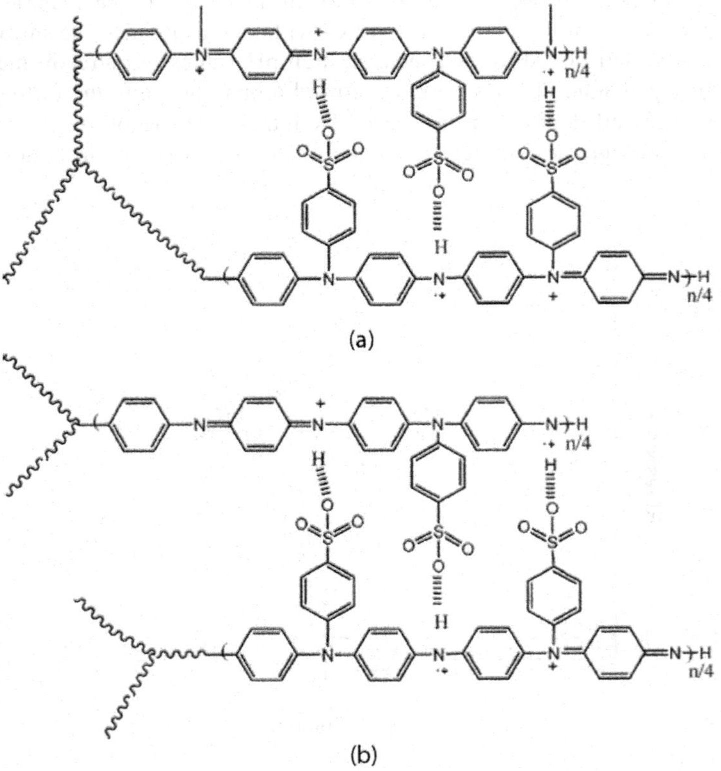

(a)

(b)

SCHEME 4.6.2 Self-doped mechanism of hyperbranched PSDA. (a) Intramolecular self-doping and (b) Intermolecular self-doping.

At pH values <7, the peak around 800 nm was red-shifted. The HCl external doping can overwhelm the self-doping because of the weak connection in the latter case. The smaller HCl molecules can easily replace the sulfonic moieties in the doping process.

The pH response of the H-PSDA conformation in aqueous solutions was investigated using AFM and SEM (see Figures 4.6.7 and 4.6.8). For the dedoped H-PSDA,

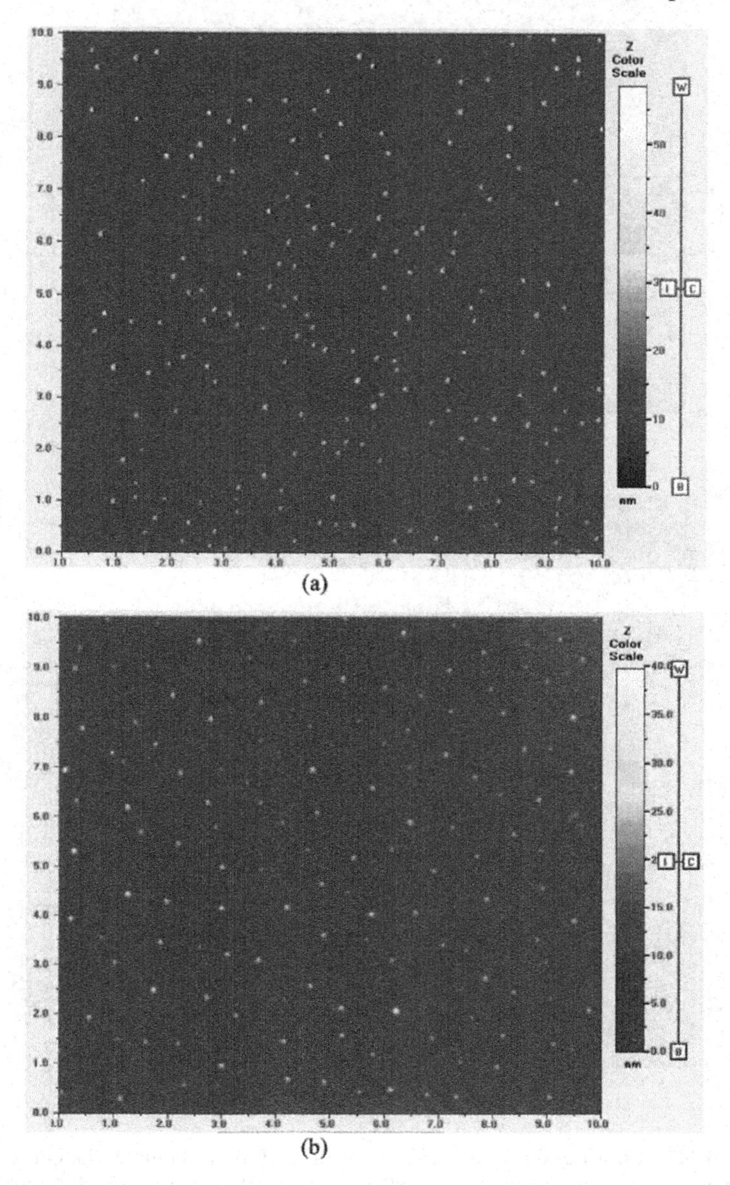

FIGURE 4.6.7 AFM images of the H-PSDA obtained via spin-coating on mica plates with a concentration of 0.1 g/L at various pH values: (a) dedoped PSDA at pH = 7, (b) dedoped PSDA at pH = 10. *(Continued)*

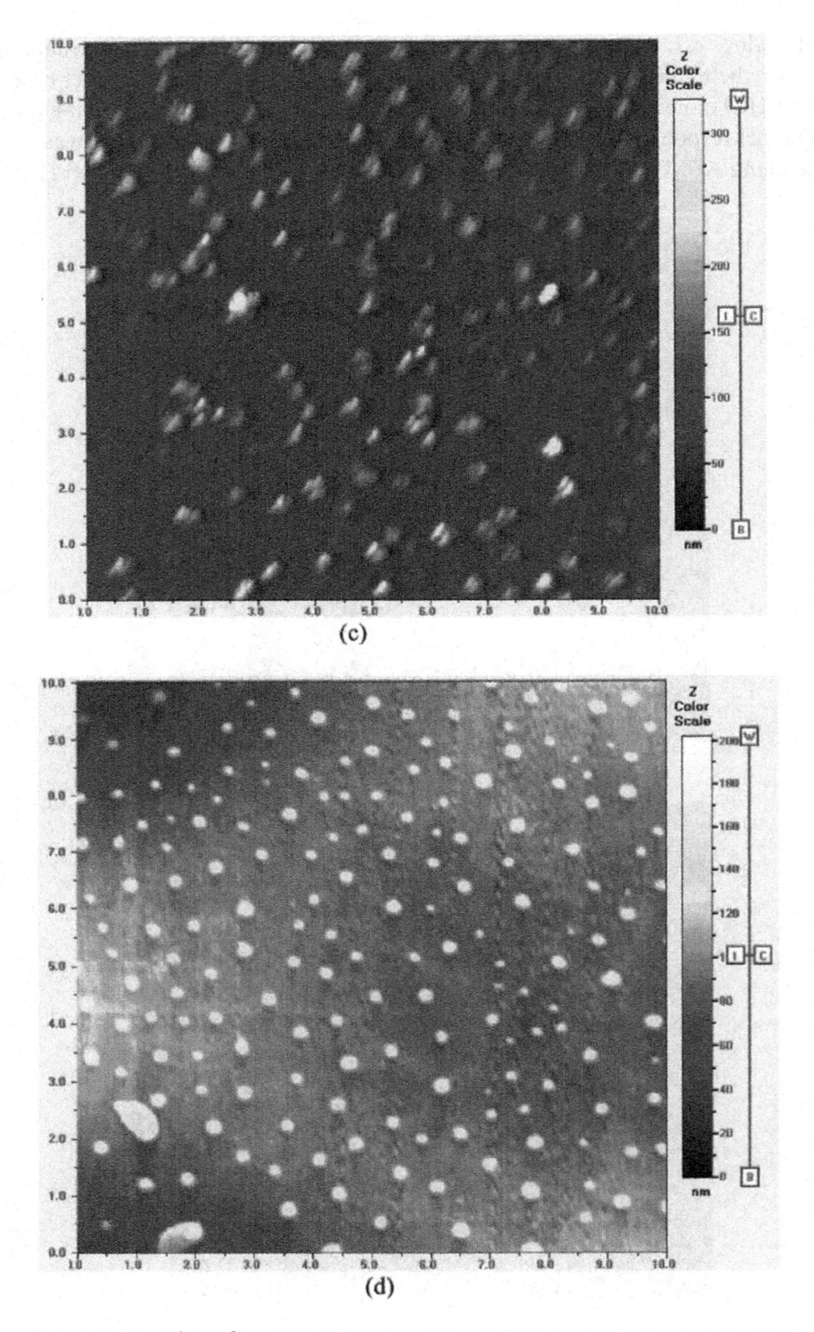

FIGURE 4.6.7 (Continued) AFM images of the H-PSDA obtained via spin-coating on mica plates with a concentration of 0.1 g/L at various pH values: (c) self-doped PSDA at pH = 7, and (d) externally doped PSDA at pH = 3.

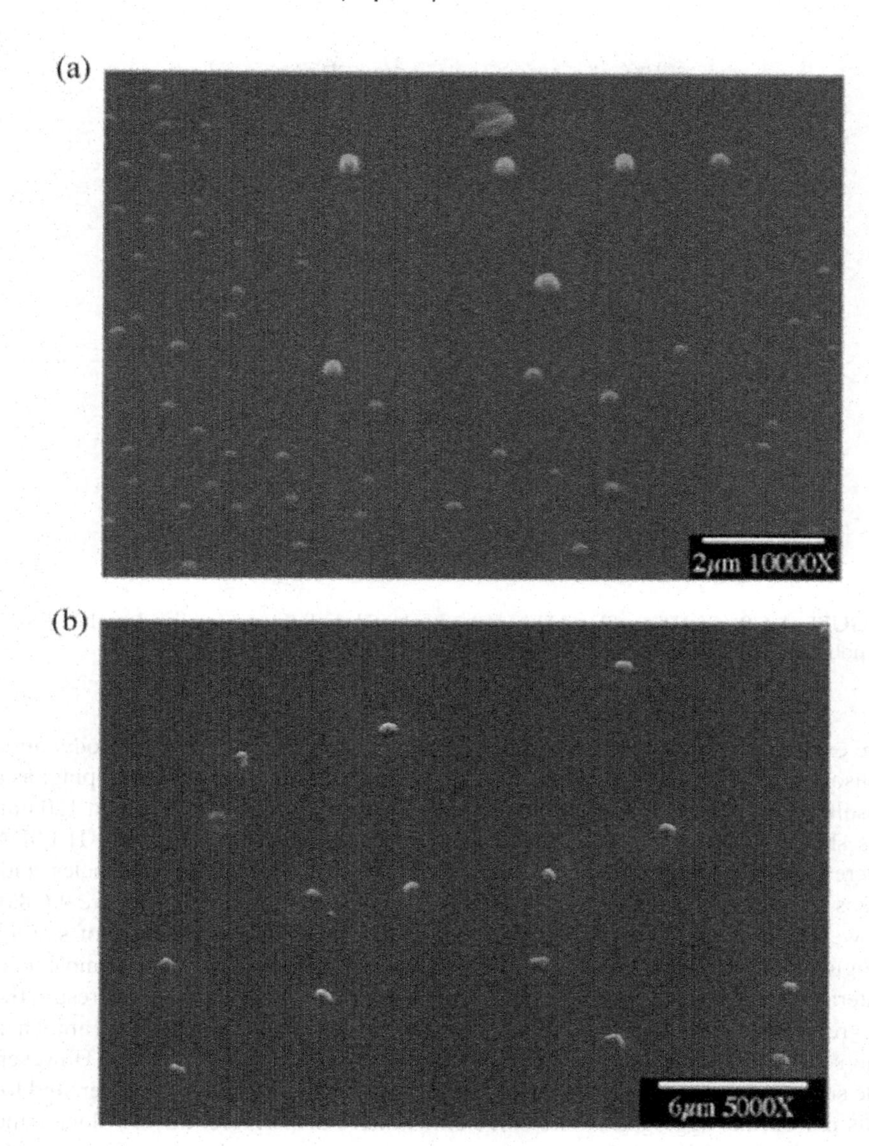

FIGURE 4.6.8 SEM pictures of self-doped H-PSDA obtained via spin-coating on mica plates with a concentrations of 0.1 g/L at pH = 7 as a function of the storage time: (a) 1 day of storage and (b) one week of storage.

the particle size decreased slightly with increasing pH from 7 to 10 from about 40 to about 28 nm (see Figure 4.6.7a and b). However, during the self-doping at pH = 7, the particle sizes became nonuniform, attaining an average of about 250 nm (Figure 4.6.7c), because the self-doping involves intermolecular interactions (Scheme 4.6.2) which lead to aggregation. When the pH was decreased to 3,

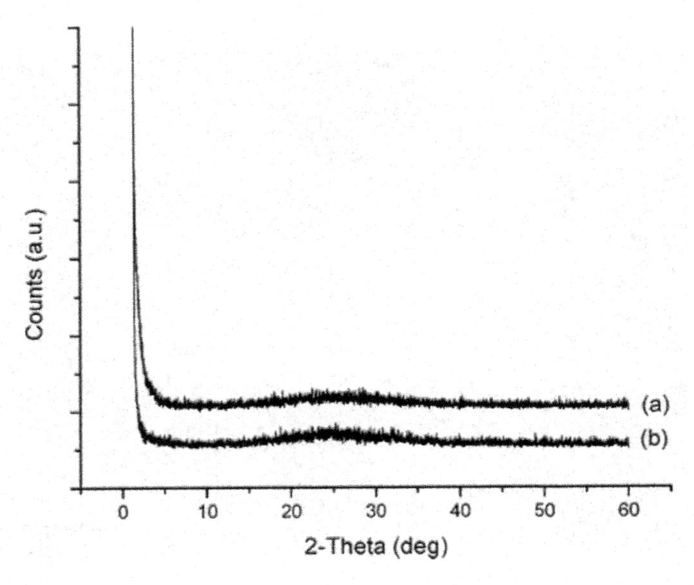

FIGURE 4.6.9 XRD profiles of the (a) self-doped and (b) externally doped H-PSDA copolymers.

the external doping by the HCl molecules could easily replace, for thermodynamic reasons (the free energy becoming smaller), the sulfonic acid moieties' doping; as a result, the aggregation decreased from an average of about 250 nm to about 120 nm. As shown in Figure 4.6.8, at pH = 7, the aggregates of the self-doped H-PSDA increased with storage time, as also observed via SEM. Irregular particles with sizes from 120 to 300 nm were identified after 1 day of storage (see Figure 4.6.8a); however, larger particles of about 350 nm were formed after one week of storage (Figure 4.6.8b). As shown in Scheme 4.6.2, self-doping occurs via intermolecular interactions between the sulfonic acid moieties and the nitrogens, and as a result, the aggregates are large. The aggregates were found to decrease to around 40 nm when the solution was neutralized with 0.1 N ammonium hydroxide solution. However, the sulfonic acid moieties which were not doped remained exposed to water, and for this reason, the aggregates were soluble in water. Furthermore, the self-doped and externally doped H-PSDAs were in an amorphous state, as shown in Figure 4.6.9. The pH response of the H-PSDA was reversible. Therefore, this pH sensitivity can have potential applications in biosensors.

Surface Elemental Analysis by XPS. Figure 4.6.10 presents the wide-scale XPS spectra of the HCl-doped, self-doped, and dedoped copolymers. They exhibit binding energies around 168, 284, 398, and 533 eV, which can be assigned to S 2p, C 1s, N 1s, and O 1s, respectively. The integral area ratios between two peaks, which are due to S 2p and N 1s, respectively, are about 1:2, 1:1, and 1:1 for the dedoped, self-doped, and externally doped copolymers, respectively. The peaks due to Cl 2p at 199 eV could be hardly identified in the dedoped and self-doped copolymers,

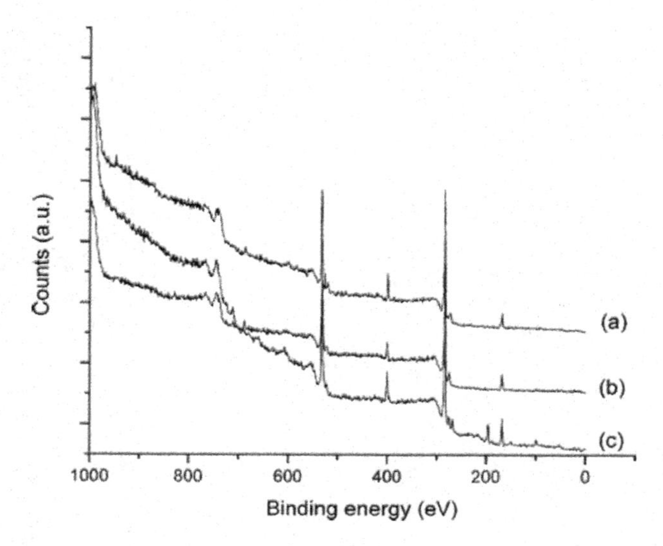

FIGURE 4.6.10 XPS spectra of H-PSDA copolymers: (a) dedoped, (b) self-doped, and (c) externally doped.

indicating that the chloride ions have been completely removed. Of course, the chloride peak appeared for the externally doped copolymer because of HCl doping. The interpretation for the C 1s and O 1s is not pursued because the surface contamination during XPS experiments cannot be avoided.

The N and S cores can be present upon doping or dedoping in several energetic states involving different components.[23,28] The oxidation and doping levels of the copolymers can be determined by analyzing the N 1s and S 2p core levels (see Table 4.6.1 and Figure 4.6.11a–c and a′–c′). The N 1s core has four

TABLE 4.6.1
Elemental Analysis of a Typical H-PSDA Copolymer in Various Doping States by XPS

Element (orbital)	Possible Components	Binding Energy (eV)	fwhm(eV)[a], Component Concentration (%)		
			Dedoped	Self-Doped	Externally Doped
N 1s	=N–	398.20	2.4, 45.6	1.1, 5.5	1.2, 2.6
	–NH–	399.41	1.6, 54.4	1.68, 52.1	1.81, 38.1
	$=N^+–$	400.83		1.49, 39.8	2.53, 45.0
	$–N^+H–$	402.31		0.92, 2.6	2.15, 14.3
S 2p	$–SO_3^-$	167.88	2.41, 94.2	1.55, 43.4	1.16, 20.7
	$–SO_3H$	168.80	1.14, 5.8	1.52, 56.6	2.11, 79.3

[a] fwhm means the full widths at half-maximum of the fitted peaks.

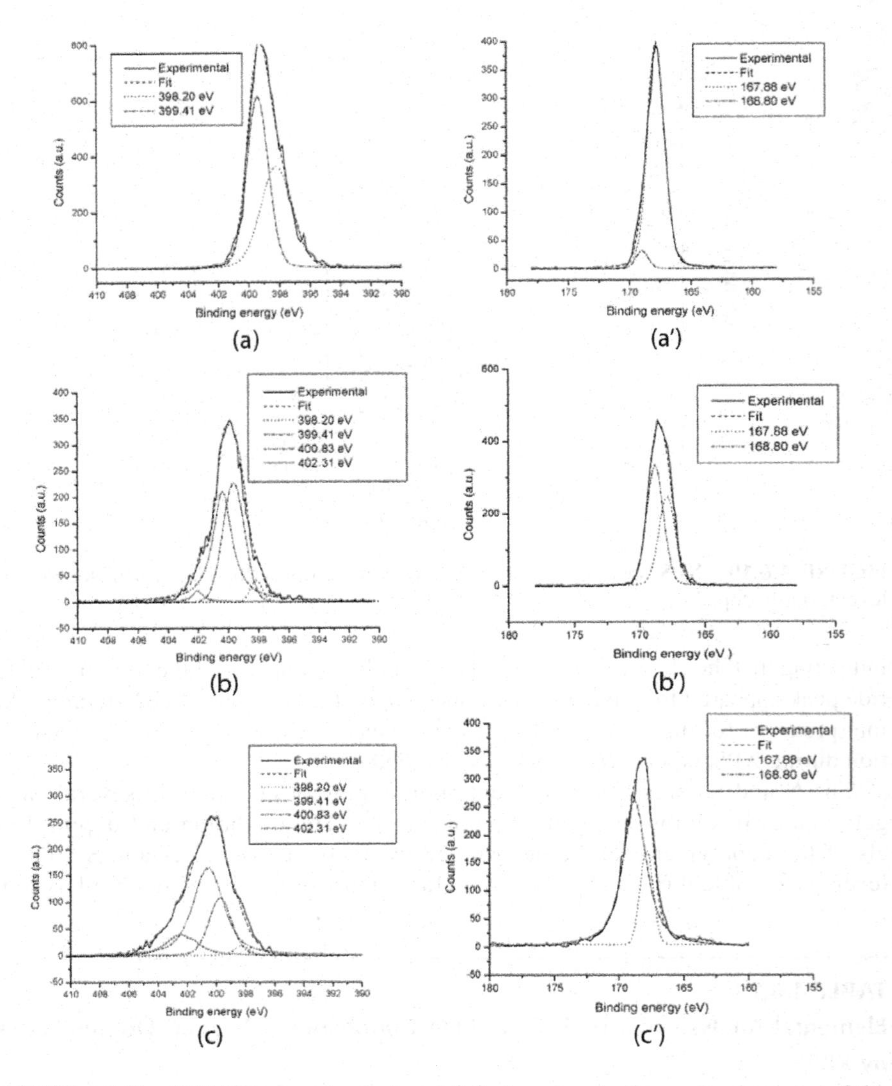

FIGURE 4.6.11 XPS N 1s (spectra a–c) and S 2p (spectra a′–c′) core level spectra of the H-PSDA copolymers in different doping states: (a) dedoped, (b) self-doped, and (c) externally doped.

fitted components with binding energies (BEs) at 398.20, 399.41, 400.83, and 402.31 eV, respectively (see Figure 4.6.11a–c). They can be assigned to the undoped imine and amine, doped imine, and positively charged amine, as listed in Table 4.6.1. The oxidation levels for the self-doped and externally doped copolymer are 45.3% and 47.6%, respectively, close to that for the copolymer in the dedoped state (its total oxidation level being about 45.6%). The doping levels attained in the former two cases are 42.3% and 59.3%, respectively, the higher value for the externally doped copolymer, being due to the additional doping by HCl on the nitrogens of the amine moieties.

The S 2p spectra could be fitted using two components with BEs of 167.88 and 168.80 eV, respectively (see Figure 4.6.11a′–c′). The former, with the lower BE, originates from the sulfur of the sulfonic acid moieties that protonates the nitrogens of the amine or the imine moieties. The latter, with the higher BE, can be assigned to the sulfur in the $-SO_3H$ moieties because the sulfur in the $-SO_3^-$ has a higher electron density than that in the $-SO_3H$. In the dedoped state, 94.2% of the sulfur cores were present in the $-SO_3^-$ anions. In the self-doped state, most of the ammonium moieties could be replaced with H^+ via the ion exchange. Consequently, there are 43.4% of $-SO_3H$ moieties protonating the nitrogens of the amine or imine moieties via inter- or intramolecular doping interaction. However, that value decreased to 20.7% after external doping. As mentioned above, a doping level as high as 59.3% was attained in the case of HCl doping. The external doping even partially replaced the self-doping.

Conductivity and Electromagnetic Shielding Effectiveness (SE). The conductivities determined on various compressed H-PSDA pellets by the four-point probe method were below 10^{-1}S/cm, much lower than that of PANI. This can be attributed to the sulfonic withdrawing effect and to the steric constrain of the aromatic substituent at the N sites. The conductivities of the self-doped and externally doped H-PSDA were 1.2×10^{-2} and 3.7×10^{-2}S/cm, respectively, which are slightly larger than those of the linear PSDAS reported in a previous paper.[29] For the conductivities being in the semiconducting range, 10^{-1}–10^{-4}S/cm, the obtained materials can be used for EMI shielding.[30] The SE values of the self-doped copolymer over a frequency range from 10 to 10^3 MHz, presented in Figure 4.6.12, are above 10 dB. The hyperbranched copolymers and their nanoscale-size aggregates have potential applications as EMI materials.

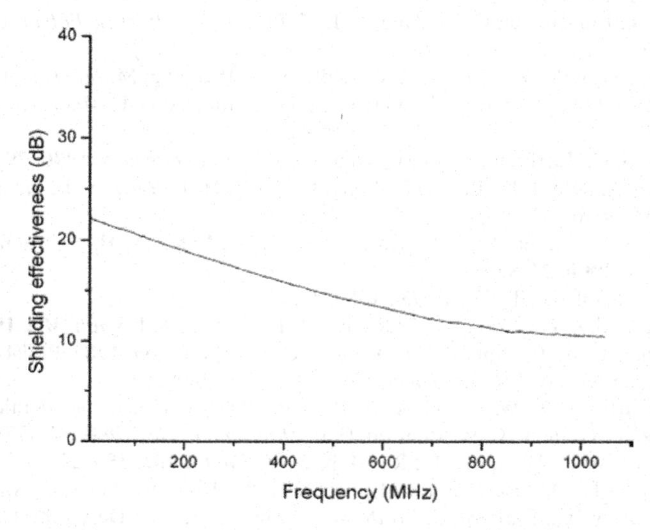

FIGURE 4.6.12 The electromagnetic shielding effectiveness (EMISE) of the self-doped H-PSDA as a function of frequency.

4.6.4 CONCLUSION

A synthesis methodology to prepare a hyperbranched sulfonated polydiphenylamine as a novel conducting polymer was developed. The obtained hyperbranched sulfonated polydiphenylamine (H-PSDA) is water soluble. The morphology of the H-PSDA in aqueous solutions is dependent on the pH. In the dedoped state, at pH values above 7, the aggregates of the polymeric particles had almost a uniform diameter around 40 nm. In the self-doped sate at pH = 7, the aggregates acquired nonuniform sizes with an average of about 250 nm because of the intermolecular doping interactions. However, at lower pH values (<5), the external HCl doping became dominant and the aggregates acquired smaller sizes of about 120 nm.

REFERENCES

1. Tomalia, D. A.; Naylor, A. M.; Goddard, W. A., III. *Angew. Chem., Int. Ed. Engl.* **1990**, *29,* 138.
2. Frechet, J. M. J. *Science* **1994**, *263,* 1710.
3. Frechet, J. M. J.; Henmi, M.; Gitsov, I.; Aoshima, S.; Leduc, M.; Grubbs, R. B. *Science* **1995**, *269,* 1080.
4. Litvinenko, G. I.; Simon, P. F. W.; Mueller, A. H. E. *Macromolecules* **1999**, *32,* 240.
5. Gaynor, S. G.; Edelman, S.; Matyjaszewski, K. *Macromolecules* **1996**, *29,* 1079.
6. Hawker, C. J.; Frechet, J. M. J.; Grubbs, R. B.; Dao, J. *J. Am. Chem. Soc.* **1995**, *117,* 10763.
7. Simon, P. F. W.; Radke, W.; Muller, A. H. E. *Macromol. Rapid Commun.* **1997**, *18,* 865.
8. Sunder, A.; Hanselmann, R.; Frey, H.; Mulhaupt, R. *Macromolecules* **1999**, *32,* 4240.
9. Hawker, C. J.; Hedrick, J. L. *Macromolecules* **1995**, *28,* 2993.
10. Wang, J. S.; Matyjaszewski, K. *J. Am. Chem. Soc.* **1995**, *117,* 5614.
11. Chiefari, J.; Chong, Y. K.; Ercole, F.; Kretina, J.; Jeffery, J.; Le, T. P. T.; Mayadunne, R. T. A.; Meijs, G. F.; Moad, C. L.; Moad, G.; Rizzardo, E.; Thang, S. H. *Macromolecules* **1998**, *31,* 5559.
12. Hua, F. J.; Liu, B.; Hu, C. P.; Yang, Y. L. *J. Polym. Sci., Part A: Polym. Chem.* **2002**, *40,* 1876.
13. (a) Matyjaszewski, K.; Gaynor, S. G.; Kulfan, A.; Podwika, M. *Macromolecules* **1997**, *30,* 5194. (b) Matyjaszewski, K.; Gaynor, S. G.; Muller, A. H. E. *Macromolecules* **1997**, *30,* 7034.
14. Gaynor, S. G.; Edelman, S.; Matyjaszewski, K. *Macromolecules* **1996**, *29,* 1081.
15. Wang, Z. M.; He, J. P.; Tao, Y. F.; Yang, L.; Jiang, H. J.; Yang, Y. L. *Macromolecules* **2003**, *36,* 7446.
16. Cheng, G. L.; Simon, P. F. W.; Hartensteinl, M.; Muller, A. H. E. *Macromol. Rapid Commun.* **2000**, *21,* 846.
17. Oyama, N.; Ohaka, T. *Synth. Met.* **1987**, *18,* 375.
18. MacDiarmid, A. G.; Chiang, J. C.; Richter, A. F.; Epstein, A. J. *Synth. Met.* **1987**, *18,* 285.
19. McManus, P. M.; Cushman, R. J.; Yang, S. C. *J. Phys. Chem.* **1987**, *91,* 744.
20. Yue, J.; Epstein, A. J. *J. Am. Chem. Soc.* **1990**, *112,* 2800.
21. (a) Kilmartin, A. P.; Wright, G. A. *Synth. Met.* **1997**, 88, 153; 163. (b) Kinlen, P. J.; Liu, J.; Ding, Y.; Graham, C. R.; Remsen, E. E. *Macromolecules* **1998**, *31,* 1735. (c) Roy, B. C.; Gupta, M. D.; Bhoumik, L.; Ray, J. K. *Synth. Met.* **2002**, *130,* 27.
22. (a) Cataldo, F.; Maltese, P. *Eur. Polym. J.* **2002**, *38,* 1791. (b) Planes, G. A.; Morales, G. M.; Miras, M. C.; Barbero, C. *Synth. Met.* **1998**, *97,* 223. (c) DeArmitt, C.; Armes, S.; Winter, J.; Uribe, F. A.; Gottesfeld S.; Mombourquette, C. *Polymer* **1993**, *34,* 158. (d) Nguyen, M.; Diaz, A. *Macromolecules* **1994**, *27,* 7003.

23. (a) Yue, J.; Wang, Z. H.; Cromack, K. R.; Epstein, A. J.; MacDiarmid, A. G. *J. Am. Chem. Soc.* **1991**, *113,* 2665. (b) Yue, J.; Gordon, G.; Epstein, A. J. *Polymer* **1992**, *33*, 4409. (c) Wei, X. L.; Wang, Y. Z.; Long, S. M.; Bobeczko, C.; Epstein, A. J. *J. Am. Chem. Soc.* **1996**, *118,* 2545. (d) Wei, X. L.; Fahlman, M.; Epstein, A. J. *Macromolecules* **1999**, *32,* 3114.
24. Ito, S.; Murata, K.; Teshima, S.; Aizawa, R.; Asako, Y.; Takahashi, K.; Hoffman, B. M. *Synth. Met.* **1998**, *96,* 161.
25. ***N-tert*-butoxylcarbonyl-4-aminophenol (4-AP-^{t}BOC).** 4-Aminophenol (5.4 g) and 11.0 g of di*tert*-butyldicarbonate were dissolved into a 55 mL mixture of THF/DMF (1:1, v/v) in a 100 mL flask. The mixture was cooled at 0°C. Triethylamine (5.1 g) was introduced dropwise within 1 h. The system was kept at 0°C for 4 h, the temperature was raised to room temperature, and the reaction was continued for 12 more h. Further, the solvent was evaporated under vacuum, 300 mL of DCM and 100 mL of water were added, and two layers were formed. The pH of the water layer was carefully adjusted with a 1.0 N HCl aqueous solution until neutral and extracted twice with 50 mL of DCM each time. The combined DCM solution was washed with water three times (50 mL each) and dried using Na_2SO_4. The DCM was removed using a rotary evaporator under vacuum. A light-yellow solid, 4-AP-^{t}BOC, was obtained with a yield of 86 wt%. FTIR (cm^{-1}): 3550–3200 (–NH and –OH), 2900 (phenyl), 1750–1670 (–C=O in ^{t}BOC). ^{1}H NMR (ppm): 7.18 (d, 2H), 6.71 (d, 2H), 6.34 (s, 1H), 1.53 (m, 9H). ESI MS $(Na + M)^+$: 232.1.
26. **4-(4-Vinylbenzoxy1)(*N-tert*-butoxycarbonyl)phenylamine) (4-VBPA-^{t}BOC).** 4-AP-^{t}BOC (2.09 g), 1.38 g of K_2CO_3 and 1.52 g of 4-vinylbenzyl chloride were mixed with 15 mL of acetone in a 50 mL flask. The system was heated with intensive stirring under a slight reflux and allowed to react under a nitrogen atmosphere. The reaction was followed using TLC plates. After the 4-AP-^{t}BOC was consumed, 1.0 g of triethylamine was introduced and the reaction continued at 60°C. After the 4-vinylbenzyl chloride (or 4-choromethylstyrene, CMS) was consumed, 300 mL of DCM and 100 mL of water were added and two liquid layers were generated. The water layer was extracted twice with 50 mL of DCM each time. The combined DCM phases were washed three times with 50 mL of water each time until neutral and dried using Na_2SO_4. Further purification was achieved by passing through an inhibitor remover disposable column (Aldrich, Catal. 30631–2) to eliminate the excess of 4-AP-^{t}BOC (which inhibits the radical homopolymerization of 4-VBPA-^{t}BOC). DCM was removed using a rotary evaporator under vacuum. A light-red solid, 4-VBPA-^{t}BOC, was obtained with a yield of 83 wt%. FTIR(cm^{-1}): 3300–3150 (–NH), 2900 (phenyl), 1750–1670 (–C=O in ^{t}BOC), 957 ($-CH=CH_2$). ^{1}H NMR (DMF-d_7, ppm): 7.3–6.8 (m, 8H), 6.67 (m, 1H), 6.41 (s, 1H), 5.6 (m, 2H), 5.2 (s, 2H), 1.56 (m, 9H). ESI MS $(Na + M)^+$: 348.0.
27. (a) Chen, S.; G. Hwang. *J. Am. Chem. Soc.* **1994**, *116*, 7939. (b) Chen, S.; Hwang, G. *J. Am. Chem. Soc.* **1996**, *117,* 10055. (c) Chen, S.; Hwang, G. *Macromolecules* **1996**, *29,* 3950.
28. Hua, F. J.; Ruckenstein E. *Macromolecules* **2004**, *37,* 6104.
29. Hua, F. J.; Ruckenstein, E. *J. Polym. Sci., Part A: Polym. Chem.* **2004**, *42,* 1429.
30. Duke, C. B.; Gibson, H. W. *Kirk-Othmer: Encyclopedia of Chemical Technology*; John Wiley and Sons: New York, 1982; Vol. 18, p 755.

5 Preparation and Modification of Conductive Surface

CONTENTS

A broad range of commonly used materials, ranging from conventional polymers to glass, are not electrically conductive. On the other hand, conductive polymers generally would not possess adequate processability and strong mechanical properties. Therefore, incorporation of conducting polymers on the surface of nonconductive materials can obtain new materials with significant application potentials. Modification of the conductive surface layers can further endow the layers with improved comprehensive properties. This chapter is devoted to the preparation and modification of conductive polymer-based surface layers or coatings on polymeric or other types of substrates, and three general approaches were employed in the corresponding studies.

The first approach is to blend conductive polymer with non-conductive polymer via solutions to modify material properties (Section 5.1). Blended films of conductive polyaniline and amphiphilic Pluronic copolymer are prepared by slow solvent evaporation of solutions of the two polymers in *N*-methylpyrrolidinone (NMP). Surface modification of polyaniline film with Pluronic polymer is accomplished by swelling of film surface with a NMP solution of Pluronic copolymer, followed by rapid surface deswelling by water treatment. As compared to pure polyaniline, the Pluronic polymer-modified polyaniline are more hydrophilic and possess improved anti-biofouling properties, therefore, may be more useful for biosensor applications.

The second approach generates conductive surface layers by polymerization in the surface layers swollen by monomer of conductive polymer (Section 5.2). Crosslinked poly(butyl acrylate) or polyurethane films are swollen with pyrrole or 3-methylthiophene, and then immersed into ferric chloride solution to induce oxidative polymerization. As a result, polypyrrole or poly(3-methylthiophene)-containing surface layers are formed on the films. Remarkable conductivity (up to 42 S/cm) is observed for these surface layers.

The third approach is based on surface treatment of inorganic surfaces and graft polymerization to form conductive polymer surface grafts (Sections 5.3–5.5). Glass or (patterned) silicon surfaces are hydroxylated, and then reacted with 3-bromopropyltrichlorosilane or phenyltrichlorosilane. The resulting surfaces are further converted to aniline-primed surfaces, which are subsequently used for graft polymerization of aniline to form polyaniline grafts covalently attached on the surfaces. Significant surface conductivity values (up to 23 S/cm) of grafted polyaniline are achieved, which are higher than the typical value for polyaniline homopolymer film (~1 S/cm).

5.1 Improved Surface Properties of Polyaniline Films by Blending with Pluronic Polymers without the Modification of the Other Characteristics*

Z.F. Li and Eli Ruckenstein

Department of Chemical Engineering, State University of New York at Buffalo, Buffalo, NY 14260, USA

ABSTRACT Films of conductive polyaniline and amphiphilic Pluronic (P105) copolymer blends were prepared by dissolving the two polymers in *N*-methylpyrrolidinone (NMP) followed by a slow solvent evaporation at 55°C. The characteristics of both doped and undoped films were determined by X-ray photoelectron spectroscopy (XPS), scanning electron microscopy (SEM), water droplet contact angles, differential scanning calorimetry (DSC), thermal gravimetry analysis (TGA), wide-angle X-ray diffraction (WAXD), and tensile strength measurements. The surface of the blends became more hydrophilic than that of the hydrophobic PANI film, but the other properties of the blends did not change appreciably for Pluronic content lower than 50 wt%. Compared to PANI films, the more hydrophilic surfaces decreased the amount of bovine serum albumin protein adsorbed. By preventing biofouling, the polyaniline-Pluronic blends can become more useful as biosensors than the polyaniline films.

* *Journal of Colloid and Interface Science* 2003, 264, 362–369.

5.1.1 INTRODUCTION

In the past decades, conducting polymers have had numerous technological applications. Among conducting polymers, polyaniline (PANI) has been of particular interest because of its easier processability, its good environmental stability, the high and reversible nature of its conductivity, and its interesting redox properties. Polyaniline has the general formula $[(–B–NH–B–NH–)_y (–B–N=Q=N–)_{1–y}]_x$, in which B and Q denote C_6H_4 rings in benzenoid and quinoid forms, respectively [1,2]. The intrinsic oxidation state of PANI ranges from the fully oxidized pernigraniline ($y = 0$) through the 50% oxidized emeraldine ($y = 0.5$) to the fully reduced leucoemeraldine ($y = 1$). PANI can reach a highly conductive state either through the protonation of the imine nitrogens (=N–) in its emeraldine state or through the oxidation of the amine nitrogens (–NH–) in its fully reduced leucoemeraldine state [2]. Usually, PANI is synthesized in the emeraldine base form, which can be processed readily into thin films, fibers, and elastomers [1,3–5]. It has great potential in numerous applications, including rechargeable batteries, light-emitting diodes, molecular sensors, corrosion protection of metals, and gas separation membranes [6–10].

The poly(ethylene oxide)–poly(propylene oxide)–poly(ethylene oxide) (PEO–PPO–PEO) triblock copolymers, commercially available as Poloxamers, Pluronics, or Synperonics, are high-molecular-weight nonionic surfactants, with the general formula $(PEO)_x–(PPO)_y–(PEO)_x$. The PPO segment is lipophilic and the PEO segments are hydrophilic. The amphiphilicity of the block copolymers leads to their self-assembly, behavior resembling that of low-molecular-weight surfactants. In addition, the lengths of the PPO and PEO segments can be easily adjusted, thus ensuring a large variety of physicochemical properties, with wide applications ranging from medical and pharmaceutical products to photographic and plastic ones [11–14]. Due to their surface-active nature and biocompatibility, these water-soluble, nontoxic polymers have found applications as biomaterials [15,16] in drug delivery and gene therapy [17,18]. The hydrophilic segments of the Pluronic polymer have very low interfacial tension with respect to water and adsorb low amounts of proteins [19,20], cells [21,22], and platelets of blood [23,24], thus avoiding biofouling. They have been used for the biofunctionalization of other conventional polymers through blending or grafting to make the latter more hydrophilic and biocompatible [19–24].

Conductive polymers can be employed as analytical biological sensors [25–27], biomembranes [28,29], and enzyme-based logic gates [30], in biochemistry [31], and as substrates for mammalian cell growth [32]. Consequently, the combination of Pluronic polymers with conductive ones can improve the biocompatibility of the latter and make them more useful in their applications in the biological and biochemical fields. In addition, Zelikin et al. [33] proposed to use "erodible conductive polymers," which "erode slowly under physiological conditions and support the growth, proliferation, and differentiation of primary human cells in vitro cell culture assays." Water-soluble, nontoxic, and biocompatible Pluronic polymers blended with PANI can also be useful in this direction.

The purpose of this paper is to modify the surface of PANI films by preparing blends with the amphiphilic Pluronic copolymer through solvent blending. The surface properties of the blends, such as the hydrophilicity and the surface composition will be investigated by water contact angle measurements and X-ray photoelectron spectroscopy (XPS). In addition, the blended films will be characterized by scanning electron microscopy (SEM), differential scanning calorimetry (DSC), thermal gravimetry analysis (TGA), wide angle X-ray diffraction (WAXD), and tensile strength measurements, in order to investigate the changes of their characteristics with the improvement of their surface characteristics. A few comparative experiments regarding the adsorption of bovine serum albumin on the surface of the film are also included in the paper to demonstrate their increased biocompatibility.

5.1.2 EXPERIMENTAL SECTION

5.1.2.1 Materials

The Pluronic P105 (PEO–PPO–PEO block copolymer, 50 wt% PEO and 50 wt% PPO), with MW 6500, was donated by BASF Corp. (Mount Olive, NJ). The protein used in the adsorption test, bovine serum albumin (BSA), was purchased from Sigma (MO, USA). The aniline, ammonium persulfate, and *N*-methylpyrrolidinone (NMP) and the other chemicals (all of reagent grade) were purchased from the Aldrich/Sigma Chemical Co. and used as received.

5.1.2.2 Oxidative Polymerization of Aniline

The emeraldine (EM) salt was prepared by the oxidative polymerization of aniline for about 5 h with ammonium persulfate in a 1 M HCl solution, at ice-water temperature. The particles were separated and converted to the EM base by their treatment with an excess of 0.5 M NaOH, and then washed thoroughly with deionized water until neutral. Finally, the EM base powder was dried under vacuum.

5.1.2.3 Preparation of PANI/Pluronic (P105) Blend Films

The PANI/Pluronic blends were prepared by dissolving 0.5 g of a mixture of P105 and finely ground freshly prepared PANI powder in a sufficiently large amount of NMP for the system to be completely dissolved (15 mL NMP). The total weight of the PANI/P105 mixture was kept constant but the weight ratio was varied. The polymer solution was stirred at room temperature for 48 h, poured onto a Teflon dish, and then heated in an oven at 55°C to evaporate the NMP slowly until the weight of the sample remained constant (about 2 days). The thickness of the obtained films was 10–20 μm. The doped state of the films was obtained by keeping them overnight (about 12 h) in a 0.1 M H_2SO_4 solution.

5.1.2.4 Characterization of the Blend Films

The blend films were characterized by X-ray photoelectron spectroscopy (XPS), scanning electron microscopy (SEM), water contact angle measurements, protein adsorption, differential scanning calorimetry (DSC), thermal gravimetric analysis (TGA), wide-angle X-ray diffraction (WAXD), and tensile strength measurements. Their conductivity was determined by the standard four-probe method. The contact angle measurements were carried out using an NRL CA Goniometer, Model 100-00(115), from Ramehart Inc. A telescope with a magnification power of 23× was equipped with a protractor of 1° gradation. The angles reported are accurate to within ±3°. For each substrate, at least five measurements on different locations were averaged. The XPS measurements were carried out on a Surface Science Model SSX-100 Small Spot ESCA, using an Al$K\alpha$ monochromatized X-ray source (1.48-KeV photons). The pressure in the analysis chamber was maintained at 10^{-9} Torr or lower during the measurements. To compensate for the surface charging effects, all binding energies were referenced to the C1.*s* hydrocarbon peak at 284.6 eV. In the peak analysis, the linewidths (the widths at half-maximum) of the Gaussian peaks were kept constant for the components of a particular spectrum. The surface elemental compositions were determined from the peak area ratios and were accurate to within 1%. The SEM experiments were carried out using a Hitachi S-4000 FESEM instrument operated at 20 kV with the specimens mounted on double-sided adhesive carbon tape. The samples were sputter-coated with carbon before being examined under the microscope. The test for protein adsorption was carried out by dipping the PANI or blend films in 0.2 mg/mL of BSA in a NaAc/HAc buffer solution (pH 7.1). The adsorption was allowed to proceed at 4°C for 24 h and then the film was removed from the solution. The amount of bovine serum albumin left in the solution was determined using the Bradford standard protein assay method [34]. The absorbance at 595 nm was measured with a UV-vis Beckman spectrophotometer. The amount adsorbed was calculated by comparing with the initial concentration of the protein solution. The conductivity of the polymer films was determined by the four-probe method, using a Hewlett-Packard Model 3478A digital multimeter. For each of the conductivities reported, at least three measurements were averaged. The X-ray diffraction (WAXD) patterns of the samples were obtained using a SIEMENS D500 diffractometer and Cu$K\alpha$ radiation of wavelength 1.5406 Å. The diffraction data were recorded for 2θ angles between 1° and 40°, with a resolution of 0.02°. Thermogravimetry analysis (TGA) was carried out on a Perkin-Elmer TGA-7 thermogravimetric analyzer, using high-purity nitrogen as purging gas. Differential scanning calorimetry (DSC) thermograms were obtained with a Perkin–Elmer DSC-7 instrument, using a standard aluminum pan. Nitrogen was used as sweeping gas, and the heating rate was 25°C/min. The mechanical properties were determined by measuring the tensile strength of the film at breakage on an Instron (Model 1000) mechanical test machine. The films were cut into the standard shape and the values plotted are averages for at least six samples.

5.1.3 RESULTS AND DISCUSSION

5.1.3.1 Surface Property and Morphology

5.1.3.1.1 XPS Analysis and Water Contact Angle Measurements

The surface composition of the films was determined by XPS and their surface hydrophilicity by contact angle measurements. Figure 5.1.1 presents the XPS

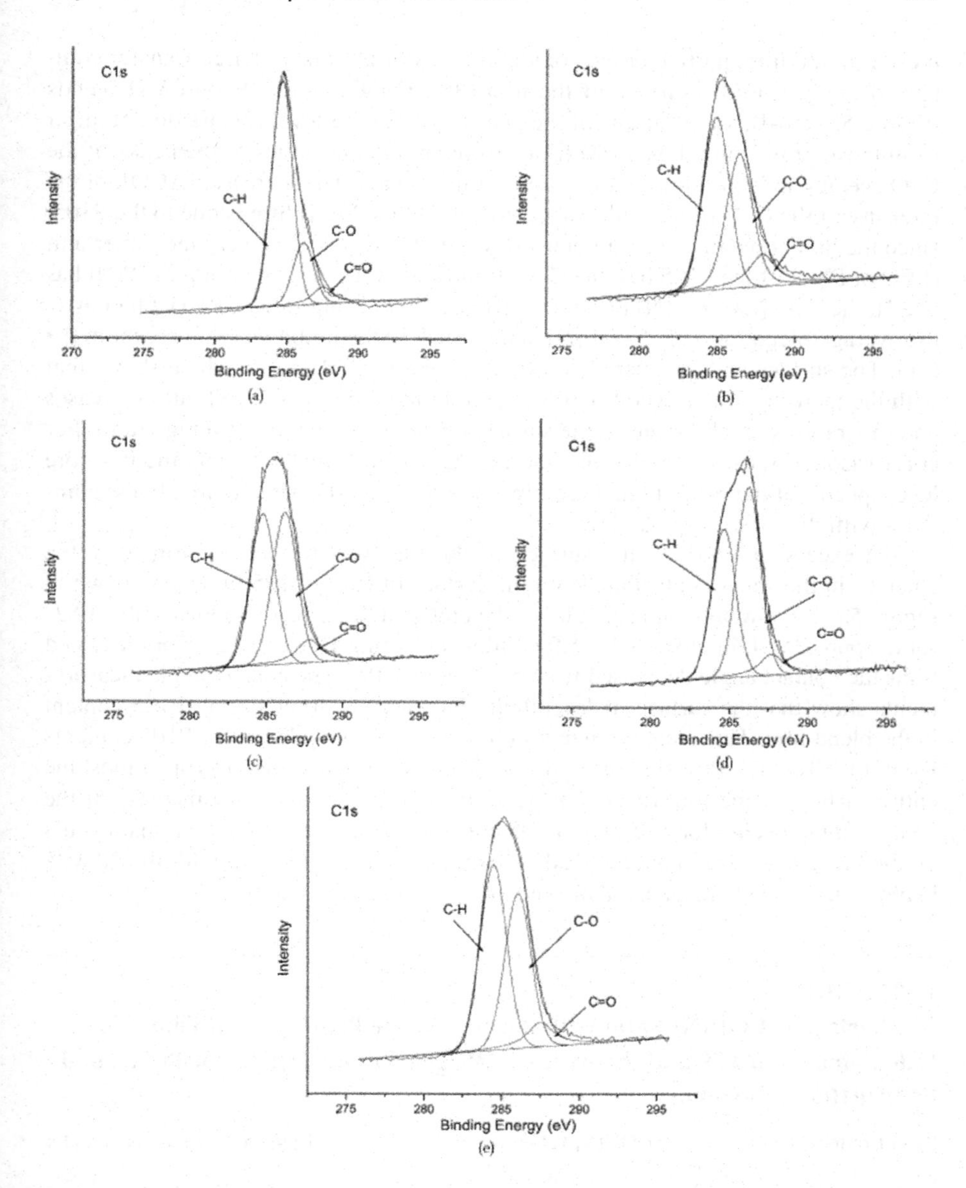

FIGURE 5.1.1 XPS C1*s* core-level spectra of blend films for various P105 contents: (a) doped pure PANI film; (b) doped 10 wt% P105; (c) doped 50 wt% P105; (d) doped 70 wt% P105; and (e) undoped 10 wt% P105.

curve-fitting of Cl *s* core-level spectra of the doped emeraldine (EM) film and of its blends with P105. The C1*s* core-level spectra of all the samples can be deconvoluted into three peaks at 284.6, 286.2, and 288.7 eV, which correspond to CH and/or CN, C–O, and C=O species, respectively [35]. For the pure doped EM film (Figure 5.1.1a), the C–O and C=O peaks can be associated with the residual NMP present in the film, partial surface oxidation, and/or a weakly charge-transfer complexed oxygen on the surface of the film [36]. For the doped P105/PANI blends (Figure 5.1.1b–d), the positions of the peaks of C1*s* core-level spectra do not differ from those of the pure doped EM film. However, the intensities of the peaks of the C–O species are substantially increased for the blends. The increase, [ΔCO], of the intensity of the C–O peaks with respect to the pristine PANI film is due to the P105, since the latter contains large amounts of C–O moieties. Since every monomer unit (PEO or PPO) of the P105 has one C–O group and every monomer unit of PANI has one N, the ratio [ΔC–O]/[N] provides the relative concentration of P105. Of course, the average blend ratio of [C–O]/[N] is provided by the initial weight percent in the film. The surface and the average ratios are listed in Table 5.1.1, which shows that with the increase of the overall P105 concentration the [ΔC–O]/[N] ratio increases and this ratio is much larger on the surface of the blend than the average ratio. For comparison purposes, we also include Figure 5.1.1e, which presents the Cl*s* core level spectra of the emeraldine base film for 10 wt% P105. This figure almost coincides with that of the doped film.

The excess of P105 on the surface of the film was further confirmed by the changes in the water contact angle on the surface of PANI/P105 films. As shown in Figure 5.1.2, the water contact angle for the undoped films decreased from about 62° for the pure PANI film to about 22° for a film containing 10 wt% P105. For the doped films the contact angle decreased from 53° to about 19°. The contact angle measurements show that the surface hydrophilicity increases with increasing P105 content in the blend film. This increase in hydrophilicity is more moderate for P105 contents larger than 10 wt%. Probably the hydrophilic chains of P105 are covering almost the entire surface of the film for a content of about 10 wt%. A sudden increase of the contact angle occurs for a 70 wt% P105 content. The phase segregation that occurs on the film surface, revealed by SEM experiments in the next section for the 70 wt% blend, is most likely responsible for the increased hydrophobicity.

TABLE 5.1.1
The Surface [ΔCO]/[N] Ratio with Respect to the Pristine PANI Film Calculated from XPS and the Average Weight Ratio [CO]/[N] for the Doped PANI/P105 Blend Film

P105 Content (wt%)	[ΔCO]/[N] Value from XPS	[CO]/[N] Value from Weight Ratio
0	0	0
10	3.6	0.20
30	5.1	0.79
50	8.2	1.84
70	11.8	4.26
100	–[a]	–

[a] No N present.

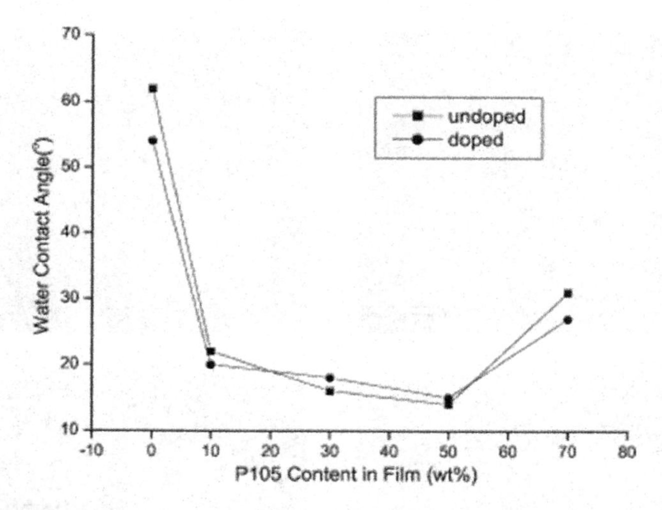

FIGURE 5.1.2 Water drop contact angle on the blend films for various P105 contents.

5.1.3.1.2 SEM

The SEM micrographs (Figure 5.1.3) show that the doped EM film cast from NMP had a rather smooth and featureless surface and that some small but uniformly dispersed aggregates appeared on the surface after the addition of P105. As the P105 content increased to 50 wt%, the surface aggregates became larger. However, no obvious phase separation occurred and this indicates good compatibility of the two polymers. When the content of P105 increased to 70 wt%, the film became very rough and large particles were present on the surface, probably because of the segregation of the two polymers.

5.1.3.1.3 Protein Adsorption

For materials used in blood-contacting applications, protein adsorption may trigger the coagulating sequence on their surface, if they are not biocompatible. The commonly used bovine serum albumin was selected by us to test the adsorption of blood protein in vitro. As shown in Table 5.1.2, it was found that on undoped PANI films, the amount of BSA adsorbed was 2.6 $\mu g/cm^2$, while on the undoped 10 wt% Pluronic film, the amount adsorbed decreased to 0.53 $\mu g/cm^2$. For Pluronic contents of 30 and 50 wt%, the adsorbed amounts decreased to 0.39 and 0.27 $\mu g/cm^2$, respectively. It is therefore obvious that the blending of PANI with the Pluronic polymer efficiently reduces the protein adsorption. For comparison, the protein adsorption on films in doped state was also determined and the results are listed in Table 5.1.2. The amounts of protein adsorbed are lower on the doped films than on the undoped ones. This occurred because the interfacial tension between the PANI films and water is decreased by both the Pluronic molecules and doping and this increases the biocompatibility [37]. However, the blending with the Pluronic copolymer is much more efficient in decreasing the amount adsorbed than the doping.

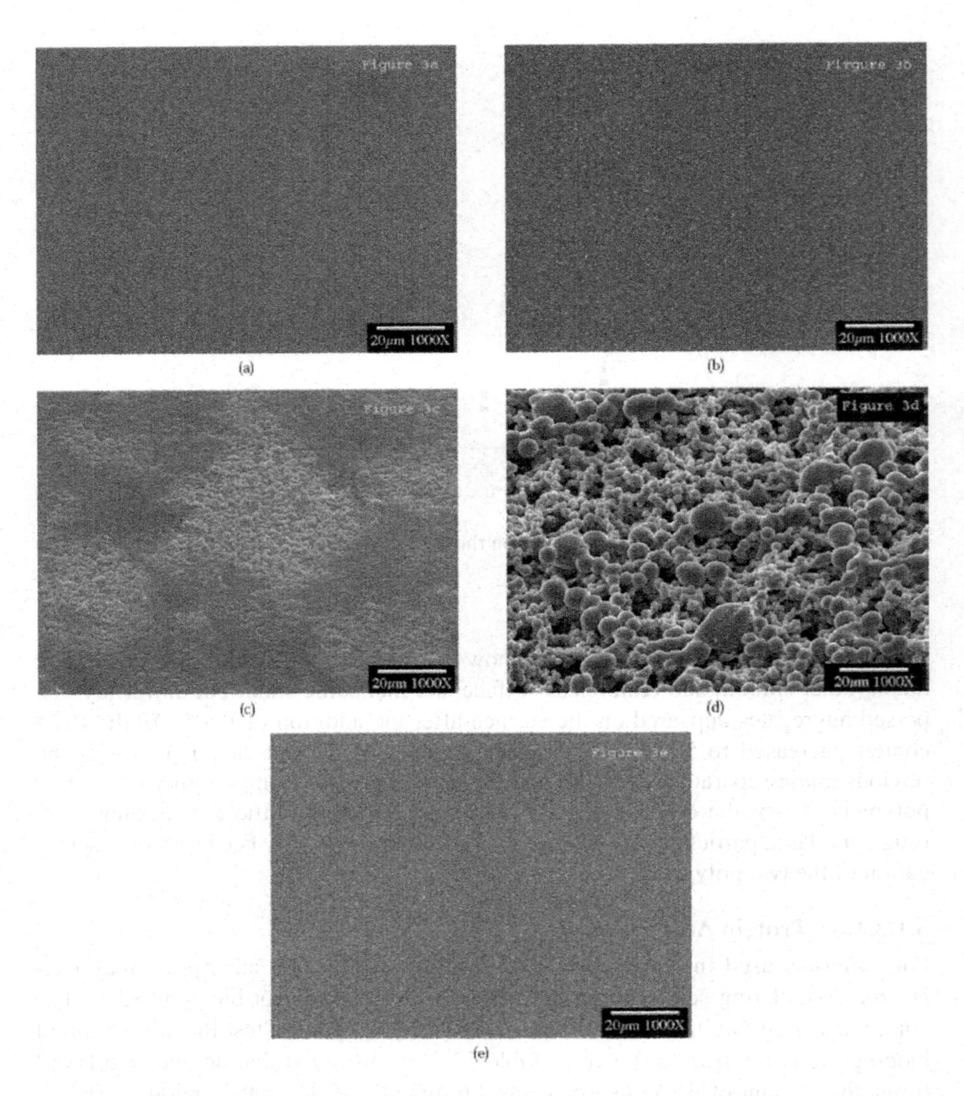

FIGURE 5.1.3 SEM micrographs of doped blend films for various P105 contents: (a) pure PANI film; (b) 10 wt% P105; (c) 50 wt% P105; (d) 70 wt% P105; and (e) undoped 10 wt% P105.

TABLE 5.1.2
Protein Adsorption on the PANI and Its Modified Films

Films	Undoped State (μg/cm²)	Doped State (μg/cm²)
PANI	2.6	1.5
10 wt% Pluronic in blend	0.53	0.41
30 wt% Pluronic in blend	0.39	0.32
50 wt% Pluronic in blend	0.27	0.22

5.1.3.2 Thermal Properties

5.1.3.2.1 Differential Scanning Calorimetry

As reported also by other authors [38], the DSC traces do not provide glass transition temperatures (T_g) or melting points for the polyaniline EM base and EM salt; they exhibit only broad exothermic peaks. Nevertheless, the thermal transitions and the corresponding heat effects of the Pluronic polymer in the PANI/P105 blends can be used to evaluate the level of polymer–polymer interactions. As shown in Figure 5.1.4a, the pure P105 exhibits a glass transition temperature at −74°C. The 70 and 50 wt% P105 blends also exhibit glass transition temperatures near that of

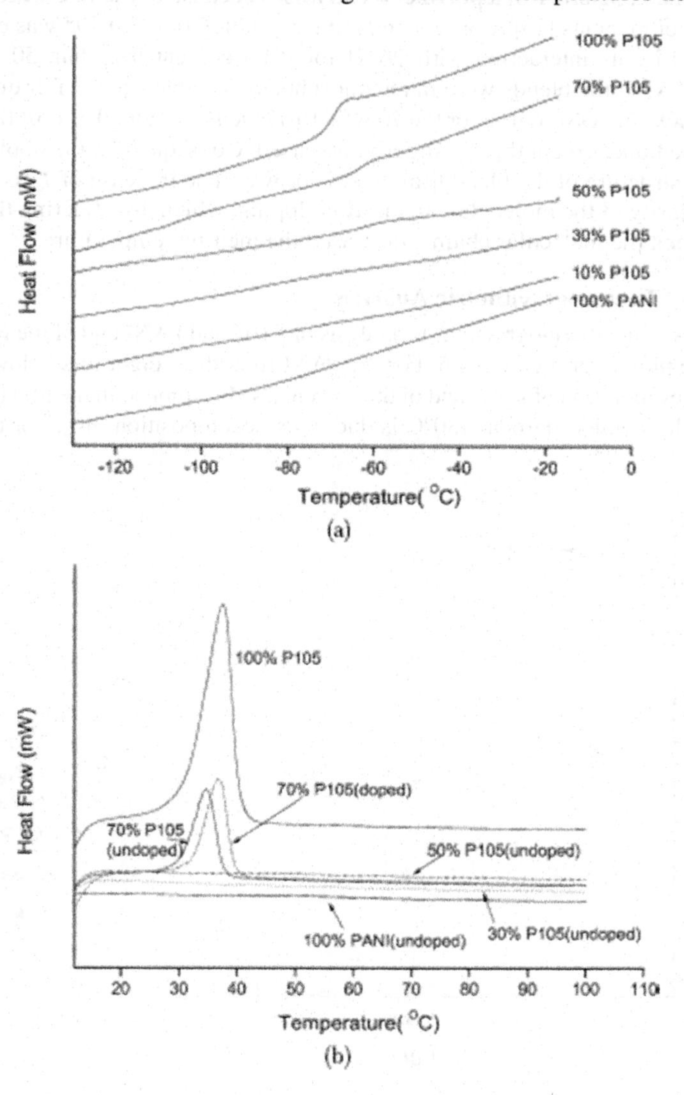

FIGURE 5.1.4 (a,b) DSC traces for various P105 contents.

the pure P105. In contrast, for the blends with P105 content below 50 wt%, no T_g could be detected. This disappearance of the T_g revealed that, for sufficiently small amounts of P105 (<50 wt%), there is total miscibility between PANI and the Pluronic polymer. Figure 5.1.4b presents the DSC traces of PANI/P105 blends during heating scans above the room temperature. It shows that the pure Pluronic P105 has a melting point T_m near 39°C, and that there is no melting point for the PANI film. For the blend with a P105 content of 70 wt%, T_m decreased from 39°C to 35°C and the heat of fusion per gram of P105 decreased from 66.1 J/g for the pure P105 to 45.3 J/g for the 70 wt% P105. For the blends with P105 content of 50 wt% or lower, no melting peak is present. The miscibility in a polymer blend arises because of the interactions among different components [39]. Consequently, the crystallization of P105 was completely suppressed by its interaction with PANI for P105 content less than 50 wt%, and, like pure PANI, the blends were in an amorphous, miscible state. After doping with 0.1 M H_2SO_4, the DSC curve for the 70 wt% P105 blend remained almost the same as the undoped one, except that T_m slightly increased. Consequently, the doping did not affect the structure of the blend film in a major way. The increase of T_m is due to the higher polarity of the molecular chains after doping, which, by affecting the interaction between the molecular chains, increased the melting temperature.

5.1.3.2.2 Thermogravimetric Analysis

The results of the thermogravimetric analysis of P105 and PANI and of the PANI/P105 blends are plotted in Figure 5.1.5. For the PANI film, the weight loss below 400°C is due to the evaporation of water and of the solvent (NMP) trapped in the film (~20 wt%), while the loss which starts at 480°C is due to its decomposition [40]. For the blends,

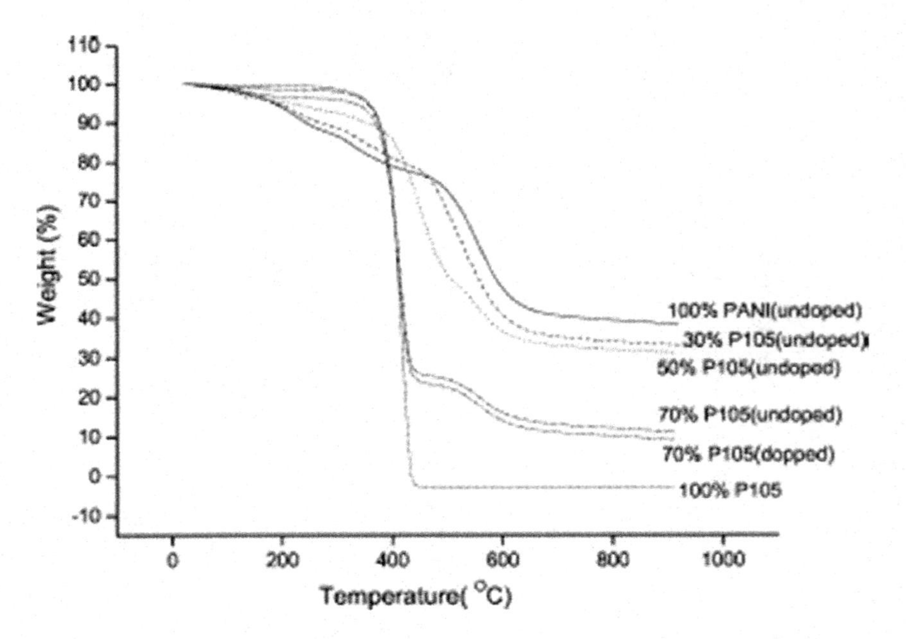

FIGURE 5.1.5 Thermogravimetric analysis for various P105 contents.

there are two major steps of degradation, one around 370°C–390°C for P105 and the other one around 480°C for PANI. For the pure P105, the onset decomposition temperature is 370°C, which increases to 385 and 393°C after its blending with 30 and 50 wt% PANI, respectively. Figure 5.1.5 reveals that the thermal stability of P105 can be improved through blending due to its interaction with PANI. In contrast, the decomposition temperature of PANI did not change through its blending with P105. The doping had little effect on the weight curve of the 70 wt% P105 blend (Figure 5.1.5).

5.1.3.3 Wide Angle X-ray Diffraction Patterns (WAXD)

Figure 5.1.6 presents the X-ray diffraction patterns of the pure PANI and P105, as well as of their doped and undoped blends, after the casting and drying of the films. For the pure PANI, there is only one broad amorphous single peak at $2\theta = 20°$. It was reported previously [40] that the residual NMP in the PANI film can act as a plasticizer and that the interaction of the C=O groups of NMP with the NH groups of PANI leads to an isotropic PANI film. The peak noted above for the PANI film cast from NMP is consistent with their results. In contrast, the WAXD pattern of the pure P105 exhibits two narrow peaks at $2\theta = 19.356°$ and 23.652°, corresponding to d spacings of 4.582 and 3.758 Å, respectively, which indicate a crystalline structure. For the 70 wt% P105 blends, the crystalline structure of P105 is still present for both the doped and undoped films. For P105 contents smaller than 70 wt%, the two diffraction peaks disappeared completely, being replaced by a weak broad

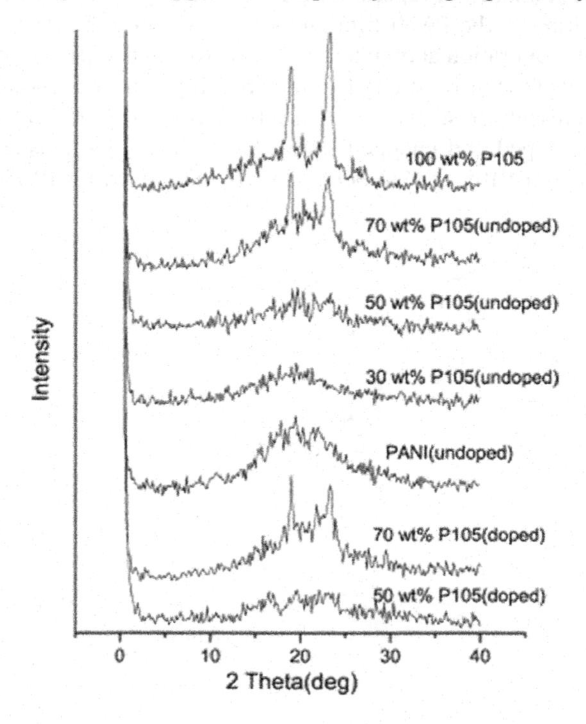

FIGURE 5.1.6 XRD patterns of doped and undoped blend for various P105 contents.

peak centered around $2\theta = 20°$ of the amorphous PANI. The above observations are consistent with the DSC results, which revealed that the 70 wt% P105 was still crystalline. The presence of PANI molecules caused in this case a depression of T_m (see Figure 5.1.4b), but no change in the XRD patterns. The disappearance of the two WAXD peaks for P105 < 50 wt% has proven that in those cases the P105 was completely imbedded in the PANI amorphous matrix.

5.1.3.4 Mechanical Properties of the Films

The mechanical properties of the PANI films depend on various factors, such as the intrinsic properties of PANI, the solvent trapped in the film, and the plasticizing effect of the water and NMP. The hydrogen-bonding interactions in the blends and the interactions between the PANI backbones are also important. It has been reported [36] that the PANI films prepared from EM base powders in NMP formed cross-linked structures when dried at 150°C, and that this improved their tensile strength at the breaking point. It is likely that either the residual NMP cross-linked the PANI chains and/or the cross-linking occurred among the PANI backbones themselves, a process which is facilitated by elevated temperatures [41]. In the present experiments, we found that a slow evaporation (48 h) at low temperatures (55°C) generated a tensile strength at the breaking point of 110 MPa for the undoped PANI films comparable to that prepared by heating at 150°C for 6 h (115 MPa). One can therefore conclude that the evaporation time of the solvent (NMP) can play a role in the cross-linking of the PANI film and that a slow evaporation at low temperature and a rapid evaporation at high temperature provide comparable results. For the doped PANI films the tensile strength at the breaking point was 95 MPa.

Figure 5.1.7 presents the tensile strength at the breaking point as a function of P105 content for both doped and undoped states. It shows that the tensile strength does not change much for P105 content below 10 wt%. However, for P105 content above

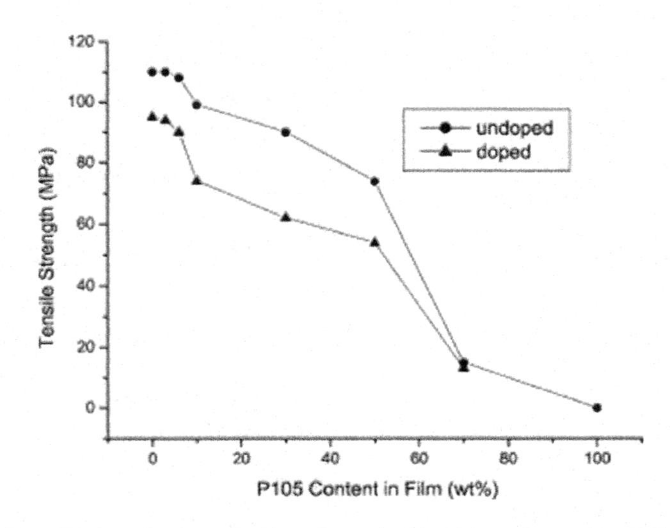

FIGURE 5.1.7 The tensile strength at breakage for blend films for various P105 contents.

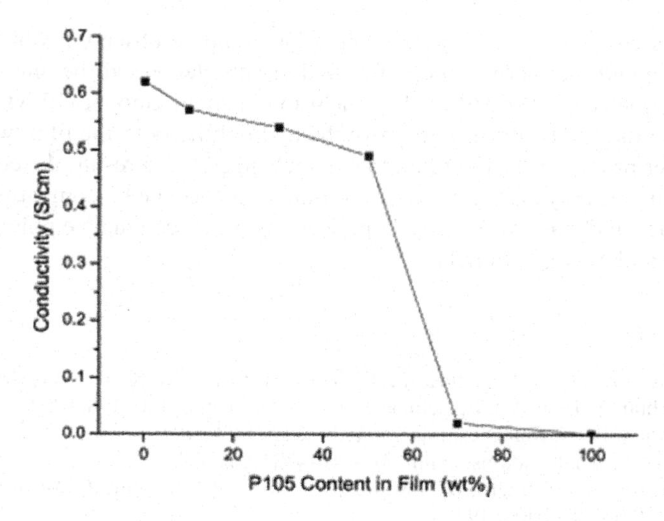

FIGURE 5.1.8 The conductivity of the blend films for various P105 contents.

10 wt% the decrease is more accentuated and becomes rapid for contents greater than about 50 wt%. Figure 5.1.7, combined with the thermal analysis results, indicates that a high content of P105 can affect the intra- and intermolecular interactions, damaging the cross-linking in the films. However, a small content of P105 in the blend does not much affect the tensile strength. This may be due to the preference of P105 for the surface of the film and the small amount of P105 which remains in the matrix.

5.1.3.5 Conductivity

The effect of P105 loading on the room temperature conductivity of doped blends is presented in Figure 5.1.8, which shows that the conductivity is smaller than that of the doped PANI film. However, the decrease is small for low contents of P105 and becomes large only for contents higher than about 70 wt%. The substantial decrease of the conductivity may be due to the phase separation in the blend, which was revealed by both the thermal analysis and the SEM investigations.

5.1.4 CONCLUSIONS

Blended films of conductive polyaniline and amphiphilic Pluronic (P105) copolymers were prepared by solventcasting from a common solvent (NMP). The Pluronic molecules prefer to be located at the surface, which thus becomes more hydrophilic than that of the PANI films. This reduces the protein adsorption on the surface of the blends compared to that on the pure PANI surface films. The DSC and WAXD spectra showed that there was no phase separation for P105 contents below 50 wt% and that the hydrophilicity of the surface was greatly improved. In addition, the thermal stability, the tensile strength at the breaking point, and the conductivity of the blend films did not change much for Pluronic contents below 50 wt%. However, for a P105 content of 70 wt%, phase separation could be observed, which changed the structure and the thermal properties and greatly decreased the tensile strength and the conductivity of the films. In conclusion, a small amount (10 wt%) of P105 blended into the PANI film can improve the hydrophilicity of the film, while keeping the other properties of the film almost unchanged. As a result, the combination of the conductive polyaniline with the Pluronic improves the biocompatibility of the PANI surface and can provide novel applications of the conductive polymers to the biomedical and biomaterial fields.

REFERENCES

1. M. Angelopoulos, G.E. Asturias, S.P. Ermer, A. Ray, E.M. Scherr, A.G. MacDiarmid, M. Akhtar, Z. Kiss, A.J. Epstein, *Mol. Cryst. Liq. Cryst.* 160 (1988) 151.
2. J.C. Chiang, A.G. MacDiarmid, *Synth. Met.* 13 (1986) 193.
3. A.P. Monkman, P. Adams, *Synth. Met.* 41–43 (1991) 627.
4. H.L. Wang, R.J. Romero, B.R. Mattes, Y. Zhu, M.J. Winokur, *J. Polym. Sci. Part B Polym. Phys.* 38 (2000) 194.
5. J.E. Fischer, Q. Zhu, X. Tang, E.M. Scherr, A.G. MacDiarmid, V.B. Cajipe, *Macromolecules* 27 (1994) 5094.
6. M.R. Anderson, B.R. Mattes, H. Reiss, R.B. Kaner, *Science* 252 (1991) 1412.
7. W.B. Liang, C.R. Martin, *Chem. Mater.* 3 (1991) 390.
8. G. Gustafsson, Y. Cao, M. Treacy, F. Klavetter, N. Colaneri, A.J.Heeger, *Nature (London)* 357 (1992) 477.
9. G. Grem, G. Leditzky, B. Ullrich, G. Leising, *Adv. Mater.* 4 (1992) 36.
10. N. Ahmad, A.G. MacDiarmid, *Synth. Meth.* 78 (1996) 103.
11. K. Mortensen, *Colloids Surf. A Physicochem. Eng. Asp.* 183–185 (2001) 277.
12. I.R. Schmolka, *Am. Perfum. Cosmet.* 82 (1967) 25.
13. M. Lawrence, *Chem. Soc. Rev.* (1994) 417.
14. M.W. Edens, in: V.M. Nace (Ed.), *Nonionic Surfactants: Polyoxyalkylene Block Copolymers*, Dekker, New York, 1996.
15. J.S. Tan, D.E. Butterfield, C.L. Voycheck, K.D. Caldwell, J.T. Li, *Biomaterials* 14 (1993) 823.
16. C.R. Deible, E.J. Beckman, A.J. Russell, W.R. Wagner, *J. Biomed. Mater. Res.* 41 (1998) 251.
17. A.V. Kabanov, E.V. Batrakova, V.Y. Alakhov, *J. Controlled Release* 82 (2002) 189.
18. P. Lemieux, N. Guerin, G. Paradis, R. Proulx, L. Chistyakova, A. Kabanov, V. Alakhov, *Gene Ther.* 7 (2000) 986.
19. S.I. Jeon, J.H. Lee, J.D. Andrade, P.G. DeGennes, *J. Colloid Interface Sci.* 142 (1991) 149.
20. J.H. Lee, J.H. Kopecek, J.D. Andrade, *J. Biomed. Mater. Res.* 23 (1989) 351.
21. V.A. Liu, W.E. Jastromb, S.N. Bhatia, *J. Biomed. Mater. Res.* 60 (2002) 126.
22. E. Detrait, J.B. Lhoest, P. Bertrand, P.V. de Aguilar, *J. Biomed. Mater. Res.* 45 (1999) 404.
23. D. Anderson, T. Nguyen, P.K. Lai, M. Amiji, *J. Appl. Polym. Sci.* 80 (2001) 1274.
24. K. Park, H.S. Shim, M.K. Dewanjee, N.L. Eigler, *J. Biomater. Sci. Polym. Ed.* 11 (2000) 1121.
25. L.J. Nagels, E. Staes, *Trac. Trend Anal. Chem.* 20 (2001) 178.
26. X.Y. Cui, V.A. Lee, Y. Raphael, J.A. Wiler, J.F. Hetke, D.J. Anderson, D.C. Martin, *J. Biomed. Mater. Res.* 56 (2001) 261.
27. M.M. Castillo-Ortega, D.E. Rodriguez, J.C. Encinas, M. Plascencia, F.A. Mendez-Velarde, R. Olayo, *Sensors Actuat. B Chem.* 85 (2002) 19.

28. S.V. Ermolaev, N. Jitariouk, A. Le Moel, *Nucl. Instrum. Meth. B* 185 (2001) 184.
29. J.M. Pernaut, J.R. Reynolds, *J. Phys. Chem. B* 104 (2000) 4080.
30. N. Lotan, G. Ashkenazi, S. Tuchman, S. Nehamkin, S. Sideman, *Mol. Cryst. Liq. Cryst.* 234 (1993) 635.
31. L.L. Miller, *Mol. Cryst. Liq. Cryst.* 160 (1988) 297.
32. V. Shastri, PhD dissertation, Rensselaer Polytechnic Institute, 1995.
33. A.N. Zelikin, D.M. Lynn, J. Farhadi, I. Martin, V. Shastri, R. Langer, *Angew. Chem. Int. Ed.* 41 (2002) 141.
34. M.M. Bradford, *Anal. Biochem.* 72 (1976) 248.
35. D. Briggs, in: *Surface Analysis of Polymers by XPS and Static SIMS*, Cambridge University Press, Cambridge, UK, 1998, p. 65.
36. Z.F. Li, E.T. Kang, K.G. Neoh, K.L. Tan, *Synth. Met.* 87 (1997) 45.
37. E. Ruckenstein, S. Gourisankar, *J. Colloid Interface Sci.* 101 (1984) 430; E. Ruckenstein, S. Gourisankar, *J. Colloid Interface Sci.* 109 (1986) 557.
38. R.A. Basheer, A.R. Hopkins, P.G. Rasmussen, *Macromolecules* 32 (1999) 4706; S.H. Goh, H.S.O. Chan, C.H. Ong, *J. Appl. Polym. Sci.* 68 (1998) 1839.
39. N.P. Chen, L. Hong, *Polymer* 43 (2002) 1429.
40. S.A. Chen, H.T. Lee, *Macromolecules* 26 (1993) 3254.
41. R. Mathew, D. Yang, B.R. Mattes, M.P. Espe, *Macromolecules* 35 (2002) 7575.

5.2 Synthesis of Surface Conductive Polyurethane Films*

Eli Ruckenstein and Yue Sun

Department of Chemical Engineering, State University of New York at Buffalo, Buffalo, NY 14260, USA

ABSTRACT Polyurethane films containing 3-methylthiophene were first prepared by the condensation of poly(propylene glycol) diol, poly(propylene glycol) triol and toluene 2,4-diisocyanate in the presence of a catalyst and 3-methylthiophene, at room temperature, overnight. The immersion of the films in a suitable organic solution of ferric chloride led to their rapid coating with conductive layers of poly(3-methylthiophene) via the diffusion-oxidative reaction of 3-methylthiophene and ferric chloride. The effects of the oxidative reaction time, the concentration of ferric chloride, the type of polyurethane, the solvent used for ferric chloride and the weight ratio of 3-methylthiophene to polyurethane were investigated. The coating layer was of the order of tens of micrometers and its conductivity was as high as 42 S/cm.

5.2.1 INTRODUCTION

In the last two decades, the conductive polymers, such as poly pyrrole (PPY), polyaniline (PANI), polythiophene (PT) and their derivatives, have received considerable attention due to their high conductivity, high thermal and chemical stability and, particularly, their potential applications in batteries, catalysis, biosensors, chemical detectors, actuators, electrochromic devices, electromagnetic shielding and antistatic coatings [1–7]. However, their commercial applications are somewhat limited because they are usually brittle and unprocessable, by both solution and melting methods.

Much research was carried out to prepare composites containing a conductive polymer and an insulating but flexible polymer, especially in the form of films. Various anionic soluble polymers could be used as dopants in the electrochemical polymerization of pyrrole to prepare polyelectrolyte–PPY composites at

* *Synthetic Metals* 1995, 75, 79–84.

the anode [8]. Similarly, latex particles with anionic surfaces together with electrochemically polymerized PPY were precipitated on an anode to generate composite films [9]. Most conductive composite films were electrochemically prepared by coating an anode with a polymer film and then immersing them in a solution containing an electrolyte and a monomer. The monomer, which imbibes the polymer, was electrochemically polymerized, thus generating a composite film. Poly(3-methylthiophene)–poly(methyl methacrylate) (PMT–PMMA) [10], PPY–poly(vinyl alcohol) [11], PPY–poly(tetrafluoroethylene) [12] and PPY–poly(vinyl chloride) [13] were prepared by this method. However, the electrochemical method cannot provide large films and is also expensive.

The chemical polymerization method can overcome these disadvantages. Several procedures to obtain conductive composite films have been developed. PPY–poly(vinyl alcohol) films were prepared by exposing poly(vinyl alcohol) that contained ferric chloride to pyrrole and water vapors [14]. PPY–poly(ethylene terephthalate) (PPY–PET) was prepared by immersing the pyrrole-swollen PET polymer into an aqueous solution of ferric chloride [15]. PPY–PMMA films were obtained by spreading a water-insoluble solvent solution of pyrrole and PMMA over the surface of an aqueous solution of ammonium persulfate [16,17]. Composite films containing PPY were also prepared by introducing polymer films at the interface between an aqueous solution of ferric chloride and a toluene solution of pyrrole [18]. Insulating plastics and fibers can be coated with thin films of conductive polymers by polymerizing monomers directly onto the substrates. For example, by immersing fluoropolymer sheets into a mixture of monomer solution and oxidant solution, one can coat the polymer with PPY, PANI, poly(*N*-methylpyrrole) or PMT films [19].

In this paper, a simple method for preparing PMT-coated polyurethane (PU) films is proposed. The oligomers of poly(propylene glycol) diol and poly(propylene glycol) triol, toluene 2,4-diisocyanate, a catalyst and 3-methylthiophene (3-MT), after being well mixed, were poured on a glass plate where they formed a 3-MT-PU film, via the condensation polymerization at room temperature. The film was peeled away from the glass and immersed into an organic solution of $FeCl_3$. The PU film was thus coated with a conductive film containing PMT via the oxidative polymerization of 3-MT. The preparation conditions, such as the weight ratio of 3-MT to PU, the reaction time, the concentration of $FeCl_3$, the nature of PU and that of the solvent used in the oxidative polymerization, were carefully studied.

Since the PUs have good mechanical properties, the PU with conductive surfaces could be employed in new fields, for example, as keyboards or buttons for calculators and computers. Due to the shining smooth surface of PMT, which easily dissipates electricity, the conducting surface has the advantage of avoiding dust deposition from air. In contrast, the conductive rubber, such as the carbon rubber, accumulates the dust from the air because of the presence of the insulating rubber at the surface of the material. The PU coated with PMT also constitutes a good material for electromagnetic shielding. The preparation procedure is very simple and the amount of 3-MT required small (as small as 0.056 weight ratio of 3-MT/PU).

5.2.2 EXPERIMENTAL

5.2.2.1 CHEMICALS

3-MT (Aldrich, 99+%), iron chloride (Aldrich, 97%), PPG725 and PPG2000 (poly(propylene glycol) diol, Aldrich, mol. wt. 725 and 2000, respectively), PPGT1500 (poly(propylene glycol) triol, Aldrich, mol. wt. 1500), PB 1200 and PB2800 (polybutadiene, hydroxyl functionalized, Aldrich, mol. wt. 1200 and 2800, respectively), PB4900 (polybutadiene, hydroxyl functionalized, Scientific Polymers Products, Inc., mol. wt. 4900), toluene 2,4-diisocyanate (Aldrich, 80%, tech), dibutyltin dilaurate (Aldrich, 95%), anhydrous ethyl acetate (Aldrich, 99+%), butyl acetate (Aldrich, 99%), methyl alcohol (Aldrich, 98%), *N, N'*-dimethyl formamide (Aldrich, 99.8%) were used as received. Distilled and deionized water was employed.

5.2.2.2 PREPARATION OF PU COATED WITH PMT

In a typical experiment, 0.8 g poly(propylene glycol) triol (PPGT1500), 1.0 g poly(propylene glycol) diol (PPG725), 0.52 g toluene 2,4-diisocyanate, 0.60 g 3-MT and two drops of butyltin dilaurate (catalyst) were mixed in a tube. Then, the liquid mixture was poured on a horizonal glass plate with an area of about 40 cm^2, which was subsequently covered with a box, whose four open edges were sealed to the surface of a desk (Figure 5.2.1), to avoid the vaporization in free air. After the condensation polymerization was allowed to proceed for 18 h, the PU film obtained containing 3-MT was peeled away from the glass. Then the film was immersed in a solution of 24.5 g $FeCl_3$ in 70 mL ethyl acetate, at room temperature, for 0.5–2 h. The surface of the film turned black rapidly. The coated film was washed in methyl alcohol and water and finally dried in vacuum overnight. In some experiments, PPG725 was replaced by PPG2000 or by PB 1200, PB2800 or PB4900.

5.2.2.3 THE MEASUREMENT OF THE THICKNESS OF THE COATING LAYER

A thin slice cut vertically from the coated film was magnified with a microscope (ZEISS AXIOVERT 35) and its picture taken with a SONY Video Print UP-930. The thickness of the coating layer was calculated from its thickness in the picture taking into account the magnification of the microscope. One of the pictures is presented in Figure 5.2.2.

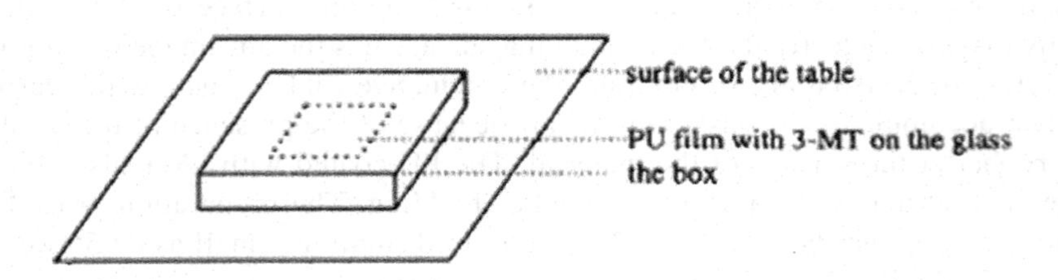

FIGURE 5.2.1 Apparatus used to prepare 3-MT-PU films.

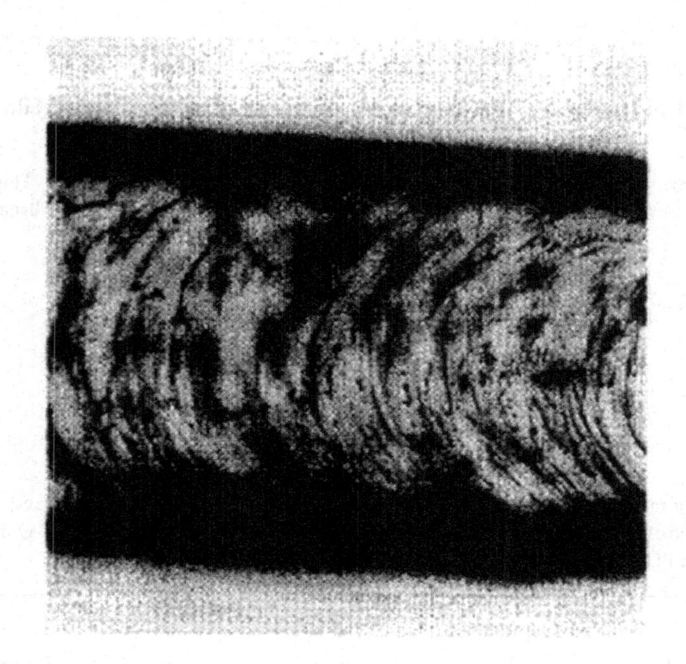

FIGURE 5.2.2 Picture of a coated PU film.

5.2.2.4 Measurement of the Electrical Resistance and Conductivity

The coated PU film was cut to rectangular shape (2.5 × 1.5 cm). The electrical resistance was measured by the four-point method. The conductivity of the coating layer was calculated from the electrical resistance and the thickness of the coating layer.

5.2.2.5 The Tensile Testing

The sample sheet was cut to the size required by the ASTM D.638-58T. The tensile testing was performed with an Instron Testing Instrument (model 1000) at room temperature with a cross-head speed of 20 mm/min.

5.2.3 RESULTS AND DISCUSSION

Table 5.2.1 presents the effect of the oxidative reaction time on the properties of the coated PU film. With increasing reaction time, the electrical resistance of the film decreases while the thickness of the coating layer increases. For an oxidative polymerization of about 5 min, the PU film has a coating layer with a conductivity as high as 10 S/cm. During the immersion of the 3-MT-PU film in the ethyl acetate solution of $FeCl_3$, several processes take place: (i) the swelling of the 3-MT–PU film by the organic solution; (ii) the diffusion of $FeCl_3$ and ethyl acetate into the 3-MT–PU film and the diffusion of 3-MT from inside towards the surface of the film; (iii) the oxidative polymerization of 3-MT by $FeCl_3$. Generally speaking, a diffusion-oxidative

TABLE 5.2.1
The Effect of the Reaction Time on the Properties of the Coated Film[a]

Reaction Time (min)	Electrical Resistance (Ω)	Thickness of Coating Layer (μm)	Conductivity of Coating Layer (S/cm)	Tensile Strength (MPa)	Elongation at Break Point (%)
0	∞	0	0	3.7	598
2	1.2×10^3	4.4	0.6	2.7	363
5	25	13	10	2.3	306
10	31	18	6.0	1.9	274
30	15	18	12	1.5	206
60	9.3	18	20	1.7	186
120	17	26	7.5	1.3	170
1440	17	35	5.6	1.6	75.0

[a] The weight ratio of PPGT1500/PPG725 was 0.8; the molar ratio of isocyanate/hydroxyl was 1.1; the weight ratio of 3-MT/PU was 0.25; the concentration of $FeCl_3$ in ethyl acetate was 0.35 g/mL; the thickness of the PU film was about 0.45 mm.

polymerization process takes place. As soon as the amount of PMT becomes large enough, a network is generated in the neighborhood of the surface of the film. The surface conductivity of the film increases from 0 to 10 S/cm when the reaction time increases from 0 to 5 min. The network forms a shield near the PU surface which prevents the counter diffusion of 3-MT and $FeCl_3$, and hence prevents additional oxidative polymerization. As a result, when the reaction time increases from 10 to 60 min, the thickness of the coating layer hardly changes. For very long reaction times, for example, 24 h, some $FeCl_3$ still penetrates through the surface layer and reacts with 3-MT behind the coating layer, increasing somewhat the thickness of the conductive layer. However, the amount of the additionally produced PMT is small and appears not to be included in the network surface layer, since the electrical resistance does not change appreciably even though the apparent thickness of the coating layer measured by microscopy increases. The conductivity decreases because the measured film thickness used in the calculations is larger than the effective conductive thickness.

Regarding the mechanical properties of the coated PU film, the elongation at the break point decreases with increasing reaction time because of the stress concentration in the coated PU caused by the brittle network of PMT. Table 5.2.1 demonstrates that only a short time is needed to coat the PU film with a highly conductive layer of PMT.

Table 5.2.2 illustrates the effect of the concentration of $FeCl_3$ on the properties of the coated PU film. With decreasing $FeCl_3$ concentration, both the electrical resistance and the thickness of the coating layer increase, while the conductivity decreases. The conductivity becomes large at a critical concentration between 0.30 and 0.35 g/mL. These observations can be explained as follows: Firstly, the oxidation potential of Fe^{3+}/Fe^{2+} increases with increasing $FeCl_3$ concentration and only

TABLE 5.2.2
The Effect of the Concentration of $FeCl_3$ in Ethyl Acetate on the Properties of the Coated Film[a]

Concentration of $FeCl_3$ (g/mL)	Electrical Resistance (Ω)	Thickness of Coating Layer (μm)	Conductivity of Coating Layer (S/cm)	Tensile Strength (MPa)	Elongation at Break Point (%)
0.50	6.0	13	42	1.7	198
0.40	6.7	18	28	1.4	149
0.35	9.3	18	20	1.7	186
0.30	10.1	35	9.4	1.9	131
0.25	0.21×10^3	40	0.4	1.4	160
0.15	1.0×10^3	44	0.08	1.3	227

[a] The reaction time was 1 h. Other preparation conditions were as in Table 5.2.1.

at a suitable oxidation potential can 3-MT be oxidized to PMT. Secondly, at higher concentrations, more $FeCl_3$ diffuses in the PU film in a short time and the PMT layer, which retards the further reaction, is formed more rapidly. As a result, the conductivity is high (42 S/cm) and the coating layer is thin (13 μm). For low $FeCl_3$ concentrations, the diffusion into the PU film is slower. It takes therefore a longer time for a PMT network to be formed and $FeCl_3$ can penetrate deeper into the film. This leads to both thicker coating layers and lower conductivities. A concentration of $FeCl_3$ of 0.35 g/mL was employed in the experiments presented below.

Table 5.2.3 shows the effect of the weight ratio of 3-MT/PU on the properties of the coated PU film. With increasing weight ratio, the electrical resistance and the

TABLE 5.2.3
The Effect of the Weight Ratio of 3-MT/PU on the Properties of the Coated Film[a]

Feeding Ratio of 3-MT/PU (%)	Electrical Resistance (Ω)	Thickness of Coating Layer (μm)	Conductivity of Coating Layer (S/cm)	Tensile Strength (MPa)	Elongation at Break Point (%)
2.0	2.0×10^6	35	4.7×10^{-5}	1.8	324
4.0	0.145×10^3	31	0.7	1.5	127
5.6	23.2	25	5.7	1.8	132
10.0	20.1	18	9.2	1.5	161
15.0	8.1	18	23	1.1	140
25.0	9.3	18	20	1.7	186
50.0	7.5	18	25	1.4	149

[a] The reaction time was 1 h. Other preparation conditions were as in Table 5.2.1.

thickness of the coating layer decrease, while the conductivity increases. This is as expected since, for higher weight ratios, the content of 3-MT near the surface of the film is higher. Therefore, in a short time, more 3-MT participates in the reaction with $FeCl_3$ and forms a conductive network and also a reaction-preventing shield. When the content of 3-MT is low, the slow diffusion of 3-MT plays an important role in its reaction with $FeCl_3$. In the latter case, $FeCl_3$ penetrates deep into the film. As a result, the coating layer is thick, but its conductivity is low because of the loose network of PMT generated from the smaller amount of 3-MT. Table 5.2.3 shows that for a relatively low weight ratio of 3-MT/PU, namely, 0.056, the conductivity of the coating layer is already 5.7 S/cm.

Table 5.2.4 presents the effect of the thickness of the PU film on the properties of the coated PU film. For a thickness of 0.25 mm, the coated film is an insulator; when it increases to 0.32 mm, the conductivity becomes 4.8 S/cm; when it increases further from 0.43 to 0.88 mm, the conductivity moderately increases. An explanation is as follows: In the 18 h process of condensation polymerization of PPG725 and PPGT1500 to form the PU film, the 3-MT from the film diffuses into the air of the box and therefore its actual concentration in the film is much lower. As a result, when the PU film is too thin, for example 0.25 mm, the content of 3-MT in the film decreases so much that a conductive network can no longer be generated; hence, the conductivity is zero. When the film becomes thicker, for example, 0.32 mm, even though the content of 3-MT on the surface of the film is low, it is still possible for a conductive network to be generated because the 3-MT present inside can diffuse to the surface and react. After the formation of the coating layer, the surplus of 3-MT seldom takes part in the reaction. This is why the thickness of the coating layer increases very little when the thickness of the PU film increases from 0.32 to 0.88 mm. Under the preparation conditions used in this paper, only PU films thicker than 0.32 mm can be well coated with a conductive layer. Decreasing the volume of the box of Figure 5.2.1 and introducing a small container containing 3-MT in the box to supply 3-MT vapor may prevent the diffusion of the 3-MT from the PU in the

TABLE 5.2.4
The Effect of the Thickness of the Film on the Properties of the Coated Film[a]

Thickness of PU Film (mm)	Electrical Resistance (Ω)	Thickness of Coating Layer (μm)	Conductivity of Coating Layer (S/cm)	Tensile Strength (MPa)	Elongation at Break Point (%)
0.25	∞	8.8	0	1.8	240
0.32	53	13	4.8	1.7	125
0.43	25.5	18	7.3	1.6	139
0.52	25.5	15	8.7	2.0	238
0.65	20.2	18	9.2	2.0	221
0.88	19.7	18	9.4	2.0	210

[a] The reaction time was 0.5 h and the weight ratio of 3-MT/PU was 0.5. Other preparation conditions were as in Table 5.2.1.

TABLE 5.2.5
The Nature of PU on the Properties of the Coated Film[a]

Diol	Weight Ratio of PPGT1500/ PPG (or PB)	Electrical Resistance (Ω)	Thickness of Coating Layer (μm)	Conductivity of Coating Layer (S/cm)	Tensile Strength (MPa)	Elongation at Break Point (%)
PB1200	0.8	0.53×10^3	76	0.08	2.9 (3.1)[b]	26
PB2800	0.8	0.25×10^3	48	0.3	3.1 (3.7)[b]	37
PB4900	0.8	0.11×10^3	79	0.4	3.2 (4.8)[b]	23
PPG2000	0.8	5.6	26	23	1.4	73
PPG725	0.4	5.8	26	22	2.4	194
PPG725	0.6	7.5	22	20	1.9	193
PPG725	0.8	9.3	18	20	1.7	186
PPG725	1.2	15	8.8	25	2.5	190

[a] The reaction time was 1 h. Other preparation conditions were as in Table 5.2.1.
[b] The value in parentheses is the tensile stress at the yield point.

atmosphere and, thus, allow the decrease of the minimum thickness of the PU film needed to obtain surface conductive PU films.

Table 5.2.5 presents the effect of the nature of PU on the properties of the coated film. The electrical resistance of the PU containing PB (1200, 2800 or 4900) is one to two orders of magnitude larger than that of PU based on PPG (725 or 2000), the coating layer is thicker and its conductivity lower. It is worth mentioning that the stress–strain curve of the coated PU based on PB has a yield point, while those of coated PU containing PPG do not have one (Figures 5.2.3 and 5.2.4). Since both PB and PMT have double bonds in their molecular chains, while PPG does not, the PB and PMT chains have higher affinity for one another than those of PPG and PMT; hence, they mix better. For this reason, at the beginning of tensile testing, the motion of the PB chains is restricted by the PMT domains; when the stress increases, the network of PMT is broken and then the sample exhibits the rubber-like mechanical behaviour of PU based on PB. At the point at which the latter process takes place, the yield point occurs. Concerning the coated PU based on PPG, the motion of the PPG chains is not completely restricted by the PMT domains, because they do not mix well, and the stress–strain curve is rubber-like.

Table 5.2.5 also shows that the weight ratio of PPGT1500/PPG725 affects somewhat the properties of the coated PU film. When it increases, the electrical resistance of the coated PU increases, while the thickness of the coating film decreases. This happens because the increased crosslinking of the PU film caused by the increased ratio of PPGT1500/PPG725 decreases the swelling ability of PU and retards the diffusion process. The conductivity remains, however, almost the same, about 20 S/cm.

The effect of the nature of the solvent used for $FeCl_3$ was also investigated. When acetone, *N, N′*-dimethyl formamide, methyl alcohol or ethyl alcohol was employed instead of ethyl acetate, the PU films remained insulators. The reason is that $FeCl_3$

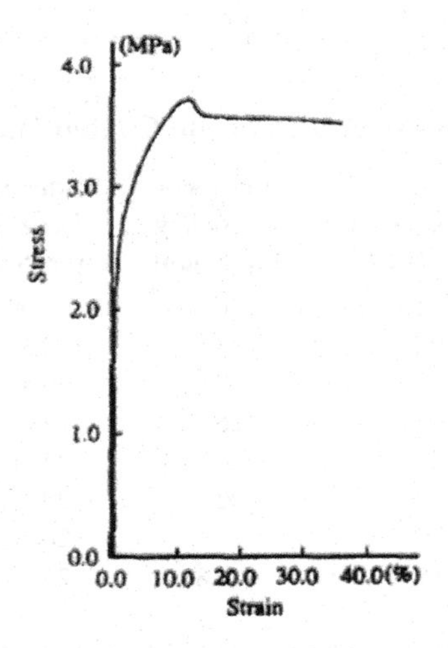

FIGURE 5.2.3 Stress–strain curve of a coated PU film containing PB 1200. The preparation conditions were as in Table 5.2.5.

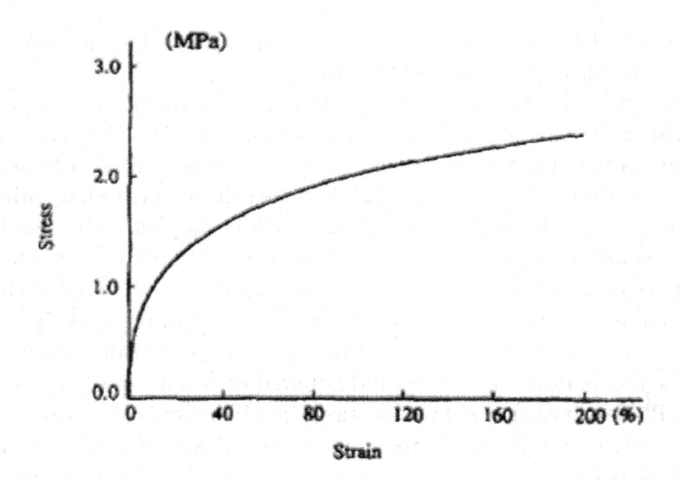

FIGURE 5.2.4 Stress–strain curve of a coated PU film containing PPG725 with a weight ratio of PPGT1500/PPG725 = 0.8. The preparation conditions were as in Table 5.2.5.

in these solvents does not have suitable oxidation potentials to polymerize the 3-MT. When an aqueous solution of $FeCl_3$ was employed, separated black spots appeared on the surface of the PU films. An aqueous solution of $FeCl_3$ is able to oxidize 3-MT; however, because PU does not swell in water, no diffusion of 3-MT and $FeCl_3$ in the PU film can occur and, hence, the polymerization cannot proceed further. In contrast, ethyl acetate and butyl acetate are suitable solvents. When butyl acetate replaced ethyl acetate, the electrical resistance increased from 17 to 106 Ω, and the thickness and the conductivity of the coating layer decreased from 26 μm and 7.5 S/cm to 4.4 μm and 7.2 S/cm, after 2 h of reaction (for the conditions of Table 5.2.1).

Concerning the electrical stability, the conductivity of the coated PU film decreased by about 50% after one month.

5.2.4 CONCLUSIONS

Surface conductive PU films can be successfully prepared by the diffusion-oxidative polymerization method proposed in this paper. First, a mixture of the precursors of PU containing 3-MT was cast and transformed into a film by condensation polymerization. Secondly, the film containing 3-MT was immersed into a ferric chloride solution. As a result, in a short time, the PU film was coated with a conductive layer near its surface. The preparation conditions, such as the oxidative reaction time, the concentration of ferric chloride, the weight ratio of 3-MT/PU, as well as the thickness and the nature of the PU film are factors that affect the properties of the coated PU film. In this method, even after a short reaction time of 5 min and for the low weight ratio of 3-MT/PU of 0.056, PU films could be coated with shighly conductive PMT layers near the surface. The highest conductivity of the coating layer was 42 S/cm.

REFERENCES

1. N.C. Billingham and P.D. Calvert, *Electrically Conducting Polymers —A Polymer Science View Point, Advances in Polymer Science*, Vol. 90, Springer, Berlin, Germany, 1989.
2. H.S.O. Chan, S.C. Ng and S.H. Seow, *Synth. Met., 66* (1994) 177.
3. Q. Pei and O. Inganäs, *Synth. Met., 55–57* (1993) 3730.
4. K. Yoshino, K. Kaneto and Y. Inuishi, *Jpn. J. Appl. Phys., 22* (1983) 157.
5. M.G. Kanatzidis, *Chem. Eng. News, 68* (1990) 36.
6. P.N. Bartlett and P.R. Birkin, *Synth. Met., 61* (1993) 15.
7. F. Janos and L. Schrader, *Synth. Met., 63* (1994) 187.
8. N. Bates, M. Cross, R. Lines and D. Walton, *J. Chem. Soc., Chem. Commun.,* (1985) 871.
9. S.J. Jasne and C.K. Chiklis, *Synth. Met., 15* (1986) 175.
10. J. Roncali and F. Gamier, *J. Chem. Soc., Chem. Commun.,* (1986) 783.
11. S.E. Lindsey and G.B. Street, *Synth. Met., 10* (1984) 67.
12. R.M. Penner and C.R. Martin, *J. Electrochem. Soc., 33* (1986) 310.
13. O. Niwa, M. Kakuchi and T. Tamamura, *Synth. Met., 18* (1987) 677.
14. T. Ojio and S. Miyata, *Polym. J., 18* (1986) 95.
15. C. Li and Z. Song, *Synth. Met., 40* (1991) 23.
16. M. Moreta, I. Hashida and M. Mishimura, *J. Appl. Polym. Sei., 36* (1988) 1639.
17. H.S.O. Chan, T.S.A. Hor, P.K.H. Ho, K.L. Tan and B.T.G. Tan, *J. Macromol. Sei., Chem., 27* (1990) 1081.
18. V. Bocchi, G.P. Gardini and S. Rapi, *J. Mater. Sei. Lett., 6* (1987) 1283.
19. L.S.V. Dyke, C.J. Brumlik, W. Liang, J. Lei, C.R. Martin, Z. Yu, L. Li and G.J. Collins, *Synth. Met., 62* (1994) 75.

5.3 Conductive Surface via Graft Polymerization of Aniline on a Modified Glass Surface*

Z.F. Li and Eli Ruckenstein

Department of Chemical Engineering
Clifford C. Furnas Hall, State University of New York at Buffalo, SUNY—Buffalo Box 604200, 14260-4200 Amherst, NY, USA

ABSTRACT A functionalization with an aniline monolayer of a modified glass surface, followed by the surface grafting of polyaniline (PANI), was used to prepare a conductive surface. A bromopropylsilane monolayer was first generated by reacting a hydroxylated surface with 3-bromopropyltrichlorosilane under an inert atmosphere. This layer was functionalized by its reaction with aniline, which substituted the bromide atoms of the silane chain. Further, the tethered aniline molecules were used as active sites for the graft polymerization of PANI on the surface. The composition and microstructure of the PANI-grafted glass surfaces were examined by X-ray photoelectron spectroscopy (XPS), and ultraviolet-visible (UV-VIS) spectroscopy, as well as by contact angle measurements. The surface conductivity of the modified glass surface-grafted with PANI was of the order of 10 S/cm, hence, larger than the usual value (~1 S/cm) of the bulk PANI.

5.3.1 INTRODUCTION

Owing to its unique electro-optical properties, polyaniline (PANI) has been extensively investigated in the field of conductive polymers [1]. Among the research directions, the chemical deposition of PANI as a thin layer on various materials, such as plastics, glass, metals, as well as on micro- and nano-porous materials generated a new area of applications [2]. The coating of various materials with a layer of

* *Synthetic Metals* 2002, 129, 73–83.

conductive polymer was achieved by using several methods, such as the spreading of a solution of conductive polymer on the surface of the material followed by the evaporation of the solvent, the electropolymerization of the monomers on an electrode, or the chemical polymerization and deposition on the surfaces of various materials immersed in the polymerization solution.

In recent years, a great deal of attention was paid to coating the surface of glass and related inorganic substrates with a layer of conductive polymers to prepare electrochromic displays, liquid crystal displays and "smart windows" [3]. On a smooth surface of glass, the adhesion of a conductive polymer is usually poor. The increase of the surface adhesion was achieved through a chemical pre-treatment, by spraying $SOCl_2$ on the surface of glass at 400°C, thus binding Cl to the glass surface; this was followed by the grafting of a polymer [3]. This method generated a strong chemical bond between glass and PANI. Some attempts have been also made to generate a self assembly (SAM) monolayer or a patterned SAM monolayer, using an alkoxysilane containing an aniline or a pyrrole group, followed by the physical, chemical or electrochemical deposition of conductive PANI or polypyrrol [4–7]. However, the above SAM methods provided incomplete coverage, and there were even difficulties in the preparation and purification of the silane compounds bearing an aniline or a pyrrole group [8,9].

In this paper, we employed the aniline monomer substitution reaction on a SAM silane monolayer, followed by the grafting of PANI on the surface. The method involves the initial formation of a stable silane monolayer through its reaction with the hydroxyl groups of the glass surface. This was followed by the functionalization of the SAM monolayer through aniline substitution, and further, by the surface oxidative graft polymerization of aniline on the modified glass surface via the covalently immobilized aniline sites. This method offers the advantage that one can select a suitable silane coupling agent which can yield a homogenous, reproducible, stable monolayer, using only commercially available, high purity reagents, that are free of the side reactions [9,10] encountered with other more exotic silane monolayers. Furthermore, the method can be used for SAM monolayers that can be photo-patterned with high resolution, thus, allowing the patterning with conductive polymers after the photolithographic process.

5.3.2 EXPERIMENTAL SECTION

5.3.2.1 MATERIALS

The microscopic glass slides were purchased from Aldrich/ Sigma Chemical Co., and were sliced into rectangular strips of about 1.0 cm × 2.0 cm in size. Aniline monomers, 3-bromopropyltrichlorosilane, ammonium persulfate and hydrazine were obtained also from Aldrich/Sigma and were used as received. The anhydrous toluene was dried over 4 Å molecular sieves prior to use. The solvents, such as acetone, ethanol, and *N*-methylpyrrolidinone (NMP), and other chemicals were of reagent grade and were also purchased from the Aldrich/Sigma Chemical Co. and used as received.

5.3.2.2 Substrate Pre-treatment

To remove the organic residues from the surface, the glass slides (1 cm × 2 cm) were first soaked in a soap solution, sonicated for 5 min, and then rinsed with a large amount of distilled water. The substrates were then immersed in a "piranha" solution (a mixture of 70% volume concentrated sulfuric acid (98 wt.%) and 30% volume of a hydrogen peroxide solution (30 wt.%) and boiled for about 50 min. The cleaned glass slides were then washed with a large amount of distilled water and dried under reduced pressure for subsequent surface treatment. The contact angle of water droplets on the glass surface was low, about 12°, over the entire surface, revealing the high cleanliness and uniformity of the surface.

5.3.2.3 Silane Treatment to Produce SAM

The pre-treated hydrophilic glass slide was placed at room temperature into a solution of 3-bromopropyltrichlorosilane (25 μL in 25 mL of dried toluene), under a nitrogen atmosphere for 24 h, and then washed twice with dried toluene in the same nitrogen atmosphere. Finally, the glass was removed from the nitrogen atmosphere and cleaned in an ultrasonic bath in toluene for 1 min, rinsed again successively with toluene, acetone, and ethanol, and finally dried.

5.3.2.4 Aniline Monomer Functionalization Through Substitution Reaction

The SAM-glass substrate was immersed in aniline contained in a Pyrex tube, for various time durations ranging from 1 to 48 h, and various temperatures, ranging from room temperature to 80°C. After the covalent substitution of aniline to the silane group, the glass substrate was washed thoroughly with a large amount of NMP to remove the unreacted aniline. The residual NMP present on the aniline-SAM-glass substrate was removed by washing with a large amount of ethanol.

5.3.2.5 Surface Oxidative Polymerization of Aniline

The oxidative graft polymerization of aniline to the aniline-SAM-glass substrate was as that usually employed for the oxidative homopolymerization of aniline to the conductive emeraldine (EM) salt [11,12]. It was carried out in a 1 M HCl solution containing 0.1 M aniline and 0.1 M $(NH_4)_2S_2O_8$. The reaction was allowed to proceed at 0°C for 5 h. The EM salt grafted on the glass surface was converted to the neutral EM base by the immersion and equilibration of the glass slide in a large amount of doubly distilled water. The surface-modified glass was subsequently immersed in a large volume of NMP for at least 48 h in order to remove the physically adsorbed EM base. During the washing process, the NMP solvent was changed several times during the first hour and then changed every 12 h. The PANI-grafted glass surface was further washed with alcohol to remove the residual NMP before being dried under reduced pressure. The EM state of the grafted PANI on the glass surface was reduced to the leucoemeraldine (LEB) state by immersion in hydrazine for 2 h, followed by a thorough rinsing with distilled water, before drying under reduced pressure. Further, the PANI-glass surface with PANI in the LEB state was immersed in a palladium nitrate solution for 10 min. After removal from the Pd nitrate solution, the Pd deposited glass substrate was rinsed thoroughly with double distilled water before being dried under reduced pressure.

5.3.2.6 Characterization of the Surface-Modified Glass Slide

The graft-modified glass surfaces were characterized by contact angle, XPS, UV–VIS spectrum, and four-probe conductivity measurements. The contact angle measurements were carried out using a NRL CA Goniometer Model 100-00(115) from Ramehart Inc. The telescope with a magnification power of 23X was equipped with a protractor of 1° gradation. The angles reported are accurate within ±3°. For each substrate, at least three measurements on different locations were averaged. The XPS measurements were carried out on a Surface Science Model SSX-100 Small Spot ESCA, possessing an Al Kα monochromatized X-ray source (1.48 keV photons). The samples were mounted on standard sample studs and the core-level spectra were obtained at a photoelectron takeoff angle of 45°, measured with respect to the surface of the film. The pressure in the analysis chamber was maintained at 10^{-9} Torr or lower during the measurement. To compensate for the surface charging effects, all binding energies were referenced to the C 1s hydrocarbon peak at 285 eV. In the peak analysis, the line widths (the width at half-maximum) of the Gaussian peaks were kept constant for the components in a particular spectrum. The surface elemental compositions were determined from the peak area ratios and were accurate to within 1%. The UV–VIS absorption measurements were carried out using a Beckman Model DU650 scanning spectrophotometer. The conductivity of the graft-modified glass surface was determined by the four-probe method, using a Hewlett-Packard Model 3478A digital multimeter. For each of the conductivities reported, at least three measurements were averaged.

5.3.3 RESULTS AND DISCUSSION

The strategy for the PANI functionalization of glass hydroxyl-terminated surfaces consists of three basic steps, depicted in Scheme 5.3.1: (1) the formation of a well-defined SAM through the reaction of 3-bromopropyltrichlorosilane with the hydroxyls of the glass surface; (2) the debromolization of this monolayer by substitution with aniline, resulting in the formation of an aniline-glass substrate; and (3) the reaction of the aniline sites with an aniline-oxidant solution to produce the desired graft polymerization of aniline on the surface. Each process is described below.

5.3.3.1 Formation of 3-Bromopropylsilane on the Glass Surface

In order to generate a well-defined functionalized surface by the method outlined in Scheme 5.3.1, a dense and uniform 3-bromopropylsilane monolayer must initially

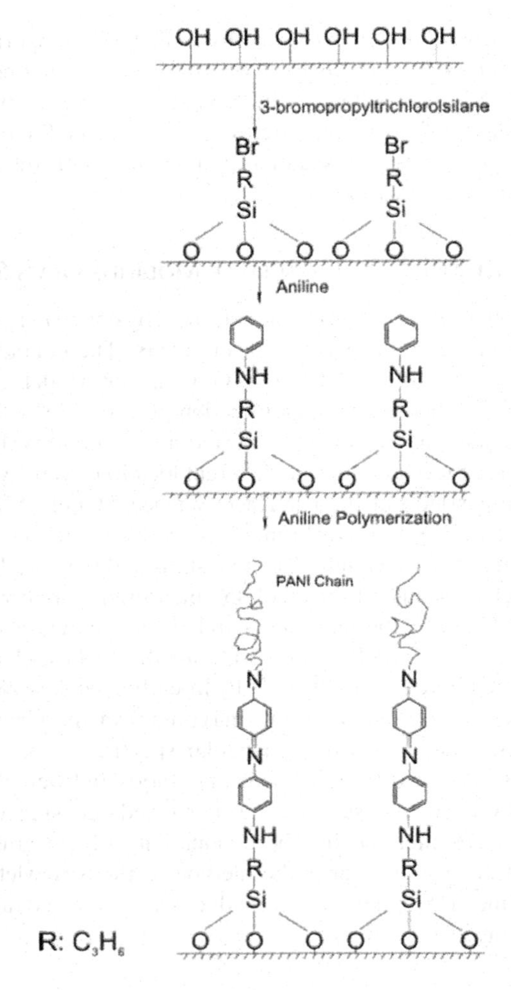

SCHEME 5.3.1 Chemical strategy for the surface graft polymerization of PANI on a glass substrate.

be formed, by a reproducible procedure, on the surface. For this reason, a maximally hydroxylated surface was first generated by the immersion of the substrate in a boiling piranha solution for at least 30 min, but < 1 h [13,14]. The contact angle of the surface so treated was the same (12°) over the entire surface, indicating that the substrate was uniformly covered with hydroxyl groups. Further, the hydroxylated surface was exposed at room temperature to a 3-bromopropyltrichlorosilane toluene solution for about 24 h; thus, the surface became covered with a uniformly dense 3-bromopropylsilane film. The increase in the wetting angle of the glass from 12° after hydroxylation to 53° after silanization demonstrates that the

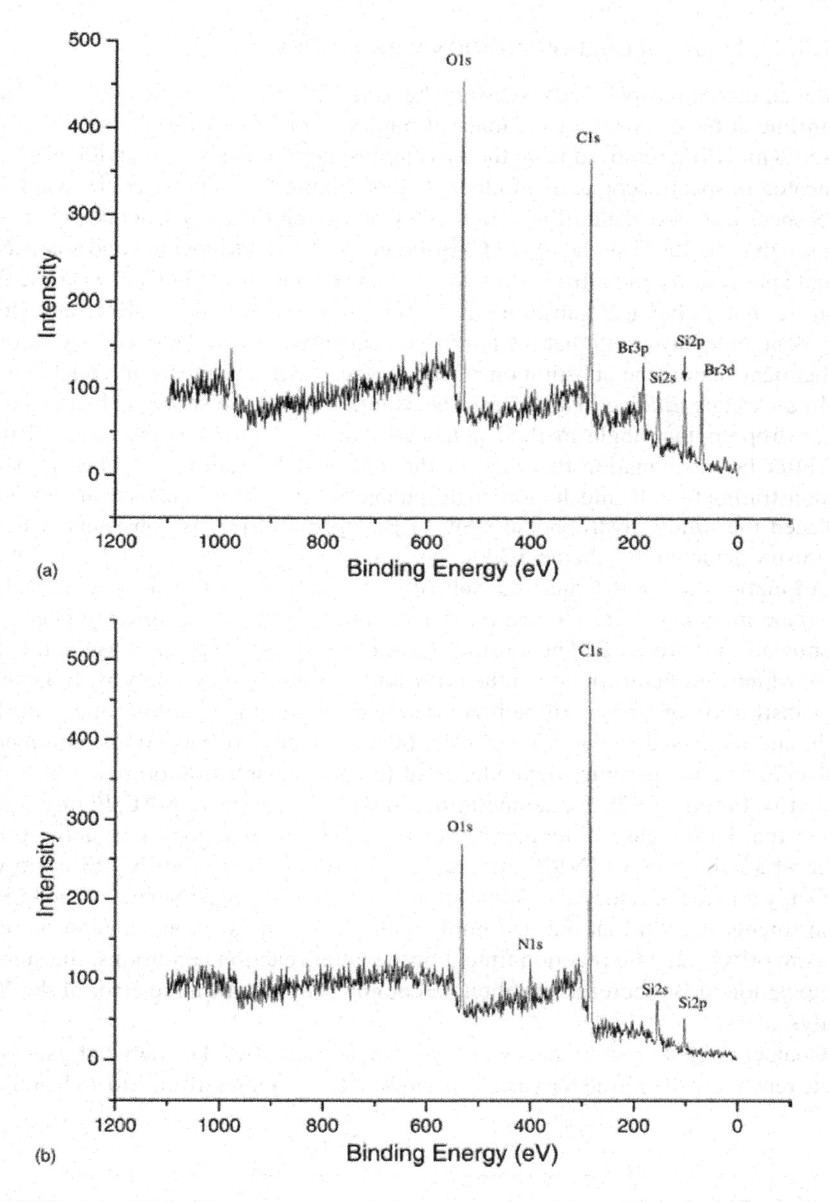

FIGURE 5.3.1 XPS wide scan spectra of glass substrates with monolayers of (a) $\equiv SiC_3H_6Br$ and (b) after the substitution reaction of Br by aniline.

hydrophobic $\equiv SiC_3H_6Br$ layer has replaced the hydrophilic hydroxyl layer. The contact angle of the $\equiv SiC_3H_6Br$ layer on glass varied by <3° between different locations, indicating the uniformity of the monolayer. The appearance of a Br peak in the wide scan XPS analysis of the surface (Figure 5.3.1a) indicates the presence of $\equiv SiC_3H_6Br$ on the surface.

5.3.3.2 Surface Functionalization with Aniline

Under an inert atmosphere, the substrate bearing a $\equiv SiC_3H_6Br$ monolayer was placed in aniline at 60°C. After a time interval ranging from 1 to 48 h, the substrate was rinsed with NMP, removed from the inert atmosphere, washed again with NMP and sonicated in spectroscopically grade methanol. Figure 5.3.1b presents the wide scan XPS spectrum after the aniline treatment. Comparing Figure 5.3.1a and 5.3.1b, one can see that the Br 3d signal of the C–Br bond (70.3 eV) disappeared and a new N 1s signal appeared for the nitrogen connected to a benzene ring (399.3 eV). These data indicate that a chemical substitution reaction occurred between aniline and Br on the silane monolayer and that the aniline was anchored to the glass surface through a chemical bond. The substitution of the surface-bound C–Br by the nucleophilic aniline was quantitative in all cases. The examination of the substrate by the sessile water drop contact angle method indicated that the contact angle increased from 53° after bromosilanation to 77° after the aniline substitution of Br (Table 5.3.1), demonstrating that the much more hydrophobic SiC_3H_6–NH–benzene ring layer has replaced the mildly hydrophobic $\equiv SiC_3H_6Br$ layer, and this is consistent with the chemistry depicted in Scheme 5.3.1.

A kinetic study of the chemical substitution reaction of Br by aniline in the bromosilane monolayer was carried out for various reaction times and temperatures, as shown in Figure 5.3.2. The aniline substitution level is expressed as the [N]/[Si] ratio, which was determined via the XPS surface composition analysis. In general, the substitution of Br by aniline increased with increasing reaction time from 1 to 48 h, and increased slowly after 24 h at 60°C, because Br was almost completely replaced. The temperature dependence of the aniline substitution reaction is presented in Figure 5.3.2b for temperatures ranging from 25 to 80°C Figure 5.3.2b shows that an increase in temperature accelerated the reaction up to about 60°C, after which the ratio of [N]/[Si] decreased. The decrease of the [N]/[Si] ratio was probably due to the removal of some silane sites from the glass surface. The kinetic experiments suggest that the optimum conditions for maximum substitution reaction are 60°C and 24 h reaction time. For the latter reaction conditions, the surface composition of Br decreased to about 0.4 atom%, value within the error of the XPS analysis (1%).

Contact angle measurements were also carried out after the silanized glass substrate reacted with aniline for various periods of time ranging from 1 to 48 h at 60°C.

TABLE 5.3.1
Water Contact Angles after Surface Treatment

	Contact Angle (°)
Glass surface	65
Glass surface after the treatment with the piranha solution	12
Surface after reaction with 3-bromopropyltrichlorosilane	53
Surface after 48 h of substitution reaction of Br by aniline	77

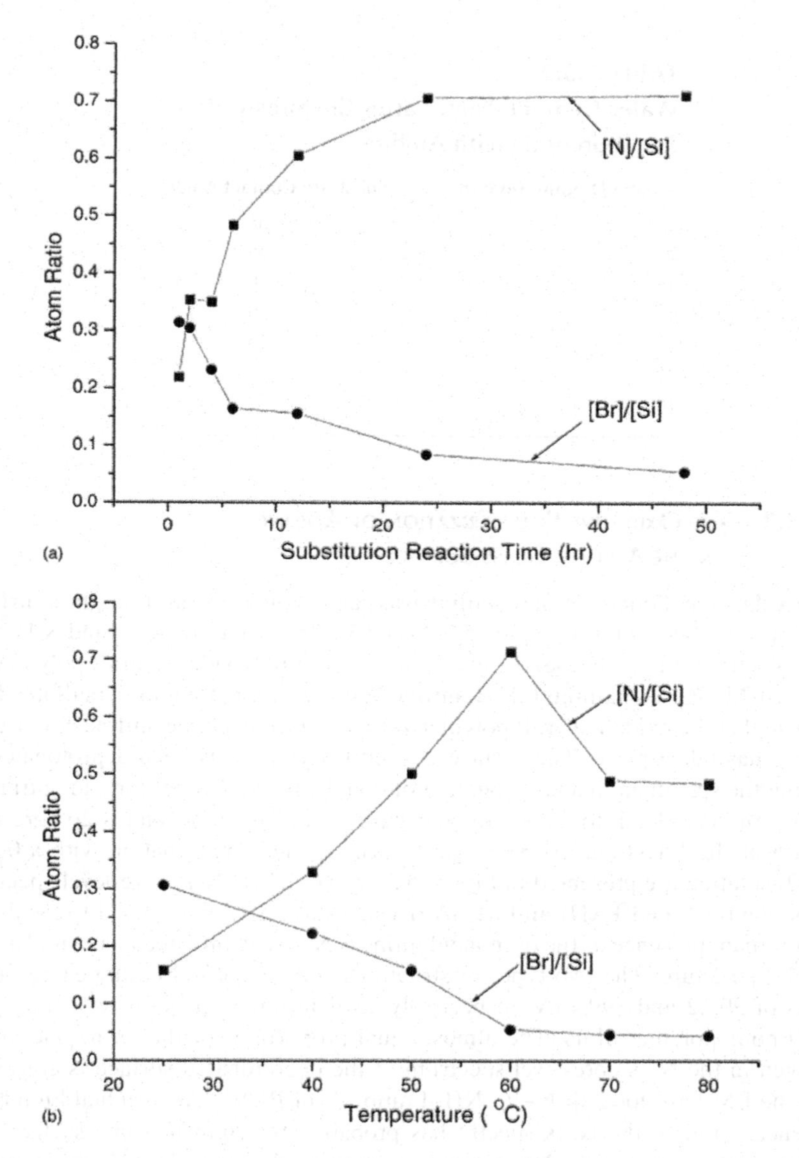

FIGURE 5.3.2 The effect of (a) substitution reaction time; and (b) temperature on the [N]/[Si] ratio and [Br]/[Si] ratio of the glass surface.

The results are listed in Table 5.3.2, which shows that the contact angle increased with increasing reaction time and leveled off for reaction times longer than about 24 h. These results confirm that the hydrophobic benzene ring was tethered to the top surface of the substrate by replacing a Br atom. The surface density of aniline bound to the substrate is expected to affect the amount of PANI chains grafted to the substrate, as will be shown in Section 3.3.

TABLE 5.3.2
Water Contact Angles after the Substitution Reaction of Br with Aniline

Aniline Dipping Time (h)	Substrate Contact Angles (°)
1	58
2	66
4	70
6	73
12	75
24	77
48	77

5.3.3.3 The Oxidative Polymerization of Aniline on the Aniline-Glass Surface

The oxidative polymerization of aniline was carried out using aniline-glass surfaces with various [N]/[Si] ratios. Figure 5.3.3a and 5.3.3b presents the C 1s and N 1s core-level spectra of the aniline-SAM-glass surface after the oxidative graft polymerization in 1 M HCl containing 0.1 M aniline and 0.1 M ammonium persulfate. They confirm that the oxidative graft polymerization of aniline on the aniline-SAM-glass surface has taken place. The surface-grafted PANI salt has been deprotonated by dipping the specimen in a large amount of double distilled water for a long time, to convert the PANI salt to its neutral EM base form. The N 1s, and S 2p core-level spectra of the PANI-aniline-SAM-glass surface after reprotonation with a 0.5 M H_2SO_4 solution are presented in Figure 5.3.3c and d. The N 1s core-level spectrum of the deprotonated PANI-aniline-SAM-glass surface (Figure 5.3.3b) reveals the predominant presence of the quinonoid imine (=N–) structure and benzenoid amine (–NH–) structure. The two types of structures correspond to binding energy (BE) peaks of 398.2 and 399.4 eV, respectively, as determined previously [12] also for PANI homopolymer films. The almost equal proportions of the imine and amine nitrogen in the N 1s core-level spectrum of the deprotonated surface is consistent with the EM base form ([=N–]/[–NH–] ratio ~1) of PANI. The residual high binding energy tail in the N 1s spectra has probably its origin in some surface oxidation products, or weak charge-transfer complexed oxygen [15]. Comparing the N 1s core-level spectrum (Figure 5.3.3b) to that of the reprotonated PANI-aniline-SAM-glass surface (Figure 5.3.3c), one can conclude that the grafted EM base on the glass surface was effectively reprotonated by H_2SO_4, since the –N= groups were replaced by the positively charged nitrogens. The protonation of the EM base form of the aniline homopolymer also occurs preferentially at the imine group [11]. Figure 5.3.3d, which reveals the presence of the SO_3^- group, provides additional evidence for doping. As the homopolymer, the EM base of the grafted PANI on the glass slide surface can readily be reduced by hydrazine to the LEB state and re-oxidized with a palladium nitrate solution. Figure 5.3.4a and 5.3.4b presents the C 1s and N 1s core-level spectra of the PANI-glass surface after its reduction with hydrazine. The N 1s core-level spectrum of the reduced PANI-glass surface is mostly dominated by the –NH– peak at a BE of 399.4 eV (Figure 5.3.4b). The residual imine group at 398.2 eV and the residual high BE above 400 eV can be attributed to the incomplete reduction of the imine nitrogen of EM by NH_2–NH_2 to the amine nitrogen, or may have resulted, at least in part, from some surface oxidation

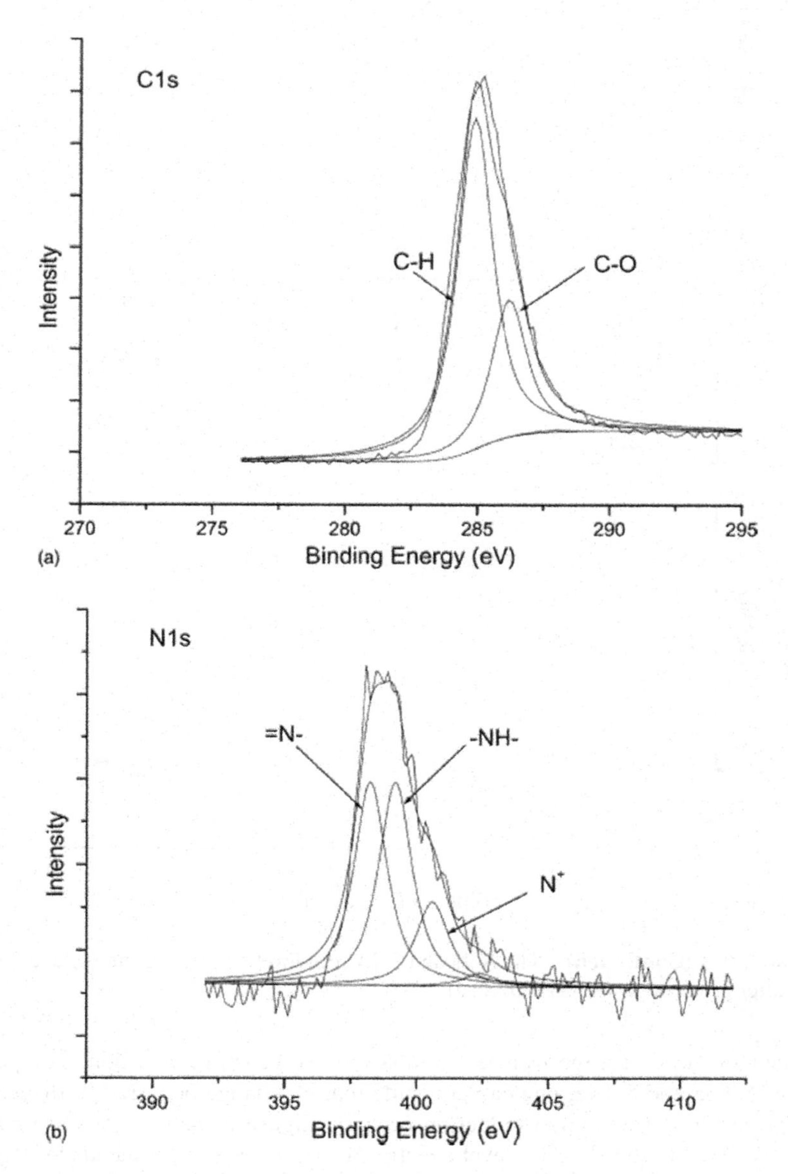

FIGURE 5.3.3 C 1s and N 1s core-level spectra of PANI-grafted on the glass substrate (a, b).
(Continued)

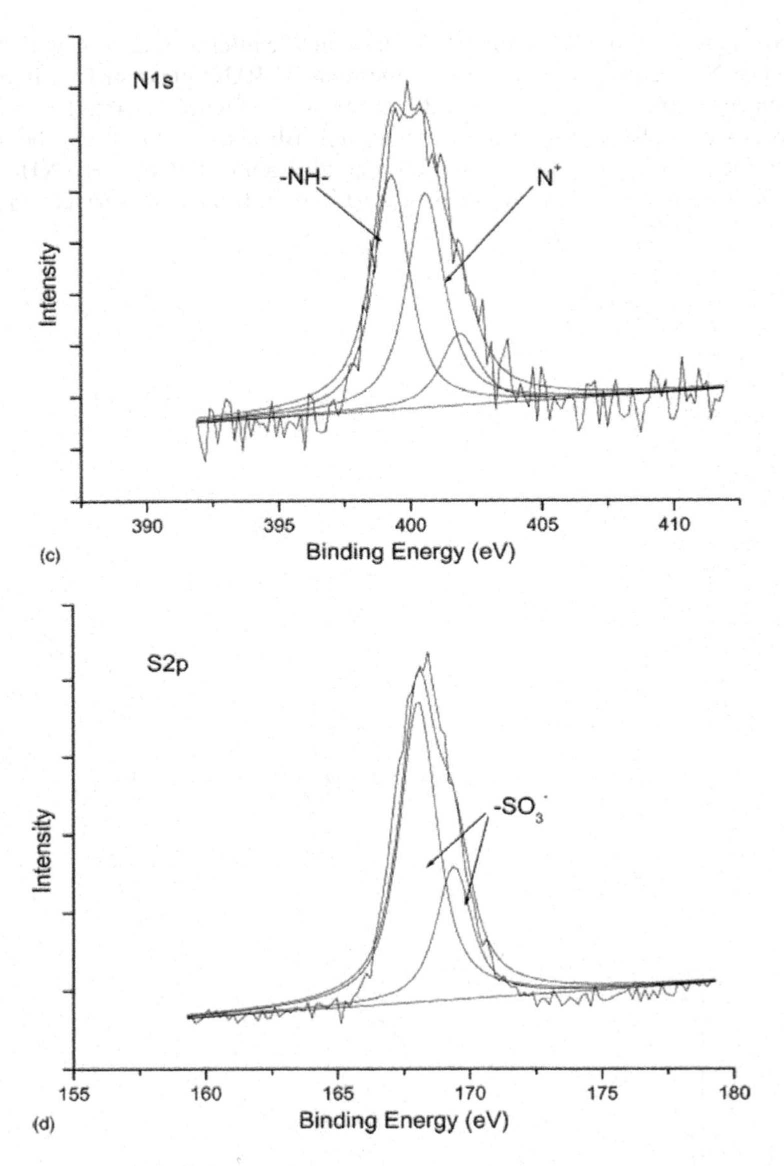

FIGURE 5.3.3 (Continued) The N 1s and S 2p core-level spectra for the surface-grafted PANI after protonation with H_2SO_4 (c, d).

products or weak charge-transfer complexed oxygens. In addition, comparing Figures 5.3.3a and 5.3.4a, one can conclude that no change occurred in the carbon spectra during the oxidation-reduction process. Figure 5.3.4c and 5.3.4d presents the N 1s and Pd 3d XPS core-level spectra, respectively, of a reduced PANI-glass surface reoxidised with a palladium nitrate solution, having an initial content of 0.02 wt.% of Pd, for about 10 min. The comparison of Figure 5.3.4b and 5.3.4c indicates that the proportion of imine nitrogen increases after oxidation, hence, that the intrinsic oxidation state of the grafted PANI in its LEB state is increased upon reduction. The Pd 3d core-level spectrum (Figure 5.3.4d) can be deconvoluted into two major BE peaks, Pd 3d5/2 and Pd 3d3/2 at about 335 and 340 eV, respectively, which can be assigned to the Pd(0) species [16]. The behavior of the glass grafted PANI is very similar to that of the PANI homopolymer [12,17].

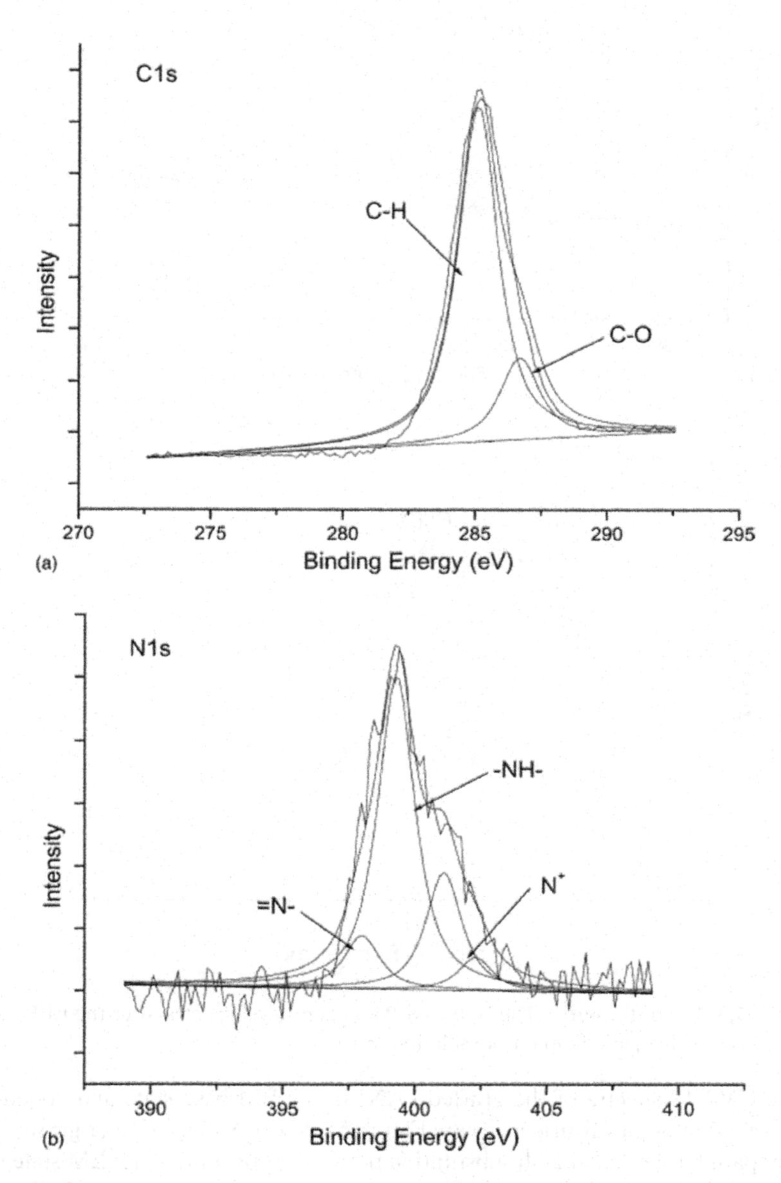

FIGURE 5.3.4 C 1s and N 1s core-level spectra of LEB state of grafted PANI on the glass substrate (a, b). *(Continued)*

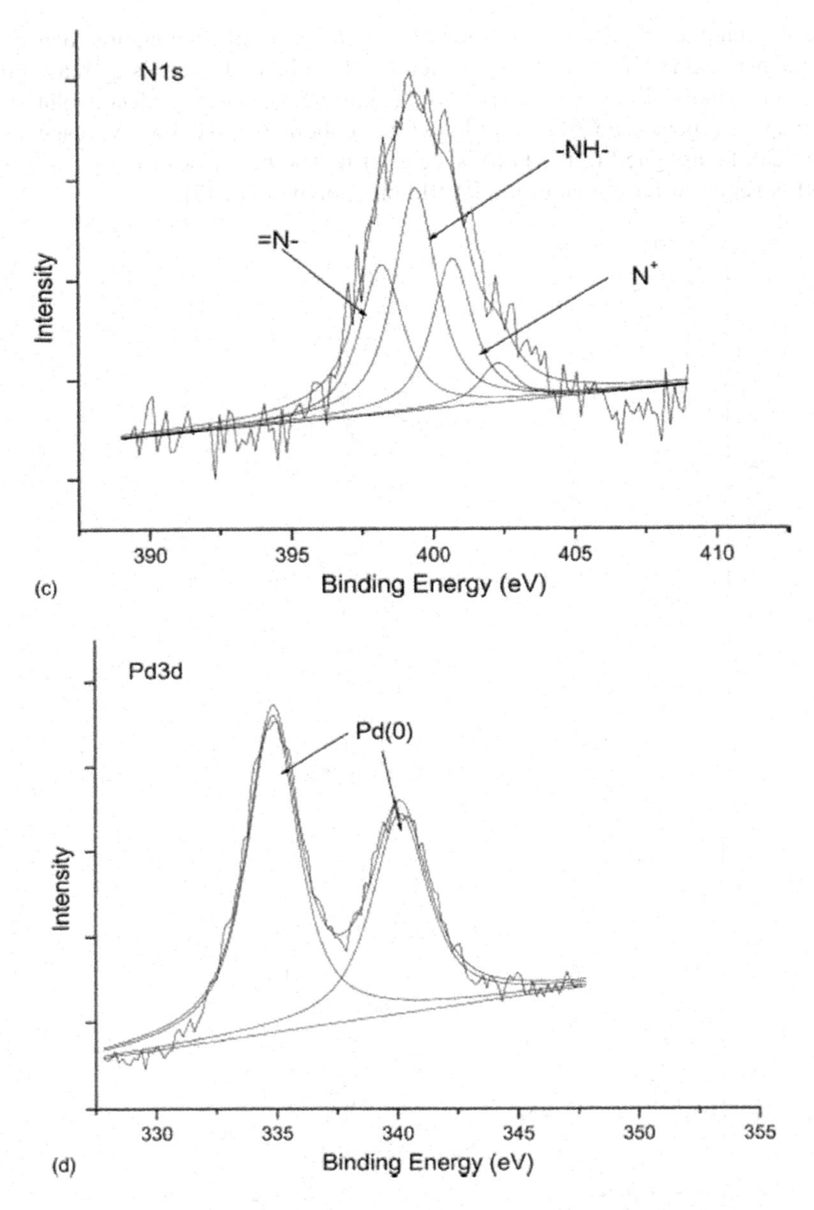

FIGURE 5.3.4 (Continued) The N 1s and Pd 3d core-level spectra after the LEB surface was oxidized with a palladium nitrate solution (c, d).

The UV-VIS spectra of the grafted PANI in the EM base state and doped state (after treatment with sulfuric acid) are Figure 5.3.5. UV-VIS spectra of grafted samples prepared using 6 and 24 h substitution times of Br by aniline: (a) EM state of the grafted PANI film on the glass substrate and (b) grafted PANI protonated with 0.5 M H_2SO_4 solution presented in Figure 5.3.5 for grafted samples prepared using 6 and

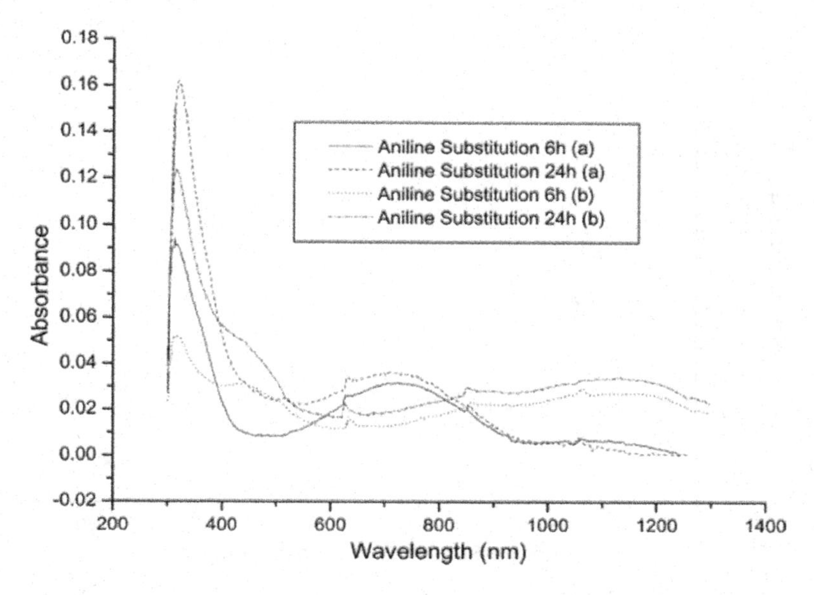

FIGURE 5.3.5 UV–VIS spectra of grafted samples prepared using 6 and 24 h substitution times of Br by aniline: (a) EM state of the grafted PANI film on the glass substrate and (b) grafted PANI protonated with 0.5 M H_2SO_4 solution.

24 h substitution times of Br by aniline. Similar to the PANI homopolymer [18,19], the grafted PANI in the EM base state (Figure 5.3.5a) exhibits the typical absorption peak at 310 nm (π–π* transition of the benzenoid rings) and the localized quinoid exciton transition band at ~640 nm. After dipping in acid solution for a short time, the color of the glass changed from light blue to light green, which indicated that PANI was doped. In the doped state of PANI (Figure 5.3.5b), a significant red shift occurred for wavelengths larger than 800 nm [18,19] due to the protonation of the imino sites and excitations to the polaron band. The absorbance provides an indirect qualitative measure of the thickness of the film, because it depends upon the amount of aniline bound to the surface. In addition, it should be mentioned that the PANI-grafted glass kept its transparency.

The successful grafting of the PANI chains was further confirmed by the significant increase in the surface conductivity. Figure 5.3.6 presents the surface resistance of the PANI-grafted glass after protonation with 0.5 M H_2SO_4 against the substitution time of Br by aniline. As expected, the surface resistance decreases with increasing substitution time from 4.85×10^6 to 2.1×10^4 Ω/area (where the area is 1 cm wide and 0.2 cm long). This again demonstrates that the aniline sites generated through substitution on the surface of the glass constitute the sites that trigger the further graft polymerization. From XPS depth profile analysis of the samples, one could evaluate the thickness of the PANI layer grafted on the glass surface. We obtained in this manner values between 7 and 10 nm for a substitution time of 48 h at 60°C. Considering a film thickness of 10 nm, one obtains a surface conductivity of 11 ± 3 S/cm. For comparison purposes, we also prepared protonated PANI films and determined their

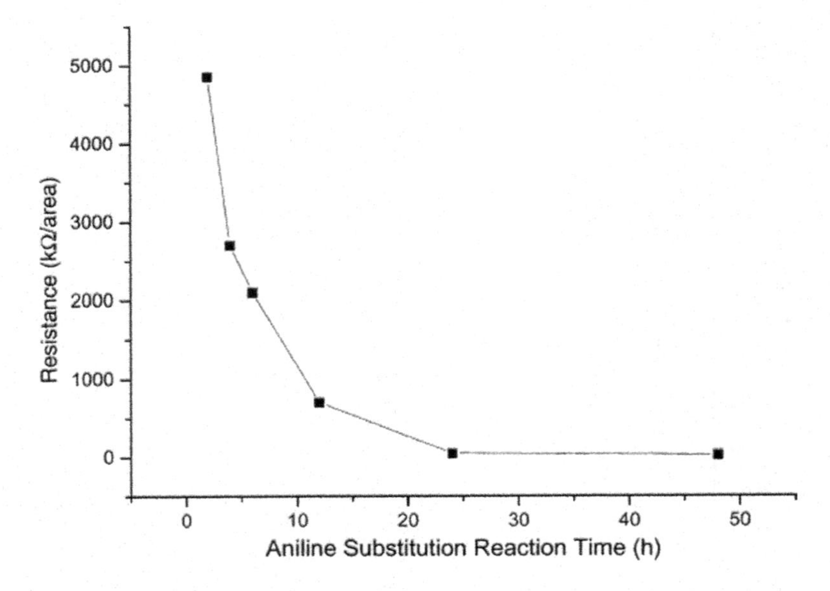

FIGURE 5.3.6 The surface resistance of the PANI-grafted glass after protonation with 0.5 M H_2SO_4 against the substitution time of Br by aniline.

bulk conductivities, which were of the order of 1 ± 0.2 S/cm. The different values can be attributed to the ordered and evenness of the grafted layer generated via a SAM method. The relatively high conductivity of the grafted PANI makes the present surface modification technique potentially valuable for antistatic applications and in the photolithographic industry. Finally, it is appropriate to emphasize that the grafted PANI chains are covalently bound to the glass surface and cannot be easily removed by extraction with organic solvents or by adhesive tape peeling.

5.3.4 CONCLUSION

A SAM of a 3-bromopropylsilane monolayer can be formed under an inert atmosphere on a glass substrate by the reaction between 3-bromopropyltrichlorosilane and the hydroxyl groups of the surface. It can be further functionalized by the substitution of the brom atoms at the end of silane chains with aniline molecules. The aniline molecules chemically tethered to the glass surface can act as active sites for surface grafting of PANI. The microstructure of the aniline monolayer and grafted PANI were characterized by XPS, UV-VIS and contact angle measurements. The extent of substitution of Br by aniline increased with increasing temperature and reaction time. The density of aniline tethered to the glass via substitution affected the density of the PANI finally grafted on the glass surface. The grafted PANI kept many of the characteristics of PANI homopolymer. The surface conductivity (~11 S/cm) was, however, higher than that of the bulk homopolymer films (~1 S/cm). The strong adhesion and smoothness of the ultra-thin grafted PANI layers may have potential in the pattern industry.

REFERENCES

1. E.M. Genies, A. Boyle, M. Lapkowski, C. Tsintavis, *Synth. Met.* 36 (1990) 139.
2. A. Malinauskas, *Polymer* 42 (2001) 3957.
3. R.M.G. Rajapakse, A.D.L. Chandani, L.P.P. Lankeshwara, N.L.W.L. Kumarasiri, *Synth. Met.* 83 (1996) 73.
4. F. Faverolle, A.J. Attias, B. Bloch, P. Audebert, C.P. Andrieux, *Chem. Mater.* 10 (1998) 740.
5. C.G. Wu, J.Y. Chen, *Chem. Mater.* 9 (1997) 399.
6. X. Li, X. Zhang, Q. Sun, W. Lu, H. Li, J. *Electroanalytical Chem.* 492 (2000) 23.
7. Z. Huang, P.O. Wang, A.G. MacDiarmid, Y. Xia, G. Whitesides, *Langmuir* 13 (1997) 6480.
8. A. Ulman, *Chem. Rev.* 96 (1996) 1533.
9. M.D.K. Ingall, C.H. Honeyman, J.V. Mercure, P.A. Bianconi, R.R. Kunz, *J. Am. Chem. Soc.* 121 (1999) 3607.
10. C.S. Dulcey, J.H. Georger, V. Krauthamer, D.A. Stenger, T.L. Fare, J.M. Calvert, *Science* 252 (1991) 551.
11. A. Ray, G.E. Asturias, D.L. Kershner, A.F. Ritchter, A.G. MacDiarmid, A.J. Epstein, *Synth. Met.* 29 (1989) E145.
12. Z.F Li, E.T. Kang, K.G. Neoh, K.L. Tan, *Synth. Met.* 87 (1997) 45.
13. E.U. Thoden van Velzen, J.FJ. Engbersen, D.N. Reinhoudt, *J. Am. Chem. Soc.* 116 (1994) 3597–3598.
14. M.G.L. Petrucci, A.K. Kakkar, *J. Chem. Soc. Chem. Commun.* (1995) 1577–1578.
15. D.C. Trivedi, in: H.S. Nalwa (Ed.), *Handbook of Organic Conductive Molecules and Polymers*, Vol. 2, John Wiley & Sons, Chichester, UK, 1997, p. 505.
16. J.F. Moulder, W.F. Stickler, P.E. Sobol, K.D. Bomben, in: J. Chastain (Ed.), *Handbook of X-Ray Photoelectron Spectroscopy*, Perkin-Elmer, Eden Prairie, MN, 1992, pp. 86 and 119.
17. Y.P. Ting, K.G. Neoh, E.T. Kang, K.L. Tan, *J. Chem. Technol. Biotechnol.* 59 (1994) 31.
18. F.L. Wuld, F.M. Nowak, A.J. Heeger, *J. Am. Chem. Soc.* 108 (1986) 8311.
19. S. Stafstrom, J.L. Bredas, A.J. Epstein, H.S. Woo, D.B. Tanner, W.S. Huang, A.G. MacDiarmid, *Phys. Rev. Lett.* 59 (1987) 1464.

5.4 Patterned Conductive Polyaniline on Si(100) Surface via Self-Assembly and Graft Polymerization*

Z.F. Li and Eli Ruckenstein

Department of Chemical Engineering, State University of New York at Buffalo, Buffalo, New York 14260

ABSTRACT A combination of surface graft polymerization of aniline and photopatterned self-assembly monolayer (SAM) was used to generate a well-defined pattern of conductive polyaniline on a Si(100) surface. A self-assembly of phenylsilane monolayer was first generated by reacting a hydroxylated silicon surface with phenyltrichlorosilane under a dry inert (N_2) atmosphere. The formed SAM layer has been photopatterned under an UV laser at 263 nm through a lithographic mask. The patterned SAM was reacted with triflic acid (HOTf) under a dry inert atmosphere to remove the benzene rings from the SAM layer. The OTf groups of the triflated SAM have been substituted with aniline under a dry inert atmosphere to generate an aniline-primed substrate which was further used for the graft polymerization of aniline to prepare a patterned conductive polyaniline (PANI) layer. The composition, microstructure, and morphology of PANI grafted silicon surfaces were examined by X-ray photoelectron spectroscopy (XPS), atomic force microscopy (AFM), scanning electron microscopy (SEM), four probe conductivity, and contact angle measurements. The surface conductivity of grafted PANI free of patterning was 23 S/cm and through the patterned wires was 21 S/cm (for the surface fraction grafted), which are larger than the usual value of the homopolymer PANI films (~1 S/cm). Microscopy images revealed a compact grafted PANI and a high edge acuity of the pattern. The present method provides a new strategy for the generation of a pattern of conductive polymers via graft polymerization.

* *Macromolecules* 2002, 35, 9506–9512.

5.4.1 INTRODUCTION

In the past years, a great deal of research was concerned with the patterning of polymer films which can be used as components in molecular electronics,[1] optical devices,[2] etch resists,[3,4] and biosensors[5,6] and as scaffolds for tissue engineering and fundamental studies in cell biology.[7–9]

Among the polymers, the conjugated organic polymers are especially attractive since they offer several advantages over the metals and the conventional inorganic semiconductors, such as more facile processing and ease of adjusting the conductivity in a wide range by changing the dopant or the doping level. They can be considered as potential alternatives to the metals and semiconductors as connecting wires and conductive channels, which can be used as active materials in optoelectronics,[10] microelectronics,[11] microelectromechanical systems (MEMS),[12] sensors,[13] and related areas.[14]

The patterning of conjugated polymers on various substrates was achieved by using a variety of techniques, such as the deposition by scanning electrochemical microscopy,[15] screen printing,[16] micromolding in capillaries (MIMIC),[17] ink-jet printing,[18] photochemical patterning by photolithography,[19] and microcontact printing through self-assembly monolayer (SAM).[20] The above methods employed the area-selected electropolymerization, the nonreactive patterning, or directly the photochemical patterning of conjugated polymers.[21] In another method, a combination of the microcontacting printing and the area-selective deposition method was used for the physical deposition of polyaniline (PANI) and polypyrrole (PPY) on a patterned SAM.[22]

It is generally believed that the "self-assembly" monolayer (SAM) provides the best control at the molecular level and constitutes a potential technique for the preparation of advanced materials.[23] In previous work,[24] we employed the aniline monomer neucleophilic substitution reaction of bromine atoms bound to a SAM silane monolayer on the surface of a glass substrate, followed by the grafting of polyaniline. The method involved the initial formation of a stable bromine silane monolayer through the reaction of a hydroxylated glass surface with 3-bromopropyltrichlorosilane. This was followed by the functionalization of the SAM monolayer through the aniline substitution of the bromine atoms and further by the surface oxidative graft polymerization of aniline on the covalently immobilized aniline sites. This method offered the advantage of flexibility in the selection of a suitable silane coupling agent, which could generate a homogeneous, stable monolayer without the use of synthetically challenging or unstable silanes. The surface conductivity of the PANI layer grafted to glass was higher than that of the homopolymer PANI films.

It has been shown that the alkyl/aryl trichlorosilanes are excellent molecules for the formation of monolayers through the condensation reaction with substrates that bear a surface layer of hydroxyl groups.[25] It was also reported[26,27] that the phenyltrichlorosilane is a reagent which does not undergo unfavorable side reactions. Phenylsilane monolayers can be formed on glass, silicon, metal oxides, and other ceramics, on which surface hydroxyls can be generated. These SAM monolayers are homogeneous and stable and do not have the drawbacks of other more exotic silane monolayers, such as irreproducibility, incomplete coverage of the substrate, and selfcondensation of silanes in solution.[26] The phenylsilane monolayers can be photopatterned with high resolution, and for this reason they can be used in photolithographic processes. Furthermore, the dearylation of the phenylsilyne groups of the silyne network with the trifluoromethanesulfonic acid (triflic acid, HOTf) results in the generation of reactive silyl cation sites on the backbones, which can react with suitable monomers to form functionalized polysilynes.[27]

We have previously demonstrated that aniline can be bound to a SAM monolayer generating functional groups which allow the further grafting of polyaniline (PANI). The goal of the present paper is (a) to employ a SAM monolayer of phenylsilane generated by reacting phenyltrichlorosilane with the OH groups of a silicon surface, which are obtained by treating the surface with a "piranha" solution (a mixture of a concentrated sulfuric acid solution and a hydrogen peroxide solution), (b) to pattern this monolayer using an UV laser, (c) to dearylate the phenylsilane by its reaction with HOTf and formation of a weak bond between the latter acid and silane, (d) to substitute the –OTf group of the triflic acid with aniline, and finally (e) to graft polymerize aniline to the aniline-primed surface. In essence, a patterned PANI will be prepared through the combination of a chemical graft polymerization and patterned SAM technique. With the advent of nanolithography, such as the AFM lithography, this method can provide nanometer size features for patterns of conductive polymers.

5.4.2 EXPERIMENTAL SECTION

5.4.2.1 Materials

Single-crystal undoped Si(100) wafers, 76 mm in diameter, polished on one side were purchased from Gelest Inc., Morrisville, PA, and sliced into pieces of about 1.0 × 1.0 cm in size. The aniline, phenyltrichlorosilane, ammonium persulfate, and trifluoromethanesulfonic acid (triflic acid, HOTf) were provided by Aldrich/Sigma and used as received. The anhydrous toluene, aniline, and hexane were dried over 4 Å molecular sieves prior to use. The solvents, acetone, ethanol, methanol, and *N*-methylpyrrolidinone (NMP), and other chemicals were of reagent grade and were also purchased from the Aldrich/Sigma Chemical Co. and used as received.

5.4.2.2 Substrate Pretreatment

The freshly sliced silicon substrates (1 × 1 cm) were first soaked in a soap solution and sonicated for 5 min to remove the organic residues from their surface and then rinsed with a large amount of distilled water. The substrates were then immersed in a "piranha" solution [a mixture of 70% volume concentrated sulfuric acid (98 wt%) and 30% volume of a hydrogen peroxide solution (30 wt%)] and boiled for about 50 min to generate OH groups. The silicon substrates were then washed with a large amount of nanopure water and dried under reduced pressure for subsequent surface treatment. The contact angle of water droplets on the silicon surface was low and almost constant over the entire surface, revealing that the surface was uniformly hydroxylated.

5.4.2.3 Silane Treatment to Prepare SAM

The pretreated hydrophilic silicon substrate was placed into a solution of phenyltrichlorosilane in dried toluene (25 μL in 25 mL), under a nitrogen atmosphere, for 12 h. Further, the substrate was washed twice with dried toluene in the same nitrogen atmosphere. Finally, the silicon substrate was removed from the nitrogen atmosphere and cleaned in an ultrasonic bath in toluene for 1 min, rinsed successively with acetone and ethanol, and finally dried.

5.4.2.4 Photolithography to Pattern a SAM Monolayer

The photolithography was conducted using a Spectra-Physics model 3960-LIS Ti:sapphire laser. Pulses of about 90 fs in length at a wavelength of 790 nm were generated and directed into a frequency tripler (U-Oplaz Technologies model TP-1B) to obtain pulses of UV light at a wavelength of 263 nm. The as-obtained 263 nm UV laser was directed to a modified silicon surface through a photomask. The total irradiation energy on the sample was about 60 J/cm^2.

5.4.2.5 Triflation Procedure and Aniline Monomer Functionalization Through a Substitution Reaction

SAM-Si substrates were immersed in 99% triflic acid under a nitrogen atmosphere and allowed to stay for 3 days. Each substrate was then individually rinsed twice successively in dried toluene and hexane and placed in dried aniline located in a Pyrex tube at room temperature, under a nitrogen atmosphere, for 24 h. After the covalent substitution of aniline to the triflated surface, the silicon substrate was washed thoroughly with large amounts of dried toluene and NMP to remove the unreacted aniline. The residual NMP present on the aniline-SAM-Si substrate was removed by washing with a large amount of methanol.

5.4.2.6 Surface Oxidative Polymerization of Aniline

The oxidative graft polymerization of aniline to an aniline-SAM-silicon substrate was carried out in an 1 M HCl solution containing 0.3 M aniline and 0.3 M $(NH_4)_2S_2O_8$ (as the oxidative homopolymerization of aniline to the conductive emeraldine (EM) salt[28]). The reaction was allowed to proceed at 0°C for 12 h. The surface-modified silicon was subsequently immersed in a large volume of NMP for at least 24 h in order to remove the physically adsorbed EM salt. The polyaniline (PANI) grafted silicon surface was further washed with alcohol to remove the residual NMP. The EM salt grafted on the silicon surface was converted to the neutral EM base by the immersion of the silicon substrate for 24 h in a large amount of doubly distilled water before being dried under reduced pressure.

5.4.2.7 Characterization of the Surface-Modified Silicon Wafer Surface

The graft-modified silicon surfaces were characterized on their polished side by contact angle, X-ray photoelectron spectroscopy (XPS), atomic force microscopy (AFM), scanning electron microscopy (SEM), and four-probe conductivity measurements. The contact angle measurements were carried out using a NRL CA goniometer model 100-00-(115) from Ramehart Inc. The telescope with a magnification power of 23× was equipped with a protractor of 1° gradation. The angles reported are accurate within ±3°. For each substrate, at least three measurements on different locations were averaged. The XPS measurements were carried out on a Surface Science model SSX-100 Small Spot ESCA, possessing an Al Kα monochromatized X-ray source (1.48 keV photons). The pressure in the analysis chamber was maintained at 10^{-9} Torr or lower during the measurements. To compensate for the surface charging, all binding energies were referenced to the C 1s hydrocarbon peak at 285 eV. In the peak analysis, the line widths (the width at half-maximum) of the Gaussian peaks were kept constant for the components of a particular spectrum. The surface elemental compositions were determined from the peak area ratios and were accurate to within 1%. The morphologies of the pristine and surface-modified silicon wafers were examined by a Quesant Resolver atomic force microscope (AFM), using the tapping mode at a scanning rate of 1.0–4.0 Hz. The thickness profile of the film on the Si surface and the root-mean-square surface roughness (rms) of the film were directly evaluated from the AFM images. The ellipsometric thickness was determined with a Gaertner Ellipsometric from Gaertner Scientific Co., Chicago, IL. The patterned images of PANI on the silicon wafers were obtained by SEM using a Hitachi S-4000 FESEM instrument operated at 20 kV with the specimen mounted on a double-sided adhesive carbon tape. The conductivity of the graft-modified silicon surface was determined by the four-probe method, using a Hewlett-Packard model 3478A digital multimeter. For each of the conductivities reported, at least three measurements were averaged.

5.4.3 RESULTS AND DISCUSSION

Two kinds of experiments have been performed. Most of the characteristics of the system have been determined by carrying out experiments free of patterns. In addition, experiments have been performed that involved patterning. The surface modification of the hydroxyl-terminated silicon substrate by phenyl silanization with phenyltrichlorosilane, the dearylation of the benzene ring monolayer with the triflic acid to generate a reactive Si-OTf surface, the functionalization of the surface through the aniline nucleophilic substitution reaction of –OTf, and finally the oxidative graft polymerization of aniline are sketched in Scheme 5.4.1. Each of the above steps is described in detail below.

5.4.3.1 Formation of Phenylsilane on the Silicon Surface

Before generating an uniform functionalized surface by the method outlined in Scheme 5.4.1, a hydroxylated silicon surface was formed by immersing the substrate

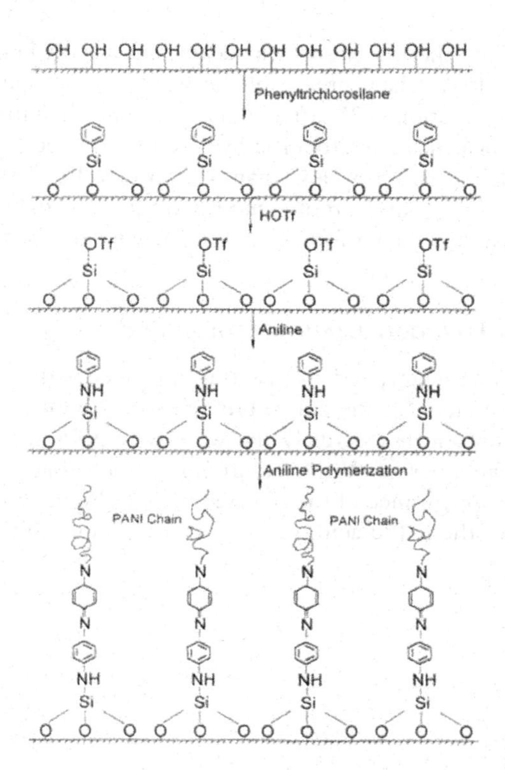

SCHEME 5.4.1 Strategy for surface graft polymerization of polyaniline (PANI) on a silicon substrate

in a boiling piranha solution for at least 30 min, but less than 1 h.[27b,29,30] The contact angle of the surface so treated changed from 51° for the pristine substrate to 20° (Table 5.4.1) and was the same all over the surface; this revealed that the substrate was uniformly covered with hydrophilic hydroxyl groups. Further, the hydroxylated silicon surface was exposed, in a dry N_2 atmosphere, to a solution of phenyltrichlorosilane in dried toluene at room temperature for about 12 h, during which the trichlorosilane groups of the phenyltrichlorosilane reacted with the hydroxyl groups

TABLE 5.4.1
Water Contact Angles after Surface Treatment

	Contact Angle (deg)
Pristine silicon wafer	51
Silicon surface after treatment with the piranha solution	20
Surface after reaction with phenyltrichlorosilane	76
Surface after treatment with triflic acid	36
Surface after substitution reaction by aniline	73

of the Si surface. Thus, the silicon surface became chemically bound to a uniformly dense phenylsilane film. The increase of the water contact angle of the surface from 20° after hydroxylation to 76° after silanization indicated that the hydrophobic ≡SiPh layer has replaced the hydrophilic hydroxyl layer. The contact angle of the ≡SiPh layer on silicon varied by less than 3° between different locations, which together with a small value of the root-mean-square surface roughness (rms = 3.1Å) of the film, demonstrated the uniformity of the phenylsilane monolayer.

5.4.3.2 Surface Functionalization with Aniline

The ≡SiPh layer was dearylized by reacting the phenylsilane sites with HOTf to produce reactive sites for the substitution reaction with aniline. Parts a and b of Figure 5.4.1 present respectively the wide-scan XPS spectra of the silicon substrate with monolayers of ≡SiPh before and after triflation followed by solvent washing. The appearance of the fluorine peak in Figure 5.4.1b indicates that the –OTf groups of the triflic acid were covalently bound to the silicon surface.

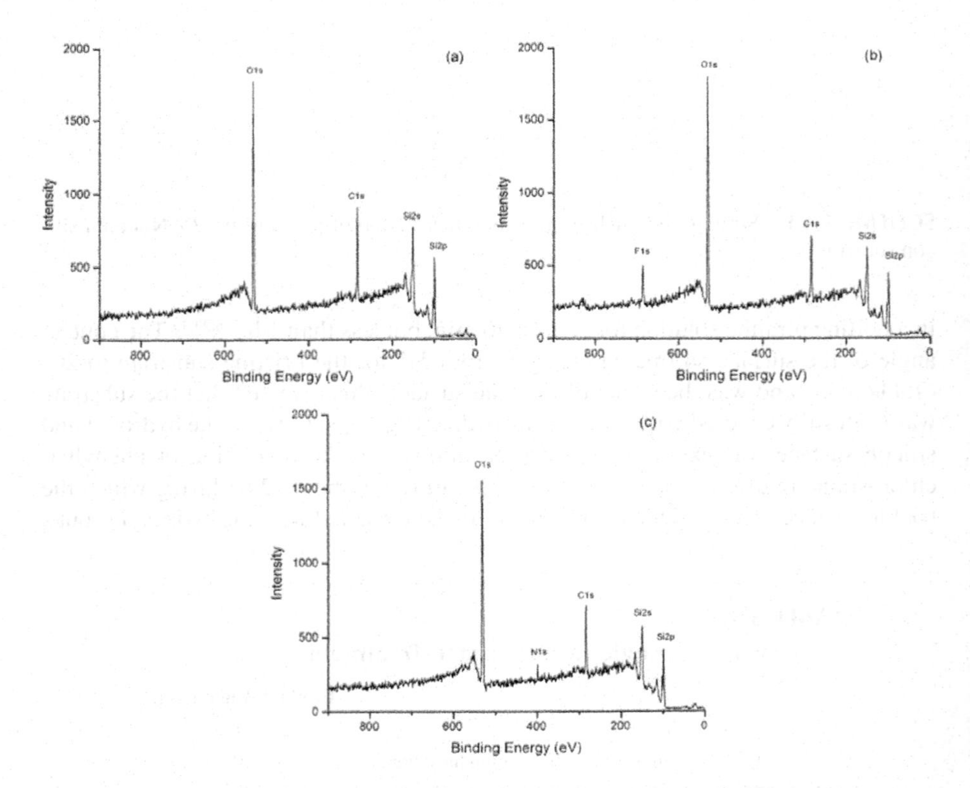

FIGURE 5.4.1 XPS wide-scan spectra of silicon substrates with monolayers of (a) ≡SiPh, (b) after the treatment with triflic acid, and (c) substitution reaction of –OTf with aniline.

Further, the triflated silicon substrate bearing a ≡SiOTf layer was placed under a dry N_2 atmosphere in a well-dried aniline, at room temperature, for 24 h. Then the substrate was rinsed with dried toluene, washed again with NMP, and sonicated in spectroscopic grade methanol. Figure 5.4.1c presents the wide-scan XPS spectrum after the surface functionalization with aniline. Comparing parts b and c of Figure 5.4.1, one can see that the F 1s signal of the –OTf group (688.5 eV) from Figure 5.4.1b was replaced by a new N 1s signal in Figure 5.4.1c. The binding energy (BE) of N 1s, 399.3 eV, indicated that a NH group was covalently connected to a benzenyl ring.[31] These results show that a chemical substitution reaction occurred between aniline and –OTf on the silane monolayer and that aniline was anchored to the silicon surface through a chemical bond. The examination of the substrate by the sessile water drop contact angle method indicated that the contact angle increased from 36° after triflation to 73° after the aniline substitution (Table 5.4.1). The above changes demonstrated that the more hydrophobic SiNH-benzyl ring layer has replaced the hydrophilic ≡SiOTf layer, and this is consistent with the aniline substitution.

5.4.3.3 Graft Polymerization of Aniline on the Aniline–Silicon Surface

The graft oxidative polymerization of aniline was carried out via the conventional method. Parts a and b of Figure 5.4.2 present the wide-scan XPS and the N 1s core-level spectra of the aniline–SAM-silicon surface after the oxidative graft polymerization in a 1 M HCl solution containing 0.3 M aniline and 0.3 M ammonium persulfate. After polymerization, the surface-grafted polyaniline (PANI) salt was first washed thoroughly with NMP and then deprotonated by dipping the specimen in a large amount of double distilled water for a long time, to convert the PANI salt to its neutral emeraldine (EM) base form. The increase in the intensity of N 1s and C 1s signal in the XPS wide-scan spectrum (Figure 5.4.2a) indicated that the oxidative graft polymerization of aniline, which brought more carbon and nitrogen on the silicon surface, has taken place. Figure 5.4.2a also shows that no Si was detected; this means that the surface was completely covered by PANI. The small sulfur signal is most likely due to the impurities accumulated during grafting, which involved ammonium persulfate. The curve-fitted N 1s core-level spectrum of the EM state of PANI-aniline-SAM-silicon surface (Figure 5.4.2b) revealed the predominant presence of two types of structures corresponding to the binding energy (BE) peaks of 398.2 and 399.4 eV, which can be assigned to the quinonoid imine (=N–) and benzenoid amine (–NH–) structures,[31] respectively. The almost equal proportions of the imine and amine nitrogen in the N 1s core-level spectrum of the deprotonated surface are consistent with the EM base form ([=N–]/[–NH–] ratio ~1) of PANI. The residual high binding energy tail (>400 eV) in the N 1s spectra is probably due to some surface oxidation products or a weakly charged transfer complexed oxygen.[32] The above N 1s spectra are similar to those of the polyaniline homopolymer films, and this indicates a successful grafting of PANI to the surface of silicon.

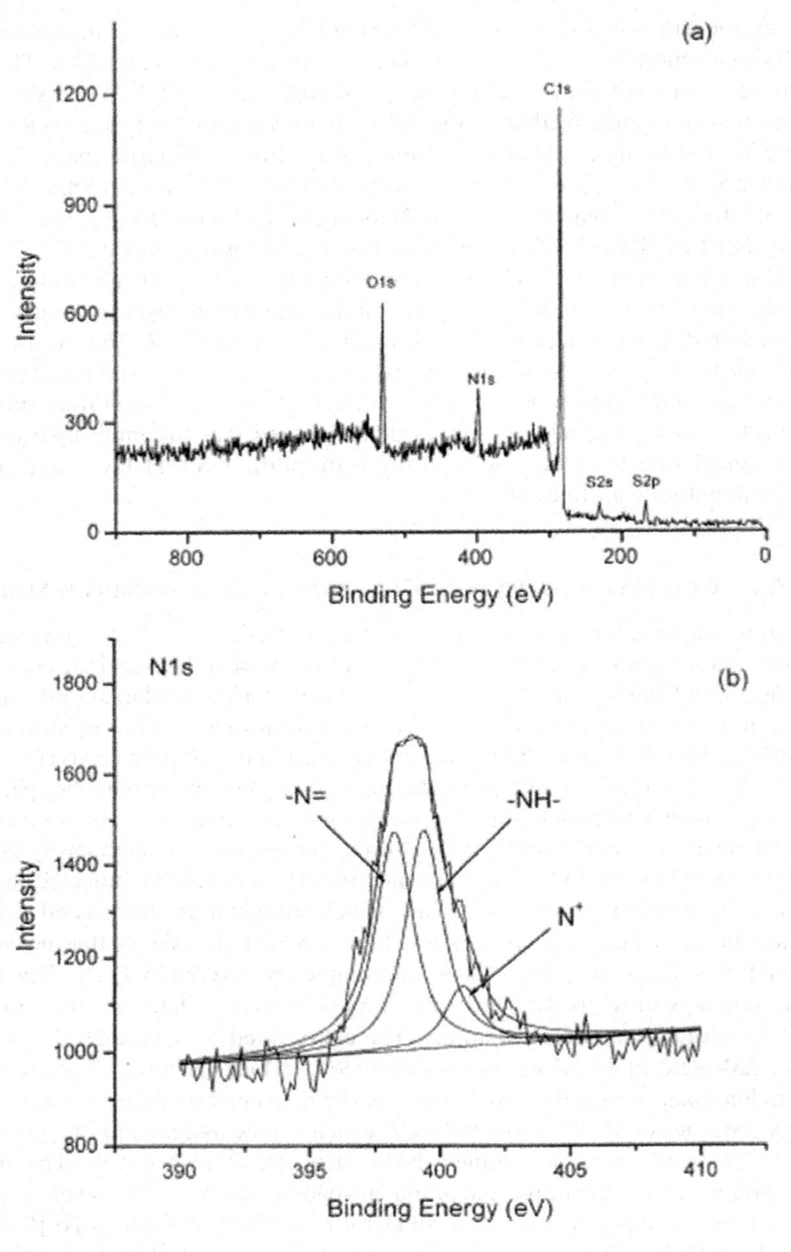

FIGURE 5.4.2 (a) XPS wide-scan spectra and (b) N 1s core-level spectra of PANI grafted on a silicon substrate.

5.4.3.4 AFM Images

The AFM images obtained are presented in Figures 5.4.3a–c, which reveal differences in morphologies. The pristine silicon wafer substrate is imaged by AFM in Figure 5.4.3a. It has a smooth surface, and its root-mean-square roughness (rms) was 1.6 Å. The assembly of a phenylsilane monolayer, the triflation by HOTf, and the substitution reaction by aniline (Figure 5.4.3b) increased the roughness, producing many fine aggregates distributed evenly on the silicon surface. While the rms roughness changed from 1.6 to 4.3 Å, the surface remained smooth. In Figure 5.4.3c, the area is well covered with PANI, and the roughness can be more easily observed. The sizes of most PANI aggregates are between 150 and 200 nm in diameter, and the aggregates are closely packed. The PANI layer has a mean height of 32 nm, which coincides with the ellipsometry result, and a surface roughness rms = 3.87 nm. It should be emphasized that a much higher regularity in the morphology was observed for the PANI grafted using the present method than for the physically deposited PANI films, the latter having a rms of 10.2 nm for a film thickness of 34 nm.

5.4.3.5 Protonated State of the Grafted PANI on the Silicon Surface

The successful grafting of the PANI chains was additionally confirmed by the significant increase in the surface conductivity and changes of XPS core-level spectra after protonation with a 0.5 M H_2SO_4 solution. Like the homopolymer, the EM base of PANI-aniline-SAM-silicon surface can be reprotonated with an acidic solution. Figure 5.4.4a, b presents the XPS analysis of N 1s and S 2p core-level spectra of

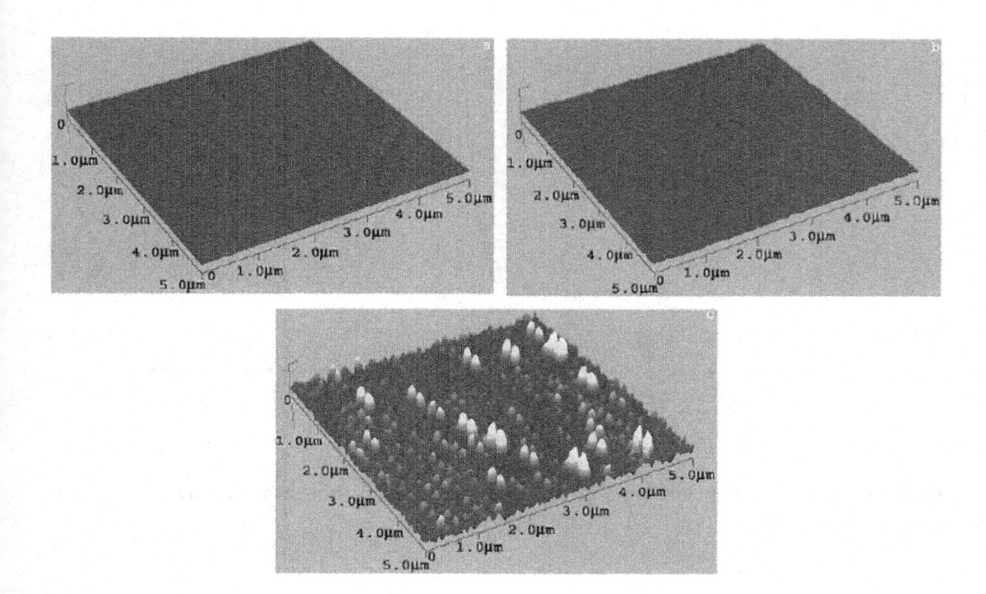

FIGURE 5.4.3 AFM image for the (a) pristine, (b) aniline primed, and (c) PANI grafted Si(100) surface.

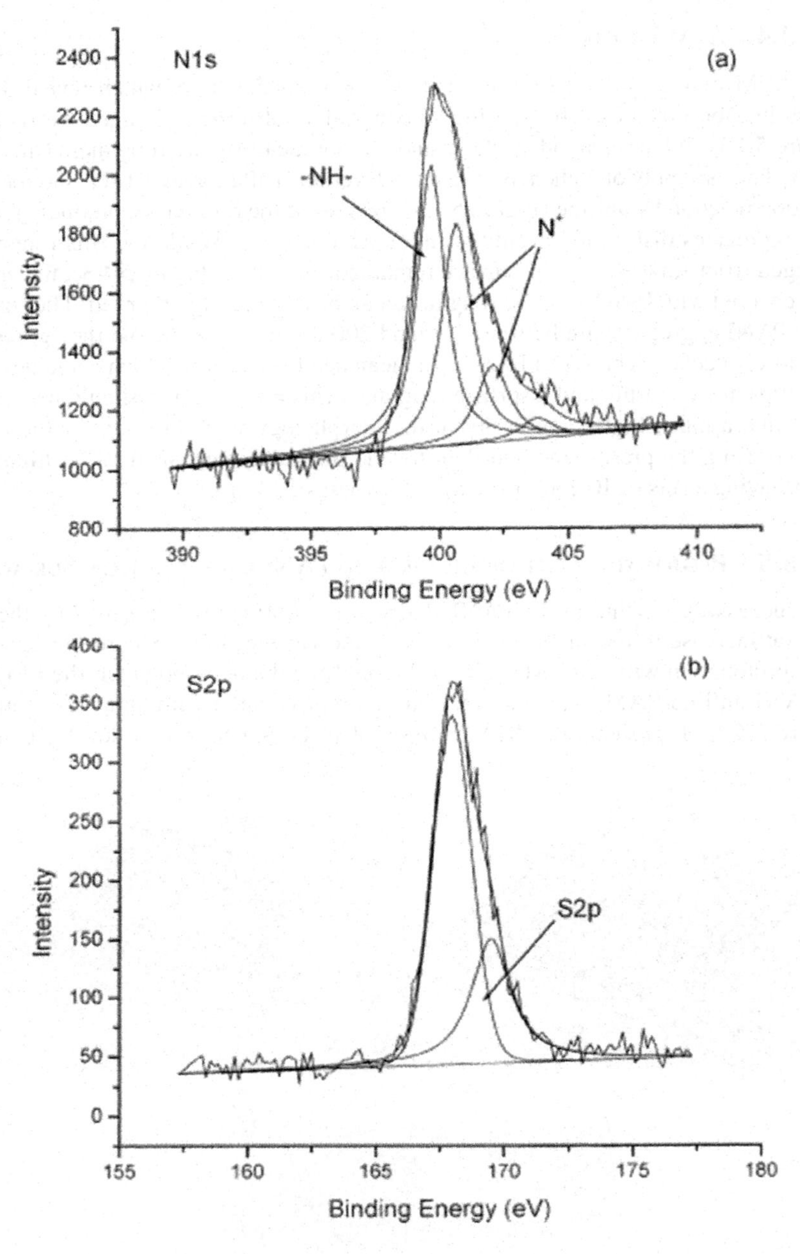

FIGURE 5.4.4 (a) N 1s and (b) S 2p core-level spectra for the surface grafted PANI after protonation with H_2SO_4.

PANI-aniline-SAM-silicon surface after doping with H_2SO_4. Comparing the N 1s core-level spectrum (Figure 5.4.2b) to that of the reprotonated PANI-aniline-SAM-silicon surface (Figure 5.4.4a), one can notice that the –N= groups (BE = 398.2 eV) were replaced after protonation by positively charged nitrogens (BE = 400.9 and 402.2 eV). Similarly, the protonation of the EM base form of the aniline homopolymer also occurred preferentially at the imine group.[28] The above change in the curve-fitted N 1s spectra, combined with the strong S 2p signal, allowed one to conclude that the EM base grafted on the silicon surface was effectively reprotonated by H_2SO_4 and that the reprotonation of the grafted PANI chains is not unlike that of the homopolymer. Regarding the conductivity of the grafted PANI in the protonated state, the surface resistance measured by the four-probe method changed from the insulation state of the pristine silicon wafer to a surface resistance of 2.96 kΩ/area (with the area 1 cm wide and 0.2 cm long) for the grafted PANI after protonation. Both the ellipsometry measurements and the AFM surface morphology analysis provided an average thickness of 35 nm for the PANI grafted on the silicon surface. On the basis of this value, one can obtain a surface conductivity of 23 S cm. The bulk conductivity of the protonated PANI films is, in general, of the order of 1 S/cm. As shown by the AFM images, the different values can be attributed to the ordered, high-density structure of the surface layer in which the PANI chains are tightly packed. The relatively high conductivity of the grafted PANI makes the present surface modification technique potentially valuable for surface graft polymerization of conductive polymers to various substrates. Finally, it is appropriate to emphasize that the thin layer of grafted PANI chains are covalently bound to the silicon surface and cannot be easily removed by extraction with organic solvents or by peeling tests with Scotch tape.

5.4.3.6 Patterned Grafted Polyaniline

Until now, the experiments have been carried out with unpatterned PANI. As shown in Scheme 5.4.2, patterned PANI can be formed by graft polymerization on preselected areas. It was generated by first patterning a phenylsilane SAM monolayer by the photolithographic technology. Some authors[26a,27b] have shown that a phenylsilane monolayer can be patterned with a deep UV laser or UV light through a lithographic mask. The UV laser/light could efficiently cleave the Si–R bond on the surface and generate SiOH groups in contact with air. We employed this method by first patterning a phenylsilane monolayer with a 263 nm laser through a predesigned lithographic mask. In this case, the Si–phenyl bonds were removed from the irradiated areas, leaving a monolayer of unreacted Si–phenyl in the unexposed regions for further functionalization. Then the reaction of the unexposed phenylsilane with triflic acid and the further exposure to the aniline resulted in a patterned aniline-primed surface. The graft oxidative polymerization of aniline could be achieved by adding an oxidant and aniline solution for polymerization on the aniline-primed areas. The grafted PANI thus formed constitutes a negative of the lithographic mask. Figure 5.4.5a presents the SEM image of the patterned PANI after graft polymerization. The white strips represent the grafted PANI area, whereas the black strips the nongrafted areas. A relatively clear boundary between the patterned and nonpatterned areas can be

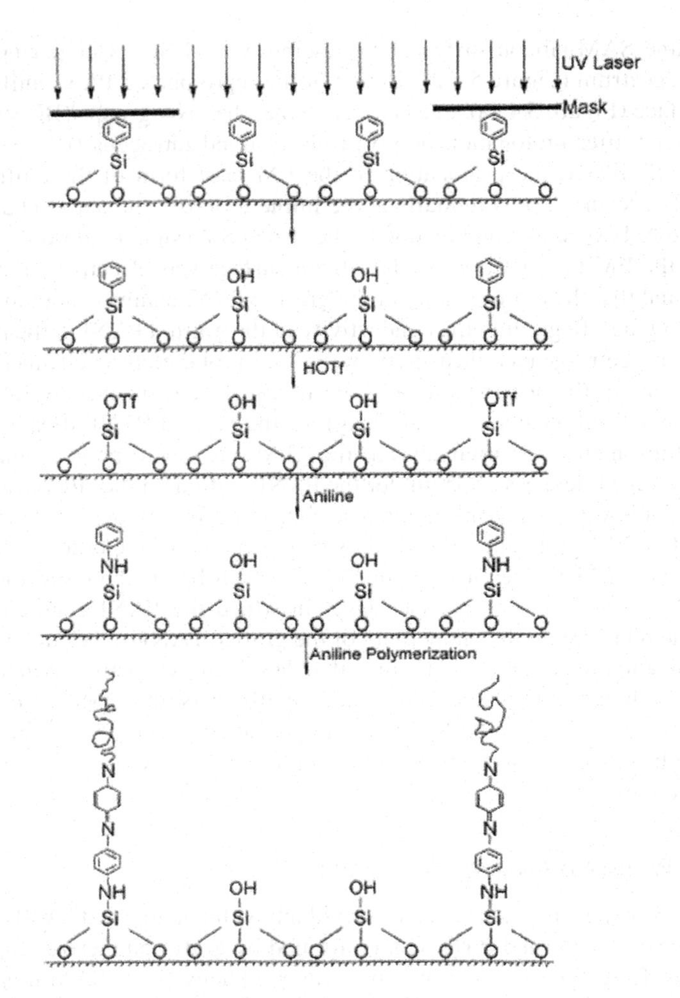

SCHEME 5.4.2 Strategy for patterned graft polymerization of polyaniline (PANI) on a silicon substrate

observed in the SEM image. A closer look at the boundaries between the two can be obtained under a higher magnification by AFM (Figure 5.4.5b), which demonstrates that this method provides a high line edge acuity of the patterned PANI. With a rich available chemistry for reactions with specified SAMs and the advent of new nanotechnological methods to pattern SAM monolayers at the nanometer scale, this method can provide a strategy for patterning conductive polymers at nanoscale arrays without the need of a photoresist.

The surface conductivity through the patterned wires was 21 S/cm (for the surface fraction grafted), a value very near to that determined for the nonpatterned case (23 S/cm). This is not surprising because the width of the wires was large (about 50 μm). As expected, the surface conductivity in the direction normal to the wires was negligible.

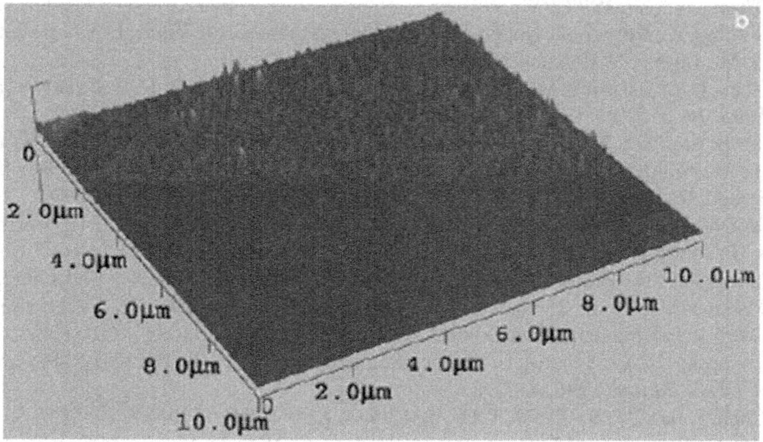

FIGURE 5.4.5 (a) SEM image of the patterned PANI film on a silicon substrate and (b) selected edge area of (a) under AFM.

5.4.4 CONCLUSION

A phenylsilane monolayer was first generated by reacting phenyltrichlorosilane with a hydroxylated silicon surface through the self-assembly method, under an inert atmosphere. The SAM layer was photopatterned under an UV laser at 263 nm through a lithographic mask. The as-patterned SAM was allowed to react with triflic acid (HOTf) in order to remove the benzene ring from the SAM layer. The –OTf groups of SAM were substituted with aniline to form an aniline-primed substrate.

The aniline sites allowed the further graft polymerization of aniline to generate a patterned conductive polyaniline (PANI). The conductivity through the patterned wires was as large as 21 S/cm, much larger than the usual value of the bulk PANI films (~1 S/cm). The microscopy experiments revealed that the patterned PANI had a compact structure and a good edge acuity of the patterns. The combination of surface graft polymerization of aniline and photopatterned self-assembly monolayer (SAM) provides a new strategy for patterning the conductive polymers with potential use in the electrical and biomedical fields.

REFERENCES

1. MacDonald, S. A.; Willson, C. G.; Frechet, J. M. J. *Acc. Chem. Res.* **1994**, *27*, 151.
2. Healy, B. G.; Foran, S. E.; Walt, D. R. *Science* **1995**, *269*, 1078.
3. Zamborini, F. P.; Crooks, R. M. *Langmuir* **1997**, *13*, 122.
4. Jeon, N. L.; Choi, I. S.; Whitesides, G. M.; Kim, N. Y.; Laibinis, P. E.; Harada, Y.; Finnie, K. R.; Girolami, G. S.; Nuzzo, R. G. *Appl. Phys. Lett.***1999**, *75*, 4201.
5. Crooks, R. M.; Ricco, A. J. *Acc. Chem. Res.* **1998**, *31*, 219.
6. Mrksich, M.; Whitesides, G. M. *Trends Biotechnol.* **1995**, *12*, 228.
7. Gosh, P.; Amirpour, M. L.; Lackowski, W. M.; Pishko, M. V.; Crooks, R. M. *Angew. Chem., Int. Ed.* **1999**, *38*, 1592.
8. Singhvi, R.; Kumar, A.; Lopez, G. P.; Stephanopoulos, G. N.; Wang, D. I. C.; Whitesides, G. M.; Ingber, D. E. *Science* **1994**, *264*, 696.
9. Chen, C. S.; Mrksich, M.; Huang, S.; Whitesides, G. M.; Ingber, D. E. *Biotechnol. Prog.* **1998**, *14*, 356.
10. (a) Braun, D.; Brown, A.; Staring, E.; Meijer, E. W. *Synth. Met.* **1994**, *65*, 85. (b) Burroughes, J. H.; Bradley, D. D. C.; Brown, A. R.; Marks, R. N.; Mackay, K.; Friend, R. H.; Burns, P. L.; Holmes, A. B. *Nature (London)* **1990**, *347*, 539.
11. (a) Dodadabalapour, A.; Torsi, L.; Katz, H. E. *Science* **1995**, *268*, 270. (b) Burroughes, J. H.; Jones, C. A.; Friend, R. H. *Nature* (*London*) **1988**, *335*, 137.
12. (a) Inganäs, O.; Pei, Q. *Adv. Mater.* **1992**, *4*, 277. (b) Baugman, R. H.; Shacklette, L. W.; Elsenbaumer, R. L.; Plichta, E. J.; Becht, C. In *Molecular Electronics*; Lazarev, P. I., Ed.; Kluwer Academic Publishers: Amsterdam, the Netherlands, 1991; p 267.
13. Lonergan, M. C.; Severin, E. J.; Doleman, B. J.; Beaber, S. A.; Grubbs, R. H.; Lewis, N. S. *Chem. Mater.* **1996**, *8*, 2298.
14. Chen, L. H.; Jin, S.; Tiefel, T. H. *Appl. Phys. Lett.* **1993**, *62*, 2440.
15. Kranz, C.; Ludwig, M.; Gaub, H. E.; Schuhmann, W. *Adv. Mater.* **1995**, *7*, 38.
16. Garnier, F.; Hajlaoui, R.; Yassar, A.; Srivastava, P. *Science* **1994**, *265*, 1684.
17. Beh, W. S.; Kim, I. T.; Qin, D.; Xia, Y. N.; Whitesides, G. M. *Adv. Mater.* **1999**, *11*, 1038.
18. Chang, S. C.; Bharathan, J.; Yang, Y.; Helgeson, R.; Wudl, F.; Ramey, M. B.; Reynolds, J. R. *Appl. Phys. Lett.* **1998**, *73*, 2561.
19. Drury, C. J.; Mutsaers, C. M. J.; Hart, C. M.; Matters, M.; de Leeuw, D. M. *Appl. Phys. Lett.* **1998**, *73*, 108.
20. Bjornholm, T.; Greve, D. R.; Reitzel, N.; Hassenkam, T.; Kjaer, K.; Howes, P. B.; Larsen, N. B.; Bogelund, J.; Jayaraman, M.; Ewbank, P. C.; McCullough, R. D. *J. Am. Chem. Soc.* **1998**, *120*, 7643.
21. Holdcroft, S. *Adv. Mater.* **2001**, *13*, 1753.
22. Huang, Z.; Wang, P. O.; MacDiarmid, A. G.; Xia, Y.; Whitesides, G. *Langmuir* **1997**, *13*, 6480.
23. Ulman, A. *J. Mater. Educ.* **1989**, *11*, 205.

24. (a) Li, Z. F.; Ruckenstein, E. *Synth. Met.* **2002**, *129*, 73. (b) Li, Z. F.; Ruckenstein, E. *J. Colloid Interface Sci.* **2002**, *251*, 343.
25. (a) Wasserman, S. R.; Tao, Y.-T.; Whitesides, G. M. *Langmuir* **1989**, *5*, 1074. (b) Tripp, C. P.; Hair, M. L. *Langmuir* **1991**, *7*, 923.
26. (a) Dulcey, C. S.; Georger, J. H.; Krauthamer, V.; Stenger, D. A.; Fare, T. L.; Calvert, J. M. *Science* **1991**, *252*, 551. (b) Plueddemann, E. P. In *Silane Coupling Agents*; Plenum Press: New York, 1991. (c) Murray, R. W. In *Introduction to the Chemistry of Molecularly Designed Electrode Surfaces*; Murray, R. W., Ed.; Techniques of Chemistry Series; John Wiley & Sons: New York, 1992.
27. (a) Smith, D. A.; Freed, C.; Bianconi, P. A. *Chem. Mater.* **1993**, *5*, 245. (b) Ingall, M. D. K.; Honeyman, C. H.; Mercure, J. V.; Bianconi, P. A.; Kunz, R. R. *J. Am. Chem. Soc.* **1999**, *121*, 3607. (c) Hrkach, J.; Ruehl, K.; Matyjaszewski, K. *Polym. Prepr. (Am. Chem. Soc., Div. Polym. Chem.)* **1988**, *29*(2), 112. (d) Matyjaszewski, K.; Chen, Y. L. *J. Organomet. Chem.* **1988**, *340*, 7. (e) Matyjaszewski, K.; Chen, Y. L.; Kim, H. K. In *Inorganic and Organometallic Polymers*; Zeldin, M., Wynne, K. J., Allcock, H. R., Eds.; ACS Symposium Series No. 360; American Chemical Society: Washington, DC, 1988; pp 78–88.
28. Ray, A.; Asturias, G. E.; Kershner, D. L.; Ritchter, A. F.; MacDiarmid, A. G.; Epstein, A. J. *Synth. Met.* **1989**, *29*, E145.
29. Thoden van Velzen, E. U.; Engbersen, J. F. J.; Reinhoudt, D. N. *J. Am. Chem. Soc.* **1994**, *116*, 3597.
30. Petrucci, M. G. L.; Kakkar, A. K. *J. Chem. Soc., Chem. Commun.* **1995**, 1577.
31. Snauwaert, P.; Lazzaroni, R.; Riga, J.; Werbist, J. J.; Gonbeau, D. *J. Chem. Phys.* **1990**, *92*, 2187.
32. Trivedi, D. C. In *Handbook of Organic Conductive Molecules and Polymers*; Nalwa, H. S., Ed.; John Wiley & Sons: Chichester, UK, 1997; Vol. 2, p 505.

5.5 Luminescent Silicon Nanoparticles Capped by Conductive Polyaniline through the Self-Assembly Method*

Z.F. Li, M.T. Swihart, and Eli Ruckenstein

Department of Chemical and Biological Engineering,
State University of New York at Buffalo, Buffalo, New York 14260

ABSTRACT Graft polymerization has been used, for the first time, to prepare a dense conductive polymer coating on free-standing luminescent silicon nanoparticles. The silicon nanoparticles maintained their photoluminescence and crystallinity after surface modification. The nanoparticles were first surface hydroxylated and then reacted with (3-bromopropyl)trichlorosilane to form a dense bromopropylsilane monolayer. This was further reacted with aniline, which displaced the bromine atoms. The surface-bound aniline molecules were then used as active sites for the graft polymerization of polyaniline (PANI). The composition, structure, morphology, and other physical properties of the PANI-capped Si nanoparticles were examined by X-ray photoelectron spectroscopy, Fourier transform infrared spectroscopy, X-ray diffraction, and transmission electron microscopy. The silane self-assembled monolayer effectively protected the silicon particles against photoluminescence quenching and degradation in basic solutions that rapidly quench the photoluminescence of unprotected particles. The PANI coating further enhanced this protection, even in its nonconducting emeraldine base state. The electrical conductivity of the HCl-doped (emeraldine salt) PANI-capped Si nanocomposite exceeded 10^{-2} S/cm, which is 6 orders of magnitude higher than that of the bare Si nanoparticles. However, there was negligible change in the photoluminescence spectrum or lifetime upon addition of the PANI layer, suggesting that the charge carriers responsible for the luminescence remained confined within the Si nanoparticles.

* *Langmuir* 2004, 20, 1963–1971.

5.5.1 INTRODUCTION

Nanoparticles of many materials have received great attention lately, as researchers have become increasingly aware that reducing the size of materials to the nanometer scale can change their properties in fundamental ways, thus generating attractive electronic, optical, magnetic, and/or catalytic properties associated with their nanoscale or quantum-scale dimensions.[1] Following the discovery in 1990 of visible red light emission upon UV excitation of electrochemically etched nanoporous silicon by Canham and co-workers,[2] a great deal of research has been performed on producing nanosized Si with visible photoluminescence and on characterization of its structure and optoelectronic properties.[3–8] This luminescence from nanosized Si has been attributed to radiative recombination of carriers confined in Si nanoparticles, and its color can be modified from blue to red by changing the nanoparticle size. One benefit of exploiting the optical properties of silicon nanoparticles over the other materials is the potential of silicon nanoparticles to be integrated within existing silicon technologies in order to create nanoscale optoelectronic devices. These applications include their use as chemical sensors,[9] optoelectronic devices,[10] electroluminescent displays,[11] photodetectors,[12] and as a lasing material for photopumped tunable lasers.[13] In addition, the silicon nanoparticles' good biocompatibility,[14] high photoluminescence quantum efficiency,[15] and stability against photobleaching,[16] make them an ideal candidate for replacing fluorescent dyes in many biological assays and fluorescence imaging techniques.

Because of the relative ease of preparing luminescent porous silicon films by etching silicon wafers, there have been many studies of surface modification of porous silicon films by various organic compounds ranging from alkyl chains to macromolecules, to develop new building blocks for optical, electrical, and biomedical applications. This includes free radical initiation to form alkyl monolayers,[17,18] derivatization with alcohols through electrochemical and thermal reactions,[19] derivatization with halogens,[20] photoelectrochemical esterification reactions with carboxylic acids and with alcohols,[21] and reactions with chlorosilanes.[22] In addition, modification of the porous silicon film by polymer deposition was also achieved to improve its functionality for use in sensors, electroluminescent devices, and other hybrid devices.[23–27]

Several methods have been developed to produce freestanding silicon nanoparticles that are nonporous and free from a substrate. These include ultrasonic dispersion of porous silicon,[28] solution synthesis,[29] gas-phase decomposition of silanes,[30] laser-vaporization controlled condensation,[31] organosilane decomposition in supercritical organic solvents,[32] and laser-driven decomposition of silane.[33] However, the above methods cannot produce more than a few milligrams per day of luminescent Si nanoparticles that are free from a substrate. This has hindered research on these free-standing silicon nanoparticles, and relatively few studies have been presented on surface functionalization of free-standing silicon nanoparticles with polymers or other organic moieties. Recently, free silicon nanoparticles that exhibit bright visible photoluminescence have been successfully produced in macroscopic quantities (hundreds of milligrams per day).[34] These are prepared in a two-step process in which silicon nanoparticles are synthesized by CO_2 laser heating of SiH_4–H_2–He mixtures followed by etching of the nanoparticles with a hydrofluoric acid/nitric acid mixture to reduce their size and passivate their surface. This technique is a valuable tool for producing large amounts of free luminescent silicon nanoparticles in a controlled and reproducible way. The availability of free silicon nanoparticles prepared by the above method allows us to prepare and study free-standing polymer/Si nanocomposites that can potentially exhibit unique optical, chemical, and mechanical properties.

Among the methods to produce nanocomposites, modification and functionalization of nanoparticle surfaces have become an area of intense interest, since one of the requirements for realizing functional nanomaterials is the availability of suitably modified nanoparticle building blocks.[35] Recently, nanocomposites of conductive polymers and inorganic particles have received much attention.[36] As one-dimensional semiconductors, some conductive polymers have the advantage of being easy to process into large-area devices, and their energy gaps and ionization potentials can be tuned by chemical modification of the polymer chains.[37] Among them, polyaniline (PANI) is distinguished by its high electrical conductivity, redox properties, environmental stability, ease of processing, and low cost.[38] It has extensive applications in areas including Li ion batteries,[39] light-emitting devices,[40,41] corrosion protection,[42] radio frequency and microwave absorption,[43] and electromagnetic interference shielding and antistatic films and coatings.[44] Several types of inorganic nanoparticles have been encapsulated in conductive polymers, and such nanocomposites have shown various interesting characteristics, particularly in their dielectric properties, energy storage capability, catalytic activity, and magnetic susceptibility.[45]

The coating of polyaniline onto silicon nanoparticles is of particular interest because of potential applications in optical emitters and detectors. Because of its large work function, PANI has been used as a hole-transporting material in polymeric LEDs.[46] The work function of PANI in its emeraldine base and emeraldine salt forms has been estimated for different samples and by various methods to be from 4.3 to 4.7 eV,[47] about 4.8 eV,[48] or more than 5 eV.[46] In addition, the work function and conductivity of PANI can be changed by varying the counterion used for doping. It appears that work function can be tuned independently of conductivity.[49] The energy band structures for the different forms of PANI are presented in a paper by Huang and MacDiarmid.[50] For bulk Si, the electron affinity is 4.05 eV and the band gap is 1.12 eV, so the top of the valence band is 5.17 eV below vacuum. This is comparable to, but probably greater than, the PANI work function. In silicon nanoparticles, the band gap widens, but it is not clear how the band alignment will change. Thus, PANI is a reasonable candidate for a hole transport material for injection of positive charge carriers into Si nanocrystals for hybrid inorganic/organic light-emitting devices. Quantum-dot photodetectors[51] based on photoconductive PANI/silicon nanoparticle nanocomposites are another interesting possibility. In these, positive charge carriers would be optically excited from the Si nanoparticles into the PANI, increasing the conductivity of the nanocomposite upon illumination.

To construct a highly dense polymer coating on the surface of the nanoparticles, the so-called "grafting from" techniques are preferable to methods that attempt to link existing polymer chains to the surface.[52] In "grafting from" methods, as used here, the small monomer molecules can freely access the active initiation sites and the ends of the growing polymer chains. In linking existing polymer chains to a surface, there is inevitably a steric barrier to incoming polymers imposed by chains that are already attached, and this limits the density of attachments of the polymer chains to the surface.[53] The surface-initiated polymerization in conjunction with a self-assembly (SAM) technique is among the most useful synthetic routes to precisely design and functionalize the surfaces of various solid materials by well-defined polymers and copolymers.[52] A method of surface graft polymerization of aniline on photopatterned self-assembled phenylsilane monolayers to generate well-defined patterns of conductive polyaniline on a planar Si(100) surface was previously reported.[54] This method generated a compact grafted PANI with high edge acuity of the pattern, which provided a new strategy for the generation of a pattern of conductive polymers via graft polymerization. In the present work, we use (3-bromopropyl) trichlorosilane to first form a self-assembled monolayer on the surface of silicon nanoparticles. Then, after functionalization of the SAM monolayer through aniline substitution, further surface oxidative graft polymerization forms a dense layer of polyaniline. This work presents a detailed characterization of conductive polymer coated Si nanoparticles. After this surface modification and acid doping of the PANI, the coated particles exhibit a bulk electrical conductivity near 10^{-2} S/cm. The photoluminescence (PL) of the bare Si nanoparticles can be quenched by exposure to various substances, as has been widely reported for porous silicon. The resulting lack of stability of the PL could limit many applications of this novel material. In contrast, silicon nanoparticles coated with PANI using the self-assembly method are much more robust and maintain their photoluminescence after long exposures to solvents and basic solutions that would otherwise degrade the PL properties. In addition, experiments regarding the electromagnetic interference (EMI) shielding of the PANI grafted Si nanoparticles are also included in the paper to demonstrate the surface conductive effect of grafted PANI. With increasing chip density of transistor electronics, the problem of electromagnetic and radio frequency interference and electrostatic dissipation to avoid electrostatic shock becomes especially acute and sometimes can damage the brittle circuit elements.[55] This first reported grafting of a conductive polymer to silicon nanoparticles shows one way that the polymer and nanoparticle properties can be combined. The graft polymerization strategy used here can also be adapted to other polymers. The ability to prepare these hybrid organic/inorganic nanoparticles will broaden the potential applications of luminescent Si nanoparticles.

5.5.2 EXPERIMENTAL SECTION

5.5.2.1 Materials

The chemical reagents, aniline, (3-bromopropyl)trichlorosilane, and ammonium persulfate were obtained from Aldrich/Sigma and were used as received. The anhydrous toluene was dried over 4 Å molecular sieves prior to use. Other solvents, such as acetone, ethanol, *N*-methylpyrrolidinone (NMP), and hexylamine, as well as other chemicals were of reagent grade and were also purchased from the Aldrich/Sigma Chemical Co. and used as received.

The nanosized (around 3–5 nm) crystalline silicon powder was synthesized as described in detail previously.[34] The particles were generated by laser-induced heating of a flowing mixture of SiH_4, He, and H_2 to produce nonluminescent nanoparticles. These were etched with a mixture of HNO_3 and HF to reduce their size and passivate their surfaces. The particles were washed with water and methanol and finally collected by filtration. The resulting Si nanoparticles exhibit bright red or orange photoluminescence under UV excitation and have predominantly Si–H bonds on their surface.

5.5.2.2 Surface Pretreatment

The Si nanoparticles having predominantly Si–H bonds on their surface were first immersed in a "Piranha" solution (a mixture of 70 vol% concentrated sulfuric acid (98 wt%) and 30 vol% of a hydrogen peroxide solution (30 wt%)), sonicated for a few minutes and boiled for about 30–60 min to generate a high coverage of Si–OH groups on the surface of the nanoparticles. The particles were then washed with a large amount of distilled water and dried under vacuum for subsequent surface treatment.

5.5.2.3 Silane Treatment to Produce SAM

After the treatment by "Piranha" solution, the nanoparticles were placed into a solution of (3-bromopropyl)trichlorosilane in dried toluene (25 μL in 25 mL) under a nitrogen atmosphere, sonicated for 15 min, held at room temperature for 24 h, and then washed with dried toluene in an ultrasonic bath to remove the unreacted (3-bromopropyl)trichlorosilane. Finally, the particles were removed from the nitrogen atmosphere, collected by filtration, rinsed successively with toluene, acetone, and ethanol, and finally dried.

5.5.2.4 Aniline Functionalization and Graft Polymerization

The SAM silicon nanoparticles were immersed in aniline contained in a Pyrex tube for 48 h at room temperature. After the covalent substitution of the Br atoms of the SAM by aniline, the particles were washed with spectroscopic-grade methanol and then added to a 1 M HCl solution containing 0.1 M aniline and sonicated for 15 min. After that, a 0.1 M$(NH_4)_2S_2O_8$/1 MHCl solution was added and the mixture was allowed to polymerize in an ultrasonic bath for 30 min. In this step polyaniline was formed in its emeraldine (EM) salt state. The EM salt grafted on the nanoparticle surface was converted to the neutral EM base by adding a 0.1 M NaOH solution and then washing with a large amount of doubly distilled water. The surface-modified nanoparticles were subsequently washed several times in a large volume of NMP in order to remove any physically adsorbed EM base. Finally, the solution was centrifuged and the nanoparticles were obtained by filtration.

The PANI–Si nanoparticles were washed with alcohol followed by deionized water to remove any residual NMP before being dried under reduced pressure.

5.5.2.5 Characterization

XPS measurements were carried out on a Surface Science model SSX-100 Small Spot ESCA, possessing an Al Kα monochromatized X-ray source (1.48 keV photons). The pressure in the analysis chamber was maintained at 10^{-9} Torr or lower during the measurements. To compensate for surface charging, all binding energies were referenced to the C 1s hydrocarbon peak at 285 eV. In the peak analysis, the line widths (the width at half-maximum) of the Gaussian peaks were kept constant for the components of a particular spectrum. The surface elemental compositions were determined from the peak area ratios and were accurate to within 1%. The X-ray diffraction (XRD) patterns of the samples were obtained using a SIEMENS D500 diffractometer with Cu Kα radiation of 1.5406 Å wavelength. The diffraction data were recorded for 2θ angles from 15 to 60°, with a resolution of 0.02°. Transmission electron microscopy (TEM) investigations were carried out using a JEM-2010 electron microscope equipped with a tungsten gun operating at an accelerating voltage of 200 kV. Before the TEM measurements, the specimens were ultrasonically dispersed in methanol and dropped onto carbon-coated Cu grids. The Fourier transform infrared (FTIR) spectra were obtained using a Perkin-Elmer 1760 spectrometer with a resolution of 4 cm^{-1}. KBr pellets were employed for all the powders. Photoluminescence spectra were recorded with a SLM model 8100 spectrofluorimeter possessing a 420 nm emission cutoff filter. The excitation wavelength was set at 380 nm (for reasons discussed later). The average experimental error in determining the intensity of the PL was 7%. For time-resolved measurements, the PL was excited by ~300 fs pulses at a wavelength of 400 nm. These were generated by frequency doubling the 800 nm output from a Coherent RegA 9000 regenerative amplifier. The time-resolved photoluminescence (TRPL) data were collected and recorded using a spectrometer and a C4334 Hamamatsu streak camera.

The conductivity of the PANI-Si nanoparticles was determined by the four-probe method using compressed pellets and a Hewlett-Packard model 3478A digital multimeter. For each of the conductivities reported, at least three measurements were averaged. The electromagnetic shielding effectiveness (EMI SE) was determined over a frequency range from 10 to 1000 MHz using the coaxial cable method. The setup consisted of an Elgal (Israel) SET 19A shielding effectiveness tester with its input and output connected to a Hewlett-Packard (HP) 8510A network analyzer. An HP APC-7 calibration kit was used to calibrate the system. The EMI SE values expressed in decibels were calculated from the ratio of the incident to transmitted power of the electromagnetic wave using the equation

$$SE = 10 \log([P_1/P_2]) \ (\mathit{decibels},\ \mathit{dB})$$

where P_1 and P_2 are the incident power and the transmitted power, respectively.

5.5.3 RESULTS AND DISCUSSION

The strategy for the grafting of PANI on the Si nanoparticles through the self-assembly (SAM) method is depicted in Scheme 5.5.1 and consists of the following steps: (1) the formation of Si–OH groups on the Si nanoparticle surface; (2) the formation of a well-defined SAM through the reaction of (3-bromopropyl)trichlorosilane with the hydroxyls of the Si–OH surface; (3) the debromination of this monolayer by substitution with aniline; and (4) the further reaction of the aniline sites with an aniline–oxidant solution to produce polyaniline grafted on the surface. Each step is discussed below in detail.

5.5.3.1 Formation of Si–OH on the Nanoparticle Surface

A freshly prepared Si nanoparticle surface is covered predominantly with hydrogen (Si–H_x) after etching with an HF/HNO_3 mixture and subsequent washing with 15% HF. To generate a well-defined functionalized surface by the method outlined in Scheme 5.5.1, a dense and uniform 3-bromopropylsilane monolayer must initially be formed, by a reproducible procedure, on the surface. For this reason, a highly hydroxylated surface was first generated by dispersing the particles in a boiling piranha solution for at least 30 min, but less than 1 h. Figure 5.5.1 presents the FTIR spectrum of the freshly prepared HF-etched Si nanoparticles before and after the treatment with piranha solution. Curve a in Figure 5.5.1 is a typical FTIR spectrum of the freshly prepared Si nanoparticles. It displays the surface Si–H_x (x = 1–3) stretching vibration modes around 2200 cm^{-1} and scissors vibration mode at 860 cm^{-1}. An absorption peak is also present at 1100 cm^{-1} due to the presence of surface silicon

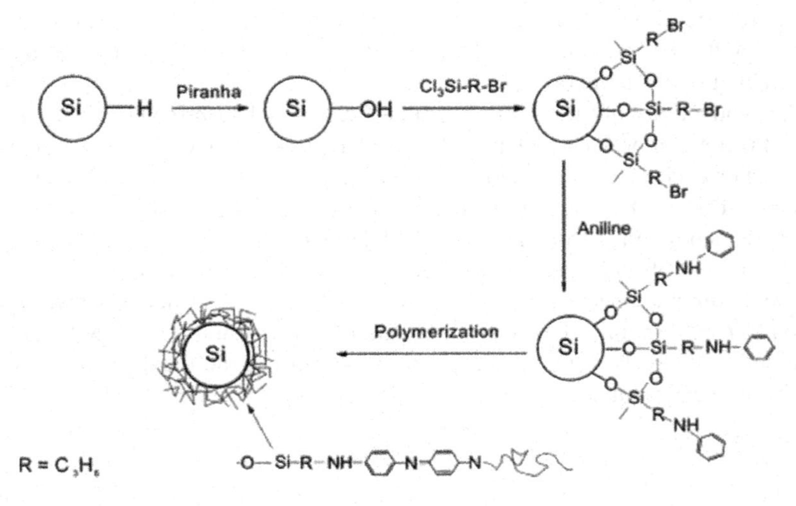

SCHEME 5.5.1 Strategy for surface graft polymerization of polyaniline (PANI) on a silicon substrate.

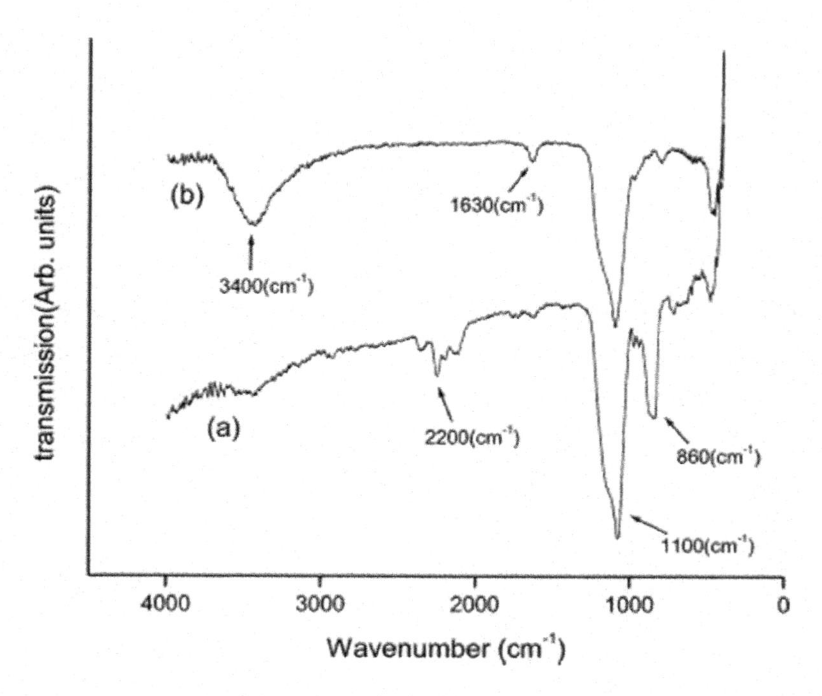

FIGURE 5.5.1 FTIR spectra of (a) freshly prepared Si nanoparticles that have Si–H on the surface (after HF treatment) and (b) Si nanoparticles bearing Si–OH on the surface after treatment with Piranha solution.

oxide (Si–O–Si bending vibrations) on the Si nanoparticles. Curve b in Figure 5.5.1 is the FTIR spectrum of the particles after treatment with piranha solution. Compared with that of the freshly prepared Si nanoparticle in curve a, both the Si–H_x stretching mode at 2200 cm^{-1} and scissors vibration mode at 860 cm^{-1} have disappeared and an intense Si–OH stretching mode at 3400 cm^{-1} and a bending mode at 1630 cm^{-1} have appeared.[56] The above results indicate that the Si–H bonds were replaced by Si–OH bonds in this step.

5.5.3.2 Formation of a Self-Assembled Bromopropylsilane Monolayer and Further Surface Functionalization with Aniline

The Si nanoparticles with hydroxylated surfaces were exposed at room temperature, in a dry N_2 atmosphere, to a (3-bromopropyl)trichlorosilane solution in anhydrous toluene for about 24 h, during which the trichlorosilane groups of the (3-bromopropyl)trichlorosilane reacted with the hydroxyl groups of the Si nanoparticle surface. The particles were then rinsed with toluene, removed from the inert atmosphere, washed again with toluene, and sonicated in spectroscopic grade methanol. After being rinsed, the particles were left with uniformly dense (3-bromopropyl) silane films covalently linked to the surface. Parts a and b of Figure 5.5.2 present the wide scan XPS spectra of Si nanoparticles with hydroxylated surfaces and

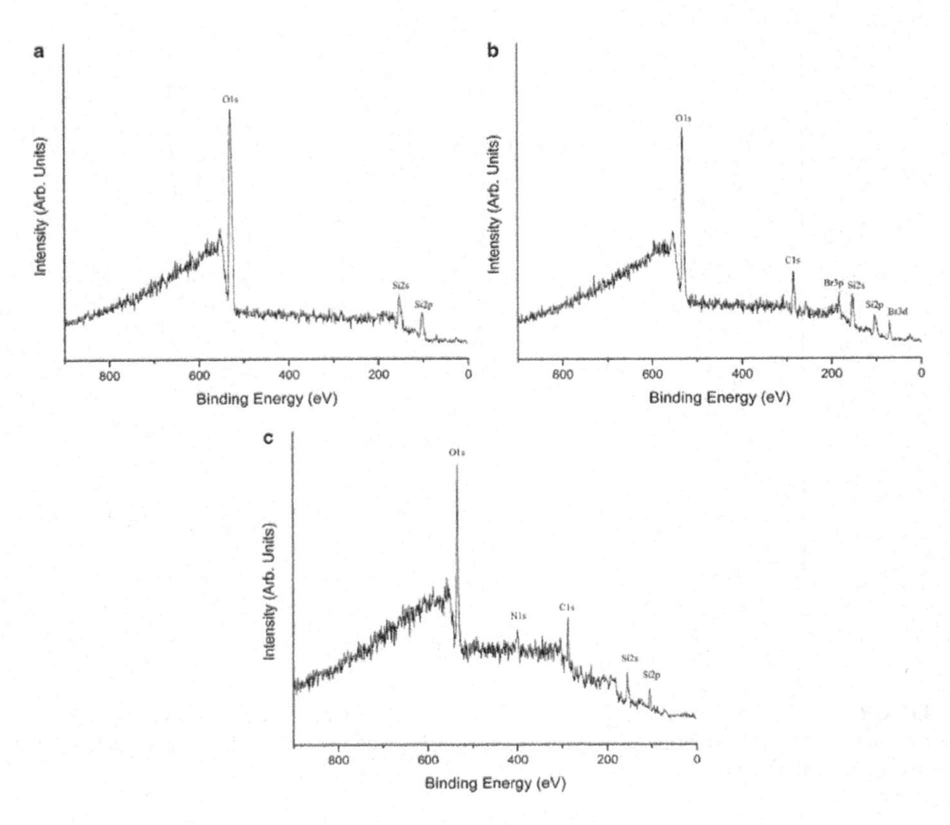

FIGURE 5.5.2 XPS wide scan spectra of silicon nanoparticles (a) after treatment with piranha solution, (b) after formation of the bromopropylsilane SAM, and (c) after substitution reaction of Br with aniline.

of particles covered by a self-assembled $–SiC_3H_6Br$ monolayer, respectively. The appearance of the Br and C peaks in Figure 5.5.2b indicates that $–SiC_3H_6Br$ was covalently bound to the silicon surface. The atomic ratio of C/Br is 3.2, which is consistent with the stoichiometry of bromopropylsilane. The silanized Si nanoparticles bearing $–SiC_3H_6Br$ monolayers were then placed in aniline at room temperature for 48 h. In this step the bromine atoms were substituted by aniline. Then the particles were rinsed with dried toluene, sonicated in spectroscopic grade methanol, and then dried. Figure 5.5.2c presents the wide scan XPS spectrum of these particles. Comparing parts b and c of Figure 5.5.2, one can see that the Br 3d signal of the $–SiC_3H_6Br$ (at 70.3 eV) from Figure 5.5.2b was replaced by a new N 1s signal in Figure 5.5.2c. The binding energy (BE) of N 1s, 399.3 eV, indicates that the signal is from an NH group covalently connected to a benzenyl ring. After reaction with aniline, the surface Br atom fraction obtained from the XPS measurements was reduced from 4.3% to 0.7%, which is within the ~1% uncertainty of the XPS measurements of atomic composition. This indicates that the displacement of Br by aniline was nearly complete, though there may still be small peaks corresponding to Br in the spectrum of Figure 5.5.2c. These results show that a chemical substitution reaction occurred between aniline and –Br on the silane monolayer and that aniline was anchored to the silicon surface through a chemical bond.

5.5.3.3 The Graft Polymerization of Aniline on the Aniline-Silicon Surface

The graft oxidative polymerization of aniline was carried out via the conventional method, by introducing the aniline-primed Si nanoparticles in aniline/HCl aqueous solution for polymerization in an ultrasonic bath. Figure 5.5.3 presents the wide scan XPS and the N 1s core-level spectra of the silicon nanoparticles after the oxidative graft polymerization in a 1 M HCl solution containing 0.1 M aniline and 0.1 M ammonium persulfate. After polymerization, the surface-grafted polyaniline in the emeraldine (EM) salt state was converted to its neutral EM base form as described in the previous section. Compared with Figure 5.5.2c, the increase in the intensities of the N 1s and C 1s signals in the XPS wide scan spectrum of Figure 5.5.3a indicates that the oxidative graft polymerization of aniline, which brings carbon and nitrogen to the silicon surface, has taken place. The curve-fitted N 1s core-level spectrum of the PANI on the Si nanoparticles (Figure 5.5.3b) reveals the predominant presence of two types of structures corresponding to the binding energy (BE) peaks at 398.2 and 399.4 eV. These can be assigned to the quinonoid imine (=N–) and benzenoid amine (–NH–) structures, respectively. The almost equal proportions of the imine and amine nitrogen in the N 1s core-level spectrum of the particle surface are consistent with the EM base form ([=N–]/[–NH–] ratio ~1) of PANI. The residual high binding energy tail (>400 eV) in the N 1s spectrum is probably due to some surface oxidation products, or a weakly charged-transfer complexed oxygen. The above N 1s spectrum is similar to that of the polyaniline homopolymer.[57] This confirms the successful grafting of the PANI to the surface of the Si nanoparticles.

The FTIR absorption spectra of PANI homopolymer powder and PANI grafted on the Si nanoparticles are compared in Figure 5.5.4. For PANI homopolymer powders the main peaks at 1590 and 1500 cm^{-1} can be assigned to the stretching vibrations of quinone and benzene rings, respectively. The peak at 1310 cm^{-1} can be assigned to the C–N stretching vibration of a secondary aromatic amine. The peak at 1160 cm^{-1} corresponds to the aromatic C–H in-plane bending mode. The out-of-plane bending of C–H in the 1,4-disubstituted benzene ring is reflected in the 830 cm^{-1} peak. The similarity between the PANI grafted on the Si nanoparticles and the PANI homopolymer powder further confirms that PANI was successfully grafted on the surface of the Si nanoparticles. In addition, the PANI-capped Si nanocomposites exhibit a strong peak at 1100 cm^{-1}, which can be attributed to Si–O–Si at the Si/PANI interface. Finally, as shown by Figure 5.5.4, after the PANI-capped Si nanoparticles were immersed in a 0.1 M NaOH solution for 24 h, the FTIR of the sample showed no change in the spectrum. This shows that PANI grafted on the Si particles surface was stable under these conditions and can protect the Si particles from degradation by basic solutions (vide infra).

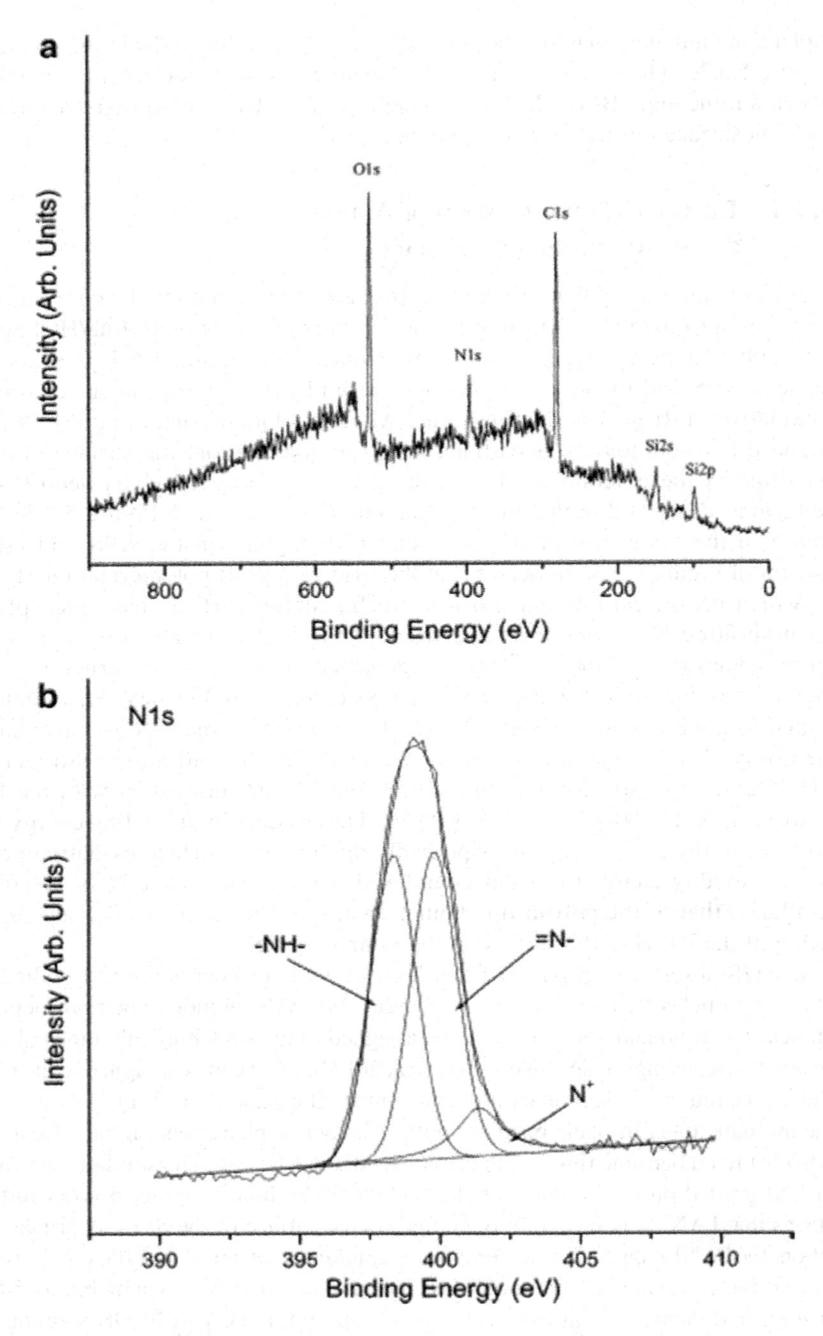

FIGURE 5.5.3 (a) XPS wide scan spectrum and (b) N 1s core-level spectrum of silicon nanoparticles grafted with PANI.

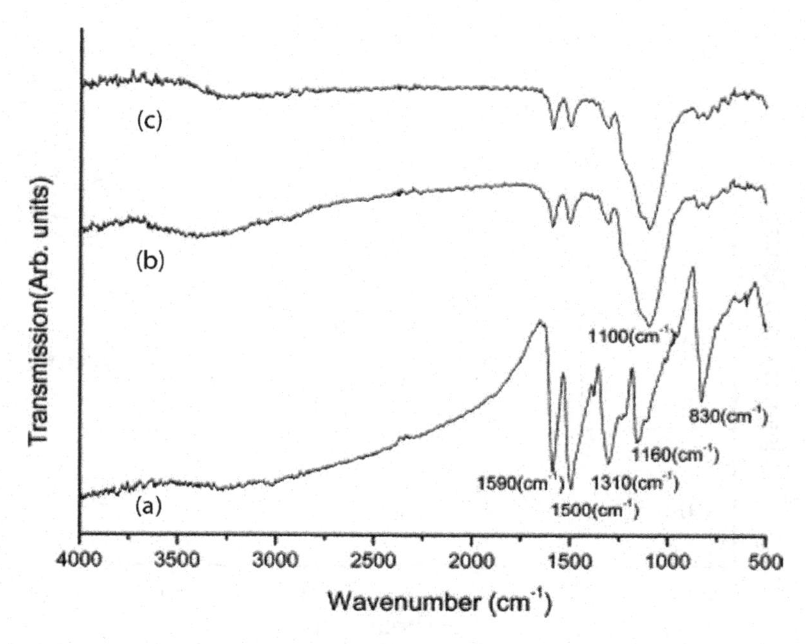

FIGURE 5.5.4 FTIR spectra of (a) EM base homopolymer, (b) PANI-capped Si nanoparticles, and (c) PANI-capped Si nanoparticles after immersion in 0.1 M NaOH for 24 h.

5.5.3.4 XRD Analysis

The crystallinity of Si nanoparticles after grafting of PANI on their surface was also investigated. Figure 5.5.5 presents the XRD patterns of the freshly prepared Si nanoparticles and of PANI-coated particles. Before the PANI coating, there are three main sharp peaks at $2\theta = 28.1°$, 47.4°, and 56.2° (Figure 5.5.5a), which are characteristic of crystalline silicon.[34] These three main peaks are still present after the graft polymerization of PANI on the particle surface (Figure 5.5.5b), but an additional broad amorphous peak is present around 20°. The additional peak corresponds to the XRD pattern of PANI and is similar to the PANI XRD results reported previously.[58] The presence of the three crystalline Si peaks and the persistence of the photoluminescence of the PANI coated Si nanoparticles, vide infra, confirm that the graft polymerization of PANI did not affect the crystallinity of the Si nanoparticles. The inset in Figure 5.5.5 shows the $2\theta = 47.4°$ peak for both samples, normalized and overlaid for comparison. For the PANI-coated particles, the peak is slightly broader, possibly indicating a slight decrease in crystal size. However, peak broadening is also affected by strain, and it is reasonable to expect that the SAM formation and PANI coating would change the strain state of the particles. Thus, one cannot conclude, based on the slight XRD peak broadening alone, that there is any significant decrease in particle size during the coating process.

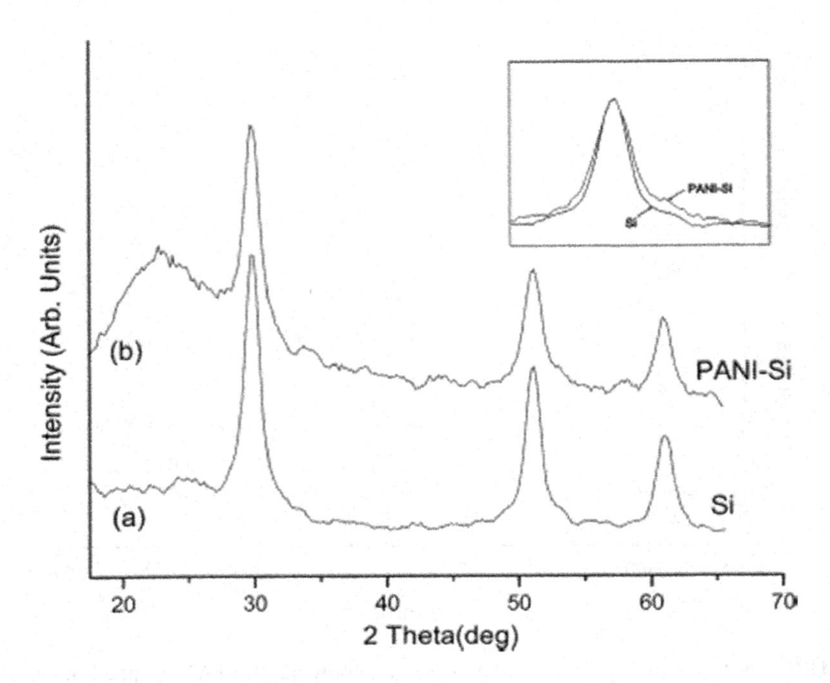

FIGURE 5.5.5 XRD patterns of (a) freshly prepared Si nanoparticles after etching and (b) PANI-capped Si nanoparticles. The inset shows the $2\theta = 47.4°$ peak for both samples, overlaid for comparison of the peak width.

5.5.3.5 TEM

Samples of uncapped and PANI-capped Si nanoparticles were dispersed and sonicated in methanol. The dispersion was dropped onto holey carbon-coated Cu TEM grids and then dried at room temperature. Panels a and b of Figure 5.5.6 present micrographs of uncapped and PANI-capped Si nanoparticles, respectively. Figure 5.5.6a shows that the uncapped particles are not well dispersed and that they aggregated significantly during solvent evaporation from the TEM grid. However, one can still observe (in Figure 5.5.6a) that most of the Si nanoparticles were in the range of 3–5 nm.[59] For the PANI-capped Si nanoparticles in Figure 5.5.6b, the dispersibility of the nanoparticles was slightly improved and one can observe that most of the particles have diameters in the range of 8–10 nm. The core–shell structure observed in Figure 5.5.6b indicates that the PANI coating was around 2–3 nm thick.

5.5.3.6 Photoluminescence and Stability of the PANI–Si Nanoparticles

Neither the hydrogen surface termination of freshly etched silicon particles nor the thin oxide that forms upon exposure of these particles to air for a few hours protects the particles from quenching of their photoluminescence by molecules in solution.

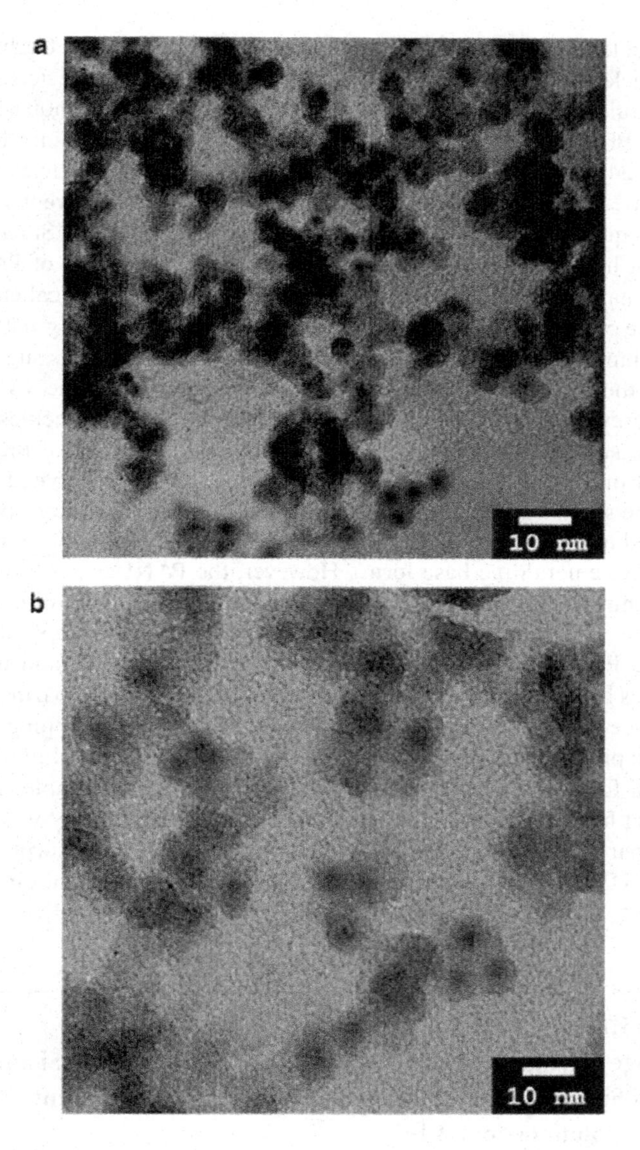

FIGURE 5.5.6 Transmission electron micrographs of (a) uncapped Si nanoparticles and (b) PANI-capped Si nanoparticles.

This restricts the potential use of Si nanoparticles in the fabrication of devices. In our experiments, it was found that the PL of freshly prepared Si nanoparticles was quenched on a time scale of seconds to minutes after immersion in ammonium hydroxide or other basic solutions and by many nitrogen-containing organic solvents such as NMP and amines. Quenching of silicon nanocrystal luminescence by amines is a well-known phenomenon.[60] The PL stability of the PANI-capped

Si nanoparticles was examined and compared with that of the freshly prepared Si nanoparticles after their immersion in a series of acidic or basic solutions and various organic solvents. We found that the PANI-capped Si nanoparticles maintained their PL even after immersion in NH_4OH or NaOH solutions, hexylamine, or NMP for 24 h. In contrast, the PL of the uncapped Si nanoparticles was quickly quenched after their immersion in the above solutions and solvents. Table 5.5.1 summarizes quantitatively the PL change of the freshly prepared Si nanoparticles, of those that had been silanized but not coated with PANI, and of PANI-capped silanized Si nanoparticles after their immersion in various chemical environments for 24 h. One can see that the PL stability of Si nanoparticles was greatly improved after their silanization by the self-assembly method. These results suggest that the coverage of the nanoparticle surface by a dense bromopropylsilane SAM stabilizes the PL by preventing intimate contact between quenching molecules and the Si nanoparticle surface. It seems that there is no dramatic effect of the conductivity of PANI on the PL intensity and stability since the PANI-coated Si particles behave in the same way as silane-coated particles in acidic solutions (where PANI is protonated and conductive) and basic solutions (where PANI is in its relatively nonconductive emeraldine base form). However, the PANI-coated SAM layer has made the nanoparticles much more robust toward nitrogen-containing organic solvents such as NMP and hexylamine. As a more detailed example, Figure 5.5.7 presents the PL spectra of the PANI-capped Si nanoparticles and uncapped Si nanoparticles before and after immersion in NMP for 24 h. Comparing parts a and b, one can see that the PL stability was enhanced greatly by capping with PANI. For PANI-capped Si nanoparticles, the PL intensity decreased slightly after 24 h in NMP, but for the uncapped particles the PL disappeared completely. It should be noted that before immersing in NMP the absolute PL intensity was lower from the coated particles than the uncoated particles due to some absorption by PANI of both the UV light used to excite the PL and the visible light emitted by the nanoparticles. The UV-vis absorption spectrum of both the PANI-coated particles

TABLE 5.5.1
Percentage Change of the Photoluminescence Intensity of Si and Modified Si Nanoparticles after Their Immersion in Different Chemical Solutions for 24 h[a]

Nanoparticles	0.1 M HCl	0.1 M NaOH	0.1M NH_4OH	NMP	Hexylamine
Si	120[b]	20	0	0	0
Silane–Si	100	75	65	78	82
PANI–silane–Si	100	75	70	95	93

[a] Percentage given is 100 times the ratio of the peak PL intensity after 24 h to that immediately after dispersion into the solution or solvent.

[b] The PL intensity increased for uncapped Si particles after its immersion in 0.1 M HCl and a blue shift by about 15 nm occurred.

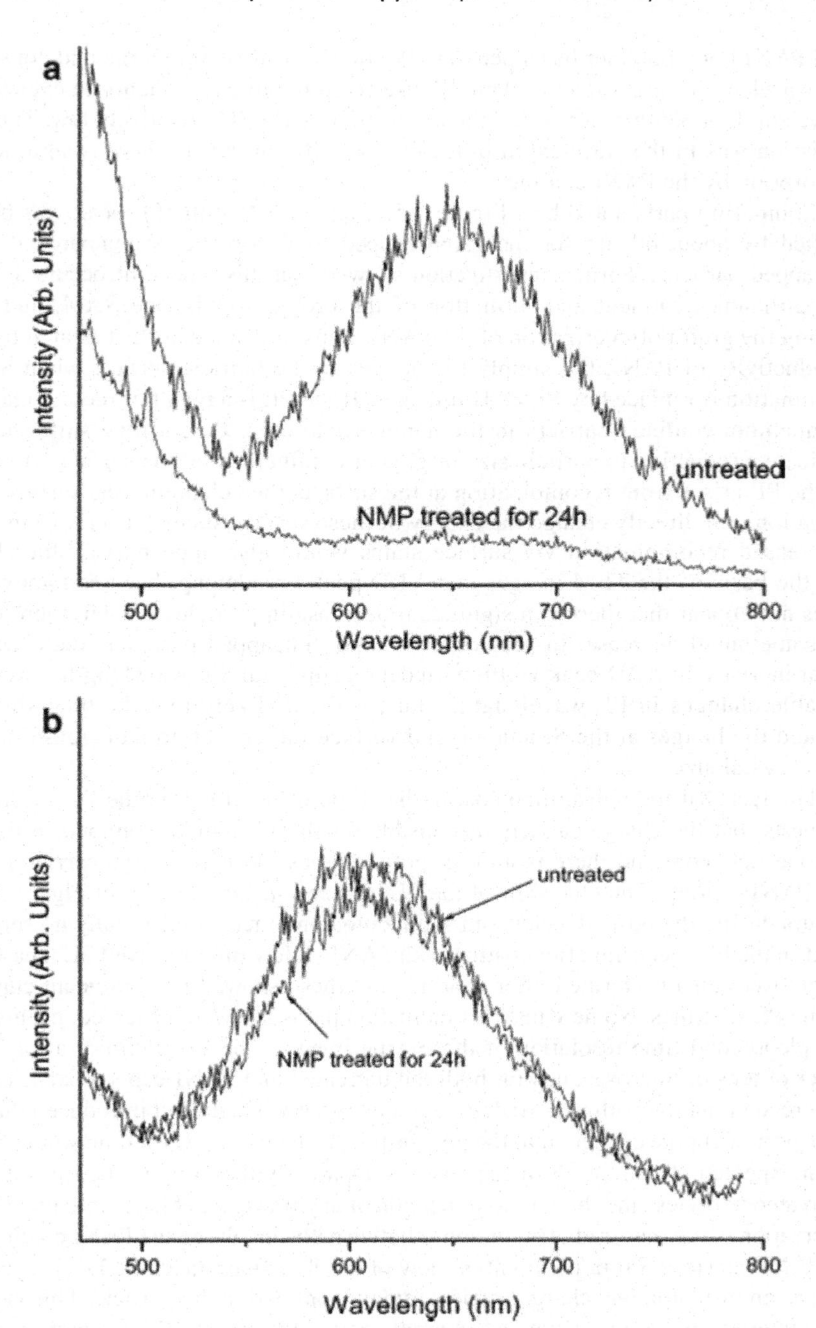

FIGURE 5.5.7 Normalized photoluminescence spectra of (a) freshly prepared Si nanoparticles before and after immersion in NMP solution for 24 h and (b) PANI-capped Si nanoparticles before and after immersion in NMP solution for 24 h.

and PANI homopolymer has a peak at 300 nm and is relatively large and constant for wavelengths larger than 600 nm. We selected in our experiments an excitation wavelength of 380 nm, for which the absorption by PANI is relatively low. The PL emission was in the wavelength range of 620–670 nm, where there is significant absorption by the PANI coating.

Comparing parts a and b of Figure 5.5.7, one sees that the PL peak was blue-shifted by about 40 nm for the PANI-capped Si nanoparticles compared to the uncapped particles. Further investigation showed that this blue shift occurs during the pirhanha treatment and formation of the bromopropylsilane SAM, and not during the graft polymerization of the PANI. Thus, it is not an effect related to the conductivity of PANI, but simply to changes in the particle surface when Si–H termination is replaced by Si–O–H linkages. If the PL is a result of recombination of quantum-confined carriers in the nanoparticle core, then a blue shift should indicate a reduction in particle size or greater confinement at the surface. If some of the PL arises from recombination at the surface, then changing the surface termination will directly change the energy of these surface states (or remove them). Decreased recombination via surface states would also appear as a blue shift. On the basis of the TEM images and XRD peak broadening discussed above, it does not appear that there is a significant decrease in particle size, but there may be some small decrease. In previous work[34] on uncapped particles, there was a clear increase in XRD peak width with decreasing peak PL wavelength for comparable changes in PL wavelength. Thus, it seems likely that the blue shift is related to changes at the Si nanocrystal surface rather than to a decrease in the nanocrystal size.

The fact that the polyaniline coating has little or no effect on the PL spectrum suggests that the charge carriers responsible for the PL remain confined in the Si nanoparticle core and there is no transport of holes (positive charge carriers) into the PANI coating. The blue shift of the PL between spectra a and b of Figure 5.5.7 occurs during the pirhanha etch and silanization and there is essentially no further shift in the PL spectrum after coating with PANI. A few time-resolved PL measurements were also performed after coating, and these showed no significant change in the PL lifetimes. No new fast recombination pathways were observed, even with the picosecond time resolution of these experiments. The PL lifetimes are of the order of tens of microseconds for both the untreated and PANI-capped particles. If new recombination pathways (radiative or nonradiative) had been introduced due to transport of charge carriers into the polyaniline, a change in PL lifetime would have been expected. Thus, the PL mechanism is apparently the same in the treated and untreated particles, and there is no indication of any transport of holes into the PANI or creation of any new radiative or nonradiative paths involving the PANI coating or Si/PANI interface. From the point of view of the Si nanoparticle, the PANI coating acts as an insulator, and charge carriers remain confined in the particle. This is perhaps interesting, because from an "external" point of view, the PANI is quite a good conductor, as detailed below. Apparently, either the work function of PANI is sufficiently high that holes generated in the Si nanoparticle cannot leave the particle or the propylsilane linking groups at the interface provide a sufficient barrier to prevent carrier transport from the silicon to the PANI.

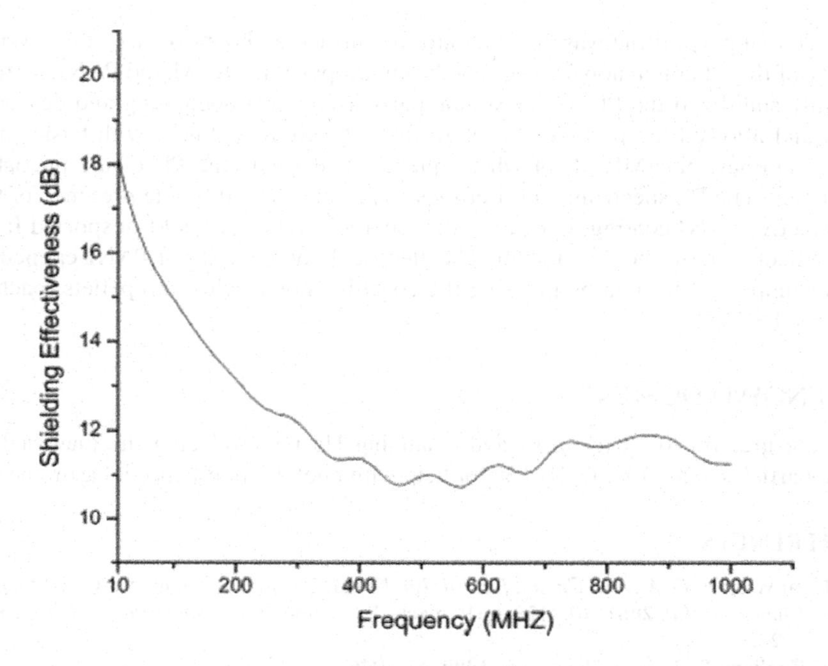

FIGURE 5.5.8 The electromagnetic shielding effectiveness of PANI-capped Si nanoparticles.

5.5.3.7 Conductivity and Electromagnetic Shielding Properties

The electrical conductivity of pellets pressed from the doped PANI-capped Si nanoparticles was as high as 2.7×10^{-2} S/cm, which is 6 orders of magnitude larger than that of pellets pressed from the freshly prepared Si nanoparticles (~10^{-8} S/cm). The relatively high conductivity rendered by PANI can provide the Si/PANI nanoparticles and nanocomposites prepared from them with good electrostatic dissipation and electromagnetic interference shielding properties. Figure 5.5.8 provides information about the electromagnetic shielding of PANI-capped Si nanoparticles. The EMI shielding effectiveness (SE) measurements were carried out using the coaxial transmission line method in the frequency range of 10–1000 MHz. Figure 5.5.8 indicates that the SE at high frequencies is around 11 dB, and at lower frequencies it is larger than 18 dB. For the freshly prepared Si nanoparticles without capping by PANI, the SE value is around zero; there is no EMI shielding effect.

5.5.4 CONCLUSION

Free-standing luminescent crystalline Si nanoparticles were successfully capped with polyaniline through reaction with a self-assembled bromopropylsilane monolayer. The composition, structure, morphology, and other physical properties of the PANI-capped Si nanoparticles were examined by X-ray photoelectron spectroscopy, Fourier transform infrared spectroscopy, and X-ray diffraction, which proved that polyaniline was grafted onto the surface without affecting the crystallinity of the silicon nanoparticles. The bromopropylsilane SAM and PANI coating greatly stabilized the PL of the Si nanoparticles against quenching and degradation and allowed the particles to retain their PL after treatment with basic solutions, amines, or NMP, all of which quench or degrade the PL of the uncoated particles. The PL spectrum and lifetimes were not affected by the presence of the conductive PANI coating, indicating that charge carriers are not transported from the silicon core to the PANI shell. The electrical conductivity of PANI-capped Si nanocomposites formed by pressing the coated nanoparticles into pellets reached 2.7×10^{-2} S/cm.

ACKNOWLEDGMENT

We are grateful to Xuegeng Li and Yuanqing He for producing the unetched Si nanoparticles, and to W. D. Kirkey for help with photoluminescence measurements.

REFERENCES

1. (a) Wegner, G. *Angew. Chem., Int. Ed. Engl.* **1981**, *20*, 361. (b) Niemeyer, C. M. *Angew. Chem., Int. Ed.* **2001**, *40*, 4128. (c) Remacle, F.; Levine, R. D. *Chem. Phys. Chem.* **2001**, *2*, 20.
2. Canham, L. T. *Appl. Phys. Lett.* **1990**, *57*, 1046.
3. Heath, J. R. *Science* **1992**, *258*, 1131.
4. Lu, Z. H.; Lockwood, D. J.; Baribeau, J.-M. *Nature* (*London*) **1995**, *378*, 258.
5. Vinciguerra, V.; Franzo, G.; Priolo, F.; Iacona, F.; Spinella, C. *J. Appl. Phys.* **2000**, *87*, 8165.
6. Wakayama, Y.; Tagami, T.; Inokuma, T.; Hasegawa, S.; Tanaka, Sh. *Res. Dev. Cryst. Growth* **1999**, *1*, 83.
7. Rinnert, H.; Vergnat, M.; Marchal, G. *Mater. Sci. Eng., B* **2000**, *69–70*, 484.
8. Huisken, F.; Kohn, B. *Appl. Phys. Lett.* **1999**, *74*, 3776.
9. (a) Lin, V. S. Y.; Motesharei, K.; Dancil, K. P. S.; Sailor, M. J.; Ghadiri, M. R. *Science* **1997**, *278*, 840. (b) Harper, J.; Sailor, M. J. *Anal. Chem.* **1996**, *68*, 3713.
10. Hamilton, B. *Semicond. Sci. Technol.* **1995**, *10*, 1187.
11. Doan, V. V.; Sailor, M. J. *Science* **1992**, *256*, 1791.
12. Sailor, M. J.; Heinrich, J. L.; Lauerhaas, J. M. *Semiconductor Nanoclusters;* Kamat, P. V., Meisel, D., Eds.; Elsevier Science: New York, **1996**; Vol. 103.
13. Canham, L. T. *Appl. Phys. Lett.* **1993**, *63*, 337.
14. Yoshinobu, T.; Ecken, H.; Ismail, A. B. M.; Iwasaki, H.; Luth, H.; Schoning, M. J. *Electrochim. Acta.* **2001**, *47*, 259.
15. Delerue, C.; Allan, G.; Lannoo, M. *Phys. Rev. B* **1993**, *48*,11024.
16. Bruchez, M., Jr.; Moronne, M.; Gin, P.; Weiss, S.; Alivisatos, A. P. *Science* **1998**, *281*, 2013.
17. (a) Buriak, J. M.; Stewart, M. P.; Geders, T. W.; Allen, M. J.; Choi, H. C.; Smith, J.; Raftery, D.; Canham, L. T. *J. Am. Chem. Soc.* **1999**, *121*, 11491. (b) Buriak, J. M. *Chem. Rev.* **2002**, *102*, 1271.
18. (a) Linford, M. R.; Chidsey, C. E. *J. Am. Chem. Soc.* **1993**, *115*, 12631. (b) Linford, M. R.; Fenter, P.; Eisenberger, P. M.; Chidsey, C. E. *J. Am. Chem. Soc.* **1995**, *117*, 3145.
19. (a) Warntjes, M.; Vieillard, C.; Ozanam, F.; Chazalviel, J.-N. *J. Electrochem. Soc.* **1995**, *142*, 4138. (b) Krawiec, B. S.; Cassagneau, T.; Fendler, J. H. *J. Phys. Chem. B* **1999**, *103*, 9524.

20. Lauerhaas, J. M.; Sailor, M. J. *Science* **1993**, *261*, 1567.
21. (a) Lee, E. J.; Ha, J. S.; Sailor, M. J. *J. Am. Chem. Soc.* **1995**, *117*, 8295. (b) Lee, E. J.; Bitner, T. W.; Shane, M. J.; Ha, J. S.; Sailor, M. J. *J. Am. Chem. Soc.* **1996**, *118*, 5375.
22. Anderson, R. C.; Muller, R. C.; Tobias, C. W. *J. Electrochem. Soc.* **1993**, *140*, 1393.
23. Bakker, J. W. P.; Arwin, H.; Wang, G.; Jarrendahl, K. *Phys. Status Solidi A* **2003**, *197*, 378.
24. Nguyen, T. P.; Le Rendu, P.; Cheah, K. W. *Physica E* **2003**, *17*, 664.
25. Lakehal, M.; Nguyen, T. P.; Le Rendu, P.; Joubert, P.; Destruel, P. *Synth. Met.* **2001**, *121*, 1631.
26. Moreno, J. D.; Marcos, M. L.; Agullo-Rueda, F.; Guerrero-Lemus, R.; Martin-Palma, R. J.; Martinez-Duart, J. M.; Gonzalez-Velasco, J. *Thin Solid Films* **1999**, *348*, 152.
27. Monastyrskii, L.; Lesiv, T.; Olenych, I. *Thin Solid Films* **1999**, *344*, 335.
28. (a) Heinrich, J. L.; Curtis, C. L.; Credo, G. M.; Kavanagh, K. L.; Sailor, M. J. *Science* **1992**, *255*, 66. (b) Bley, R. A.; Kauzlarich, S. M.; Davis, J. E.; Lee, H. W. H. *Chem. Mater.* **1996**, *8*, 1881.
29. (a) Wilcoxon, J. P.; Samara, G. A.; Provencio, P. N. *Phys. Rev. B* **1999**, *60*, 2704. (b) Mayeri, D.; Phillips, B. L.; Augustine, M. P.; Kauzlarich, S. M. *Chem. Mater.* **2001**, *13*, 765. (c) Bley, R. A.; Kauzlarich, S. M. A. *J. Am. Chem. Soc.* **1996**, *118*, 12461. (d) Heath, J. R. *Science* **1992**, *258*, 1131.
30. (a) Wilson, W. L.; Szajowski, P. J.; Brus, L. *Science* **1993**, *262*, 1242. (b) Ostraat, M. L.; De Blauwe, J. W.; Green, M. L.; Bell, L. D.; Brongersma, M. L.; Casperson, J.; Flagan, R. C.; Atwater, H. A. *Appl. Phys. Lett.* **2001**, *79*, 433. (c) Fojtik, A.; Weller, H.; Fiechter, S.; Henglein, A *Chem. Phys. Lett.* **1987**, *134*, 477. (d) Littau, K. A.; Szajowshki, P. J.; Muller, A. J.; Kortan, A. R.; Brus, L. E. *J. Phys. Chem.* **1993**, *97*, 1224.
31. Carlisle, J. A.; Dongol, M.; Germanenko, I. N.; Pithawalla, Y. B.; El-Shall, M. S. *Chem. Phys. Lett.* **2000**, *326*, 335.
32. (a) English, D. S.; Pell, L. E.; Yu, Z. H.; Barbara, P. F.; Korgel, B. A. *Nano Lett.* **2002**, *2*, 681. (b) Holmes, J. D.; Ziegler, K. J.; Doty, R. C.; Pell, L. E.; Johnston, K. P.; Korgel, B. A. *J. Am. Chem. Soc.* **2001**, *123*, 3743.
33. Ehbrecht, M.; Ferkel, H.; Smirnov, V. V.; Stelmakh, O.; Zhang, W.; Huisken, F. *Surf. Rev. Lett.* **1996**, *3*, 807. (b) Ehbrecht, M.; Kohn, B.; Huisken, F.; Laguna, M. A.; Paillard, V. *Phys. Rev. B* **1997**, *56*, 6958. (c) Ledoux, G.; Gong, J.; Huisken, F.; Guillois, O.; Reynaud, C. *Appl. Phys. Lett.* **2002**, *80*, 4834.
34. Li, Xuegeng, He Yuanqing, Talukdar, S. S.; Swihart, M. T. *Langmuir* **2003**, *19*, 8490.
35. (a) Sailor, M. J.; Lee, M. J. *Adv. Mater.* **1997**, *9*, 78. (b) Kamat P. V. *J Phys. Chem. B* **2002**, *106*, 7729.
36. (a) Colvin, V. L.; Schlamp, M. C.; Alivisatos, A. P. *Nature* **1994**, *370*, 354. (b) Cassagneau, T.; Mallouk, T. E.; Fendler, J. H. *J. Am. Chem. Soc.* **1998**, *120*, 7848 (c) Gao, M.; Richer, B.; Kirstein, S. *Adv. Mater.* **1997**, *9*, 802. (d) Gangopadhyay, R.; De, A. *Chem. Mater.* **2000**, *12*, 608.
37. Grem, G.; Leditzky, G.; Ullrich, B.; Leising, G. *Adv. Mater.* **1992**, *4*, 36. (b) Greenham, N. C.; Morratti, S. C.; Bradley, D. D. C.; Friend, R. H.; Holmes, A. B. *Nature* **1993**, *365*, 628.
38. (a) Singh, R.; Tandon, R. P.; Chandra, S. *J. Appl. Phys.* **1991**, *70*, 243. (b) Joo, J.; Epstein, A. J. *Appl. Phys. Lett.* **1994**, *65*, 2278.
39. MacDiarmid, A. G.; Kanner, R. B. In *Handbook of Conducting Polymers;* Skotheim, T. A., Ed.; Marcel Dekker: New York, 1986; Vol. 1, p 687.
40. Burroughes, J. H.; Bradley, D. D. C.; Brown, A. R.; Marks, R. N.; Friend, R. H.; Burns, P. L.; Holmes, A. B. *Nature* **1990**, *347*, 539.
41. (a) Halliday, D. P.; Gray, J. W.; Adams, P. N.; Monkman, A. P. *Synth. Met.* **1999**, *102*, 877. (b) Halliday, D. P.; Holland, E. R.; Eggleston, J. M.; Adams, P. N.; Cox, S. E.; Monkman, A. P. *Thin Solid Films* **1996**, *276*, 299.

42. Racicot, R.; Brown, R.; Yang, S. C. *Synth. Met.* **1997**, *85*, 1263.
43. Dhawan, S. K.; Singh, N.; Venkatachalam, S. *Synth. Met.* **2002**, *129*, 261.
44. Makela, T.; Pienimaa, S.; Taka, T.; Jussila, S.; Isotalo, H. *Synth. Met.* **1997**, *85*, 1335.
45. (a) Sukeerthi, S.; Contractor, A. Q. *Chem. Mater.* **1998**, *10*, 2412. (b) Flitton, R.; Johal, J.; Maeda, S.; Armes, S. P. *J. Colloid Interface Sci.* **1995**, *173*, 135. (c) Gangopadhyay, G.; De, A. *J. Appl. Phys.* **2000**, *87*, 2363. (d) Cho, G.; Fung, B. M.; Glatzhofer, D. T.; Lee, J. S.; Shul, Y. G. *Langmuir* **2001**, *17*, 456.
46. Higgins, R. W. T.; Zaidi, N. A.; Monkman, A. P. *Adv. Functional Mater.* **2001**, *11*, 407.
47. Liess, M.; Chinn, D.; Petelenz, D.; Janata, J. *Thin Solid Films* **1996**, *286*, 252.
48. Halliday, D. P.; Eggleston, J. M.; Adams, P. N.; Pentland, I. A. Monkman, A. P. *Synth Met.* **1997**, *85*, 1245.
49. Polk, B. J.; Potje-Kamloth, K.; Josowicz, M.; Janata, J. *J. Phys. Chem. B* **2002**, *106*, 11457.
50. Huang, W. S.; MacDiarmid, A. G. *Polymer* **1993**, *34*, 1833.
51. Sergeev, A.; Mitin, V.; Srtrocio, M. *Physica B* **2002**, *316*, 369.
52. (a) Jordan, R.; West, N.; Ulman, A.; Chou, Y.-M.; Nuyken, O. *Macromolecules* **2001**, *34*, 1606. (b) Nuss, S.; Böttcher, H.; Wurm, H.; Hallensleben, M. L. *Angew. Chem, Int. Ed.* **2001**, *40*, 4016. (c) Mandal, T. K.; Fleming, M. S.; Walt, D. R. *Nano Lett.* **2002**, *2*, 3–7. (d) Weck, M.; Jackiw, J. J.; Rossi, R. R.; Weiss, P. S.; Grubbs, R. H. *J. Am. Chem. Soc.* **1999**, *121*, 4088.
53 (a) Prucker, O.; Rühe, J. *Macromolecules* **1998**, *31*, 592. (b) Prucker, O.; Rühe, J. *Macromolecules* **1998**, *31*, 602.
54. Li, Z. F.; Ruckenstein, E. *Macromolecules* **2002**, *35*, 9506.
55. Ellis, J. R. *Handbook of Conductive Polymers*; Skotheim, T. A., Eds.; Marcel Dekker: New York, 1986; Vol. 1, p 505.
56. Beckmann, K. H. *Surf. Sci.* **1965**, *3*, 314.
57. Li, Z. F.; Kang, E. T.; Neoh, K. G.; Tan, K. L. *Synth. Met.* **1997**, *87*, 45.
58. Chen, S. A.; Lee, H. T. *Macromolecules* **1993**, *26*, 3254.
59. Covalent attachment of other small organic molecules (octadecene, undecylenic acid, octadecyltrimethoxysilane) to the silicon nanoparticles allows them to form stable dispersions in a variety of nonpolar and semipolar solvents. These well-dispersed particles do not agglomerate during solvent evaporation when cast on a TEM grid. For those particles, high-resolution TEM imaging clearly shows individual silicon nanoparticles 3–5 nm in diameter. Details of this will be presented in a separate article (Li, X.; He, Y.; Swihart, M. T. *Langmuir* **2004**, *20*, 4720).
60. Harper, J.; Sailor, M. J. *Langmuir* **1997**, *13*, 4652.

6 Miscellaneous Topics

CONTENTS

Research investigations on several different polymer-based topics are presented in this chapter, including (1) the synthesis of polymer and polymer-containing nanocomposite by coordinated polymerization, (2) the preparation of dendritic polymers and dendrimer-containing hybrids, and (3) the toughing and compatibilization or brittle polymers.

Coordinated polymerization is a crucial technology of polymer synthesis, with major industrial applications in the production of high-density polyethylene (PE), linear low-density PE, and isotactic polypropylene (iPP). Synthetic studies on the preparation of syndiotactic polystyrene (sPS) by coordinated polymerization are important (Section 6.1), because of the significant properties of sPS, including high crystallinity and melting point (275°C), superior heat and chemical resistance, etc. Coordinated polymerization can also be used to prepare polymer-containing nanocomposites by using inorganic material-supported catalysts (Section 6.2). The resulting PE/palygorskite nanocomposites with uniformly dispersed palygorskite nanofibers in PE matrix exhibit enhanced mechanical properties as compared to the PE/palygorskite nanocomposites obtained by melt blending approach.

Dendritic polymers are an important class of highly branched polymers that possess higher solubility and miscibility with other materials, as well as more terminal groups, than linear or lightly branched polymers. A special type of dendritic polymers is prepared by self-condensing cationic polymerization of 1-[(2-vinyloxy) ethoxy]ethyl acetate (Section 6.3). Hybrids consisting of SiO_2 and poly(amidoamine) (PAMAM) dendrimer are synthesized via a chemical reaction approach (Section 6.4). These hybrids not only maintain a high metal ion complexing capacity due to their PAMAM contents, but demonstrate much higher thermal stability than the PAMAM dendrimer.

The enhancement of toughness of a brittle plastic by toughing and compatibilization approaches can greatly increase the applicability of the plastic. As a widely used biodegradable polymer, polylactide is relatively brittle. Polyurethane-toughened polylactide, prepared through solution blending followed by crosslinking, leads to an increase of the maximum toughness by one magnitude as compared to pure polylactide (Section 6.5). Neither ethylene–vinyl acetate copolymer (EVA) nor ethylene–acrylic acid copolymer (EAA) is compatible with nylon 6, a commonly used plastic. However, the combination of EVA and EAA can effectively toughen nylon 6, and the effects of EVA and EAA in toughening and compatibilization are cooperative (Section 6.6). The resulting trinary blends exhibit a much higher impact strength than nylon 6.

6.1 Syndio-Specific Polymerization of Styrene Using Fluorinated Indenyltitanium Complexes*

Guangxue Xu and Eli Ruckenstein

Department of Chemical Engineering, State University of New York at Buffalo, Buffalo, New York 14260

ABSTRACT Nine new fluorinated half-sandwich titanocene complexes **(1b–9b)** based on substituted alkylindenes were synthesized, by reacting Me_3SnF with the corresponding chloride species, and employed as catalyst precursors for the syndio-specific polymerization of styrene. When activated with methylaluminoxane (MAO), the new precursors **1b–9b** exhibited increased activities by factors of 15–40 compared with the corresponding chlorinated compounds and provided improved syndiotacticity, enhanced melting temperature, and higher polymer molecular weights. The activities of indenyl and methyl- or phenyl-substituted indenyl complexes were found to be higher by factors of 4–12.5 than those of $CpTiF_3$ and Cp^*TiF_3. More importantly, the amount of MAO can be reduced to an Al: Ti molar ratio of 300 in the temperature range of 10°C–90°C. It is likely that Ti–F, more polarized than the Ti–Cl bond in the half-sandwich titanocenes, allows the formation of more active and stable active sites of Ti(III) complexes needed for the syndio-specific polymerization of styrene. Evidence in this direction is brought via the electron paramagnetic resonance (EPR) spectrum and redox titration. The higher activity and syndio-specificity of the fluorinated catalysts are attributable to a greater number, more stable Ti(III) active sites, and/or higher propagation rate constant.

* *Journal of Polymer Science: Part A: Polymer Chemistry* 1999, 37, 2481–2488.

6.1.1 INTRODUCTION

The control of stereoregularity is of practical importance in the development of new polymers or tailor-made polymers as well as in the control of polymer properties. The discovery of metallocene catalysts opened up the possibility of stereochemical control in olefin polymerization at the molecular level, thus producing polymers with new microstructures (e.g., isotactic, syndiotactic, and stereoblock) via the design of the catalyst.[1] Consequently, extensive studies concerning the metallocene catalysts have been carried out. Most remarkable is the recent development of the syndiospecific styrene polymerization. In contrast to the well known isotactic polystyrene (iPS), which has a very low crystallization rate and therefore is useless for practical applications,[2] the syndio-tactic polystyrene (sPS) has a fast crystallization rate, with more than an order of magnitude higher than that of iPS. In addition, it has a high melting point (275°C) and crystallinity, superior heat and chemical resistance, and unique mechanical properties similar to those of some expensive engineering plastics. Consequently, sPS provides technologically important characteristics for the electronic and engineering industries.[3]

Since Ishihara and coworkers reported the synthesis of sPS by using titanium/methylaluminoxane (MAO) catalysts, there has been interest in the synthesis of homogeneous organometallic complexes able to provide an efficient and stereoregular polymerization of styrene.[4] The polymerization mechanism[5] and the structure of the active site[4g,4i,6] as well as the structural characterization and technical applications of sPS were investigated.[3] A variety of titanium, zirconium, and hafnium complexes have been evaluated as catalysts for the syndio-specific polymerization of styrene.[4–6] The half-sandwich titanocenes of the type $CpTiCl_3$, $IndTiCl_3$, and substituted $IndTiCl_3$, activated with MAO, showed the highest activities. Recently, Kaminsky et al.[7] reported that the fluorinated catalysts of the type $CpTiF_3$ and Cp^*TiF_3 have much higher activities and produce polymers with higher molecular weight than the chlorinated catalysts. However, compared with the polymerization of olefins, the activity for styrene polymerization was much lower. To this end, intensive studies have been carried out to develop new and better catalysts for the syndio-specific polymerization of styrene.

In the present study, we synthesize a series of fluorinated indenyl titanium complexes as catalyst precursors for the syndio-specific polymerization of styrene. Also, we determine the oxidation states of titanium by redox titration and by electron paramagnetic resonance (EPR). The findings show that the catalytic activity and polymer molecular weight are dependent on the electronic and steric effects of ligands in half-sandwich metallocenes, and that the higher activity and syndiospecificity of the fluorinated catalysts is caused by a greater number, more stable Ti(III) active sites, and/or a higher propagation rate constant.

6.1.2 EXPERIMENTAL

6.1.2.1 General Procedures

All experiments were performed under a dry nitrogen atmosphere using either standard Schlenk techniques or a dry box. Reagent grade hexane, THF, and methylene chloride were distilled prior to use from calcium hydride under nitrogen. Toluene and all other solvents were purified by refluxing over Na-K alloy/benzophenone ketyl under nitrogen for at least a week followed by distillation. Styrene was purchased from Aldrich and dried over calcium hydride for 1 week at room temperature and distilled under reduced pressure. Trimethylsilyl chloride (Aldrich) was distilled from CaH_2 and dichlorodimethylsilane from quinoline. MAO and all other reagents were purchased from Aldrich and used without further purification. The Me_3SnF was synthesized by using a published method[8] and sublimed at 100°C/1.32 × 10^{-2} Pa prior to use. ^1H- and ^{19}F-NMR spectra were recorded on an INOVA-500 spectrometer. The $CpTiCl_3$, $CpTiF_3$, Cp^*TiCl_3, Cp^*TiF_3, $(MeCp)TiCl_3$, $(MeCp)TiF_3$, $(MeCp)_2TiCl_2$, and $(MeCp)_2TiF_3$ were prepared according to the literature.[4,7]

6.1.2.2 Indenyltrifluorotitanium (1b)

To a suspension of Me_3SnF (5.49 g, 30.0 mmol) in toluene (30 mL), a solution of $IndTiCl_3$ (**1a**)[9a–b] (2.69 g, 10 mmol) in toluene (50 mL) was added. The resulting mixture was stirred at room temperature overnight. The solvent and Me_3SnCl were removed in vacuum, and the residue was recrystallized from THF/hexane to yield 2.05 g (93%) of the orange-yellow $IndTiF_3$ (**1b**).

^1H-NMR ($CDCl_3$): δ 7.81 (m, 2H, arom), 7.48 (m, 2H, arom), 7.10 (d, 2H, H–C_5(3) and H–C_5(1), J = 10.4 Hz), 7.03 (t, 1H, H–C_5(2), J = 10.4 Hz). ^{19}F-NMR ($CDCl_3$): δ 111.9. MS (EI): m/z 220 (M^+).

Anal. calcd. for $C_9H_7F_3Ti$: C, 49.13; H, 3.21. Found: C, 49.03; H, 3.15.

6.1.2.3 1-Methylindenyltrifluorotitanium (2b)

The same procedure as that described for **1b** was used. The 1-methylindenyltrichlorotitanium [1-(Me)$IndTiCl_3$] (**2a**)[9b] (2.83 g, 10.0 mmol) and Me_3SnF (5.49, 30 mmol) were used to give 2.13 g (91%) of the orange-yellow 1-(Me)$IndTiF_3$ (**2b**).

^1H-NMR ($CDCl_3$): δ 7.85–7.60 (m, 2H, arom), 7.45 (m, 2H, arom), 7.05 [d, 1H, H–C_5(3), J = 4.1 Hz], 6.81 [d, 1H, H–C_5(2), J = 4.1 Hz)], 2.65 (s, 3H, CH_3). ^{19}F-NMR ($CDCl_3$): δ 109.5. MS (EI): m/z 234 (M^+).

Anal. calcd. for $C_{10}H_9F_3Ti$: C, 51.31; H, 3.88. Found: C, 51.25; H, 3.85.

6.1.2.4 1-Ethylindenyltrifluorotitanium (3b)

Compound **3b** was prepared according to the method described for **lb**. 1-(Et)$IndTiCl_3$ (**3a**)[9b] (2.98 g, 10.0 mmol) and Me_3SnF (5.49 g, 30.0 mmol) yielded 2.28 g (92%) of 1-(Et)$IndTiF_3$ (**3b**) after recrystallization from THF/hexane.

^1H-NMR ($CDCl_3$): δ 7.65 (m, 2H, arom), 7.40 (m, 2H, arom), 7.03 [d, 1H, H–C_5(3), J = 3.5 Hz], 6.89 (d, 1H, H–C_5(2), J = 3.5 Hz], 3.15 (q, 2H, CH_2, J = 7.4 Hz), 1.36 (t, 3H, CH_3, J = 7.4 Hz). ^{19}F-NMR ($CDCl_3$): δ 118.5. MS (EI): m/z 248 (M^+).

Anal. calcd for $C_{11}H_{11}F_3Ti$: C, 53.25; H, 4.48. Found: C, 53.15; H, 4.46.

6.1.2.5 1-*tert*-Butylmdenyltrifluorotitamum (4b)

Following the procedure described for **1b**, 1-(Me_3C)$IndTiCl_3$ (**4a**)[9b] (3.26 g, 10.0 mmol) and Me_3SnF (5.49 g, 30.0 mmol) gave 1-(Me_3C)$IndTiF_3$ (**4b**) (2.48 g, 90%).

^{1}H-NMR ($CDCl_3$): δ 8.00–7.81 (m, 2H, arom), 7.58–7.41 (m, 2H, arom), 7.05 [d, 1H, H–C_5(3). J = 3.3 Hz], 6.85 (d, 1H, H–C_5(2), J = 3.3 Hz], 1.45 (s, 9H, 3CH_3). ^{19}F-NMR ($CDCl_3$): δ 108.5. MS (EI): *m/z* 276 (M^+).

Anal. calcd for $C_{13}H_{15}F_3Ti$: C, 56.54; H, 5.49. Found: C, 56.56; H, 5.50.

6.1.2.6 1-Trimethylsilylindenyltrifluorotitanium (5b)

A suspension of Me_3SnF (5.49 g, 30.0 mmol) and 1-(Me_3Si)Ind$TiCl_3$ (**5a**)[9b] (3.42 g, 10.0 mmol) was stirred for 12 h in toluene (80 mL) at room temperature. The solvent was removed under vacuum and the residue was sublimed at 105°C/0.001 mmHg to yield 2.54 g (87%) of air- and moisture-sensitive orange-yellow 1-(Me_3Si)IndTiF_3 (**5b**) after recrystallization from THF/hexane.

^{1}H-NMR ($CDCl_3$): δ 7.80 (m, 2H, arom), 7.52 (m, 2H, arom), 7.31 [d, 1H, H–C_5(3), J = 1.5 Hz], 7.15 [d, 1H, H–C_5(2), J = 1.5 Hz], 1.56 (s, 9H, 3CH_3). ^{19}F-NMR ($CDCl_3$): δ 120.5. MS (EI): *m/z* 282 (M^+).

Anal. calcd for $C_{12}H_{15}F_3SiTi$: C, 49.32; H, 5.18. Found: C, 49.25; H, 5.16.

6.1.2.7 1-Isoprylindenyltrifluorotitanium (6b)

Compound **6b** was synthesized according to the method described for **5b**, 1-(iPrInd)$TiCl_3$ (**6a**)[9b] (3.12 g, 10.0 mmol) and Me_3SnF (5.49 g, 30.0 mmol) gave 1-(iPrInd)TiF_3 (**6b**) (2.49 g, 95%).

^{1}H-NMR ($CDCl_3$): δ 7.75–7.48 (m, 4H, arom), 7.05 (d, J = 3.5 Hz, 1H), 6.98 (d, J = 3.4 Hz, 1H), 3.65 (m, 1H), 1.50 (d, J = 7.4 Hz, 3H), 1.28 (d, J = 7.4 Hz, 3H). ^{19}F-NMR ($CDCl_3$): δ 118.5 (s). MS (EI): *m/z* 262 (M^+).

Anal. calcd for $C_{12}H_{13}F_3Ti$: C, 54.98; H, 5.01. Found: C, 54.54; H, 5.06.

6.1.2.8 1,3-Dimethylindenyltrifluorotitanium (7b)

Following the procedure described for **5b**, 1,3-(Me_2Ind)$TiCl_3$ (**7a**)[9c] (2.98 g, 10.0 mmol) and Me_3SnF (5.49 g, 30.0 mmol) produced 1,3-(Me_2Ind)TiF_3 (**7b**) (2.27 g, 91.5%).

^{1}H-NMR ($CDCl_3$): δ 7.85–7.52 (m, 4H, arom), 6.85 [s, 1H, H–C_5(2)], 2.75 (s, 6H, CH_3). ^{19}F-NMR ($CDCl_3$): δ 120.9. MS (EI): *m/z* 248 (M^+).

Anal. calcd for $C_{11}H_{11}F_3Ti$: C, 53.25; H, 4.48. Found: C, 53.20; H, 4.45.

6.1.2.9 1-Phenylindenyltrifluorotitanium (8b)

Following the procedure described for **1b**, 1-(PhInd)$TiCl_3$ (**8a**)[4j] (3.45 g, 10.0 mmol) and Me_3SnF (5.49 g, 30.0 mmol) gave 1-(PhInd)TiF_3 (**8b**) (2.75 g, 93%).

^{1}H-NMR ($CDCl_3$): δ 7.52–8.30 (m, 9 H, arom), 7.45 (d, 1H, H–C_5(2), 7.25 (dd, 1H, H–C_5(2). ^{19}F-NMR ($CDCl_3$): δ 121.5. MS (EI): *m/z* 296 (M^+).

Anal. calcd for $C_{15}H_{11}F_3Ti$: C, 60.84; H, 3.75. Found: C, 60.55; H, 3.70.

6.1.2.10 1,3-Diphenylindenyltrifluorotitanium (9b)

Following the procedure described for **5b**, 1,3-(Ph_2Ind)$TiCl_3$ (**9a**)[4j] (4.22 g, 10.0 mmol) and Me_3SnF (5.49 g, 30.0 mmol) yielded 1,3-(Ph_2Ind)TiF_3 (**9b**) (3.31 g, 89%) as dark green crystals.

^{1}H-NMR ($CDCl_3$): δ 7.45–8.40 (m, 14H, arom), 7.38 [s, 1H, H–C_5(2)]. ^{19}F-NMR ($CDCl_3$): δ 119.2. MS (EI): *m/z* 372 (M^+).

Anal. calcd for $C_{21}H_{15}F_3Ti$: C, 67.76; H, 4.07. Found: C, 67.68; H, 4.05.

6.1.2.11 Polymerization and Analytical Procedures

A 100-mL glass reactor equipped with a magnetic stirrer was attached to a high-vacuum line and then sealed under a nitrogen atmosphere. Freshly distilled toluene (20 mL) was introduced through a syringe, followed by addition of styrene (20 mL) and of the appropriate amount of MAO. The bottle was placed in a bath at the desired temperature and stirred for 10 min. The preactivated titanocene compound (2.5 μmol) in toluene then was added, and the mixture was stirred for selected reaction times. The reaction mixture subsequently was quenched with 10% HCl in methanol, filtered, and dried in a vacuum oven at 80°C. The polymer then was extracted with 2-butanone for 48 h in a Soxhlet extractor to remove any atactic polymer. The syndiotactic polymer was determined as the amount of polymer insoluble in 2-butanone.

The molecular weight was determined from intrinsic viscosity in *o*-dichlorobenzene at 135°C[10] as

$$[\eta] = 1.38 \times 10^{-4} Mw^{0.7}$$

DSC thermograms were recorded with a Thermal Analyst 2100 (Du Pont Instruments) at a heating rate of 10 K/min. The melting temperature of the polymers was determined from the second heating scan. ^{1}H- and ^{19}F-NMR spectra were recorded on an INOVA-500 spectrometer. The Ti oxidation states[6a] and the concentration of active species[4c] were determined according to the literature.

6.1.3 RESULTS AND DISCUSSION

6.1.3.1 Synthesis of Catalyst Precursors

Indenyltrichlorotitanium (**1a**),[9a–b] 1-methylinde-nyltrichlorotitanium (**2a**),[9b] 1-ethylindenyltrichlorotitanium (**3a**),[9b] 1-tert-butylindenyltrichlorotitanium (**4a**),[9b] 1-trimethylsilylindenyltrichlorotitanium (**5a**),[9b] 1-isoprylindenyltrifluorotitanium (**6a**),[9b] 1,3-dimethylindenyltrichlorotitanium (**7a**),[9c] 1-phenylindenyltrichlorotitanium (**8a**),[4j] and 1,3-diphenylindenyltrichlorotitanium (**9a**)[4j] were synthesized as indicated in the literature. The new catalyst precursors **1b–9b** were prepared with excellent yields from the corresponding chloride species via their reaction with Me_3SnF, by using the procedures employed for the analogous cyclopentadienyltitanium complexes $Cp'TiF_3$ ($Cp' = C_5Me_5$, C_5Me_4Et, C_5Me_4H, or C_5H_5)[8] (Scheme 6.1.1). The recrystallization of the complexes from THF/hexane mixtures resulted in **1b–9b** with 87%–95% yields. Compounds **1b–9b** were purified further by sublimation in vacuum. They are soluble in polar solvents such as toluene or THF without decomposition. The compounds **1b–9b** are air and moisture sensitive. The attempt to synthesize **1b–9b** complexes by reacting the corresponding chloride species with AsF_3 was unsuccessful because of the difficulty to isolate and purify the target compounds

Me_3SnF, toluene

1a : $R_1 = R_2 = R_3 = H$
2a : $R_1 = Me, R_2 = R_3 = H$
3a : $R_1 = Et, R_2 = R_3 = H$
4a : $R_1 = t-Bu, R_2 = R_3 = H$
5a : $R_1 = Me_3Si, R_2 = R_3 = H$
6a : $R_1 = iPr, R_2 = R_3 = H$
7a : $R_1 = R_3 = Me, R_2 = H$
8a : $R_1 = Ph, R_2 = R_3 = H$
9a : $R_1 = R_3 = Ph, R_2 = H$

1b : $R_1 = R_2 = R_3 = H$
2b : $R_1 = Me, R_2 = R_3 = H$
3b : $R_1 = Et, R_2 = R_3 = H$
4b : $R_1 = t-Bu, R_2 = R_3 = H$
5b : $R_1 = Me_3Si, R_2 = R_3 = H$
6b : $R_1 = iPr, R_2 = R_3 = H$
7b : $R_1 = R_3 = Me, R_2 = H$
8b : $R_1 = Ph, R_2 = R_3 = H$
9b : $R_1 = R_3 = Ph, R_2 = H$

SCHEME 6.1.1

from the dark reaction mixtures; however, the latter method was successful in preparing the analogous cyclopentadienyltitanium complexes.[11]

6.1.3.2 Syndio-Specific Polymerization of Styrene

As shown in Table 6.1.1, the fluorinated titanocenes are much more active than their chlorinated counterparts. Depending on the ligand structure, the polymerization activity of the fluorinated catalyst is 10–30 times higher than that of the chlorinated one at the relatively low Al: Ti ratio of 300. In addition, the fluorinated catalysts provide higher molecular weight and melting temperature sPS. Compared with the precursors $CpTiF_3$ and Cp^*TiF_3 employed by Kaminsky,[7] the Ind-TiF_3 was found to have much higher activity. Under identical polymerization conditions, the activity of $IndTiF_3$ catalyst was 4.5 times higher than that of $CpTiF_3$ and 13.3 times higher than that of Cp^*TiF_3. The increased activity of $IndTiF_3$ may be attributed to the stimulation of propagation rate and to an increase of the number and/or the stability of the active species by the higher electron-donating ability of the indenyl than of the Cp moiety.[12] Cp* is a stronger electron donor than Cp and the indenyl ring but, because of the greater steric hindrance it generates, the polymerization activity of Cp^*TiF_3 is much smaller. In addition, the $IndTiF_3$ provides a molecular weight and a melting temperature higher than those of the $CpTiF_3$ catalyst and comparable with those of Cp^*TiF_3. The two Cp ring catalysts hinder the insertion of the styrene monomer, and for this reason their activity is the lowest.

From Table 6.1.1 one can see that the methyl-substituted catalysts affect not only the activity, which is higher, but also the syndiotacticity and the molecular weight, which are also higher. These results suggest that the ligand structure has a significant effect on the syndio-specific polymerization of styrene. Therefore, the effect of other substitutions into indenyltitanium trifluorides also was investigated (Table 6.1.2). Table 6.1.2 compares the catalytic performances of indenyl-, methylindenyl-,

TABLE 6.1.1
Syndio-Specific Polymerization of Styrene Using Methylaluminoxane and Various Titanium Compounds[a]

Compound	Activity[b]	Syndiotactic Index (%)[c]	T_w (°C)[d]	10^{-5} *Mw*[e]
$CpTiCl_3$	198	75.3	258	0.5
$CpTiF_3$	2800	86.2	265	1.2
$(MeCp)TiCl_3$	370	82.5	263	—
$(MeCp)TiF_3$	3600	88.7	269	—
Cp^*TiCl_3	40	97.2	271	1.8
Cp^*TiF_3	940	99.4	277	7.9
$IndTiCl_3$	400	94.3	268	1.0
$IndTiF_3$	12,500	98.1	275	5.2
$(MeInd)TiCl_3$	600	94.2	268	1.4
$(MeInd)TiF_3$	17,500	99.0	276	6.5
$(MeCp)_2TiCl_2$	0.02	—	260	2.1
$(MeCp)_2TiF_2$	0.50	—	268	—

[a] Polymerization temperature (T_p) = 50°C, polymerization time (t_p) = 10 min, total volume (styrene + toluene) = 40 mL, styrene concentration = 4.38 mol/L, [Ti] = 6.25×10^{-5} mol/L, Al: Ti = 300: 1.

[b] Activity = kilograms of PS/(mol of Ti × h).

[c] Determined by ^{13}C-NMR.

[d] Melting point determined by DSC at a heating rate of 10 K/min.

[e] Molecular weights determined by intrinsic viscosity.

ethylindenyl-, trimethylsilyl-, propylindenyl-, butylindenyl-, phenylindenyl-, dimethylindenyl-, and diphenylindenyltitanium trichlorides and trifluorides at a polymerization temperature of 50°C. In all the cases, the fluorinated indenyltitanium catalysts are over 20 times more active than the chlorinated ones and provide higher syndiotacticity, melting temperature, and molecular weight. Among the methyl to butyl substituted compounds, only the methyl ones, $(MeInd)TiF_3$ and $(Me_2Ind)TiF_3$, have a significantly higher activity than the unsubstituted ones and produce polymers with higher molecular weight. This indicates that only the stronger electron-donating and less bulky substituents are beneficial and increase the rate of propagation without increasing the rate of chain termination by β-hydrogen abstraction. An increase in the steric constraints interferes with the styrene coordination and migratory insertion, thus leading to a lower activity and to a reduction of stereochemical control, reflected in lower syndiotacticity and melting temperature. Although the ethyl to butyl moieties in the corresponding substituted compounds have some electron-donating abilities, their catalytic effect is lower because it is dominated by the steric constraints induced by their bulkiness. The phenyl in the corresponding substituted indene is effective as an electron donor because the electron is delocalized into the phenyl ring.[4,13] Compared with the unsubstituted catalyst, $(PhInd)TiF_3$ provides a

TABLE 6.1.2
Comparison of Chlorinated and Fluorinated (RInd)TiX$_3$ for Syndio-Specific Polymerization of Styrene at 50°C[a]

	X = Cl				X = F			
Catalyst	Activity[b]	sPS (%)[c]	10^{-5} *MW*[d]	*Tm* (°C)[e]	Activity[b]	sPS (%)[c]	10^{-5} *MW*[d]	*Tm* (°C)[e]
IndTiX$_3$/MAO	400	94.3	1.0	268	12,500	98.5	5.2	272
(MeInd)TiX$_3$/MAO	600	94.2	1.4	268	17,500	99.1	6.5	273
(Me$_2$Ind)TiX$_3$/MAO	570	96.3	2.0	270	15,000	99.7	9.5	277
(EtInd)TiX$_3$/MAO	450	93.0	1.4	267	10,200	96.5	4.9	270
(iPrInd)TiX$_3$/MAO	150	85.2	1.2	264	3,200	90.0	4.6	267
(Me$_3$CInd)TiX$_3$/MAO	80	83.0	1.3	253	2,000	87.5	5.4	266
(Me$_3$SiInd)TiX$_3$/MAO	100	83.4	1.5	260	3,500	90.5	6.0	265
(PhInd)TiX$_3$/MAO	510	75.6	1.8	264	14,000	94.6	6.2	268
(Ph$_2$Ind)TiX$_3$/MAO	420	64.2	2.2	260	12,000	85.2	6.8	265

[a] Polymerization time (t_p) = 10 min, total volume (styrene + toluene) = 40 mL, styrene concentration = 4.38 mol/L, [Ti] = 6.25 × 10^{-5} mol/L, Al: Ti = 300: 1.
[b] Activity = kilograms of PS/(mol of Ti × h).
[c] sPS% = (grams of polymer insoluble in 2-butanone)/(grams of total polymer) × 100.
[d] Molecular weights determined by intrinsic viscosity.
[e] Melting point determined by DSC at a heating rate of 10 K/min.

somewhat higher catalytic activity and molecular weight, whereas (1,3-Ph$_2$Ind)TiF$_3$ provides comparable activity but higher molecular weight. In the latter cases, the effect of the electron-donating ability is somewhat higher than that of the steric constraints. The steric effect of the phenyl substitution causes a reduction of the stereochemical control, and this is reflected in the lower syndiotacticity and melting temperature provided by (PhInd)TiF$_3$ and (1,3-Ph$_2$Ind)TiF$_3$ as compared with those of the unsubstituted IndTiF$_3$.

Table 6.1.3 summarizes the results obtained regarding the styrene polymerization catalyzed by MAO-activated chlorinated and fluorinated (MeInd)TiX$_3$ catalysts for various polymerization conditions. At all temperatures examined, the fluorinated compound is more active than its chlorinated counterpart. The chlorinated compounds are, in addition, less stable, as one can see from the significant decrease in activity from 50°C to 90°C. In contrast, the maximum activity of the fluorinated complex is reached at about 70°C, without any significant decrease at 90°C. These findings indicate that the fluorinated complex is more stable than the chlorinated one even at high polymerization temperatures. On the other hand, the activity of the fluorinated catalyst is dependent on the polymerization time. At 50°C, a maximum activity is reached after 5 min and there after decreases, similar to the polymerization behavior of the chlorinated catalyst. However, the activity of the fluorinated catalyst is still 10 times higher than that of the chlorinated one at longer reaction times. A time

TABLE 6.1.3
Comparison of Activities (kg of sPS)/(mol of Ti X h) of Chlorinated and Fluorinated (MeInd) TiX_3 for the Syndio-Specific Polymerization of Styrene at Different Polymerization Conditions[a]

T_p (°C)	t_p(min)	Al/Ti (mol/mol)	Activity[b]	
			X = Cl	X = F
10	10	300	50	1,400
30	10	300	250	7,000
50	10	300	600	17,500
70	10	300	400	20,000
90	10	300	120	19,000
50	5	300	850	22,000
50	10	300	600	17,500
50	30	300	—	8,000
50	60	300	200	2,000
50	10	300	600	17,500
50	10	600	750	16,000
50	10	1000	1300	13,000
50	10	2000	2500	11,000
50	10	4000	3000	6,500

[a] T_p = polymerization temperature, t_p = polymerization time, total volume (styrene + toluene) = 40 mL, styrene concentration = 4.38 mol/L, [Ti] = 6.25×10^{-5} mol/L.

[b] Activity = kg of PS/(mol of Ti × h).

dependence of activity was observed in the MAO-activated $CpTiF_3$ and (MeCp) TiF_3 catalysts.[7] In addition, Table 6.1.3 shows that the activity is affected by the Al: Ti ratio. High activities of the fluorinated catalysts are reached at the relatively low Al: Ti ratio of 300 and decrease for larger ratios. In contrast, the chlorinated counterparts are more active only at the high Al: Ti ratio of 2000–4000. As a result, the fluorinated systems are very important from an industrial point of view because they exhibit a high activity, provide high molecular weight, and, very importantly, employ a low Al: Ti molar ratio, thus reducing the costs of the catalysts and sPS industrial process.

From the above findings it can be seen that the catalyst performances are very sensitive to the nature of the metal Ti center and of the counteranion. The electronegativity of fluorine in the fluorinated compounds is much stronger than that of chlorine in the chlorinated complexes. As a result, the former has a stronger polarizing effect. This stronger polarization of the Ti–F bond facilitates the replacement of the F ligand by MAO at the electrophilic Ti metal, thus generating more active sites and enhancing the catalytic activity.

To get some insight into the effect of the ligands F and Cl in the titanocenes on the polymerization activity, the oxidation states distribution of the Ti species for $(MeInd)TiCl_3$ and $(MeInd)\text{-}TiF_3$ was investigated by redox titration[6a] under the conditions of the styrene polymerization. The distribution of the oxidation states of Ti in the $(MeInd)TiCl_3/MAO$ (Al/Ti = 300) catalyst was found to be Ti(IV): Ti(III): Ti(II) = 76.2: 20.5: 3.3%. In contrast, the distribution of the oxidation states of Ti in the $(MeInd)TiF_3/MAO$ (Al/Ti = 300) catalyst was Ti(IV): Ti(III): Ti(II) = 15.2: 72.5: 12.3%. For Al/Ti = 2000, the distribution of the oxidation states was Ti(IV): Ti(III): Ti(II) = 40.0: 36.2: 23.8% for $(MeInd)TiCl_3$, but 18.2: 67.5: 14.3% for $(MeInd)TiF_3$. It is clear that the amount of Ti(III) complex in the fluorinated system is over two times greater than that of Ti(III) in the chlorinated counterparts. The EPR spectra provide Ti(III) = 85% for the $(MeInd)TiF_3/MAO$ (Al/Ti = 300) catalyst, but 38.5% for $(MeInd)TiCl_3$ (Al/Ti = 300).

The EPR spectra also reveal that the amount of Ti(III) in the fluorinated catalysts is much higher than that in the chlorinated ones. Figure 6.1.1 shows the EPR spectrum of MAO activated $(MeInd)TiX_3$ (X = Cl, F) catalyst systems. The main EPR signal for the $(MeInd)TiF_3/MAO$ catalyst (Al/Ti = 300) is a doublet at g = 1.987 attributable to a Ti(III) species with ^{H}a = 7.2 G (Figure 6.1.1b). Hyperfine splitting of about 7.2 G for the ^{47}Ti (I = 5/2) and ^{49}Ti (I = 7/2) isotopes also can be seen. The integrated intensity of this spectrum gave a Ti(III) of 95%. Although the EPR spectrum

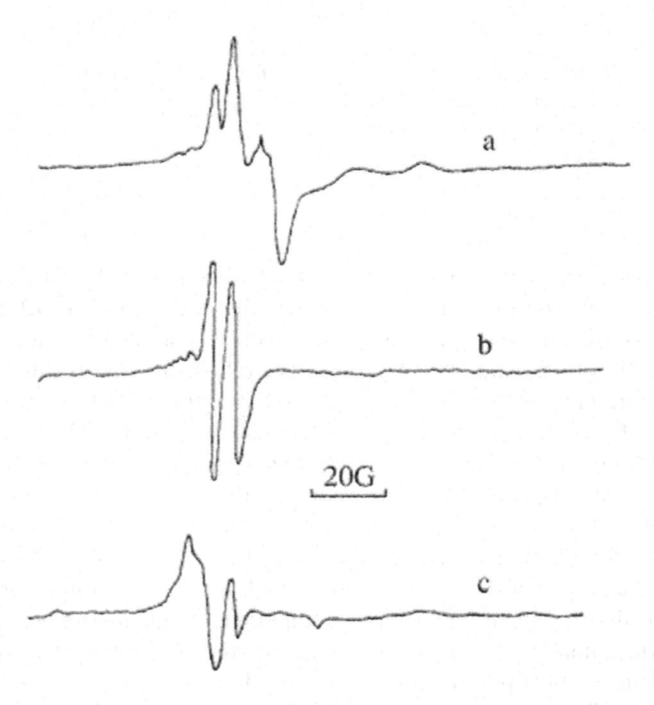

FIGURE 6.1.1 EPR spectra of (a) $(MeInd)TiCl_3/MAO$, Al/Ti = 300; (b) $(MeInd)TiF_3/MAO$, Al/Ti = 300; (c) $(MeInd)TiF_3/MAO$, Al/Ti = 2000.

(Figure 6.1.1a), obtained for the (MeInd)$TiCl_3$/MAO catalyst (Al/Ti = 300), is different from that in Figure 6.1.1b, there are additional resonances at g = 1.966 and g = 1.960. The amount of Ti(III) in this figure is 46%. One cannot correlate Ti(III) values obtained with redox titration and EPR, but these findings clearly indicate that the highly polarized Ti–F bond is alkylated more easily by MAO via the replacement of F, and the MAO-alkylated titanocene is reduced more easily from the Ti(IV) to the Ti(III) complex. The higher the Ti(III) content, the higher the catalytic activity. Hence, in agreement with previous studies,[4,6a] the Ti(III) complexes constitute the active species for the syndio-specific polymerization of styrene. Moreover, one can notice that the spectrum of Figure 6.1.1c, obtained for the (MeInd)TiF_3/MAO system at the Al/Ti ratio of 2000, is almost similar to that in Figure 6.1.1b (Al/Ti = 300) but has a lower Ti(III) of 78%. Both the EPR spectrum and the redox titration are consistent with the polymerization results as shown in Table 6.1.3; the polymerization activity of fluorinated systems is highest at the Al/Ti ratio of 300, but decreases with either higher or lower Al/Ti ratios. In contrast, the activity and the Ti(III) percentage increase with the Al/Ti ratio for the chlorinated catalyst systems.

The molecular weight of the polymer is controlled by the relative rates of two competing processes, the β-hydride elimination and the styrene insertion.[4] This means that the difference in the molecular weight (Mw) provided by the fluorinated complexes and their chlorinated counterparts is a result of disparities in the two rates. In addition, the molecular weight of the polymer also is dependent on the life of the active sites. The active sites of the fluorinated catalysts are more stable than those of the chlorinated catalysts (*vide supra*), thus increasing the polymer molecular weight. The higher stability probably is provided by the greater electronegativity of the F counterion, which together with MAO and Ti generate a more stable complex.

The fluorinated catalysts are highly syndio-specific. This high syndio-specificity is attributable to the large number of syndio-specific active species, [C_s*], and their high propagation rate constant ($k_{p,s}$); the number of atactic catalytic species, [C_a*], and their rate constant ($k_{p,a}$) are both much smaller. For example, in the (MeInd)TiF_3/MAO catalyst system, these quantities are [C_s*] = 90% and $k_{p,s}$ = 14.8 $(Ms)^{-1}$ as compared with [C_a*] = 8% and $k_{p,a}$ = 1.8 $(Ms)^{-1}$. It was found that the (MeInd)$TiCl_3$/MAO catalyst has [C_s*] = 38% and $k_{p,s}$ = 1.5 $(Ms)^{-1}$, [C_a*] = 6% and $k_{p,a}$ = 1.2 $(Ms)^{-1}$. These differences of [C_s*] and $k_{p,s}$ values between the fluorinated complex and chlorinated one are in good agreement with the differences in the catalytic activity and PS syndiotacticity.

REFERENCES

1. (a) Sinn, H.; Kaminsky, W. *Adv Organomet Chem* 1980, 18, 99; (b) Ewen, J. A.; Jones, R. L.; Razavi, A.; Ferrara, J. D. *J. Am Chem Soc* 1988, 110, 6255; (c) Yang, X.; Stern, C. L.; Marks, T. J. *J. Am Chem Soc* 1994, 116, 10015; (d) Shapira, P. J.; Cotter, W. D.; Schaefer, W. P.; Labinger, J. A.; Bercaw, J. E. *J Am Chem Soc* 1994, 116, 4623; (e) Coates, G. W.; Waymouth, R. M. *Science* 1995, 267; (f) Xu, G. X. *Macromolecules* 1998, 31, 2395; (g) Xu, G. X.; Ruckenstein, E. *Macromolecules* 1998, 31, 4724.
2. (a) Natta, G. *J Polym Sci* 1955, 16, 143; (b) Longo, P.; Grassi, A.; Oliva, L.; Ammendola, P. *Makromol Chem* 1990, 191, 237; (c) Xu, G. X.; Lin, S. A. *Macromol Chem, Rapid Commun* 1994, 15, 873; (d) Xu, G. X.; Lin, S. *A Chin Polym Bull* 1994, 2, 67.

3. (a) Ishihara, N. *Macromol Symp* 1995, 89, 553; (b) Po, R.; Cardi, N. *Prog Polym Sci* 1996, 21, 47; (c) Cartier, L.; Okihara, T.; Lotz, B. *Macromolecules* 1998, 31, 3303; (d) Ishihara, N.; Kuramoto, M.; Uoi, M. *Eur Pat Appl* 224096, 1986; (e) Campbell, R. E., Jr.; Hefner, J. G. Int. Pat. Appl WO 88-10276, 1988.
4. (a) Ishihara, N.; Seimiya, T.; Kuramoto, M.; Uoi, M. *Macromolecules* 1986, 19, 2464; (b) Zambelli, A.; Oliva, L.; Pellechia, C. *Macromolecules* 1989, 22, 2129; (c) Chien, J. C. W.; Salajka, Z. *J Polym Sci, Part A: Chem* 1991, 29, 1253; (d) Soga, K.; Yu, C. H.; Shiono, T. *Macromol Chem, Rapid Commun* 1988, 9, 351; (e) Kuncht, A.; Kuncht, H.; Barry, S.; Chien, J. C. W.; Rausch, M. D. *Organometallics* 1993, 12, 3075; (f) Kaminsky, W.; Lenk, S. *Macromol Chem Phys* 1994, 195, 2093; (g) Pellecchia, C.; Pappalardo, D.; Oliva, L.; Zambelli, A. *J Am Chem Soc* 1995, 117, 6593; (h) Grassi, A.; Zambelli, A. *Organometallics* 1996, 15, 480; (i) Wang, Q.; Quy-oum, R.; Gillis, D. J.; Tudoret, M. J.; Jeremic, D.; Hunter, B. K.; Baird, M. C. *Organometallics* 1996, 15, 693; (j) Foster, P. J.; Chien, J. C. W.; Rausch, M. D. *Organometallics* 1996, 15, 2404; (k) Xu, G. X. *Macromolecules* 1998, 31, 586; (l) Xu, G. X.; Lin, S.A. *Macromolecules* 1997, 30, 685.
5. (a) Zambeli, A.; Oliva, L.; Pellecchia, C.; Grassi, A. *Macromolecules* 1987, 20, 2035; (b) Longo, P.; Grassi, A.; Proto, A.; Ammendola, P. *Macromolecules* 1988, 21, 24.
6. (a) Chien, J. C. W.; Salajka, Z.; Dong, S. H. *Macromolecules* 1992, 25, 3199; (b) Pellecchia, C.; Immirzi, A.; Grassi, A.; Zambelli, A. *Organometallics* 1993, 12, 4473; (c) Xu, G. X.; Lin, S. A. *Acta Polym Sin* 1997, 3, 380; (d) Grassi, A.; Pellecchia, C.; Oliva, L. *Macromol Chem Phys* 1995, 196, 1093.
7. Kaminsky, W.; Lenk, S.; Scholz, V.; Roesky, H. W.; Herzog, A. *Macromolecule* 1997, 30, 7647; (b) Kaminsky, W. *Macromol Chem Phys* 1996, 197, 3907.
8. Herzog, A.; Liu, F. Q.; Roesky, H. W.; Demsar, A.; Keller, K.; Noltemeyer, M.; Pauer, F. *Organometallics* 1994, 13, 1251.
9. (a) Ready, T. E.; Day, R. Q.; Chien, J. C W.; Rausch, M. D. *Macromolecules* 1993, 26, 5822; (b) Ready, T. E.; Chien, J. C. W.; Rausch, M. D. *J Organomet Chem* 1996, 519, 21; (c) Kim, Y.; Koo, B. H.; Do, Y. *J Organomet Chem* 1997, 527, 155.
10. (a) Dawkins, J. V.; Maddock, J. W.; Coupe, D. *J Polym Sci Part A: Polym Chem* 1970, 8, 1803; (b) Smith, W. V. *J Appl Polym Sci* 1974, 18, 3685.
11. (a) Liinas, G. H.; Mensa, M.; Palacios, F.; Royo, P.; Serrano, R. *J Organomet Chem* 1988, 340, 37; (b) Sotoodeh, M.; Leichtweis, Z.; Roesky, H. W.; Noltemeyer, M.; Schmidt, H. G. *Chem Ber* 1993, 126, 913.
12. Gassman, P. G.; Winter, C. H. *J Am Chem Soc* 1988, 110, 6130.
13. (a) Bordwell, F. G.; Satish, A. V. *J Am Chem Soc* 1992, 114, 10173; (b) Bordwell, F. G.; Drucker, G. E. *J Org Chem* 1980, 45, 3325.

6.2 Polyethylene-Palygorskite Nanocomposite Prepared via *In Situ* Coordinated Polymerization*

Junfeng Rong, Miao Sheng, and Hangquan Li

School of Materials Science and Technology, Beijing University of Chemical Technology, Beijing 100029, China

Eli Ruckenstein

Department of Chemical Engineering, State University of New York at Buffalo, Amherst, New York 14260

ABSTRACT A polyethylene/palygorskite nano-composite (IPC composite) was prepared via an *in situ* coordinated polymerization method, using $TiCl_4$ supported on palygorskite fibers as catalyst and alkyl aluminum as co-catalyst. These composites were compared with those prepared by melt blending (MBC composites). It was found that in the IPC composites, nano-size fibers of palygorskite were uniformly dispersed in the polyethylene matrix. In contrast, in the MBC composites, the palygorskite was dispersed as large clusters of fibers. Regarding the mechanical properties of the IPCs, the tensile modulus increased and the elongation at break decreased with increasing fiber content, while the tensile strength passed through a maximum. The tensile strength and elongation at break were much smaller for the MBC composites. The final degree of crystallinity of the IPC composites decreased with increasing palygorskite content. Regarding the kinetics of crystallization, the ratio between the degree of crystallinity at a given time and the final one was a universal function of time. It was found that large amounts of gel were present in the IPC composites and much smaller amounts in the MBC composites.

* *Polymer Composites* 2002, 23, 4.

6.2.1 INTRODUCTION

The field of organic/inorganic nano-composites has enjoyed a rapid development. The high stiffness, the dimension and thermal stability of inorganic materials and the toughness, processability and dielectric properties of polymers can be combined in these nano-composites. Because of the large surface areas of the nano-particles and fibers, the inorganic components can interact strongly with the polymer matrix, resulting in superior properties (1,2).

In order to prepare nano-composites, several approaches have been used. In the simplest, which was employed for reinforcing elastomers (3–5), an organometallic material was absorbed into a crosslinked network, and the swollen elastomer was placed into water containing a catalyst. The hydrolysis resulted in the formation of particles with diameters in the range of 20–30 nm and a narrow size distribution.

Polyethylene (PE) is one of the most widely used polymers. It possesses several valuable characteristics such as high electrical resistance, high ductility, nontoxicity and bio-compatibility (6), in addition to its low density, low cost, recyclability and good processability. However, neither its stiffness nor its low temperature toughness is satisfactory. It is expected that reinforcing PE with nano-size fillers could overcome these shortcomings. Several methods have been employed to disperse montmorillonite in the matrices of various polymers. Montmorillonite was dispersed in polymer solutions (7–10), polymer melts (11,12), monomers or emulsions (13–15). The technique of preparation of polymer/montmorillonite nanocomposites via *in situ* polymerization has been commercialized (16,17). However, montmorillonite can be dispersed only in polar liquids. Because ethylene, the monomer of polyethylene, is gaseous, and both the melt and solutions of polyethylene are nonpolar, none of the methods mentioned above could be applied to polyethylene.

A novel methodology, *in situ* coordinated polymerization (18), was employed in this paper to prepare PE reinforced with nano-size palygorskite fibers. The major component of palygorskite is hydrous magnesium silicate, which is present as fibers with diameters less than 100 nm and lengths ranging from hundreds nm to several μm (Figure 6.2.1). Each Mg atom has two vacancies associated with it, which by reacting with $TiCl_4$, generate bridges between Mg and $TiCl_4$:

$$\text{Mg}\genfrac{}{}{0pt}{}{\square}{\square} + \text{TiCl}_4 \longrightarrow \text{Mg}\genfrac{}{}{0pt}{}{\text{Cl}}{\text{Cl}}\text{Ti}\genfrac{}{}{0pt}{}{\text{Cl}}{\text{Cl}}$$

The catalyst was supported on palygorskite fibers, and used to polymerize ethylene. As the ethylene polymerized, the fibers became covered with polymers and a polyethylene/palygorskite nano-composite was generated. In this paper, the relationships among the preparation method, structure and properties of the obtained nanocomposites were investigated. For comparative purposes, the composites of PE and palygorskite prepared by introducing the latter into a PE melt were also examined in parallel.

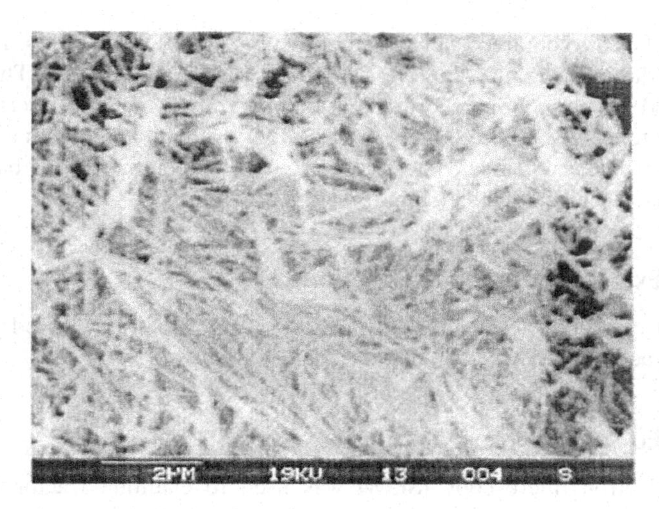

FIGURE 6.2.1 SEM micrograph of palygorskite.

6.2.2 EXPERIMENTAL

6.2.2.1 Materials

Ethylene of polymerization grade was purchased from Yanshan Petrochemical Co., China, and purified over a 4-Å molecular sieve; palygorskite was bought from Jiashan, Anhui, China, the dekalin from Shanghai chemical reagent plant (99%) and the ultra-high molecular weight polyethylene (UHMWPE, $MW = 1{,}200{,}000$) from Yanshan Petrochemical Co., China.

6.2.2.2 Preparation of the Supported Catalyst

Palygorskite fibers were soaked for 6 hours in a Muffle at a temperature of 600°C, and subsequently introduced into a sufficiently concentrated $TiCl_4$ solution in an organic solvent (hexane, heptane, or octane) to adsorb 0.3 wt% Ti. After palygorskite reacted with $TiCl_4$, the fibers were washed three times with hexane at 40°C and dried in N_2 at a temperature of 60°C. It was noticed that the Ti content was not affected by washing, which indicated a strong attachment of $TiCl_4$ to the surface of palygorskite.

6.2.2.3 Preparation of the Composites

The activated palygorskite was placed in a 500 mL flask equipped with a stirrer, into which 200 mL heptane and $Al(iBu_3)$ (Al/Ti = 15/1 atomic ratio) were introduced. Ethylene was supplied under a pressure of 1 atm to carry out the polymerization for 30 min at 45°C. The ethylene polymerized on the surface of the palygorskite. The polyethylene generated wrapped the fibers, resulting in a composite *in situ*. The polymerization was terminated by adding an inhibitor (acid solution in ethanol).

The composites thus obtained were dried in vacuum at 60°C for 6 hours. The content of palygorskite in the composites was varied between 2 and 30 wt%. The composites prepared via the *in situ* coordinated polymerization method described above will be denoted as IPC. The composites prepared via melt blending on a Brabender twin-screw extruder will be denoted as MBC. The temperature of the barrel of the extruder was 180°C and the blending time lasted 10 min.

6.2.2.4 Extraction Test

The solvent extraction test of PE/palygorskite composites was carried out with a Soxhlet extractor at 135°C for 20 hours using dekalin as solvent.

6.2.2.5 Electron Microscopy Observations

Scanning electron microscopy (SEM) was used to examine fractured surfaces obtained by breaking the sample bars in liquid nitrogen. The fractured surfaces were sputter-coated with gold before examination with an electron microscope instrument (Cambridge S250, UK).

Transmission electron microscopy (TEM) was used to examine the morphology of the composites. The specimens were prepared by super-thin slicing, and examined with a Hitachi H-800, Japan instrument operated at 100 kV.

6.2.2.6 Tensile and Impact Tests

The tensile and impact bars were prepared by heat molding. The molding temperature was 180°C, and the molding pressure 8 MPa. The tensile test was performed according to ASTM D638-81 at a strain rate of 50 mm/min with a universal testing instrument (Instron 1185, UK). The Izod impact test was performed according to GB/T 1043-93 with an impact tester XJJ-5, Japan instrument. The tensile test speed was 25 mm/s. Both tests were carried out at room temperature.

6.2.2.7 Thermal Analysis

Thermal gravimetric (TG) analysis was carried out with a Perkin-Elmer TGS-2 thermal balance in a N_2 flow at a heating rate of 20°C/min. The DSC analysis was carried out on a Perkin-Elmer DSC-2C instrument in a N_2 flow at a heating rate of 10°C/min. The degree of crystallinity of the samples was calculated using the equation $X\% = \left(\Delta H_f / \Delta H_f^0\right) \times 100$, where ΔH_f *is* the heat of fusion determined by DSC, and ΔH_f^0 is the heat of fusion of pure PE, which has the value 293 J/g (19).

The rates of isothermal crystallization of the samples were determined by DSC. A sample was first heated rapidly to 250°C, and kept at that temperature for 20 min to eliminate the effect of the specimen's thermal history; subsequently the temperature was rapidly reduced to a particular crystallization temperature, and kept at that temperature for isothermal crystallization. The computer recorded the change of crystallization heat (ΔH_C) with time.

6.2.3 RESULTS AND DISCUSSION

6.2.3.1 Morphology of the Composites

Figures 6.2.2 and 6.2.3 present the TEM micrographs of IPC and MBC, respectively. The white areas represent the PE and the black areas the palygorskite. One can see that nano-composites could be prepared via the *in situ* coordinated polymerization. Because the catalyst was implanted onto the surface of the palygorskite through Ti-Cl-Mg bridges, individual fibers of palygorskite could be well separated by the growing chains. As the polymerization proceeded, the inorganic fibers were wrapped completely by the polymer chains. As a result, the fibers became well dispersed, as shown in Figure 6.2.2. In contrast, the palygorskite could not be dispersed as individual fibers via mechanical mixing, but only as large clusters in the PE matrix (Figure 6.2.3).

The quality of the dispersion of palygorskite in PE is also revealed by the SEM micrographs of the fractured surfaces of samples. For the IPC, the fractured surfaces are relatively fine and smooth, without large holes and knots (Figure 6.2.4). This indicates not only a good dispersion of palygorskite, but also a good adhesion between PE and palygorskite. In contrast, for the MBC (Figure 6.2.5), the surface shows a poor combination between palygorskite and PE. The material appears to be composed of domains loosely packed together, some being clusters of palygorskite and others aggregates of PE.

6.2.3.2 Interaction Between PE and Palygorskite in the Composites

The strength of the interactions between palygorskite and PE can be evaluated from solubility measurements. After washing with a good solvent for PE (dekalin) in an extractor, all the soluble species were removed, leaving only the palygorskite and the insoluble PE. The results of the solubility measurements are listed in Table 6.2.1,

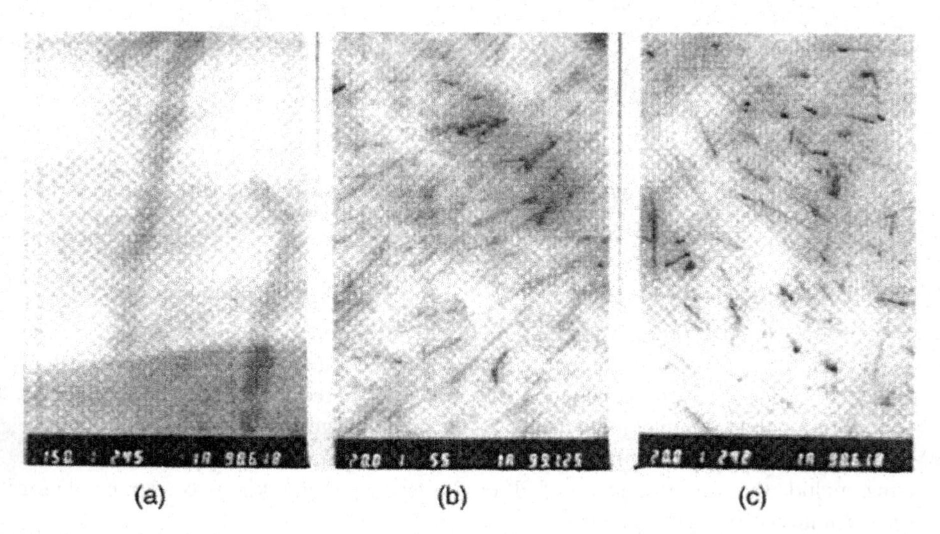

FIGURE 6.2.2 TEM micrographs of the PE/palygorskite nano-composites prepared via in situ coordinated polymerization (magnification: a: 150,000; b, c: 20,000).

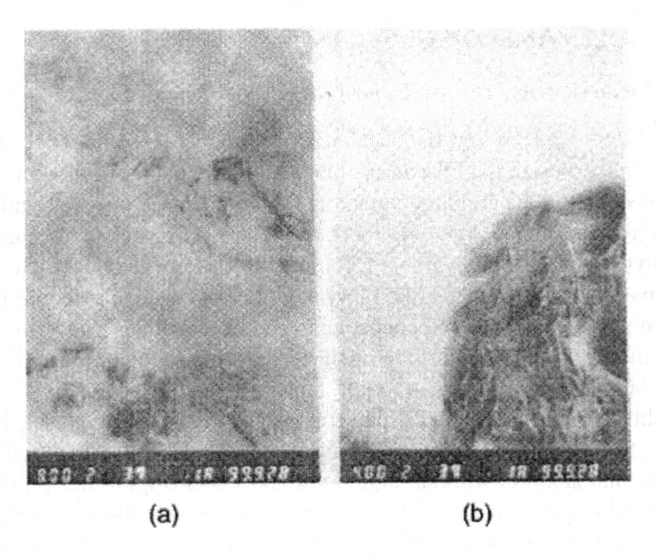

FIGURE 6.2.3 TEM micrographs of the composites prepared via melt blending (magnification: a: 4000; b: 8000).

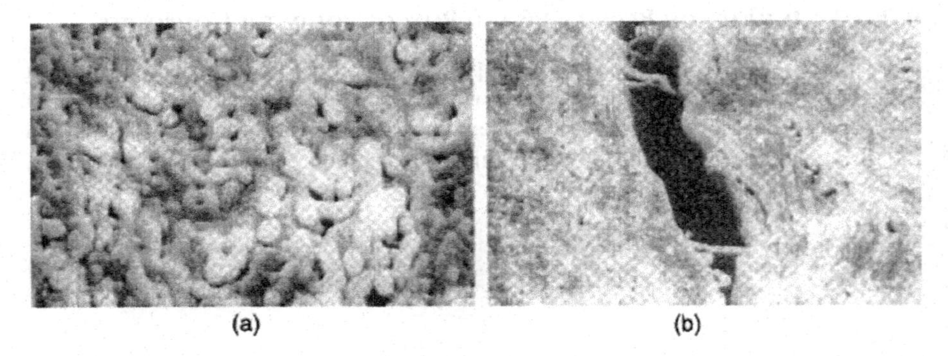

FIGURE 6.2.4 SEM micrographs of impact fractured surface of IPC (magnification: a: 1000; b: 25).

which shows that the preparation method greatly affected the gel content. Indeed, the gel contents of the MBCs were below 5 wt%, whereas those of IPCs were higher than 50 wt%. For the IPCs, the gel content was greatly affected by the content of palygorskite. For a palygorskite content above 20 wt%, the PE gel content exceeded 90 wt%. Since no mechanisms leading to crosslinking can occur during polymerization, one can conclude that the attachment of PE chains to the palygorskite was responsible for the formation of the gel species.

In the IPC, the PE chains grew directly from the surface of the palygorskite via coordinated polymerization; consequently, a large number of chains were rooted

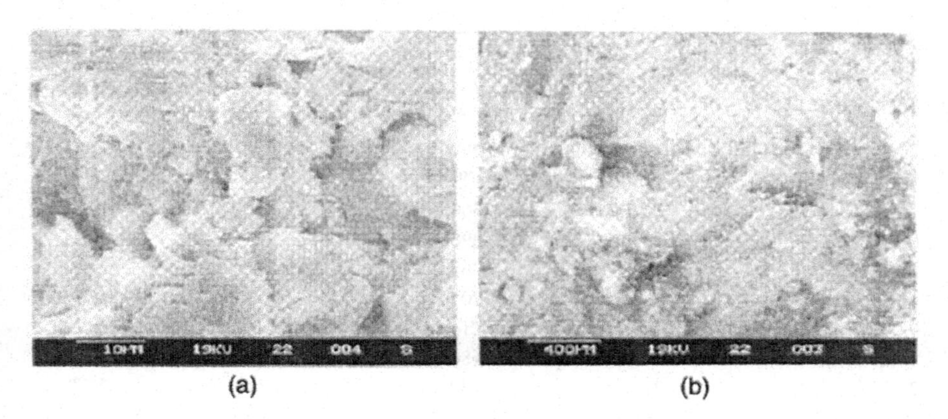

FIGURE 6.2.5 SEM micrographs of impact fractured surface of MBC (scale bar: a: 10 μm; b: 400 μm).

TABLE 6.2.1
The Result of the Dekalin Extraction Test

Sample	Content of the Samples before Extraction (wt%)		Percent of PE Extracted (wt%)	Gel Content (wt%)
	PE	Palygorskite		
Pure PE	100	0	100	0
MBC-1	97.0	3.0	96.9	3.1
MBC-2	92.0	8.0	97.6	2.4
MBC-3	84.0	16.0	99.4	0.6
MBC-4	75.0	25.0	100.0	0
IPC-1	95.6	5.4	46.2	53.8
IPC-2	90.1	9.9	31.0	69.0
IPC-3	79.4	20.6	5.0	95.0
IPC-4	72.0	28.0	6.0	94.0

on one fiber, forming a large cluster. Furthermore, the long chains formed on the same fiber or on different fibers were partly crystallized and entangled together. The tangled chains could be detached only with great difficulty from the inorganic fibers, and this probably explains the high gel content of the IPCs.

6.2.3.3 Mechanical Properties

The mechanical properties of the IPCs are listed in Table 6.2.2, which shows that the tensile modulus and yield strength increase, and the elongation at break decreases in most cases with increasing palygorskite content. It is well known that the elongation of polymer is a result of the movement of the segments. Because the chains are anchored on the fibers and entangled, their motion was greatly inhibited. As a result, the increase in the palygorskite content had negative effects for the elongation at break but positive

TABLE 6.2.2
Mechanical Properties of the IPCs[a]

Palygorskite Content (wt%)	Tensile Strength (MPa)	Elongation at Break (%)	Tensile Modulus (MPa)	Yield Strength (MPa)
0.1	26.9	366.3	322.0	19.7
0.5	28.9	396.8	355.1	20.4
3.1	40.9	266.9	448.6	21.9
5.5	36.9	205.8	545.9	22.4
9.9	31.8	214.2	850.0	25.6
18.7	33.7	180.5	878.5	26.4
24.5	27.2	80.8	1167.6	26.3

[a] The molecular weight of the dekalin extracted PE was between 4 and 5×10^6 and was determined from intrinsic viscosity measurements using the expression $[\eta] = 6.77 \times 10^{-4} MW^{0.67}$. (Ref. 20)

for the yield strength. The effect of the content of inorganic fiber on the tensile strength is more complex. One may notice that the tensile strength exhibits a maximum as the content of palygorskite increases. For a polymeric material, the tensile strength depends on the cohesive energy of the material. When the palygorskite content was relative low, the tensile strength increased with increasing palygorskite content because of the strong interactions generated. However, at large palygorskite contents, too much strain was generated in PE, and the tensile strength decreased with increasing palygorskite content. Regarding the tensile modulus, it is clear that the presence of the reinforcer is responsible for its increase with increasing palygorskite content.

The results the MBCs are presented in Table 6.2.3. As described in the Experimental section, the MBCs listed in Table 6.2.3 were prepared using a twin-screw extruder. Other preparation methods, such as millers and single-screw extruders, were also used to prepare MBCs. It was found that the twin-screw extruder was best at mechanically dispersing the fibers, though the dispersion could not achieve the nanoscale. One may notice that both the tensile strength and the elongation at break are much lower than those of IPCs. This can be easily attributed to the presence of clusters of palygorskite in the PE matrix, which acted as defects in the material. However, the tensile moduli of

TABLE 6.2.3
Mechanical Properties of the MBCs

Palygorskite Content (wt%)	Tensile Strength (MPa)	Elongation at Break (%)	Tensile Modulus (MPa)
3.0	14.0	5.2	572.7
8.0	23.4	33.8[a]	400.6
16.0	20.2	5.1	850.0
25.0	19.8	5.1	816.0

Note: The samples did not exhibit a yield point.

[a] The heterogeneity of the material may explain the large differences in the elongation at break.

the MBCs are of the same order as those of the IPCs, since the modulus is dependent on the chemical composition and not on the microstructure.

The Izod impact strength of the composites could not be determined because both the IPCs and the MBCs contain ultra-high molecular weight polyethylenes, which do not break in the impact test at room temperature.

6.2.3.4 Crystallization Behavior of PE

The temperature employed to study the thermal behavior of the composites was well below the melting point of palygorskite, which is above 1500°C. For this reason, only the degree of crystallinity, the melting point and the crystallization rate of PE were of concern.

The final degree of crystallinity of the IPC composites is plotted against the palygorskite content in Figure 6.2.6 which shows that the higher the palygorskite content, the lower the crystallinity of PE. Intuition suggests that palygorskite fibers provide sites for nucleation, which promote crystallization. However, as already mentioned, the fibers are wrapped by chains and are not accessible to segments from outside layers. The segments that are directly anchored on the fibers are not able to fold to form regular crystallites because they do not possess enough mobility. They, are present in an amorphous state, as illustrated in Figure 6.2.7. For this reason, the fibers are not promoters, but rather inhibitors, of crystallization. Consequently, not only the crystallinity decreased, but also the melting point of PE decreased with increasing palygorskite content (Figure 6.2.8). The anchoring of the PE chains interfered with crystallization and thus became responsible for the lowering of the melting point.

The kinetics of crystallization can be represented using the equation:

$$\theta = \frac{X(t)}{X(t_\infty)} = \frac{\int_0^t \frac{dH(t)}{dt}\,dt}{\int_0^\infty \frac{dH(t)}{dt}\,dt}$$

where $dH(t)/dt$ is the rate of heat flow, $X(t)$ is the crystallinity at time t and $X(t_\infty)$ is the final crystallinity after a sufficiently long time. The DSC experiments provided

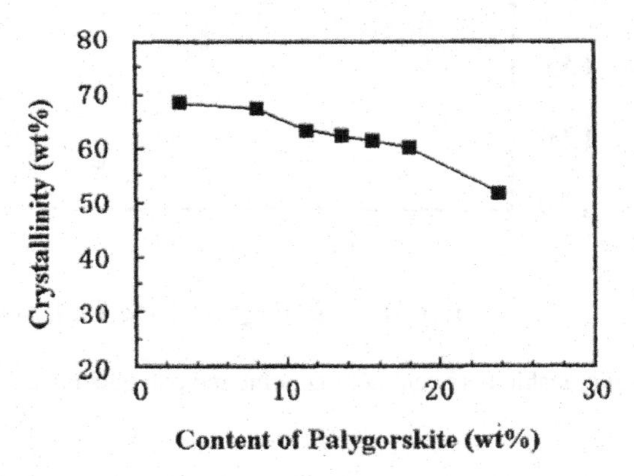

FIGURE 6.2.6 Relationship between crystallinity and palygorskite content for the IPC.

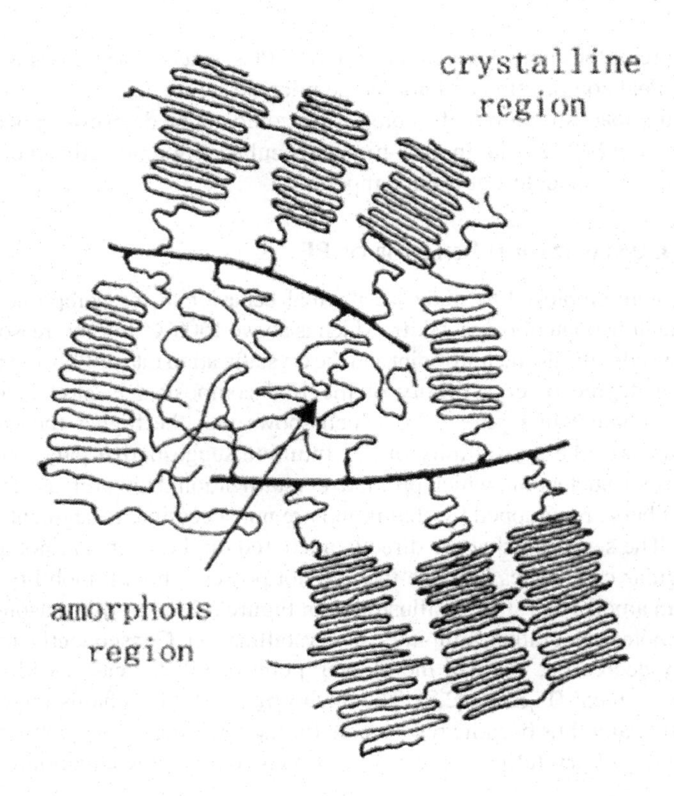

FIGURE 6.2.7 A possible structure of the IPC.

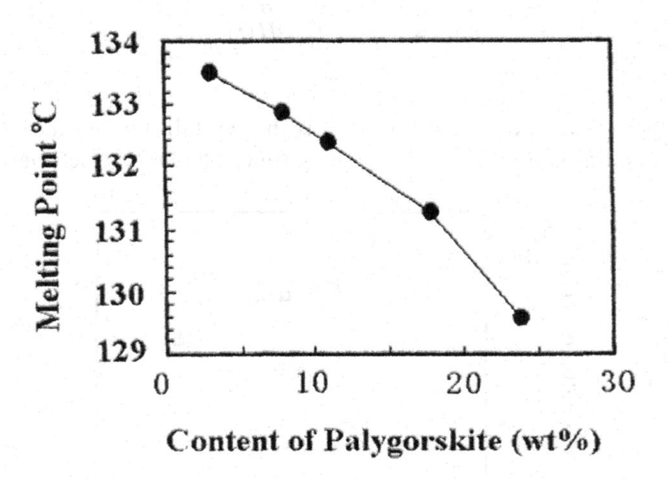

FIGURE 6.2.8 Relationship between melting point and palygorshite content for the IPC.

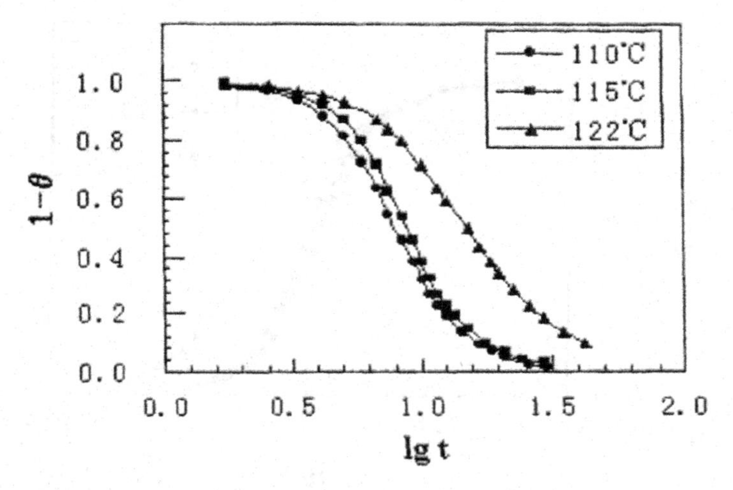

FIGURE 6.2.9 Plot of (1–θ) vs.lgt for IPC (palygorskite content 3 wt%).

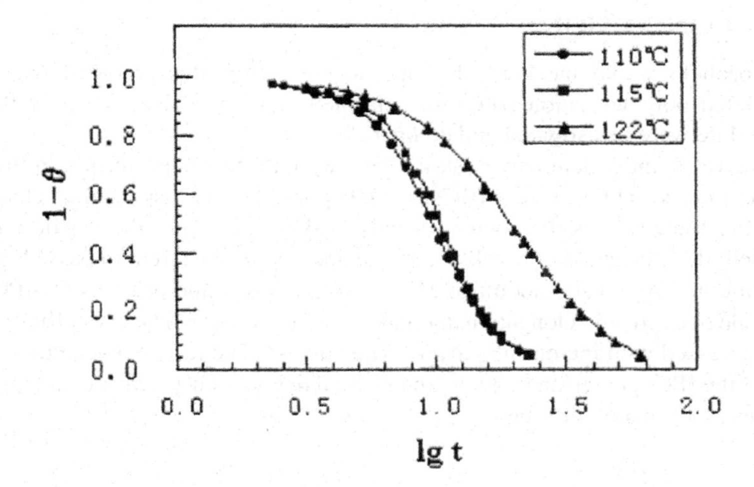

FIGURE 6.2.10 Plot of (1–θ) vs. lgt for IPC (palygorskite content 8 wt%).

the values of $dH(t)/dt$. θ is plotted in Figures 6.2.9 through 6.2.11 as a function of ln$t$. Comparing the three figures, one can conclude that they almost coincide, and hence that θ is a universal function of the time t, independent of palygorskite content. Of course, $X(t)$ is dependent via $X(t_\infty)$ on palygorskite content.

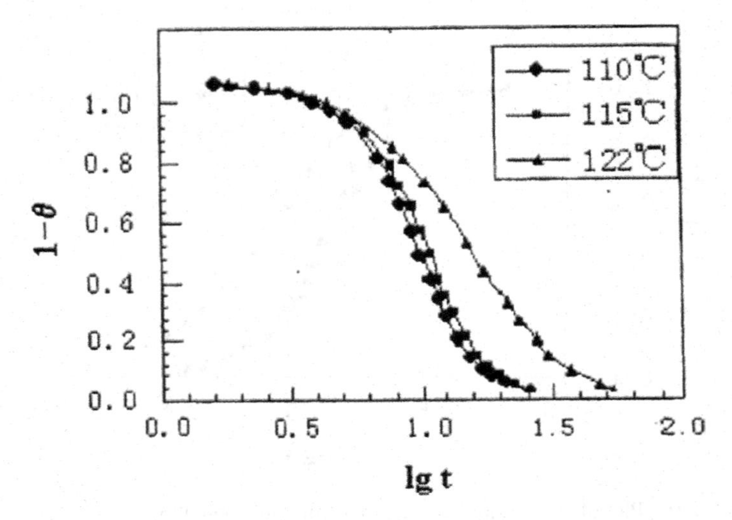

FIGURE 6.2.11 Plot of (1–θ) vs. lgt for IPC (palygorskite content 18 wt%).

6.2.4 CONCLUSION

The morphology and mechanical properties of composites prepared via *in situ* coordinated polymerization (IPC) and via melt blending (MBC) of polyethylene/palygorskite were investigated and compared.

In the IPCs, individual palygorskite fibers were dispersed uniformly in the polyethylene matrix, while in the MBCs, the palygorskite was dispersed as clusters of fibers. For the IPCs, as the content of palygorskite increased, the tensile modulus increased, the elongation at break decreased and the tensile strength passed through a maximum. The tensile moduli of MBCs were of the same order as for IPCs, but the tensile strength and elongation at break were much lower. The crystallinity of the IPCs decreased with increasing palygorskite content. The ratio between the crystallinity of the IPCs at a given moment and its final degree of crystallinity at long times is a universal function of time.

REFERENCES

1. G. Camino, A. Ya. Polishchuk, and M. P. Luda, *Polymer Composites*, **21**, 821 (2000).
2. P. Ingran, *Makromol Chem.*, **267**, 111 (1968).
3. J. E. Mark and B. Erman, *Rubber-like Elasticity, A Molecular Primer*, Wiley-Interscience, New York (1988).
4. J. E. Mark, *Chemtech*, **19**, 230 (1989).
5. J. E. Mark and D. W. Schaefer, *Polymer-Based Molecular Composites*, Materials Research Society, Pittsburgh, PA (1990).
6. P. Blais, *Science*, **121**, 42 (1966).
7. R. A. Vaia, H. Ishii, and E. P. Giannelis, *Chem. Mater.*, **5**, 632 (1993).
8. A. K. Carrado, P. Thiyagarajan, and L. D. Elder, *Clays Clay Miner.*, **14**, 506 (1996).
9. T. Lan and T. J. Pinanavaia, *Chem. Mater.*, **6**, 573 (1994).

10. J. Wu and M. Lemer, *Chem. Mater.*, **5**, 835 (1993).
11. R. A. Vaia, K. D. Jandt, and E. P. Giannelis, *Macromolecules*, **28**, 8086 (1995).
12. R. A. Vaia and E. P. Giannelis, *Macromolecules*, **30**, 7990 (1997).
13. J. E. Pillion and M. E. Thompson, *Chem. Mater.*, **3**, 777 (1991).
14. P. B. Messersmith and S. I. Stupp, *J. Mater. Res.*, **7**, 2599 (1992).
15. M. G. Kanatzids, C. G. Wu, and H. O. Maracy, *Chem. Mater.*, **2**, 222 (1990).
16. G. Lagaly, *Appl Clay Sci.*, **15**, 1 (1999).
17. P. C. Lebaron, Z. Wang, and T. J. Pinnavaia, *Appl Clay Sci.*, **15**, 11 (1999).
18. J. F. Rong, H. Q. Li, Z. H. Jing, X. Y. Hong, and M. Sheng, *J. Appl Polym. Sci*, **82**, 1829 (2001).
19. B. P. Wang, PhD dissertation, University of Massachusetts, Boston, MA (1989).
20. J. Brandrup and E. H. Immergut, eds., *Polymer Handbook*, 3rd Ed., John Wiley & Sons, New York (1989).

6.3 Dendritic Polymers from Vinyl Ether*

Hongmin Zhang and Eli Ruckenstein

Chemical Engineering Department,
State University of New York at Buffalo,
Amherst, New York 14260, USA

ABSTRACT 1-[(2-Vinyloxy)ethoxy]ethyl acetate (1) was prepared by the addition reaction between ethylene glycol divinyl ether and acetic acid. 1 contains both a cationically polymerizable C=C double bond and a dormant initiating moiety for cationic polymerization. It can, therefore, undergo self-condensing cationic polymerization in the presence of a Lewis acid activator, such as zinc chloride. Using this procedure, a novel dendritic polymer consisting of vinyl ether was prepared and its hyperbranched molecular structure confirmed by FT-IR and ^{1}H NMR spectra.

6.3.1 INTRODUCTION

The dendritic polymers have received increased attention in recent years, because of their quite different macromolecular architecture compared to that of the traditional linear polymers, and of their unusual properties (1–4). The special hyperbranched structure and unique properties of these polymers can be exploited to design novel polymer materials which possess unique viscoelastic properties (5,6), can form unusual blends (7,8) and can be employed as molecular carriers (9,10).

The dendritic polymers were generally prepared step by step using both divergent and convergent methods (11,12). However, these complex multistep procedures limit the capability for preparing the dendritic polymers. Recently, Fréchet and his coworkers (13,14) developed a convenient method, called self-condensing vinyl polymerization, to prepare dendritic polymers with broad molecular weight distributions. They employed a vinyl monomer, 3-(l-chloroethyl)-ethenylbenzene, which, containing both a polymerizable C=C double bond and a dormant initiating moiety, can undergo self-condensing in the presence of an activator ($SnCl_4$) to generate a dendritic polymer. Several styrene type dendritic polymers were thus prepared (13,14).

* *Polymer Bulletin* 1997, 39, 399–406.

Higashimura and coworkers (15,16) found that the adduct between an alkyl vinyl ether and an acetic acid derivative (RCOOH, R = CF_3, $CC1_3$, $CHC1_2$, CH_2C1, CH_3) can be used as initiator for the living cationic polymerization of vinyl ether in the presence of $ZnCl_2$. This suggested to us to prepare a novel dendritic polymer based on vinyl ether. The vinyl monomer, l-[(2-vinyloxy)ethoxy]ethyl acetate (1; in Scheme 6.3.1), which is an adduct between ethylene glycol divinyl ether and acetic acid, was first prepared. 1 contains both a cationically polymerizable C=C double bond and an initiating moiety, whose structure is similar to that of the initiating moiety in the cationic polymerization earned out by Higashimura et al. (15,16). The initiating moiety of 1 can be activated by a Lewis acid ($ZnCl_2$) to give a partly dissociated carbocation which is able to initiate the cationic polymerization of the vinyl groups. There are two possible routes for the polymerization of 1. As shown in Scheme 6.3.1, route a provides a linear polymer (2). In this case, 1 acts just as a monomer. According to route b, 1 undergoes a self-condensing polymerization. One initiating moiety reacts with the double bond of another 1 molecule to form a dimer (3) which, having one double bond and two reactive initiating moieties, can further react with the initiating moieties and/or the double bonds of other molecules. If it just reacts with the two double bonds of 1, a tetramer (4) will be produced. Further condensations involving the activated 1, dimers and oligomers, will generate a highly branched dendritic polymer. In this paper, the polymerization of 1 was proved to proceed according to route b.

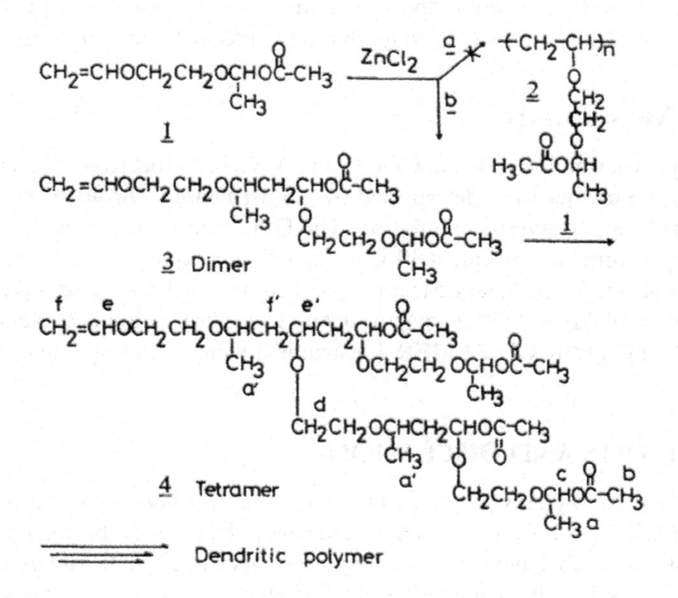

SCHEME 6.3.1

6.3.2 EXPERIMENTAL

6.3.2.1 MATERIALS

Toluene (solvent) was washed with concentrated sulfuric acid, then with water, dried over $MgSO_4$ and distilled twice over calcium hydride just before use. Et_2O was dried with CaH_2 and distilled in the presence of $LiAlH_4$. $ZnCl_2$ (Aldrich, 1.O M solution in diethyl ether) was diluted with purified Et_2O.

6.3.2.2 SYNTHESIS OF 1-[(2-VINYLOXY)ETHOXY]ETHYL ACETAE (1)

1 was prepared through the reaction between ethylene glycol divinyl ether (EGDE; Aldrich, 97%) and acetic acid (AA; Aldrich, 99.8%). AA (22.2 g, 0.37 mol) was drop-wise added to EGDE (45.7 g, 0.39 mol) with magnetic stirring at 60°C and kept at this temperature for 8 h. The pure 1 was obtained as a colorless oil by careful distillation under reduced pressure.

6.3.2.3 POLYMERIZATION PROCEDURE

The polymerization was carried out in a 100 mL round-bottom glass flask, under nitrogen, with magnetic stirring. The reaction was initiated by adding an Et_2O solution of $ZnCl_2$ to a solution of 1 with a dry syringe, at a selected temperature. The polymerization was terminated with methanol (ca. 2 mL) containing a small amount of ammonia. The quenched reaction mixture was washed with 10 wt% aqueous sodium thiosulfate solution and then with water, evaporated to dryness under reduced pressure, and vacuum dried to obtain the product polymer.

6.3.2.4 MEASUREMENTS

^1H NMR spectra were recorded in $CDC1_3$ on a VXR-400 spectrometer. The $\bar{M}_n$ and $\bar{M}_w/\bar{M}_n$ of the polymer were determined by gel permeation chromatography (GPC) on the basis of a polystyrene calibration. The GPC measurements were carried out with tetrahydrofuran as solvent, at 30°C, using two polystyrene gel columns (Waters, Linear) connected to a Waters 515 precision pump. FT-IR spectra were recorded on a PERKIN-ELMER 1760-X spectrometer. The thermogravimetric analysis was carried out with a PERKIN-ELMER-7 instrument using a heating rate of 10°C/min.

6.3.3 RESULTS AND DISCUSSION

1-[(2-Vinyloxy)ethoxy]ethyl acetate (1) is an adduct between ethylene glycol divinyl ether (EGDE) and acetic acid (AA). However, the reaction between EGDE and AA generates not only 1, but also a by-product between one EGDE and two AA molecules, as well as some unreacted EGDE. In the polymerization of 1, the presence of the unreacted EGDE results in gelation because EGDE ($CH_2{=}CHOCH_2CH_2OCH{=}CH_2$) has two double bonds. This was proved by mixing 1 with EGDE in the proportion 9/1 wt/wt. The polymerization of this mixture in the presence of $ZnCl_2$ produced a gel which could not be dissolved in any solvent. It is, therefore, necessary to separate 1 from the reaction mixture carefully. This was carried out by many successive distillations under reduced pressure. Just prior to polymerization, 1 was again doubly distilled in the presence of CaH_2. A 40% yield based on AA was achieved (bp: 41°C at 0.1 mmHg; purity > 99%). Comparing the FT-IR spectrum of the prepared 1 (B, Figure 6.3.1) with that of EGDE (A, Figure 6.3.1), one can note the presence of the carbonyl stretching band at 1739 cm^{-1}. As shown in Figure 6.3.2A, the chemical shifts and their intensities in the ^1H NMR spectrum of the prepared 1 are consistent with its molecular structure.

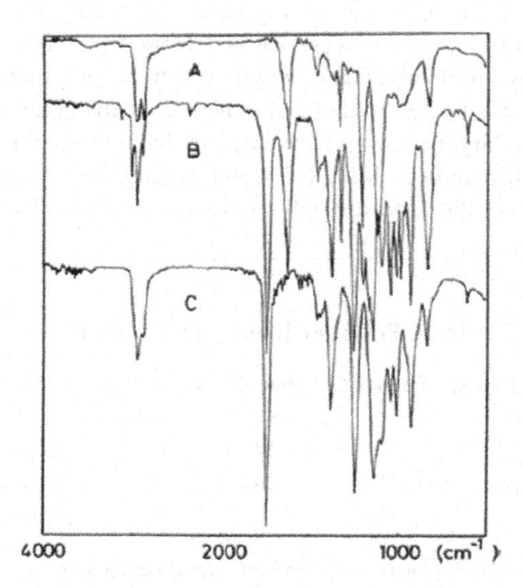

FIGURE 6.3.1 FT-IR spectra of EGDE (A), 1 (B) and the dendritic polymer of 1 (C; no. 1 in Table 6.3.1).

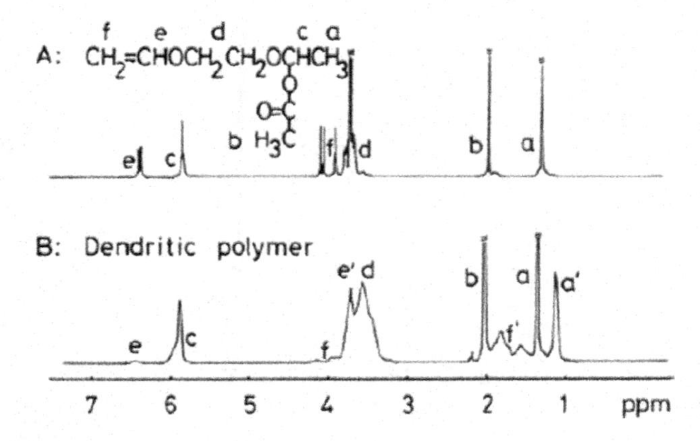

FIGURE 6.3.2 ^1H NMR spectra of 1 (A) and the dendritic polymer of 1 (B refers to the molecular structure of tetramer in Scheme 6.3.1). The polymerization was carried out in toluene at 0°C for 60 min. [1] = 0.57 M, [$ZnCl_2$] = 6.3 mM.

The polymerization of 1 was carried out in the presence of $ZnCl_2$ in toluene at 0 or 20°C. After adding $ZnCl_2$ to the toluene solution of 1, the polymerization system was transparent. As the polymerization proceeded, a red color appeared sometimes gradually, depending on the concentration of $ZnCl_2$ and the polymerization temperature. For instance, when $[ZnCl_2]$ = 25.0 mM at 0°C (no. 1, Table 6.3.1), the red color emerged after 5 h. At the same temperature, but for $[ZnCl_2]$ = 12.5 mM (no. 2, Table 6.3.1), it took 12 h. When the polymerization temperature was raised to 20°C (no. 3, Table 6.3.1), the red color appeared after 8 h. If the concentration of $ZnCl_2$ was too low (no. 4, Table 6.3.1), no color change was observed during polymerization.

As depicted in the GPC traces of Figure 6.3.3, during the initial period, low molecular weight oligomers (b) were formed from 1 (a). Then, the condensation generated gradually high molecular weight dendritic polymers (c). At the same concentration of 1 (no. 1 to 3, Table 6.3.1), the higher the concentration of $ZnCl_2$ or the temperature, the larger the weight-average molecular weight and the broader the molecular weight distribution. Table 6.3.1 and Figure 6.3.3 show that the molecular weight distributions of the dendritic polymers are broad $\left(\bar{M}_w/\bar{M}_n = 5.4\text{–}9.8\right)$. This is

TABLE 6.3.1
Preparation of a Dendritic Polymer from 1 in Toluene

No.	$[1]_0$ (M)	$[ZnCl_2]$ (mM)	Temp. (°C)	Time (h)	Conv. (%)	$10^{-3}\bar{M}_w$ [a]	$10^{-3}\bar{M}_n$ [a]	$\bar{M}_w/\bar{M}_n$ [a]
1	0.57	25.0	0	24	96	44.3	4.51	9.8
2	0.57	12.5	0	24	95	23.4	4.32	5.4
3	0.57	12.5	20	24	99	27.3	3.50	7.8
4	0.22	2.0	0	42	91	32.6	3.80	8.6

[a] Determined from GPC measurements based on a polystyrene calibration.

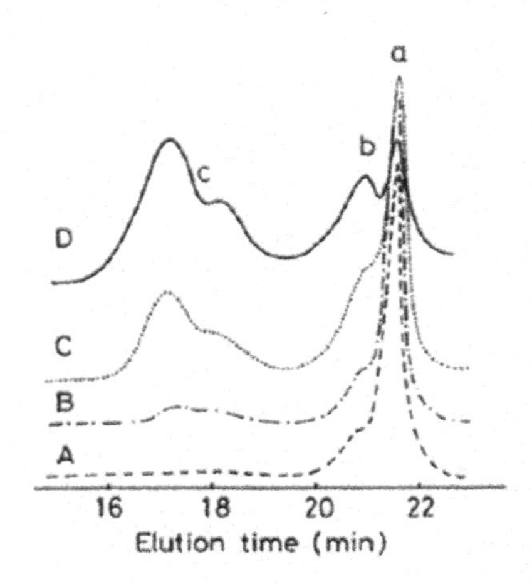

FIGURE 6.3.3 GPC traces of the oligomers and dendritic polymers of 1 at various reaction times. Polymerization in toluene at 0°C for 5 min (A), 45 min (B), 3.5 h (C) and 23 h (D), $[1]_0$ = 0.22 M, $[ZnCl_2]$ = 2.0 mM. Peak a: 1, peak b: oligomers, peak c: dendritic polymers.

mainly due to the self-condensing vinyl polymerization (13), which because of its complex propagation provides dendritic polymers with broad molecular weight distributions. In addition, the broad molecular weight distribution can be also related to the structure of 1. According to Higashimura et al. (15,16), in the cationic polymerization of vinyl ethers initiated by $CH_3CH(OiBu)OCOR$, a broad molecular weight distribution is obtained when $R{=}CH_3$. The effect of the monomer structure on the molecular weight distribution of the dendritic polymer is now investigated by using the monomer $CH_2{=}CHOCH_2CH_2OCH(CH_3)OCOCF_3$.

The dendritic polymer obtained is soluble in methanol, acetone, acetic acid, ethyl ether, tetrahydrofuran, benzene, toluene, dichloromethane, chloroform, carbon tetrachloride, *N,N′*-dimethylformamide and methyl sulfoxide, but is insoluble in water, 1-butanol, hexane and cyclohexane. As shown in Figure 6.3.1, the absorption of C=C double bond in 1 (B) at 1638 cm^{-1} almost disappeared in the IR spectrum (C) of the dendritic polymer, indicating that only a trace amount of C=C double bonds remained unreacted. This trace amount arises because each molecule of the dendritic polymer possesses one double bond (see Scheme 6.3.1). This result is consistent with the ^1H NMR measurements. As shown by Figure 6.3.2, very weak absorptions of $CH_2{=}CH$ double bonds (f, e) can still be detected. The obvious evidence for the formation of the dendritic polymer is that the absorption at 1.37 ppm (a) of methyl in 1 was split into two peaks a (1.37 ppm) and a′ (1.14 ppm) after polymerization, corresponding to the methyls (a′) inside the molecules and those (a) at the periphery of the dendritic polymer, respectively (see the molecular structure of the tetramer in Scheme 6.3.1). It was, furthermore, observed that the intensity of the absorption a′ increased gradually as the polymerization proceeded. If the polymerization would have proceeded by route a (Scheme 6.3.1), the peak a′ would have had to be absent in the ^1H NMR spectrum. These results indicate that the polymerization of 1 proceeded according to route b (Scheme 6.3.1) and that a dendritic polymer with hyperbranched structures was indeed obtained.

Figure 6.3.4 presents the thermogravimetric curve (B) of the dendritic polymer. For comparison, the thermogravimetric curve (A) of poly(vinyl ethyl ether) (PVEE) is also depicted. The decomposition of PVEE exhibits one step from 421°C to 558°C, corresponding to the decomposition of the carbon-carbon main chain. However,

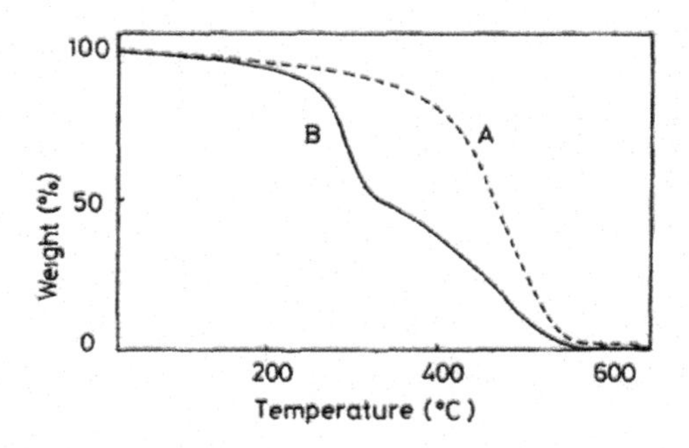

FIGURE 6.3.4 Thermogravimetric curves of poly(VEE) (A; Aldrich, Catalog No. 18265-6) and the dendritic polymer of 1 (B; no. 1 in Table 6.3.1).

the decomposition of the dendritic polymer is much more complicated. The first step from 259°C to 314°C can be attributed to the groups ($CH_3CHOCOCH_3$) at the periphery (see the molecular structure of the tetramer in Scheme 6.3.1). The second step from 314°C to 545°C can be attributed not only to the carbon-carbon, but also to the carbon-oxygen chains, and for this reason the decomposition temperature is lower than that of the linear PVEE. These results also confirm the complicated molecular structure of the dendritic polymers obtained from 1.

REFERENCES

1. Fréchet JMJ (1994) *Science* 263: 1710.
2. Tomalia DA (1994) *Adv Mater* 6: 529.
3. Newkome GR (1993) *Advances in Dendritic Macromolecules*. JAI Press, Greenwich, CT.
4. Kim YH, Webster OW (1990) *J Am Chem Soc* 112:4592.
5. Mourey TH, Turner SR, Rubenstein M, Fréchet JMJ, Hawker CJ, Wooley KL (1992) *Macromolecules* 25:2401.
6. Hawker CJ, Farrington P, Mackay M, Fréchet JMJ, Wooley KL (1995) *J Am Chem Soc* 117:6123.
7. Kim YH, Webster OW (1992) *Macromolecules* 25: 2501.
8. Massa DJ, Shriner KA, Turner SR, Voit BI (1995) *Macromolecules* 28: 3214.
9. Jansen JF, de Brabander van den Berg EM, Meijer EW (1994) *Science* 266: 1226.
10. Jansen JF, Meijer EW, de Brabander van den Berg EM (1995) *J Am Chem Soc* 117: 4417.
11. Tomalia DA, Durst HD (1993) *Top Curr Chem* 165: 193.
12. Hawker CJ, Fréchet JMJ (1990) *J Am Chem Soc* 112: 7638.
13. Fréchet JMJ, Henmi M, Gitsov I, Aoshima S, Leduc Mr, Grubbs RB (1995) *Science* 269:1080.
14. Hawker CJ, Fréchet JMJ, Grubbs RB, Dao J (1995) *J Am Chem Soc* 117: 10763.
15. Kamigaito M, Sawamoto M, Higashimura T (1991) *Macromolecules* 24: 3988.
16. Aoshima S, Higashimura T (1989) *Macromolecules* 22: 1009.

6.4 SiO_2–Poly(amidoamine) Dendrimer Inorganic/Organic Hybrids*

Eli Ruckenstein and Wusheng Yin

Department of Chemical Engineering, State University of New York at Buffalo, Buffalo, New York 14260

ABSTRACT SiO_2-poly(amidoamine) (PAMAM) dendrimer hybrids were synthesized via (1) a Michael addition reaction between the dendrimer and 3-(trimethoxysilyl) propyl acrylate, (2) the dissolution of the formed compound in methanol, and (3) the mixing of the latter solution with a methanol solution of partly hydrolyzed tetraethylorthosilicate (TEOS) and its casting on a glass substrate. ^1H NMR analysis indicated that in the first step, 77% of the secondary amines were converted into tertiary amines when the fourth-generation dendrimer was employed and 46% were converted when the second-generation dendrimer was used. The final SiO_2-PAMAM dendrimer hybrids were obtained via the hydrolysis and condensation of the compound obtained via the Michael addition and the methanol solution of partly hydrolyzed TEOS. The compartmentalized structure of the hybrids due to the compartments of the dendrimers could be controlled by changing the dendrimer and the amount of TEOS. Scanning electron microscopy and transmission electron microscopy micrographs provided information about the structure of the hybrids. Like the PAMAM dendrimer, the SiO_2-PAMAM dendrimer hybrids exhibited a high metal ion complexing capacity because of the presence of the compartments of the dendrimer; they can be, however, much more easily handled, and, as demonstrated by thermogravimetric experiments, have much higher thermal resistance.

* *Journal of Polymer Science: Part A: Polymer Chemistry* 2000, 38, 1443–1449.

6.4.1 INTRODUCTION

To obtain mechanically and thermally improved polymer materials, hybrids, consisting of an organic polymer matrix and inorganic oxide, have been prepared and investigated.[1–8] A low-temperature sol-gel process of an organometallic precursor, such as silicate, titanate, or aluminate, has been a convenient technique for the synthesis of the inorganic constituent of these hybrids. Tetraethylorthosilicate (TEOS) is the most common precursor employed because it yields a glassy silica network by hydrolysis and condensation under mild conditions. Some of these hybrids prepared by *in situ* hydrolysis and polycondensation of silicon alkoxide in the presence of an organic polymer exhibited high improvements in mechanical properties.[9,10] Recently, some interesting hybrids of silicon oxide and conjugated polymers have been reported that exhibited improved electroluminescence efficiency.[11,12] When silicon oxide was employed in the preparation of light-emitting-diode devices, their durability was improved.[11,12]

The dendrimers have well-defined, branched and compartmentalized structures in the nanosize range and thus exhibit some unique properties. A dendrimer was used to generate spherical or cylindrical polymers by its grafting onto poly(methyl acrylate).[13] Organic/inorganic hybrids have been prepared with hyperbranched polymers as the organic constituent and cubic silsesquioxanes as the inorganic component.[14,15] In this study, we designed and synthesized an inorganic/organic SiO_2–dendrimer hybrid by imbedding the dendrimer into a SiO_2 network, thus incorporating the compartmentalized structure of the dendrimer into the hybrid. This approach allows one to incorporate a variety of well-defined, nanosize dendrimers into SiO_2 networks and to develop some functional inorganic/organic hybrids that might be employed as membranes for separation processes and as catalytic substrates. In this article, we report the synthesis of SiO_2-poly(amidoamine) (PAMAM) dendrimer hybrids and their characterization. Such hybrids are employed to extract Cu^{2+} ions from aqueous solutions.

6.4.2 EXPERIMENTAL

6.4.2.1 MATERIALS

Calcium hydride (CaH_2), methanol solutions of PAMAM dendrimers [20 wt% Generation 2 (G2) and 10 wt% Generation 4 (G4)], methanol, tetrahydrofuran (THF), TEOS, dodecylbenzene sulfonic acid (DBSA), and 3-(trimethoxysilyl) propyl acrylate (TPA) were purchased from Aldrich. Prior to use, THF was dehydrated over CaH_2, whereas the other chemicals were used as received.

6.4.2.2 MICHAEL ADDITION OF TPA TO THE PAMAM DENDRIMER

All glassware was dried before use. A rubber-septum-capped reaction vessel was first subjected to a high vacuum (20 mbar), after which high-purity nitrogen was allowed to flow. A 20 wt% G2 PAMAM dendrimer solution in methanol (1 g of solution, 0.06 mmol of dendrimer, 0.98 mmol of $-NH_2$), 0.62 g of TPA (2.4 mmol), and 5 mL of anhydrous THF or a 10 wt% G4 PAMAM dendrimer solution in methanol (1 g of solution, 0.007 mmol of dendrimer, 0.45 mmol of $-NH_2$), 0.274 g of TPA (1.1 mmol), and 5 mL of anhydrous THF were introduced into the reaction vessel with dry syringes. The reaction was carried out with magnetic stirring in an oil bath at 50°C for 24 h. On the completion of the reaction, the solvent was removed with dry nitrogen, which was followed by vacuum (about 30 mbar).

6.4.2.3 Synthesis of the TEOS-Based Precursor

The $\equiv SiOCH_3$ group formed via the addition reaction can be more easily hydrolyzed than the $\equiv SiOCH_2CH_3$ group of TEOS. To ensure a higher polycondensation of the $\equiv SiOCH_3$ with TEOS, the latter was first partly hydrolyzed. TEOS (5 g), 10 g of methanol, 0.14 g of H_2O, and 0.01 g of DBSA were introduced into a reaction vessel. The mixture was refluxed for 1 h, and then the vessel was capped with a rubber septum. The following reactions can be considered to take place:

$$\begin{aligned} &\equiv SiOCH_2CH_3 + H_2O \\ &\rightarrow \equiv SiOH + HOCH_2CH_3 \end{aligned} \tag{6.4.1}$$

$$\begin{aligned} &\equiv SiOH + \equiv HOCH_2CH_3 \\ &\rightarrow \equiv SiOSi \equiv + SiOCH_2CH_3 \end{aligned} \tag{6.4.2}$$

$$2 \equiv SiOH \rightarrow \equiv SiOSi \equiv + H_2O \tag{6.4.3}$$

6.4.2.4 Preparation of the SiO_2-PAMAM Dendrimer Hybrids

A 10 wt% solution in methanol of the addition product was mixed with the TEOS-based precursor in selected ratios. After mixing, the solution was cast onto a glass substrate and dried at room temperature for 3 days to obtain the SiO_2-PAMAM dendrimer hybrid.

6.4.2.5 Measurements

^1H NMR spectra of the addition products were recorded on an INOVA-500 NMR instrument with $CDCl_3$ as the solvent. The Fourier transform infrared (FTIR) spectrum of the SiO_2-PAMAM dendrimer hybrid was recorded on a Perkin-Elmer (Model 1760-X) instrument. The morphological characteristics of the samples were examined by scanning electron microscopy (SEM). The specimens were frozen in liquid nitrogen, fractured, mounted on a sample holder, and coated with gold. They were then examined with a HITACHI S-800 instrument. The samples for transmission electron microscopy (TEM) were prepared via the dipping of copper grids into dilute solutions of the addition product, followed by drying in air for 1 day. They were then examined with a GEOL, model JEM-20100 instrument. The thermal degradation was investigated with a Perkin-Elmer 7 thermogravimetric analyzer (TGA). A 10-mg film was heated in air from 30°C to 500°C at a heating rate of 10.0°C/min.

6.4.2.6 Complexing Capacity of the SiO_2-PAMAM Dendrimer Hybrids

The SiO_2–PAMAM dendrimer hybrid (44 mg) was soaked into 40 mL of a 0.1 M $CuBr_2$ aqueous solution. UV spectroscopy was employed to determine the time dependence of the Cu^{2+} ion complexing by the hybrid.

6.4.3 RESULTS AND DISCUSSION

6.4.3.1 Synthesis of the Addition Product

In the first step, TPA was allowed to react with the PAMAM dendrimer with anhydrous THF as the solvent at 50°C for 24 h (Scheme 6.4.1). Even though a 1.5 molar excess of TPA was employed, the Michael addition of TPA to the primary amine group of the dendrimer still yielded two adducts (as indicated by ^1H NMR), namely, Compound A with a secondary amino group and Compound B with a tertiary amino group (Scheme 6.4.2). The ^1H NMR spectrum of the addition product of the G4 dendrimer is presented in Figure 6.4.1. The peaks can be assigned as follows: the multi-absorption peak at 0.78 ppm to the protons of $CH_2\underline{CH_2}Si$, the peak at 1.8 ppm to the protons of $\underline{CH_2}CH_2Si$, the multipeaks between 2.3 and 3.8 ppm to the protons of the PAMAM dendrimer,[16] the single peak at 3.45 ppm to the protons of $-Si(O\underline{CH_3})_3$, the peak at 4.18 ppm to the protons of $-COO-\underline{CH_2}-$, the peak at 4.38 ppm to the protons of $CH_2{=}CH-COO-\underline{CH_2}$, and the peaks between 5.8 and 6.4 ppm to the protons of $\underline{CH_2}{=}\underline{CH}-COO-CH_2$.[17] On the basis of the proton peak intensity ratio of $-COO-\underline{CH_2}$ to $CH_2{=}CH-COO-\underline{CH_2}$, one can conclude that 12 mol% unreacted

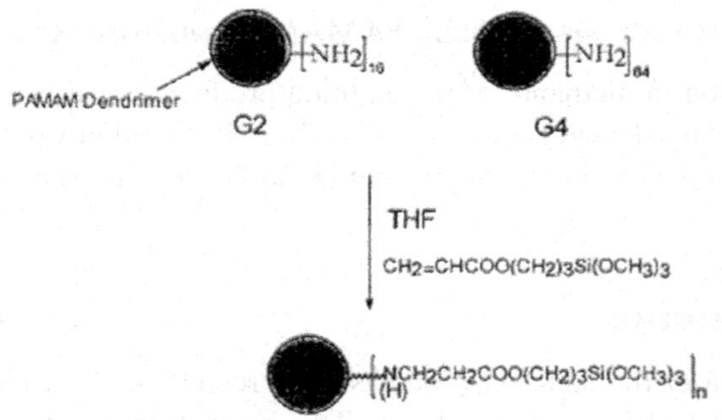

SCHEME 6.4.1 As indicated by ^1H NMR, n is 24 for G2 and 114 for G4.

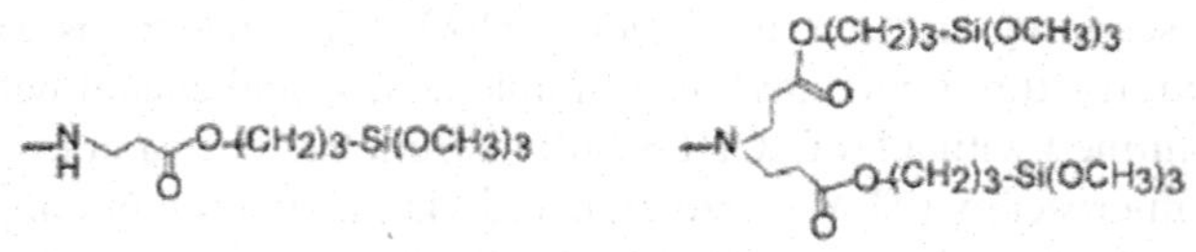

SCHEME 6.4.2

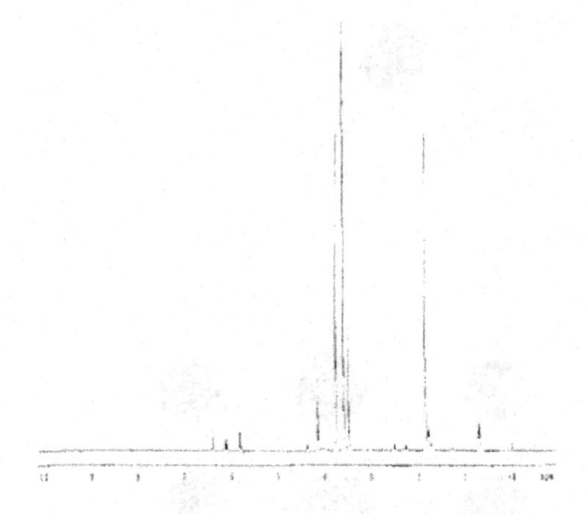

FIGURE 6.4.1 ^1H NMR spectrum of the addition product between the fourth-generation PAMAM dendrimer and TPA.

TPA was left in the addition product involving the G4 dendrimer and 77% of the secondary amino groups (>NH) were converted into tertiary amino groups (>N—); no NH_2 groups were detected by NMR). The average molecular weight (evaluated starting from ^1H NMR data) of the addition product was 40,700. Similarly, 14 mol% TPA was left in the addition product based on the G2 dendrimer, 46% of the secondary amino groups were converted into tertiary groups, and the average molecular weight was 8700.

6.4.3.2 SiO_2-PAMAM Dendrimer Hybrid

The SiO_2-PAMAM dendrimer hybrid film was obtained from the mixed solution of the addition compound and TEOS by casting onto a glass substrate; this was followed by drying in air for several days (Scheme 6.4.3). The hybrid film was insoluble in methanol or THF, whereas the addition product could be dissolved in either of them. Figure 6.4.2 presents the FTIR spectrum of the SiO_2–PAMAM dendrimer hybrid G4Si. The absorption peak at 1745 cm^{-1} can be assigned to the C═O stretching vibration of the $-CH_2CH_2\underline{CO}O(CH_2)_3Si(OCH_3)_3$ segment, and the peak at 1650 cm^{-1} can be assigned to the C═O stretching vibration inside the PAMAM dendrimer segment of the addition product.[17] These results agree with the structure identified by ^1H NMR. The formation of the SiO_2 network was proven by the absorption peak at about 1030 cm^{-1}, which can be assigned to the Si—O—Si stretching vibration.[17]

The SEM micrographs of the SiO_2–PAMAM dendrimer hybrids prepared with various ratios of the addition product to TEOS are presented in Figure 6.4.3. Without TEOS, the addition product was crosslinked by the TPA moieties only. Two-phase structures are observed in that figure. Of course, these structures are present because of the incorporation of the compartmentalized PAMAM dendrimer in the SiO_2 network. The size of the dispersed phase for G2Si (about

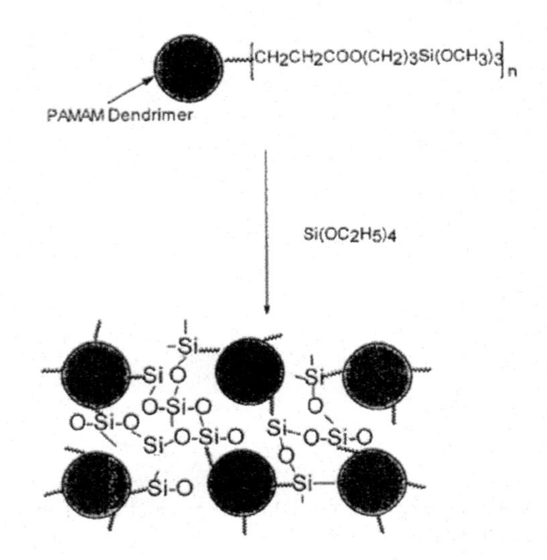

SCHEME 6.4.3 SiO_2-PAMAM inorganic/organic hybrid: schematic presentation of the formation of the hybrid.

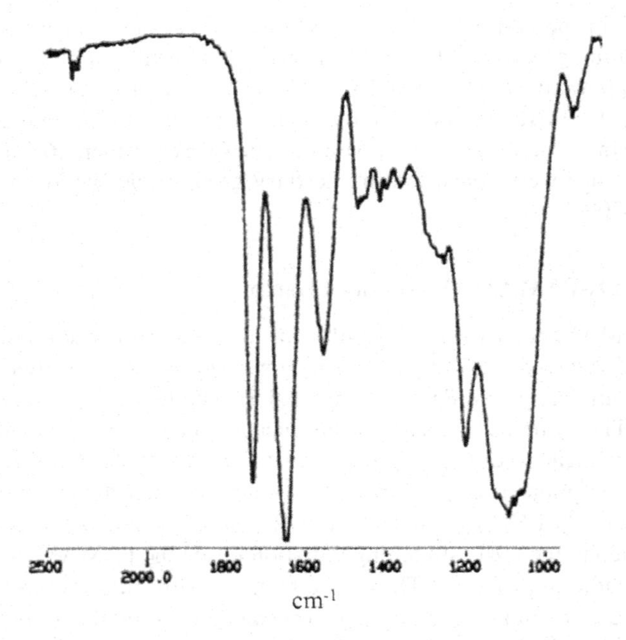

FIGURE 6.4.2 FTIR spectrum of the SiO_2–PAMAM dendrimer hybrid (G4Si).

0.2 μm) is smaller than that of G4Si (3 μm). The increase in the amount of TEOS is expected to affect the domains of the dispersed phase. With increasing amounts of TEOS, the size becomes smaller for the fourth-generation dendrimer, and the number of dispersed domains decreases for the second-generation dendrimer. During the hydrolysis and condensation of TEOS with the addition product, the

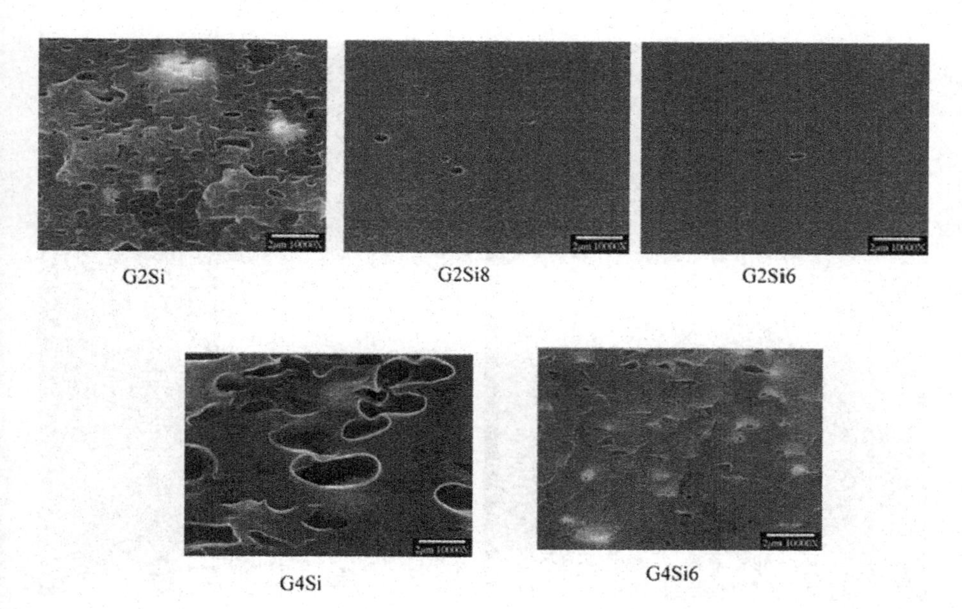

FIGURE 6.4.3 SEM of the SiO_2–PAMAM dendrimer hybrids prepared with various ratios of methanol solutions of 33 wt% TEOS to a 10 wt% solution of the addition product (v/v): G2Si (0/1), G2Si8 (1.1/0.8), G2Si6 (1.1/0.6), G4Si (0/1), and G4Si6 (1.1/0.6).

volume of the film contracted a lot, and this caused the decrease in size or even the elimination of some domains of the dispersed phase. At high amounts of TEOS, hybrid films were generated that were almost free of the dispersed phase when the second-generation PAMAM dendrimer was used, and only powders were obtained when the fourth-generation dendrimer was employed. The condensation of TEOS in the compartments of the PAMAM dendrimer might be responsible for this behavior.

The morphology of G2Si was also investigated with a TEM instrument, and the micrographs are presented in Figure 6.4.4. From Figure 6.4.4a, one can see that light spots are dispersed in the hybrid film. The light spots are likely due to PAMAM dendrimers because of the compartmentalized structure. One of these spots at a higher magnification is presented in Figure 6.4.4b. The sizes of the spots were larger than those of the dendrimers because of their aggregation.

The PAMAM dendrimers were recently employed to cluster Cu ions into nanoparticles of Cu in their compartments.[18,19] It is, however, difficult to work with the dendrimer. For this reason, we carried out experiments in which the SiO_2–PAMAM dendrimer hybrids were employed for the clustering of Cu ions. The metal ion absorption capacity of the SiO_2–PAMAM-dendrimer hybrids from $CuBr_2$ aqueous solutions was investigated by UV spectroscopy (the $CuBr_2$ aqueous solution has an absorption peak at 814 nm). Figure 6.4.5 shows that for the hybrid of the fourth-generation PAMAM dendrimer, the peak intensity of $CuBr_2$ decreased with increasing time and remained unchanged after about 76 h. Similarly, when the hybrid based

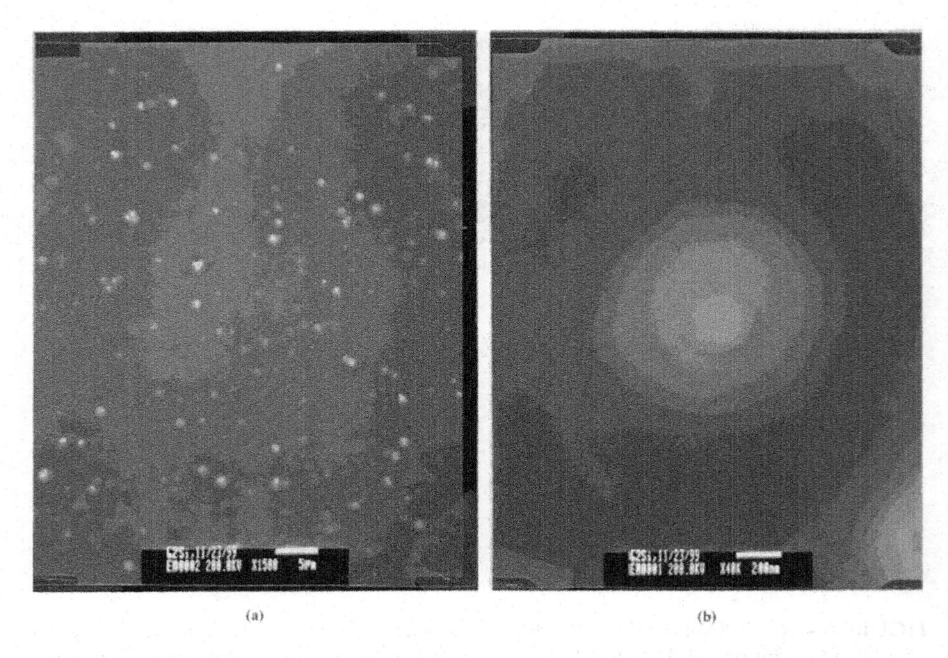

FIGURE 6.4.4 TEM micrographs of G2Si (magnification: a: 1500; b: 40000).

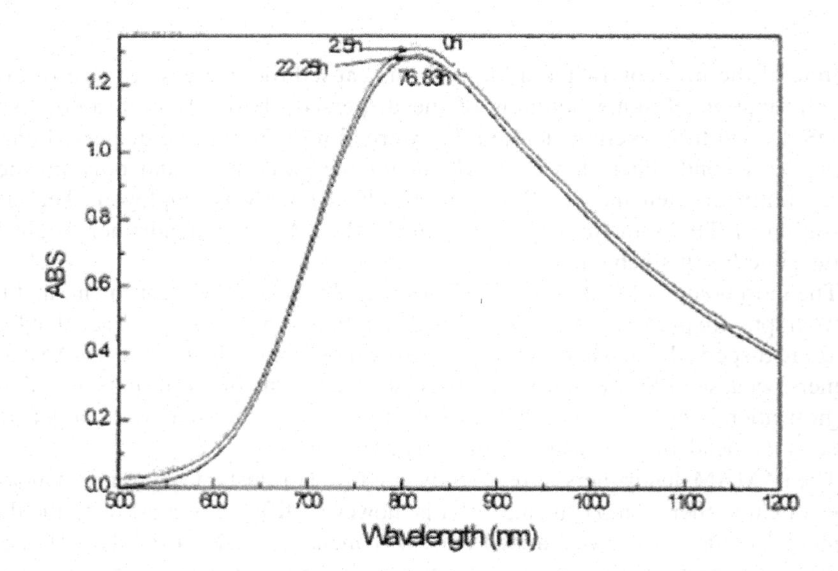

FIGURE 6.4.5 UV spectra of a 0.1 M $CuBr_2$ aqueous solution soaked by a G4Si hybrid for various lengths of time (ABS = absorption).

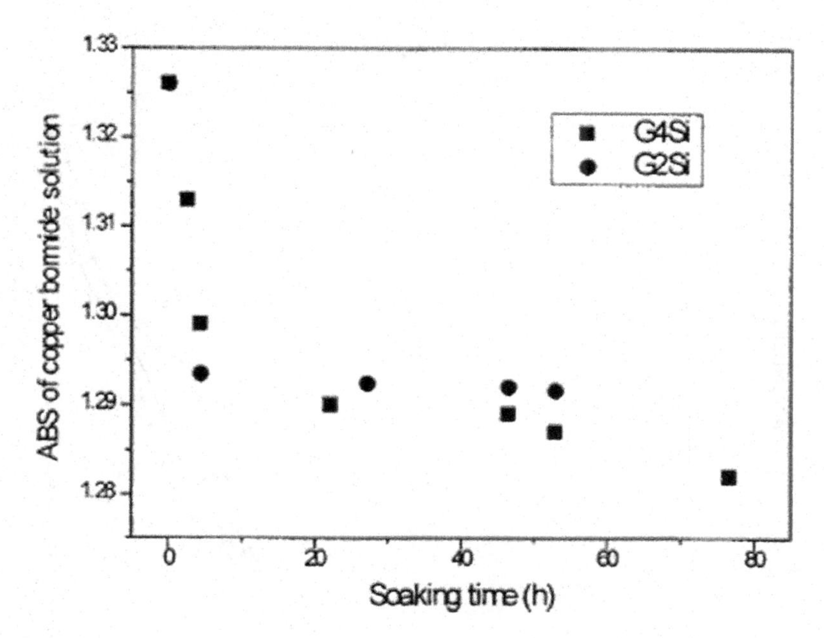

FIGURE 6.4.6 Absorption intensity of a $CuBr_2$ aqueous solution to the soaking time for G4Si and G2Si.

on the second-generation PAMAM dendrimer was employed, the absorption intensity of the $CuBr_2$ aqueous solution exhibited the same trend but remained unchanged after about 28 h. After the light-blue hybrids were taken out from the $CuBr_2$ solutions, their color remained unchanged on soaking in distilled water.

From Figure 6.4.6, one can obtain that (1) the capacities of the second- and fourth-generation PAMAM-dendrimer-based hybrids were 0.058 mol Cu^{2+}/g (3.69 g Cu^{2+}/g) and 0.075 mol Cu^{2+}/g (4.80 g Cu^{2+}/g), respectively, and (2) the absorption intensity for the former hybrid decreased more rapidly with time than that for the latter. The hybrid of the higher generation dendrimer has a denser outer shell, which restricted the diffusion of metal ions, but has a larger number of compartments, which favored a higher capacity.

Figure 6.4.7 presents the TGA curves of the fourth-generation PAMAM dendrimer-based hybrids prepared with various addition-product/TEOS ratios. They show that the hybrids began to decompose at about 185°C. The G4Si sample lost 65% of its weight when the temperature reached 500°C. With the addition of TEOs, the thermal stability of the hybrids was greatly improved, and the weight loss was reduced to only 30% when the temperature reached 500°C.

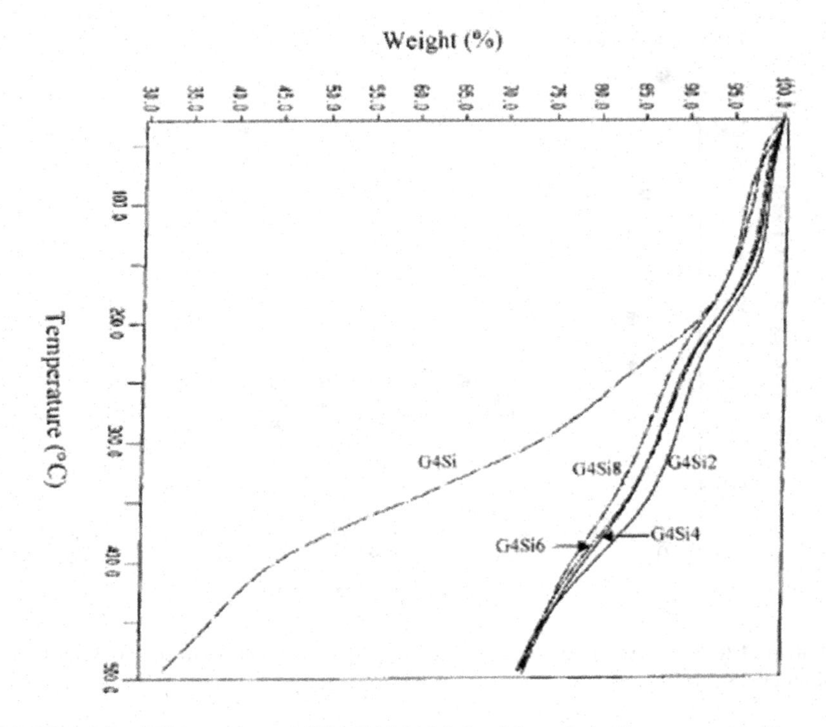

FIGURE 6.4.7 TGA curves of SiO_2–PAMAM dendrimer hybrids prepared with various ratios of TEOS to the addition product indicated in Figure 6.4.3.

6.4.4 CONCLUSION

A silicon-containing PAMAM dendrimer was first synthesized via a Michael addition reaction between PAMAM dendrimer and TPA. This compound was further reacted with TEOS, and a hybrid was generated by casting onto a glass substrate. The SiO_2–PAMAM dendrimer hybrid has a compartmentalized structure because of the presence of the PAMAM dendrimer compartments. The hybrids have a two-phase structure, and the size of the domains of the dispersed phase is dependent on the size of the PAMAM dendrimer and the density of the SiO_2 network. The Cu metal ion complexing capacity of the SiO_2–PAMAM dendrimer hybrids was investigated and related to the size of the PAMAM dendrimer. The thermal stability of the hybrid increased with increasing amounts of TEOS in the feed. In contrast to the pure PAMAM dendrimer, the hybrid has a higher thermal stability and can be more easily handled.

REFERENCES

1. Yilgor, I.; McGrath, J. *Adv Chem Ser* 1990, 224, 207.
2. Saegusa, T.; Chujo, Y. *Makromol Chem Macromol Symp* 1992, 64, 1.
3. Novak, B. M. *Adv Mater* 1993, 5, 422.
4. Matejka, L.; Dusek, K.; Plestil, J.; Kriz, J.; Ledicky, F. *Polymer* 1998, 40, 171.
5. Novak, B. M.; Davies, C. *Macromolecules* 1991, 24, 5481.
6. Haraguchi, K.; Usami, Y.; Yamamura, K.; Matsumoto, S. *Polymer* 1998, 39, 6243.
7. Yin, W. S.; Liu, H. W.; Yin, B.; Gu, T. *Polymer* 1997, 38, 5173.
8. Haddad, T. S.; Lichtenhan, J. D. *Macromolecules* 1996, 29, 7302.
9. Haraguchi, K.; Usami, Y. *Chem Lett* 1997, 1997, 51.
10. Morikawa, A.; Iyoku, Y.; Kakimoto, M.; Imai, Y. *J Mater Chem* 1996, 2, 679.
11. Komada, S.; Fujihana, K.; Osaka, T. *J Electrochem Soc* 1998, 145, 1126.
12. Ho, P. K. H.; Thomas, D. S.; Friend, R. H.; Tessler, N. *Science* 1999, 285, 233.
13. Percec, V.; Ahn, C.-H.; Ungar, G.; Yeardley, D. J. P.; Moller, M.; Sheiko, S. S. *Nature* 1998, 391, 161.
14. Srinivasan, S. A.; Twieg, R.; Hedrick, J. L.; Hawker, C. J. *Macromolecules* 1996, 29, 8543.
15. Hedrick, J. L.; Hawker, C. J.; Miller, R. D.; Twieg, R.; Srinivasan, S. A.; Trollsas, M. *Macromolecules* 1997, 30, 7607.
16. Dvonic, P. R.; Jallouli, A. M.; Swanson, D.; Owen, M. J. U.S. Patent 5, 739, 218, 1998.
17. Pavia, P. R.; Lampman, G. M., Jr.; Kriz, G. S. *Introduction to Spectroscopy*; Saunders: Philadelphia, PA, 1979; p. 20–80.
18. Balogh, L.; Tomalia, D. A. *J Am Chem Soc* 1998, 120, 7355.
19. Zhao, M.; Jun, L.; Crooks, R. M. *J Am Chem Soc* 1998, 120, 4877.

6.5 Polyurethane Toughened Polylactide*

Yumin Yuan and Eli Ruckenstein

Department of Chemical Engineering, State University of New York at Buffalo, Buffalo, New York 14260, USA

ABSTRACT Brittle polylactide (PLA) was toughened by introducing 5 wt% of a poly(ε-caprolactone)(PCL) diol- and triol-based polyurethane (PU) network. The extent of crosslinking of the PU was varied by changing the ratio between diol and triol. The effects of the PU content and its crosslink density on the mechanical properties and the toughness of PU/PLA blends were investigated. Maximum toughness of PU/PLA blends, an order of magnitude higher than that of pure PLA, could be achieved by the use of a proper amount of PU and a proper extent of cross-linking.

6.5.1 INTRODUCTION

Polylactide (PLA) is biodegradable and can be hydrolyzed to nontoxic, water soluble or metabolic products. It has been used successfully for drug delivery, in medical devices, and as an absorbable material.[1–3] However, owing to its brittleness, wider applications of PLA to replace the non-degradable polystyrene and poly(vinyl chloride) are restricted.

The toughening of brittle thermoplastics and thermosets has been investigated for several decades.[4–6] Such materials can be toughened significantly by introducing a discrete rubber phase in the brittle matrix, a typical example being high-impact polystyrene, in which the polystyrene is toughened with grafted poly(butadiene) rubber. A number of factors, such as the entanglement density of the matrix, the rubber content, the rubber particle size and size distribution, the interfacial tension and the phase behavior of the matrix and the rubber, affect the toughening of a brittle material.[7,8]

Poly(ε-caprolactone) (PCL) is a biodegradable rubber that has been used to toughen brittle PLA plastics by copolymerization of PCL diol with lactide monomer. Grijpma et al.[9] improved the impact strength of PLA by synthesizing PCL-PLA-PCL

* *Polymer Bulletin* 1998, 40, 485–490.

triblock copolymers. Hiljanen-Vainio et al.[10] toughened PLA via direct polycondensation of L-lactic acid and ε-caprolactone followed by chain extension through urethane linkages of the polyester to poly(ester-urethane).

In the present paper, PLA (96 mol% L-isomer) was toughened using a small amount of a semi-interpenetrating polyurethane (PU)-PLA network (SIPN) whose extent of crosslinking was varied by changing the ratio between PCL diol and triol. A maximum toughness of 18 MJ/m^3 was achieved by blending the PLA with 5 wt% PU based on an OH mole ratio between diol and triol of 9 to 1; this represents an order-of-magnitude increase over the 1.6 MJ/m^3 toughness of pure PLA.

6.5.2 EXPERIMENTAL

6.5.2.1 Materials

Polylactide (PLA, 96 mol L-isomer, M_n = 112,000), poly(ε-caprolactone) (PCL) diol and triol (M_n = 1250 and 900, respectively), toluene-2,4-diisocyanate, dibutyltin dilaurate and anhydrous toluene were purchased from Aldrich and used as received.

6.5.2.2 Preparation of PU/PLA Blends

Three kinds of PU were used with an OH mole ratio between diol and triol of 10/0, 9/1 and 7/3. The corresponding PUs will be denoted as PU-0, PU-1 and PU-3, respectively. PU/PLA blends were prepared by solution blending in toluene, followed by interchain cross-linking. A typical preparation of a PU-1/PLA blend proceeded as follows: 1.125 g of pre-dried PCL diol was dissolved in 20 mL anhydrous toluene, to which 0.191 g toluene-2,4-diisocyanate along with a drop of dibutyltin dilaurate were added. Reaction was allowed to proceed under a nitrogen atmosphere at 90°C for 1 h and at 60°C for an additional 1 h. Then 0.06 g of pre-dried PCL triol was added to the solution containing NCO-terminated prepolymer molecules, to promote partial cross-linking. Finally, solutions containing 0.1 g and 0.2 g of PU-1 were removed from the flask, just before the gelation point was reached, and mixed with toluene solutions containing 1.9 g and 1.8 g PLA, respectively. Further, inter-chain cross-linking of PU-1 was carried out for 1 h at 60°C to complete the formation of a semi-interpenetrating PU-PLA network (SIPN). The resulting polymer blend solutions were poured into aluminum pans and the solvent was allowed to evaporate slowly overnight. The rough films obtained, with PU-1 contents of 5 and 10 wt%, respectively, were dried thoroughly in a vacuum oven at 60°C, and thin semitransparent blend films were prepared by a short hot pressing at 180°C, followed by rapid air cooling to room temperature.

6.5.2.3 DSC Characterization

The thermal behavior of the sample was examined with a DuPont 910 differential scanning calorimeter (DSC), under nitrogen. The samples were first heated to 175°C, with a heating rate of 10°C/min and then held at 175°C for 1 min to remove any previous thermal history. The samples were then rapidly quenched in liquid nitrogen. The second scanning was carried out from 0°C to 180°C, also with a heating rate of 10°C/min. The data were collected during the second scanning.

6.5.2.4 Determination of Mechanical Properties

Thin films of 0.10–0.15 mm thickness were cut into the dumbbell-shaped form indicated by ASTM D.638-58T. The tensile testing was performed at room temperature, using an Instron Universal Testing Instrument (Model 1000), with an elongation rate of 10 mm/min. The yield and tensile strengths were calculated based on the initial cross-sectional area of the specimen. The toughness is defined as the energy needed to break a sample of unit area and unit length (J/m^3); it is given by the area under the stress-strain curve. The data regarding the mechanical properties represent averages for at least five specimens.

6.5.3 RESULTS AND DISCUSSION

Figure 6.5.1 presents the stress-strain curves of PLA and PU/PLA blends containing 5 wt% PU, in which the crosslink density was varied by changing the OH mole ratio between diol and triol. PLA (curve a) is a brittle thermoplastic material, which exhibits a yield point and a low elongation of about 6%. When the linear PU-0 was introduced into the PLA matrix (curve b), neither the tensile strength, nor the elongation were improved. In contrast, the elongation of the PU-1/PLA blend (curve c) increased

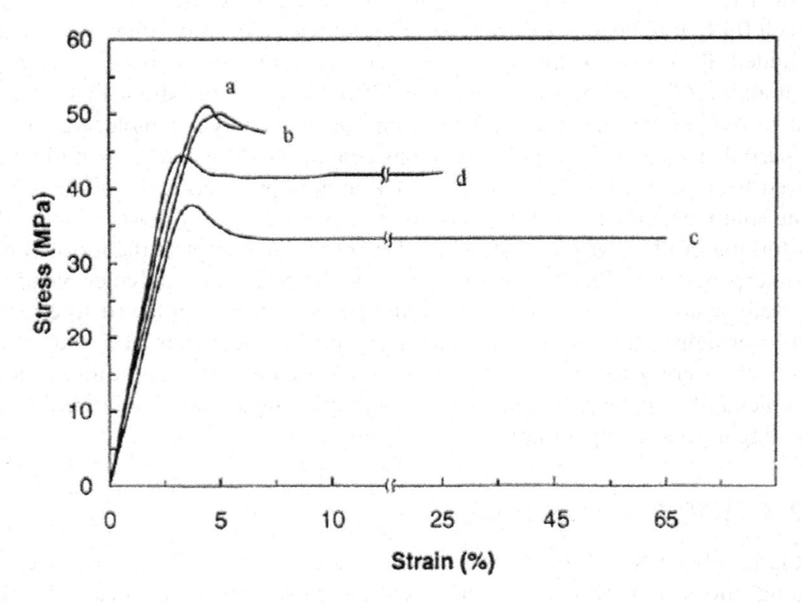

FIGURE 6.5.1 The stress-strain curves of PLA and PU (5 wt%) PAL blends a: Pure PLA; b: PU-0/PLA; c: PU-1/PLA; d: PU-3/PLA.

TABLE 6.5.1
The PLA and PU/PLA Blends: Mechanical Properties

PU Content wt%	Yield Strength (MPa)	Tensile Strength (MPa)	Young's Modulus (MPa)	Elongation (%)	Toughness (MJ/m³)
0 (PLA)	51	48	1480	6	1.6
Linear PU-0					
5	49	47	1340	7	2
10	/	38	1160	5	1.4
PU-1: diol/triol = 9/1					
5	37.8	33.5	1160	60	18
10	35	30.5	1000	35	10.5
PU-3: diol/triol = 7/3					
5	44	41	1490	20	13.5
10	31.4	29	1100	15	4

up to 60%, and the yield and the tensile strengths decreased somewhat, indicating that the PLA was significantly toughened by the PU-1-PLA SIPN. However, as the crosslink density was further increased (PU-3/PLA blend, curve d), the elongation decreased to 25%, and the yield and tensile strengths acquired values between those of PLA and the PU-1/PLA blend, demonstrating that the greater cross-linking in the PU-3 SIPN yielded less toughness. The effects of the PU content and the cross-link density on the mechanical properties of PU/PLA composites are summarized in Table 6.5.1.

Due to its low elongation (6%), PLA has low toughness, about 1.6 MJ/m³. When blended with 5–10 wt% linear PU-0 (free of PCL triol), the strengths and moduli decreased, but since the elongation has barely improved, the toughness of the blend remained as low as that of the pure PLA. This occurs because the PU-0 is more polar than PLA, and strong hydrogen-bonding in the former stimulates self-aggregation. Consequently, the adhesion between the polymers is poor. When, however, the PLA was mixed with a properly crosslinked PU-1, the toughness of the material significantly improved. With only 5 wt% of PU-1 in the blend, the elongation increased to about 60% and the toughness increased by one order of magnitude (to 18 MJ/m³). This occurs because some of the PLA interpenetrates the PU-1 networks, generating PU-PLA semi-interpenetrating networks which are more compatible with PLA. With a further increase of the crosslink density (PU-3/PLA blend), the toughness decreases compared to the PU-1/PLA blend. The lower toughness of the PU-3/PLA blend is due to the increased stiffness of the semi-interpenetrating PU-PLA network. This results in less intermingling between the PU-PLA networks and the PLA.

The specimens containing 5 or 10 wt% PU-1 and 5 wt% PU-3 exhibit a well defined yield point followed by the appearance of a stress whitening during the tensile testing. The surfaces of the broken samples have a fibrous structure. This indicates that a large amount of energy is dissipated during the drawing of the fibrils, which enhances the toughness of these materials.

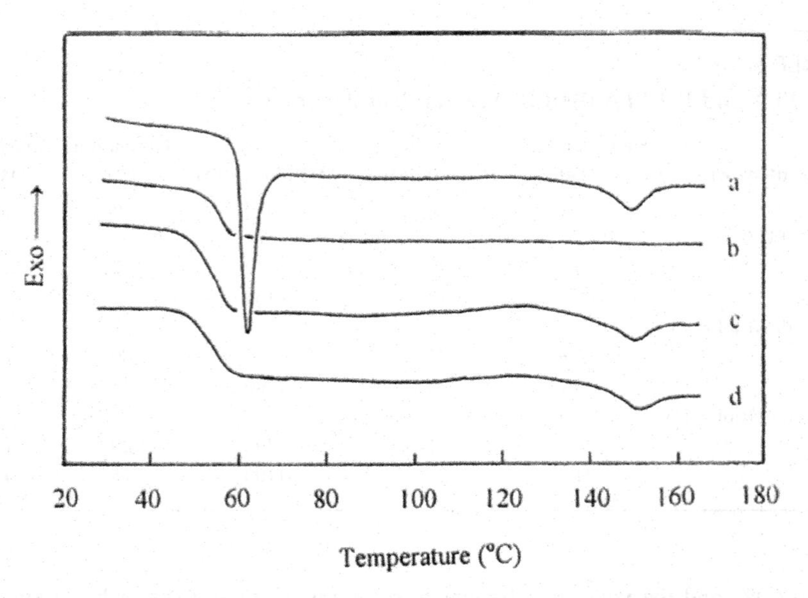

FIGURE 6.5.2 DSC traces of PLA and PU (5 wt%)/PLA blends a: Pure PLA, first scanning; b: pure PLA, second scanning; c: PU-l/PLA blend, second scanning; d: PU-3/PLA blend, second scanning.

Figure 6.5.2 presents the DSC traces of pure PLA and PU/PLA blends. For pure PLA, the T_g is about 56°C and the T_m is 150°C. After being quenched in liquid nitrogen, the PLA became completely amorphous, since only its T_g (but not its T_m) was detected during the second scanning. The T_gs and T_ms of the PU/PLA blends were not affected by the small amount of PU. However, for the quenched PU/PLA samples, a small, wide exothermic cool crystallization temperature at about 130°C, and an endothermic melting peak corresponding to the crystalline PLA, appeared during the second scanning. This occurs because of the higher mobility of the PLA chains in the presence of the flexible PU chains.

6.5.4 CONCLUSION

A maximum toughness of PLA can be achieved by introducing 5 wt% of a properly crosslinked polyurethane network (having an OH mole ratio of diol to triol of 9 to 1). The toughness thus achieved (18 MJ/m^3) is an order of magnitude higher than that of pure PLA. Lower toughness occurs when the polyurethane is linear or overly crosslinked. Optimum toughness is a result of a balance between the compatibility of the semi-interpenetrating polyurethane-PLA network with PLA and the stiffness of this network.

ACKNOWLEDGMENT

This work was supported by the National Science Foundation.

REFERENCES

1. Leenslag JW and Pennings AJ (1987) *Macromol Chem* 188:1809.
2. Holland SJ, Tighe BJ and Gould PLJ (1986) *J Control Release* 4:155.
3. Li MS, Garreau H and Vert M (1990) *J Mater Sci Mater Med* 1:198.
4. Bucknall CB (1977) *Toughen Plastics*, Applied Science Publisher, London, UK.
5. Wu S (1990) *Polym Int* 29:29.
6. Ruckenstein E and Li H (1997) *Polym Composite* 18(3):320.
7. Wu S (1985) *Polymer* 26:1855.
8. Joziasse CAP, Topp MDC, Veenstra K, Grijpma DW and Pennings AJ (1994) *Polym Bull* 33:599.
9. Grijpma, Van Hofslot RDA, Super K, Nijenhuis AJ and Pennings AJ (1994) *Polym Eng Sci* 34:1674.
10. Hiljanen-Vainio M, Kylma J, Hiltunen K and Seppala JV (1997) *J Appl Polym Sci* 63:1335.

6.6 Cooperative Toughening and Cooperative Compatibilization: *The Blends of Nylon 6, Ethylene-co-Vinyl Acetate, and Ethylene-co-Acrylic Acid**

Xiaodong Wang and Hangquan Li

School of Materials Science and Engineering,
Beijing University of Chemical Technology,
Beijing 100029, People's Republic of China

Eli Ruckenstein

Department of Chemical Engineering,
State University of New York at Buffalo,
Buffalo, New York 14260-4200, USA

ABSTRACT A polymer blend consisting of nylon 6 and ethylene–vinyl acetate copolymer (EVA) was compatibilized with an ethylene-acrylic acid copolymer (EAA). Neither EVA nor EAA are compatible with nylon 6; however, the combination of the two resulted in a toughened nylon 6. The compatibilization was revealed by the dramatic increase in impact strength, and the smaller particle size and finer dispersion of EVA in the nylon 6 matrix in the presence of EAA. The degree of toughening was evaluated through its effect on the mechanical, morphological and rheological properties, by changing the proportion of the components in the nylon 6/EVA/EAA blends. Because EAA is a compatibilizer for nylon 6/EVA and EVA is a compatibilizer for nylon 6/EAA, both the toughening and the compatibilization are cooperative.

* *Polymer* 2001, 42, 9211–9216.

6.6.1 INTRODUCTION

The blending with suitable elastomeric materials has become one of the important means for the improvement of the toughness of a brittle plastic. There is general agreement that the elastomer particle size and the elastomer matrix adhesion are the important factors that determine the toughness of the plastic/elastomer blends [1–3]. These two factors are, however, inter-related, since changing one of them changes the other one also. Therefore, when studying the two factors, they must be carefully controlled.

The toughening of nylon 6 with elastomers has attracted a great deal of attention [4–6]. However, the hydrocarbon elastomers do not have sufficient affinity for the polar polymer molecules such as nylon 6 [7,8]. This difficulty was avoided by incorporating functional groups into the elastomer that can react with the amine groups of nylon 6 [9,10]. The functional groups provided the adhesion needed and dramatically increased the dispersion of the elastomer resulting in improved toughness [11–13]. Among the hydrocarbon elastomers employed one can list the ethylene–propylene–diene copolymer (EPDM), the styrene–butadiene–styrene block copolymer (SBS), the styrene–ethylene–butadiene-styrene block copolymer (SEBS), etc. [14–17].

The blending with core-shell particles provided another pathway for the toughening of nylon 6. ABS possesses a core of styrene-butadiene copolymer and an incomplete shell of styrene-acrylonitrile copolymer. To improve the adhesion between nylon 6 and the styrene–acrylonitrile copolymer, the common practice was to graft glycidal methacrylate onto the shell of ABS, or acrylamide onto the nylon 6 chains [18–21]. Particles with the flexible core of poly(butyl acrylate) (PBA) and the glassy shell of poly(methyl methacrylate) (PMMA) were recently employed to toughen nylon 6. Since PMMA and nylon 6 are not miscible, epoxy resins were used as compatibilizers [22–25].

In this paper, results are presented regarding the toughening of nylon 6 by a copolymer of ethylene and vinyl acetate, which is suggested as a novel impact modifier of nylon 6. While the ethylene–vinyl acetate copolymer (EVA) is a flexible and easily available compound, it is not compatible with nylon 6. An ethylene–acrylic acid copolymer (EAA) developed by DuPont was therefore employed by us as a compatibilizer of the nylon 6/EVA blends. The goal of this paper is to examine the mechanical properties, morphology and rheological behavior of the ternary system nylon 6/ EVA/EAA. It will be shown that EAA is a compatibilizer for nylon 6/EVA, and that EVA is a compatibilizer for nylon 6/EAA. Hence EAA and EVA are cooperative compatibilizers and tougheners of nylon 6.

6.6.2 EXPERIMENTAL

6.6.2.1 MATERIALS

Nylon 6 (commercial grade: 1013B, number average molecular weight: 25,000) was supplied by UCB Chemical, Japan. The ethylene-vinyl acetate copolymer (EVA), with a content of 24 wt% vinyl acetate, was supplied by Beijing Organic Chemical, China. The ethylene–acrylic acid (EAA) (commercial name: Nucrel) copolymer with a content of 3.5 wt% acrylic acid was purchased from DuPont Chemical, USA.

6.6.2.2 Preparation of the Blends

Nylon 6 was dried for 12 h and kept in an airtight aluminum-polyethylene package before use. The blends were mixed using a WP 35 mm twin-screw extruder (*L/D* = 35). All the ingredients were tumble-blended and fed through the throat of an extruder. The barrel and die temperatures were increased from 205°C to 24°C, and the rotation speed of the screw was 180 rpm. The blends passed through a cooling water bath and were finally pelletized. The extrusion parameters were changed very little from one composition to another.

6.6.2.3 Mechanical Properties Test

The tensile properties were determined at room temperature with an Instron Universal Testing Machine (Model 1130) according to the ASTM D638. The notched Izod impact strength was determined with a SUMITOMO impact tester according to the ASTM D256. The thickness of the Izod impact specimens was 1/8 in. Five determinations were carried out for each data point.

6.6.2.4 Scanning Electron Microscopy

The sample bars were fractured in liquid nitrogen. The fractured surface was etched for 2 h with a boiling mixture of toluene and methyl-ethyl-ketone at a weight ratio of 60/40, then coated with an Au/Pd alloy, and subsequently subjected to observation under a scanning electron microscope (Cambridge S250). The micrographs of SEM were analyzed with an image analyzer (IBAS 1/2).

6.6.2.5 Rheological Measurements

The torque of the blend samples was determined with a Brabender mixer (Plasticorder model PLE 330) at 240°C for 20 min, and recorded as a function of time. The apparent viscosities at various shear rates were determined with a capillary rheometer possessing a capillary with an *L/D* ratio of 43/1 (TOYOSEIKI Mode 1B) at 240°C. The melt flow index was determined according to ASTM D1238 at 250°C.

6.6.3 RESULTS AND DISCUSSION

6.6.3.1 Mechanical Property

The incompatibility of the nylon 6/EVA blends is reflected in the stress–strain curves of Figure 6.6.1. While the pure nylon 6 possesses a tensile strength of about 60 MPa and an elongation at break of over 160%, the nylon 6/EVA blend at a weight ratio of 80/20 provided lower values for both properties. The decrease in tensile strength occurred partly because of the weakening of the inter- and intra hydrogen bonding of the nylon 6 molecules by the segments of EVA, and more

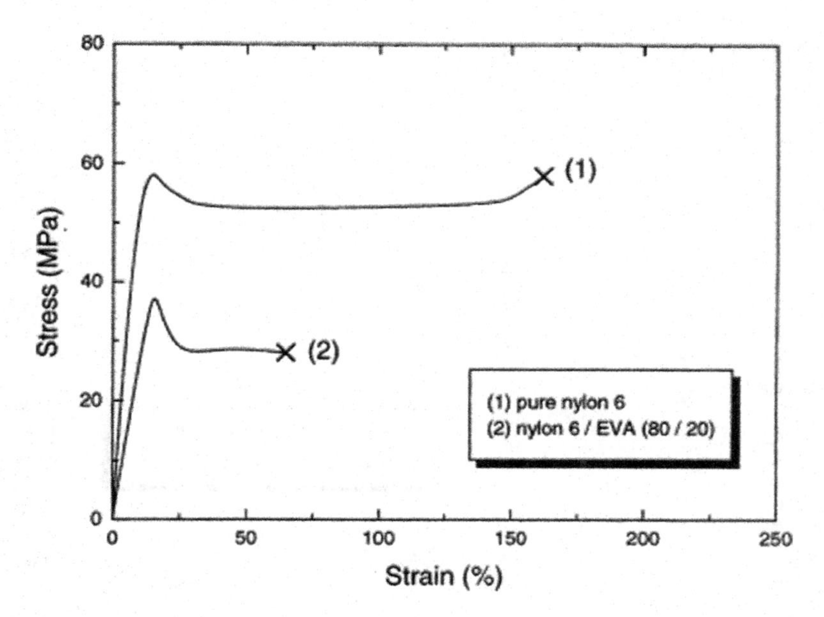

FIGURE 6.6.1 The stress-strain curves of nylon 6 and nylon 6/EVA (80/20 weight ratio) binary blend.

importantly, because of the incompatibility between the polar segments of nylon 6 and the nonpolar ones of EVA. For the latter reason, the elongation at break also decreased. The compatibility could be, however, improved by adding EAA, a copolymer of ethylene and acrylic acid, to the system. Because of the similarity of the chain structures, EAA and EVA are miscible in all proportions. The acrylic acid moieties generated H-bonds and/or reacted with the amine groups of nylon 6, which became thus compatibilized with EVA. Indeed, as shown in Figure 6.6.2, the tensile strengths of nylon 6/EVA blends were improved by the addition of EAA, and the elongation at break became as large as 200% when a sufficiently large amount of EAA was added. This suggests that adhesion occurred between the segments of EVA and those of EAA grafted on nylon. Figure 6.6.2 shows that the tensile strength increased from 40 MPa for the uncompatibilized to 54 MPa for compatibilized blends.

A high improvement in mechanical properties was achieved regarding the toughness (Figure 6.6.3). The notched impact strength of pure nylon 6 is about 19 J/m [26,27], while the nylon 6/EVA blend possessed at a weight ratio of 80/20 a notched impact strength of about 60 J/m. Although EVA alone increased the toughness of nylon 6 by a factor of 3, which represents a moderate improvement, the blend behaved as a brittle material at impact.

Because of incompatibility, EVA could not be finely dispersed in the nylon 6 matrix, resulting only in a moderate toughening. However, the compatibilizer, EAA, generated bridges between the two components and decreased the particle size

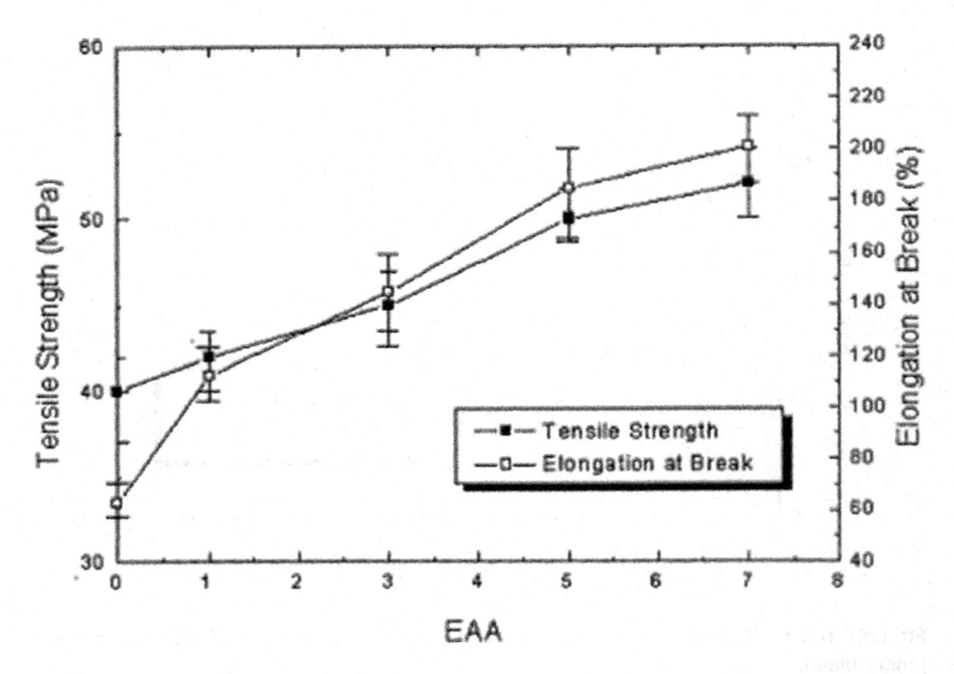

FIGURE 6.6.2 The effect of EAA content [g/(100 g nylon 6 + EVA)] on the tensile strength and elongation at break of nylon 6/EVA/EAA ternary blends (weight ratio nylon 6/EVA = 80/20).

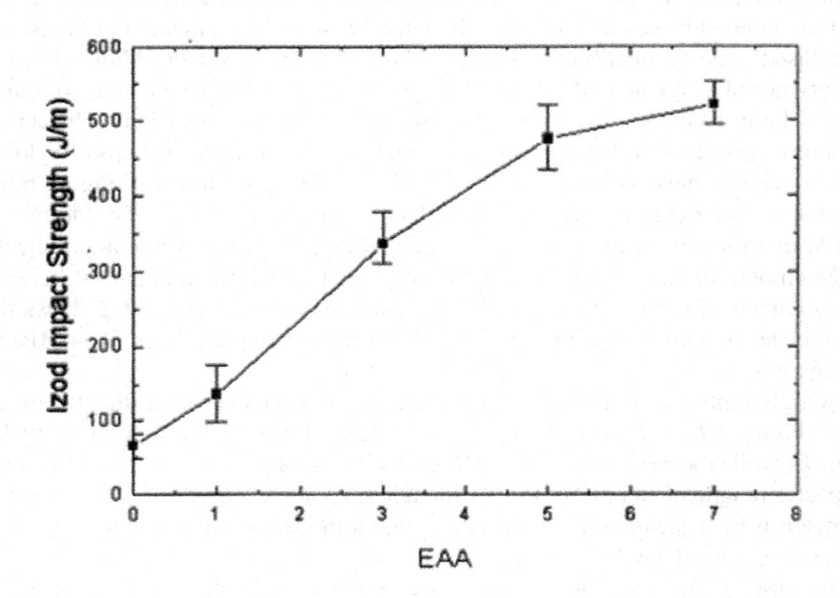

FIGURE 6.6.3 The effect of EAA content [g/(100 g nylon 6 + EVA)] on the notched impact strength of nylon 6/EVA/EAA ternary blends (weight ratio nylon 6/EVA = 80/20).

of EVA (more details later). Major increases in the notched impact strengths were obtained as a result of compatibilization. For a sufficiently large amount of EAA, the sample achieved the high notched impact strength of 520 J/m, which represents a tough behavior. Wu [27] found that for the nylon 6/SEBS-*g*-maleic anhydride system the tough-brittle transition was very sharp, the impact strength increasing sharply from below 200 to above 500 J/m as the particle size became smaller than a threshold value. It is interesting to notice that in the present system the tough–brittle transition was rather a gradual one. The toughness increased with increasing content of compatibilizer (Figure 6.6.3) and the impact behavior of the sample changed from brittle to semi-tough to fully tough.

6.6.3.2 Morphology

The morphology of the blends was investigated by scanning electron microscopy (SEM). All the samples were first subjected to brittle fracture at low temperature, and the fractured surface was etched with a mixture of toluene and methyl-ethyl-ketone to remove the EVA and/ or EAA species. The holes left on the fractured surface of the nylon 6 matrix reflect the morphology of the dispersed phase (Figure 6.6.4). The average diameter of the holes, determined from the SEM micrographs using an image analyzer, is plotted in Figure 6.6.5. Figure 6.6.4a presents the fractured surface of a nylon 6/EVA blend at a weight ratio of 80/20 and shows that the holes are large and non-uniform, ranging from 0.5 to 2.4 μm. It is now accepted that for a nylon 6/elastomer blend to be well toughened, the size of the elastomer particles should be smaller than 0.7 μm [13–27]. For non-uniform EVA sizes of 0.5–2.4 μm, the tensile properties were poor, and only a moderate toughening could be achieved. As shown in Figures 6.6.4 and 6.6.5, the interaction between the two components was improved by introducing EAA. For 1 wt% EAA, the size of the holes decreased under 1.2 μm. As the EAA content increased, the size of the holes became increasingly smaller and more uniform. For an EAA content of 7 g/(100 g nylon + EVA), the size of the dispersed domains decreased under 0.4 μm. Such a fine dispersion can be attributed to the compatibilization and resulted in a well-toughened blend.

It is of interest to examine the morphology of the nylon 6/EAA blends (Figure 6.6.6). Because the acrylic acid moieties of EAA can react with the amine groups of nylon 6, it was expected the two to be compatible and that EAA would be well dispersed in the nylon 6 matrix. However, the micrographs of Figure 6.6.6 indicate poor dispersions, the sizes of the EAA domains being as large as 1–2 μm and non-uniform. The mechanical properties of the blends, which are presented in Figure 6.6.7, also indicate that these two components are not compatible. Although a reaction can occur between the acrylic acid moieties and the amine groups, the poly ethylene segments of the EAA were strongly repulsed by nylon 6, resulting in phase segregation. When EVA was added to the system, a well-toughened nylon 6/EAA blend was obtained. This happened probably because the segments of poly(vinyl acetate) have a polarity between those of polyethylene and nylon 6.

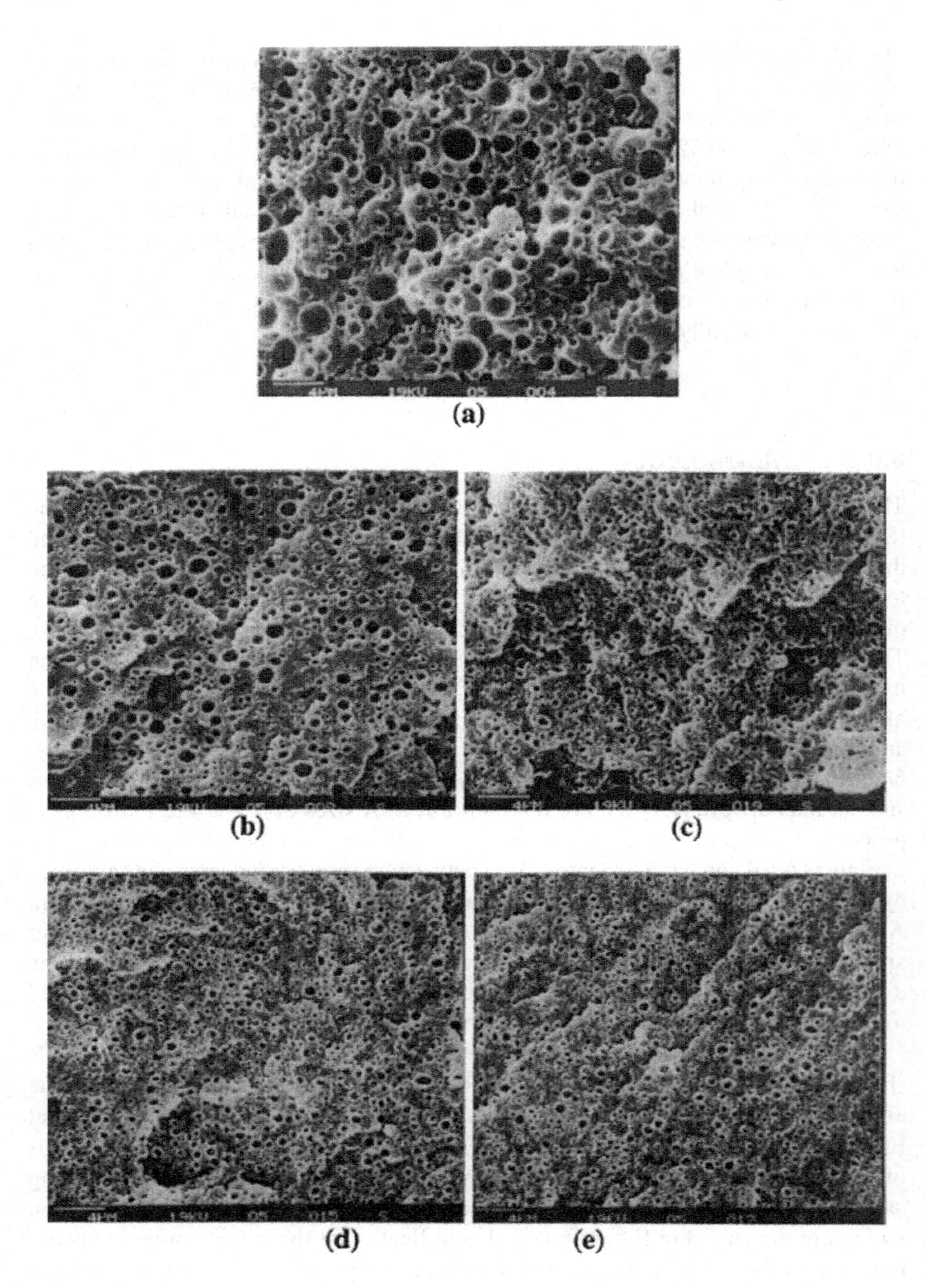

FIGURE 6.6.4 SEM micrographs of the fractured surface of the nylon 6/EVA blends (80/20 weight ratio) with EAA contents [g/(100 g nylon 6 + EVA)]: (a) 0, (b) 1, (c) 3, (d) 5, (e) 7.

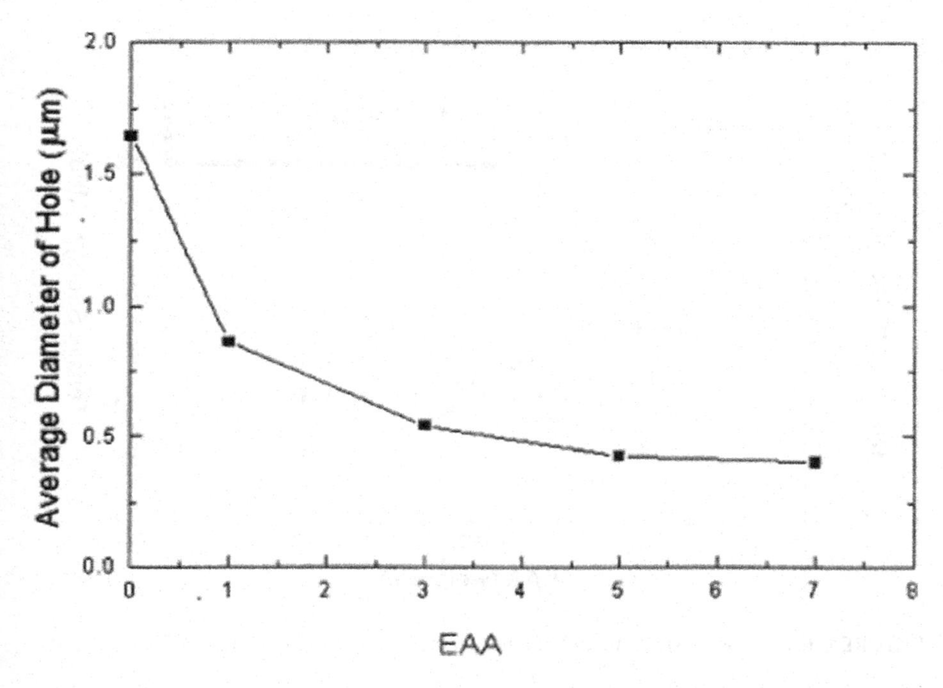

FIGURE 6.6.5 Average hole size on SEM micrographs of the fractured surface vs. EAA content [g/(100 g nylon 6 + EVA)]. Weight ratio nylon 6/EVA = 80/20.

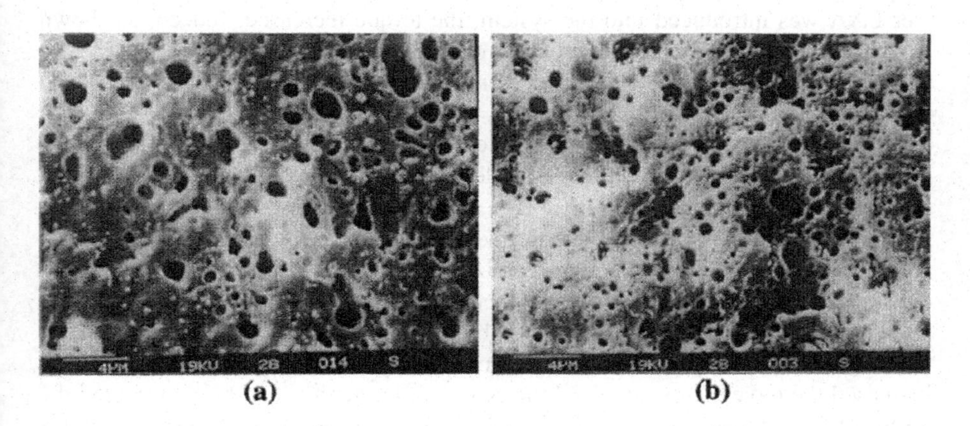

FIGURE 6.6.6 SEM micrographs of the fractured surface of the nylon 6/EAA binary blends for the weight ratios of: (a) 90/10, and (b) 80/20.

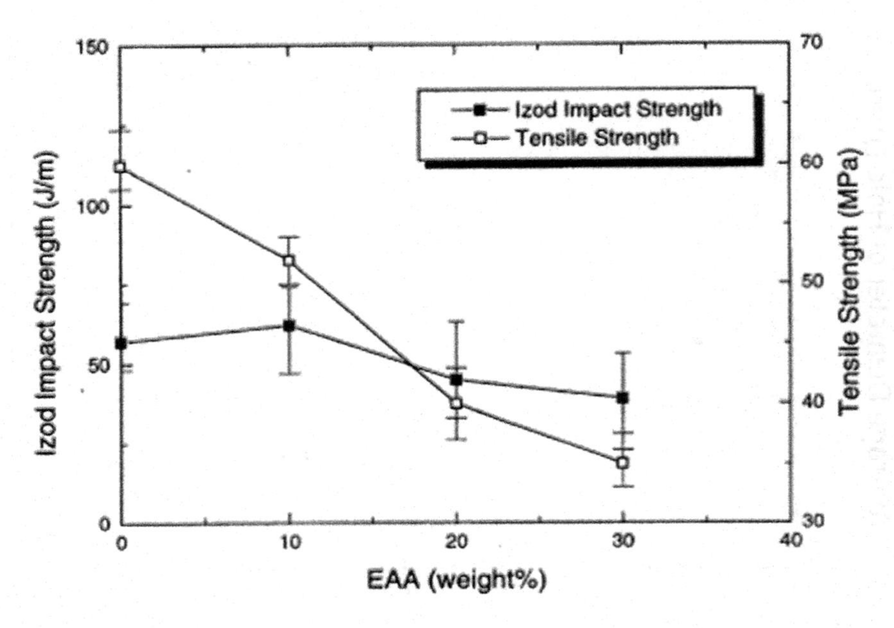

FIGURE 6.6.7 Tensile strength and notched impact strength of nylon/EAA binary blends.

6.6.3.3 Rheological Behavior

In Figure 6.6.8 the mixing torque of the melt is plotted against the mixing time for systems with various compositions. The pure nylon 6 exhibited a low torque, and the torque of the blend of nylon 6/EVA was even lower. When, however, the compatibilizer EAA was introduced into the system, the torque increased. Indeed, as shown by Figure 6.6.8, the greater the content of EAA, the higher was the torque. The decrease of the torque when EVA was blended with nylon 6 constitutes evidence for incompatibility.

Besides the volume of the chain, the resistance to the flow of a polymer melt is mainly due to the entanglement of its molecules. The chains of nylon 6 are flexible, and they are entangled in the melt. The introduction of EVA disentangled some nylon 6 chains and, as a result, the torque was lowered. However, when EAA was added, nylon 6 and EVA were compatibilized. This increased the interaction among the segments, and thus the torque was increased. However, there is another mechanism for the increase of the torque caused by EAA that involves the reaction between the acrylic acid moieties of EAA and the amine groups of nylon 6. This reaction increased the molecular weight and the degree of branching, and both increased the torque of the compatibilized blends. Similar changes occurred in the melt flow index of the blends (Figure 6.6.9).

The melt viscosity at various shear rates is plotted in Figure 6.6.10 as a function of the EAA content. It was found that the melt viscosity of nylon 6/EVA blend without compatibilizer was lower than that of pure nylon 6; however, the greater the EAA content, the higher became the viscosity. The change in the melt viscosity confirms

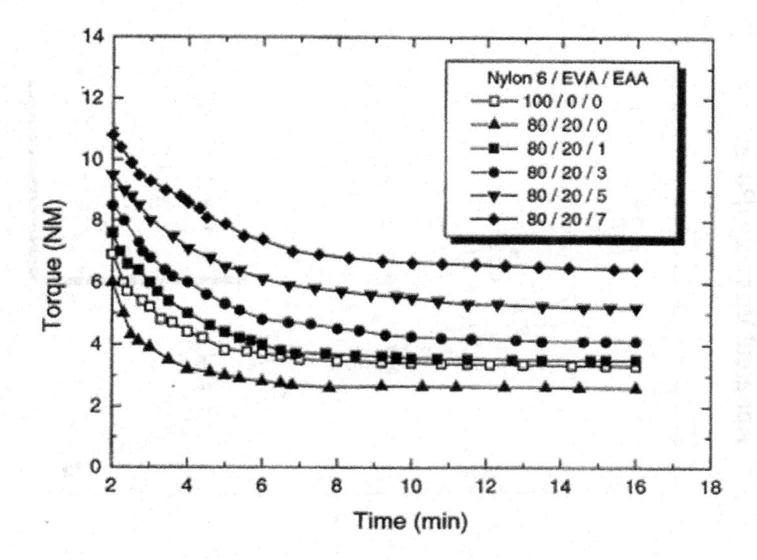

FIGURE 6.6.8 The mixing torque vs. mixing time for nylon 6 and its blends for various EAA contents [g/(100g nylon 6 + EVA)], weight ratio nylon nylon 6/EVA = 80/20.

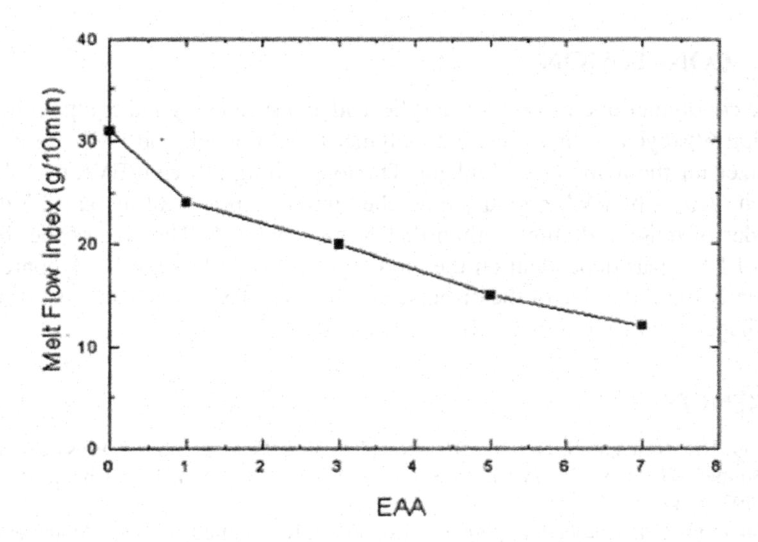

FIGURE 6.6.9 The effect of EAA content [g/(100 g nylon 6 + EVA)] on the melt flow index of nylon 6/EVA/EAA ternary blends (wt ratio nylon 6/EVA = 80/20).

the explanation provided in the preceding paragraph for the changes in the torque. Figure 6.6.10 also shows that the higher the shear rate, the lower was the viscosity. This can be considered to be a result of increased disentanglement of the chains with increasing shear rate.

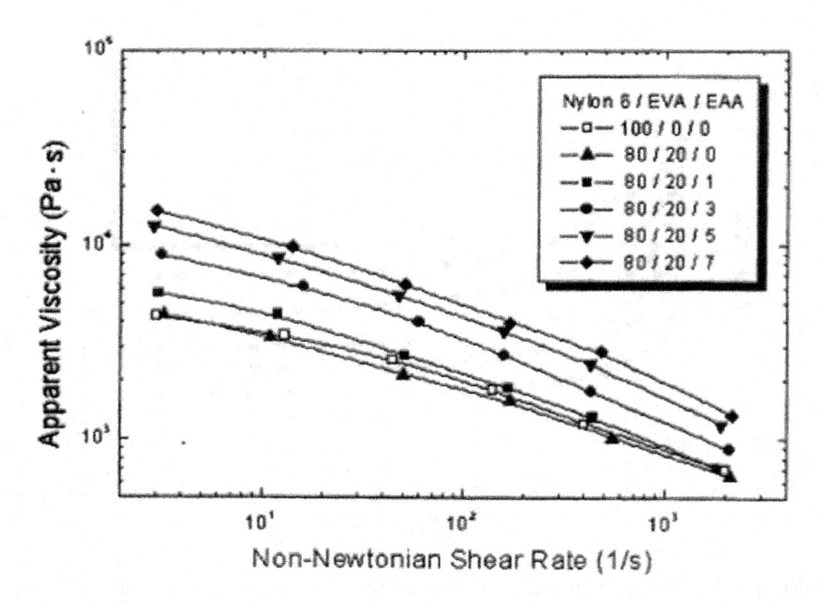

FIGURE 6.6.10 The effect of EAA content [g/(100 g nylon 6 + EVA)] on the melt viscosity of nylon 6/EVA/EAA ternary blends (weight ratio nylon 6/EVA = 80/20).

6.6.4 CONCLUSION

EAA, a copolymer of ethylene and acrylic acid, constitutes a good compatibilizer for the nylon 6/Ethylene-vinyl acetate copolymer (EVA) blend, and EVA a good compatibilizer for the nylon 6/EAA blend. The toughening of nylon/EVA by EAA and of nylon 6/EAA by EVA provides a notched impact strength of nylon 6/EVA/EAA one order of magnitude higher than that of pure nylon 6. The size of the disperse phase of EVA was dependent on the amount of EAA. When the EAA content was sufficiently high, the size of the dispersed domains of EVA was below 0.4 μm. Such a fine dispersion resulted in a well-toughened blend.

REFERENCES

1. Paul DR, Newman S. *Polymer Blends*, vol. 2. New York: Academic Press, 1978. p. 35.
2. Bonner JG, Hope PS. *Polymer Blends and Alloys*. London, UK: Chapman & Hall, 1993. p. 45.
3. Bucknall CB. *Toughening Plastics*. London, UK: Applied Science Publishers, 1977. p. 19.
4. Shi Z, Yang W, Tang L. *Handbook of Polyamide Resins*, vol. 65. Beijing, China: Chinese Petrochemical Industry Press, 1994.
5. Dagli SS, Xanthos M. *Polym Eng Sci* 1994;34:1720.
6. Holsti RM, Seppala JV. *Polym Eng Sci* 1994;34:395.
7. Takeda Y, Keskkula H, Paul DR. *Polymer* 1992;33:3173.
8. Datta S, Lohse DJ. *Polymeric Compatibilizers*, vol. 271. Munich, Germany: Hanser Publishers, 1996.
9. Wu J, Kuo JF, Chen CY. *Polym Eng Sci* 1993;33:1329.

10. Dijkstra K, Laoe L, Gaymans RJ. *Polymer* 1995;35:315.
11. Majumdar B, Keskkula H, Paul DR. *Polymer* 1994;35:1386.
12. Lu M, Keskkula H, Paul DR. *Polymer* 1987;28:1073.
13. Wu S. *J Polym Sci, Polym Phys Ed* 1988;26:807.
14. Willis M, Favis BD. *Polym Eng Sci* 1988;28:1416.
15. Borggreve RJM, Gaymans RJ, Eichenwald HM. *Polymer* 1990;31:971.
16. Gaymans RJ, Werff JW. *Polymer* 1994;35:3658.
17. Jha A, Bhowmick AK. *Rubber Chem Technol* 1997;70:799.
18. Muratoglu OK, Argon AS, Cohen RE. *Polymer* 1995;36:921.
19. Majumdar B, Keskkula H, Paul DR. *Polymer* 1994;35:4263.
20. Majumdar B, Keskkula H, Paul DR. *Polymer* 1994;35:5453.
21. Majumdar B, Keskkula H, Paul DR. *Polymer* 1994;35:3164.
22. Li D, Yee AF. *Polymer* 1993;34:4471.
23. Lu M, Keskkula H, Paul DR. *Polym Eng Sci* 1994;34:33.
24. Soh YS. *J Appl Polym Sci* 1992;45:1831.
25. Soh YS. *Polymer* 1994;35:2764.
26. Oshinski AJ, Keskkula H, Paul DR. *Polymer* 1992;33:268.
27. Wu S. *Polymer* 1985;26:1885.

Index